Food, Pharmaceutical and Bioengineering Division 2013

Core Programming Area at the 2013 AIChE Annual Meeting: Global Challenges for Engineering a Sustainable Future

San Francisco, California, USA
3-8 November 2013

ISBN: 978-1-63439-041-5

Printed from e-media with permission by:

Curran Associates, Inc.
57 Morehouse Lane
Red Hook, NY 12571

Some format issues inherent in the e-media version may also appear in this print version.

Printed by Curran Associates, Inc. (2014)

AIChE
3 Park Avenue
New York, NY 10016-5991

Phone: (203) 702-7660
Fax: (203) 775-5177

www.aiche.org

Additional copies of this publication are available from:

Curran Associates, Inc.
57 Morehouse Lane
Red Hook, NY 12571 USA
Phone: 845-758-0400
Fax: 845-758-2634
Email: curran@proceedings.com
Web: www.proceedings.com

TABLE OF CONTENTS

(6a) Design, Synthesis and Characterization of Transthyretin Mimetics for Inhibition of Beta-Amyloid Toxicity

Murphy, R. M., University of Wisconsin-Madison
Cho, P. Y., University of Wisconsin-Madison
Yang, D. T., University of Wisconsin-Madison
Originally presented on: 11/3/2013 15:30:00 - 15:48:00

Neuronal damage and destruction in Alzheimer's disease has been linked to soluble oligomers of beta-amyloid. Transthyretin is a transport protein present in both serum and cerebrospinal fluid that protects against beta-amyloid toxicity, both in vitro and in vivo. Previously we showed that beta-amyloid oligomers bind to two regions on TTR: strand G (particularly L110) and the EF helix/loop (L82). We also showed that mutation of either leucine to alanine (L82A or L110A) significantly reduced beta-amyloid binding, and that these mutants were ineffective at protecting neurons against beta-amyloid toxicity. Here we present our efforts to design a peptidyl compound based on strand G that can mimic TTR's ability to bind to beta-amyloid and prevent its toxicity. Beginning with the 16-mer corresponding to TTR residues 102-117, we used SPOT peptide arrays to determine the minimum length and to identify critical residues. 6- and 8-mers were too short to bind beta-amyloid; the minimum length required was 10 residues in our screen. Residues Ile107, Leu110, Leu111, Tyr114 and Tyr116 were identified as critical, because mutation led to loss of beta-amyloid binding. We synthesized several 12-mer and 16-mers, including a scrambled sequence peptide and a control in which Leu110 was mutated to alanine, and tested their ability to bind to beta-amyloid using Trp fluorescence, proteolytic fragmentation, and PICUP crosslinking assays. We also examined the role of charged residues at the N- and C-terminus on binding. We observed that active peptides self-assembled in the absence of beta-amyloid, and speculate that self-assembly may be necessary for binding activity. We tested for specificity by examining whether the strand G-based peptide also interacted with other amyloidogenic proteins, and evaluated their effect on beta-amyloid aggregation kinetics. Finally, we observed that our optimized peptide, but not the LtoA control mutant, was able to protect neurons against beta-amyloid toxicity. Thus, transthyretin's binding and protective capability can be retained in a small designed compound.

(6b) A Novel Computer-Aided Molecular Approach Using the Signature Molecular Descriptor to Design Potentially New Non-Intuitive Amyloid-Î² Aggregation Inhibitors

Kayello, H., The University of Akron
Visco, D. P. Jr., The University of Akron
Tseng, J. H., University of South Carolina
Soto-Ortega, D., University of South Carolina
Suo, C., University of South Carolina
Gao, J., University of South Carolina
Chastain, S., University of South Carolina
Murphy, B. P., University of South Carolina
Lim, M., University of South Carolina
Xie, F., University of South Carolina
Chapman, J., University of South Carolina
Wang, Q., University of South Carolina
Moss, M., University of South Carolina
Originally presented on: 11/3/2013 15:48:00 - 16:06:00

The 'amyloid cascade hypothesis', linking aggregation of the amyloid-β protein (Aβ) to the pathogenesis of Alzheimer's disease (AD), has led to the emergence of inhibition of Aβ aggregation as a therapeutic strategy for this currently unpreventable and devastating disease. The identification of Aβ aggregation inhibitors has proceeded primarily through the screening of drug-like molecules, leading to the discovery of hundreds of aggregation inhibitors belonging to multiple classes of compounds. However, only two small-molecule amyloid aggregation inhibitors have proven effective enough to advance to late clinical trials. Albeit proven effective, identifying drugs from specific classes with optimal inhibition effect can be incremental, time consuming, and expensive where improved inhibitors are often developed by small "refinements". Such an approach cannot identify novel and non-intuitive inhibitors with desired properties, however.

Alternatively, a computer-aided molecular design (CAMD) strategy that uses an inverse quantitative structure property relationship (I-QSPR) is proposed for the *design* of drug-like molecules. The CAMD technique is a powerful tool that applies information about how known substances behave in a given application, encodes their molecular structure using what are referred to as "molecular descriptors," relates the descriptors to its performance, then uses that information to generate new molecules with optimal predicted performance without the need for detailed information about drug target. Such an approach has the potential to accelerate the discovery of drug-like molecules.

In this work, 21 molecules that belong to the classes "proven effective" including naphthalimides, coumarin analogs, dihydropyridines, and polyphenols were evaluated for their effect upon aggregation of Aβ monomer. Aggregation was initiated by incubating monomeric protein in the presence of NaCl and under continuous agitation. Progression from monomer to aggregates was evaluated using thioflavin T, a fluorescent dye that yields a shifted and enhanced fluorescence when bound to the β-sheet structure of amyloid aggregates. Uninhibited aggregation exhibited a lag time, characteristic of nucleation, followed by a period of rapid growth, and culminating with a plateau as monomer and aggregate species reached equilibrium. Compounds were characterized for both their ability to extend nucleation, indicated by a fold-increase in the lag time, and their ability to reduce the quantity of aggregate formed, indicated by a decrease in the fluorescence of the equilibrium plateau. After deconstructing all evaluated structures into their corresponding building blocks, or *molecular descriptors*, two quantitative structure property relationship (QSPR) models were created and refined in order to score the over 200 million inverse solutions, which are potential structures. Both QSPR models employed advanced machine-learning techniques. For the lag extension time, a classifier was developed where solutions with a positive lag extension time greater than 2-fold were kept. For the decrease in equilibrium plateau, a support vector regression model was created to filter out solutions with percent decreases less than 95%. The final step of refinement ensured drug-likeness of model outputs where solutions were screened via Lipinski's Rule. Optimal inverse solutions were then converted into structures by employing an enumeration algorithm and structures with a certain molecular stability were retained. The next step was to identify commercially available compounds or synthesize optimal candidates and test them for efficacy. Screening the PubChem chemical database in order to identify structures *similar* to the CAMD output based on a specific predefined similarity metric could expedite the discovery of potentially new drugs if structure synthesis was inevitable.

(6c) Engineering Picomolar Affinity Into a 5 Kda Scaffold for Tumor Targeting

Kruziki, M. A., University of Wisconsin Madison
Bhatnagar, S.,
Hackel, B. J., University of Minnesota
Originally presented on: 11/3/2013 16:06:00 - 16:24:00

Engineered protein scaffolds are conserved stable frameworks upon which amino acid diversity can be introduced to develop novel function. Scaffolds serve as strong platforms for ligand discovery towards the multitude of clinically, scientifically, and industrially impactful targets. Of particular interest here are small protein scaffolds capable of improved permeability and distribution for targeting nascent tumors. We aim to develop a platform for engineering a very small (5 kDa), stable, scaffold capable of efficient evolution to picomolar affinity requisite for sustained delivery in vivo. Towards this end, we analyzed the secondary structure, stability, phylogenetic sequence diversity, and the accessible surface area and shape of the potential paratope of proteins throughout characterized proteomes to identify candidate scaffolds for engineering molecular recognition. We identified a 45 amino acid domain consisting of both beta strand and alpha helical content including two exposed loops amenable to mutation. We diversified twelve amino acids in these two loops using complementarity bias and length diversity to generate a combinatorial library for ligand discovery. Directed evolution was performed using yeast surface display and magnetic and fluorescence selections. Binders to three targets were isolated with affinities as strong as 160 pM. Biophysical characterization, including circular dichroism and thermal stability, will be presented. In vivo biodistribution, via molecular imaging and excised tissue activity analysis of tumor xenografted mice, will be shown. Robust, efficient engineering of small (5 kDa), stable ligands has strong potential for both diagnostic and therapeutic tumor targeting.

(6d) Transcripts for Combined Synthetic Microrna and Gene Delivery

Kashyap, N., Center for Systems Biology, The University of Texas at Dallas
Pham, B.,
Bleris, L., University of Texas at Dallas
Xie, Z., Tsinghua University
Originally presented on: 11/3/2013 16:24:00 - 16:42:00

MicroRNAs (miRNAs) are a class of short noncoding RNAs which are endogenously expressed in many organisms and regulate gene expression by binding to messenger RNA (mRNA). MicroRNAs are either produced from their independent transcription units in intergenic regions or lie in intragenic regions. Intragenic miRNAs and their host mRNAs are produced from the same transcript by the microprocessor and the spliceosome protein complex respectively. The details and exact timing of the processing events have implications for downstream RNA interference (RNAi) efficiency and mRNA stability. Here we engineer and study in mammalian cells a range of synthetic intragenic miRNAs co-expressed with their host genes. Furthermore, we study transcripts which carry the target of the miRNA, thereby emulating a common regulation mechanism. We perform fluorescence microscopy and flow cytometry to characterize the engineered transcripts and investigate the properties of the underlying biological processes. Our results shed additional light on miRNA and pre-mRNA processing but importantly provide insight into engineering transcripts customized for combined delivery and use in synthetic gene circuits.

(6e) The Design of Peptide Amphiphile Micelles for Diagnostic Applications in Atherosclerosis

Chung, E. J., University of Chicago
Drews, L. B., University of California, Berkeley
Tirrell, M., University of Chicago
Originally presented on: 11/3/2013 16:42:00 - 17:00:00

Peptide amphiphiles self-assemble into dynamic micellar structures and provide a multivalent display of targeting peptides which can be developed for use in a wide range of diagnostic and therapeutic applications. Recently, fibrin-binding peptide amphiphile micelles were reported to be suitable for detecting blood clots, a characteristic of late staged atherosclerosis[1]. Although visualization of a clot within a plaque provides a powerful, molecular imaging tool, fibrin expression represents a progressed state of atherosclerotic disease, limiting the system to identifying the outcome of plaque rupture vs. diagnosing the onset of plaque vulnerability for preventative measures.

The presence of monocytes is a known marker of atherosclerotic plaque vulnerability. Specifically, activated endothelial cells secrete cytokines which recruit monocytes in large quantities. Therefore, in order to intervene before plaque rupture, we are engineering monocyte-targeting, peptide amphiphile micelles through the incorporation of the cytokine binding motif. The binding peptide sequence or Cy-7 was conjugated to a lipid tail and mixed to form fluorescently-labeled, monocyte-targeting peptide amphiphile micelles. The critical micelle concentration was determined to be approximately 1 μM with an average diameter of approximately 15 nm, confirmed via dynamic light scattering and transmission electron microscopy. These spherical micelles were stable at 37°C in cell culture media for up to 30 minutes and bound to murine monocytes when evaluated via flow cytometry and confocal microscopy. Moreover, micelles were confirmed to be biocompatible and biofunctional using Live/Dead and chemotaxis assays, respectively.

A molecular imaging tool that binds specifically to monocytes has the potential to meet the sensitivity and accuracy standards necessary to detect vulnerable plaques and therefore provide the time and the opportunity for the patient to treat and manage them. Future studies in Apo/E knockout mice will further delineate the full potential of these novel, peptide amphiphile micelles for atherosclerosis diagnostics. Notably, since peptide micelles can be synthesized with multiple functionalities (i.e. targeting element, imaging dye, therapeutic, etc.), our investigations will lay the ground work for peptide amphiphile micelle-mediated drug delivery at the site of atherosclerotic plaques in future studies.

1. Peters, D. *et al.* Targeting atherosclerosis by using modular, multifunctional micelles. Proceedings of the National Academy of Sciences **106**, 9815-9819, 2009.

(6f) Discovery of Antibody Biomarkers Present in Pre-Eclampsia

Elliott, S. E., University of California, Santa Barbara
Soffici, A. R., Cottage Health System
Daugherty, P. S., University of California, Santa Barbara
Originally presented on: 11/3/2013 17:00:00 - 17:18:00

Discovery of biologic molecules specific to a diseased state, or biomarkers, can lead to diagnostic development, therapeutic target identification, and improved understanding of disease pathogenesis. Antibodies remain an attractive class of biomarkers given their amplification by the immune system, stability, and current clinical use. In this work, we demonstrate the utility of bacteria-displayed peptide libraries in conjunction with fluorescence activated cell sorting (FACS) to identify the presence of novel antibody biomarkers by quantitatively screening for peptides with enhanced binding to antibodies present in a diseased state over normally present antibodies. We applied these methods in an effort to identify molecular diagnostics for pre-eclampsia (PE), a condition with unknown etiology that affects 5-8% of pregnancies. Previous studies have implicated a pathologic role for autoantibodies against a specific epitope on the angiotensin II AT_1 receptor in PE patients [1][2]. However, these antibodies remain difficult to detect, vary in prevalence amongst studies, and most importantly, lack specificity. We hypothesized that additional PE specific antibody biomarkers may exist, and this complex disease could serve as a model for developing methodologies to identify novel antibody biomarkers using the known epitope as a control. Therefore, we expressed the seven amino acid epitope on the surface of *E. coli* to determine the cross-reactivity and specificity in a set of 45 PE and 48 normal-outcome pregnancies (N), and we applied two distinct screening strategies to a fully randomized 15 amino acid bacteria-displayed peptide library with a subset of this cohort. At an optimized cutoff, the cell-surface expressed AT_1 epitope achieved 78% sensitivity (35/45 PE) and 56% specificity (27/48 N). For screening, we used either antibody enriched fractions or unprocessed, diluted plasma. Although both screens resulted in peptides with significantly ($p < 0.05$) higher PE reactivity than N and diagnostic potential in a validation set, sequence analysis yielded a stronger consensus family with three subgroups from the dilute plasma screen. Based upon this consensus, we created and screened a focused library to further evolve the motif and enhance sensitivity. This work demonstrates the utility of bacteria-displayed libraries to identify novel diagnostic reagents for antibody biomarker detection in PE. Furthermore, the improved consensus motif may enable identification of antigens mimicked by these antibody binding peptides. The present methodology may thus improve understanding of disease pathogenesis and guide therapeutic development.

[1] Wallukat, G. *et al.* Patients with Preeclampsia Develop Agonistic Autoantibodies against the Angiotensin AT_1 Receptor. *The Journal of Clinical Investigation* **103**,945–952 (1999).

[2] Zhou, C. C. *et al.* Angiotensin Receptor Agonistic Autoantibodies Induce Pre-eclampsia in Pregnant Mice. *Nature Medicine* **14,** 855–862 (2008).

(6g) Universal Cell Detection By Immobilized Antimicrobial Peptide Plantaricin

Albayrak-Guralp, S., University of Michigan
Gubbuk, I. H., Selcuk University
Kucukkolbasi, S., Selcuk University
Gulari, E., University of Michigan
Originally presented on: 11/3/2013 17:18:00 - 17:36:00

Antimicrobial peptides (AMPs) are an essential part of innate immune response of many organisms and they have demonstrated direct antimicrobial activity against a large spectrum of microorganisms. These peptides exert their activity by binding to surface components of target cells and causing membrane disruption. The semi-selective cell recognition and binding property of AMPs make them potentially valuable detection agents. Plantaricin-423 is a Class-IIa bacteriocin and display bactericidal activities against several foodborne pathogens and spoilage gram positive bacteria. The goal of this study was to investigate the potential of using immobilized plantaricin for capture of a broad variety of microbial species in a biosensor format. The peptides used for immobilization consist of N-terminal domain of Pln-423 (first 18 amino acids) and an additional cysteine residue (Pln-18C) to form a disulfide bond between thiol groups of the silanized glass surface and the peptide. The capture capability of peptide-immobilized slides was tested with *Bacillus coagulans*, *Enterococcus hirae*, *Escherichia coli*, *Staphylococcus epidermidis*, *Listeria innocua* and *Mycobacterium smegmatis* cells representing both Gram positive and Gram negative bacteria families. Our results showed that Pln-18C is able to recognize and capture cells of all seven species that were tested. The cell binding was observed to occur within five minutes of incubation and the binding levels were similar for all cell types except *M. smegmatis* which was lower compared to other species. The detection limit for *L. innocua* 33090 was determined to be $5x10^4$ to $1x10^5$cells/ml depending on the immobilized peptide concentration. We have also evaluated the viability of surface-bound *L. innocua* 33090 and *E. coli* JE5505 cells and observed that immobilized Pln-18C maintained its anti-listerial activity, however, it did not cause any membrane disruption of *E. coli* cells as expected. Our results demonstrate the potential of utilizing antimicrobial peptides, specifically Pln-18C, in biosensors for pathogen detection and creating antimicrobial surfaces against *Listeria* species. Moreover, in combination with other cell-binding AMPs, a higher number of different target species, from food-borne pathogens to biodefense agents, can be captured on a more stable, economic and robust platform.

(6h) Multivalent Ligands to Control Stem Cell Fate

Conway, A., University of California, Berkeley
Schaffer, D., University of California, Berkeley
Originally presented on: 11/3/2013 17:36:00 - 17:54:00

There is currently significant interest in designing nanostructured materials that can interact with cells and regulate key cellular functions. We demonstrate for the first time the design of potent multivalent conjugates that can organize stem cell receptors into nanoscale clusters on the cell membrane and control stem cell fate, both in vitro and in vivo. The ectodomain of ephrin-B2, normally an integral membrane protein ligand, was conjugated to a soluble biopolymer to yield multivalent nanoscale conjugates that potently induced signaling in neural stem cells and promoted their neuronal differentiation both in culture and within the adult brain. Furthermore, synthetic multivalent conjugates of ephrin-B1 strongly enhanced human embryonic and induced pluripotent stem cell differentiation into functional dopaminergic neurons. Multivalent bioconjugates thus represent a powerful tool for controlling stem cell fate in vitro and in vivo. These results also highlight the potential of nanoscale therapeutics that interact with target cells and elicit desired cellular responses.

(7a) Multifunctional DNA Nanostructures for Therapeutic Delivery
Charoenphol, P., University of Michigan
Bermudez, H., University of Massachusetts Amherst
Originally presented on: 11/3/2013 15:30:00 - 15:48:00

DNA nanostructures offer several advantages over conventional therapeutic carriers such as liposomes and polymeric particles, including biocompatibility, biodegradability, and low immunogenicity. Furthermore, a simple one-pot self-assembly fabrication process results in nearly monodisperse particle sizes with controllable spatial location of ligands. We have previously shown that these DNA nanostructures are resistant to nuclease degradation and can be used to deliver antisense DNA to cells. However, these first-generation carriers lacked target specificity and demonstrated limited cell uptake.

In this work, self-assembled DNA tetrahedron nanostructures bearing multivalent DNA aptamers are shown to selectively target HeLa cancer cells and trigger efficient intracellular uptake, as determined by flow cytometry. The improved uptake relative to naked DNA nanostructures and linear DNA relies on aptamer multivalency and is concentration-dependent. In addition, the specific aptamer used in this work has therapeutic activity of its own: it is known to cause the down-regulation of Bcl-2, a protein regulator of apoptosis.

We are currently exploring strategies to induce cancer cell apoptosis through combined antisense and aptamer delivery via DNA tetrahedrons. Our in vitro results demonstrate a synergistic effect due to antisense and aptamer co-delivery and suggest that multifunctional DNA nanostructures could have great potential as therapeutic delivery vehicles.

(7b) Genetically Encoded Gas Nanostructures As Reporters for Non-Invasive Molecular Imaging

Shapiro, M. G., UC Berkeley

Originally presented on: 11/3/2013 15:48:00 - 16:06:00

Many important biological processes – ranging from simple metabolism to complex cognition – take place deep inside living organisms, yet our ability to study them in this context is very limited. Technologies such as fluorescent proteins and optogenetics enable exquisitely precise imaging and control of cellular function in small, translucent specimens using visible light, but are limited by the poor penetration of such light into larger tissues. In contrast, most non-invasive technologies such as magnetic resonance imaging (MRI) and ultrasound – while based on energy forms that penetrate tissue effectively – lack the needed molecular precision. Our work attempts to bridge this gap by engineering new molecular technologies that connect penetrant energy to specific aspects of cellular function *in vivo*.

Here, we descrbe the development of highly sensitive molecular reporters for MRI[1] and ultrasound[2] based on genetically encoded gas nanostructures from buoyant microorganisms. Gas vesicles (GVs) are gas-filled protein-shelled compartments ~200 nm in size that exclude water but are permeable to gas. We show that GVs produce ultrasound contrast based on the echogenicity of their hollow interiors, enabling them to be sensitively detected *in vitro* and *in vivo*. Whereas the physics of conventional microbubble contrast agents dictates sizes larger than one micron and half-lives shorter than a few hours, GVs are nano-sized and inherently stable, enabling a broader range of biological applications. Furthermore, the unique mechanical properties and genetic diversity of GVs enable multiplexed imaging and dynamic biosensing.

In addition, we have adapted GVs as the first genetically encoded reporters for hyperpolarized MRI, a highly sensitive technique that uses non-proton nuclei such as the biocompatible noble gas xenon to achieve orders of magnitude greater molecular sensitivity than conventional ^{1}H-MRI. We show that the gas permeability of GVs enables them to interact with dissolved hyperpolarized xenon and alter its magnetic properties. As a result, GVs can be detected at picomolar concentrations, a 10,000-fold improvement over comparable reporter genes for conventional MRI. We have shown that GVs can be used to detect tumor cells and act as reporters of gene expression. As with ultrasound, we have also taken advantage of the genetic diversity of GVs to enable multiplexed imaging.

Our findings establish a promising new class of molecular reporters for two non-invasive imaging modalities and demonstrate the biochemical engineering potential of a unique family of naturally evolved nanostructures.

The presented work was conducted by the Shapiro group in collaboration with the Schaffer, Pines and Conolly groups at UC Berkeley.

1. Shapiro MG*, Ramirez ME, Sperling LJ, Sun G, Pines A , Schaffer DV, Bajaj VS*. Genetically encoded reporters for hyperpolarized MRI. *In revision.* 2. Shapiro MG*, Goodwill PW, Neogy A, Schaffer DV, Conolly SM. Genetically encoded gas nanostructures as ultrasonic molecular reporters. *In revision.* (*corresponding)

(7c) Nano-Biosensors for Fast, High-Throughput Detection of Antimicrobial Resistance

Casamada, J.,
Nagpal, P., University of Colorado Boulder
Chatterjee, A., University of Colorado Boulder
Originally presented on: 11/3/2013 16:06:00 - 16:24:00

Antibiotics have been an important weapon in the fight against infections for over half a century. However, the extensive use and misuse of antibiotics has resulted in selection and worldwide spread of antibiotic resistant bacteria. Bacteria and other microorganisms that are known to cause infection are remarkably resilient and can develop antibiotic resistance as an outcome of evolution via natural selection. Bacteria acquire antibiotic resistance via spontaneous or induced mutations, or horizontally transferring resistance-conferring genes from one bacterium to the other. Detection and identification of DNA sequences encoding antimicrobial resistance is the key to medical intervention. However, often resistance-causing sequences exist at trace concentrations *in vivo* within a microbial population, and hence their detection is a challenging pursuit. While second generation deep and ultra-deep sequencing methods, and single copy PCR amplification methods are capable of detecting single nucleotide polymorphisms, however, these methods are not applicable to the clinical laboratory setting because of cost and technical complexity. Previously, Scanning tunneling microscopy and spectroscopy (STM and STS respectively) has long been used in physical sciences to study electron tunneling, over angstrom scale distances, to measure electronic density of states and use them as fingerprints in surface science to identify adsorbed species. Recently, several groups have started exploring the use of STM for fast, high-throughput sequencing of nucleotides in the field of biophysics. While this technique can be very useful for fast, single molecule detection and simultaneous sequencing of fast mutating pathogens, notorious for developing drug resistance, only Guanine (G) nucleotides have been identified using this technique so far. Here, we present first demonstration of nano-electronic detection of single DNA strands and sequencing of all bases (A or G,T,C), with nanometer scale precision. As a proof of concept on the DNA sequencing potential using STM and STS, we partially sequenced one copy of single stranded DNA molecule of ampicillin resistance gene with over 95% success rate.

(7d) Integrated Micro Nanotechnology for Sensitive Detection of Cancer Cells in Circulation

Yoon, H. J., University of Michigan
Kim, T. H., University of Michigan
Zhang, Z., University of Michigan
Ebrahim, A.,
Paoletti, C.,
Lin, J.,
Ramnath, N.,
Wicha, M.,
Hayes, D.,
Simeone, D.,
Nagrath, S., University of Michigan

Originally presented on: 11/3/2013 16:24:00 - 16:42:00

Circulating tumor cells (CTCs) are shed from the primary tumor into the peripheral blood. CTCs are emerging as important biomarkers with high clinical relevance. Enumeration of CTCs may have several clinical uses, including determination of prognosis in patients with established malignancy, or even detection of previously undiagnosed cancer. However, due to the limitation of sensitivity and specificity of current technologies for CTC isolation, the full potential of CTCs has yet to be realized. Emerging microfluidic technologies are promising for isolating CTCs with a high yield; however, these platforms often have three-dimensional structures, thus lacking the advantages of traditional planar surface and limiting further characterization and expansion of cells on the chip. Here we describe a new approach to more effectively isolate CTCs incorporating the nanomaterial, graphene oxide (GO). Using self-assembly of GO on a patterned gold surface, we create islands of nano-arms for sensitive CTC capture without the aid of three dimensional posts or structures. The capture is truly planar and we demonstrate the isolation of cells with high sensitivity even at low frequency of target cells (75.22% ±29.46 at 5 cells/mL spiked in blood). This 2D planar GO-chip provides a maximum recovery rate of 95% and a stable mean capture efficiency of 82% up to 3 mL/hr sample processing rate. The cells are specifically captured on the planar islands of nano-arms and are able to proliferate after capture when placed in culture conditions in the chip. We further show that the GO-chip is able to capture CTCs from blood samples collected from pancreatic, breast and lung cancer patients. In summary, we present here a novel integrated nano microfluidic technology using GO for sensitive planar capture of CTCs.

(7e) Single-Cell Level Study of Precise siRNA Delivery and Dynamic in Acute Myeloidleukemia Cells By Nanochannel Electroporation System
Gao, K., The Ohio State University
Originally presented on: 11/3/2013 16:42:00 - 17:00:00

Single-Cell Level Study of Precise siRNA Delivery and Dynamic in Acute Myeloidleukemia Cells by Nanochannel Electroporation System

Keliang Gao[a], Xiaomeng Huang[a,b], Lei Li[a], Lingqian Chang[a], Xinmei Wang[a], Guido Marcucci[b], and L. James Lee[a,c *]

[a] Nanoscale Science and Engineering Center for Affordable Nanoengineering of Polymeric Biomedical Devices

[b] Department of Internal Medicine

[C] Department of Chemical and Biomolecular Engineering

The Ohio State University, Columbus, Ohio 43210, USA

[*]Corresponding Author's E-mail: lee.31@osu.edu

A micro/nano-fabrication process of a nanochannel electroporation (NEP) biochip and its application for precise delivery of siRNA for non-viral gene transfection was described to study the dynamic of siRNA in leukemia cells. A novel dip coating process was optimized to produce DNA nanowires across a microridge array patterned on the polydimethylsiloxane (PDMS) surface with a yield up to 95%. Molecular imprinting based on a low viscosity resin, 1,4-butanediol diacrylate (1,4-BDDA), was adopted to convert the microridge-nanowire-microridge array into a microchannel-nanochannel-microchannel array. Secondary machining by the femtosecond laser ablation was applied to shorten one side of microchannels from 3000 to 50 μm to facilitate cell loading and unloading. The biochip was then sealed in a packaging case with reservoirs and microfluidic channels to enable cell and plasmid loading, and to protect the biochip from leakage and contamination. The package case could be opened for cell unloading after NEP to allow for the follow-up cell culture and analysis. To demonstrate its application, NEP was applied to deliver precise dosage of Mcl-1 siRNA to single acute myeloid leukemia cells. Various amount of siRNA was injected into the cells through different NEP conditions, and critical thereshold was established by checking cell viability. Mcl-1 mRNA was down-regulated by siRNA, shown by molecular beacon detection. Single cell PCR was adopted to illustrate intracytoplasmic siRNA copy number. Two different kinds of leukemia cells, KG1a as wild type FLT3 and MV4-11 as FLT3-ITD, were studied for comparison. The results established different thereshold for wild type and mutant cells due to apoptosis resistance, and showed that more siRNA were required for apoptosis of aggressive FLT3-ITD cells. Dynamic analysis was revealed with different dosage of siRNA, and illustrated the intracytoplasmic mechanism of down-regulation and recovery.

(7f) Nanoscale Roughness and Surface Charge of Functionalized Halloysite Nanotubes Control Selectin-Mediated Adhesion of Malignant and Non-Malignant Cells Under Flow

Mitchell, M., Cornell University
Castellanos, C., Cornell University
King, M. R., Cornell University
Originally presented on: 11/3/2013 17:00:00 - 17:18:00

Nanoscale Roughness and Surface Charge of Functionalized Halloysite Nanotubes Control Selectin-Mediated Adhesion of Malignant and Non-Malignant Cells Under Flow

Michael J. Mitchell[1], Carlos A. Castellanos[1], Michael R. King[1]

[1]Department of Biomedical Engineering, Cornell University, Ithaca, New York USA

Introduction: The separation of circulating tumor cells (CTCs) in large numbers and at high purity levels from patient blood could lead to the development of effective personalized medicine regimens for those with metastatic cancer. Our lab has recently developed microscale flow devices containing nanostructured surfaces of halloysite nanotubes (HNTs) and recombinant human E-selectin (ES) to both isolate CTCs from patient blood and deliver targeted chemotherapeutics to cancer cells (1-3). Improvement of current CTC isolation purity levels is challenged by the fact that both CTCs and leukocytes possess ligands for ES (4). Herein, we summarize our study of the role of nanoscale roughness and surface charge of functionalized HNTs in differentially recruiting and capturing CTCs and leukocytes from flow.

Materials and Methods: HNTs were functionalized by incubation with sodium dodecanoate (NaL) or decyltrimethylammonium bromide (DTAB). Untreated, NaL- and DTAB-functionalized HNTs were immobilized within microscale flow devices, followed by incubation with ES. HNTs and immobilized HNT surfaces were characterized using dynamic light scattering, AFM, SEM, fluorescence microscopy, and contact angle goniometry. Cancer cell lines (COLO 205 and MCF7) and human neutrophils isolated from peripheral blood were perfused through flow devices via a motorized syringe pump and exposed to physiologically relevant wall shear stresses. Micrographs and videos of adhesion phenomena were acquired to quantify cell rolling velocity, rolling flux, and the number of cells captured.

Results and Discussion: Adsorption of NaL surfactant into the HNT inner lumen increased the net negative HNT charge, which enhanced the electrostatic repulsion and thus the colloidal stability of the HNT solution. NaL treatment altered the HNT average negative zeta potential from -25.1 to -103.4 mV (Fig. 1A), compared to the average negative zeta potential for untreated HNTs. Conversely, the HNT average negative zeta potential was reduced to -4.0 mV upon treatment with DTAB. The number of cancer cells captured from flow increased on surfaces coated with NaL-functionalized HNTs and ES (Fig. 1B). Interestingly, the number of leukocytes captured from flow decreased on identical surfaces (Fig. 1B). Cancer cell capture significantly increased on nanostructured surfaces of increasing net negative charge, while leukocyte capture significantly decreased (Fig. 1C).

Conclusions: We have shown that surfaces of nanoscale roughness and altered surface charge significantly affect the differential adhesion of cancer cells and leukocytes to ES under flow. While both cancer cells and leukocytes possess adhesion receptors for ES, an increase in the net negative HNT charge acts to repel leukocytes from interaction with ES, while *simultaneously* enhancing cancer cell ES-mediated adhesion. The combination of both nanoscale roughness and HNT surface charge can serve to increase both the number and purity of CTCs isolated from patient blood, which can enable the development of effective personalized cancer therapies.

Acknowledgements: The work described was supported by the Cornell Center on the Microenvironment and Metastasis through Award Number U54CA143876 from the National Cancer Institute.

Figure 1. (A) Schematic of HNT functionalization and immobilization within a microscale flow device. (B) Reduced adhesion of leukocytes and enhanced adhesion of cancer cells from flow using a microscale flow device coated with NaL-functionalized HNTs and ES. Scale bar = 150 um. (C) Number of leukocytes and COLO 205 cancer cells captured from flow per 180,000 um^2 using functionalized HNT surfaces coated with ES. ***P<0.001. NS: not significant.

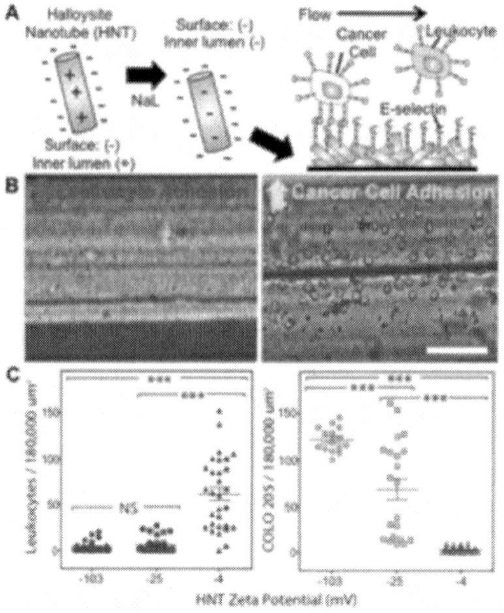

References:

1. Hughes AD, Mattison J, Western LT, Powderly JD, Greene BT, King MR. Microtube Device for Selectin Mediated Capture of Viable Circulating Tumor Cells from Blood. *Clinical Chemistry*. 2012 Feb 16;58:846–53.

2. Mitchell MJ, Chen CS, Ponmudi V, Hughes AD, King MR. E-selectin liposomal and nanotube-targeted delivery of doxorubicin to circulating tumor cells. *Journal of Controlled Release*. 2012 Jun;160:609–17.

3. Hughes AD, King MR. Use of Naturally Occurring Halloysite Nanotubes for Enhanced Capture of Flowing Cells. *Langmuir*. 2010 Jul 20;26(14):12155–64.

4. Mitchell MJ, Castellanos CA, King MR. Nanostructured Surfaces to Target and Kill Circulating Tumor Cells While Repelling Leukocytes. *Journal of Nanomaterials*. 2012;2012(3):1–10.

(7g) Nanoparticles As Autophagy Activators

Song, W., Rice University
Lee, S. S., Rice University
Savini, M.,
Segatori, L., Rice University
Originally presented on: 11/3/2013 17:18:00 - 17:36:00

Autophagy is the main catabolic pathway in mammalian cells that mediates degradation of cytoplasmic material through the lysosomes. Deficiencies in this fundamental homeostatic process result in accumulation of cellular waste material, which is associated with development of a range of human diseases. A number of nanomaterials induce autophagy—most likely as a response to the presence of what the cell perceives as foreign of aberrant material. While activation of autophagy may lead to enhanced clearance of waste material, it can also induce activation of cell death programs. Moreover, we found that nanoparticles of specific charge and composition can activate the autophagic response but also impair cellular components that mediate degradation, ultimately blocking autophagic flux. Thus, effective strategies aimed at enhancing cellular clearance should not only activate autophagosome formation to compartmentalize toxic material, but also promote autophagosome clearance by the lysosomes. Evidence of co-regulated roles of lysosomes and autophagosomes emerged from the recent discovery of an overarching regulatory gene network (CLEAR, Coordinated Lysosomal Expression and Regulation) and its master regulator, the transcription factor EB (TFEB). Here, we propose a strategy aimed at activating autophagy and promoting cellular clearance using nanoparticles that could potentially impact a number of human diseases associated with autophagic deficiencies. We investigated the effect of physicochemical properties of a number of nanoparticles, namely geometric parameters, charge, and composition, on downstream effects associated with activation of autophagy, namely clearance of toxic autophagic substrates. In summary, this study provides a mechanistic understanding of the interaction of nanomaterials with the lysosome-autophagy system and identifies the design rules to engineer nanoparticles that enhance clearance of toxic cellular waste for therapeutic applications.

(7h) Dendrimer-Based Therapeutic Delivery for Cerebral Palsy: Mechanism of Nanoparticle Uptake

Kannan, R., Johns Hopkins School of Medicine
Nance, E.,
Zhang, F.,
Mishra, M., Johns Hopkins School of Medicine
Balakrishnan, B., Johns Hopkins School of Medicine
Kannan, S., Johns Hopkins School of Medicine
Originally presented on: 11/3/2013 17:36:00 - 17:54:00

Cerebral palsy is a chronic childhood disorder that currently has no effective cure. Injury to the developing brain that occurs either in utero or soon after birth can result in motor, sensory and cognitive deficits as seen in cerebral palsy, with neuroinflammation playing a key role in the pathophysiology. Nanoparticles such asdendrimers provide opportunities for the targeted delivery of multiple drugs that can mitigate pathways involved in injury and can be delivered selectively to cells that are responsible for neuroinflammation and injury. We have previously demonstrated that a systemic, postnatal therapy with dendrimer-drug nanodevices in rabbit kits with CP,can cross the disrupted BBB and localize in activated microglia and astrocytes, resulting in an improvement in motor deficits (1).

Understanding factors affecting uptake will allow for better design and more efficient delivery of dendrimer platforms to diseased cells, further increasing therapeutic efficacy. In this work, we used a combination of in vivo and ex vivo methods to study the effect of dendrimer size (generation 4 (G4), ~4nm in size) and surface charge (-OH terminated neutral vs -NH2 terminated cationic dendrimers) on the ability to cross an impaired BBB, diffuse within the brain parenchyma, and selectively uptake in microglia cells in rabbit kits exposed to endotoxin in utero. In brain slices from endotoxin kits, the neutral G4-OH-Cy5 dendrimer could diffuse through the brain parenchyma, preferentially accumulating in microglia, whereas the cationic G4-NH2-FITC dendrimers did not diffuse in slices. Following intracranial administration into an endotoxin day 1 kit, G4-OH dendrimer was found distributed in the parenchyma at 4 hrs and 24 hrs, and localized in microglial cells at 24 hrs. In age-matched healthy controls, minimal uptake in microglia was seen in either white matter or cortical regions; however, dendrimer was spread throughout the parenchyma at 24 hours confirming the diffusive nature of these particles. Following intravenous (IV) administration, G4-OH-Cy5 dendrimer penetrate a disrupted BBB in endotoxin kits and were found within the parenchyma at 4 hours, and within microglial cells at 24 hours. Polystyrene nanoparticles (20nm) did not cross the BBB or selectively uptake into microglia cells. Dendrimer was not found in the brain in age-matched healthy controls following systemic administration. These results suggest that appropriate tailoring of the physicochemical properties of the dendrimer nanodevice to optimize targeting and localization can lead to improvement in drug delivery to the brain.

1. S Kannan, H Dai, RS Navath, B Balakrishnan, A Jyoti, J Janisse, R Romero, RM Kannan (2012). 'Dendrimer-based postnatal therapy for neuroinflammation and cerebral palsy in a rabbit model'. *Science Translational Medicine*, 4(130), p. 130ra46. Highlighted in *Nature*, *Science*, *Nature Review Drug Discovery*.

(9a) Selective Recovery From Interstitial Spaces of Leaf Tissue

Kingsbury, N., University of California, Davis
McDonald, K. A., University of California, Davis

Originally presented on: 11/3/2013 15:30:00 - 15:48:00

Title: Selective Recovery of Apoplast-Targeted Recombinant Proteins from Plant Leaf Tissue

Nathanial J. Kingsbury and Karen A. McDonald

Department of Chemical Engineering and Materials Science

University of California, Davis, CA 95616

Transient production of recombinant proteins in plant leaf tissue using vacuum infiltration with recombinant *Agrobacterium tumefaciens*has advantages over other approaches including speed, scalability, safety and cost-effectiveness. This method is being considered for large scale production of vaccines, biodefense agents and industrial enzymes and other products requiring a eukaryotic host. Recovery of functional protein from leaf tissue such as tobacco using traditional methods of homogenization and extractioncan be challenging due to high amounts of fiber, phenolics and proteases. Targeting the product to the apoplast has advantages in terms of product quality, particularly glycoform type and uniformity, as well as ease of product recovery, since the secreted product can be removed from the leaf tissue without tissue disruption. To recover transiently expressed apoplast-targeted recombinant proteins, a vacuum infiltration-centrifugation methodis applied to *Nicotiana benthamiana* and *Nicotiana tabacum* leaves. Leaves are submerged in a buffer solution and a moderate vacuum is applied for several minutes and then quickly released, allowing the buffer solution to infiltrate the interstitial spaces of the leaf. The leaf is then centrifuged to remove the liquid which is enriched with components from the apoplast and cell wall matrix. The collected fluid, called the apoplast wash fluid (AWF), is highly concentrated up to 16-fold and purified as much as 50-fold compared to a homogenate extract. Multiple rounds of vacuum infiltration-centrifugation, applied to further remove secreted protein from the residual tissue, can result in recovery ofover 90% of the secreted protein of interest. The ratio of air phase to liquid phase in the interstitial space, called the dilution factor, is calculated and related to the rate of recovery from each round of vacuum infiltration-centrifugation.

Leaves subjected to the vacuum infiltration-centrifugation method showed no signs of damage after recovery of AWF and continued producing the recombinant protein of interest. For example, when leaves expressing a recombinant cellulase enzyme, endoglucanase catalytic domain (E1cd) from *Acidothermus cellulolyticus*, were subjected to daily rounds of vacuum infiltration-centrifugation starting at two days post-agroinfiltration, the overall amount of E1cd collected fromthe leaves after six days post-agroinfiltration was as much as four times greater than in leaves where the E1cd was allowed to accumulate in the apoplast unperturbed. Possible mechanisms for this improvement in expression level are improved stability of the E1cd in the AWF compared to the leaf apoplast or improved secretion kinetics upon diluting the E1cd concentration in the apoplast and by extension the endomembrane system as a whole. The results therefore demonstrate a novel technique for boosting expression levels in transgenic leaf tissue.

Out of the model proteins used these studies, those with smaller molecular weights were recovered with superior yields, suggesting that the porosity of the cell wall matrix may be limiting removal by the vacuum infiltration–centrifugation method. Pretreatment of the leaves with low concentrations of pectinase enzyme improved the yield and concentration of an 82kDa recombinant protein, human butyrylcholinesterase, four-fold while improving its protein purity two-fold in the AWF. These results demonstrate that even relatively large proteins can be efficiently recovered using the vacuum infiltration-centrifugation method.

(9b) Control of Intein-Mediated Recombinant Protein Purification In a Mammalian Cell Expression System

Han, T. C., The Ohio State University
Originally presented on: 11/3/2013 15:48:00 - 16:06:00

Control of Intein-mediated Recombinant Protein Purification

In a Mammalian Cell Expression System

Tzu-Chiang Han

Dr. David W. Wood

Inteins are genetic elements that can self-excise from various host proteins in a post-translational protein splicing process. This activity can be modified by a simple mutation to yield isolated, controllable cleaving activity at either the N- or C-terminus of the intein. By combining self-cleaving inteins with conventional affinity tags, self-cleaving affinity tags have been developed. Once the tagged target protein is purified, the affinity tag and intein components are removed easily through the controlled self-cleavage of the intein without adding any protease. Although this efficient tag-removal method has been successfully applied in prokaryotic cell systems in our previous studies, the *in vivo* premature cleaving of the precursor protein has been a serious drawback for this method in eukaryotic cell systems. Here we present our most recent work in developing a next-generation intein for use in mammalian cells. In particular, we have engineered a new intein with enhanced control, where the control mechanism is compatible with our mammalian cell culture. This intein has been tested with several conventional affinity tags in mammalian cell culture, and the results will be presented. By controlling the *in vivo* premature cleaving in a mammalian cell culture context, we have generated a potential platform technology for the manufacture of non-antibody glycoprotein therapeutics at large scale. Additionally, this intein has been modified for high-throughput applications in drug and target discovery and development.

(9d) A Potential Platform Technology for Antibody Purification: Mixed-Mode Chromatography

Lin, D. Q., Zhejiang University
Yao, S. J., Zhejiang University
Zhang, Q. L., Zhejiang University
Tong, H. F., Zhejiang University
Wang, R. Z., Zhejiang University
Lu, H. L., Zhejiang University
Originally presented on: 11/3/2013 16:24:00 - 16:42:00

Mixed-mode chromatography (MMC) is a novel technology for bioproduct separation, especially antibodies. Mixed-mode ligands typically contain a combination of multiple binding modes like hydrophobic interaction, electrostatic forces, hydrogen bonding, etc.. High capacity, salt-tolerance, good stability and relatively low cost are the major advantages of MMC for direct capture processes. However, compared with Protein A capture, the separation selectivity of MMC for antibodies is still limited.

Based on series of tests with MMC using IgG, IgY and IgY(ΔFc), it was found that the control of loading pH or salt addition in the buffer was critical to improve antibody purification from albumin containing feedstocks. The control of caprylate addition might be another effective way for improving antibody purification. High purity and recovery can be obtained after the optimization of operation conditions. To better understand the ligand-protein interactions, a molecular simulation approach was established with combinative methods of molecular docking and dynamic simulation. The results indicated that at neutral conditions the ligand can bind onto some special sites of IgG surface which showed strong hydrophobicity. When the pH changed, the ligand departed quickly from the protein surface due to electrostatic repulsion. In addition, the ligand-net was constructed to simulate the protein adsorption on the porous surface of chromatographic resins. The MM/PBSA method was used to acquire the binding energy map which was used to scan the most suitable binding orientations of IgG on the ligand-net. The effects of ligand structure, ligand cluster, gel matrix and liquid conditions were also studied. The simulation results could provide useful information on the process development and the design of novel ligands for MMC.

For biopharmaceuticals such as monoclonal antibodies (mAbs), downstream processes require highly selective and robust technologies to achieve extremely high purity. The complexity of the interactions between the mixed-mode ligand and the desired protein might be considered as a challenge to improve the adsorption selectivity. Due to the flexibility of operation development and the innovative design of novel ligands, MMC is likely to become a potential platform for antibody purification with high efficiency and low cost.

* This work was supported by National Natural Science Foundation of China and the Zhejiang Provincial Natural Science Foundation of China.

(9e) Modeling the Synergies of Saline and Organic Modifiers On Rpc Separation

Johansson, K., Lund University
S ndergaard Frederiksen, S.,
Degerman, M., Lund University
Breil, M. P., Technical University of Denmark
Mollerup, J.,
Nilsson, B., Lund University
Originally presented on: 11/3/2013 16:42:00 - 17:00:00

During the last decades, great advances have been made regarding both modeling and understanding of preparative ion exchange chromatography (IEC). The Langmuir mobile phase modifier (MPM) [1] and the steric mass action (SMA) [2] models each give an accurate description of the effect of the salt concentration and recent developments in high throughput screening [3] as well as molecular dynamics simulations [4] have greatly increased the knowledge of the underlying mechanisms. In comparison, the chromatographic modes based on hydrophobic interactions, i.e. hydrophobic interaction chromatography (HIC) and reversed-phase chromatography (RPC), have been left in the shade. The complexity of the interaction between fatty ligands and hydrophobic patches on proteins has made it elusive. Although some serious attempts have been made, e.g. using a thermodynamic approach [5], many aspects of these hydrophobic phenomena are still not well understood.

As a step towards a better understanding of RPC and a model that applies for both HIC and RPC, an RPC separation of three insulins has been studied. This is a continuation of a project which was initiated with a study of the corresponding HIC separation. In both these cases, the effects of salt as well as organic modifiers have been investigated. This study included two different resins with C4 and C18 ligands, respectively, both on silica backbone.

The linear range of the isotherm was mapped by isocratic pulse experiments, performed at various concentrations of salt and organic modifier. The concentration of organic modifier was limited by nature; either by the lack of retention or by the salt solubility at the higher end, and by low UV signal-to-noise ratio, due to very high retention, at the lower end. For either resin, a series of pulse experiments at varying load within the non-linear range was performed to investigate the capacity properties.

Based on the experimental data, a mechanistic model of the system is developed. Since the resin particles are rather small and since the adsorption kinetics are generally rate-controlling for RPC, the column is described by a reaction dispersive model. The adsorption is described by a combination of thermodynamics [5] and the Langmuir MPM model, with adjustments which account for the observed effects of the modifier concentrations. The model parameters related to these effects as well as Henry's constant were calibrated against the retention data from the linear range experiments. The model parameters concerning capacity and kinetics were tuned by hand to make the peak shape fit the overloaded and linear range experiments, respectively.

The final model can be used as a tool to design new separation processes, as well as to optimize already existing ones. It is also useful in robustness studies and for filing purposes. Apart from the obvious practical benefits, it provides a new step of insight into the underlying phenomena and gives a deeper understanding of the process.

References

1. Melander, W.R., S. El Rassi, and C. Horváth, *Interplay of hydrophobic and electrostatic interactions in biopolymer chromatography: Effects of salts on the retention of proteins.* J. Chromatogr., 1989. **469**: p. 3-27.

2. Brooks, C.A. and S.M. Cramer, *Steric Mass-Action Ion Exchange: Displacement Profiles and Induced Salt Gradients.* AIChE J., 1992. **38**(12): p. 1969-1978.

3. Nfor, B.K., et al., *Rational and systematic protein purification process development: the next generation.* Trends Biotechnol., 2009. **27**(12): p. 673-679.

4. Holstein, M.A., et al., *Effects of Urea on Selectivity and Protein−Ligand Interactions in Multimodal Cation Exchange Chromatography.* Langmuir, 2012. **29**(1): p. 158-167.

5. Mollerup, J.M., *A Review of the Thermodynamics of Protein Association to Ligands, Protein Adsorption, and Adsorption Isotherms.* Chem. Eng. Technol., 2008. **31**(6): p. 864-874.

(9f) Novel Thermo-Responsive Fucose Binding Ligands for Glycoprotein Purification By Affinity Precipitation

Arnold, L., Georgia Tech
Originally presented on: 11/3/2013 17:00:00 - 17:18:00

Carbohydrates are the most abundant organic molecules found in nature, and they are metabolized and synthesized by nearly all organisms. Recently, focus in understanding the structure and function of glycoproteins has expanded into the rapidly growing field of glycobiology. However, the complexity and microheterogeneity of these oligosaccharide chains have made identification, isolation, and analysis very difficult. To aid in this demanding field, sugar binding proteins, or lectins, have emerged as critical tools in glycobiology. Lectins and lectin-containing biomaterials have been used in separation and identification, such as lectin affinity chromatography. However, while amenable at bench scale for glycoprotein separation, scale up of this chromatographic process is notably inefficient and expensive. Column fouling, diffusion limitations, and flow rate constraints are just some of costly disadvantages of this time consuming process.

To address these problems, we designed a recombinant fusion protein that potentially could be used for affinity precipitation purification of glycoproteins without chromatography. This protein has dual functionality: a fucose binding bacterial lectin from *Ralstonia solanacearum* (RSL) for carbohydrate affinity, and an elastin-like protein (ELP) thermo-responsive biopolymer for precipitation. Affinity separation will be achieved by simple inverse temperature cycling (ITC), which consists of temperature oscillation, followed by recovery of the precipitate via filtration or centrifugation. Separation may also be achieved by fluctuation of the salt concentration, maintaining the integrity of possible target enzymes. Furthermore, a newly developed extracellular secretion method for high expression level in *E. coli* is utilized to establish a cost effective process for production and isolation of this protein, achieving a secretion yield of 35 mg/L over a 48 hour cultivation period.

The presentation will detail the design, overexperession and characterization of several lectin-ELP fusions. The successful application of glycoprotein purification using these novel proteins will be demonstrated with a model protein, horseradish peroxidase (HRP), which contains α1,3-fucosylated glycans. Isolation of this target glycoprotein from a protein mixture in an affinity precipitation process resulted in a pure glycoprotein and a recovery yield of 88%.

(9g) Elastin-Like Polypeptide Tag Length Effect On Protein Expression and Purification

Shakalli Tang, M., The Ohio State University
Coolbaugh, M. J., The Ohio State University
Lease, R. A., Ohio State University
Wood, D. W., the Ohio State University
Originally presented on: 11/3/2013 17:18:00 - 17:36:00

Recombinant protein technology has enabled the expression of fusion proteins that can be purified in bacterial systems with the use of purification tags. In particular, the elastin-like polypeptide (ELP) precipitation tag in the presence of a self-cleaving intein protein domain allows the non-chromatographic purification of a target protein with only a change in salt concentration and pH. Shortening the ELP tag length has the potential to increase expression by freeing the energy that would be required to synthesize the large ELP repetitive protein sequence. An increase in production yield of a given protein would thus be expected when decreasing the ELP tag length. Furthermore, the optimal ELP tag length needed may be affected by the size and solubility of the protein product fused with the tag. In this study, the effect of five different ELP tag sizes in product yield was examined in the expression and purification of several proteins, including β-galactosidase and β-lactamase. Sodium dodecyl sulfate polyacrylamide gel electrophoresis (SDS-PAGE) qualitatively suggested the success of each purification. The corresponding activity and Bradford assays were then used to confirm these results quantitatively. Finally, studies including a wider range of salt concentrations and a wider selection of target proteins with varying hydrophobicity can help elucidate the relationship between the target protein fused to the ELP-intein system and the ELP tag length required for its successful purification.

(9h) High-Throughput Bioprocess Development for Polymer-Polymer Aqueous Two-Phase Systems

Sorensen, E., University College London
Bracewell, D. G., University College London
Patel, N.,
Originally presented on: 11/3/2013 17:36:00 - 17:54:00

The market for therapeutic proteins is currently increasing at a remarkable pace. In many cases, therapeutic proteins, such as monoclonal antibodies, must be administered in relatively large dosages. This large dosage requirement (> 100 mg $dose^{-1}$) (Shire et al., 2004), coupled with the high demand for these drugs, has resulted in an increased pressure on pharmaceutical companies to deliver high quality drugs at sufficient quantities. As a result, there has been an increasing amount of research & development (R&D) aimed at improving upstream efficiency, which has resulted in increased cell titres during the cell culture/fermentation stage of protein production. To date, improvements in upstream technology have, however, not been matched by improvements in downstream technology. Consequently expensive chromatography columns must be scaled accordingly to cope with the increasing amounts of material generated, which is proving to be a process bottleneck due to their capacity and cost, and has highlighted the need to re-evaluate current downstream separation methodologies.

Aqueous two-phase extraction (ATPE) has been shown to be a promising alternative downstream separation technology (Rito-Palomares, 2004). Systems can be formed from combining two incompatible polymers or a polymer and salt above a critical composition. Although having many advantages such as low surface-tension, easy scalability and a gentle environment for biopharmaceuticals, there is still a lack of widespread use of such systems. This can be attributed in part to the difficulty involved in characterising the large experimental space required for optimisation and effective process development. Factors such as charge, hydrophobicity, solute affinity, polymer molecular weight and concentration all play an important role on partitioning in ATPE (Asenjo and Andrews, 2011) and characterising all of the above factors manually is very resource intensive.

The high-throughput characterisation of polymer-salt aqueous two-phase systems has been explored (Oelmeier et al., 2011, Bensch et al., 2007) however, polymer-polymer aqueous two-phase systems behave somewhat differently and have so far not been considered although they have a clear potential for high separation efficiency. To help address this issue, we present an approach that utilises automated liquid handling systems for the high-throughput determination of polymer-polymer phase diagrams and protein partition behaviour. Such an approach allows for rapid exploration of the large experimental space provided by such systems, which in turn will allow for easier and more rapid development of optimised separation systems. Conventional techniques used to determine the binodial curve can be cumbersome and time consuming. An automated approach for polymer-polymer aqueous two-phase systems based on the determination of cloud point by light obscuration measurements is presented, and demonstrated for a system of PEG and Dextran. In addition, phase diagram tie-lines are generated by integration of automated phase volume determination with the lever arm rule. Measurement of protein partitioning is determined by measurements of UV absorbance at 280 nm with an appropriate control system as a blank.

The developed methodology will allow for screening of the wide variety of factors known to affect aqueous two-phase systems, such as polymer molecular weight, salt type, salt concentration, pH etc., thereby allowing the potential benefits of ATPE to be more readily explored. Future work would include the integration of more sophisticated assays into the high-throughput methodology to characterise components in complicated feed material such as impurities and products of interest. In addition, the integration of the experimental methodology with mechanistic models via parameter estimation is of particular interest as this would further speed up process development.

Acknowledgement:

We gratefully acknowledge the part-funding and support from BioMarin.

References:

Asenjo, J. A. & Andrews, B. A. 2011. Aqueous two-phase systems for protein separation: A perspective. *Journal of Chromatography A,* 1218, 8826-8835.

Bensch, M., Selbach, B. & Hubbuch, J. 2007. High throughput screening techniques in downstream processing: Preparation, characterization and optimization of aqueous two-phase systems. *Chemical Engineering Science,* 62, 2011-2021.

Oelmeier, S. A., Dismer, F. & Hubbuch, J. 2011. Application of an Aqueous Two-Phase Systems High-Throughput Screening Method to Evaluate mAb HCP Separation. *Biotechnology and Bioengineering,* 108, 69-81.

Rito-Palomares, M. 2004. Practical application of aqueous two-phase partition to process development for the recovery of biological products. *Journal of Chromatography B-Analytical Technologies in the Biomedical and Life Sciences,* 807, 3-11.

Shire, S. J., Shahrokh, Z. & Liu, J. 2004. Challenges in the development of high protein concentration formulations. *Journal of Pharmaceutical Sciences,* 93, 1390-1402.

(12a) Transcription Activator-Like Effector Hybrids for Conditional Control and Rewiring of Chromosomal Transgene Expression

Li, Y., The University of Texas at Dallas
Moore, R., The University of Texas at Dallas
Guinn, M.,
Bleris, L., University of Texas at Dallas
Originally presented on: 11/3/2013 15:30:00 - 15:48:00

The ability to conditionally rewire pathways in human cells holds great therapeutic potential. Transcription activator-like effectors (TALEs) are a class of naturally occurring specific DNA binding proteins that can be used to introduce targeted genome modifications or control gene expression. Here (1) we present TALE hybrids engineered to respond to endogenous signals and capable of controlling transgenes by applying a predetermined and tunable action at the single-cell level. Specifically, we first demonstrate that combinations of TALEs can be used to modulate the expression of stably integrated genes in kidney cells. We then introduce a general purpose two-hybrid approach that can be customized to regulate the function of any TALE either using effector molecules or a heterodimerization reaction. Finally, we demonstrate the successful interface of TALEs to specific endogenous signals, namely hypoxia signaling and microRNAs, essentially closing the loop between cellular information and chromosomal transgene expression.

1. Yi Li, Richard Moore, Michael Guinn, Leonidas Bleris. Transcription activator-like effector hybrids for conditional control and rewiring of chromosomal transgene expression. Sci Rep. 2012;2:897. doi: 10.1038/srep00897. Epub 2012 Nov 28.

(12b) Exploiting Post-Transcriptional Regulatory Interactions for in Vivo RNA Characterization

Sowa, S., University of Texas at Austin
Vazquez-Anderson, J., University of Texas at Austin
Contreras, L. M., University of Texas at Austin
Originally presented on: 11/3/2013 15:48:00 - 16:06:00

"Exploiting Post-transcriptional Regulatory Interactions for *in vivo* RNA Characterization"

Sowa, S.*; Vazquez-Anderson, J.*; Contreras, L.

McKetta Department of Chemical Engineering, The University of Texas at Austin, 200 E. Dean Keeton Blvd., CO400 Austin, Texas 78712-1589

Microbiology Graduate Program, University of Texas at Austin, 100 E. 24th Street, STOP A6500 Austin, Texas 78712

*These authors contributed equally to these work

RNA structure has been extensively studied using a variety of *in vitro* and *in silico* techniques (e.g. crystallography, NMR, footprinting, comparative analysis, etc.). However, to date, only a few RNA structural characterization techniques have been used within an intracellular environment. In addition, although *in vitro* and *in vivo* RNA folding share some basic features, is still unclear how RNA folding *in vitro* compares to *in vivo* (Schroeder, 2002). In this study, we exploit molecular interactions that trigger activation of GFP to engineer a controllable riboswitch-inspired system capable of detecting structural perturbations in RNAs that result from sequence mutations. In this reporter system, expression of GFP is tuned by controlling local structural changes in an RBS and a cis-blocking element (cb) containing-loop. We explore different riboswitch conformations in the context of their interactions with the *tetrahymena* ribozyme Group I Intron (GI). Flow cytometry data indicates the potential of this method to detect differences between RNA variants and RNA conformers. We will discuss various design features of the system, and design algorithms that we have now established for optimal sensitivity. We will also discuss potential mechanisms that describe specific interactions between the target RNA and the designed RNA responsive elements that leads to detection of structural changes.

(12c) The Riboswitch Calculator: De Novo Design of Synthetic Cis-Acting Riboswitches From Ligand-Binding Aptamers

Espah Borujeni, A., Pennsylvania State University
Mishler, D.,
Gallivan, J. P., Emory University
Salis, H., Pennsylvania State University
Originally presented on: 11/3/2013 16:06:00 - 16:24:00

Riboswitches are structured 5' untranslated regions (UTRs) that regulate gene expression in response to changing ligand concentrations. As cellular sensors, riboswitches have the ability to detect toxic chemicals, trigger the in vivo production of therapeutics, or report the cell's internal metabolic state for metabolic engineering efforts. Currently, the majority of engineered riboswitches respond to theophylline or similarly structured ligands; however, thousands of ligand-binding mRNAs, called aptamers, have been generated through in vitro selection methods. The great challenge of riboswitch engineering has been to convert these aptamers into riboswitches capable of regulating gene expression with practically useful dynamic ranges.

We have developed and experimentally validated a computational design method, the Riboswitch Calculator, that designs synthetic riboswitches capable of binding to diverse chemical ligands (3+) and activating translation rates by 10- to 100-fold. We combine a predictive statistical thermodynamic model of translation initiation -- the RBS Calculator v1.2 -- with known aptamer sequences to identify synthetic 5' UTRs that will undergo targeted RNA shape changes to maximally regulate translation initiation in response to ligand binding events. Importantly, no additional parameterization was necessary to predict translation rates in response to ligand binding events.

We have employed both in vitro and in vivo expression assays to measure changes in translation rates and to identify the key rate-limiting steps in ligand-dependent riboswitch activation. We have identified the key reasons why some riboswitches fail to regulate gene expression, and developed quantitative criteria to discard such sequences during computational optimization. We also identify instrinsic limitations to utilizing certain aptamer sequences. The Riboswitch Calculator provides a quantitative and falsifiable approach to engineering synthetic riboswitches with predictable behaviors. It has also provided an fruitful avenue for further advancing our understanding of RNA biophysics inside cells.

(12d) Developing RNA-Based Genetic Control Systems to Quantitatively-Program Functions in Engineered Metabolic Pathways

Carothers, J.,

Originally presented on: 11/3/2013 16:24:00 - 16:42:00

Naturally-occurring metabolic systems have evolved control circuitry that minimizes the accumulation of toxic intermediates and maintains homeostasis. At present, the process of engineering even rudimentary control circuitry for synthetic biological systems remains very difficult. We have successfully formulated design-driven approaches that use mechanistic modeling and kinetic RNA folding simulations to engineer RNA-regulated genetic devices with quantitatively-predictable activities that can control gene expression in metabolic pathways producing industrial chemicals in *E. coli* (Carothers et al. & Keasling. *Science* 2011). We have also demonstrated that models and simulation tools can be used to inform the design of microbial dynamic sensor-regulator systems (DSRS) engineered to produce fatty acid-based chemicals and fuels (Zhang, Carothers, & Keasling. *Nature Biotech.* 2012). We are currently designing RNA-based systems to solve canonical control problems in a 15 gene pathway engineered to produce substituted styrenes, components of polymer composites with optical and mechanical properties favorable for advanced applications in photonics, photolithography and biomedicine. Here, I will present these results and discuss the potential for using the conceptual and experimental frameworks that we have established to create full-fledged design platforms.

(12e) Engineering a Synthetic Malonyl-CoA Controller for Dynamic Tuning of Metabolic Flux in E. Coli

Xu, P., Rensselaer Polytechnic Institute
Wang, W., Beijing University of Chemical Technology
Li, L., Rensselaer Polytechnic Institute
Zhang, F., Rensselaer Polytechnic Institute
Koffas, M. A. G., Rensselaer Polytechnic Institute
Originally presented on: 11/3/2013 16:42:00 - 17:00:00

Natural biological systems use a lot of negative and positive feedback controls to coordinate gene expression and achieve optimal biological functions. As heterologous pathways become larger and more complicated, it becomes increasingly difficult to optimize them with static regulatory control. An ideal approach would be to use dynamic regulatory networks to control gene expression so that the host cells can adjust their metabolic functions when the environmental condition changes. Here we report engineering both the positive and negative feedback controls for dynamic tuning of metabolic flux in *E. coli*. We have identified a dual transcriptional regulator which can act either as an activator or a repressor for two different promoters. The level of activation or repression is dependent on the level of intracellular malonyl-CoA. As a proof of concept, we demonstrated that the expression of two reporter proteins can be exclusively switched between the on and off state. By engineering this synthetic malonyl-CoA controller, we envision that both the malonyl-CoA source pathway and the malonyl-CoA sink pathway can be dynamically modulated so that carbon flux can be efficiently redirected to synthesize our target compounds. Implementation of this dynamic control will maintain the intracellular malonyl-CoA at the optimal level and improve both the productivity and yield of value-added metabolites in *E. coli*.

(12f) Fine-Tuning of Gene Expression in Saccharomyces Cerevisiae Through RNA Interference

Si, T., University of Illinois at Champaign-Urbana
Zhao, H., University of Illinois at Urbana-Champaign
Originally presented on: 11/3/2013 17:00:00 - 17:18:00

Beyond gene overexpression or knockout, fine-tuning of gene expression is often required to optimize a target phenotype. This requirement results from the need to balance the trade-off between catalytic activity and metabolic burden, or to avoid accumulation of toxic metabolic intermediates. RNA interference (RNAi) is an effective gene silencing mechanism through specific degradation of target messenger RNAs (mRNAs). But there is no functional RNAi pathway in the eukaryotic model *S. cerevisiae* until recently, when a heterologous RNAi pathway from *Saccharomyces castellii* was reconstituted. Here we report the design of different RNAi expression constructs to achieve a continuum of gene knockdown levels in *S. cerevisiae*. By varying the length and position of the anti-sense RNAs or the double-stranded RNAs expressed under the control of strong constitutive type II promoters, 10% to 80% gene silencing levels were achieved for green fluorescence protein (GFP). We further modulated the gene expression of *ptc6* by RNAi constructs for improved acetic acid (HAc) tolerance, a key trait for microbial production of chemicals and fuels. An optimal knockdown level was identified which resulted in the highest acetic acid tolerance among the knockout, knockdown and wild type strains. This is the first example in *S. cerevisiae* where RNAi is employed as a gene regulatory tool for metabolic engineering. Our results also provided important insights on the design principles of RNAi in yeast.

(12g) Identification of High Capacity Terminators for Improved Control of Gene Expression and Metabolic Engineering in Saccharomyces cerevisiae

Curran, K., The University of Texas at Austin
Karim, A., The University of Texas at Austin
Gupta, A., The University of Texas at Austin
Alper, H., The University of Texas at Austin
Originally presented on: 11/3/2013 17:18:00 - 17:36:00

Control of gene and protein expression of both endogenous and heterologous genes is a key component of metabolic engineering. While a large amount of work has been published characterizing promoters for this purpose, less effort has been exerted to elucidate the role of the 3' regulatory region of genes, known as the terminator. In this study, we characterize over 30 terminators for use in metabolic engineering applications in *Saccharomyces cerevisiae* and identify mRNA half-life changes to be the major cause of the varied protein and transcript expression level. We demonstrate that the difference in transcript level can be over 6.5-fold even when paired with high strength promoters. The influence of terminator selection is magnified when coupled with a low-expression promoter, with a maximum difference in protein expression of 11-fold between a high-capacity terminator and the parent plasmid terminator and over 35-fold difference when compared with a no-terminator baseline. Next, we demonstrate the utility of terminator selection for metabolic engineering by using a mutant xylose isomerase gene as a proof-of-concept. Through pairing a high-capacity terminator with a low-expression promoter, we were able to achieve the same phenotypic result – equal growth rates on xylose – as with a promoter an order of magnitude higher in strength. Moreover, we can further boost the growth rate on xylose by pairing the high-strength promoter with a high-capacity terminator. Finally, we demonstrate that minimal synthetic terminators (less than 75 bp long) can be as effective as native, full-length high-capacity terminators. These synthetic versions possess several advantages including shorter length and less homology when compared to traditionally used elements. Together, this work demonstrates that terminators will be an important part of heterologous gene expression and metabolic engineering in the future.

(12h) Using a Synthetic Gene Network to Model and Understand the Effects of Shuttling On Gene Expression Patterns

Jermusyk, A., North Carolina State University
Reeves, G. T., North Carolina State University
Originally presented on: 11/3/2013 17:36:00 - 17:54:00

Using a Synthetic Gene Network to Model and Understand the Effects of Shuttling on Gene Expression Patterns

Gene Regulation Engineering

Session 15C23

Ashley A. Jermusyk and Gregory T. Reeves

In multicellular organisms, cellular signaling events are crucial for patterning tissues, as well as for maintaining healthy adult tissues, while improper signaling can lead to disease states, such as cancer. Therefore, cellular signaling processes must be tightly regulated. A complex system of gene regulatory circuits controls this signaling process and acts to buffer this system against noise, thereby minimizing mistakes in gene expression and preventing patterning defects or disease states. Despite their importance to patterning and development, hypotheses regarding these gene regulatory circuits have been difficult to test experimentally due to their complexity and high connectivity. Therefore, to better understand the fundamental processes involved, we created a synthetic gene network in the fruit fly *Drosophila melanogaster*. This approach has the advantages that (1) the gene network is orthogonal to native *Drosophila* biology, and (2) the network is designed. These two aspects imply the connectivity of the network is understood, and thus hypotheses regarding the designed network motif can be experimentally tested with this system.

As a multicellular organism, the fruit fly is an ideal system specifically for testing hypotheses regarding multicellular phenomena, such as tissue patterning. The fundamental pattern in developing tissues is established by morphogens. Morphogens determine the domains of gene expression in developing tissue by activating gene expression based on concentration thresholds. Morphogen gradients are formed by localized protein production and subsequent diffusion and degradation. A second protein can bind to the morphogen to facilitate transport and thus broaden the morphogen gradient. It has been proposed that this ?shuttling? mechanism can increase the robustness of gene expression (Eldar et al., 2003; Haskel-Ittah et al., 2012), but this hypothesis is difficult to test experimentally. Therefore, we have created a synthetic network to test the effects of shuttling utilizing genes from yeast and *E. coli*, namely, *gal4*, *gal80*, and *lacZ*. We expressed *gal4* in a graded fashion along the anterior-posterior axis of the embryo, mimicking the intracellular diffusion of an endogenous transcription factor and morphogen, Bicoid.Â As seen below in Figure 1, the Gal4 protein activates expression of *UAS*-linked *gal80* and *lacZ*. Gal80 binds to Gal4, preventing Gal4 from binding to *UAS* and activating expression of *gal80* and *lacZ*, and this sequestration should create a negative feedback loop in our system. However, due to the spatial dynamics of the system, Gal80 instead acts to facilitate the diffusion of Gal4, thereby creating a shuttling mechanism. The effects of this shuttling mechanism on the network can be determined by looking at lacZ expression, specifically changes in expression due to the addition of Gal80 to the network. These genes were chosen since they are endogenous to yeast (and therefore not to *Drosophila)*, so all interactions in this network are fully understood. Our goal is to measure the effect of Gal80-mediated Gal4 shuttling on the robustness of the location of the *lacZ* domain. This system provides a direct experimental test of the effects of shuttling in cellular signaling events, specifically whether this shuttling can lead to increases in robustness of the system. It is this robustness that is important for combating diseases and defects in development and maintenance of expression in the organism.

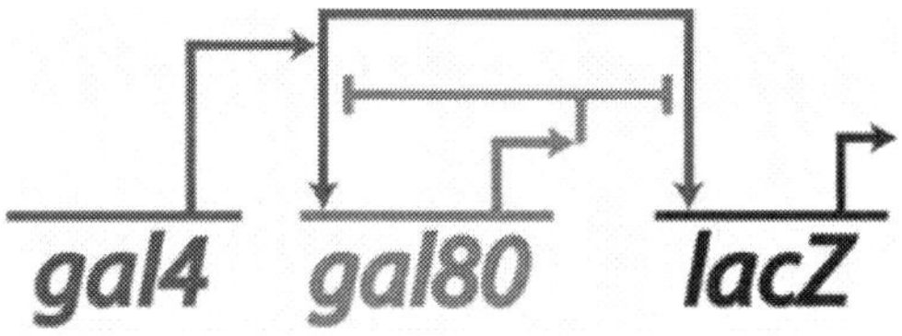

Figure 1. Network Diagram

Representation of the interactions in the studied network where arrows represent activation and flat arrowheads denote repression by binding of Gal80 to Gal4.

References

Eldar A, Rosin D, Shilo B, Barkai N. 2003. Self-enhanced ligand degradation underlies robustness of morphogen gradients. Dev Cell 5(4):635-46.

Haskel-Ittah M, Ben-Zvi D, Branski-Arieli M, Schejter ED, Shilo B, Barkai N. 2012. Self-organized shuttling: Generating sharp dorsoventral polarity in the early drosophila embryo. Cell 150(5):1016-28.

(15a) Use of 23S Ribosomal RNA Circular Permutants to Investigate E. Coli Ribosome Biogenesis in Vivo

Carlson, E., Northwestern University
Orelle, C., University of Illinois at Chicago
Szal, T., University of Illinois at Chicago
Mankin, A., University of Illinois at Chicago
Jewett, M. C., Northwestern University
Originally presented on: 11/3/2013 15:30:00 - 15:48:00

The ribosome is an amazingly complex and intricate molecular machine, responsible for producing the cell's proteome. *Escherichia coli* ribosome synthesis and assembly (biogenesis) is a tightly regulated and ordered process with complex—and not fully understood—folding and assembly steps. Understanding the intricacies and regulation of this process is key for ribosome engineering efforts. During rRNA transcription, co-transcriptional folding occurs as local secondary structure begins to emerge. Additionally, ribosomal proteins interact with the nescient rRNA as binding domains are transcribed. As different regions of the rRNA are exposed, tertiary interactions form as well. One aspect of understanding is the order in which the rRNA is transcribed. To study this, a plasmid based ribosomal operon is mutated such that the native 5' and 3' ends of the 23S rRNA are covalently linked, and a circular permutant (CP) is created by opening up new 5' and 3' ends at an internal helix. By observing effects on cell growth, a more complete understanding of transcription order effects on functional ribosome formation can be obtained. The current data set in this space is severely limited; CPs at only two of the over 100 23S helices are reported. Towards testing the remaining circular permutations, we utilize a parallel and efficient construction method to synthesize CPs of 23S rRNA at 90 helices on a plasmid based operon. Then CP23S constructs are evaluated on their ability to support cell growth in a strain lacking all 7 genomic copies of ribosomal operons. In preliminary work, we identified several new mutations that support cell growth with limited reduction in growth rate. Several mutants cannot support cell growth alone, but cells are viable in their presence. Finally, several mutations are toxic when expressed in *E. coli*. In this presentation, I will discuss our efforts to catalogue all possible 23S rRNA CPs. This work will provide a more thorough understanding of ribosome biogenesis *in vivo* and inform ongoing efforts to control gene regulation with orthogonal ribosomes.

(15b) Mechanism-Based Design of Highly Active Sirnas

Angart, P., Michigan State University
Walton, S. P., Michigan State University
Originally presented on: 11/3/2013 15:48:00 - 16:06:00

Short interfering RNAs (siRNAs) provide a method of post-transcriptional, targeted gene knockdown that is of interest as a therapeutic strategy for the treatment of a variety of diseases including metabolic disorders, cancers, viral infections, and degenerative diseases. Gene knockdown or silencing occurs through interactions of the siRNA with the native eukaryotic RNA interference (RNAi) pathway. siRNA specificity derives from the Watson-Crick base pairing interactions between the bases of the siRNA and its complementary target. The active silencing complex, known as RISC, operates as a multiple turnover enzymatic complex capable of decreasing target transcript levels sufficiently to initiate the therapeutic effect. However, siRNA activity can vary greatly depending upon its sequence and the reasons for this variability are not completely understood, hindering the progression of siRNAs to use as therapeutics. Our goal is to establish the mechanism-based guidelines for selecting siRNA sequences that are most likely to be highly active and specific for a given target.

Structurally, siRNAs are double-stranded RNAs with 19 central base pairs, two-nucleotide 3' overhangs, and 5' phosphates. Central to the function of siRNAs is the ability of the RNAi pathway to discriminate which strand of the structurally symmetric siRNA is to be active (guide strand) and which strand is to be degraded (passenger strand). Failure to incorporate the correct strand can cause both off-target effects as well as competition for the protein Argonaute 2, the core protein of RISC. This loading step represents a significant event in the success of an siRNA and must occur properly for maximal siRNA function. In this presentation, we will discuss our findings on the contributing factors that lead to siRNA strand selection and its influence on siRNA activity.

(15c) A Fluorescent Complementation Assay for Investigating RNA-Protein Interactions in Vivo

Gelderman, G.,
Contreras, L. M., University of Texas at Austin
Originally presented on: 11/3/2013 16:06:00 - 16:24:00

A fluorescent complementation assay for investigating RNA-protein interactions *in vivo*

Grant Gelderman and Lydia Contreras

McKetta Department of Chemical Engineering, University of Texas at Austin, 200 E. Dean KeetonÂ St., Stop C0400, Austin TX 78712

Small non coding RNAs are known as powerful cellular regulators of translation, capable of promoting or inhibiting translation initiation by binding mRNAs changing their structure, promoting or catalyzing mRNA degradation, or by binding other regulatory proteins and affecting their activity [1]. The interactions of these RNAs and proteins are often part of a large regulatory pathway responsible for complex forms of genetic regulation. Many methods to study RNA-protein interactions are limited largely to *in vitro* studies such as, alanine scanning, mobility assays, SELEX, electron microscopy, X-ray diffraction, etc. While these methods are powerful analytical tools and are useful for determining and characterizing binding domains in the protein and RNA sequence, the findings of these studies do not always extrapolate to the interactions occurring *in vivo*.

In this work, we present a method for studying RNA-protein interactions *in vivo* using fluorescent protein complementation [2] and flow cytometery. An advantage of this system is that it allows the interaction to occur *in vivo* and can elucidate how a complex system responds to stimuli from the environment by allowing single cell resolution to determine the average conditions of cell populations. We will demonstrate the proof of concept using a protein and an RNA from the carbon storage regulator (Csr) elements of *E. coli* (CsrA and CsrB, respectively). We evaluate the effect that mutations to CsrA and CsrB have on the fluorescent activity in order to probe the contribution of specific amino acid and nucleic acid residues in the RNA-protein interaction. Additionally, we present a rationale to screen for residues involved in RNA-protein interactions using random mutagenesis and flow cytometery.

Our method for detecting and quantifying RNA-protein interactions can give useful insights into RNA-protein interactions *in vivo*. Additionally, this method can be used to provide a rationale for targeting specific RNA-protein regulatory elements and tuning their interaction in order to optimize cellular conditions for metabolic engineering.

[1] Storz, G., Vogel, J. & Wassarman, K. Regulation by small RNAs in bacteria: expanding frontiers. *Molecular cell* **43**, 880?91 (2011).

[2] Shyu, Y. & Hu, C.-D. Fluorescence complementation: an emerging tool for biological research. *Trends in biotechnology* **26**, 622?30 (2008).

(15d) Exploring the Specificity of Nuclear Pore Transport Using AFM and QCM-D

Sorci, M., Rensselaer Polytechnic Institute
Hayama, R., Rockefeller University
Chait, B. T., The Rockefeller University
Rout, M., The Rockefeller University
Belfort, G., Rensselaer Polytechnic Institute
Originally presented on: 11/3/2013 16:24:00 - 16:42:00

The Nuclear Pore Complex (NPC) is the sole mediator of exchange between the nucleus and the cytoplasm in all eukaryotic cells[1]. Its architecture is well understood and described in the literature,[2,3] yet the molecular transport mechanism remains unclear. Transport across the NPC is fast, energy-dependent (to give directionality) and often receptor-mediated. While small molecules pass through the NPCs unchallenged, large macromolecules (>40 kDa) are excluded unless chaperoned across by transport factors collectively termed Karyopherins (Kaps). The translocation of the complexes of Kaps and their cargo proteins/RNAs occurs through the specific affinity and binding between Kaps and particular nuclear pore complex proteins (nucleoporins) called FG-Nups, which share a degenerate multiple-repeated "Phe-Gly" motif. Because FG-Nups are the major component of the selective gating mechanism, we first investigated the nanomechanical properties of cysteine-modified Nsp1 using the volume force mapping technique of atomic force microscopy (AFM). From single molecule AFM on a sparse Nsp1 surface, we estimated structural parameter as persistence length and contour length. In an attempt to better understand the transport and the selective process under crowding conditions, we then used quartz crystal microbalance with dissipation (QCM-D). Nsp1 and truncated variations of it were immobilized on QCM-D sensors. The binding and unbinding of Kap95, other binding proteins, as well as control proteins, was studied in order to investigate specificity, kinetics rate constants, effect of competitive binding. Inspired by Nature, we aim to gain sufficient understanding of the molecular scale engineering principles behind nuclear transport to allow us to design the next generation of synthetic selective nanosorters capable of purifying any protein that we desire.

1. Grünwald, Singer and Rout, *Nature* **2011**, 475, 333.

2. Alber et al., *Nature* **2007**, 450, 683

3. Alber et al., *Nature* **2007**, 450, 695

(15e) Capturing Binding-Induced Conformational Changes in Protein Engineering Calculations

Kuroda, D.,
Gray, J. J., Johns Hopkins University
Originally presented on: 11/3/2013 16:42:00 - 17:00:00

Protein-protein interactions are often targets of protein engineering, and binding is often associated with conformational changes. Such changes can affect affinities and specificities, and the mechanisms of allosteric proteins or molecular switches are often based on binding-associated conformational changes. Despite this importance, it is still very difficult to capture protein backbone conformational changes in protein-protein docking and design calculations. In this study, we use a set of known bound and unbound protein structures to test methods to computationally capture protein backbone conformational change related to binding. We address the following questions: (1) How well can we sample protein backbone conformations using existing computational models? (2) How much closer must unbound proteins need to be to the bound-state proteins for successful high-resolution protein-protein docking? (3) How do induced-fit and conformational selection mechanisms balance in protein-protein interactions? We test various Rosetta-based sampling techniques and coarse-grained normal mode analysis. Based on principal component analysis of computational protein ensembles, we find that protein motions made by each method show better correlations with interface regions than entire protein folds. Although the magnitude of motions in each protein ensemble are smaller in computational models, the direction of motions overlap with the actual conformational change between unbound and bound-state proteins, suggesting that the principal components can be used to extend motions to biologically relevant magnitudes. We demonstrate how the essential dynamics extracted from each computational model can be used in protein-protein docking simulations. Finally, we present backbone and docking energy landscapes that delineate the scope of induced-fit and conformational selections molecular recognition mechanisms, suggesting strategies for improving protein engineering approaches.

(15f) Force-Dependent Interactions Between Actin Filaments and a Minimal Adherens Junction Complex

Buckley, C. D., The Ohio State University
Tan, J., Stanford University
Pruitt, B. L., Stanford University
Weis, W. I., Stanford University
Nelson, W. J., Stanford University
Dunn, A. R., Stanford University
Originally presented on: 11/3/2013 17:00:00 - 17:18:00

Introduction

Adherens junctions (AJs) mediate adhesion between neighboring cells and are essential for coordinating cell-cell adhesion and cytoskeletal dynamics. Disruption of cytoskeletally generated tension in living cells impairs AJ formation, suggesting that localized mechanical stress may be required for AJ assembly. Previous attempts to measure the interaction of the core AJ components E-cadherin, beta-catenin, and alpha-catenin with actin *in vitro* have been unsuccessful, perhaps due to the absence of a mechanical load in bulk biochemical assays. Here we describe a new optical trap-based single-molecule assay that allows us to directly probe the interaction of the E-cadherin/catenin complex with filamentous actin as a function of applied mechanical load (Fig. 1A). Optical traps (also known as optical tweezers) utilize highly focused laser light to capture and exert force on micron-sized dielectric objects. The instrument we have constructed consists of a dual-beam optical trap that can apply up to ~20 piconewtons (pN) of force while measuring the location of a trapped 1 mm polystyrene bead with nanometer and millisecond resolutions.

Results

Our assay works as follows: two streptavidin-coated polystyrene beads are trapped simultaneously using the dual-bead optical trap and a biotinylated actin filament is stretched between them. This filament is then lowered over a platform bead immobilized on a coverslip and functionalized with the reconstituted E-cadherin/beta-catenin/alpha-catenin complex. Oscillation of the piezo-actuated microscope stage moves the platform bead and attached E-cadherin/catenin complexes back and forth along the axis of the actin filament while the optical traps remain stationary. Load is applied when the immobilized complexes bind to the filament as the stage continues moving and one of the two beads is pulled away from the trap center. As the bead is pulled away, the force gradient produced by the focused laser beam attempts to pull the bead back to the center of the optical trap. The farther the bead is pulled, the greater the restoring force it and the attached protein complex experience. The microscope stage velocity and the stiffness of the optical traps thus determine the force loading rate experienced by the E-cadherin/catenin complexes. Using high concentrations of the E-cadherin/catenin complex (~10 to 100 nM), we observe binding events that likely involve the attachment of multiple complexes to a single actin filament. Under these conditions, we observe displacement events of over 100 nm with loading rates of ~45 pN/s that can sustain over 15 pN of force. Detachment occurs via a stepwise release mechanism that may reflect sequential detachment of individual complexes (Fig. 1B).

Conclusions

Our data indicate that several E-cadherin/catenin complexes can together form a robust, load-resistant connection to an actin filament under tension. This observation is consistent with a hypothesized role for alpha-catenin in linking the adherens junction to the actin cytoskeleton. Ongoing work tests the possibility that the E-cadherin/catenin complex acts as a mechanosensitive element whose affinity for actin is directly modulated by mechanical load.

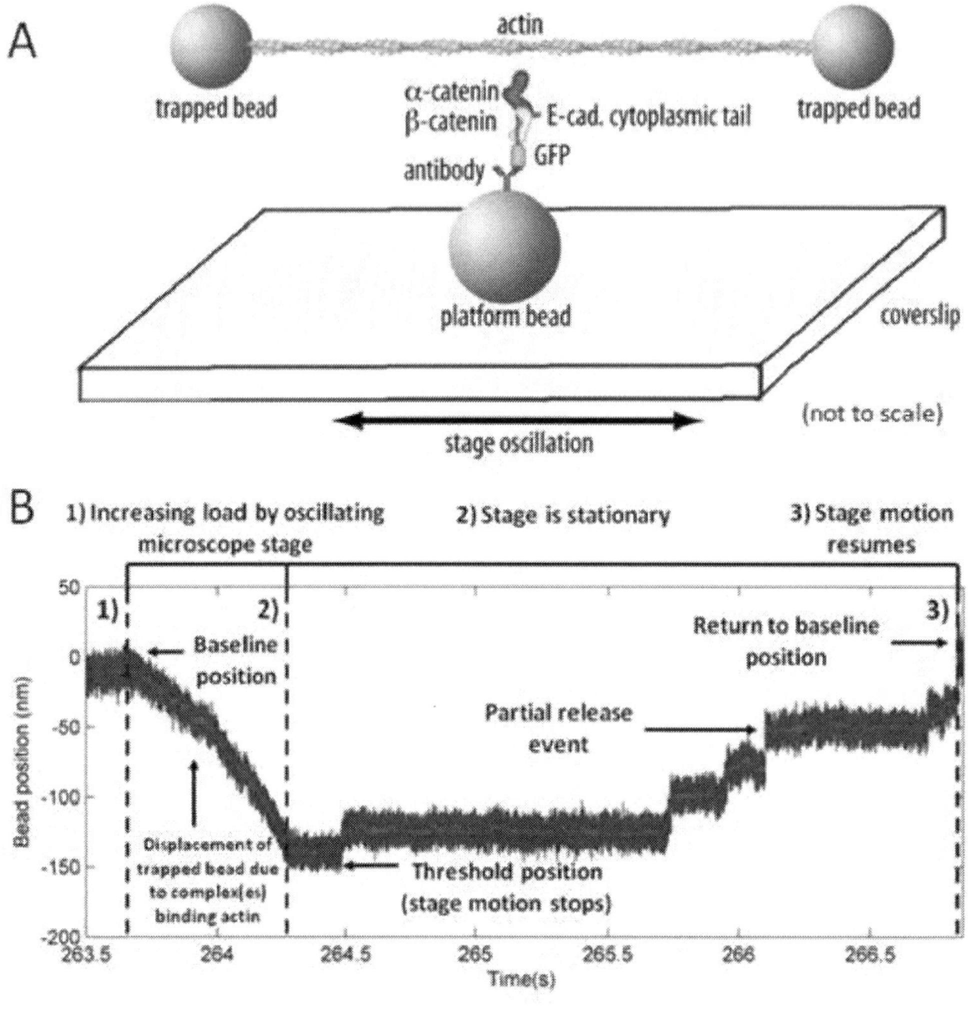

Figure 1. (A) Three-bead optical trap assay for characterizing force-dependent binding interactions between the E-cadherin/catenin complex and actin filaments. **(B)** Example trace showing the position of one of the two optically trapped beads during a binding event.

(15g) Understanding the Antagonistic Mechanism of a De Novo Designed Peptide With Experimental HIV Fusion Inhibition Via Molecular Simulation

Khoury, G. A., Pennsylvania State University-University Park
Tamamis, P., Princeton University
Smadbeck, J., Princeton University
Szekely, Z., Rutgers University
Sinko, P. J., Rutgers University
Floudas, C. A., Princeton University
Originally presented on: 11/3/2013 17:18:00 - 17:36:00

A noble goal in protein design is to predict the potency and understand in detail the mechanism of inhibition of a candidate antagonist. Previous *de novo* design has produced several HIV fusion inhibitors targeting HIV-gp41[1]. These inhibitors were experimentally confirmed to inhibit viral entry. Here, we probe, through computational studies based on molecular dynamics simulations the mechanism of inhibition of SQ175, a design among the best in antagonistic activity.

Multiple independent explicit solvent (TIP3) molecular dynamics simulations of SQ175 bound to the trimeric coiled-coil are performed in the AMBER[2] MD package. Using the snapshots from the trajectories, interaction maps were constructed identifying key SQ175:gp41 hydrophobic and electrostatic interactions. Binding free energy calculations of SQ175 confirm its affinity for the trimeric coiled-coil of gp41. Subsequently, in CHARMM[3] and using generalized Born implicit solvent[4], simulations of the C-terminus binding fragment bound to the trimeric coiled-coil of gp41 were performed. These simulations aimed at evaluating the native interactions between the C-terminus binding fragment:trimeric core and to construct a map denoting the minimum Cα-Cα distances the C-terminus binding fragment residues experience with respect to the gp41 trimeric coiled-coil. Next, multiple independent docking simulations of the C-terminus binding fragment starting 10-12 Å away from its fusion intermediate structure were performed, where after the MD, the C-terminus binding fragment was found to spontaneously dock within 1.5 Å of its conformation in the crystal structure in one simulation, indicating the system behaves well. Then, in the presence of SQ175, 12 independent docking simulations were performed starting from 12 different initial coordinates to mimic the C-terminus binding fragment approaching gp41. No restraints were implemented to drive them together. The SQ175 peptide was observed to dramatically reduce the number of native C-terminus binding fragment:gp41 contacts of which several are necessary for HIV fusion and subsequent entry, as observed by minimum Cα-Cα distances. The antagonist SQ175 was observed to stably bind to the trimeric coiled-coil of gp41 and physically block a large fraction of the native C-terminus binding fragment:gp41 contacts. Based on these results, new insights for future generations of fusion inhibitors are described.

References

1. Bellows, M. L.; Taylor, M. S.; Cole, P. A.; Shen, L.; Siliciano, R. F.; Fung, H. K.; Floudas, C. A., Discovery of Entry Inhibitors for HIV-1 via a New De Novo Protein Design Framework. *Biophysical Journal* **2010,** *99* (10), 3445-3453.

2. Weiner, P. K.; Kollman, P. A., AMBER: Assisted model building with energy refinement. A general program for modeling molecules and their interactions. *Journal of Computational Chemistry* **1981,** *2* (3), 287-303.

3. MacKerell, J., A. D., Brooks, B., Brooks, III, C.L., Nilsson, L., Roux, B., Won, Y., and Karplus, M., CHARMM: The Energy Function and Its Parameterization with an Overview of the Program. In *The Encyclopedia of Computational Chemistry*, al., P. v. R. S. e., Ed. John Wiley & Sons: Chichester: 1998; Vol. 1, pp 271-277.

4. (a) Chen, J.; Im, W.; Brooks, C. L., 3rd, Balancing solvation and intramolecular interactions: toward a consistent generalized Born force field. *Journal of the American Chemical Society* **2006,** *128* (11), 3728-36; (b) Im, W.; Lee, M. S.; Brooks, C. L., 3rd, Generalized born model with a simple smoothing function. *J Comput Chem* **2003,** *24* (14), 1691-702.

(15h) Fibronectin Increases the Rate of Fibrin Clot Polymerization and Alters Matrix Morphology

Ramanathan, A., Illinois Institute of Technology
Karuri, N. W., Illinois Institute of Technology
Originally presented on: 11/3/2013 17:36:00 - 17:54:00

Plasma fibronectin is essential for the formation of thrombi, a provisional fibrin matrix that arrests bleeding in damaged tissues. The initial rate of fibrin polymerization and the steady state absorbance for fibrin clots with varying concentrations of full-length plasma fibronectin were monitored via light absorbance. At higher fibrin concentrations, the initial rate of polymerization and steady state absorbance increased with increasing fibronectin concentration. Similar behaviors were seen at a lower fibrin concentration, but less pronounced. Three dimensional fibrin clot structure was examined by confocal microscopy z-stacks. Fibrin clots formed in the absence of fibronectin appeared to have straight, pin-like fibers and clots formed with high concentrations of fibronectin had finer, densely packed fibers. When incubated in denaturing buffer, fibronectin was not released from the fibrin clot, but the entire clot was soluble in reducing buffer; indicating possible disulfide bonding between fibrin and fibronectin. These results provide insight into the mechanism in which thrombi form *in vivo* and indicate that plasma fibronectin could be a useful target for clot destabilizing therapeutics.

(41a) Comparison of Non-Natural Pathways for Phenol Biosynthesis in Engineered Escherichia Coli

Thompson, B., Arizona State University
Pugh, S., Arizona State University
McKenna, R., Arizona State University
Nielsen, D. R., Arizona State University
Originally presented on: 11/4/2013 8:30:00 - 8:48:00

The non-renewable petrochemical phenol is used as a precursor to produce numerous fine and commodity chemicals (e.g., bisphenol A, caprolactam, and salicylic acid) as well as materials such as phenolic resins. A renewable route to phenol from glucose-derived tyrosine via tyrosine phenol lyase has previously been demonstrated using engineered *Pseudomonas putida* as the biocatalyst. In contrast, here we present on the development of a novel pathway that converts chorismate to phenol via the concerted action of isochorismate synthase, isochorismate pyruvate lyase, and salicylate decarboxylase. Whereas the former route involves fewer heterologous steps, it is derived from an endogenous precursor further removed from the central metabolism. Although both pathways have identical theoretical yields, comparison of the two routes provides interesting insight into pathway design strategies for non-natural chemical production. Both routes were systematically constructed in *Escherichia coli* following screening studies to identify the most robust pathway enzymes resulting in initial shake flask titers of 25 mg/L. Current strategies to increase final product titers through both routes include pathway de-bottlenecking by development of optimal expression strategies and re-engineering of the host metabolism to maximize production and availability of respective precursors.

(41b) Biosynthesis of a Branched C6 Fatty Alcohol Via a De Novo Pathway in Escherichia Coli

Sheppard, M., MIT
Wenck, S., Massachusetts Institute of Technology
Prather, K. L. J., Massachusetts Institute of Technology
Originally presented on: 11/4/2013 8:48:00 - 9:06:00

The exploration of microbial biofuel production has expanded in recent years beyond ethanol to a variety of heavier alcohols and fatty acid derived biodiesel. Synthesis of branched alcohols in the six to seven carbon range are an appealing target because their properties, including energy density and improved octane rating, approach those of primary components of gasoline. In addition these compounds have the potential to reduce energy costs of downstream separations. We have developed a *de novo* pathway to synthesize one such compound, 4-methyl-pentanol, by coupling *E. coli* primary metabolism to exogenous fatty acid and secondary metabolism. The overall redox neutrality of ethanol fermentation is a key to achieving near theoretical yields and we have maintained redox neutrality in design of our novel 4MP pathway. Our design is also modular, enabling testing and screening of alternative genes for individual segments of the overall pathway.

(41c) Modular Pathway Design for Combinatorial Biosynthesis of Designer Bioesters

Layton, D. S., University of Tennessee
Trinh, C. T., University of Tennessee, Knoxville

Originally presented on: 11/4/2013 9:06:00 - 9:24:00

Sensation profiles are observed all around us which are composed of many chemicals such as esters. We can mimic these profiles in everyday items for their uses in foods, beverages, cosmetics, solvents, and perfumes. Here, we present an organic way to derive these products via a combinatorial biosynthesis in *Escherichia coli*. The bioester-producing pathway is composed of two different modules for precursor formation of alcohols and acyl-CoAs that are then condensed by an alcohol acyltransferase to produce bioesters. We employ metabolic engineering and synthetic biology principles and techniques to design distinct bioester-producing pathways that can be coupled with a universal optimal modular cell in a plug-and-play design for enhanced bioester production. We will demonstrate the combinatorial production of a bioester library derived from short-chain alcohols and acyl CoAs that can be used for flavorings, fragrances, and solvents.

(41d) Engineering Auto-Regulatory Time-Dependent Genetic Circuits for Improved Biochemical Production in Escherichia Coli

Lo, T. M., Nanyang Technological University
Hwang, I. Y., Nanyang Technological University
Tan, M. H., Nanyang Technological University
Chang, M. W., Nanyang Technological University
Originally presented on: 11/4/2013 9:24:00 - 9:42:00

Escherichia coli, a common industrial workhorse, can be engineered to perform biocatalysis of lignocellulosic feedstock into valuable products. However, issues such as cell viability, substrate toxicity and product release often plague the efficiency of bioconversion. In this study, to overcome these challenges, we aimed to develop a range of novel genetic circuits that enable auto-regulatory time-dependent production of biochemicals in *E. coli*, To improve cell viability, we implemented the control theory in protein expression through the introduction of bio-sensor with time-delay element. The time-delay element ensures that the host cells would not be induced at early stage of growth. This time-delay biosensor not only eliminates the need for costly inducers such as arabinose and IPTG, but it also allows optimal performance in dynamic biological system through corrective actions. As proof of concept, we designed and installed a time-delay biosensor to enable *E. coli* to detect hydroxycinnamic acid and convert it to valuable biochemicals. Further, to counter the problem of substrate toxicity, we performed strain optimization using hydroxycinnamic acid as our test substrate, in which the improved strain was capable of tolerating up to 2.5-fold of the previously accepted concentration. Lastly, to aid in the extraction of biochemical products and simplify the product harvest process from *E. coli*, we developed a novel genetic circuit that conferred cell density-dependent auto-regulatory lysis. We envision that our auto-regulatory time-dependent approach can be readily extended to improving production of a range of biochemicals in *E. coli*.

(41e) Engineering Biosynthetic Pathways and Improving Solvent Productivities in Clostridium Acetobutylicum Toward the Production of Biodiesel Blendstocks and High-Value Chemicals

Kuchenreuther, J. M., UC Berkeley
Baer, Z., UC Berkeley
Bormann, S., UC Berkeley
Sreekumar, S., UC Berkeley
TostÃ©, D., University of California
Blanch, H. W., University of California, Berkeley
Clark, D. S., University of California

Originally presented on: 11/4/2013 9:42:00 - 10:00:00

Advancements in metabolic engineering have facilitated the biological conversion of carbohydrates into various jet and diesel fuel compounds, albeit at low yields and titers that hinder their production at the industrial level. Recently, our team developed a chemocatalytic technology to efficiently condense electrophilic alcohols with nucleophilic ketones – specifically, the acetone, n-butanol, and ethanol products from *Clostridium acetobutylicum* – for creating jet fuel and biodiesel blendstocks. Several features of Clostridia including their flexibility towards fermentable substrates, high solvent titers, and yields near the theoretical maxima have contributed to the resurgence of these microbes and the ABE fermentation for producing biofuels at the industrial scale. In this work, we describe the engineering of *Cl. acetobutylicum* via the recombinant co-expression of both native and heterologous proteins. Specifically, we discuss our progress with alleviating product toxicity, increasing solvent titers and productivities in batch and fed-batch fermentations, and introducing non-native biosynthetic pathways for making new metabolic products. We also discuss how our engineered strains, in combination with the chemocatalytic technology that we have developed, demonstrate a promising step toward establishing industrially relevant microbial processes for the production of jet fuel and diesel blendstocks as well as other high-value products.

(41f) Ethanol Tolerance in Yeast Is Determined By Membrane Gradients

Lam, F., Massachusetts Institute of Technology
Fink, G., Whitehead Institute
Stephanopoulos, G. N., Massachusetts Institute of Technology
Originally presented on: 11/4/2013 10:00:00 - 10:18:00

Ethanol made from the yeast *Saccharomyces cerevisiae* will continue to be the dominant mass-market biofuel in the near to intermediate term. One of the primary bottlenecks to higher bioethanol production is the toxicity of ethanol itself, with cell death occurring at concentrations greater than 15-20% (v/v). Increasing ethanol tolerance could improve both product yield and the viability of yeast for reuse in subsequent fermentations. However, its biological basis remains elusive: whole genome studies have shown that ethanol tolerance is influenced by many genes, and thus far, none have produced significant improvement when manipulated alone. An alternative approach, which would not necessitate any genetic manipulation of strains, is to uncover extracellular conditions that would desensitize cells to high ethanol concentrations. To this end, we have identified several micronutrients which, when added to fermentation culture, can increase ethanol production in *S. cerevisiae* by >50% in the very high sugar and high cell density conditions that mimic industrial fermentation. Moreover, we find that these additives confer increased resistance to propanol, collectively suggesting the augmentation of general cellular tolerance to solvents. Dissecting the component effects of these additives led us to uncover a novel mechanism of ethanol toxicity based on the perturbation of membrane gradients. Indeed, by using yeast mutants that specifically strengthen or weaken these gradients, we can elicit corresponding increases or decreases to ethanol performance. The strengthening of membrane gradients by these culture additives is not unique to special strains of *S. cerevisiae*: this effect raises ethanol yields (to nearly 100%) universally among a sampling of genetically divergent industrial and laboratory strains. Moreover, it is not specific to glucose: these media supplements also raise ethanol yields in a strain engineered to ferment xylose. Taken together, these results point to an unanticipated method for increasing the economic potential of cellulosic biomass and reveal the fundamental role that membrane gradients play in determining ethanol tolerance.

(41g) Engineering Yarrowia Lipolytica for Fuels and Oleochemicals Applications

Liu, L., University of Texas at Austin
Hill, A., University of Texas at Austin
Blazeck, J., University of Texas at Austin
Alper, H., The University of Texas at Austin
Originally presented on: 11/4/2013 10:18:00 - 10:36:00

Concerns about energy security, the global petroleum supply and climate change have increased interest in the production of sustainable and renewable biofuels. The oleaginous yeast *Yarrowia lipolytica* naturally possesses moderate lipid (biodiesel precursor) production and grows on different kinds of biomass (and organic waste). However, production from native, un-engineered strains is not sufficient for an industrial process. Here, we report on a rational and combinatorial metabolic engineering approach to establish *Y. lipolytica* as a premier platform for industrial-level, high lipid production. Specifically, several rational gene targets were combined to uncover potential synergistic genetic influencing lipogenesis. This study resulted in the largest collection of genetically modified strains of *Y. lipolytica*. The lipid content in the best engineered strain exceeded 80% of its dry weight. In parallel with these efforts, an inverse, combinatorial metabolic engineering approach was used to isolate improved lipid production strains. Whole genome re-sequencing of isolated strains revealed a novel lipid enhancer element. Further improvements of lipid production were achieved by combining these two approaches. Through this effort, we enabled extremely high lipid production in *Y. lipolytica*. This project helps elucidate the mechanism of lipogenesis and establish *Y. lipolytica* as an oleochemicals platform strain.

(41h) Exploiting Issatchenkia Orientalis SD108 As a New Platform Organism for Organic Acids Production

Xiao, H.,
Shao, Z., Iowa State University
Jiang, Y., University of Illinois at Urbana-Champaign
Dole, S.,
Zhao, H., University of Illinois at Urbana-Champaign
Originally presented on: 11/4/2013 10:36:00 - 10:54:00

The toxicity of organic acids presents a key challenge to the development of an economically viable fermentation process. Therefore, a platform organism that not only produces a high level of organic acids but also tolerates the associated toxicity is highly desirable. *Issatchenkia orientalis* strain SD108 was found to be tolerant to low pH and high concentration of organic acids. It can ferment fructose as efficiently as glucose, while no growth on other sugars (xylose, arabinose, cellobiose, galactose and sucrose) was observed. Tolerance assays were subsequently carried out with acetic acid, succinic acid, itaconic acid, and adipic acid. SD108 could grow even when the concentrations of succinic acid and adipic acid approached or exceeded their aqueous solubility limits and demonstrated a superior tolerance towards all four of these organic acids compared to *I. orientalis* ATCC 24210 and Classic Distiller's Turbo yeast. The draft genome sequence of SD108 was provided, which offers a basis for our future functional genomics studies on its tolerance mechanisms and a source for engineering individual organic acid biosynthetic pathways. As proof of concept, metabolic engineering strategies were further applied in SD108 for succinic acid production. By exploiting genome information, four gene expression cassettes related to the reductive TCA cycle were assembled and integrated into the genome of uracil auxotroph of SD108. The final engineered strain was able to reach a titer of 11.63 g/L, yield of 0.12 g/g, and productivity of 0.11 g/L·h of succinic acid.

(51a) Oral Delivery of Therapeutic Proteins Exhibiting High Isoelectric Points in pH-Responsive Hydrogels

Koetting, M. C., The University of Texas at Austin
Peppas, N. A., The University of Texas at Austin
Originally presented on: 11/4/2013 8:30:00 - 8:48:00

Oral Delivery of Therapeutic Proteins Exhibiting High Isoelectric Points in pH-Responsive Hydrogels

Michael C. Koetting and Nicholas A. Peppas, The University of Texas at Austin, Department of Chemical Engineering, Austin, TX

Introduction: Therapeutic delivery of proteins via intravenous injection is an invasive and often painful method which has been shown to cause low patient compliance with doctor-recommended treatment. Delivery via the oral route is strongly preferred to avoid the pain and embarrassment of injection and therefore increase patient compliance and improve quality of life. Environmentally-responsive hydrogel systems have shown great promise as drug delivery vehicles to overcome the obstacles of oral delivery. Our group has previously developed copolymer hydrogel delivery vehicles?notably poly(methacrylic acid) grafted with poly(ethylene glycol), or P(MAA-*g*-EG)?for the oral delivery of insulin. These hydrogels protect the protein through the gastrointestinal tract and release the protein in the small intestine by swelling at the higher pH found in the intestine.

Our current work is focused on creating a pH-responsive copolymeric hydrogel delivery platform for high isoelectric point-exhibiting proteins, specifically salmon calcitonin which is used to treat osteoporosis and Paget's disease. Calcitonin displays very different properties than insulin, primarily a high isoelectric point of 8.86 compared to 5.39 for insulin. Since roughly half of the human proteome consists of proteins with similarly high isoelectric points, the ability to orally deliver these many compounds will be a significant accomplishment toward making protein therapy cost effective, minimally invasive, and clinically desirable. In this work, we investigate potential hydrogel delivery systems to develop an effective oral delivery platform for achieving high bioavailability with salmon calcitonin.

Methods and Results: pH-Responsive hydrogels were synthesized by UV-initiated free radical polymerization in 0.7 mm thick films.Â The hydrogels were composed of either methacrylic acid (MAA) or itaconic acid (IA) copolymerized with either N-vinyl pyrrolidone (NVP) or poly(ethylene glycol) methyl ether methacrylate (PEGMMA). The hydrogels' swelling characteristics were determined in equilibrium and dynamic swelling studies, with all gels demonstrating a collapsed network at low pH (pH 2.0) and a significantly swelled network at neutral pH (pH 7.4) that is suitable for protection of encapsulated protein through the acidic, proteolytic stomach environment and for diffusive release in the neutral environment of the small intestine. Protein delivery capabilities were tested *in vitro* by loading 90-150 μm microparticles (from crushing/sieving of dried films) with salmon calcitonin by imbibition in solutions of varying ionic strength. Calcitonin was then released in a pH 7.4 solution of 0.150 M PBS.Â Loading and release levels were determined by HPLC or MicroBCA assay as appropriate. Hydrogels displayed significantly improved delivery capability when loaded in low ionic strength solutions, achieving up to 15.8 times as much calcitonin released per unit of hydrogel compared to when loaded in standard 0.150 M PBS. Additionally, hydrogels based on methacrylic acid, specifically poly(methacrylic acid-*co*-N-vinyl pyrrolidone) or P(MAA-*co*-NVP), achieved improved delivery levels compared to gels based on itaconic acid.Â Together, the platform achieves delivery levels that would suggest its suitability as a therapeutic oral delivery platform for salmon calcitonin.

Conclusions: pH-Responsive hydrogels were synthesized and screened as candidates for oral delivery of the therapeutic protein salmon calcitonin, as well as similar high isoelectric point-exhibiting proteins. Loading and release studies demonstrated that hydrogels based on methacrylic acid as the pH-responsive moiety offer greater delivery capabilities than itaconic acid-based hydrogels. Importantly, it was also shown that the ionic strength of the loading solution strongly affects both loading and release levels, with low ionic strength loading solutions offering significantly improved delivery capabilities *in vitro*. Therefore, we have shown improved delivery capability achieving levels sufficient for clinical therapy.

Acknowledgment: This work was supported by grant 1R01-EB000246-19 from the National Institutes of Health.

(51b) Delivering Nanoparticles to Lungs and Brain Via Hitchhiking On Red Blood Cells

Anselmo, A. C., University of California Santa Barbara
Gupta, V., University of California Santa Barbara
Mitragotri, S., University of California Santa Barbara
Originally presented on: 11/4/2013 8:48:00 - 9:06:00

Nanoparticulate drug delivery systems are one of the most widely investigated approaches for development of novel therapeutic strategies for a variety of diseases. However, rapid clearance by the reticuloendothelial system (RES) and lack of targeting to sites of interest prevent the clinical translation of nanoparticle therapies. Here, the unique flexibility, circulation times and vascular mobility of red blood cells (RBCs) are utilized to circumvent these roadblocks (Cellular Hitchhiking). 200nm nanoparticles non-covalently anchored onto RBC surface have been shown to increase circulation time and significantly reduce clearance by RES [1, 2]. In this work, Cellular Hitchhiking is used to reduce nanoparticle accumulation in RES organs (liver and spleen) and increase accumulation in all other organs, notably lungs and brain. RBC-hitchhiked nanoparticles exhibited ~7-fold higher accumulation in lungs and over 20-fold higher accumulation in the brain. Surprisingly, this accumulation was higher than that achieved by coating the nanoparticles by commonly used targeting antibodies for both lungs and brain. Further, we show that hitchhiking and antibody-mediated targeting approaches can be combined to further enhance tissue accumulation past that of either hitchhiking or antibody targeting alone. Parameters defining mechanisms for enhanced lung and brain accumulation, and its therapeutic potential will be discussed. These results highlight the efficacy of Cellular Hitchhiking in providing a new platform for targeted delivery of therapeutic agents for treatment of various cardiopulmonary, respiratory and neurodegenerative diseases.

1. Chambers, E. and S. Mitragotri, *Long circulating nanoparticles via adhesion on red blood cells: mechanism and extended circulation.* Exp Biol Med (Maywood), 2007. **232**(7): p. 958-66.

2. Chambers, E. and S. Mitragotri, *Prolonged circulation of large polymeric nanoparticles by non-covalent adsorption on erythrocytes.* J Control Release, 2004. **100**(1): p. 111-9.

(51c) Simulation of In-Vitro Gastrointestinal Digestion for Understanding Food Effect On Oral Drug Delivery

Rezhdo, O.,
Carrier, R. L., Northeastern University
Originally presented on: 11/4/2013 9:06:00 - 9:24:00

Oral delivery is preferred by the pharmaceutical industry, in large part due to patient convenience and compliance, but is not always possible due to low oral bioavailability. It has been observed that the presence of food can positively or negatively affect oral drug absorption, but these effects are not currently amenable to quantitative prediction. To understand food-drug interactions, it is crucial to develop an in-vitro system that simulates the digestion process in the stomach and intestine. Many in-vitro models have been developed that mimic digestion events in the gastrointestinal (GI) tract. However, they fail to accurately represent this dynamic environment, to a large extent because the majority of these models only take into account intestinal digestion, disregarding the fact that a significant portion of food is first digested as it passes through the stomach. Little information is available for gastric digestion. Hence, it has either been ignored or simulated via an extremely simplified in-vitro gastric digestion process that does not nearly match in-vivo events. Here we present an effort to simulate gastric digestion of milk as a relatively simple representative food model containing major components (protein, carbohydrate, fat) of a normal diet. We have designed model milk containing 33 g/L triolein, 32 g/L casein, and 53 g/L lactose. The simulated gastric fluid contains a mixture of phospholipids, specifically phosphatidylcholine at 0.133 g/L, and salts, specifically NaCl (67 mM), KCl (14 mM), and $CaCl_2$ (0.6 mM). Moreover, it also contains the two major gastric enzymes, pepsin and lipase, the concentrations of which are determined based on reported in-vivo activity data. Gastric digestion is mimicked via concurrent lipolysis and proteolysis while dropping pH stepwise every 5 min over a 1h timeframe from pH 6.5 to 1.7. The extent of digestion achieved (~30%) agrees with reported in-vivo gastric digestion. Some protein precipitation occurs as the pH of the solution approaches the isoelectric point of caseins (pI = 4.6). Dynamic light scattering analysis of particle size reveals the formation of relatively large (~900 nm – 1 um) emulsions in solution as well as micellar size (~1-5 nm) particles which are believed to include phospholipids. As the colloidal structures present during digestion serve as carriers for lipophilic drugs, impacting dissolution and absorption, analysis of structure, composition, and kinetics of partitioning of compounds within them will be useful for mechanistic and predictive models of the impact of food on bioavailability.

(51d) Pollen Grains for Oral Vaccination

Atwe, S., Texas Tech University
Ma, Y., Texas Tech University
Gill, H. S., Texas Tech University
Originally presented on: 11/4/2013 9:24:00 - 9:42:00

The oral route remains a highly desirable method of vaccination. Here we propose the use of pollen grains (PGs) as a novel material for oral vaccine delivery. PGs have the unique property that their outer shell is highly resistant that remains intact even in the acid- and enzyme-rich environment of the stomach. Furthermore, PGs could perhaps be persorbed as intact particles across the intestinal epithelium, enabling PGs to be used as microshuttles for antigen transport. PGs have been used in traditional medicine and as nutritive material in various civilizations providing anecdotal evidence to the safety of PGs for human consumption. We present the first study demonstrating the potential of PGs for oral vaccination.

Lycopodium clavatum spores (LSs) were investigated as a representative PG. Ovalbumin (OVA) was used as the model antigen. LSs were chemically treated to remove native proteins, lipids and other biomolecules. Systemic immune response was compared between the LS, cholera toxin (CT) and its B subunit (CTB). Balb/c mice were immunized orally using the different formulations at days 0 and 28. Blood was collected at days 0, 28 and 56 to determine serum immunoglobulin G (IgG) levels and Th1 and Th2 response using ELISA.

Chemical treatment of LSs was effective in removal of native plant material. Scanning electron micrographs of LSs before and after treatment provide a visual indication that processed LSs are devoid of plant biomolecules. OVA was successfully loaded into the spore interior as demonstrated by the confocal micrograph of LSs. ELISA for serum showed that mice immunized with 100 μg OVA+5mg LSs had IgG levels even higher than the group immunized with 5 mg OVA+5μg cholera toxin (CT). This is exciting data because CT is the current gold standard in oral mucosal vaccination and our study shows that LSs are even superior to CT. The strong adjuvant effect of spores has been reproduced by us in multiple independent studies. In conclusion, we present the first study demonstrating the adjuvant effect of LSs for oral vaccination.

(51e) Surfactant-Enhanced Post-Deposition Spreading of Aerosols With Application to Pulmonary Drug Delivery

Khanal, A., Carnegie Mellon University
Sharma, R., Carnegie Mellon University
Corcoran, T., University of Pittsburgh
Garoff, S., Carnegie Mellon University
Przybycien, T., Carnegie Mellon University
Swanson, E., Centre College
Tilton, R. D., Carnegie Mellon University
Originally presented on: 11/4/2013 9:42:00 - 10:00:00

Altered ventilation patterns in cystic fibrosis and other chronic obstructive pulmonary diseases cause inhaled drugs to deposit non-uniformly in the lungs. To overcome the limitations of aerodynamic delivery in these situations, a means of spreading aerosol droplets after they deposit on the airway surface liquid is required. We have shown previously that the addition of surfactants to microliter aqueous drops deposited on complex aqueous subphases that mimic the airway surface liquid results in significantly enhanced spread extents relative to saline controls. Marangoni stresses initiate spreading and escape of the surfactant from the spreading drop drives subphase surface movement at distances far from the spreading drop. Spreading stops when a capillary force balance is re-established among the subphase/air surface tension modified by the escaped surfactant monolayer, the very low interfacial tension between the aqueous subphase and the drop, and the drop/air surface tension. Every surfactant-subphase combination we have examined shows enhanced spreading relative to saline controls (cationic, anionic, nonionic, small molecule or polymeric surfactant types; aqueous entangled subphases containing porcine gastric mucins, nonionic polyacrylamide or pulmonary mucus); the drop-subphase systems behave as though they were immiscible over the time scale of the spreading event. We have translated this work to the post-deposition spreading of sub-picoliter aerosol droplets generated using vibrating mesh and nebulizers on complex subphases. While we have found that the individual surfactant-laden droplets spread more after deposition than do saline controls, importantly, the escaping surfactant from individual newly deposited droplets repels previously deposited droplets, significantly amplifying the spread area. We have demonstrated this enhanced spreading behavior in systems mimicking expected therapeutic aerosol deposition fluxes at the $2^{nd}/3^{rd}$ airway generation under shallow breathing and at the $12^{th}/13^{th}$ airway generation under moderate breathing. In both cases, the spread extent is on the order of several centimeters, a very large enhancement relative to the saline control, and may be limited only by the finite extent of the subphases we have used in our experimental systems. These results suggests new aerosol delivery strategies that could maximize the extent of drug spreading throughout the lungs of patients afflicted by obstructive lung diseases.

(51f) Planar Bioadhesive Microdevices for Enhanced Oral Drug Delivery

Chirra, H. D., University of Kentucky
Shao, L.,
Ciaccio, N., University of California San Francisco
Desai, T., University of California, San Francisco
Originally presented on: 11/4/2013 10:00:00 - 10:18:00

Oral delivery is a preferred route of administration as it is less invasive, provides patient compliance, and less costly. However, the harsh physiological conditions of the gastrointestinal tract, along with an array of intestinal enzymes and high shear fluid flow reduce the therapeutic efficacy of oral drugs. Although several delivery paradigms including microparticles have been developed, their non-uniform size distributions and symmetrical radial-directional drug release limit oral bioavailability. To overcome these issues, we demonstrate the use of microfabricated thin, planar, asymmetric, poly(methyl methacrylate); PMMA biocompatible microdevices to facilitate the delivery of oral drugs across the intestinal epithelium. Multiple reservoirs were introduced into the PMMA microdevices for an independent and controlled release of differentially tagged model drugs (BSA) across an *in vitro* caco-2 monolayer. Intestinal epithelia-specific targeting protein, tomato lectin was also conjugated to the microdevice surface to provide enhanced bioadhesion. An *in vivo* mouse model was used to study the bioadhesion, retention, and delivery of a poorly absorbable and poorly soluble model drug Acyclovir across the intestine. The use of unidirectional, flat microdevices, with multiple drug loading capacity can prove to be a simple yet powerful method to greatly enhance the therapeutic efficacy of oral drugs.

(51g) Application of Novel Ionic Liquids for Cutaneous Drug Delivery

Zakrewsky, M., University of California Santa Barbara
Lovejoy, K., Los Alamos National Laboratory
Kern, T.,
Miller, T.,
Del Sesto, R.,
Fox, D.,
Mitragotri, S., University of California Santa Barbara
Originally presented on: 11/4/2013 10:18:00 - 10:36:00

The skin is the largest organ in the body. It provides a compliant interface for needle-free drug delivery, while avoiding major degradative pathways associated with the GI tract. These result in improved patient compliance and sustained and controlled release compared to other standard delivery methods. Concurrently, for the treatment of skin related diseases (e.g. bacterial infection, skin cancer, psoriasis, atopic dermatitis, etc.) cutaneous application provides targeted delivery to the diseased site, allowing the use of more potent therapeutics with fewer systemic side effects. Unfortunately, the outer layer of the skin – the stratum corneum (SC) – presents a significant barrier to most foreign material. This is particularly true for large hydrophilic molecules (>500Da), which must partition through tortuous lipid channels in the SC to penetrate deep tissue layers where the majority of skin-related diseases reside.

Interestingly, over the last few decades ionic liquids (ILs) have emerged as a burgeoning class of designer solvents. Indeed, ILs have been proven beneficial for use in industrial processing, catalysis, pharmaceuticals, and electrochemistry to name a few. The ability to modulate either the cation or anion individually presents an advantageous framework for tuning secondary characteristics without sacrificing the primary function of the IL. We report, for the first time, the use of novel ILs for cutaneous drug delivery. The ILs employed were rationally designed by incorporating known chemical penetration enhancers (CPEs), and retained tunability of potential secondary properties (e.g. viscosity, hydrophobicity/hydrophilicity, antibacterial/antibiofilm activity, thermal stability, etc.) through manipulation of ion pairing. The work presented shows the potential of ILs for cutaneous drug delivery.

(51h) Zeolite Microneedle: Design for Drug Delivery

Poon, H. Y., The Hong Kong University of Science and Technology
Wong, W. K.,
Chau, L. Y., The Chinese University of Hong Kong
Han, W., The Hong Kong University of Science and Technology
Kwan, S. M., Hong Kong University of Science and Technology
Sim, V., The Hong Kong University of Science and Technology
Lee, M. H., The Hong Kong Polytechnic University
Chow, H. L., The Chinese University of Hong Kong
Yeung, K. L., The Hong Kong University of Science and Technology
Originally presented on: 11/4/2013 10:36:00 - 10:54:00

The most common routes for drug delivery are by oral administration, hypodermic injection and transdermal delivery. For transdermal delivery, the rigid stratum corneum is an effective barrier against the effective delivery of drugs. Different innovative strategies were developed to breach this barrier including iontophoresis, phonophoresis, skin ablation using chemicals, laser or radiofrequency and microneedle. Microneedle offers a simple, painless and less invasive transdermal delivery of drug that can be inexpensively manufactured and implemented. Unlike hypodermic injection, less medical wastes are produced and it poses a smaller risk for disease transmission.

This work presents for the first time the use of porous zeolite as material for microneedle for use in drug delivery. The study includes the fabrication of zeolite microneedle, mechanical properties of zeolite microneedles and its use for transdermal delivery of insulin and influenza vaccine in rodents (i.e. rats and mice animal models). Pure silica zeolite that has similar chemical composition as glass but porous, was grown on needle template to create an array of free standing hollow zeolite microneedles. The needle lumen and length were designed for efficient drug delivery, while maintaining good mechanical strength for skin insertion. The zeolite growth conditions were adjusted to obtain different inter- and intra-crystalline pore size for controlled drug delivery and strength that has a high safety margin comparable to metal microneedles.

The zeolite microneedles did not cause noticeable skin irritation or redness during their use in the animal study. The *in vitro* and *in vivo* studies of insulin delivery using zeolite microneedles were carried out. The blood glucose and insulin levels in diabetic rats were monitored during the insulin study for the *in vivo* study. Influenza vaccination using zeolite microneedle was done on mice, and the IgG level in blood serum was measured to determine the effectiveness of microneedles for vaccination.

(78a) Fairytale: A High-Throughput TAL Effector Synthesis Platform

Liang, J., University of Illinois at Urbana-Champaign
Chao, R., University of Illinois at Urbana-Champaign
Abil, Z., University of Illinois at Urbana-Champaign
Bao, Z.,
Zhao, H., University of Illinois at Urbana-Champaign
Originally presented on: 11/4/2013 8:30:00 - 8:48:00

Recombinant transcription activator-like effectors (TALEs) have been effectively used for genome editing and gene regulation applications. Due to their remarkable modularity, TALEs can be tailored to specifically target almost any user-defined DNA sequences. Here, we introduce fairyTALE, a liquid phase high-throughput Golden Gate TALE synthesis platform capable of producing TALE-nucleases, activators, and repressors that recognize DNA sequences between 14 and 31 bp. It features a highly efficient reaction scheme, a flexible functionalization platform, and fully automated robotic liquid handling that enable the production of hundreds of TALEs within a single day with over 98% assembly efficiency at a reagent cost of just $5 per TALE – from desired targets to expression-ready plasmids. As proof of concept, we synthesized and tested 90 TALEs, each recognizing 27 bp, without restrictions on their sequence composition. 96% of these TALEs were found to be functional, while sequencing confirmation revealed that the non-functional constructs were all correctly assembled. FairyTALE offers an unprecedented combination of efficiency, fidelity, speed, and cost. We believe it has the potential to revolutionize the use of TALEs for genome scale applications.

(78b) Gradient Diversity Enriches Combinatorial Protein Library Design

Woldring, D., University of Minnesota
Hackel, B. J., University of Minnesota
Originally presented on: 11/4/2013 8:48:00 - 9:06:00

Efficient evolution of molecular recognition reagents is vital to clinical targeting and biotechnology. We examined the hypothesis that the most efficient evolution is achieved with a combinatorial library exhibiting a gradient of diversity from extensive diversity in the paratope hot spot to full conservation in the framework. Importantly, this gradient includes moderate diversity, with structural bias, within the paratope interfacing with target yet peripheral to the hot spot. Moreover, more mild diversity is included adjacent to the interfacial residues to yield optimal intramolecular contacts with the newly identified paratope. We assessed this approach to library design within the context of two distinct protein topologies: the beta sandwich type III fibronectin domain with the paratope in three loops and the three-helix bundle affibody domain with the paratope on one surface of two helices. A range of diversities was introduced into sites throughout the paratope and its periphery. The resultant naïve combinatorial libraries were screened for specific binding to four different protein targets using yeast surface display and magnetic and fluorescence binding selections. Thorough sequencing of binding ligands quantitatively revealed the evolutionary efficacy of each diversity option at each site. The proposed evolutionary benefit of gradient diversity was evident from the sequence analysis. Current work on second-generation libraries, and continued expansion of the target pool, will fine tune the optimal diversity for each site within both topologies. These results improve the evolutionary efficiency of two validated ligand scaffolds and provide a generalizable roadmap for library design in other scaffolds, which is facilitated by the current ease and power of high throughput selection and sequencing.

(78c) A Survey of Epistasic Effects in TEM-1 Beta-Lactamase

Gonzalez, C., Johns Hopkins University
Firnberg, E., Johns Hopkins University
Ostermeier, M., Johns Hopkins University

Originally presented on: 11/4/2013 9:06:00 - 9:24:00

Epistasis is the interaction between two mutations such that the effect of the mutations together differs from what one would expect based on the effect of the individual mutations. Models and recent experiments suggest that epistatic interactions between mutations play a crucial role in molecular evolution. We exploited recent advancements in mutagenesis techniques and deep-sequencing technology to extensively test the pervasiveness and types of epistasis in *TEM-1* beta-lactamase, a convenient model protein for the study of protein evolution and the fitness effects of mutations. *TEM-1* confers high resistance to penicillin antibiotics such as ampicillin (Amp). Thus, when *E. coli* cells bearing *TEM-1* are challenged to grow in the presence of Amp, gene fitness can be directly connected to organismal fitness. Using PFunkel mutagenesis, we constructed a comprehensive codon mutagenesis library of *TEM-1* beta-lactamase designed to contain all possible single codon subsitutions in *TEM-1*. We used a synthetic gene circuit to partition this library based on relative fitness for Amp resistance and performed deep sequencing on the resulting sublibraries to determine the fitness effects of 96% of the amino acid mutations in *TEM-1*. We then introduced into this library random second codon substitutions using PFunkel mutagenesis and measured the fitness of a set of several thousand double mutants in an analogous manner. The extent and types of epistasis observed will be presented, and the implications for natural and applied evolution will be discussed.

(78d) In Vitro Ribosome Construction for Ribosome Engineering

Fritz, B. R., Northwestern University
Jewett, M. C., Northwestern University
Originally presented on: 11/4/2013 9:24:00 - 9:42:00

The ribosome is a complex macromolecular machine responsible for synthesis of proteins with diverse biological functions. The extraordinary synthetic capability of the ribosome has long made the isolation of engineered variants with altered functions an attractive prospect. For example, limited studies have shown that modifying ribosomes can enhance tolerance of unnatural amino acid substrates. Unfortunately, constraints imposed by living cells limit efforts to make engineered ribosomes *in vivo*. A simple, purely *in vitro* strategy for ribosome construction would offer a powerful solution for studying and exploiting modified ribosomes. To that end, we have developed an integrated synthesis, assembly, and translation technology (termed iSAT) for the *in vitro* construction of *Escherichia coli* ribosomes in crude S150 extracts. iSAT allows for the simultaneous synthesis of ribosomal RNA (rRNA), assembly of rRNA with purified ribosomal proteins, and translation of a reporter protein (*e.g.* luciferase) as a measure of ribosome activity in the same compartment. Here we report the improvement of iSAT activity by ~45-fold as compared to the previous state-of-the-art by tuning the transcriptional balance of rRNA and mRNA production. Specifically, modifications to the 3' rRNA gene sequence to increase transcriptional efficiency, optimization of plasmid and polymerase concentrations, and use of native rRNA operon genes for stoichiometrically balanced rRNA synthesis and native rRNA processing resulted in increases in iSAT activity. With enhanced iSAT activity, we are now developing and will also report on a novel *in vitro* ribosome evolution strategy for ribosome engineering. Specifically, we are using iSAT to generate a library of ribosomes with rRNA mutations and selecting for functionality using ribosome display. Looking forward, we believe our platform will enable *in vitro* evolution to select for ribosomes that have enhanced functions or altered chemical properties.

(78f) High-Throughput Screening of Biological Hydrogen Production

Koo, J., Stanford University
Schnabel, T., Stanford University
Shiigi, S., Stanford University
Swartz, J. R., Stanford University
Originally presented on: 11/4/2013 10:00:00 - 10:18:00

Hydrogen (H_2) has the potential to become a sustainable fuel depending on its production process. One possible process involves biological H_2 evolution using enzymes known as hydrogenases. The maximum catalytic activity of [FeFe] hydrogenases can produce H_2 at a turn over frequency (TOF) in the order of 10,000 s^{-1}. However, these hydrogenases are highly sensitive to oxygen (O_2), and it is not currently possible to combine [FeFe] hydrogenases with solar water splitting where O_2 is a byproduct. For these reasons, there is great interest in using protein engineering to evolve O_2-tolerance in these enzymes. Previous work has discovered a mutant of the *Clostridium pasteurianum* [FeFe] hydrogenase (CpI) with enhanced O_2-tolerance during H_2 consumption. However, tolerance was not observed when the mutant was producing hydrogen. We therefore sought to develop a high-throughput assay to screen for oxygen tolerance during H_2 production. It has been previously shown that hydrogen molecules can undergo a series of reactions in thin films of Pd-Pt/WO_3 to reduce the WO_3 layer. The reduction is accompanied by a change in color. Our screen uses such a sensor to monitor hydrogen production in the individual wells of a 96-wells plate. A custom fabricated elastomeric gasket provides a seal between the sensor plate and a 96-well plate. A standard CCD camera provides time-lapse images and image analysis software gives a quantitative assessment of the rate of color formation. The device is also designed so that gases of known H_2 partial pressure can be injected below the 96-wells plate for calibration purposes. We will describe the device and show results from initial evaluation of mutated hydrogenases.

(78g) Engineering Molecular Escape Artists - Human Lysozyme Variants Evade Pathogen-Derived Inhibitory Proteins

Griswold, K. E., Dartmouth
Dostal, S.,
Fang, Y.,
Originally presented on: 11/4/2013 10:18:00 - 10:36:00

Lysozymes are lytic hydrolases that break the b-(1,4) glycosidic bond between N-acetylmuramic acid and N-acetylglucosamine in bacterial peptidoglycan. These enzymes catalytically degrade bacterial cell walls, ultimately causing lysis and death. Lysozymes constitute a central element of innate immunity in the animal kingdom, and their role in protecting against bacterial pathogens has been experimentally confirmed in various animal models. In humans, the conventional or C-type lysozyme is broadly distributed throughout various tissues, secretions, and cell types, and therapeutic administration of recombinant human lysozyme represents an interesting possibility for drug-resistant infections. One barrier to such an approach is the fact that some pathogenic bacteria have evolved resistance mechanisms to this powerful antibacterial agent. Of particular importance to this talk is the recent discovery of proteinaceous lysozyme inhibitors among numerous Gram-negative bacteria. For example, *Escherchia coli, Pseudomonas aeruginosa, Burkholderia cepacia* and other pathogens produce a periplasmic protein termed "Inhibitor of Vertebrate Lysozyme", or Ivy. In *E. coli*, Ivy represents a potent inhibitor of C-type lysozymes having low nanomolar affinity and aiding bacterial survival in biological fluids such as hen egg white and human saliva. Here, we describe our efforts to engineer variants of human lysozyme able to escape the *E. coli* Ivy inhibitor. We employed a molecular model to identify key lysozyme residues at the binding interface, and we subjected these sites to combinatorial mutagenesis creating millions of mutants. To screen the library, we co-encapsulated yeast expression hosts with bacterial target cells in 50 micron hydrogel droplets. These gel microdroplets were incubated in yeast induction media supplemented with high concentrations of the inhibitory Ivy protein. Hydrogel droplets encasing performance-enhanced yeast clones were selectively stained with a fluorogenic viability probe and then isolated by fluorescence activated cell sorting. Two engineered enzymes were ultimately found to retain high inherent activity while also demonstrating 10- to 100-fold reduced affinity for the *E. coli* Ivy inhibitor. Our results provide new and unexpected insights into molecular recognition between human lysozyme and pathogen-derived Ivy proteins, and they suggest that further molecular engineering could yield novel lysozymes with clinical utility.

(78h) Evolution of Inteins for Mammalian Cell Expression

Coolbaugh, M. J., The Ohio State University
Wood, D. W., The Ohio State University

Originally presented on: 11/4/2013 10:36:00 - 10:54:00

The lack of a platform technology for the purification of non-mAb proteins remains a major concern in the biopharmaceutical industry. The implementation of the Protein A platform for the purification of monoclonal antibodies has revolutionized the industry. However, no analogous purification platform exists for non-antibody products. Purification tags present one intriguing possibility for the development of such a platform. The use of purification tags is ubiquitous at small scale; however, the cost of tag removal becomes prohibitive upon scale-up. To address this, we have developed self-cleaving purification tags based on intein technology. This intein based system has been well established for expression and purification of proteins in *E. coli*. The major limitation of the technology is due to premature cleavage; intein cleavage during expression leads directly to product losses. Due to premature cleavage, the technology is currently limited to bacterial expression systems. The goal of the current work is to develop inteins that are suitable for use in mammalian cell expression systems.

The current work describes the development of inteins suitable for mammalian cell expression via directed evolution. We have developed a high throughput screen for intein evolution based on yeast surface display. We have combined this screen with a targeted library of intein mutants to select for improved variants. We report progress toward the development of inteins with a shifted pH optimum for cleaving and increased sensitivity to small molecule inhibition that can be used in mammalian cell expression systems. This would represent a significant advance in the purification of non-mAb proteins for the biopharmaceutical industry.

(84a) Synchronized Delivery of Vascular Endothelial Growth Factor With Protease Cues Sustains Activation Toward Vessel Formation

Zhu, S., University of California Los Angeles
Lu, Y., University of California Los Angeles
Segura, T., University of California, Los Angeles
Originally presented on: 11/4/2013 8:30:00 - 8:48:00

Synchronized delivery of vascular endothelial growth factor with protease cues sustains activation toward vessel formation

Suwei Zhu, Yunfeng Lu, Tatiana Segura

Chemical & Biomolecular Engineering Department, University of California Los Angeles

Over 6.5 million patients annually in the United States are affected by wound-related health problems resulting from impeded healing processes. Following cutaneous wounding, dermal angiogenesis -- the formation of new blood vessels to bring in nutritive substances and oxygen -- is important for the reconstruction of the normal aspects of skin physiology and yet to be improved in tissue regeneration. More specifically, the signal protein of vascular endothelial growth factor (VEGF), which enhances the survival, proliferation and migration behaviors of endothelial cells, needs to be presented at the wound site in a cell-responsive and sustainable fashion. We have designed a *nanocapsule delivery system of VEGF (nVEGF)* allowing cargo to be liberated only in the presence of cell-secreted proteases. Unlike conventional protein therapeutic delivery approaches that resort to multi-dose administrations or matrix-assisted retention of growth factors, nVEGF act as a reservoir to sustain modulated release of active VEGF, based on a nanocapsule platform that can be applied alone or within versatile scaffolds.

In our strategy, a nanogel shell comprised of acrylamide based monomers and plamin-sensitive peptide (KNRVK) crosslinker was weaved around VEGF to synthesize protein nanocapsules (nVEGF), which were characterized by transmission electron microscopy (TEM) to be 15~30 nm in diameter, dynamic light scattering (DLS) and enzyme-linked immunosorbent assay (ELISA) for targeted release (Fig. 1). In the aspect of molecular biology, nVEGF altered the phosphorylation profile of VEGF receptor-2 on human umbilical endothelial cells (HUVECs) and downstream signalings into a prolonged activation state significantly different than naked VEGF did (Fig. 2). A three-dimensional *in vitro* assay of HUVEC sprouting and lumen formation showed that one dose of plasmin-degradable nVEGF supported such behaviors of HUVECs while naked VEGF was soon depleted and supplemental additions were required (Fig. 3). *Ex vivo* application of nVEGF on the chicken chorioallantoic membrane stimulated the formation of capillaries (Fig. 4). In mouse skin wound healing model, the application of nVEGF within fibrin matrices promoted a faster wound closure with a significantly higher density of infiltrated endothelial cells and capillaries (Fig. 5).

Our results demonstrated plasmin-degradable VEGF nanocapsules sustained the activity of VEGF (preventing clearage and degradation) and achieved the same degree of angiogenesis as re-dosing with exogenous VEGF. Here enzymatically-triggered release of VEGF from the nanocapsule reservoir synchronized the dosage with tissue endogenous protease levels. By reducing the frequency of therapeutic administration, such applications of growth factor nanocapsules would be clinically relevant and practically feasible to continuous deliver active growth factors in regions with proteases and at times of demand.

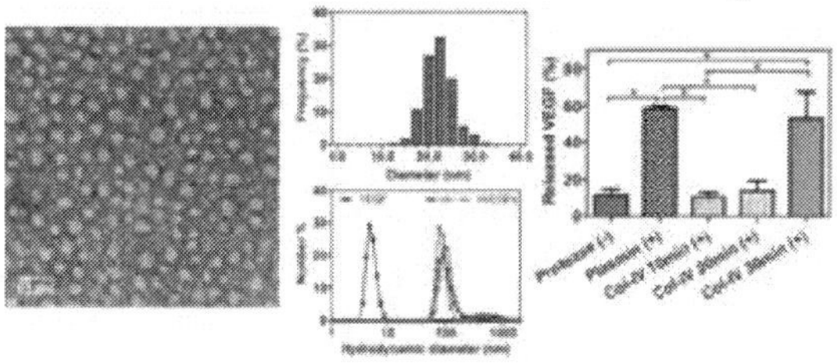

Figure 1 Characterization of the size of nVEGF and the degradability of its nanogel shell to release cargo VEGF. TEM image of purified nVEGF showing spherical particle morphology (left) and measured distribution of nanocapsule diameters (N = 322) (upper center). Hydrodynamic sizes of different batches of nVEGFs and naked VEGF (lower center). Release of VEGF from nVEGF upon protease degradation of nanogel shells (n = 3; mean ± SEM) (right).

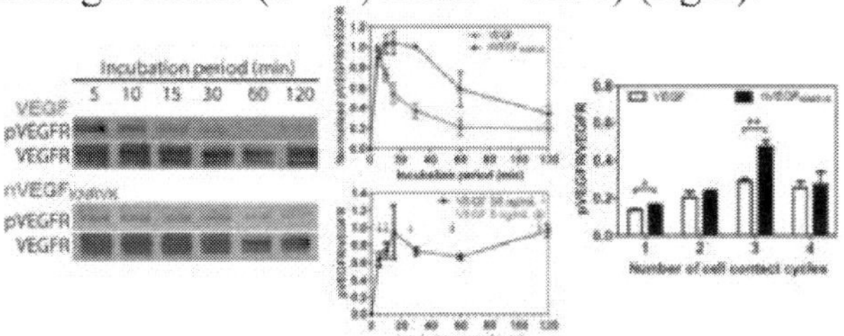

Figure 2 nVEGF prolongs the phosphorylation of VEGFR-*2* (Tyr 1175) due to VEGF release by cell-produced proteases. Western blot analysis from HUVECs exposed to native VEGF and nVEGF for different lengths of time (left) and normalized phosphorylation profiles (n = 3; mean Â± SEM) (upper center). Phospho-ELISA analysis of HUVECs exposed for 30 min to consecutively reapplied VEGF and nVEGF (right) and the prolonged phosphorylation profile reproduced by additions of a small amount of VEGF in the last 5 min of each incubation period (lower center) (n = 3; mean Â± SEM).

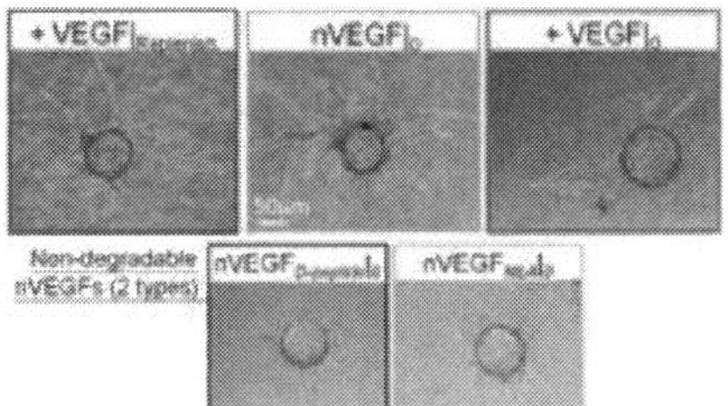

Figure 3 Single dose plasmin-degradable nVEGF supports the sprouting of HUVECs in fibrin matrices over 7 days in the bead assay.

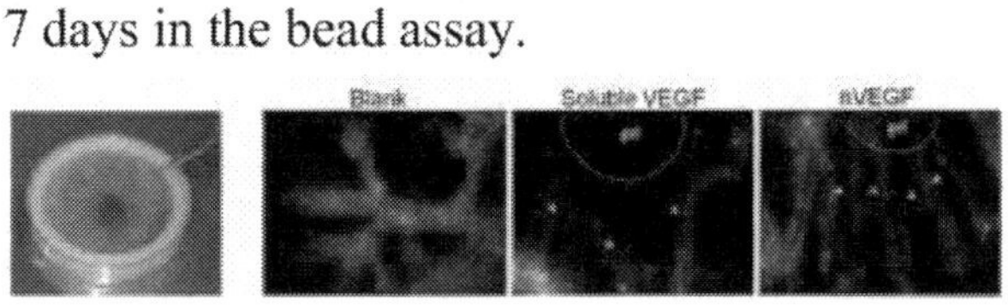

Figure 4 *Ex vivo* implantation of fibrin matrices with VEGF or nVEGF induces capillaries in chicken chorioallantoic membrane assay.

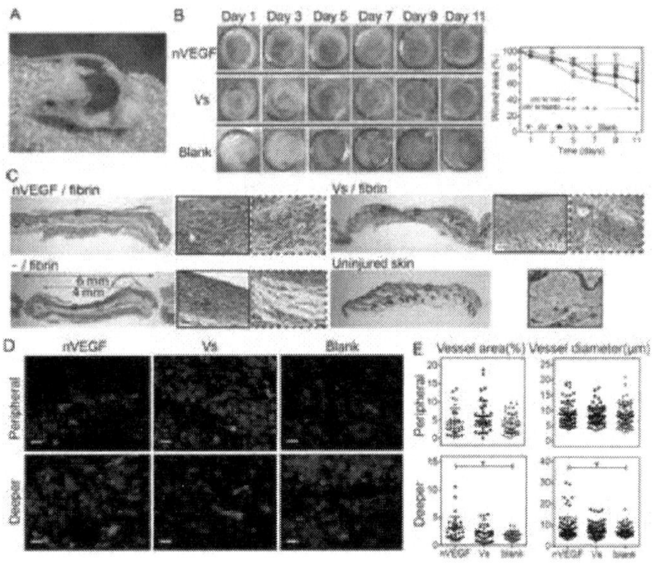

Figure 5 Mouse wound healing model and angiogenesis.

(84b) Gradient Biomaterials Accelerate Vascularization in Vivo

Akar, B., Research Service, Edward Hines, Jr. V.A. Hospital, Hines
Jiang, B., Illinois Institute of Technology
Somo, S., Illinois Institute of Technology
Appel, A., Illinois Institute of Technology
Larson, J., Illinois Institute of Technology
Mehdizadeh, H., Illinois Institute of Technology
Cinar, A., Illinois Institute of Technology
Brey, E. M., Illinois Institute of Technology
Originally presented on: 11/4/2013 8:48:00 - 9:06:00

Controlled angiogenesis is essential for the clinical translation of tissue engineering. Blood vessel formation *in vivo* is influenced by gradients of soluble growth factors. The use of microfluidic systems has enabled the evaluation of cell behaviors in controlled, small volume applications. Studies have shown the importance of gradients of soluble factors on the directed and accelerated growth and migration of cells and vessels. The complex nature and small volume of the microfluidic systems used precludes their application *in vivo*. Recently, we have developed a method for generating gradients of soluble factors within our porous hydrogels which can be easily applied to increase vascular invasion depth *in vivo*.

Composite hydrogels of poly(ethyelene glycol) (PEG) and fibrin were generated using a particulate leaching method. These hydrogels support vascularized tissue formation but at levels insufficient for clinical application.[1]Gradients of soluble factors were generated through the placement of a growth factor source within the hydrogel system distal to the tissue bed. The source used in this case was poly(lactic-co-glycolic acid) (PLGA) microspheres prepared using a double emulsion procedure. When placed in the distal region of the hydrogel system the released proteins diffuse to the tissue bed generating a gradient within the hydrogel from the microspheres to the tissue that promotes vessel invasion.

Platelet derived growth factor BB (PDGF-BB) was chosen for the following studies as it has been shown to stimulate directed growth of multiple cell types and plays an important role in vessel maturation and stabilization. Cell and tissue response depends, in part, on the magnitude of the gradient. The release of PDGF-BB was varied based on the composition of the PLGA microspheres. Using this technique, persistent gradients (greater than 50 days) were generated within the hydrogel scaffolds. The gradient, formed by the transport of PDGF-BB through the hydrogel, was modeled following Fick's second law, which assumes that diffusion is the only mechanism of transport and that the proteins do not interact (react) with the porous structure as they diffuse. The gradient in the first 5 days is higher due to the burst release exhibited by the microspheres. From 10 days the concentration profile gradually decreases as the microspheres approach zero order (constant) release. The model predictions were verified by measuring transport from the gel using radiolabeled proteins.

In order to further optimize our release system prior to analysis in a murine subject, an agent-based model of angiogenesis in engineered tissues was extended to enable testing the influence of biomaterial structures and growth factor release rates on vascularization.[2] Parameters, including the magnitude of the concentration gradient, pore size and pore interconnectivity were varied and the depth and extent of vascularization evaluated *in silico*. These studies suggest that under optimized hydrogel conditions, gradients of soluble factors can increase the rate at which vessels invade the engineered tissues.

A murine subcutaneous implant model was used to examine the influence of gradients and the magnitude of gradients on vascularization. PEG/Fibrin composite hydrogels with increasing concentrations (0, 1, 10, and 100 μg/ml) of PDGF-BB were prepared and implanted. The concentration was varied to investigate the role of gradient magnitude on vascular invasion. The hydrogels were harvested at 1, 3, and 6 weeks. Vascularization was evaluated using lectin staining and confocal imaging. Tissue structure was evaluated with histological stains.

The gradient hydrogels resulted in significantly higher vascular density within the hydrogels after 1 and 3 weeks of implantation in comparison to controls. At the highest concentration (100μg/ml), the depth of vessel invasion was 1.5X greater than any other conditions we have studied in the same model, including multiple growth factor systems described in other studies. These results show that we have developed a novel method for generating persistent gradients within porous hydrogel systems that can be implanted *in vivo*. The gradients rapidly accelerate vascularization into the porous hydrogels. We are currently examining vascularization in gradient hydrogels at longer time points.

References:

1. Jiang B. *et al.*, Fibrin Loaded Porous Poly(ethylene glycol) Hydrogels Induce Vascular Network Formation. *Tissue Engineering Part A*. 19(1-2):224-34, 2013.

2. Mehdizadeh H. *et al.*, Three-Dimensional Modeling of Angiogenesis in Porous Biomaterial Scaffolds. *Biomaterials*. 34(12):2875-87, 2013.

(84c) Engineering a Three-Dimensional Model of Vascularized Human Tumor Tissue

Ehsan, S. M., University of California, Irvine
Alonzo, L. F., University of California, Irvine
Waterman, M. L.,
Hughes, C. C. W., University of California, Irvine
George, S. C., University of California, Irvine
Originally presented on: 11/4/2013 9:06:00 - 9:24:00

The increased awareness that 3D culture systems provide a more physiologically relevant environment compared to 2D monolayers has stimulated interest in creating multicellular tumor spheroids to study cancer *in vitro*. While there have been significant advances in "tumor engineering", few systems are able to capture the subtlety of complex *in vivo* processes such as angiogenesis and neovascularization. Here, we introduce a novel 3D *in vitro* model of tumor angiogenesis that features cellular heterogeneity, stromal driven vessel formation, and direct cell-cell contact of cancer cells and endothelial cells. Co-cultures of endothelial colony forming cell-derived endothelial cells (ECFC-ECs) from cord blood and human epithelial-based tumor cells of breast (MDA-MB-231), colon (SW620), or lung (A549) origin were formulated into multicellular spheroids. The spheroids were then embedded in a fibrin gel distributed with a normal stromal cell (primary human lung fibroblasts). After 7 days of culture in either 20% or 5% O_2 conditions, the resulting vessel networks were imaged using confocal microscopy. The ECFC-ECs exhibited robust sprouting angiogenesis from the spheroid, and also formed a distinct and contiguous vessel network within the spheroid itself. The model was amenable to all four different cell lines, each demonstrating a characteristic influence on the resulting vessel networks. Interestingly, the model cell line SW620 also exhibited marked intravascular tumor cell migration. Individual tumor cells were tracked as they migrated within the lumens of the microvessels. Tissues cultured in 5% O_2 exhibited significantly enhanced intravascular cell migration compared to 20% O_2, consistent with epithelial to mesenchymal transition. Our results demonstrate the utility of a model system that may be used to study how different cancer cell types communicate with normal endothelial cells to direct microvessel formation, and intraluminal cell migration. This flexible and reproducible experimental platform has the potential to improve the clinical relevance of 3D *in vitro* culture systems by not only providing a more accurate simulation of *in vivo* tumor angiogenesis, but also enabling the delivery of anticancer agents via an engineered vascular network.

(84d) Modular Design of Microfluidic Hydrogel Scaffold for Bone Marrow Bioengineering

Lee, J., University of Michigan, Ann Arbor
Kohl, N.,
Shanbhag, S., University of Michigan
Yarmush, M., Massachusetts General Hospital, Harvard Medical School
Parekkadan, B., Massachusetts General Hospital, Harvard Medical School
Originally presented on: 11/4/2013 9:24:00 - 9:42:00

There is a critical need to develop micro-physiological human bone marrow tissue models to facilitate fundamental understanding of bone marrow biology and to improve the predictive power of bone marrow toxicity testing. Various scaffolds and microfluidic designs have been developed to simulate complex and dynamic bone marrow microenvironment; however, these platforms fall short of capturing full-spectrum anatomical, physical and hydrodynamical features of the bone marrow niche in a single platform that is essential for mechanistical studies. Here we introduce a modular design of microfluidic hydrogel scaffold that systematically increases the bone marrow tissue complexity combining a marrow-mimicking hydrogel scaffold with a microfluidic device. Hydrogel scaffolds prepared using colloidal crystals as a template closely emulated the physical and anatomical features of bone marrow to scale. The hydrogel scaffold was then integrated into a PDMS-based fluidic chamber by taking advantage of dramatic and reversible volume change (~78%) between a hydrated and a dehydrated state. The relative flow rate of a scaffold-chamber was ~90% of an empty chamber and its perfusion profile recapitulated both vascular convection and tissue diffusion in the bone marrow sinusoids because of the nature of the biomaterial design. Primary human bone marrow stromal cells were effectively seeded into a hydrogel scaffold in a chip and maintained high viability under perfusion culture condition. Finally as a proof of concept we demonstrated the dynamic interaction between floating human Nalm-6 leukemic cells and adhered human bone marrow stromal cells in the scaffold-chamber. We envision that the presented modular design of marrow-mimicking microfluidic scaffolds would provide a new preclinical testing dimension for various bone marrow related in vitro studies with advanced analytical and predictive power.

(84e) An Interactive Engineered Protein Hydrogel: Controlling and Responding to Neurite Growth

Lampe, K. J., Stanford University
Heilshorn, S. C., Stanford University
Originally presented on: 11/4/2013 9:42:00 - 10:00:00

Neural regeneration within the central nervous system (CNS) is a critical unmet challenge as brain and CNS disorders continue to be the leading cause of disability nationwide. Common tissue engineering goals seek to customize cell-biomaterial interactions and guide cell behavior. Here we have developed a material that is both cell instructive and cell responsive, creating a dynamic interplay between cells and their engineered extracellular matrix (ECM). Using recombinant protein technology, we engineered a family of elastin-like protein hydrogels with multiple independently tunable properties. The concentration of adhesive ligands such as the fibronectin-derived sequence RGD can be precisely controlled without altering the mechanical properties of the hydrogel. When dorsal root ganglion neurons were cultured in 3D in these gels, RGD ligands at 1.9 x10^7 ligands/μm^3 promoted neuron-specific growth, and more than doubled the rate of neurite extension compared to hydrogels without the adhesive sequence. Crosslinking density was tuned to create scaffolds with elastic moduli from 0.5-2.1 kPa with constant adhesive site densities. The most compliant gels led to the greatest outgrowth from encapsulated DRGs with neurites extending over 1800 μm by day 7. In contrast, the stiffest gels permitted far fewer extensions and limited outgrowth to a maximum of 600 μm over the same time frame.

We have designed the proteins with a second set of bioactive sequences that specifically respond to changes in cell phenotype. Neural stem cells (NSCs) undergoing differentiation may change their production of the protease urokinase plasminogen activator (uPA), which has previously been found at the growth cones of extending neurites. By incorporating cell-mediated degradable subunits into the elastin-like proteins, we are able to mimic the natural remodeling of the ECM. We engineered multiple uPA target sites with different degradation kinetics into the elastin-like protein to allow neural cell-mediated control of the scaffold degradation dynamics. DRGs encapsulated in uPA degradable scaffolds extended neurites at a faster rate than in gels of similar RGD density and initial stiffness without the uPA degradable sequence.

These crosslinked scaffolds are useful for directing the growth and differentiation of multiple cell types including clinically relevant NSCs. The tunable scaffolds are responsive to neuronal cells which may be able to specifically self-modulate the release of multiple bioactive factors while undergoing differentiation. This work demonstrates the versatility and responsiveness of our modularly-designed protein hydrogels for neural cell culture and encourages continued development as a biomaterial tissue construct for treating spinal cord injury.

(84f) Anisotropic Chitosan-Based Hydrogels for Cartilage Regeneration

Walker, K. J., Oklahoma State University
Madihally, S. V., Oklahoma State University
Originally presented on: 11/4/2013 10:00:00 - 10:18:00

Cartilage is a tough, semi-transparent, elastic tissue that assists in reducing friction facilitating bone movement. Cartilage could be ravaged by various diseases and regular physical activities; osteoarthritis is the most common joint disorder requiring significant medical aid. Tissue engineering or regeneration techniques offer alternative strategies, but require seeding immuno-compatible cells to colonize biodegradable scaffolds to circumvent immune rejection. Before cells can be seeded to promote the regeneration of the tissue, scaffolds made of biomaterials with biomimetic properties must be formulated. Both synthetic and natural biomaterials have been explored and processed differently to regenerate tissues both *in vivo* and *in vitro*. Further, different processing strategies have led to the development of novel surgical procedures. More recently, hydrogels have been desired for cartilage scaffolding because the hydrogel offers an environment similar to physiological conditions. Also, injectable hydrogels have been extensively explored because they offer a minimally invasive alternative to arthroscopic surgeries and ease of incorporation of cells and biologically active agent. The objectives of this study are to investigate the bio-mimicry of an anisotropic chitosan-based hydrogel that represents the multilayered architecture and properties of native cartilage.

Hydrogels have been formulated isotropically to match three of the four layers of articular cartilage. The hydrogels are chitosan based and each combined with a component that is found in native cartilage. The additional components for the layers are gelatin, hyaluronic acid (HA), and beta tri-calcium phosphate (β-TCP) superficial, radial, and calcified cartilage zones. The transitional zone is formed by the anisotropy of the superficial and radial zones. Mechanical, rheological, and physical characteristics were assessed on the chitosan-gelatin/ HA/ β-TCP hydrogels of different concentrations at the physiological conditions. After isotropic analysis of the hydrogels, anisotropy of the hydrogels was tested to establish a hydrogel morphology that is consistent with cartilage.

The compressive modulus was significantly improved by increasing the chitosan concentration. Further, increasing the chitosan concentration of the hydrogels increased the rate of gelation as well as the structural integrity at physiological diameters. Cyclical tests demonstrated repeatable strength and durability with no sign of failure in the cycle range. Anisotropic gels were formed with different gradations of chitosan-based gels.

(84g) 3-Dimensional Osteochondral Tissue Engineering

Rivera, A. L., Case Western Reserve University
Mulcahy, B.,
Chou, C. L., Case Western Reserve University
Sakai, T., Cleveland Clinic Lerner Research Institute
Caplan, A. I., Case Western Reserve University
Welter, J. F., Case Western Reserve University
Baskaran, H., Case Western Reserve University
Originally presented on: 11/4/2013 10:18:00 - 10:36:00

Osteoarthritis affects millions of individuals and is characterized by joint pain caused by articular cartilage wear. Cartilage, due to its avascular nature, has poor self-repair capabilities, which result in mechanically inferior repair tissue. One of the more successful current clinical techniques involves replacing the damaged cartilage and part of the underlying subchondral bone with osteochondral grafts. However, the availability of such grafts is limited and if taken from the patient, results in another defect site. Therefore, tissue engineering has been investigated as a promising solution for cartilage defects. In this study, we developed a novel gradient generating bioreactor that was used to investigate tissue engineering of osteochondral tissues by using molecular gradients to control human mesenchymal cell (MSC) differentiation. Within the bioreactor, a flow channel with two inlets allows for the flow of molecular gradients across 3-dimensional tissue constructs (**Fig. 1** and **Fig. 2**). Using computational fluid dynamics simulations, we designed the reactor and showed that these flow established gradients can be created and maintained indefinitely (**Fig. 2**). Further, through fluorescent tracer experiments, we demonstrated the generation and maintenance of the gradients of bioactive molecules (**Fig. 3**). To test the effect of molecular gradients on chondrogenesis, we exposed human MSCs within 3-dimensional collagen scaffolds to gradients of transforming growth factor beta 1 (TGF-β1). Chondrogenesis was visualized via toluidine blue staining for glycosaminoglycans and was only observed within the construct regions exposed to higher TGF- β1 concentrations (**Fig. 4**). To develop osteochondral constructs, the tissue constructs were exposed to opposing gradients of chondrogenic and osteogenic media. The bioreactor designed here allows for the production of stable molecular gradients that can be controlled by adjusting various parameters including the inlet flow rates, inlet molecular concentrations, and tissue construct position within the reactor (distance from inlets). The reactor can also serve as a platform to study the effect of molecular gradients on other composite tissue constructs.

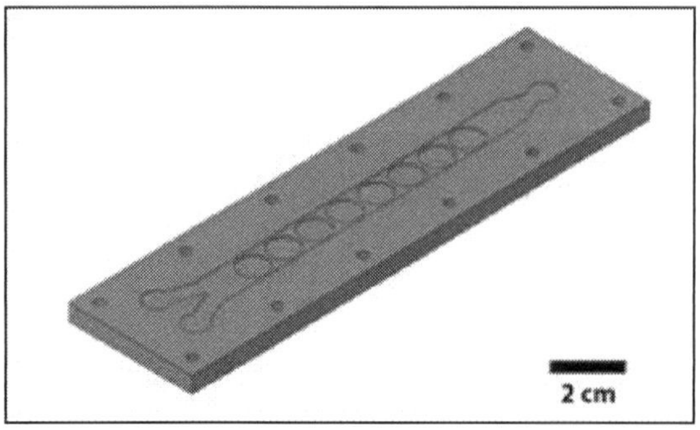

Figure 1: Bioreactor Design. Two inlets combine into a single flow channel to form a continuous molecular gradient across the width of the channel. Three dimensional tissue constructs are placed inside circular wells below the flow channel.

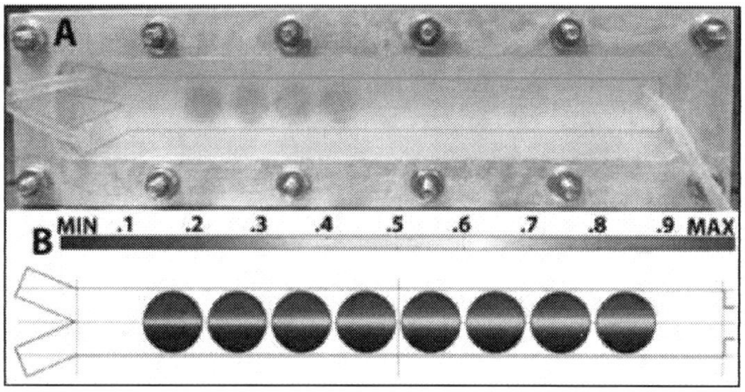

Figure 2: Bioreactor and Gradient Simulation. A. Assembled bioreactor containing MSC-collagen sponge constructs within the first 4 wells. B. Gradient simulation across each of the wells of the bioreactor.

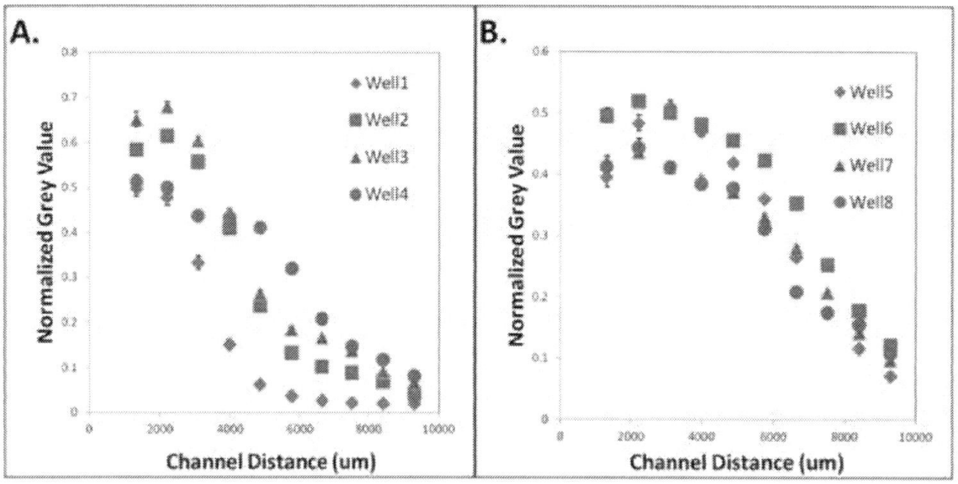

Figure 3: Fluorescent Gradient Characterization. Fluorescent intensity (normalized grey value) was measured across the width of the flow channel at each well containing collagen sponges. Wells are numbered starting from the well closest to the inlets (well 1). A. Wells 1-4. B. Wells 5-8.

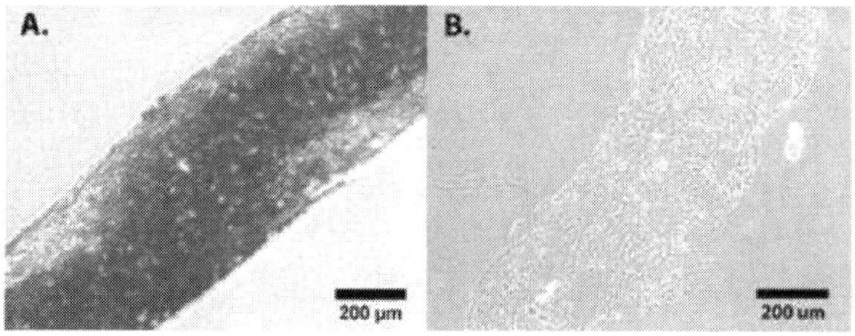

Figure 4: Chondrogenesis of MSC-Collagen Sponge Constructs. Toluidine blue staining for MSC-collagen sponge constructs exposed to A. higher TGF-β1 concentrations and B. lower TGF-β1 concentrations along the gradient.

(84h) Phenotypic Control of Valvular Interstitial Cells Using Self-Assembled Monolayers

Hedberg-Dirk, E. L., University of New Mexico, Center for Biomedical Engineering
Rush, M. N., University of New Mexico
Originally presented on: 11/4/2013 10:36:00 - 10:54:00

Current research in heart valve tissue engineering focuses on the use of valvular interstitial cells (VICs) for repair and regeneration of healthy, living valve tissue, with the goal of synthesizing replacement tissue for diseased valves. As myofibroblasts, VICs exhibit phenotypic plasticity in response to their micro-environment. However, the specific cues that regulate phenotypic expression and ECM production are not well understood. This work aims to isolate the effects of material surface chemistry on VIC behavior *in vitro* using well-defined, homogeneous self-assembled monolayers (SAMs). Due to the large quantities of negatively charged molecules present during valvogenesis, it was hypothesized that a negatively charged environment would induce VICs to an active, ECM producing phenotype. To examine the influence of physiologically relevant functional groups on attachment, proliferation, and differentiation of VICs, SAMs presenting CH_3 (hydrophobic), OH (hydrophilic), COO^- (negative at physiological pH), and NH_2^+ (positive at physiological pH) end groups were fabricated. Surfaces were consistent with previous literature as determined by static sessile drop contact angle and x-ray photoelectron spectroscopy (XPS). Initial seeding of VICs showed uniform attachment on all surfaces except CH_3, which exhibited a 50% reduction in cellular binding. Over seven days in culture, VICs on the negatively charged COO^- surfaces exhibited the greatest rate of proliferation (MTT assay) and alpha-smooth muscle actin expression (immunocytochemical staining, qPCR) among the treatment groups indicating a healthy, activated phenotypic state. Amine (NH_2^+) surfaces also showed increased proliferation and aSMA expression but exhibited indications of differentiation to an osteobastic-like, diseased state with changes in morphology, formation of calcified nodules, and production of osteocalcin. These results support our hypothesis that the negatively charged surface of the carboxyl terminated SAMs leads to an environment that induces the healthy, active VIC phenotype most suited for use in tissue engineering applications. Further studies are currently being conducted to examine integrin expression as well as tissue production.

(107a) Enhanced Biofuel Production Through Coupled Acetic Acid and Xylose Consumption By Engineered Yeast

Wei, N., University of Illinois at Urbana Champaign
Quarterman, J. C., University of Illinois at Urbana-Champaign
Kim, S. R., University of Illinois at Urbana-Champaign
Cate, J. H. D., University of California at Berkeley
Jin, Y. S., University of Illinois at Urbana-Champaign
Originally presented on: 11/4/2013 12:30:00 - 12:48:00

The anticipation for substituting conventional fossil fuels with cellulosic biofuels fuels is growing in the face of increasing demand for energy and rising concerns of greenhouse gas emissions. However, commercial production of cellulosic biofuel has been hampered by inefficient fermentation of xylose and toxicity of acetic acid released from hemicellulose in the plant cell wall. While constituting substantial portions of cellulosic biomass, xylose and acetate cannot be fermented to ethanol by *Saccharomyces cerevisiae*. Here we use a redox balancing strategy to enable efficient xylose fermentation and simultaneous *in situ* detoxification of plant cell wall-derived feedstocks. By combining an NADH-consuming acetate consumption pathway and an NADH-producing xylose utilization pathway, engineered yeast converted cellulosic sugars and toxic levels of acetate together into ethanol under anaerobic conditions. The engineered yeast strain showed increased ethanol yields, decreased byproduct accumulation, and improved cell growth during co-utilization of xylose and acetic acid compared to the reference strain without the anaerobic acetate-consuming pathway. The results demonstrate a breakthrough in making efficient use of carbon compounds in cellulosic biomass and present an innovative strategy for metabolic engineering whereby an undesirable redox state can be exploited to drive desirable metabolic reactions, even improving productivity and yield.

(107b) Biologically Upgrading Enzymatically Free Hydrolysate Derived From Lignocellulosic Biomass

Rand, J., University of Wisconsin

Originally presented on: 11/4/2013 12:48:00 - 13:06:00

Chemicals can be produce in a renewable manner from engineered microorganisms from simple sugar feedstocks; however, for these processes to ever be economically viable the feedstocks must be attainable at inexpensive cost. Currently, sugar feedstocks are obtained through the degradation of renewable lignocellulosic biomass through pretreatment and enzymatic hydrolysis steps, but the costs and inefficiencies associated with the pretreatments and enzymes necessary for deconstruction present hurdles to widespread application. The opportunity to bypass enzymatic involvement and yet still microbially produce chemicals from biomass would expand the potential of this renewable method.

Recent work on biomass deconstruction has identified enzymatically free routes that bypass the necessity of enzymatic hydrolysis in the production of potential substitute feedstocks. One route has been discovered that is capable of generating the biomass byproduct levulinic acid (LA) without the use of enzymes. Recently, the Dumesic Lab of University of Wisconsin has developed a novel technique that utilizes the solvent γ-valerolactone (GVL) to deconstruct biomass into the simple sugars glucose and xylose. This talk will discuss the efforts made towards biologically upgrading these enzymatically free feedstocks through identifying an LA metabolic route and evaluating the toxicity of these new feedstocks.

For LA to be considered as a substitute feedstock, it must have the capability to support growth as the sole carbon source. It has been demonstrated that some organisms can grow on LA, such as *Pseudomonas putida*; however, the metabolic pathway remains unknown. In order to elucidate the potential genes involved, a library of *P. putida* mutants was screened and the disrupted genes identified. Current work is involved in isolating the genes that are necessary and sufficient for growth and determining their function. The goal is to use these genes to engineer alternative organisms for LA growth capability.

(107c) Cell Free Metabolic Engineering for Generation of Biohydrogen From Cellulosic Biomass

Lu, F.,
Swartz, J. R., Stanford University
Originally presented on: 11/4/2013 13:06:00 - 13:24:00

Cell Free Metabolic Engineering for Generation of Biohydrogen from Cellulosic Biomass

Franklin Lu, James Swartz

Unpurified cell extracts can be easily created from engineered cell lines through cell lysis while maintaining the majority of cellular proteins in their active forms. Consequently, these extracts have the capability of rapidly producing a large number of enzymes in a cost effective manner for *in vitro* catalysis. One application for these unpurified extracts is the conversion of cellulosic biomass to hydrogen, which has already been demonstrated *in vitro* using purified enzymes. While unpurified extracts are much cheaper to produce, they share similar problems with in vivo metabolic engineering in that side reactions can negatively impact product yield. However, they also differ in that the cell growth and maintenance requirements are removed. Therefore, cell free metabolic engineering can provide unique opportunities as well as challenges.

We have applied a cell free metabolic approach to inactivate competing reactions and improve the yield of a biological method to convert glucose to hydrogen. By generating a library of linear DNA templates, expressing mutants using cell free protein synthesis, and screening with spectroscopic activity assays, we selected mutant enzymes from key metabolic junctions with stable protease insertion sites. These enzymes were inserted into the genome of *E.coli* for *in vivo* expression in cell lines transformed with a plasmid for expression of the corresponding protease into the periplasm. After a growth phase, the cells were lysed, mixing the active protease and mutant enzymes, and thereby selectively inactivating undesirable metabolic pathways. These unpurified extracts were used in conjunction with a three enzyme synthetic pathway to produce hydrogen and improve the conversion yield of glucose to hydrogen.

(107d) Characterization of Modular Designer Xylanosomes for Enhanced Hemicellulose Hydrolysis in Yeast

Srikrishnan, S., University of California, Irvine
Chen, W., University of Delaware
Da Silva, N. A., University of California, Irvine
Originally presented on: 11/4/2013 13:24:00 - 13:42:00

In nature, cellulosomes are displayed on the surface of anaerobes as highly ordered enzyme-complexes that enhance the catalytic efficiency based on enzyme-substrate-microbe synergy. The goal of this research was to functionally assemble and characterize five modular, designer "xylanosomes" on the surface of *Saccharomyces cerevisiae* for the enhanced saccharification of birchwood xylan. An endoxylanase (*Thermomyces lanuginosus)*, a β-xylosidase (*Aspergillus niger*), and an acetylxylan esterase (*Aspergillus awamori*) were successfully displayed on the yeast cell surface from scaffoldins that were designed with or without carbohydrate binding modules (family 22 or family 3 CBMs). The combinatorial effects of enzyme-enzyme synergy (1.6-fold) within the scaffoldin and enzyme-substrate synergy (2.1-fold) on xylan hydrolysis were decoupled, and the influence of choice and position of the CBM was also investigated. The addition of a xylan binding domain from *Thermotoga maritima* improved hydrolysis by 3.3-fold over free enzymes, with a xylose productivity of 105 mg g^{-1} substrate after 72 hours hydrolysis. The results of this study demonstrate the applicability of designer xylanosomes towards hemicellulose saccharification in yeast, and the importance of choice and position of the CBM in these modular constructs. In addition, recent work on similarly constructed cellulosomes for cellulose saccharification, for the production of ethanol and biorenewable chemical precursors will be presented.

(107e) Development of E. Coli Cell Factories for Consolidated Bioprocessing of Hemicellulose to Biosuccinic Acid Through Metabolic Engineering and Cell-Surface Display

Zhao, X. M., Tianjin University
Chen, T., Tiamjin University
Originally presented on: 11/4/2013 13:42:00 - 14:00:00

Bio-based succinate is predicted to be one of the future platform chemicals that can be derived from renewable resources, but it is still faced with the challenge of becoming cost competitive against petrochemical-based alternatives. Consolidated bioprocessing (CBP), incorporating enzyme-generating, biomass-degrading and bioproduct-producing capabilities into a single microorganism, could potentially avoid the cost of the dedicated enzyme generation in the process of hemicellulosic materials utilization. In this study, drawing on the basic principles of metabolic engineering, synthetic biology and cell-surface display, we engineered *Escherichia coli* strains capable of exporting three hemicellulases to the broth for the succinate production directly from beechwood xylan.

Xylanases were extracellular environment-directed by fusing with OsmY. Subsequently, twelve variant OsmY fused endoxylanase-xylosidase combinations were characterized and tested. The combination of XynC-A from *Fibrobacter succinogenes* S85 and XyloA from *Fusarium graminearum* which appeared to have optimal enzymatic properties was identified as the best choice for xylan hydrolysis.

To improve the hydrolysis efficiency and for better meet the needs of the CBP production, further improvements of hemicellulases secretion were investigated by *lpp* deletion, *dsbA* overexpression, *blc* surface displaying. Three xylanases (β-xylosidase, endoxylanase, and α-arabinofuranosidase) were expressed on the surface of the *E. coli*. The xylanases were displayed on the surface of the cells by fusing with anchor proteins, Blc.

The hemicellulolytic capacity was further integrated with a succinate-producing strain to demonstrate the production of succinate directly from xylan without externally supplied hydrolases and any other organic nutrient. The resulting *E. coli* was able to produce 0.37 g/g succinate from xylan anaerobically equivalent to 76% of that from xylan acid hydrolysates. This engineered *E. coli* expressing and secreting three hemicellulases demonstrated a considerable succinate production on the released monosaccharides from xylan. Because a cost-effective process cannot be realized until all three of titer, rate (or productivity), and yield (TRY) have been optimized, so it will need much more work to improve it to the point that it can support a commercial process.

(107f) Fungal Processes for Direct Bioconversion of Cellulose to Hydrocarbons By Ascocoryne Sarcoides

Mallette, N., Montana State University
Hunt, K. A., Montana State University
Strobel, G., Montana State University
Carlson, R. P., Montana State University
Peyton, B., Montana State Univeristy
Originally presented on: 11/4/2013 14:00:00 - 14:18:00

Beyond characterization of cellulolytic fungal enzymes, very little research has examined the potential role of fungi in renewable fuel production. *Ascocoryne sarcoides* (NRRL 50072) is an endophytic fungus that can utilize cellulose as its sole carbon source. This fungus excretes "myco-diesel", a mixture of straight and branched chain hydrocarbons of C_5-C_{12} chain length, in the range of gasoline and aviation fuel, including heptane, 2-pentene, octane, 1-methyl-cyclohexene, 3,5-octadiene, and cyclodecene (Griffin et al. 2010). Experimental results from shake flask and 5 liter reactor runs have verified hydrocarbon compound production under a wide array of growth conditions. Biomass yields in liquid culture have improved from 0.05 g/L to 4.8 g/L through medium improvements. The pH tolerance of *A. sarcoides* is in the acidic range, and optimal temperature is between 16-23°C. Volatiles are released by the fungus throughout its growth cycle and have reached 2 ppm using glucose as a substrate (Mallette et al. 2012). Growth on cellulose produced a higher number of fuel-related hydrocarbons including alkanes and aromatics than growth on glucose. Future research will focus on combining knowledge of genetics with metabolic modeling to develop a deeper understanding of how *A. sarcoides* and other fungi are able to produce fuel-related compounds from cellulosic substrates.

(107g) In Planta Transient Expression of Cell Wall Degrading Enzymes for Biofuel Applications

Lindenmuth, B.,
Hwang, M. S., UC Davis
Anthony, L.,
Falk, B. W., University of California, Davis
McDonald, K. A., University of California, Davis
Originally presented on: 11/4/2013 14:18:00 - 14:36:00

The United States has mandated that 16 billion gallons of cellulosic biofuels be produced annually by 2022. This production process includes the saccharification of the plant cell wall into sugars by cell wall degrading enzymes for fermentation. Currently, these enzymes contribute a significant cost to the total process. Expressing cell wall degrading enzymes recombinantly in the biomass substrate could potentially reduce the costs in this step. In recent years, it has been determined that transient expression in plants can produce high yields of heterologous proteins.

Thermostable endoglucanase E1 and endo-xylanase from *Acidothermus cellulolyticus* were transiently expressed in *Nicotiana benthamiana* utilizing several vectors and a gene silencing suppressor. Gene constructs were designed to target the enzymes to the apoplast, codon optimized for high level expression in the plant host, and synthesized. Genes were inserted into binary vector systems which incorporate the gene into the T-DNA region in *Agrobacterium tumefaciens*. Production of the enzymes using the CaMV 35S, *Tobacco mosaic virus*-based TRBO, and *Cucumber mosaic virus*-based CMVar expression systems were compared. The use of the gene silencing suppressor p19 from *Tomato bushy stunt virus* was shown to improve yields for all systems with expression levels of functional endoglucanase and xylanase reaching levels of 15.2 and 4.3 mg/kg fresh weight, respectively.

Co-expression of E1 and xylanase using the *Cucumber mosaic virus*-based CMVar system was evaluated. E1 and xylanase were co-expressed with simple mixing of *A. tumefaciens* carrying separate expression cassettes. E1 and xylanase were also co-expressed with both genes carried on the same expression cassette. Proteins expressed from the same CMVar cassette were produced in the same cell, whereas proteins expressed with mixed cassettes are produced in distinct cells. This has implications for the thermally activated hydrolysis of the enzyme-producing host tissue for cellulosic biofuel production.

Efficient transient production and thermal activation of multiple lignocellulosic enzymes within the same plant cells may be a cost-effective saccharification route for biofuel production.

(107h) Microbial Production of Renewable Monoethylene Glycol

Pereira, B., Massachusetts Institute of Technology
De Mey, M., Massachusetts Institute of Technology
Lim, C. G., University at Buffalo, the State University of New York
Zhang, H., Tufts University
Stephanopoulos, G. N., Massachusetts Institute of Technology
Originally presented on: 11/4/2013 14:36:00 - 14:54:00

Monoethylene glycol (MEG) is an important commodity chemical which is utilized in such applications as plastic packaging, polyester fabric, and antifreeze. It is derived from fossil fuels, and therefore, to reduce its environmental impact, we propose a single-step bioprocess in which plant-derived carbohydrates are converted by engineered microorganisms into renewable MEG. Toward this goal, we have engineered novel metabolic pathways for the biological production of MEG into strains of *E. coli*. In this work, we validate these pathways and demonstrate the production of MEG from various sugars, achieving titers up to 40 g/L at yields of 0.35 g-MEG/g-xylose.

(109a) Invited Speaker: Design of Synthetic Materials for Spatial, Temporal, and Active Control Over Nucleic Acid Delivery: Surface-Mediated and Redox-Based Approaches

Lynn, D. M., University of Wisconsin-Madison
Originally presented on: 11/4/2013 12:30:00 - 13:00:00

The ability to deliver nucleic acids to cells and tissues exactly when and where they are needed is critical in a broad range of fundamental and applied contexts, ranging from the development of new tools for biotechnology and biomedical research to the development of new therapies and more effective clinical interventions. This presentation will focus on the design of synthetic soft materials that can be used to provide spatial, temporal, and/or external (user-defined) control over the delivery of DNA and siRNA to cells in two different contexts. The first part of the presentation will provide an overview of our recent work on the design and fabrication of ultrathin and erodible polymer-based thin films (polyelectrolyte multilayers) that promote the localized release of DNA from surfaces, with a focus on (i) fundamental aspects of design and assembly that can be used to provide spatial and temporal control over erosion and release, and (ii) the ability of these ultrathin coatings to release or locally transfer therapeutic plasmid DNA constructs from the surfaces of interventional devices and promote beneficial physiological responses *in vivo*. The second part of the presentation will focus on the design of ferrocene-containing cationic lipids that provide redox-based control over the extent to which DNA and siRNA is (or is not) internalized by cells. The unique properties of these redox-active lipids can be exploited to design 'inactive' or dormant lipoplexes that do not enter cells, but that can be transformed into active lipoplexes that promote high levels of transfection or gene silencing by the simple extracellular addition of soluble or solid-phase reducing agents. This redox-based approach thus provides new principles for the active chemical transformation, external and remote activation, and spatial patterning of transgene expression in ways that are difficult to achieve using conventional cationic lipids.

(109b) Well-Defined Cationic Polylactides As Biodegradable Vector for Effective Delivery of Plasmid DNA

Chen, C. K., University at Buffalo
Jones, C. H., University at Buffalo
Jiang, M., University at Buffalo
Fang, L., SUNY-Buffalo
Pfeifer, B. A., Tufts University
Cheng, C., SUNY-Buffalo
Originally presented on: 11/4/2013 13:00:00 - 13:15:00

Two well-defined cationic polylactides (CPLAs) with different densities of pendant tertiary amine functionalities were synthesized by thiol-ene click functionalization of an allyl-functionalized polylactide. Well-defined chemical structures of CPLAs were characterized by ^{1}H-NMR and GPC. Additionally, the degradability of CPLAs at aqueous condtions was systematically studied using GPC as a analytic tool. Having significant hydrolytic degradability, CPLAs were utilized to complex with plasmid DNA (pDNA) to yield CPLA/pDNA polyplexes through electrostatic interaction. Biophysical properties of such polyplexes including size, surface potential and morphology were characterized by DLS and TEM. pDNA encoding luciferase was delivered to two physiological distinct cell lines (macrophage and fibroblast) via these polyplexes. The formulated polyplexes led to high levels of transfection while exhibiting low toxicity. Gene transfection efficiency was positively correlated with charge density of CPLAs, as well as polymer:pDNA weight ratios in the polyplexes. Results of this work provide helpful guidelines for the design of next-generation biodegradable biomaterials for gene delivery applications.

(109c) A Photo-Degradable Gene Delivery System for Enhanced Gene Transcription

Lee, H., Purdue University
Konieczny, S. F., Purdue University
Won, Y. Y., Purdue University
Originally presented on: 11/4/2013 13:15:00 - 13:30:00

Polymer-based gene carriers, though promising as safer alternatives to viral gene carriers, have yet been limited in large part due to their poor transfection performance. Future design and development of better polycation gene carriers will be greatly facilitated by an improved understanding of the relationship between the polycation chemistry and performance mechanism. In particular, there exists a significant gap in our understanding of how the size and compactness of polycation carriers influence their delivery performances in overcoming the most important cellular-level transport barrier, i.e., nuclear entry. In order to study this problem, we developed a photo-degradable polymer DNA carrier that is degradable upon exposure to mild UV irradiation; this carrier allows us to control the precise location of the disintegration of the PEI/DNA polyplex particles within the intracellular environment. By using this photolytic DNA carrier, we found that the exact (timing and) location of the photo degradation (i.e., cytosol, nucleus or no degradation) significantly influence the nuclear localization and the gene transfection efficiencies of the polymer/DNA complexes; the degradation of the carrier molecules impacts not only the size characteristics of the associated polyplexes which directly influence their nuclear import properties, but also the capabilities of the polyplexes to protect (against endonucleases) and release (for polymerase transcription) DNA which both impact the final gene expression levels.

(109d) Combinatorial Synthesis and Cheminformatics Modeling of Antibiotics-Based Polymers for Transgene Expression

Potta, T., Arizona State University
Zhen, Z.,
Grandhi, T. S. P., Arizona State University
Christensen, M., Arizona State University
Ramos, J., Arizona State University
Breneman, C. M., Rensselaer Polytechnic Institute
Rege, K., Arizona State University
Originally presented on: 11/4/2013 13:30:00 - 13:45:00

Gene delivery, using viral or non-viral vectors is an attractive therapeutic modality for correcting diseases of genetic origin or consequence. Non-viral vectors, especially polymers, are limited by low transgene expression levels and cytotoxicity. Consequently, new synthesis and predictive approaches that facilitate rapid discovery of effective polymeric vehicles for transgene expression are necessary. Here, we developed a library of fifty six polymers based on aminoglycosides, several of which are approved by the FDA as antibacterial drugs. Parallel screening of transgene expression activity resulted in the identification of several candidates that demonstrated lower cytotoxicities and higher efficacies than 25kDa poly(ethyleneimine), a current standard for polymer-mediated transgene delivery. We first correlated transgene expression efficacies of polymers with physicochemical properties obtained using their 'building block' structures. Predictive Quantitative Structure-Activity Relationship (QSAR) models were generated using Support Vector Machine (SVM) regression. Different methods, includin K-fold cross-validation, model evaluation metrics and Y-Scrambling, were applied during the model construction process to ensure predictive and stable QSAR models. According to our structure based QSAR modeling, polymers with smaller sized aminoglycosides, hydrophilic oxygen atoms and shorter hydrocarbon chain between ether oxygen atoms in the cross-linkers are favorable for efficient transgene delivery. For further analyzing the influencing factors of the transfection efficacies of our polycation library, we also constructed several Quantitative Structure-Properties Relationship (QSPR) models to correlate polymer structures to the their experimentally determined polymer/polyplex properties, including size, zeta-potential, molecular weight, hydrophobicity, amine content and DNA binding efficacy. Taken together, this combined combinatorial synthesis, parallel screening and cheminformatics modeling approach can led to materials with high efficacies of transgene expression, fundamental understanding into polymer physicochemical properties that facilitate this activity, and predictive approach for the rational a priori design of efficient polymers.

(109e) Invited Speaker: Combinatorial Development of Nucleic Acid Delivery Materials

Anderson, D. G., Massachusetts Institute of Technology

Originally presented on: 11/4/2013 13:45:00 - 14:15:00

The combinatorial development of small molecules has revolutionized drug discovery. Here we describe our work developing lipid-like and polymeric materials for nanoformulation of nucleic acid payloads. These materials have demonstrated efficacy in a range of tissues and animal models including non human primates. We will describe our efforts to develop new materials with improved clinical potential for a range of therapeutic and biological applications

(109f) Dual-Ligand Peptide-Targeted Liposomal Gene Delivery for Cancer Therapy

Levine, R., University of Minnesota
Pearce, T. R., University of Minnesota
Adil, M., University of Minnesota
Kokkoli, E., University of Minnesota

Originally presented on: 11/4/2013 14:15:00 - 14:30:00

Gene therapy is a powerful tool with the potential to fight cancer at the genetic level, however, effective delivery of gene therapy is a great obstacle to the clinical application of cancer gene therapy. DNA encapsulating stealth liposomes specifically targeted to cancer tissue with peptide ligands have the potential to overcome this obstacle. Moreover, dual peptide targeting can be utilized to improve both the binding efficency and specificity of delivery vehicles. In this work, a reliable DNA quantification method and optimal production conditions were developed and analyzed for DNA encapsulated stealth liposomes. Stealth liposomes were then functionalized with two peptide ligands to explore the effect of dual targeting on liposome binding to cancer cells. Dual targeting was shown to improve binding compared to single targeting with stealth liposomes *in vitro*. Several peptide concentrations and ratios were evaluated to understand the effect of ligand valency on cell binding. The optimal dual targeted stealth liposome formulation was used to deliver gene therapy showing enhanced toxicity to cancer cells compared to healthy cells *in vitro*.

(109g) Photocleavable Polyplexes As Dynamic Carriers for Controlled Nucleic Acid Delivery

Palmer, A. A., University of Delaware
Green, M. D., University of Delaware
Greco, C. T.,
Roy, R., University of Delaware
Epps, T. H. III, University of Delaware
Sullivan, M. O., University of Delaware
Originally presented on: 11/4/2013 14:30:00 - 14:45:00

Nucleic acid unpackaging has been widely identified as a key hurdle towards the successful clinical application of gene therapies. Many current delivery systems employ cationic polymers such as polyethylenimine to strongly bind and compact the nucleic acid and create stable polyplexes, which protect the nucleic acid from enzymatic degradation. However, these polyplex-based approaches offer limited control over the timing and specificity of nucleic acid delivery. Recent strategies to address this issue include the design of materials that release the nucleic acid cargo in response to specific intracellular cues such as a change in pH. An appealing alternative approach is the design of materials that are responsive to external stimuli and can be used to control the timing and specific location of delivery.

In this work, we present a novel block copolymer-based carrier that allows added tunability in the delivery system and decoupling of shielding and complexation requirements in the carrier by incorporating nonfouling and cationic blocks. A photocleavable segment in the cationic block allows for user-controlled release of the nucleic acid at a subcellular target, through light-induced hydrolysis and charge reversal. Furthermore, a poly(ethylene glycol) block provides a stealth coating during delivery that is removed in response to the light trigger. Here we report the synthesis of this photocleavable block copolymer and the assembly into stable polyplexes with nucleic acids. Additionally, we demonstrate robust cellular uptake and light-responsive cleavage of this polymer system, laying a platform for enhancing the efficiency of nucleic acid delivery.

(109h) Efficiency of siRNA Delivery By Lipid Nanoparticles Is Limited By Endocytic Recycling

Sahay, G., Massachusetts Institute of Technology

Originally presented on: 11/4/2013 14:45:00 - 15:00:00

Despite substantial efforts to understand the interactions between nanoparticles and cells, the cellular processes that determine the efficiency of intracellular drug delivery remain largely unclear. Here we examined cellular uptake of siRNA delivered in lipid nanoparticles (LNPs) using cellular trafficking probes in combination with automated high-throughput confocal microscopy as well as defined perturbations of cellular pathways paired with systems biology approaches to uncover protein-protein and protein-small molecule interactions. We show that multiple cell signaling effectors are required for initial cellular entry of LNPs through macropinocytosis, including proton pumps, mTOR, and cathepsins. SiRNA delivery is substantially reduced by ≅70% as internalized siRNA undergoes exocytosis through egress of LNPs from late endosomes/lysosomes. Niemann Pick type C1 (NPC1) is shown to be an important regulator of the major recycling pathways of LNP-delivered siRNAs. NPC1-deficient cells show enhanced cellular retention of LNPs inside late endosomes and lysosomes and increased gene silencing of the target gene. Our data suggests that siRNA delivery efficiency might be improved by designing delivery vehicles that can escape the recycling pathways.

(117a) Development of a Subnormothermic Machine Perfusion System for the Human Liver

Bruinsma, B., Center for Engineering in Medicine, Massachusetts General Hospital, Harvard Medical School, Shriners Hospital for Children
Farmer, A.,
Ozer, S., Center for Engineering in Medicine (CEM) at Massachusetts General Hospital â€“ Harvard Medical School - Shriners Hospital for Children
Berendsen, T., MGH/Harvard Medical School
Izamis, M. L., Harvard Medical School / Massachusetts General Hospital
Martins, P.,
Yeh, H.,
Uygun, K., Harvard Medical School
Originally presented on: 11/4/2013 12:30:00 - 12:54:00

The use of livers donated after cardiac death (DCD) of the donor is limited by the increased incidence of primary nonfunction and biliary complications following transplantation of these grafts. While the transplantation field continues to struggle with an ongoing shortage of donor organs, various strategies are being investigated to increase the availability of livers for transplantation, including the exploration of novel techniques to make DCD livers a safer option. Ex vivo machine perfusion holds the potential to strongly increase the quality of a donor liver and improving post-transplantation function and reducing complications. Collaborations with the New England Organ Bank (NEOB), the MGH transplant center and UMass Memorial medical center have provided us with human livers, discarded due to extended warm ischemic time, donor age, excessive steatosis or other contraindications for transplantation. Thus far, we have successfully perfused 8 discarded human livers, showing feasibility of subnormothermic machine perfusion in preserving the human liver ex vivo with active metabolism and demonstrating recovery of viability expressed in significantly improved energy status. Ongoing inclusion of human livers will allow characterization of these livers by broadening the ex vivo liver function testing, to include markers of injury and function.

Isolated (sub)normothermic organ perfusion is challenged by the absence of other homeostatic organs that are involved with maintenance of acid-base physiology (kidney, lung), energy metabolism (pancreas, intestine), fluid and electrolyte balance (kidney). These functions need to be compensated for by exogenous administration of hormones, nutrients, electrolytes and other substrates for metabolism along with a perfusion solution that supports this regulation as an artificial extracellular compartment. Controlling these functions requires a thorough understanding of both hepatic and systemic physiology and a system that supports it by allowing for real-time measurement of various parameters, easy intervention and reliable viability testing.

The subnormothermic system for the perfusion of human organs that we have developed is user friendly, but also enables advanced monitoring of hepatic function and intervention. An artificial cardiopulmonary system, including a set of pumps and hollow fiber oxygenators, drives the flow of an oxygen rich perfusion solution through the portal vein and hepatic artery of the liver. A carbogen mixture of 95% O_2/5% CO_2 creates an inflow pCO_2 of ±75 mmHg, which, in combination with exogenous bicarbonate plays a role in the carbonate buffer system. The pH of the system is periodically monitored and can be controlled by boluses of bicarbonate. Superphysiological partial oxygen pressures of >700 mmHg ensure adequate oxygenation of the tissue in the absence of hemoglobin. A commonly used hepatocyte culture medium is used a base solution for perfusion. Williams medium E is rich in nutrients and antioxidants. Supplemented with hydrocortisone and insulin it has been shown to provide a rich environment supporting hepatocyte viability in ex vivo liver perfusion and increasing survival after transplantation in the rat. Ports integrated into the system allow drawing samples for blood gas analysis, which provides periodical and on demand information about acid/base, oxygenation and electrolyte balance. Transducers proximal to both the portal and arterial inflow of the liver provide live pressure readings that can be used to calculate hepatic resistance. Samples from the perfusion solution can be analyzed for a multitude of metabolites.

Our results show that the human liver can be preserved for a minimum of three hours ex vivo in a subnormothermic setting (±21 °C), with continued metabolic function. A continuous consumption of O_2 and CO_2 output reflects active oxidative metabolism. An initially acidic perfusion solution and high lactate levels are normalized by the end of perfusion and electrolyte levels remain normal. An active biliary system is reflected in the steady production of bile. The portal venous and hepatic arterial circulation decrease in resistance, suggesting an improving circulation. The recovery of energy status of the graft is demonstrated by an increase in ATP content during perfusion. To our knowledge this is the first report of the application of subnormothermic machine perfusion in human livers and this work will provide a basis for use of this system in a clinical trial for liver transplantation.

(117b) Hemodialysis: A Sea Change Needed, Not Realized

Leonard, E. F., Columbia University
Cortell, S., St Luke's/Roosevelt Hospital
Originally presented on: 11/4/2013 12:54:00 - 13:18:00

The treatment of end-stage renal disease, ESRD, in the second decade of the 21st century involves a half-million patients in the US, for whom about $30 billion are spent each year, with fewer than 5% significantly rehabilitated. Surprisingly, after years of exotic research, the accepted most critical problem of these patients is water management, and there is not yet a way to match water removal and water generation rates, which leads to unacceptable accumulations between thrice-weekly treatments A thoroughly researched, 'simple' method for removing water continuously would be a superb sea change, but it doesn't yet work. This paper will review technical approaches, problems, and prospects for therapeutic improvement in a growing patient population surrounded by a cost-contained healthcare system. In particular the promise and new problems involved in ambulatory blood processing will be enumerated and assessed in terms of those that are fundamental, like coagulation and membrane fouling versus those that should respond to known technical innovations. All blood processing devices are nominally microfluidic but biological and artificial microfluidic systems differ in numerous important ways that will also be assessed.

(117c) An Automated Tissue Digester for Pancreatic Islet Production

Lu, Z., Auburn University
Boland, E., Techshot Inc
Todd, P. W., Space Hardware Optimization Technology, Inc
Hanley, T. R., Auburn University
Originally presented on: 11/4/2013 13:18:00 - 13:42:00

Pancreatic digestion is the primary step in preparing islets of Langerhans for clinical transplantation for restoring euglycemia in Type-1 Diabetes patients. In 1988, Ricordi et al. developed a semi-automated method in which pancreas fragments and glass balls were placed into a cylindrical vessel to form a closed-loop recirculation and dissociation system, operated by intermittent manual shaking of the vessel. However, the yield of islets from this technique tends to be variable and depends on several parameters. Collagenase-containing enzyme blends used for releasing islets from the exocrine tissue also produce variable results and can cause damage to islets. Operator function (shaking the vessel by hand) also produces variability from case to case. An automated digestion unit capable of producing reproducible operating parameters is needed for the consistent preparation of islets.

This research focuses on the incorporation of an automated horizontally rotating digester (MDR) consisting of a rotating outer cylindrical shell and a counter-rotating core both with hemispherical baffles. This configuration is designed to enhance the turbulent effect and contact area between tissue fragments and walls. In the digestion process, the rotation of the inner core is adjusted for optimum dissociation, and independently the exterior wall rotation is optimized to maintain particles in suspension without centrifuging them. Rotation speeds are also adjusted as particle size distribution is diminished during digestion. In addition, a discrete phase model in FLUENT is used to simulate the flow patterns and track the tissue particles in the MDR and Ricordi chamber. The computational fluid dynamics (CFD) analysis and comparison of flow variable contours indicate that MDR has predictable, controllable fluid shear environment and optimized differential rotation mode at 120 ml/min velocity inlet and 100 rpm rotation speed in opposite directions. The agreement between experimental and CFD predicted results is shown by analysis of viable islets under a Grasshopper® digital camera with an ultra-fast 800Mb/s digital interface. Finally, the validation of the model is performed through the comparison of simulation results with experimental data obtained from porcine pancreatic tissue digestion.

(117d) Improving the Post-Thaw Wash Process for Frozen Red Blood Cells: An Integrated Theoretical and Experimental Approach

Lusianti, R. E., Oregon State University
Higgins, A. Z., Oregon State University
Originally presented on: 11/4/2013 13:42:00 - 14:06:00

Transfusion of red blood cells (RBCs) is one of the most commonly practiced medical procedures, with approximately 15 million units transfused per year in the United States alone. Currently, blood banks and hospitals use RBCs refrigerated at 4°C in their reserve almost exclusively. Depending on the preservative, the FDA allows a storage period of 21 to 42 days for refrigerated blood. The short storage period makes it difficult to maintain a steady supply of refrigerated blood to meet the unpredictable demand for transfusion. On the contrary, freezing blood in the presence of 40% glycerol extends the FDA approved shelf life of the product to 10 years. The use of frozen blood in blood banking would enable stockpiling strategic reserves to eliminate seasonal shortages, improve availability in the event of a natural disaster, facilitate military use during the time of war, and provide overall stability in the RBC supply chain. Although logistically more advantageous, the use frozen blood in clinical therapy is uncommon due to the time consuming post-thaw wash process required to remove the glycerol before the product can be transfused. The current state of the art removal method requires approximately one hour to process a single blood unit, making short notice use of frozen blood particularly challenging. Until a more rapid method of glycerol removal is discovered, the increased use of frozen blood in blood banking remains unlikely.

In this study, modern approaches such as mathematical modeling and microfluidics were used to develop a method that expeditiously removes glycerol from thawed frozen RBCs without causing excessive cellular damage. A cell membrane transport model was used in guiding the design of several procedures for varying the extracellular solution concentration with time to theoretically remove glycerol in a rapid manner without causing cell membrane damage. Batch experiments employing a series of dilutions were then conducted to test the theoretical deglycerolization procedures, resulting in the identification of a four-step procedure capable of removing glycerol while achieving a cell recovery that meets transfusion standards in a total process time of three minutes. Direct scaled up implementation of the batch procedure would require excessive amounts of diluting solution and produce an extremely dilute cell suspension. Hence, a membrane microfluidic device was built to practically adapt the batch procedure in a continuous manner. Additionally, a model capable of predicting the glycerol removal in the membrane device was developed to assist in selecting the appropriate operational conditions to match the extracellular concentration changes in the batch procedure. Preliminary experiments using thawed frozen blood and 3.4% saline solution in a single device resulted in a reduction of glycerol mass composition of up to 55.0%±0.3% with a corresponding hemolysis of 15.1%±0.7%. The glycerol removal achieved in this experiment is comparable with the first step of the batch procedure, demonstrating partial implementation in a continuous platform. Matching the step-by-step concentration changes from the remaining steps is possible with multiple membrane devices linked together.

The operating conditions necessary to achieve such concentration changes can be predicted using the mathematical model for each device in the assembly. The use of a combination approach of mathematical modeling and microfluidics shows promise in enabling full continuous implementation of the rapid batch deglycerolization procedure. Ultimately, the findings from this study has the potential to dramatically reduce the post processing time of frozen blood, making it a more attractive option for blood banking in the future.

(117e) Polymeric Nanosensors for in Vivo Lithium Monitoring

Cash, K. J., Northeastern University
Clark, H. A., Northeastern University
Originally presented on: 11/4/2013 14:06:00 - 14:30:00

Recent years have shown the incredible versatility and utility of polymeric nanosensors. Sensors have been developed for a wide variety of ionic analytes (e.g. sodium, potassium, pH) as well as nonionic analytics (e.g. glucose) and have been applied for challenging research fields such as detection of sodium sparks in live cardiomyocytes. This field of research has yielded new knowledge on cellular metabolism and function. Additionally, these sensors can be applied in vivo, through implantation in or under the skin in animals; showing potential as a research tool for continuous monitoring of analyte concentrations without the need for direct samples to be taken from the animal. I will present research on the development and testing of *in vivo* lithium nanosensors based on this platform which are able to continuously monitor lithium concentrations for eventual application in the treatment of bipolar disorder.

This work was supported by the Research reported in this publication was supported by the National Institute Of Biomedical Imaging And Bioengineering of the National Institutes of Health under Award Number F32 EB015270, the National Institute of General Medicine of the National Institutes of Health under award number R01 GM084366 and the National Institute of Neurological Disorders and Stroke under award number R01 NS081641.

(117f) Modeling the Ability of Lubricin to Reduce Bacterial Proliferation On the Surfaces of Medical Devices

Ravi, V.,
Aninwene, G. II, Northeastern University
Webster, T. J., Northeastern University
Originally presented on: 11/4/2013 14:30:00 - 14:54:00

Abstract:

Infection of medical devices is a major clinical issue. This study used mathematical modeling to describe and quantify the ability of lubricin, an anti-adhesive glycoprotein found in synovial fluid, to inhibit the proliferation of bacteria on the surfaces of medical devices. In a previous experiment, bacterial growth trials were performed with *Staphylococcus aureus* and *Staphylococcus epidermidis* on tissue cultures with either lubricin, vitronectin, or mucin to investigate how lubricin and protein sub-regions of lubricin may reduce bacterial adhesion and proliferation. Mathematical models were created with the resulting data and were used to describe the growth dynamics of these bacteria in cultures with each of these proteins. The effect on lag time (λ), maximum specific growth rate (μ_m), and the asymptote reached during the stationary phase (A) were determined and compared. Lubricin was found to significantly increase λ of *S. aureus* and *S. epidermidis* and significantly decreased both the μ_m of *S. aureus* and A of *S. aureus* and *S. epidermidis*.

Introduction:

The attachment and proliferation of bacteria on the surface of medical devices threatens the success of these devices. It can lead to a reduction or loss of device function as well as infection, inflammation, and/or damage to the surrounding tissue[1]. *Staphylococcus aureus* and *Staphylococcus epidermidis* colonize a large percentage of the human population[2] and are two of the major culprits of biofouling, the proliferation of unwanted biological material on the surfaces of an implanted device. Following attachment to surfaces, they can form biofilms which are resistant to antimicrobials and difficult for the host's immune system to combat[3].

This study examined the ability of lubricin (LUB), a glycoprotein found in the synovial fluid that has lubricating and anti-adhesive properties, to decrease bacteria functions. LUB has a mucin-like center domain and vitronectin-like tail domains[4]. It is thought that LUB can serve as an effective non-immune opsonification agent for resisting bacteria colonization[5]. If bacteria are unable to colonize a surface and form a biofilm, it will be much easier for the immune system and antimicrobial agents to clear the bacteria before they can cause an infection.

The ability of LUB to inhibit bacterial proliferation is quantified in this study by determining the effect that LUB and protein sub-regions of LUB have on the growth curve of each bacterium. This curve is generally separated into several phases including a lag phase, during which the bacteria are adapting to the conditions of their environment and not dividing, an exponential phase during which exponential growth is seen, and a stationary phase during which the growth and death rate are equal[6]. The quantities determined by our modeling were the lag time (the length of the lag phase), the maximum specific growth rate (the greatest growth rate during the acceleration of the exponential phase), and the asymptote (a value directly correlated to the number of organisms in the stationary phase where the growth and death rates are in equilibrium). These values were determined by fitting the growth data to two sigmoid growth functions: logistic and Gompertz.

Methods:

In a previous experiment, 96-well polystyrene plates were treated with either lubricin (LUB), bovine submaxillary mucin (BSM), or bovine vitronectin (VTN). BSM was used to assess the effect of the central mucin-like domain in LUB, and VTN to assess the effect of the globular vitronectin-like tail domains[4]. LUB used in this study was extracted from bovine synovial fluid (Pel-Freez)[7]. BSM and VTN were obtained from Sigma Aldrich. LUB and BSM were used at a concentration of 200μg/mL. VTN was used at a concentration of 50μg/mL. 50μL of each protein solution were dried on the polystyrene well surface overnight.

Staphylococcus aureus (ATCC (25923)) and *Staphylococcus epidermidis* (ATCC (35984)) obtained from the ATCC (25923) were cultured in tryptic soy broth (TSB) (Sigma Aldrich) for 18 hours to reach stationary phase and were then diluted to a density of 10^7 bacteria/mL. The bacterial cell culture procedure involved inoculating approximately 3 mL of sterile TSB inoculated with one colony of desired bacteria, then incubating the solution on a shaker set to 200 rpm inside an incubator maintained at 37°C. Bacteria were seeded into the wells and were allowed to incubate for 15 minutes in a stationary incubator maintained at 37°C. After 15 minutes the bacterial solution was removed, the plates were rinsed with sterile PBS, and the wells were filled with 200μL of fresh TSB. Optical density measurements were taken every 4 minutes for 24 hours.

For this study, the growth data (optical density vs time) obtained from these bacterial trials were fit to modified forms of the logistic model and the Gompertz model using MATLAB. The model equations used were modified from their generic forms to contain coefficients for the three bacterial growth parameters lag time (λ), maximum specific growth rate (μ_m), and asymptote of the stationary phase (A) by deriving formulas to represent the mathematical parameters in terms of the bacterial growth parameters[8]. The goodness-of-fit was determined using R-square and sum of squares for error (SSE) and the models with the best fit were used to determine the value of the three parameters.

Results and Discussion:

The Gompertz and logistic models both fit the bacterial growth data very closely. The Gompertz model had a slightly better fit for 8 out of 10 data sets, based on the R-square and SSE.

S. aureus: LUB, BSM, and VTN all significantly increased the duration of the lag phase (λ) compared to the PBS soaked control. On average, LUB showed a 27% increase (52 minutes), VTN showed a 43% increase (82 minutes), and BSM showed a 29% increase (55 minutes). LUB and BSM decreased the maximum specific growth rate (μ_m) while VTN increased it. VTN and LUB also decreased the upper asymptote (A). The lag time increases seen with LUB, BSM, and VTN indicate that these proteins were making it more difficult for *S. aureus* to adapt itself to the environment and start growing exponentially. The decrease in μ_m seen with LUB and BSM shows that these proteins are reducing the acceleration of the growth phase. Though the acceleration was reduced, BSM reached a similar asymptote in the stationary phase as the PBS soaked control, while LUB also led to a much larger reduction in the asymptote. Interestingly, VTN increased the μ_m, but still led to a decrease in the asymptote. It was suggested that vitronectin can enhance adhesion in some bacteria[9], which may explain the increase in μ_m.

S. epidermidis: LUB and BSM significantly increased the lag time while VTN decreased it compared to the PBS soaked control. There was a 35% increase (105 minutes) in lag time with LUB, a 30% increase with BSM (89 minutes), and a 2% decrease (6 minutes) with VTN. LUB and BSM increased the maximum specific growth rate while VTN decreased it and increased the asymptote while VTN decreased it. Though the increase in asymptote caused by LUB is statistically significant, the difference is so small as to not indicate a biological or clinical significance. Once again, a significant increase in lag time is seen with both LUB and BSM indicating that these proteins inhibited *S. epidermidis* from adapting to the environment and slowed the entry into the exponential phase.

Conclusions:

Lubricin significantly inhibited the ability of both *S. aureus* and *S. epidermidis* to attach and proliferate on surfaces. The most striking finding was the large increase in lag time. This is particularly significant because it may give the immune system and antimicrobials more time to clear the bacteria before an infection is established.

References:

[1] a) S. D. Puckett, E. Taylor, T. Raimondo and T. J. Webster, *Biomaterials* **2010**, *31*, 706-713; b) D. J. Weber, K. L. Hoffman, R. A. Thoft and A. S. Baker, *Review of Infectious Diseases* **1986**, *8*, 12-20.

[2] S. R. Gill, D. E. Fouts, G. L. Archer, E. F. Mongodin, R. T. DeBoy, J. Ravel, I. T. Paulsen, J. F. Kolonay, L. Brinkac, M. Beanan, R. J. Dodson, S. C. Daugherty, R. Madupu, S. V. Angiuoli, A. S. Durkin, D. H. Haft, J. Vamathevan, H. Khouri, T. Utterback, C. Lee, G. Dimitrov, L. Jiang, H. Qin, J. Weidman, K. Tran, K. Kang, I. R. Hance, K. E. Nelson and C. M. Fraser, *J. Bacteriol.* **2005**, *187*, 2426-2438.

[3] J. W. Costerton, P. S. Stewart and E. P. Greenberg, *Science* **1999**, *284*, 1318-1322.

[4] D. P. Chang, N. I. Abu-Lail, F. Guilak, G. D. Jay and S. Zauscher, *Langmuir* **2008**, *24*, 1183-1193.

[5] G. E. I. Aninwene, E. Taylor, D. Hall, A. Mei, G. D. Jay and T. J. Webster, *MRS Online Proceedings Library* **2012**, *1417*.

[6] M. L. Shuler and F. Kargi, *Bioprocess Engineering: Basic Concepts (Second Edition)*, Prentice Hall PTR, Upper Saddle River, NJ, **2002**, p.

[7] G. D. Jay, D. A. Harris and C.-J. Cha, *Glycoconjugate Journal* **2001**, *18*, 807-815.

[8] M. H. Zwietering, I. Jongenburger, F. M. Rombouts and K. Vantriet, *Applied and Environmental Microbiology* **1990**, *56*, 1875-1881.

[9] B. Singh, Y. C. Su and K. Riesbeck, *Mol Microbiol* **2010**, *78*, 545-560.

(119a) Synergistic Subcellular Targeting By Shape-Engineered Nanoparticles

Barua, S., University of California, Santa Barbara
Mitragotri, S., University of California, Santa Barbara

Originally presented on: 11/4/2013 12:30:00 - 12:48:00

The efficiency of drug nanoparticles depends not only on their known biological functions, but also on their ability to bind to the target cell membrane, and being internalized into right cellular compartments without causing undesirable side effects. While it has been shown that nanoparticle properties such as size, shape and surface charge affect their binding, uptake and lysosomal rupture for effective therapeutic actions, however, insights of synergistic targeting of multiple subcellular compartments using shape specific single drug nanoparticles have not been explored. Here, we describe the preparation of rod-shape drug nanoparticles that simultaneously target cell membrane, endosome and nucleus of cancer cells to inhibit cancer cell growth synergistically. Our recent study showed how nanoparticle shape modulated cell membrane targeting and uptake by enhancing multivalent interactions of rod-shaped nanoparticles than spherical nanoparticles with the cell membrane. In this study, we investigate how synergistic targeting of cell surface, endosome and nucleus by rod-shaped drug nanoparticles regulates a balance of internalization and recycling rates, and thus enhances the therapeutic efficiency at very low drug concentrations. The results are important for the rational design of drug nanoparticles to maximize their sub-cellular distribution for effective therapeutic actions at the target site.

(119b) Layer-By-Layer Nanoparticles As Multi-Functional Delivery Systems

Morton, S., Massachusetts Institute of Technology
Deng, J., Koch Institute for Integrative Cancer Research
Shah, N., Massachusetts Institute of Technology
Shopsowitz, K.,
Quadir, M.,
Dreaden, E.,
Hammond, P. T., Massachusetts Institute of Technology
Originally presented on: 11/4/2013 12:48:00 - 13:06:00

Layer-by-Layer (LbL) assembly is a highly tunable, modular approach to surface-limited functionalization of materials with nanoscale precision over the composition and properties of the film components. This high level of control affords the capability to design systems that are multi-functional in nature, with temporal and spatial control over the release of a diverse range of materials and therapeutics of interest. Surface-mediated interactions are of particular interest in biological settings where this interface between delivery systems and cells/biological environment govern binding and internalization on a cellular level and the pharmacokinetic properties of the system at a higher level *in vivo.* Application of this approach to nanomedicine has been successful in designing hydrated, protein-resistive long-circulating nanoparticles, as well as systems that shed a hydrated shell used for enhanced persistence in the bloodstream and EPR-based accumulation in the hypoxic tumor microenvironment, whereby exposure of a positively-charged material facilitates rapid uptake by tumor cells. This has driven much interest in further characterizing these systems as drug carriers, with the means to sustain drug in complex biological settings, such as the bloodstream. Using a two-color *in vivo* imaging approach, the biological performance of drug-loaded LbL nanoparticles was evaluated, affording us the capability to screen a library of material systems to rapidly evaluate them as systemic therapeutic carriers. This approach has further allowed us to conceive of systems that provide a means for cellular recognition by incorporation of a targeting ligand, as well as ones that are capable of programmable release of synergistic combinations of drugs for enhanced efficacy in invasive cancer cell types.

(119c) Anti-PSMA Liposomes Loaded With Actinium-225 for Potential Targeted Antivascular Alpha-Particle Therapy of Cancer

Bandekar, A., Rutgers University
Sofou, S., Polytechnic Institute of NYU

Originally presented on: 11/4/2013 13:06:00 - 13:24:00

Objectives. We study targeted liposomes loaded with the alpha-particle generator Actinium-225 (^{225}Ac) to selectively kill prostate-specific-membrane-antigen-(PSMA)-expressing cells with the aim to evaluate their potential for targeted antivascular alpha-radiotherapy. This is a bottom-up design and the rationale for the chosen components is the following: The targeted moiety is PSMA which is unique to human tumor vasculature for several different types of primary human cancers. The therapeutic radionuclide, ^{225}Ac, emits a total of four alpha-particles per decay providing highly lethal and localized irradiation of targeted cells with minimal exposure of surrounding healthy tissues. And, finally, the delivery carrier is a lipid-based nanocarrier (a liposome). Liposomes can be loaded with high radioactivities, and, in terms of toxicities, liposomes can be easily engineered (by altering their size and surface characteristics) to become rapidly cleared from circulation with the aim to shift the distribution of normal organ toxicities away from the kidneys (the dose limiting organ in the case of ^{225}Ac-labeled antibodies) and toward the spleen and the liver.

Methods. Actinium-225 was loaded in preformed PEGylated liposomes encapsulating DOTA using the A23187 ionophore. Different anti-PSMA ligands (antibodies, a urea-based ligand, and A10 aptamer) were conjugated to liposomes. The targeting selectivity, extent of internalization and killing efficacy of liposomes were evaluated on monolayers of prostate cancer cells intrinsically expressing PSMA (human LNCaP and rat Mat-Lu cells) and on monolayers of HUVEC induced to express PSMA (induced HUVEC) in static incubation conditions and under flow at a 15 s^{-1} shear rate using a parallel-plate flow chamber. The effect of liposomes' surface functionalization and length of incubation on the observed killing efficacy was evaluated.

Results. The loading efficiency of ^{225}Ac into preformed liposomes ranged from 58 % to 85 % of introduced radioactivity. The conjugation reactions resulted in approximately 17 antibodies and 9 A10 aptamers per liposome. The density of approximately 300 urea-based ligands per liposome was measured to exhibit optimum behavior. Anti-PSMA-Ab-labeled liposomes display the highest levels of total specific binding to all cell lines followed by the urea-based labeled liposomes and then by the A10 aptamer-labeled liposomes. Specific cell association of targeted liposomes increases with incubation time and decreases by two-to-three fold upon introduction of the flow field. The cytotoxicity studies, which are currently in progress, demonstrate that the radiolabeled anti-PSMA-Ab-labeled liposomes exhibit LD_{50} values comparable to the corresponding radiolabeled antibody.

Conclusions. Our studies demonstrate that liposomes targeted to PSMA via various ligands and loaded with ^{225}Ac selectively bind, become internalized and kill PSMA-expressing cells including endothelial cells induced to express PSMA. These findings combined with the unique ability of liposomes to be easily tuned - in terms of size and surface modification - for optimum biodistributions suggest the potential of PSMA targeting liposomes encapsulating alpha-particle emitters for selective antivascular alpha-radiotherapy.

(119d) Polyanhydride Nanoparticle Platforms for the Prevention and Treatment of Streptococcus Pneumoniae Infections

Haughney, S., Iowa State University
King, J., University of Alabama Birmingham
Briles, D., University of Alabama Birmingham
Lueth, P., Iowa State University
Bellaire, B. H., Iowa State University
Wannemuehler, M. J., Iowa State University
Narasimhan, B., Iowa State University
Originally presented on: 11/4/2013 13:24:00 - 13:42:00

Nanoparticles and microparticles made from polyanhydrides have been demonstrated to act effectively as vaccine adjuvants and drug delivery platforms. Amphiphilic polyanhydrides are excellent candidates for vaccine delivery because of their ability to: stabilize and protect fragile antigens; provide a sustained release of antigen; and modulate the immune response. These polymers have excellent biocompatibility yielding non-toxic, non-mutagenic degradation products. We have previously demonstrated the encapsulation and release of pneumococcal surface protein A (PspA), a surface protein of *Streptococcus pneumoniae*, from two polyanhydride chemistries based on sebacic acid (SA), 1,6-bis-(*p*-carboxyphenoxy)hexane (CPH) and 1,8-bis-(*p*-carboxyphenoxy)-3,6-dioxaoctane (CPTEG). PspA is a choline-binding protein found on the surface of all pneumococcal strains and has been shown to be a protective antigen against *S. pneumoniae*. Our *in vitro* studies demonstrate the sustained release of stable, antigenic PspA and in *in vivo* studies, nanovaccine formulations containing PspA induced high antibody titers in immunized mice.

Polyanhydride nanoparticles have also previously been shown to be an effective delivery platform for antibiotics. Encapsulation of antibiotics into particles resulted in an increase in the local concentration of drug and prevented rapid clearance of the antibiotic after administration. Additionally, particle persistence and sustained release of antibiotic (or antimicrobial agent) may obviate the need for a multiple-dose course of treatment. Together, these studies lay the groundwork for a polyanhydride nanoparticle-based combination treatment against *S. pneumoniae* that addresses both pre-exposure (i.e., prophylaxis) as well as post-exposure (i.e., therapeutics).

(119e) Effect of Material Type and Particle Density On the Margination of Vascular-Targeted Carriers

Thompson, A., University of Michigan
Eniola-Adefeso, O., University of Michigan
Originally presented on: 11/4/2013 13:42:00 - 14:00:00

Effect of Material Type and Particle Density on the Margination of Vascular-Targeted Carriers

Alex Thompson[1] and Omolola Eniola-Adefeso[1]

(1) Chemical Engineering, University of Michigan, Ann Arbor, MI

Introduction: The development of vascular-targeted carriers (VTC) for the localized delivery of therapeutics could greatly improve the treatment of many human diseases. The ability to deliver a high fraction of an injected dose to a targeted diseased area is the paramount goal of vascular targeting. In order for VTCs to be effective, they must be able to marginate (localize and adhere) to diseased endothelium from blood flow. Recent research has shown that particle size and shape affect particle margination in the presence of red blood cells.[1,2,3] Neutrally buoyant microparticles (relative to blood) with equivalent spherical diameter (ESD) ≥ 2 Âµm undergo a forced localization to the red blood cell free layer (RBC-FL); analogous to the mechanism by which red blood cells force leukocytes and platelets to the RBC-FL, allowing them to effectively interact with the endothelium. Smaller, neutrally buoyant spherical particles (100-1000 nm ESD) do not benefit from this effect, therefore making them suboptimal for use as VTCs, particularly in treating diseases that occur in M/LBV, such as atherosclerosis. VTCs can be made from a variety of materials which have densities different from that of blood. The purpose of this study is to examine how the size and density of potential VTCs prescribe their ability to localize the RBC-FL *in vitro* in human blood flow under physiological conditions.

Materials and Methods: Fluorescent spherical particles of different material with different densities were conjugated with targeting molecules (sialyl Lewis a, sLe[a]) which bind to selectins at a fixed surface density. Targeted particles were tested for their margination efficiency from blood flow *in vitro* using parallel plate flow chamber (PPFC) assays. The particles explored ranged in diameter from 100-1000 nm.

vascular-targeted spheres at a fixed concentration were mixed with blood from human donors and allowed to flow over a layer of IL-1β activated human umbilical vein endothelial cells (HUVEC) lining one wall of a parallel plate flow chamber. Adhesion of VTCs from laminar, steady blood flow from a range of wall shear rates (200-1000 s^{-1}) was investigated. Channel orientation was varied in four different configurations in relation to gravity to examine how particles of different densities marginate from blood flow. Particle adhesion in recirculating blood flow, in which a sudden expansion in the PPFC causes a region of recirculating flow to occur in the channel, was also examined to see if a disturbed flow profile helps particles of different densities from blood separate from the RBC core more efficiently. The number of adherent particles under various flow conditions was imaged/quantified using brightfield and fluorescence microscopy. Also, the localization to the RBC-FL during blood flow was visualized using confocal microscopy.

Results and Discussion: **Results** from *in vitro* flow assays show that, as expected, the adhesion of neutrally buoyant polystyrene (PS) microspheres was not affected by chamber orientation. Silica (Si) microspheres with 1 Âµm diameter display roughly double the adhesion of equivalent polystyrene (PS) spheres when the chamber was oriented with the HUVEC monolayer at the bottom; however, the same Si microspheres adhere at roughly the same level as PS microspheres when the chamber is inverted so that the monolayer is at the top of the chamber. When the target site is in the downward direction, gravity favors the more dense Si particle in terms of adhesion; however when the target site is upward, perhaps the presence of the RBC core below allows similar levels of Si microparticles to remain in the RBC-FL to interact with the endothelium. In recirculation flow, our results show that at lower shear rates (200 s^{-1}) Si microparticles adhere at higher levels than PS microparticles whether the chamber is oriented with the monolayer at the top or the bottom. It appears that the flow normal to the monolayer caused by the recirculation region helps Si redistribute to the RBC-FL in this disturbed flow, resulting in an improved adhesion in this area. This finding could have implication in targeting atherosclerosis, which is known to occur in areas of the vasculature where this type of disturbed flow is found. Ongoing work is investigating the margination smaller (20-200 nm) and more dense (FeO, Au, etc) particles with the goal of identifying particle characteristics which would make them suitable for targeting specific areas of the vasculature.

Conclusions: Our study shows that particle density can affect the margination of vascular-targeted carriers, which can be particularly useful in designing carriers which target diseases, such as atherosclerosis, which occur in areas of the vasculature that exhibit unique hemodynamics.

(119f) Protein Nanocluster Dispersions With Tunable Size and Low Viscosities for Drug Delivery

Borwankar, A. U., University of Texas at Austin
Wilson, B. K., University of Texas at Austin
Dear, B. J., University of Texas at Austin
Twu, A., University of Texas at Austin
Hung, J., University of Texas at Austin
Dinin, A. K., University of Texas at Austin
Yue, J.,
Maynard, J., University of Texas at Austin
Truskett, T. M., The University of Texas at Austin
Johnston, K. P., University of Texas at Austin
Originally presented on: 11/4/2013 14:00:00 - 14:18:00

Protein solutions and dispersions at high concentrations (e.g., greater than 250 mg/ml) are of broad interest in subcutaneous drug delivery for treatment of cancer and many other diseases. At these concentrations, proteins have a tendency to undergo irreversible aggregation or gelation as a consequence of specific short ranged forces. Herein, we create highly concentrated dispersions (~ 250 mg/ml) comprising dense protein nanoclusters that have very low viscosities in the sub-100 cp range and which yield, upon dilution, monomeric protein molecules in their native state. The low viscosities result from weak interactions between the nanoclusters as characterized by reduced intrinsic viscosities as compared to monomer solutions. We observe lack of shear thinning at low Peclet number for nanoclusters of several different proteins including BSA and monoclonal antibodies in contrast with the behavior for protein solutions. These results are consistent with the nanoclusters being farther apart (on average) and more weakly interacting than the monomers, such that they do not form networks that would otherwise be deformed by equivalent amounts of shear. The nanoclusters were formed by addition of a variety of excipients with different polarities, charges and sizes to produce a short-ranged effective attraction between protein molecules. This attraction is balanced by electrostatic repulsion to yield a tunable cluster size suggesting equilibrium aspects. The protein nanocluster diameter can be reversibly tuned from 1 to 10 times the monomer diameter as shown by dynamic light scattering (DLS) and electron microscopy at a pH near their isoelectric point (pI). Upon dilution of the dispersion, the clusters rapidly dissociate into protein monomers as shown by size exclusion chromatography (SEC) and dynamic light scattering (DLS). Stable protein dispersions with low viscosities may potentially enable patient self-administration by subcutaneous injection of antibody therapeutics currently being discovered and developed.

(119g) A Novel Nano-Carrier Engineered for High Payload and Efficient Delivery of Daunomycin

Sundaram, P., Auburn University
Pradhan, S., Auburn University
Kurniawan, H., Auburn University
Lipke, E. A., Auburn University
Wower, J., Auburn University
Byrne, M. E., Auburn University
Originally presented on: 11/4/2013 14:18:00 - 14:36:00

Controlled delivery of therapeutics to target tissues constitutes a great challenge of modern medicine, especially in the field of cancer treatment. By exploiting the canonical mechanism of action of the anticancer drug daunomycin we have developed a bio-hybrid nano-carrier for controlled and extended release of daunomycin. In this nano-carrier, daunomycin is bound to DNA aptamers conjugated to 15 nm gold nanoparticle (AuNps) via single-stranded DNA anchors. Our results are the first to demonstrate that drug release can be exquisitely controlled and modulated by adjusting the strength of the molecular interactions between daunomycin and its aptamers.

We investigated the factors affecting the loading of DNA on AuNps. A maximum loading of 108 strands of anchor DNAs per AuNp was observed with 0.4M NaCl and 4μM DNA. On binding daunomycin to the aptamer-anchor DNA complex conjugated to AuNps we demonstrated a very high payload of 1157 ± 18 daunomycin molecules per AuNp. Such a large drug payload per AuNp has never been reported before. The in-vitro release of daunomycin at body temperature demonstrated controlled release of drug from the nano-carrier for an extended period of time. After 6 days of release only about 55% of the bound daunomycin was released from the nano-carrier which implies that the release can be extended much further. With this high drug payload and an ability to release it slowly for an extended period of time this nano-carrier can bring a revolution to the cancer treatments. To test the efficiency of the nano-carrier in delivering the drug to cells, we studied the cellular uptake of the nano-carrier and cell viability in MCF7 breast cancer cells. The nano-carrier displayed excellent cellular uptake, and the cell death was observed to be proportional to the concentration of the nano-carrier. At any tested concentration, cell death by the nano-carrier was significantly higher compared to the cell death by the corresponding concentration of free daunomycin. After 4 hours of incubation, the cell death by 100 nM nano-carriers was twice as much as cell death by same concentration of free daunomycin, proving that it is much more efficient in killing cancer cells compared to that of the free drug. Furthermore, to highlight the versatility of the nano-carrier, we modified the aptamers on the particles demonstrating excellent control over drug release using affinity modulation. One and two point mutants of the DNA aptamer displayed different drug binding affinity and thereby varied drug release profile. After 2 hrs of drug release, the strongest aptamer released just 20% of bound drug while the weakest released about 70%. Thus by having various combinations of two aptamers (weak and strong) on AuNps we were able to obtain any desired drug release profile. Such high variation in drug release with just one point mutation of the DNA is the ultimate control that could be achieve in a nano-scale system.

This nano-carrier with a very high payload and efficient drug delivery when equipped with molecules that allow for precise targeting of diseased cells and tissues will provide medical doctors with a powerful, yet flexible, tool for fighting cancer. Being versatile to be modified for any drug it is expected to impact a number of treatment strategies.

(119h) Nanotechnology-Based Targeted Therapies for Perinatal Brain Injury

Kannan, R., Johns Hopkins School of Medicine
Mishra, M., Johns Hopkins School of Medicine
Balakrishnan, B., Johns Hopkins School of Medicine
Kannan, S., Johns Hopkins School of Medicine
Originally presented on: 11/4/2013 14:36:00 - 14:54:00

Neuroinflammation, caused by activated microglia and astrocytes, plays a key role in the pathogenesis of cerebral palsy (CP), and other debilitating neurodegenerative disorders. Engineering and reprogramming the microglial response, to achieve targeted attenuation of neuroinflammation, can be a potent therapeutic strategy. However, drug delivery to the central nervous system is strongly restricted for most drugs by the blood-brain-barrier, making treatment of diffuse neuroinflammation a challenge. We take advantage of the unique, intrinsic, pathology-dependent, biodistribution patterns of dendrimers (with no targeting moieties) in diseases models of neurodegeneration. For example, dendrimers are transported to the periventricular region of the brain of newborn rabbit kits with cerebral palsy (CP), whereas little brain uptake is seen in healthy animals. Interestingly, they further localize selectively in activated microglia and astrocytes in animals with CP. Such selective localization in activated microglia is also seen in retinal degeneration models, upon intravitreal administration.[1] Building on these findings, we have designed and synthesized dendrimer-drug nanodevices, taking advantage of their rich surface functionality using appropriate linking chemistry. They can deliver and release the drug in the targeted tissue in a tailored and sustained manner. We show that a single intravenous dose of dendrimer-drug conjugate, administered after birth to rabbit kits with CP, results in significant improvement in motor function along with decrease in neuroinflammation and oxidative/neuronal injury, followed by improved myelination, by 5 days of age.[1] These studies suggest that attenuation of ongoing neuroinflammation, achieving by appropriate engineering of the glial response, can have significant positive consequences in these and other debilitating neurodegenerative diseases. Application of this approach to designing dendrimer-based targeted therapeutic platforms is being explored in a variety of systemic inflammation and neuroinflammation-associated disorders, especially in the perinatal period.

References:

1. S Kannan, H Dai, RS Navath, B Balakrishnan, A Jyoti, J Janisse, R Romero, RM Kannan (2012). 'Dendrimer-based postnatal therapy for neuroinflammation and cerebral palsy in a rabbit model'. *Science Translational Medicine*, 4(130), p. 130ra46. Highlighted in *Nature*, *Science*, *Nature Review Drug Discovery*.

2. R Iezzi, B Raja Guru, I Glybina, M Mishra, A Kennedy, RM Kannan (2012). 'Dendrimer-based targeted intravitreal therapy for sustained attenuation of neuroinflammation in retinal degeneration'. *Biomaterials*, 33(3), 979-988.

(126a) Basic BioChE: Agent of Change, How Harry Bungay Changed The World of Teaching Engineers

Ladisch, M., Purdue University
Tsao, G., Purdue University
Originally presented on: 11/4/2013 12:40:00 - 13:00:00

Basic BioChE, first published in 1988, was exactly what its title implied: Principles of Biochemical Engineering communicated through lessons in an interactive computer environment based on BASIC™. BASIC™ (and later visual BASIC™) was a computer code used by programmers for purposes of control, calculations, and early forms of process monitoring in an MS-DOS environment. Harry Bungay, as a Professor in the Environmental and Chemical Engineering at Rensselaer Polytechnic Institute, pioneered the use of BASIC to enable implementation of the revolutionary concept of direct access, via individual terminals, to the RPI mainframe computer in an interactive manner. The concept evolved into progams for PC's and thus further built the concept of interactive learning. Students learned principles of biochemical engineering, cellular growth, chemostats, predator/prey interactions, modeling of Monod type kinetics, and bioreactor simulation in a real-time computational environment. While the computer language and capabilities may seem "basic" by current standards, the concepts embedded in these simple programs were truly sophisticated and 25 years ahead of their time. Instructors in the RPI summer courses were fortunate to be part of Professor Bungay's vision; we learned from the students, as well as from Professor Bungay himself. His concepts evolved with the emergence of PC's in the mid-1980's, and became a valuable, but perhaps under-recognized teaching tool at the time. This talk is about a journey with Harry Bungay, from the sunset of the age of slide rules, through the early days of biochemical engineering, and to the evolution of the current era of computer-based learning and the buzz around massive online courses (MOCs). His approach anticipated the relation of computers to students in the learning process – where one serves the other. Professor Bungay's vision continues to impact both students and teachers. He forever changed the world for biochemical engineering students through Basic BioChE, and through his vision has made a lasting impact on biochemical engineering.

(126b) Polar Bear Engineering

Tanner, D. R. D., Vanderbilt University
Bungay, H. R., Rensselaer Polytechnic Institute
Originally presented on: 11/4/2013 13:00:00 - 13:20:00

Polar bears are believed to be endangered. As the Arctic continues to warm, two-thirds of the world's polar bears could disappear by mid-century. Rapid loss of sea ice is the major threat. Polar bears can't survive for long in ice water but need to hunt for seals hundreds of miles off shore.

We propose to assist them using temporary floating structures made from ice and biodegradable additives. One such additive is sawdust which when added to ice in approximately 14 per cent concentration creates a material called Pykrete. Pykrete was proposed during World War II by Geoffrey Pyke for a huge, unsinkable aircraft carrier. Pykrete is slow melting and is tough and comparable to concrete. Pykete will be compared to other ice mixtures for possible use in floating ice islands for polar bears in the summer.

Robert D. Tanner, Chemical and Biomolecular Engineering, Vanderbilt University, Nashville, Tennessee and Henry R. Bungay, Chemical and Biological Engineering, Rensselaer Polytechnic Institute, Troy, New York.

parameters vary with pressure, something that was not considered in earlier models. The goal of the present model is to investigate if increasing endothelial AQP expression – say, pharmacologically or via diet, which leads to higher intrinsic endothelial Lp_e and which reduces SI compression causing higher overall vessel wall Lp_t - can decrease the overall concentration of macromolecules in the SI and thereby slow down kinetics of its binding to SI ECM. This could have clinical relevance since such binding is believed to be the triggering event in lesion formation and subsequent pre-atherosclerotic events. Our theoretical findings predict that increasing transcellular transport by increasing AQP expression shifts the pressure regime over which Lp_{e+i} is a strong function of pressure to higher pressure values. Based on these predictions, experiments in our lab tested and confirmed these predictions. The theory further predicts that such an upregulation lowers the concentration of HRP in the sub-endothelial space and similarly affects the more biologically-relevant macromolecule LDL. By November, we hope to have data on HRP transport into vessel walls with upregulated AQPs to evaluate our predictions.

Xylose to Ethanol - Why is it still a challenge?

Originally presented on: 11/4/2013 13:20:00 - 13:40:00

It was early recognized that sugars with five carbon atoms, pentose (C5) sugars, had to be utilized efficiently when renewable lignocellulose biomass is converted to replace fossil transportation fuels. Lignocellulose biomass includes agriculture and forest residues as well as dedicated energy crops. Typical examples are corn stover, sugar cane bagasse, wheat straw, poplar, pine and switchgrass. In these raw materials the C5 content ranges from 7% DM in pine to 22% DM in switchgrass. The most investigated substitute for fossil transportation fuel is ethanol. Biomass conversion to ethanol is a very old and well established process based on sugar and starch, which contain only hexose (C6) sugars. The preferred microorganism in these processes is the yeast *Saccharomyces cerevisiae* because of its superior tolerance to osmostic pressure and to ethanol. However, this organism does not naturally ferment C5 sugars to ethanol.

C5 fermenting *S. cerevisiae* strains have been generated by applying genetic engineering first in haploid laboratory strains and later in diploid and polyploid industrial strains. Many C5 strains have then been further adapted to lignocellulose materials such as wheat straw and corn stover. An alternative approach is to select *S. cerevisiae* host strains based on their suitability for a particular lignocellulose raw material and then to introduce the xylose fermenting pathway. The latter rational engineering approch will be discussed in relation to the fermentation of different lignocellulose raw materials.

(126d) Process Development for Very-High-Gravity Ethanol Fermentation

Lin, Y. H., University of Saskatchewan

Originally presented on: 11/4/2013 13:40:00 - 14:00:00

Metabolic flux distribution may be altered by manipulating intracellular reducing equivalents. To favour ethanol synthesis by *Saccharomyces cerevisiae*, a reduced cytosolic environment is desired, otherwise biomass formation is favoured. Direct variation of intracellular $NADH/NAD^+$ is difficult, however, indirect control through measurement of fermentation redox potential is applicable. In this presentation, we first report a correlation between yeast growth pattern and fermentation redox potential profile under very-high-gravity (VHG) conditions [1]. When *S. cerevisiae* was grown under VHG conditions, it initially encounters osmotic stress due to a high glucose feed. As fermentation proceeds, ethanol built-up retards yeast propagation, leading to sudden cell death and incomplete sugar utilization. To overcome this problem, strategies for proper aeration during ethanol fermentation could be employed [2]. It is known that yeast grows 33% faster at its active state than at its inactive state. This feature was used to develop a redox potential-driven repeated-batch ethanol fermentation process. Compared to the equivalent batch and the continuous ethanol fermentation processes, the annual ethanol productivity of the repeated-batch operation is 2.4% and 13.2% greater, respectively, under 200 g feeding glucose/l conditions [3]. Incomplete glucose conversion under VHG condition is common. To overcome this problem, a fermentation process incorporating a chemostat device and an ageing vessel along with a proper operating scheme was designed and validated [4, 5].

(126e) The Impact of Metabolic Uncouplers on the Elimination Capacity of Toluene in a Gas-Phase Biofilter

Gostomski, P.,
Detchanamurthy, S.,
Originally presented on: 11/4/2013 14:00:00 - 14:20:00

A series of metabolic uncouplers were tested in a lab-scale biofilter to determine the effect on the elimination capacity (EC) of toluene. Uncouplers are lipophilic weak acids capable of carrying protons across the cell membrane, thereby influencing ATP production. Uncouplers have been used successfully in activated sludge systems to lower the sludge yield but their use has not been reported in gas phase biofilters. The hypothesis was that as biofilters (without supplemental nutrient addition) are operated in an energy excess environment driven by maintenance energy, uncouplers could potentially increase the specific removal rate per cell thereby increasing the volumetric removal rate, EC (g m^{-3} h^{-1}).

The lab-scale biofilter system combined a differential reactor design with a hanging water column to give good control over most environmental parameters including water content of the unsaturated soil. It also allowed easy addition and removal of the uncouplers to the packing material. Soil was the packing material tested at 30 ^{o}C and -10 cm_{H2O} matric potential buffered at pH 7. Toluene in air was used as the feed gas at 25 ml $min^{-1}$and the inlet load was varied by adjusting the inlet toluene concentration between 0 – 300 ppm using a diffusion tube system. The metabolic uncouplers tested included benzoic acid, pentachlorophenol (PCP), m-chlorophenol, trichlorophenol, malonic acid, CCCP and FCCP. These uncouplers were shown to have a positive influence in activated sludge systems and also did not contain nitrogen, which was the limiting nutrient in the soil system.

The system response varied between the uncouplers. The concentrations tested were based on effective concentrations in the activated sludge literature. Benzoic acid (10 mM) and m-chlorophenol (160 mM) irreversibly decreased the EC by approximately 60% and 50% respectively. CCCP (0.002 mM) reversibly decreased the EC by 30% but the effect was reversible. FCCP (0.01 mM), malonic acid (25 mM) and trichlorophenol (4.1 mM) had slight impacts on the EC. PCP (140 mM) increased the EC by 60% albeit irreversibly.

As only the CCCP response was reversible, the uncoupler mechanism in these mixed culture soil microcosms was unclear. The mechanism is being investigated in pure cultures to try to elucidate the mechanism. Also as most of these compounds are considered water pollutants, even the 60% increase in the EC for PCP is of questionable benefit in a full-scale biofilter. Even though the water release from a traditional biofilter is low, the addition of PCP would greatly complicate the water disposal process. The increase in EC would need to be much greater to justify the added complexity.

(126f) Use of a pH Auxostat to Investigate Physiological Responses of E. coli to Rapid Growth Conditions

Sharfstein, S. T., University at Albany
Originally presented on: 11/4/2013 14:20:00 - 14:40:00

Microbial cells must adapt to rapidly changing culture environments by alteration of gene expression. While regulatory mechanisms in *E. coli* have been studied extensively, many questions still remain unanswered. In particular, understanding how growth rate is regulated is critically important for effective use of microorganisms for biotechnology applications. We have explored the cellular response of *E. coli* to alterations in growth rate using a pH auxostat. An auxostat is a continuous culture system, with feedback control. In the pH auxostat, a pH set point is established slightly lower than the pH of the feed medium (for an acid-producing medium such as a glucose feed). As the cells grow and metabolize the feed, they produce organic acids (e.g. acetate), driving down the pH. To increase the pH, the feed rate is automatically increased. As the feed rate approaches the maximum growth rate of the organism, the culture begins to "wash out", resulting in less acid production and a reduction in feed rate. Thus, over time, the culture reaches a steady state based on the maximum growth rate of the organism.

Using the pH auxostat, we obtained a specific growth rate of 0.044 min^{-1} (corresponding to a doubling time of 15 minutes) for *E. coli* grown in a minimal medium with 10g/L of glucose and 2 g/L of yeast extract. To determine the changes in gene expression we performed a transcriptomic study using Affymetrix Gene chip arrays. We identified a number of proteins in the RpoS regulon that were differentially expressed. In addition, we found a significant modulation of iron transporter genes. Upregulation of the RpoS system was confirmed by western blotting and increased iron transport was verified by atomic absorption analysis and siderophore analysis. Current studies are focused on identifying the upstream regulators of RpoS in this system and a detailed proteomic analysis of cellular responses. We are also investigating whether the upregulation of RpoS is caused by mutation or simply alteration of gene expression patterns.

(143a) Engineering Flavin-Based Photoreceptors As An Emerging Class of Fluorescent Reporter Proteins

Mukherjee, A., University of Illinois at Urbana-Champaign
Walker, J., University of Illinois at Urbana-Champaign
Weyant, K. B., University of Illinois at Urbana-Champaign
Schroeder, C. M., University of Illinois at Urbana-Champaign
Originally presented on: 11/4/2013 12:30:00 - 12:48:00

We report on the development of flavin-binding photoreceptor proteins as an emerging class of fluorescent reporters with potential applications in synthetic biology and biological engineering. Flavin-based fluorescent proteins (FbFPs) were recently developed as genetically encoded reporters that are characterized by small size (≈ 120 amino acids *vs.* 240 amino acids in GFP) and oxygen-independent maturation of fluorescence, which are key advantages over the widely used green fluorescent protein (GFP) [1,2]. Furthermore, we previously demonstrated that FbFPs present additional enhanced properties compared to GFP-based probes: specifically, broad pH tolerance (4-11), thermal stability (up to 60°C), and rapid maturation of fluorescence (≈ 3 min. *vs.* ≈ 30 min. in case of GFP) [3]. In this way, FbFPs open up exciting avenues for investigating a broad class of biological systems that are poorly suited to GFP-based imaging (for example, thermophiles, anaerobic bioprocess such as fermentation, and cellular processes characterized by fast time-scales). However, FbFPs are at a nascent stage of development and are limited by modest brightness and a small library of available probes comprising only three known reporter proteins. These attributes substantially limit versatility and flexibility when selecting FbFPs for a specific application. Based on the considerable promise of FbFPs as robust fluorescent reporters under a broad range of environmental conditions, we sought to address the aforementioned issues using 1) directed evolution to enhance probe brightness and 2) mining sequenced microbial genomes to identify potential new members of the FbFP imaging toolkit.

In a first set of experiments, we applied directed evolution to develop bright mutants of FbFPs. Specifically, we mutated nine amino acids in the fluorophore (flavin) binding pocket using site saturation mutagenesis and isolated two bright FbFP mutants (F37S and F37T) [4]. The F37S and F37T mutants exhibited a nearly 2-fold increase in overall brightness of fluorescence emission. In a second set of experiments, with a view towards expanding FbFP library size, we applied genome mining techniques to identify homologous photoreceptor domains in bacterial, algal, and archaeal genomes. Specifically, sequence alignments between existing FbFPs were employed to query genome databases for homologous flavin-binding photoreceptors. Identified proteins were cloned and expressed in wild type *Escherichia coli* and screened for fluorescence using whole cell spectrofluorometry as well as fluorescence measurements on purified protein preparations. Using this approach, we discovered two entirely new FbFPs identified in blue light photoreceptors from the fresh water algae: *Chlamydomonas reinhradtii* and *Vaucheria frigida*. In particular, mutation of a critical cysteine residue in the flavin-binding pocket of the wild type (nonfluorescent) photoreceptors was found to inactivate the fluorescence-quenching photochemistry of the native photoreceptor protein and render the mutants fluorescent (quantum yields of 0.24 and 0.31 respectively). Additional improvements in the brightness of newly discovered FbFPs were sought via random mutagenesis using error-prone PCR and site directed mutagenesis.

In summary, we developed a biochemical framework to identify and engineer new and improved FbFP-variants using site saturation mutagenesis and genome mining approaches. We anticipate that our work will encourage and enable the broad application of FbFPs and engineered bright mutants as fluorescent reporters to investigate biological systems that are poorly suited to GFP-based labeling.

References

1. T. Drepper, **et al**. *Nature Biotechnology*, 25: 443:445 (2007).

2. S. Chapman, **et al**. *PNAS*, 105(50):20038:20043 (2008)

3. A. Mukherjee, **et al**. PLOS ONE, accepted

4. A. Mukherjee, **et al**. Journal of Biological Engineering, 6(1):20 (2012)

(143b) Transition State Analogue-Based Computational Enzyme Design

Grisewood, M., The Pennsylvania State University
Gifford, N., The Pennsylvania State University
Janik, M. J., Pennsylvania State University
Maranas, C. D., The Pennsylvania State University
Originally presented on: 11/4/2013 12:48:00 - 13:06:00

Despite a handful of recent successes, computational design of active enzymes remains a difficult yet important task. Enzymes are ubiquitous commercially and achieve reaction rates that are typically unattainable with inorganic catalysts. Catalysts function by effectively lowering the reaction energy barrier, namely through transition state stabilization, ground state destabilization, or both. Because the transition state structure for a reaction is challenging to resolve, our design approach focuses on the use of transition state analogue compounds as proxies for the true transition state structure. The fundamental basis of catalysis (i.e., lowering the reaction energy barrier) in coordination with the use of transition state analogue compounds as proxies for the unknown transition state structure stimulates the derivation of correlating expressions that relate enzyme catalytic properties to computationally-accessible interaction energies.

Here, we will provide a detailed description of the new design approach predicated on the use of interaction energy calculations and transition state analogues. Furthermore, we will validate the correlating expressions using experimental data from a benchmark system and apply these expressions to alter enzyme specificity from a native substrate analogue towards a non-native substrate. The validation of the design technique for both the native substrate analogue and the non-native substrate provides the basis for a new computational enzyme design method- OptZyme. OptZyme is a novel design methodology that implements a structure-based design protocol for improving activity of an enzyme with a new substrate. We will show that the application of OptZyme results in a library of mutants with an improved catalytic property for a particular substrate. We will illustrate trends observed within the libraries of engineered enzymes and provide a structural basis for some of the common occurrences. Finally, we will analyze differences between the constructed libraries of mutants, with a special emphasis on point mutations that improve catalytic activity for a non-native substrate.

(143c) High-Throughput Screening Of Combinatorial Regulatory Protein Libraries for Isolation of Novel Endogenous Molecular Biosensors

Frei, C. S., University of Houston
Gredell, J. A., University of Houston
Quan, S., University of Houston
Cirino, P. C., University of Houston
Originally presented on: 11/4/2013 13:06:00 - 13:24:00

HIGH-THROUGHPUT SCREENING OF COMBINATORIAL REGULATORY PROTEIN LIBRARIES FOR ISOLATION OF NOVEL ENDOGENOUS MOLECULAR BIOSENSORS

Christopher Frei, Joseph Gredell,Shuai Qian, Patrick C. Cirino

University of Houston, S294 Engineering Bldg 1, Houston, TX, 77204-4004, USA
T: 832-842-6306, csfrei@uh.edu

Endogenous molecular biosensors based on effector-dependent transcriptional regulatory proteins are emerging and powerful biomolecular tools for specific, sensitive, and high-throughput detection/monitoring of intracellular molecules [1]. The naturally occurring proteins couple molecular recognition with changes in gene expression. To direct the evolution of molecular recognition toward compounds of interest, we construct large libraries of regulatory protein variants and screen for ligand responsiveness using a GFP reporter system[2]. These customized molecular biosensors enable high-throughput screening of combinatorial biosynthesis libraries to improve enzymatic or microbial production of metabolites [3, 4].

Our group has shown success engineering AraC, a well-studied native *E. coli* dual-regulatory protein. By generating multiple AraC saturation mutagenesis libraries and using positive and negative FACS-based screening, we have isolated variants with altered specificity toward D-arabinose (D-ara), mevalonate (mev), and triacetic acid lactone (TAL)[2-4]. We then explored the limits of engineering AraC for altered molecule recognition. We sought variants that respond to many compounds including p-coumaric acid (pCA), theophylline, propionic acid, vanillin, malonic acid, levulinic acid, and several others. In the majority of cases AraC variants remaining after sorting showed minimal ligand responsiveness, although a notable exception was pCA-responsive variants. Improvements in biosensor properties of variants isolated from the initial saturation mutagenesis library screening (e.g. reduced leaky expression and enhanced sensitivity) were also sought by further directed evolution. Properties of select variants and mutation patterns that correlate with these properties will be described.

The isolation of responsive clones is both tedious and not guaranteed, and to streamline the process we aimed to optimize multiple parameters in the high-throughput screen. First, schemes with different combinations of positive and negative sorting were tested for a subset of target ligands, and the responses of the populations and individual variants present at the sorting endpoints were compared. The influence of reporter gene copy number and copy number variability on sorting results was also studied. As expected, stable chromosomal integration of a single reporter gene reduced expression variability. Also, replacing GFP with a variant having attenuated stability reduces the signal from leaky AraC variants, amplifies the on/off expression ratio, and could improve the accuracy of the positive sorting step.

Finally, high-throughput sequencing of individual gene populations surviving different stages of sorting was also performed to develop a more comprehensive understanding of the sequence evolution of specific clones. This sequencing study and the resulting insights into future screening and AraC protein library design will be discussed.

1. Gredell, J.A., C.S. Frei, and P.C. Cirino, *Protein and RNA engineering to customize microbial molecular reporting.* Biotechnology Journal, 2012. **7**(4): p. 477-499.

2. Tang, S.Y., H. Fazelinia, and P.C. Cirino, *AraC regulatory protein mutants with altered effector specificity.* Journal of the American Chemical Society, 2008. **130**(15): p. 5267-5271.

3. Tang, S.-Y., O. Akinterinwa, and P.C. Cirino, *A novel endogenous reporter of triacetic acid lactone (TAL) enables screening for improved TAL production by E. coli.* submitted.

4. Tang, S.-Y. and P.C. Cirino, *Design and Application of a Mevalonate-Responsive Regulatory Protein.* Angewandte Chemie International Edition, 2011. **50**(5): p. 1084-1086.

(143d) Repurposing Bacterial Quality Control for the Engineering of Proteins With Superior Traits

Boock, J. T., Cornell University
King, B. C.,
Gibson, D. M.,
DeLisa, M. P., Cornell University
Originally presented on: 11/4/2013 13:24:00 - 13:42:00

In this study, we have repurposed the intrinsic quality control of the twin-arginine translocation (Tat) pathway to enhance the traits of target proteins using directed evolution. As a proof-of-concept, we applied this method to the endocellulase Cel5A from the fungal plant pathogen *Fusarium graminearum* to improve its production and function. Our approach is based on a novel two-tiered genetic selection and screening method. First, a Tat-based genetic selection is applied that links protein translocation with resistance to beta-lactam antibiotics. Since the quality control mechanism of the Tat pathway only permits the export of folded proteins, this genetic selection allows for the rapid and high-throughput isolation of well-folded, stable Cel5A library members while eliminating those that are poorly folded. A second screening step is imposed to ensure that the proteins that pass the Tat quality control retain high activity. For Cel5A, this involves a screen of enzyme activity using the soluble cellulose substrate carboxymethyl cellulose (CMC). Following two iterations through our two-tiered selection and screen, we isolated a Cel5A variant whose production is increased 10-fold over the parent enzyme. The gain in production is achieved without any loss in activity, underscoring the value in this two-step evolution approach. Further, we have characterized several of the improved variants to determine which biophysical properties are selected by the Tat quality control mechanism. The results of these experiments are helping to shed light on the poorly understood quality control mechanism and should guide future protein engineering attempts that exploit this pathway.

(143e) Engineering Light-Activated Proteins to Control Biological Processes in the Brain

Cho, Y. K., University of Wisconsin-Madison
Chen, F.,
Chuong, A. S.,
Klapoetke, N. C.,
Boyden, E. S.,
Originally presented on: 11/4/2013 13:42:00 - 14:00:00

In nature, diverse organisms ranging from bacteria to plants have acquired the ability to use light as a ubiquitous energy and information source to survive and adapt to its environment. The main mechanism for sensing light and transducing it into physiological functions is carried out by various classes of light-activated proteins that are often modular and reversible in their ability to regulate downstream effectors. Taking advantage of these facts, engineered light-activated proteins are emerging as powerful tools to control biological processes with high spatial resolution and temporal accuracy. However, engineering light-activated proteins to improve existing tools and create novel functionality remains a challenge, especially for multifaceted molecules such as channelrhodopsins, light-activated cation channels which show complex kinetic, spectral, and ion conduction properties. In this talk, we present an automated microscopy method capable of physiological screening through libraries of mutant channelrhodopsins, optimizing them for multiple properties in parallel. We use it to find mutations in channelrhodopsin-2 (ChR2) that sculpt its ion selectivity, a physiologically important property that has been hard to engineer, while maintaining high capacity to mediate light-driven control of membrane potential. We also show that combining multiple mutations discovered through such a screen can result in additive effects on ion selectivity. Using this methodology and screening principles, we find a mutant with order-of-magnitude improved selectivity for sodium over both calcium and proton ions, which displays photocurrents larger than that of the original wild-type ChR2. Such tools with high ion selectivity may find use in scientific and prototype medical applications, and the approach utilized may be useful for engineering other difficult properties of light-activated proteins, as well as for creating novel tools.

(143f) Methodology for Redesigning the Specificity of Secreted Proteases

Guerrero, J.,
O'Malley, M. A., University of California, Santa Barbara
Daugherty, P. S., University of California, Santa Barbara
Originally presented on: 11/4/2013 14:00:00 - 14:18:00

Proteases constitute 2% of the human genome and regulate many biological processes including cell growth and migration, inflammation, blood coagulation, and programmed cell death. It is estimated that most proteases have on the order of 10-100 physiological substrates, though few studies have elucidated the molecular mechanisms by which proteases achieve a high level of substrate selectivity. A generally applicable, high-throughput strategy to engineer proteases to cleave a target substrate with high specificity and high catalytic efficiency would greatly expand the use of proteases for analytical, biotechnological, and therapeutic applications. We have developed a cell-based assay for redesigning protease selectivity by screening protease mutant libraries for cleavage of a fluorogenic peptide substrate exhibiting Förster resonance energy transfer (FRET). Our screening method relies on using fluorescence activated cell sorting (FACS) for detecting cleavage of the FRET reporter substrate, which will allow screening of large protease mutant libraries (~10^7members) to identify the activity of interest. As a model system, these novel methods were applied to the protease human kallikrein 7 (hK7) to identify variants that selectively cleave the central hydrophobic core of the amyloid beta (Aβ) peptide, involved in Alzheimer's disease pathology.

Using *Saccharomyces cerevisiae,* an expression system was designed and constructed to produce active hK7 and a FRET substrate probe intracellularly. Expression of correctly folded, active hK7 was detected in the yeast cell lysate using an hK7 FRET reporter substrate and confirmed by Western blot. FRET probe production in yeast was measured using flow cytometry and an average 8-fold increase in yellow fluorescence over background was observed after expression for 16 hours at 30°C. We optimized expression conditions to detect hK7 activity in yeast with a co-expressed Aβ8 (KLVF↓F↓AED) FRET substrate using flow cytometry. No activity was detected for yeast cells co-expressing an inactive hK7 mutant or the FRET substrate alone. hK7 displays modest activity towards the target Aβ8 substrate but prefers tyrosine (Y) at the P1 position. We hypothesize that amino acid substitutions around and in the active site may yield variants that prefer the phenylalanine (F) at P1 of Aβ8 and exclude tyrosine, thereby narrowing the specificity towards Aβ8. We randomly mutated the hK7 gene using error-prone PCR to generate two libraries. These libraries were then screened for enhanced activity and selectivity towards the therapeutically relevant target substrate Aβ8 using FACS. This methodology can be applied to engineer other human proteases for highly specific degradation of proteins implicated in a disease state.

(143g) Novel Method for Engineering Aggregation-Resistant Antibody Fragments

Perchiacca, J. M., Rensselaer Polytechnic Institute
Lee, C.,
Tiller, K., Rensselaer Polytechnic Institute
Tessier, P. M., Rensselaer Polytechnic Institute
Originally presented on: 11/4/2013 14:18:00 - 14:36:00

Protein aggregation is a significant challenge in developing safe and effective therapeutic antibodies. We are investigating how antibody fragments can be engineered to possess extremely high solubility without altering binding activity. Although we find that hydrophobic peptides can be introduced into the complementarity-determining regions (CDRs) of single-domain antibodies in a rational manner to mediate specific binding to the Alzheimer's and Parkinson's polypeptides (Perchiacca et al., *PNAS*, 2012; Ladiwala et al., *PNAS*, 2012), these hydrophobic loops also mediate antibody aggregation. Therefore, we have developed a novel approach for engineering the edges of hydrophobic CDRs with charged mutations that dramatically increases antibody solubility without reducing their binding affinity. In contrast, similar mutations within the antibody scaffold and non-CDR loops are ineffective at preventing aggregation. We are currently using this mutational strategy for engineering several high-affinity antibody fragments and full-length antibodies to improve their solubility.

(143h) Structure Guided Engineering of Monomeric Streptavidin With Improved Ligand Binding

DeMonte, D., University at Buffalo
Drake, E.,
Lim, K. H., University at Buffalo
Gulick, A.,
Park, S., University of Pennsylvania
Originally presented on: 11/4/2013 14:36:00 - 14:54:00

We recently reported the engineering of monomeric streptavidin, mSA, corresponding to a single subunit of wild type (wt) streptavidin tetramer. The monomer was designed by homology modeling, in which the streptavidin and rhizavidin sequences were combined to engineer a high affinity binding pocket containing residues from a single subunit only. Although mSA is stable and binds biotin with nanomolar affinity, its fast off rate (koff) creates practical challenges during applications. We obtained a 1.9 Ã… crystal structure of mSA bound to biotin to understand their interaction in detail, and used the structure to introduce targeted mutations to improve its binding kinetics. To this end, we analyzed the structure of a streptavidin analog, shwanavidin, which binds biotin with high affinity as a dimer. The binding pocket of shwanavidin differs from that of mSA or rhizavidin in that it contains an aromatic residue that appears to form a lid over bound biotin. This led us to hypothesize that a comparable lid would likewise enhance the binding affinity of mSA. However, the T48F mutation in mSA only resulted in a modest 20 ? 40% improvement in the measured koff. On the other hand, introducing the S25H mutation near the bicyclic ring of bound biotin increases the dissociation half life (tÂ½) by an order of magnitude. Molecular dynamics (MD) simulations suggest that the mutation stabilizes the binding loop L3,4 by interacting with a residue at the base of the binding loop, and protects key intermolecular hydrogen bonds by limiting solvent entry into the binding pocket. The latest results from a directed evolution study will also be presented. The engineered binding kinetics of the mSA mutant significantly improves the usability of the streptavidin monomer in applications that require stable ligand binding.

(149a) Dissecting the Role of Asymmetric Division in Control of Stem Cell Lineage Specification

Harris, G. M., University of South Carolina
Jabbarzadeh, E., University of South Carolina
Originally presented on: 11/4/2013 12:30:00 - 12:48:00

Human pluripotent stem cells (hPSCs) have proven to be a promising tool for biomedical researchers due to their capabilities of self-renewal, almost indefinite proliferation, and ability to give rise to all human cell types. Regarding self-renewal, stem cells also have the ability to give rise to daughter cells with a committed cell fate as well as an undifferentiated and uncommitted cell. To accomplish this feat, cells intrinsically organize polarity proteins and cell fate determinants into only one dividing daughter cell and divide asymmetrically with the cell fate determinants segregated. In observing this asymmetric division, cytoskeletal organization has proven to be a critical factor for polarization as well as many other stem cell behaviors including cellular organization and adhesion. This study will outline the role of cell polarization in hPSC function and stem cell lineage commitment. Deep ultraviolet (UV) lithography was utilized to create patterns with symmetrical and non-symmetrical features to display and promote polarization of stem cells. This process was accomplished by utilizing a photomask with microfeatures to irradiate an already passivated substrate with low wavelength UV light making is possible to attach proteins selectively to irradiated areas. These studies showed cells were able to adhere and display prominent stress fibers while certain patterns were able to show segregation of polarity protein GPSM2. Cell polarity, spindle orientation, cell division, and lineage specification of dividing and daughter cells were then quantified on adherent cells and correlated to illicit the roles of asymmetric division in hPSCs. Understanding the mechanisms by which cell shape and cell division can control lineage specification has the ability to provide great insight for the rational design of biomaterials and usage in regenerative medicine.

(149b) Microenvironment Regulation of Tissue Development From Pluripotent Stem Cells Post-Cryopreservation

Sart, S., Florida State University
Yan, Y.,
Liu, Y., Florida State University
Ma, T., Florida State University
Li, Y., FAMU-FSU College of Engineering
Originally presented on: 11/4/2013 12:48:00 - 13:06:00

Neural progenitor cells (NPCs) are usually derived from pluripotent stem cells (PSCs) through the formation of embryonic bodies (EBs), a 3-D microtissue-like structure mimicking embryonic development. PSCs could also be expanded as microtissue-like aggregates in suspension culture, which enables the large scale production in stirred bioreactors. The cryo-banking of PSCs and committed EBs is required to meet the demand of large amount of functional neural populations in view of potential clinical application for the treatment of various neurological disorders.

In this study, it is hypothesized that the microenvironment (including biochemical factors and cell configuration) of undifferentiated PSC aggregates and EBs affects their recovery and neural differentiation post-cryopreservation. To assess the influence of PSC structural environment on cell recovery, intact microtissues from PSCs and EBs (I-P and I-E) were compared to dissociated single cells (D-P and D-E). The results showed that I-P and I-E recovered better after thawing compared to D-P and D-E respectively, due to the preserved extracellular matrix (ECM) and cell organization. Moreover, the F-actin organization was maintained in I-E group while it was depolymerized in D-E group. Reactive oxygen species (ROS) and caspase expression were also investigated under various conditions. Small aggregates generated significantly less ROS and caspases compared to large ones. In the same vein, mesenchymal stem cell secretome, known to bear broad spectrum of protective factors, enhance EB recovery through anti-oxidative and anti-apoptotic effects. Finally, I-P maintained the pluripotent marker expression, and the ability to differentiate into three-germ layers. I-E and D-E showed the enhanced neuronal differentiation and D-E further enhanced astrocyte differentiation.

In summary, modulation of structural and biochemical environment is essential to promote efficient recovery and minimize ROS-mediated apoptosis for PSC tissue development. The structural microenvironment also sustained PSC differentiation potential and regulated the neural commitment post-cryopreservation.

(149c) Comparing the Effects of Mechanical Stimulation On Bone Marrow and Hair-Follicle Mesenchymal Stem Cells: Vascular Tissue Engineering

Koobatian, M. T., State University of New York at Buffalo
Liang, M. S., State University of New York at Buffalo
Swartz, D. D., University at Buffalo, The State University of New York
Andreadis, S. T., University at Buffalo, The State University of New York
Originally presented on: 11/4/2013 13:06:00 - 13:24:00

Adult stem cells such as mesenchymal stem cells (MSC) show significant proliferation and differentiation potential; however these capacities decrease significantly with donor aging. This problem is further compounded by culture senescence limiting their culture time to about 8-10 passages. This is a major concern as patients mostly in need for vascular grafts and cellular therapies are elderly.

In the present study we compared adult and neonatal ovine hair-follicle MSC (aHF,nHF) to adult and neonatal bone-marrow MSC (aBM, nBM) in 2D culture examining their susceptibility to culture senescence as well as their differentiation potential to smooth muscle cell (SMC) lineage. We concluded that aHF/nHF and nBM do not show significant cellular senescence and also demonstrate increased cell proliferation compared to aBM. This was confirmed by long term proliferation experiments and β-galactosidase activity staining. aHF/nHF also showed significantly higher clonogenic potential compared to nBM, while nBM showed significantly higher clonogenic potential compared to aBM isolations.

We then generated fibrin based Tissue Engineered Vascular constructs (TEVs) using these cell isolations and compared their vasoreactive response to receptor and non-receptor mediated pathways under static and pulsed conditions. Pulsation was induced in a dose response fashion for one week at (2%, 3.6% and 5.5% distention) after culturing the vascular constructs statically for one week. To accomplish this task a 24 well-pulsation plate (24-PP) was designed, capable of mechanically conditioning 24 TEVs at the same time. Under static conditions all aHF/nHF isolations demonstrated superior vasoreactive responses when compared to aBM/nBM isolations. However, mechanical properties and vasoreactive responses were either unchanged or diminished when aHF/nHF were pulsed, while BM isolations showed improvements. Young's Modulus increased significantly following pulsation for aBM/nBM and Ultimate Tensile Stress also increased significantly for aBM isolations following pulsation. Collagen content was measured from static and pulsed TEV constructs to determine the extent of ECM remodeling and it was found there were significant improvements for aBM/nBM isolations, while aHF/nHF isolations were unchanged or diminished. In summary we concluded that there are significant similarities and differences between HF and BM MSCs. Developing a stronger understanding on these similarities and differences will aid us in the further development and optimization for the use of MSCs in vascular tissue engineering.

(149d) Cells Tune Matrix Metalloproteinase Activity By Sensing Mechanical Inputs

Schneider, I., Iowa State University
Haage, A., Iowa State University
Originally presented on: 11/4/2013 13:24:00 - 13:42:00

One of the primary mechanisms of tissue microenvironmental remodeling is extracellular matrix (ECM) degradation. ECM degradation is highly controlled during immune response, wound healing and development. When ECM degradation is uncontrolled, pathologies like fibrosis or cancer cell invasion occur. ECM degradation is mediated in large part by the matrix metalloproteinase (MMP) family. MMP level can be controlled through expression; however MMP activity is highly regulated through post-translational modifications, limiting the usefulness of expression studies. Environmental signals, such as growth factor and ECM concentration are known to alter MMP activity. However, another important microenvironmental characteristic is the ECM crosslinking density. Highly crosslinked ECM has small pores, is stiff and constitutes a difficult matrix through which to migrate, while uncrosslinked ECM has larger pores, is softer and constitutes less of a barrier for cell migration. The fibrotic microenvironment as well as the tumor microenvironment tend to be much stiffer than in normal tissue. In the case of the tumor microenvironment this leads to enhanced invasion. How does this occur? One mechanism might include the upregulation of MMP activity in order for cells to migrate through stiff, densely crosslinked networks of ECM.

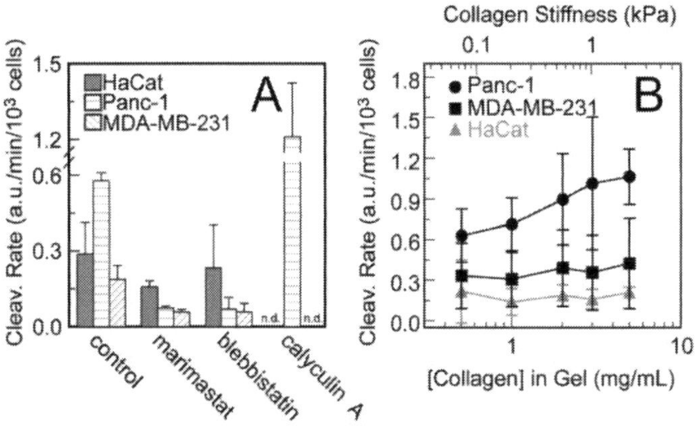

We are interested in determining if cells can sense chemical crosslinks in the ECM network through the mechanical properties of the ECM. Since cell contractility generally correlates with ECM stiffness, we are additionally examining how changes in cell contractility regulate MMP activity. We are using fluorescently quenched cleavage peptides in a high-throughput live-cell 96-well assay that can directly detect changes in MMP activity. MMP activity was higher in cells plated on high concentration collagen gels (5 mg/mL) as compared to low concentration collagen gels (0.5 mg/mL) (Fig. 1B). This response was cell-type specific and more pronounced in highly invasive cancer cells. Because these gels differ in Young's moduli by over 10-fold, we hypothesized that these changes in MMP activity were due to changes in the mechanical properties of the ECM. To test this in part, cells were treated with pharmacological inhibitors (blebbistatin) and enhancers (calyculin A) of contractility and MMP activity was measured. As predicted, when cell contractility increased with calyculin A treatment, MMP activity increased. When cell contractility decreased with blebbistatin treatment, MMP activity decreased to roughly the same level as that seen by marimastat, a pan-MMP inhibitor (Fig. 1A). Currently, we are examining ways to independently control ECM density and mechanical properties to probe MMP regulation by each. These findings suggest an interesting mechanism to sense crosslinked ECM. When cells are exposed to a highly crosslinked ECM network that requires high MMP activity, they sense high ECM crosslinking density by probing the mechanical properties of the environment and respond by activating MMPs. Insights into the mechanical control over MMP activity can guide the design of artificial tissues that can elicit a particular MMP activity in cells or therapeutic targets that lie upstream of MMP activity and cancer invasion.

(149e) Mesenchymal Stem Cell Alignment On the Pre-Stretched Surface

Liu, C., Michigan State University
Kim, J.,
Baek, S., Michigan State University
Chan, C., Michigan State University
Originally presented on: 11/4/2013 13:42:00 - 14:00:00

Mechanical cues in the cellular environment play important roles in guiding various cell behaviors, such as cell alignment, migration, proliferation, and differentiation. Numerous studies have shown that mechanical cyclic stretch can induce cells to align perpendicular to the stretch direction, while relatively fewer studies focus on static stretch. However, almost all of the previous studies of static stretch were post-stretch, which means the cells were first seeded to allow attachment and then the substrate subsequently stretched. In contrast, we create a static pre-stretched anisotropic surface in which the cells are seeded after the substrate is stretched. The results show that cells align in the direction of pre-stretch, which is induced by the anisotropy that can be predicted by the theory of finite elasticity.

The experimental results agreed with a "Cell mechano-active sensing" model, which suggests that the cells can sense and respond to surface anisotropy by orienting in the direction of maximal effective stiffness. In this study we employed a theory of "small deformation superimposed on large" to predict the anisotropy induced by the uniaxial pre-stretch. After a 10% uniaxial pre-stretch, the effective stiffness that the cells sense in the stretched direction is 1.33 times of that in the perpendicular direction. Next, we explored the impact of pre-stretch magnitude on cell orientation angle distribution. Cells (mesenchymal stem cells or primary neurons) seeded on poly-L-lysine coated PDMS membrane surfaces with 10%, 20%, or 30% pre-stretch were quantified after 4 days of culture for their cell orientation angles. The results showed that the ratio of cells that orient parallel (within $\pm 10°$) increased with the magnitude of the stretch.

In summary, we demonstrated that cells aligned on static pre-stretched anisotropic surface, and the number of cells that aligned in parallel orientation increased with the pre-stretch magnitude. Besides alignment of MSCs, we investigated the impact of pre-stretched surface on axonal growth and orientation. The results showed that axonal alignment also increased with the pre-stretch magnitude and a larger pre-stretch can promote thicker and longer axonal growth.

(149f) A Comparison of Two Dimensional Monolayer and Three Dimensional Spheroid Adipocyte Cultures

Janorkar, A. V., University of Mississippi Medical Center
Turner, P. A., University of Mississippi Medical Center
Harris, L.,
Originally presented on: 11/4/2013 14:00:00 - 14:18:00

According to the Center for Disease Control and Prevention, 72.5 million adults in the United States are classified as obese, with the annual mortality rate exceeding 300,000. In order to effectively treat this disease, obesity must be better understood at the cellular level with respect to metabolic state and environmental stress; however, current two-dimensional (2-D) *in vitro* cell culture methods do not represent the *in vivo* adipose tissue appropriately due to absence of complex architecture and cellular signaling. Conversely, 3-D *in vitro* cultures have been reported in the literature to have optimal results mimicking the adipose tissue *in vivo*. In previous studies, our laboratory created 3-D spheroid hepatocyte constructs when cells were cultured on tissue culture polystyrene (TCPS) culture plates coated with elastin-like polypeptide (ELP) coupled to polyethyleneimine (PEI). The aims of this study were to examine the efficacy of the ELP-PEI conjugate toward creating a 3-D preadipocyte culture system, and to study their differentiation and maintenance process in physiologically-relevant microenvironments when subjected to various concentrations of nutritionally relevant free fatty acids with respect to total DNA content, cell viability, and intracellular triglyceride accumulation. Our results showed that 3T3-L1 preadipocytes cultured on the ELP-PEI surface formed 3-D spheroids within 72 hours, whereas the cells cultured on unmodified TCPS remained in monolayer configuration. Significant statistical difference was discovered between the 3-D spheroid and 2-D monolayer culture with respect to the DNA content and triglyceride accumulation, indicating differences in cellular response. Results indicated that the 3-D culture may be a more sensitive modeling technique for *in vitro* adipocyte culture and provides a platform for future evaluation of 3-D *in vitro* adipocyte culture. This outcome may prove advantageous for more rapidly promoting a differentiated phenotype in adipose cell cultures for investigating the influence of exogenous drugs and nutrient treatments on a mature cell population over a shorter *in vitro* experimental period.

(149g) Development of a Mathematical Model to Describe the Release of Monocyte Chemoattractant Protein - 1 (MCP-1) From Human Aortic Endothelial Cells and the Transport Through a Three-Dimensional Collagen Matrix

Ghousifam, N., Oklahoma State University
Fahlenkamp, H. G., Oklahoma State University
Originally presented on: 11/4/2013 14:18:00 - 14:36:00

Atherosclerosis is an inflammatory disease, initiated by the accumulation of lipid substances in the subendothelial layer of major arteries, followed by adhesion and transmigration of monocytes. This adhesion and transmigration of monocytes involve several steps, mediated by bioactive molecules named chemotactic cytokines (chemokines). Monocyte Chemoattractant Protein-1 (MCP-1) is one chemokine that is expressed highly in atherosclerotic lesions and plays a role in monocyte trafficking across the endothelial cell layer. The end state of the monocyte after transendothelial migration is dependent on many factors, such as the specific tissue involved in the process, the type of stimulus, and the formation of MCP-1 concentration gradients in the tissue. Having a better understanding of the underlying mechanisms and identifying potential pro-atherogenic markers will help to target these cytokines with specific therapeutic strategies. A 3D *in vitro* vascular tissue model, consisting of human aortic endothelial cells (HAECs) grown on the surface of a collagen matrix, was used to investigate MCP-1 release within a 3D environment. The 3D tissue model provides the added dimension that is required for the creation of concentration gradients, along with cellular movement and interactions created by such gradients. During analysis of the system, a binding reaction between MCP-1 and the collagen matrix was discovered, which shows that in addition to a soluble gradient of MCP-1, a static gradient (the gradient of collagen-bound MCP-1) may be formed in the collagen matrix as well. The main objective of this study was to develop a mathematical model that can be used to predict the MCP-1 concentration gradients in the collagen matrix of the tissue model. The unsteady-state transport model includes a source term to describe MCP-1 production from the HAECs and a binding reaction term to describe the interaction of MCP-1 with the collagen matrix. The release of MCP-1 from the HAECs in the tissue model, under normal culture conditions and in response to TNF-α, was determined at various time points over a 24-hour period. The MCP-1 release profile was used to derive the source term of MCP-1 production of the HAECs in the mathematical model. The binding reaction expression and rate constant were determined experimentally by measuring the initial rates of reaction for various MCP-1 concentrations applied to a collagen matrix without cells. The mathematical model indicates that the concentration gradients of both soluble and static MCP-1 are formed inside the collagen matrix. The model predicts the increase of MCP-1 concentration (both soluble and static) with time, due to the release of MCP-1 from the HAECs in response to TNF-α activation. The concentration of static MCP-1 remains higher than that of the soluble MCP-1 after 18 hours and overcomes the soluble gradient of MCP-1 as time passes. The model further substantiates that apart from the soluble gradient of MCP-1, the static gradient of MCP-1 is another potent factor that may mediate monocyte transendothelial migration. The mathematical model was validated with experimental data. The mathematical model demonstrated a good prediction of MCP-1 concentrations in both the top and the bottom reservoirs surrounding the collagen matrix, for selected incubation time points with TNF-α. The mathematical model can be used to provide new information about the relationship between MCP-1 concentration gradients and monocyte transendothelial migration associated with inflammation.

(149h) Regulation of Cancer Stem Cells By the Tumor Microenvironment
Wokpetah, J., The Pennsylvania State University
Originally presented on: 11/4/2013 14:36:00 - 14:54:00

Regulation of cancer stem cells by the tumor microenvironment

Joseph Wokpetah and Esther W. Gomez

Departments of Chemical Engineering and Bioengineering, The Pennsylvania State University, University Park, PA

Tumors are composed of a heterogeneous population of cells surrounded by a dynamic microenvironment. A subpopulation of cells within tumors, termed cancer stem cells (CSCs), is believed to drive tumor growth and recurrence. CSCs have been implicated in the progression of a variety of cancers as their self-renewal potential is thought to be important for colonization of secondary organ sites. Likewise, CSCs exhibit increased resistance to chemotherapeutics and ionizing radiation; thus, current cancer therapies may fail due to survival of CSCs that can then regenerate new tumors. Here, we sought to understand how properties of the tumor microenvironment regulate CSC qualities. We find that the rigidity of the microenvironment is a critical determinant of CSC phenotype. Results from these studies may suggest therapeutic approaches for reducing the population of CSCs within tumors and for improving the efficacy of cancer treatments.

(161a) Metabolic Flux Redistribution for Efficient Anaerobic Production of 1,2-Propanediol and 1-Propanol in Escherichia Coli

Jain, R.,
Originally presented on: 11/4/2013 15:15:00 - 15:33:00

Metabolic flux redistribution for efficient anaerobic production of 1,2-Propanediol and 1-Propanol in *Escherichia coli*

Rachit Jain[a] and Yajun Yan[b]

[a]College of Engineering, Driftmier Engineering Center, University of Georgia, Athens, GA 30602, USA

Email: rachit@uga.edu

[b]Biochemical Engineering Program, College of Engineering, Driftmier Engineering Center, University of Georgia, Athens, GA 30602, USA

Email: yajunyan@uga.edu

Abstract

With the industrial revolution of the last century and years of rapid globalization a surge in energy production, consumption and demand has affected all countries. With most of the world relying on oil and coal as the primary source of energy, its depletion is at hand. The rapid consumption of oil for fuel and chemicals is leading to crisis in two directions – environmental and economic. The release of green house gas carbon dioxide is primarily due to oil derived fuels and coal based energy generation. At this hour the need is to gradually reduce the reliance on oil based chemicals and fuels. Over the years metabolic engineering and synthetic biology have enabled chemical and biological engineers alike to develop and construct novel metabolic pathways. Truly, the production of "green chemicals" holds paramount importance in reducing the number of chemicals manufactured from oil. In establishing biological processes for their manufacture the dependence on oil can be greatly reduced and hence the expenses on importing oil into the country can be decreased.

The manufacture of biofuels and green chemicals has been proposed by researches across the globe by constructing novel synthetic routes. Although researchers report the achievement of either high titer or high yield, only a handful of processes exist with the capability to achieve both from a single strain. In this work we first engineer an efficient method for the production of high value commodity- 1,2-propanediol and further dehydrated it to 1-propanol. To do this we hypothesized a novel fermentative approach to produce 1-propanol by dragging the carbon flux from the native glycolytic pathway in *Escherichia coli*. Using bio-prospecting approaches we evaluated eight different methylglyoxal synthases, three alcohol dehydrogenases and three dehydratases that are involved in the pathway via enzyme assays.

After the identification of the minimal set of enzymes we (i) constructed a synthetic pathway via 3 steps leading to the production of 1,2-propanediol and enhancing it from 0.28g/L to 1.52g/L in shake flask studies (ii) redirected the native carbon flux into our synthetic pathway and reduced by-products, improving the yield to about 50% of theoretical maximum (iii) demonstrated the manipulation of cellular energetics (NADH availability) to boost the titer to 2.14g/L in wild type *E. coli* (iv) developed cell adaptation strategies to improve the growth, resulting in 1,2-propanediol titer to biomass ratio of >4:1 after 48 hours and a titer of 2.7g/L (representing a yield >90% of theoretical maximum) after 72 hours from our engineered strains. This work represents the highest titer and yield of 1,2-propanediol production achieved in shake flask studies from *Escherichia coli*till date. This process holds potential for future scale up by achieving high titer and yield simultaneously from a single strain.

After the establishment of an efficient 1,2-propanediol pathway, we identified the optimal dehydratase for the production of 1-propanol and achieved its production at 0.11g/L (conversion efficiency of 11%). After establishing this novel metabolic route we proceeded to address the conversion efficiency of 1,2-propanediol to 1-propanol. We constructed a fusion diol dehydratase protein in order to regulate the expression level of individual subunits. By doing so the titer further increased to 0.62g/L. By studying the regulation of upstream pathway (1,2-propanediol) and downstream pathway (1-propanol) via operon organization and plasmid copy number we achieved the improvement in tweaking the cell to improve the conversion efficiency to about 50%.

This work represents metabolic and synthetic biology approaches and demonstrates solutions for the production of bulk chemicals like 1,2-propanediol by developing processes which achieve high titer and high yield simultaneously. This work also highlights protein based engineering and regulation of expression levels of proteins to strike a balance between upstream and downstream pathways to improve production of chemicals (1-propanol) via novel metabolic routes.

(161b) Metabolic Engineering Of P. Freudenreichii For Enhanced Propionic Acid Production

Ammar, E. M., Ohio State University
Wang, Z., The Ohio State University
Yang, S. T., Ohio State University
Originally presented on: 11/4/2013 15:33:00 - 15:51:00

Propionibacteria are widely used in industry for manufacturing of Swiss cheese, vitamin B12 and propionic acid. Little is known about their genetics and only few reports are available on their metabolic engineering. Most previous metabolic engineering work on propionibacteria aimed at improving the fermentative production of vitamin B12. Propionic acid is well known as food preservative and as an intermediate in many industrial processes. Currently, it is mainly produced from petrochemicals which are perishable and costly. Its fermentative production suffers from low yields, low productivity and byproducts formation. In this research, by engineering *P. freudenreichii* using heterologous gene from E. coli, the productivity and purity of propionate were enhanced with proper choice of the carbon source. To our best knowledge, this is the first such report and it could be a novel route to enhance propionate production from biomass using renewable feedstocks.

(161c) Investigation of Cyanobacterial Alkane Biosynthesis Pathway for Microbial Production of "Drop in" Biofuels

Lin, F., University of michigan
Das, D., University of michigan
Marsh, N., University of michigan
Lin, X., University of Michigan
Originally presented on: 11/4/2013 15:51:00 - 16:09:00

An alkane biosynthesis pathway originally from cyanobacteria has attracted great interest in the past several years for metabolic engineering for "drop in" biofuels, since its products (alkanes and alkenes) have high energy density and can be used directly as substitutes for gasoline, diesel and jet fuel without costly and energy-intensive chemical conversion. In this pathway, fatty acyl-CoA (ACP) is first converted by fatty acyl reductase (ACR) to aldehydes, which subsequently undergo decarbonylation catalyzed by aldehyde decarbonylase to yield alkanes and alkenes. To obtain knowledge of this newly-identified pathway, which will facilitate its engineering for biofuels and other biotechnology applications, over-expression, purification and initial characterization of ACR from *Synechococus elongatus* PCC7942 was performed. ACR operates by a 'ping-pong' mechanism which involves an acyl-enzyme thioester intermediate, with $k_{cat} = 0.36 \pm 0.023$ min^{-1}, K_m (stearoyl-CoA) = 31.9 ± 4.2 μM and K_m (NADPH) = 35.6 ± 4.9 μM. We are currently evaluating the performance of this pathway in recombinant *E. coli*. Synthetic protein scaffolds and other methods are being employed to improve the efficiency of alkane and alkene synthesis.

(161d) Converting 2,3-Butanediol to Butanone By Engineered Diol Dehydratase

Jacobson, G.,
Ge, X., University of California Riverside
Originally presented on: 11/4/2013 16:09:00 - 16:27:00

High titers of 2,3-butanediol are commonly present as the major end product in anaerobic fermentation in various microbes (e.g. 150g/L by *Klebsiella pneumonia*), but with an excessively high boiling point of 177°C, cost-effective separation is not feasible. Diol dehydratase (DDH, EC 4.2.1.28 or 4.2.1.30) converts a diol molecule into its corresponding ketone or aldehyde. Catalyzed by DDH, the dehydration of 2,3-butanediol yields butanone (methyl ethyl ketone or MEK), a potential advanced liquid fuel/fuel additive due to its high energy density (31.5 MJ/kg) and a low heat of vaporization (0.45 MJ/Kg). However, biosynthesis of butanone has not been demonstrated, due to (1) butanone is not a natural metabolite present in any microbial fermentation; and (2) the native substrates of all known diol dehydratases (DDHs) are 1,2-propanediol, glycerol and ethanediol. 2,3-butanediol, when used as a substrate for wild type DDHs to produce butanone, exhibits an activity 1 or 2 fold lower than the native substrates. This research aims to not only construct and optimize metabolic pathways, but also more importantly, engineer the key enzymes of rate-limiting steps to improve the specific activities and substrate/co-factor preference desired for high productivity. Here we report our recent progresses on cloning, expression and characterization of DDHs from *Lactobacillus brevis,* and the creation of engineered DDHs using a rational designed approach based on known structures of DDHs. These engineered DDHs demonstrate substrate specificity shifts with enhanced activity toward *meso*-2,3-butanediol, thus paving a path for metabolic synthesis of MEK from low cost renewable biomass.

(161e) Application of Omics Technologies in n-Biobutanol Production By Metabolically Engineered C. Tyrobutyricum

Ma, C., University of Alabama
Xu, N.,
Liu, X., The University of Alabama
Originally presented on: 11/4/2013 16:27:00 - 16:45:00

Application of Omics technologies in n-biobutanol production by metabolically engineered *C. tyrobutyricum*

Chao Ma*, Ningning Xu, and Margaret Liu

As a sustainable and environmentally friendly biofuel with similar energy intensity to petroleum gas, biobutanol is an excellent alternative for fossil fuel. The metabolically engineered acidogenic *C. tyrobutyricum* has been used to produce biobutanol through genetic engineering such as *ack* gene knockout (ACKKO) and *adhE2* gene synthesis (ACKKO-adhE2) in previous studies. To further engineer *C. tyrobutyricum*, it is essential to understand the mechanism of host cell regulation of biobutanol generation. In this study, Omics technologies (i.e. proteomics and metabolomics) were applied to investigate the key enzymes and critical metabolism activities responsible for the high yield, productivity, and selectivity of biobutanol. First, the 2-L free-cell fermentations were performed in 3-L stirred tank bioreactors to compare the biobutanol and biobutyrate productions from glucose among wild type, ACKKO mutant and ACKKO-adhE2 mutant. A high butanol titer of 14 g/L was achieved by ACKKO-adhE2, and high butyric acid titer of 36 g/L was obtained by ACKKO. Second, the global proteomics profiling and metabolomics maps of these three strains were investigated and compared. It is found that, as compared to wild type and ACKKO, the ACKKO-adhE2 had lower intracellular expression of pyruvate: ferredoxin oxidoreductase and hydrogenase in the hydrogen generation pathway, but higher bifunctional acetaldehyde-CoA/alcohol dehydrogenase in the butanol formation pathway. These findings indicated the critical role of reducing power in butanol production and demonstrated the possibility of further improvement of biobutanol production by increasing NADH accumulation in the ACKKO-adhE2 mutant of *C. tyrobutyricum*. In future, novel engineered *C. tyrobutyricum* will be rationally designed to achieve higher biobutanol production by integrating the Omics investigation with metabolic engineering.

Authors:

Chao Ma

Department of Chemical and Biological Engineering

The University of Alabama

Hauser Hall R116

301 7th Avenue

Tuscaloosa, AL 35405

(Email): cma3@crimson.ua.edu

(Phone): 205-292-8115

Ningning Xu

Department of Chemical and Biological Engineering

The University of Alabama

Hauser Hall R116

301 7th Avenue

Tuscaloosa, AL 35405

(Email): nxu2@crimson.ua.edu

(Phone): 205-348-7340

Margaret Liu, PhD

Assistant Professor

Department of Chemical and Biological Engineering

The University of Alabama

Hauser Hall R116

301 7th Avenue

Tuscaloosa, AL 35405

(Email): mliu@eng.ua.edu

(Phone): 205-348-0868

(161f) Biorefinery for Xylitol, Lactic Acid and Astaxanthin Production From Corn Cobs

Horiuchi, J. I., Kitami Institute of Technology
Konishi, M., Kitami Institute of Technology
Kanno, T.,
Takamura, H.,
Matsui, K.,
Kidachi, T.,
Originally presented on: 11/4/2013 16:45:00 - 17:03:00

A biorefinery is an integrated process to produce multiple products including fuels and chemicals from renewable resources with low environmental impact. In this research, we experimentally examined the feasibility for producing multiple products including, xylitol, lactic acid and astaxanthin from corn cobs as a feed stock. The corn cobs were first hydrolysed by the dilute acid to obtain to xylose and its residues were then enzymatically hydrolyzed to glucose. By this treatment, hydrolysates containing approximately 27 g-xylose/ *l* and 30 g-glucose/ *l* were selectively obtained from 100 g-corn cobs/ *l* . Using the hydrolysate containing xylose, approximately 18 g/*l* xylitol was successfully produced by *Candida magnolia*. The hydrolysate containing glucose was easily utilized for lactic acid production by *Lactococcus lactis*. Simultaneous enzymatic saccharification and fermentation (SFF) was also effective to produce concentrated lactic acid (approx. 90 g/L) from corn cob residues after dilute sulphuric acid hydrolysis, which mainly contain cellulose. It was also found that the hydrolysates containing xylose and glucose could be used as a favourable source for astaxanthin production using *Xanthophyllomyces dentrohous*. By incorporating the high value product such as astaxanthin into the biorefinery system, the economic feasibility of the biorefinery was greatly improved. Finally total process evaluation will be presented from the view points of profitability, energy consumption and environmental impact.

(161g) Biohydrogen Production From Untreated Lignocellulosic Biomass Using Thermophiles

Bibra, M., SDSMT
Kainth, A. S., South Dakota School of Mines and Technology
Bhalla, A., South Dakota School of Mines & Technology
Bhardwaj, R.,
Kumar, S., South Dakota School of Mines & Technology
Sani, R. K., South Dakota School of Mines & Technology
Originally presented on: 11/4/2013 17:03:00 - 17:21:00

The desire for a greener world with renewable energy source and to reduce the anthropogenic pollutant emissions has shifted the research interest toward biofuels.Â Hydrogen, being a highly energy efficient, environmentally friendly, and renewable source, has been given importance in the past few years. Â The present hydrogen production processes are limited by the substrate shortage, operation costs, hazardous chemicals usage and production, complex designs, and their high set-up costs. Â Biohydrogen production form the lignocellulosic biomass using microorganisms offers promising alternatives for the costly chemical and physical processes. Â However, pretreatment costs of the lignocellulosic biomass often make the biohydrogen production process less cost-effective.Â Therefore, the present research is focused on using the untreated biomass so as to lower the cost of hydrogen production. Â Locally available lignocellulosic biomass (Prairie Cord Grass -PCG) was used for the biohydrogen production using thermophillic microbes isolated from Thermopolis hot springs of Wyoming.Â Dark fermentation (60Â°C) was carried in the 135-ml serum bottles containing 100 ml growth medium amended with 1% untreated PCG and hydrogen was measured.Â A maximum cumulative yield of 8.68 mL/g-dry PCG was achieved under optimum conditions of temperature and pH. Utilization of untreated PCG and metabolite production (lactic and acetic acids) profiles was also compared. Â The results show that thermophillic consortium was able to convert untreated PCG into hydrogen in one step. This thermophillic consolidated bioprocess will likely impact ongoing multiple-step conversion processes of lignocellulosic biomass to hydrogen by providing a more efficient and economical process.

(161h) Immobilization of Lipase From Candida Antarctica Type B Onto Epoxy-Functionalized Poly (methyl methacrylate)

Matos, L. J. B. L. D., Universidade Federal do Cear
Pinheiro, B. B., Universidade Federal do Ceara
Santos, J. C. S. D., Universidade Federal do Cear
Goncalves, L. R. B., Universidade Federal do Cear
Originally presented on: 11/4/2013 17:21:00 - 17:39:00

Immobilization of enzymes and proteins on activated supports may be used to improve some enzyme properties in addition to allowing the simplification of reactor design. In this sense, supports containing epoxy groups may be very useful tools to stabilize proteins via uni or multipoint covalent attachment if the immobilization is properly designed. In this work, the support used for immobilization was Immobead 350-Chiralvision, which is a poly (methyl methacrylate) epoxylated with nonpolar characteristics support, diameter ranging between 300- 700μm. For the adsorption immobilization, assays were performed varying the pH between 6-8 and ionic strength, an important factor, between 100-700mM. Best immobilization results were obtained at pH 7.5, ionic strength 400 and 700 mM, where recovery activity was around 25%. For multipoint immobilization, conducted at pH 10.05, the epoxylated support was modified by introducing amino groups, since the presence of amino groups allows a very rapid ionic adsorption of the proteins onto the support at low ionic strength. The new support has a layer of epoxy groups over a layer of ethylenediamine that is covalently bound to the support. Thus, this support has a great anionic exchanger power and a high number of epoxy groups. Immobilization was conducted at low ionic strength using amino-Immobead 350, while high ionic strength using conventional monofunctional epoxy supports was required. Furthermore, immobilization is expected to be much more rapid using amino-epoxy supports than employing conventional epoxy supports. Last but not least, the performance of both biocatalysts will be investigated in the synthesis of the ethyl oleate, aiming biodiesel production by esterification of acidic substrates.

(171a) Intracellular Delivery of Antigenic Proteins for Cancer Vaccination

Sharei, A., MIT
Mao, S., MIT
Basto, P., MIT
Szeto, G., MIT
Talkar, T.,
Hartoularos, G., MIT
Irvine, D. J., Massachusetts Institute of Technology
Langer, R. S., Massachusetts Institute of Technology
Jensen, K. F., Massachusetts Institute of Technology
Originally presented on: 11/4/2013 15:15:00 - 15:33:00

Introduction

Cell-based vaccines against cancer, composed of tumor antigen-loaded antigen presenting cells (APCs), hold great therapeutic potential due to their ability to activate the patient's immune system and drive a long-lasting, tumor-specific, CD8 T cell response. These therapies, such as the recently approved Provenge® for prostate cancer, have minimal side-effects relative to chemotherapies and radiation treatment. However, one of the main barriers to developing such vaccines has been achieving efficient intracellular loading of APCs with protein antigens for presentation to CD8 T cells. Since the mechanism of cross-presentation (a process wherein endocytosed proteins can induce a CD8 response) within antigen presenting cells remains elusive, one must develop a reliable method of delivering the desired antigen directly to the cell cytoplasm to advance therapies utilizing the potent effector CD8 response. In this work, our team has adapted our recently developed microfluidic delivery system (Sharei, A. et al. *PNAS, 2013.* 110, 1082-2087) to induce selective MHC class I antigen presentation by direct delivery of the target protein to the cell cytosol. Compared to endocytotic methods, this system has demonstrated a 10-100 fold improvement in antigen-specific CD8 T cell activation and can potentially be translated to more challenging systems, such as B cells.

Materials and Methods

This work was performed using the ovalbumin (ova) antigen presentation model. In our experiments, dendritic cells were derived from the bone marrow of healthy B6 mice and treated by the aforementioned microfluidic delivery system to facilitate intracellular delivery of the ovalbumin protein. Antibody staining, ELISA, flow cytometry, and proliferation assays using OT-I and OT-II transgenic mice were used to assess antigen presentation efficiency and T cell cytokine secretion.

Results and Discussion

In our studies, we have demonstrated enhanced MHC-I presentation of the SIINFEKL epitope by antibody staining and verified that the delivery process does not interfere with DC maturation as measured by up-regulation of CD80 and CD86 in response to stimulation by lipopolysaccharide. As illustrated in **Fig. 1**, this method has also successfully induced the activation/proliferation of antigen-specific CD8 T cells (derived from OT-I mice) as measured by T cell proliferation and cytokine secretion. Moreover, preliminary results indicate that this method can induce functional antigen presentation in primary B cells ? a significant challenge for existing technologies.

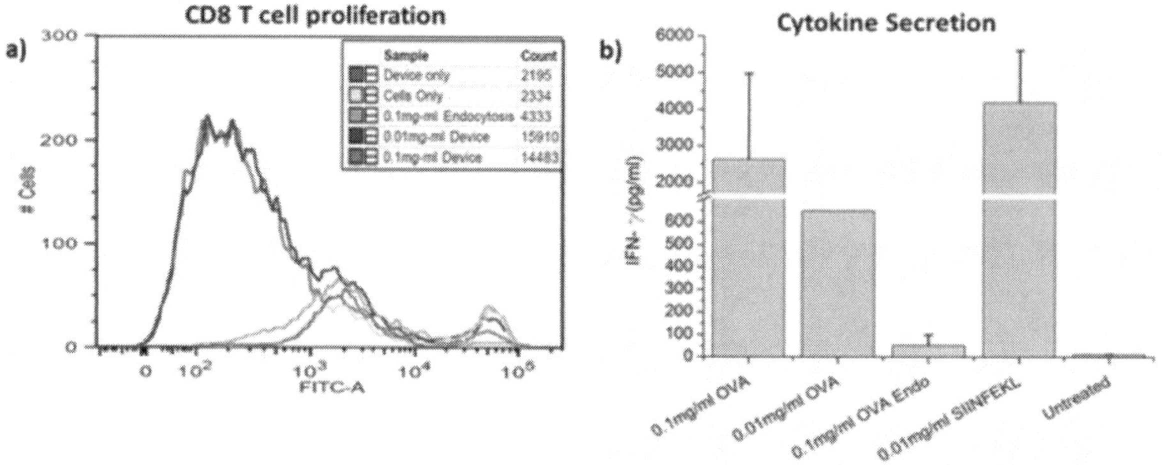

Figure 1. a) Bone marrow derived DCs were treated by the device in the presence of varying OVA concentrations and T cell proliferation was measured by CFSE staining and flow cytometry. b) Secretion of IFN-γ from CD8 T cells stimulated by treated DCs. The ?OVA Endo' condition corresponds to an endocytosis only case.

Conclusion

The described intracellular delivery technique has demonstrated enabling potential as a platform for high efficiency antigen presentation in cancer vaccine applications.

(171b) Sensing the Tumor Microenvironment: Genetically Modified Bacteria Sense and Trigger Gene Expression At the Site of Solid Tumors

Panteli, J., University of Massachussetts Amherst
Forbes, N. S., University of Massachusetts
Originally presented on: 11/4/2013 15:33:00 - 15:51:00

To improve the therapeutic specificity of bacterial cancer therapies, we have utilized chemosensory receptors coupled to the osmoporin transduction pathway to create novel a bacterial gene expression system that switches on specifically in solid tumors. We hypothesize that tumor necrosis leads to increased ribose concentrations due to the abundance of degrading DNA and RNA. It has also previously been shown that there is enhanced serine production in metastatic cancer making it a good target for specific expression within malignant tissue. A genetic circuit has been created using a previously developed fusion protein, Trz1, which couples the ribose and glucose sensing of the Trg protein to osmoporin signal transduction domain of EnvZ, from *E. coli,* and produces green fluorescent protein (GFP) via the porin promoter pOmpC. This genetic switch enables bacterial sensing of ribose and glucose and subsequent visualization, via fluorescence, of sugar gradients within the tumor microenvironment. In addition to this we have created a novel fusion protein, Tsz2, joining the serine sensing Tsr chemoreceptor to EnvZ to detect serine gradients within the tumor microenvironment. The ribose-glucose sensing bacterial probe responds to both glucose and ribose gradients. A previously developed *in vitro* 3D tumor model mimicking tumor tissue adjacent to a blood vessel was used to determine the ability of bacterial to probe the tumor microenvironment. The tumor-on-a-chip microfluidic device creates gradients in nutrients and oxygen within tumor tissue similar to *in vivo* conditions. Image analysis revealed gradients of the sugars within solid carcinoma tissue via the fluorescent bacterial sensors for specific gene expression. Bacterial colonization occurred in the regions of tissue with the highest sugar content as expected. The sugar gradients controlled GFP expression levels within the tissue based on bacterial density and depth within the tissue. These bacterial sensors have the potential to improve targeted cancer therapy by limiting expression to within tumors.

(171c) Characterization of the Transport of Live Autonomous Drug Particles (DrugBots) in Tumor Spheroids Interstitium

Traore, M. A., Virginia Tech
Sahari, A., Virginia Tech
Behkam, B., Virginia Tech
Originally presented on: 11/4/2013 15:51:00 - 16:09:00

Limited efficacy of the existing chemotherapeutic treatments of primary cancer tumors is partially attributed to the limited diffusion distance of the drug particles which prevents successful treatment of the quiescent tumor cells. For an anticancer drug to be effective against a solid primary tumor, it must be able to reach all cancerous cells within the tumor to prevent clonogenic tumor stem cells from regeneration. Current chemotherapeutic drugs have limited penetration within the tumor. The low selectivity of anti-cancer drugs with respect to cancerous tissue is also problematic due to the exposure of healthy cells to anti-cancer drugs. Thus, chemotherapy can be enhanced through both better anti-cancer drug cargo targeting towards cancerous cells. Several strains of attenuated bacteria have been identified to possess the innate ability to target and preferentially colonize tumor tissues. In this work, we demonstrate that the tumor targeting *E. coli* bacteria can be coupled with therapeutic nanoparticles to form intelligent autonomous drug carriers (DrugBots). DrugBots consists of an engineered abiotic body (i.e. drug-laden particles) and a living component (i.e. bacteria) for sensing and actuation purposes. The use of innate sensing and propulsion capabilities in DrugBots can dramatically improve delivery of chemotherapeutic to solid primary tumors. DrugBots actively sense and target tumors and utilize chemotaxis to carry drug nanoparticles deep into the tumor[1,2]. In this work, we demonstrate that DrugBots can penetrate deeper within the tumor tissue when compared to the commonly used passively diffusing chemotherapeutic nanoparticles.

For this study, tumor spheroids were grown in-vitro using human colon carcinoma cells (HT-29) in low adhesion round bottom 96-well plates[3]. To quantify the extravascular transport of the DrugBots, DrugBots suspended in DMEM culture media were incubated with the tumor spheroids and imaged using confocal microscopy.

The distribution of the DrugBots as well as passively diffusing nanoparticles was quantified by measuring the fluorescence intensity within the tumor spheroid. Our preliminary results indicate that DrugBots have the capability to transport loads significantly deeper within the tumor tissue. This study demonstrates that bio-hybrid tumor targeting systems (drug loaded nanoparticles self-assembled with tumor-targeting bacteria) can be used as live autonomous drug delivery agents with enhanced selectivity and penetration of solid tumors. As illustrated in this work, bacteria carrying particles can penetrate a solid tumor past the proliferating ring into the quiescent and necrotic core with a fairly homogeneous distribution in the tumor.

References:

1. K. Low *et al.* Nature biotech., 17:37-41, 1999
2. R. Kasinskas *et al.* Biotechnol and Bioeng., 94: 710-721, 2006
3. A. Ivascu *et al.* J Biomol Screen., 11: 922-932, 2006

(171d) Cell-Mediated Drug Delivery With Phagocytosis Resistant Polymeric Backpacks

Gilbert, J. B., Johns Hopkins University
Anselmo, A. C., University of California Santa Barbara
Mitragotri, S., University of California Santa Barbara
Rubner, M. F., Massachusetts Institute of Technology
Cohen, R. E., Massachusetts Institute of Technology
Originally presented on: 11/4/2013 16:09:00 - 16:27:00

Advances in drug delivery have been enabled by modifying the surface chemistry and, more recently, the shape of particles to target delivery of therapeutics to diseased sites. Recent studies have utilized the unique ability of certain mammalian cells to target and home to sites of inflammation, or disease, by conjugating drug carrying particles directly to the surface of these cells for targeted delivery to these diseased sites(1). The synergism between particle and cell has great potential to increase the efficiency of drug delivery in the future.

In this work we have developed a polymeric particle termed a backpack that conjugates strongly to monocytes through specific antibody interactions and has many applications in cell-mediated delivery. The backpacks are anisotropic, stratified polymeric thin films that are hundreds of nanometers thick and microns wide. We will discuss the fabrication and applications of backpacks in monocyte-mediated drug delivery. Since the backpack leaves most of the cell surface unaltered, we show that the biological functions and characteristics of monocytes are not changed upon its conjugation with the cell surface. For example, an *in vitro* transwell assay was used to show that backpack-laden monocytes still perform native functions such as extravasation through an endothelial cell layer in response to a biological stimulus. Furthermore, prior results show that the backpack, due to its unique shape, is resistant to phagocytosis by macrophages(2). The capability to resist internalization while strongly attach to the surface of immune cells such as monocytes and macrophages, uniquely position the backpack for cell-mediated extracellular therapeutic delivery in areas such as arthrosclerosis or cancer.

References:

1. Stephan MT, Moon JJ, Um SH, Bershteyn A, Irvine DJ (2010) Therapeutic cell engineering with surface-conjugated synthetic nanoparticles. *Nat Med*16(9):1035-U1135.

2. Doshi N, *et al.* (2011) Cell-based drug delivery devices using phagocytosis- resistant backpacks. *Adv Mater* 23(12):H105-H109.

(171e) Unnatural Killer Cells: TRAIL-Coated Leukocytes That Kill Cancer Cells in the Circulation

Mitchell, M., Cornell University
Rana, K., Cornell University
Wayne, E.,
Schaffer, C., Cornell University
King, M. R., Cornell University
Originally presented on: 11/4/2013 16:27:00 - 16:45:00

Unnatural Killer Cells: TRAIL-coated Leukocytes that Kill Cancer Cells in the Circulation

Michael J. Mitchell[1], Kuldeepsinh Rana[1], Elizabeth C. Wayne[1], Chris B. Schaffer[1], Michael R. King[1]

[1]Department of Biomedical Engineering, Cornell University, Ithaca, New York USA

Introduction: Hematogenous metastasis contributes to a poor patient prognosis in many types of cancer, with over 90% of cancer-related death attributed to metastasis (1). For metastasis to occur, cancer cells detach from the primary tumor, invade through the basement membrane, and intravasate into the peripheral circulation as circulating tumor cells (CTCs) (2). It is believed that adhesive interactions between selectins on the blood vessel wall and selectin ligands on the CTC surface facilitate metastatic progression, in a manner similar to the leukocyte adhesion cascade essential in the inflammatory response (3). Here, we describe a novel approach to functionalize human and murine leukocytes with tumor necrosis factor-related apoptosis-inducing ligand (TRAIL) on their surface, along with E-selectin (ES), to capture and induce cancer cell apoptosis both *in vitro* in human blood and *in vivo* in mouse circulation.

Materials and Methods: ES/TRAIL nanoscale liposomes and COLO 205 or PC-3 cancer cells suspended in buffer or spiked in peripheral human blood were subjected to shear flow in a cone-and-plate viscometer. Cancer cells and mononuclear cells were isolated, cultured overnight, and analyzed for viability using flow cytometry (FC). To determine leukocyte functionalization with ES/TRAIL, fluorescent (FL) liposomes were exposed to human blood under shear flow, and leukocytes were isolated and analyzed for ES/TRAIL attachment using FC and confocal microscopy. To examine cancer cell capture and killing *in vivo*, mice were injected with ES/TRAIL retro-orbitally, followed by injection of FL-labeled COLO 205 cells via tail vein injection. Mouse blood was removed via cardiac puncture, and viable and apoptotic cancer cells were analyzed using FC. Viable and apoptotic cancer cells lodged within the mouse lung, liver, and lymph nodes were quantified using multiphoton microscopy.

Results and Discussion: Upon exposure to shear flow, ES/TRAIL liposomes in human blood were functionalized to the surface of leukocytes, as confirmed by confocal microscopy (Fig. 1A) and FC. Exposure of functionalized ES/TRAIL leukocytes into cancer cell-spiked blood successfully targeted and killed cancer cells under shear flow, compared to controls (Fig. 1B). Mouse studies revealed that ES/TRAIL-functionalized leukocytes capture and induce apoptosis in cancer cells in the peripheral circulation *in vivo* (Fig. 1C). Cancer cells lodged within the mouse lung were apoptotic after treatment with ES/TRAIL leukocytes, as determined by multiphoton microscopy.

Conclusions: We have shown that leukocytes functionalized with ES/TRAIL can capture and induce cancer cell apoptosis *in vitro* in human blood and *in vivo* in mouse experiments. ES/TRAIL-functionalized leukocytes can form the basis of a novel method to target CTCs in the bloodstream as a means to prevent cancer metastasis. Clinically, this therapeutic strategy can serve as preventive measure upon diagnosis of metastatic hematogenous cancers.

Acknowledgements: The work described was supported by the Cornell Center on the Microenvironment and Metastasis through Award Number U54CA143876 from the National Cancer Institute.

Figure 1: (A) Functionalized ES/TRAIL leukocytes after ES/TRAIL liposome exposure to human blood under shear flow. Green = FL ES/TRAIL liposomes. Blue = leukocyte nuclei. Scale bar = 5 um. (B) Viable COLO 205 cancer cells (gated FC population) after exposure leukocytes functionalized with ES or ES/TRAIL. Untreated: viable COLO 205 control sample. (C) COLO 205 cells (yellow arrow) and functionalized leukocytes (red arrow) isolated from mouse blood after exposure to ES/TRAIL or ES liposomes, sTRAIL, or buffer injections *in vivo*. Scale bar = 20 um.

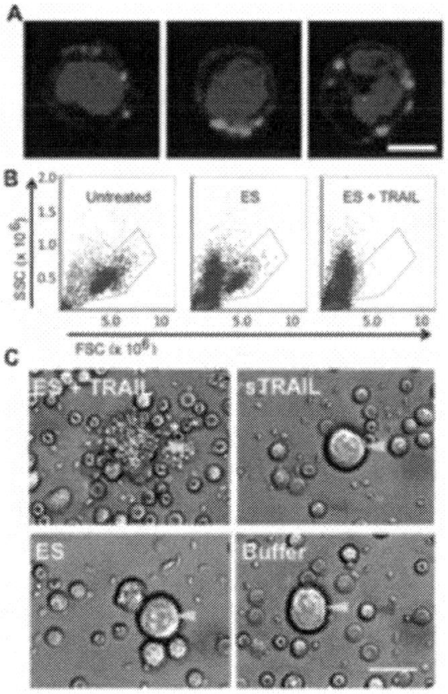

References:

1. Mitchell MJ, King MR. Computational and experimental models of cancer cell response to fluid shear stress. Frontiers in Oncology. 2013 Mar 2;:1–11.

2. Hughes AD, King MR. Nanobiotechnology for the capture and manipulation of circulating tumor cells. WIREs Nanomed Nanobiotechnol. 2011. Dec 7;4(3):291–309.

3. Geng Y, Marshall JR, King MR. Glycomechanics of the Metastatic Cascade: Tumor Cell–Endothelial Cell Interactions in the Circulation. Ann Biomed Eng. 2011 Nov 19;40(4):790–805.

(171f) Ex-Vivo Analysis of Lipid Impact On Intestinal Mucosa Barrier

Yildiz, H. M., Northeastern University
Carrier, R. L., Northeastern University
Originally presented on: 11/4/2013 16:45:00 - 17:03:00

Ex-vivo Analysis of Lipid Impact on Intestinal Mucosa Barrier

Hasan Yildiz, yildiz.h@husky.neu.edu

Rebecca L. Carrier, rebecca@coe.neu.edu

Introduction:

A heterogeneous mucus layer has evolved to protect exposed epithelial surfaces by trapping environmental microorganisms while allowing the passage of certain compounds (e.g., nutrients) across the epithelium. Our previous results indicate that exogenous lipids that may be associated with food or drug delivery systems significantly impact intestinal mucus barrier properties. However, these studies were conducted using collected (scraped) porcine intestinal mucus, and recent literature indicates that intestinal mucus is not isotropic. The main objective of this work is to analyze the impact of lipids on particle transport across intact mucus on an intestinal explant.

Materials and Methods:

Transport of model drug carriers was investigated using 3 week old wild type FVB/N mouse small intestine. Mice were euthanized via CO_2. Intestine fragments were dissected, cut open to expose the intestinal lumen, and placed into a chamber on a concavity slide. 200 nm carboxylate- modified polystyrene microspheres were diluted in maleate buffer or simulated fed state intestinal contents (?FED state?) including maleate buffer, bile salts, phospholipids (lecithin) and a lipid mixture comprised of 35 mM soybean oil, 30 mM sodium oleate and 15 mM monoglycerol for a final particle concentration of 0.0025 wt.-%. 5 Âµl of diluted particle solutions were then added on top of tissue segments and the system was equilibrated for 1 h at 25 ^{0}C prior to microscopy in a humid chamber. Particle diffusion was measured by tracking the positions of diluted microspheres using real- time multiple particle tracking technique (MPT). Positions of particle centroids were used to calculate time-averaged mean squared displacements (MSD) and effective diffusivities (D_{eff}): $MSD = [x(t+\tau)-x(t)]^2 + [y(t+\tau)-y(t)]^2$ and $D_{eff} = MSD/(4\tau)$ where x and y are positional data and τ is the time scale.

Results:

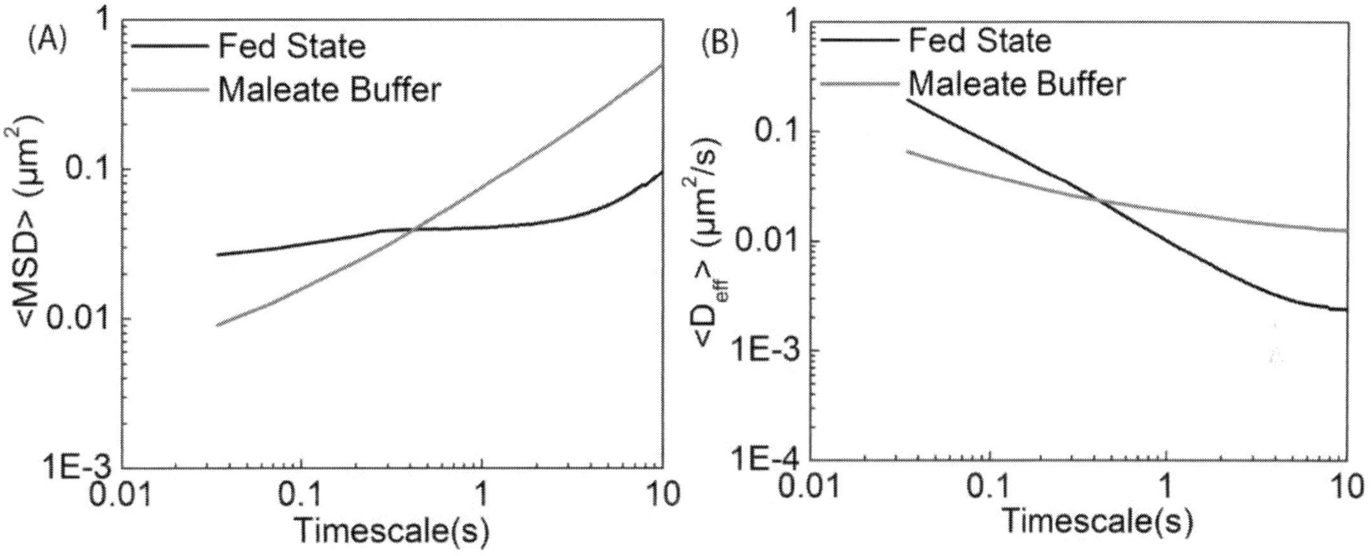

Intestinal contents characteristic of the fed state, including model bile and lipids, markedly reduced particle transport rate through intact ex vivo intestinal mucus. At a time scale of 10 s, transport rates of 200 nm carboxylate-modified particles in FED State were nearly 5-fold lower than those of same-size particles in maleate buffer. Intermolecular interactions, potentially including lipid-particle and/or lipid-mucus interactions, possibly retard particle diffusion in the gastrointestinal tract. Furthermore, effective diffusivities in maleate buffer in scraped mucus and intestinal explant were on the same order, whereas in Fed State diffusivities in explant tissue are an order of magnitude higher than collected porcine intestine mucus.

Conclusions:

Food-associated lipids strengthen the intestinal mucus barrier to particulate matter. These results motivate in vivo in-depth investigation of the impact of lipids on mucosal barriers to facilitate lipid-based delivery of drugs and potentially enable control of pathogen invasion in disease.

(171g) Photo-Activated Drug Release of a Liposomal Carrier, Comparison Between Hollow Gold Nanoparticles and Sulfide Nanoparticles

Ogunyankin, M. O., University of California, Davis
Zasadzinski, J. A., University of Minnesota
Lapotko, D., Rice University

Originally presented on: 11/4/2013 17:03:00 - 17:21:00

Liposomes, lipid bilayer capsules, have been widely evaluated as drug nanocarriers, however, they are often limited by slow and non-specific release. A novel strategy is to initiate drug release by an external agent such as near infra-red pulsed laser light adsorbed by encapsulated or tethered nanoparticles. The light energy is rapidly converted into thermal energy that generates nanobubble formation that can mechanically disrupt the liposome membranes, thereby releasing the drug with temporal and spatial control. This study presents a comparison of the photoactivated liposomal drug release induced by two different types of nanoparticles (NPs): hollow gold nanoshells (HGN) and cupper sulfide nanoparticles (CSN). Both NPs exhibit strong absorption of near infra-red (NIR) light and offer the opportunity to convert optical energy into thermal energy. The absorbance of NIR light is desirable because it causes minimal thermal injury to normal tissues with optimal light penetration. A picosecond pulsed NIR laser is used to trigger and target the release of the drug. The photothermal heating of the nanoparticles generated by short laser pulses results in large temperature gradients that leads to the formation of transient vapor nanobubbles in aqueous solution. The collapse of the nanobubbles induces the rupture of the liposome membrane and the subsequent drug release with minimal damage to the surroundings. We are particularly interested in the relative light intensity necessary to initiate nanobubble formation for the ~ 10 nm diameter CSN and the ~ 50 nm diameter HGN. We are also interested in the ease of encapsulating the smaller CSN compared to the larger HGN within 100 nm liposomes most used for drug delivery. The study will lead to a better understanding of the capabilities of multi-functional nanomaterials for NIR laser-controlled drug release in order to facilitate the development of more complex therapeutic studies.

(171h) Pulsatile Release of Model Drugs Through Layered Polymers

Gandhi, S., University of Iowa
Nuxoll, E., University of Iowa
Originally presented on: 11/4/2013 17:21:00 - 17:39:00

Barrier-Mediated Pulsatile Release (BMPR) systems offer a materials-based alternative for automated pulsatile drug delivery, without pumps, power supplies, or complex circuitry. While earlier materials-based approaches such as delayed-release microcapsules are limited to two or three pulses due to the independent nature of each pulse's timing control, BMPR systems link the timing of each pulse to the previous pulse. Each dose of drug is sequestered in its own stimuli-sensitive depot, releasing only upon contact with the stimulant. These depots are stacked with sacrificial barriers in between, each of which block the stimulant for a predetermined time. For instance, layers of soluble drug may be separated by degradable polymer layers. Water, as the stimulant, will erode the polymer layer over a fixed period of time, followed by quick dissolution and release of the underlying drug and the start of degradation for the next polymer layer. This example, however, is quickly limited by irregular polymer erosion, a single stimulant (water), and difficulty in scaling delay times.

In this presentation, we report the development of a generalized BMPR system which overcomes those limitations. Drug (methylene blue) was immobilized in pH-sensitive polymer (poly(methyl methacrylate-co-dimethylaminoethyl methacrylate)) which released only at low pH. Zinc oxide (ZnO) nanoparticles immobilized in a pH-insensitive matrix (poly(vinyl alcohol)) served as the barrier layer. The time required for acid to penetrate the barrier layer scaled with the ZnO concentration and with the square of the polymer thickness, allowing wide scaling of the delay time with only minor changes to the barrier layer. This system has demonstrated tuned release using a citric acid stimulant to produce up to ten pulses of model drug (methylene blue) over various preset timescales. This system has also demonstrated the alternate release of multiple solutes (methylene blue and methyl orange) at a regular time intervals up to five pulses from a single BMPR device. This presentation will discuss the synthesis of depot & barrier films, design of experimental BMPR device, pulse behavior of model drugs and introduce efforts to further generalize this delivery strategy by incorporating glucose as a stimulant.

(176a) Propagating and Inflating Viral Progenies Using Drop-Based Microfluidics and Biophysics-Based Forward Evolutionary Models

Rotem, A., Harvard University
Chang, C., Harvard University
Serohijos, A.,
Feldman, A.,
Zhang, H.,
Cheron, N., Harvard University
Tao, Y.,
Choi, J. M.,
Cockrell, S.,
Wobus, C.,
Pipas, J.,
Shakhnovich, E.,
Weitz, D. A., Harvard University
Originally presented on: 11/4/2013 15:15:00 - 15:33:00

The study of how viruses propagate is important for curing disease and preventing viral outbreaks, yet is challenging due to inability to access enough variants in the population, including ones that do not survive genetic drift and selection. In nature, viruses can compete with one another, and the most evolutionary fit virus usually takes over a population. Yet there exist variants in the population that can escape subjected evolutionary pressures and eventually dominate the population. Successful studies of viral epidemics *in vitro* hinges on the ability to access these variants, in order to obtain a comprehensive understanding of fitness in an evolving population and to allow the exploration of multiple virus trajectories, as well as prediction of their direction. Unfortunately, measurements of fitness trajectories can currently be observed in the laboratory only when the mutations have a large effect on fitness in the population. Accessing the richness of the background genomic diversity has only been performed upon a small number of species, since these experiments have been both labor intensive and time-consuming due to their large population sizes. Thus, there remains a need for a simple experimental method that allows greater access to the genetic diversity in a large population, in order to allow exploration of fitness space.

Here, we present the use of droplet-based microfluidics as a simple method to segregate and propagate a viral population as individual viral lineages, effectively eliminating competition between viral strains and minimizing natural selection. We show that the population obtained from isolating viruses in drops is more diverse than that from bulk culture, thus allowing access to mutations and variants that would not otherwise occur be seen from natural selection events, allowing the potential to control mutation rates and accelerate evolution *in vitro* or *in silico*. This general method of permitting smaller population sizes to propagate, similar in vein to mutation accumulation experiments, can be widely applied to other genetic systems to revolutionize our understanding of evolution and the distribution of fitness.

(176b) A Single-Cell Perspective On Signaling and Heterogeneity Maintenance in Glioblastoma Multiforme

Lam, L. K. W., Massachusetts Institute of Technology
Maire, C., Dana-Farber Cancer Institute
Francis, J., Broad Institute of MIT and Harvard
Ligon, K. L., Brigham and Women's Hospital
Meyerson, M. L., Dana-Farber Cancer Institute
Adalsteinsson, V., Pennsylvania State University
Love, J. C., Massachusetts Institute of Technology
Originally presented on: 11/4/2013 15:33:00 - 15:51:00

Glioblastoma multiforme (GBM) is the most common and lethal brain cancer in adults. The current standard-of-care involves surgical resection followed by administration of radiotherapy and temozolomide. This treatment, however, has generally proven ineffective in patients, with a median 2-year survival rate of about 26.5%. This is largely because this 'blanket' therapy fails to account for the complex inter-tumoral and intra-tumoral genomic and functional heterogeneity of the cancer, as well as its myriad interactions with the tumor microenvironment.

Inter-tumoral heterogeneity in GBM is often associated with the 2 GBM progression pathways: the de novo and progressive pathways. Although the tumors that arise via these separate pathways are histologically indistinguishable, closer inspection reveals that they harbor distinct combinations of genetic aberrations, with each disrupting different players in classic oncogenic and tumor suppressor pathways. Intra-tumoral heterogeneity, on the other hand,involves the development of distinct cellular subcompartments within GBMs that demonstrate different biological behavior, most importantly with respect to drug resistance. A classic example of this involves heterogeneity at the EGFR locus. Studies have indicated that around 50% of all GBMs have amplified levels of wtEGFR and of these, about half harbor the constitutively-active EGFRvIII mutant variant. It has been shown that EGFRvIII+ cells can enhance the proliferation of wtEGFR+ cells via IL-6 paracrine signaling. maximizing overall tumorigenicity. GBM cells are also able to co-opt their stromal compartment to their advantage via the secretion of various immunomodulatory, hypoxic and angiogenic factors. With an occurrence rate of roughly 7 in 100,000 individuals per year, it is clear that targeted therapies that are better able to handle these intricacies have to be devised to combat this disease. This will require a more intimate knowledge of the relevant intercellular pathways that lead to heightened tumorigenicity, supported by advancements in medical genetics that continue to make genomic and transcriptomic screening more economically viable. This will allow us to correlate genomics with function, allowing the design of better treatments for GBM.

This presentation will focus on elucidating links between the secretomes of distinct cell subpopulations within GBM primary samples and their mutational status at the EGFR locus, as well as their wider genomic profile. To investigate the link between genomics and function, we have used arrays of subnanoliter wells (nanowells) to study primary GBM samples with single-cell resolution. Cells are loaded onto the nanowell device and are allowed to settle via gravity before being stained with various surface markers (e.g., CD45 and EGFR), allowing the discrimination of distinct cell subpopulations via image-based cytometry. Once loaded, the process of 'microengraving' is performed to determine the cytokine secretion profiles of the cell(s) in each well: this process involves sealing a glass slide functionalized with appropriate capture antibodies against the loaded nanowell device for a set duration, allowing for the capture of cytokines that can later be detected using fluorescent dye-conjugated detection antibodies. The image-based cytometry and microengraving data are analyzed in an integrated fashion to identify cells of interest. These cells are isolated using a micromanipulator robot into microtiter plates for downstream sequencing, yielding genomic/transcriptomic data linked to the functional data obtained via microengraving.

Proof-of-concept experiments using the commercially available T98G cell line demonstrate heterogeneity in the secretion of the cytokine IL-6, with 11% of the events registering secretion over four hours. Examination of both single-cell events and multi-cell events suggests secretion occurs in an independent manner—that is, there is no synergy or antagonism of IL-6 secretion on the numbers of IL-6-secreting cells. This example illustrates how the discrete co-cultures enabled in the nanowells provides a unique experimental platform for investigating functional intercellular interactions. Experiments will also be reported for primary GBM-derived neurosphere lines in microtiter plates to investigate the effects of depletion of various exogenous factors (e.g. EGF) and drug treatments (e.g. erlotinib) on their bulk functional and genomic profiles, as well as their proliferative capacity and lineage differentiation. This systematic analysis may shed light on possible escape pathways exploited by GBMs to circumvent treatments and promote tumor relapse, as well as revealing critical players in the homeostatic maintenance of heterogeneity. Similar studies recapitulated with single-cell resolution may further reveal possible heterogeneities in response. These functional investigations, coupled with characterization of genetic heterogeneity, should begin to reveal how intra-tumoral diversity promotes survival and maintenance, and may reveal new targets for therapy.

(176c) Effects of Coupled Dose and Rhythm Manipulation of Plasma Cortisol Levels On Leukocyte Transcriptional Response to Endotoxin in Humans

Kamisoglu, K., Rutgers, The State University of New Jersey
Sleight, K., Rutgers, The State University of New Jersey
Nguyen, T. T., University of California, San Diego
Calvano, S. E., Robert Wood Johnson Medical School
Corbett, S. A., Robert Wood Johnson Medical School
Androulakis, I. P., Rutgers University
Originally presented on: 11/4/2013 15:51:00 - 16:09:00

Healthy function of the immune system is intrinsically tied to the biological rhythms running in harmony. There is a bi-directional information flow between the immune and neuroendocrine systems such that; while a wide range of immune parameters go through daily fluctuations under the control of autonomic and endocrine rhythms; immune effectors, such as cytokines, can, in turn, influence the phase of these rhythms (1). In the case of acute stress, such as trauma or infection, inflammatory response is mounted by the coordinated action of immune and neuroendocrine systems. Hypothalamus-pituitary-adrenal (HPA) axis is a crucial mediator of this action because of its control on circadian and ultradian glucocorticoid (GC) secretion patterns (2). These fluctuating patterns are hypothesized to provide the basis for a ?continuous dynamic equilibration?, fundamental to establishment of homeostasis and allostasis (3). Cortisol is the major among circulating human GCs which, through feedforward and feedback regulatory loops, maintains this dynamic state essential for proper response to stressors. Although GCs are mostly famous for their immunosuppressive and anti-inflammatory actions that have been clinically utilized for the treatment of autoimmune and inflammatory disorders for decades; due to their non-linear effects on transcription and highly complex interactions with other inflammatory mediators; their downstream reactions are highly diverse. Effect of GCs on the immune response can be so disparate, such as to permit, suppress or stimulate the response, depending on their secretion frequency, concentration and timing with respect to a stressor (4-7).

Since GCs are both major endogenous hormones regulating stress response as well as clinically used pharmacological agents; building a solid understanding on their effects on immune cells is of primary importance. With this vision in mind, this *in vivo* study in healthy human subjects seeks to assess the responses of whole blood leukocytes to coupled dose and rhythm manipulation of cortisol preceding an immune challenge induced by endotoxin (LPS) administration. Healthy human model of LPS administration has been used as a reproducible experimental platform providing mechanistic insights into how cells and organs respond to systemic inflammation. The model transiently alters many physiologic and metabolic processes in a qualitatively similar manner to those observed after acute injury and systemic inflammation (8); therefore served as a suitable platform for investigating the changes in immune response due to hormonal manipulations. Transcriptional responses of leukocytes as well as the change in total leukocyte count in blood and plasma cortisol levels were monitored over 24 hours both before and after LPS administration. Differential gene expression and clustering analyses were done separately for these two periods to allow focusing both on the response to cortisol infusion alone and on the effects of this treatment to subsequent LPS exposure.

Through continuous infusion; plasma cortisol concentration was increased to and kept constant at a level associated with major physical stress and almost twice as much as the normal daily peak. In response to this treatment, total leukocyte count in blood was increased both during cortisol infusion period and within 2 hours after LPS exposure, after which it started to decrease. Transcriptional programming of leukocytes was also significantly altered where a priming response at the end of 24hr infusion period was evident. Enhanced expression of a number of pattern recognition and cytokine receptors, receptor regulatory and signal transduction elements; as well as tuned down protein translation and mitochondrial function portrayed a phenotype sensitized against potential infectious threats. Promoter analysis within cortisol responsive genes revealed that only less than half of this transcriptional response might have been as a direct consequence of glucocorticoid receptor (GR)-DNA binding, therefore transcriptional regulation mechanisms other than DNA binding, such as tethering, may play a significant role. Furthermore, absence of any correlations between the direction of the response and different modes of transcriptional regulation underscored the complexity of mechanisms that govern the transcriptional programming through GR.

Despite these significant changes in transcriptional programming in leukocytes prior to LPS administration, response to LPS followed very similar patterns in both cortisol and saline pre-treated groups except one significantly different cluster including probesets associated with major players known to regulate the inflammatory response such as cytokines and chemokines, anti-inflammatory signal transduction elements and receptors mediating recruitment of leukocytes to sites of inflammation. Observed much more intense response in the cortisol group in this cluster indicated that although continuous cortisol infusion did not directly affect the transcription of major inflammatory players, it indirectly affected their regulation, possibly through modulating their up-stream signaling cascades.

In sum, altered natural fluctuation coupled with increased concentration of plasma cortisol levels induced a primed phenotype in the circulating leukocytes associated with sensitization of surveillance function and tuning down of energy-intensive processes which may be linked to increasing their efficiency in responding to subsequent infectious threats.

References

1. Mavroudis PD, Scheff JD, Calvano SE, Androulakis IP. Systems Biology of Circadian-Immune Interactions. J Innate Immun. 2012.

2. de Kloet ER, Sarabdjitsingh RA. Everything has rhythm: focus on glucocorticoid pulsatility. Endocrinology. 2008;149(7):3241-3.

3. Lightman SL, Conway-Campbell BL. The crucial role of pulsatile activity of the HPA axis for continuous dynamic equilibration. Nat Rev Neuroscience. 2010;11(doi:10.1038/nrn2914):710-8.

4. Scheff JD, Calvano SE, Lowry SF, Androulakis IP. Transcriptional implications of ultradian glucocorticoid secretion in homeostasis and in the acute stress response. Physiol Genomics. 2012;44(2):121-9. PMCID: 3289115.

5. Young E, Abelson J, Lightman S. Cortisol pulsatility and its role in stress regulation and health. Frontiers in Neuroendocrinology. 2004;25(2):69-76.

6. Munck A, Guyre PM. Glucocorticoid physiology, pharmacology and stress. Advances in experimental medicine and biology. 1986;196:81.

7. Sapolsky RM, Romero LM, Munck AU. How do glucocorticoids influence stress responses? Integrating permissive, suppressive, stimulatory, and preparative actions. Endocrine reviews. 2000;21(1):55-89.

8. Lowry SF. Human endotoxemia: a model for mechanistic insight and therapeutic targeting. Shock. 2005;24 Suppl 1:94-100.

(176d) Disruption of Tumor-Associated Immune Dysfunction Via Coupled Multicellular Networks

Chuang, Y., Northwestern University
Leonard, J. N., Northwestern University
Originally presented on: 11/4/2013 16:09:00 - 16:27:00

Macrophages play a critical role in maintaining the balance between homeostasis and protective inflammation by adopting either immunostimulatory (M1) or immunosuppressive (M2) phenotypes. Tumors and pathogens manipulate this balance by inducing the production of immunosuppressive stimuli, which shifts macrophages into a M2 phenotype, supporting tumor growth and pathogen survival. In addition, M2 macrophages have been reported to convert to an M1 phenotype given certain stimuli, which is termed "plasticity". Despite the centrality of these phenomena to diverse disease processes, the mechanisms regulating macrophage "decision-making" vis-à-vis polarization remain poorly understood. Here, we describe several novel systems-level insights into how such dysfunctional immune network states are established, maintained, and potentially may be disrupted.

Since most macrophage polarization studies to date involve applying coherent stimuli (either pro-M1 or pro-M2 stimuli alone), and these stimuli rarely exist independently *in vivo*, we first investigated how macrophages integrate contradictory signals to "calculate" a response. Macrophages were exposed to combinatorial variations in dose of IL-10 and IL-12 (pro-M2 and pro-M1 stimuli, respectively), and macrophages were "activated" using bacterial endotoxin. We previously observed that M2-type responses increased with IL-10 dose and were largely independent of IL-12 co-treatment. Furthermore, we also showed that the removal of IL-10 was sufficient to restore an M1-type phenotype. Here, we identify a biochemical "node" within the macrophage intracellular network that mediates IL-10-driven M2 functions and also explains the reversion of macrophage phenotype from M2 to M1 upon removal of IL-10. We also identified a novel multicellular network through which IL-12 may *indirectly* promote commitment to M1 phenotypes via the pro-inflammatory mediator, IFNγ. This novel network-level model of macrophage polarization identifies several candidate nodes that may be targeted for restoring of an immune network state that promotes immunological clearance or control of cancer.

Finally, we have observed that under some conditions, distinct M1 and M2 states are induced in different cells within a single population. The probability of polarization towards an M2 state increased with IL-10 dose and was again independent of IL-12 co-treatment. These data represent the first evidence to date that macrophage polarization may be a stochastic process. We will present novel insights into the mechanisms underpinning this heterogeneity, and we will discuss connections between these phenomena and the dynamic evolution of macrophage phenotypes we have observed during tumor progression *in vivo*. These systems-level analyses of macrophage function provides new insights into innate immune function and will help to formulate therapeutic strategies for treating the many chronic diseases characterized by local immune dysfunction.

(176e) Metabolic Rewiring Induced By Hyperactive Akt Signaling in Cancer Cells

Duckwall, C. S.,
Young, J. D., Vanderbilt University
Originally presented on: 11/4/2013 16:27:00 - 16:45:00

Metabolic Rewiring Induced by Hyperactive Akt Signaling in Cancer Cells

At a molecular level, cancer is a disease of deregulated cell growth arising from genetic mutations that result in improper cell signaling. A key enzyme in mediating cellular growth response is the Akt serine/threonine kinase. This enzyme mediates a host of proliferative and survival responses including advancing cell cycle progression, increasing rate of biosynthesis, and increasing apoptotic resistance. Hyperactivation of the Akt signaling axis is implicated in a number of cancers including prostate, breast, lung, and colon cancers. Akt is also known to exhibit profound regulation of central carbon metabolism at the transcriptional, translational, post-translational, and allosteric levels. Our group has employed metabolic flux analysis using isotopically labeled glucose to map the flow of carbon through central metabolic processes in both a wild-type pro-B-cell line and a derivative line expressing constitutively active myr-Akt. Through comparison of flux maps generated for these cell lines, we have discovered significant metabolic reprogramming in multiple central carbon pathways as a result of Akt signaling.

Prior biochemical studies of Akt have shown that its activation leads to increased expression of glycolytic enzymes including the GLUT1 glucose transporter and hexokinase, two rate-limiting enzymes in glycolysis. Consistent with these previous results, we show that constitutive Akt signaling leads to a 3-fold increase in glucose uptake and a 5-fold increase in lactate secretion. Interestingly, we show that Akt activity also has profound effects on amino acid utilization. Although extracellular exchange rates for the amino acids glycine and serine are similar, Akt is shown to rewire cells' intracellular utilization as evidenced through increased labeling enrichment of serine in cells with hyperactive Akt and through direct comparison of flux maps. Abnormalities in glycine and serine utilization have recently been linked to more aggressive, cancerous phenotypes in other cell lines. Our results also show changes in pentose phosphate activity with wild-type cells failing to send carbon through the oxidative branch, imposing interesting constraints on NADPH production that are quantified in the metabolic flux analyses. Full characterization of the metabolic changes induced by Akt to support proliferation and survival of cancer cells will lead to better therapeutic and adjuvant treatments for the host of cancer subtypes with aberrant Akt signaling.

(176f) Probing Mammalian Cell Metabolism Using Stable Isotope Tracers and Metabolic Flux Analysis

Parker, S. J., University of California, San Diego
Metallo, C. M., University of California San Diego
Originally presented on: 11/4/2013 16:45:00 - 17:03:00

Metabolism is central to virtually all cellular functions, and cells must negotiate their surrounding microenvironment to effectively utilize their surrounding nutrients. A detailed understanding of how metabolic pathways are regulated is necessary to design better culture systems and identify therapeutic targets in diseases where metabolism is deregulated. To this end we apply stable isotope tracers, mass spectrometry, and metabolic flux analysis to characterize central carbon metabolism in mammalian cells, including normal cells and tumor cell lines. We have developed unique, tracer-based methods to probe glycolysis, the pentose phosphate pathway, and the tricarboxylic acid cycle. Using these approaches we have identified the importance of reductive glutamine metabolism for cells proliferating under low oxygen tensions, whereby glutamine rather than glucose serves as the primary carbon source for lipid synthesis. We have extended these findings to further elucidate how cells obtain amino acids for cell growth and regenerate cofactors for reductive biosynthesis. These findings highlight a critical role of oxygen and reductive TCA metabolism in supporting cell growth under hypoxia.

(176g) Metabolomic and Proteomic Analysis of Metabolic Enzyme Disruption in Adipocytes

Sims, J. K., Tufts University
Choi, K., Texas A&M University
Manteiga, S.,
Russell, W.,
Jayaraman, A., Texas A&M University
Lee, K., Tufts University
Originally presented on: 11/4/2013 17:03:00 - 17:21:00

Metabolomic and Proteomic Analysis of Metabolic Enzyme Disruption in Adipocytes

Introduction: Obesity results from a chronic imbalance in caloric intake and expenditure. The excess calories are stored mainly as intracellular lipids (triglycerides, TG), leading to an expansion of body fat or adipose tissue through increases in fat cell (adipocyte, AD) size and number. One approach to controlling body fat could be to intervene in the metabolic processes of the adipose tissue that directly contribute to lipid accumulation and cell growth. This approach requires a systematic investigation of adipose tissue-specific metabolism, preferably in isolation from confounding systemic influences. The goal of the present study is to investigate the effect of targeted knockdown of lipid pathway enzymes through metabolomic and proteomic analyses.

Materials and Methods: 3T3-L1 preadipocytes (PAs) were in grown in 12-well plates and induced to differentiate into ADs using a standard hormone cocktail. Experiments were conducted in both 2D planar culture and 3D collagen culture. For 3D culture, the induced cells were detached, concentrated, and mixed into an ice-cold collagen pre-polymer solution with (coculture) or without ECs (monoculture). The collagen cell suspensions were added into 12-well plates and allowed to gel. The cells were exposed to siRNA by lipofection. Fresh siRNA was added every 48 h. Gene expression, TG content, and droplet distribution were quantified using qRT-PCR, enzymatic assay, and image analysis, respectively. Intracellular and extracellular metabolite concentrations were measured using in-house developed LC-MS/MS protocols, and changes in expression levels of metabolic enzymes were analyzed using isobaric tags (iTRAQ).

Results and Discussion: Metabolic proteins targeted for knockdown were selected from different stages of lipid accumulation: upstream synthesis (breakdown and transformation of glycolysis intermediates into precursors of fatty acid synthesis), downstream synthesis (formation of triglyceride from synthesized precursors), and droplet stability (regulation of intracellular lipases and formation of larger, more stable lipid droplets). Based on our prior work characterizing the metabolic flux distribution in differentiating ADs, pyruvate carboxylase (PCX) and isocitrate dehydrogenase (IDH) were identified as targets for upstream synthesis, diglyceride acyltransferase (DGAT) for downstream synthesis, and fat-specific protein 27 (FSP) and perilipin (PLIN1) for droplet stability. We first tested the efficiency of knockdown in 2D planar culture. Each of the targets we tested led to significant reduction in triglyceride accumulation (Fig. 1). To understand the efficacy of the lipofection in 3D culture, we first characterized the effect of PLIN1 knockdown, as PLIN1 expression is specific to ADs and correlates with lipid droplet stability. As expected, TG accumulation was reduced in both monoculture and coculture (Fig. 2A). Results of image analysis showed significant reduction in average lipid droplet size in both 3D mono- and coculture (Fig. 2B). The decrease in the average droplet size reflected a *shift* in the *distribution* of droplet sizes (Fig. 2C). The siRNA treatment did not affect the *variance* in the lipid droplet sizes (see error bars), indicating knockdown was achieved. Interestingly, PLIN1 knockdown did not significantly affect the TG content (not shown). Using metabolite uptake and release data in conjunction with a stoichiometric model of AD metabolism, we calculated the impact of the knockdowns on major metabolic pathway fluxes. The flux data were combined with enzyme concentration data to determine the rate controlling steps in TG accumulation using metabolic control analysis as a framework. Lastly, we performed pairwise knockdown of enzymes. We investigated knocking down enzymes in two different stages of accumulation (e.g. PCX and DGAT) as well as two enzymes in the same stage (e.g. PCX and IDH). As expected, we found that combined knockdown led to a greater reduction in triglyceride accumulation.

Conclusions: We report the effects of metabolic gene knockdowns in a 3D coculture system capturing the cell-cell interactions found in adipose tissue. We have shown successful knockdown of several metabolic proteins, and each lead to a reduction in accumulation of triglycerides; however, inhibition of PCX (an upstream step) had the greatest effect. Image analysis of lipid droplets provides an efficient and non-invasive method of monitoring TG accumulation and cellular hypertrophy.Â Further work is warranted to identify additional silencing targets to limit lipid accumulation and adipose tissue growth, to better understand the metabolic consequences of knockdown and to continue to improve our 3D coculture system. Furthermore, results from MFA and MCA will be used to focus future experiments on key regulatory steps and identify which proteins could be used as potential therapeutic targets.

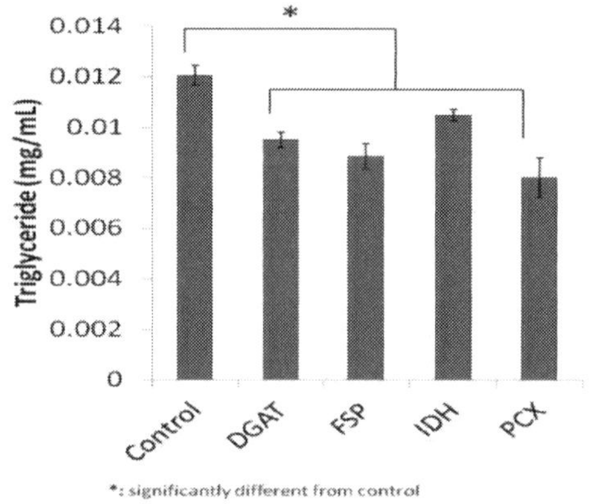

Fig 1 Identification of knockdown targets in 2D planar monoculture.

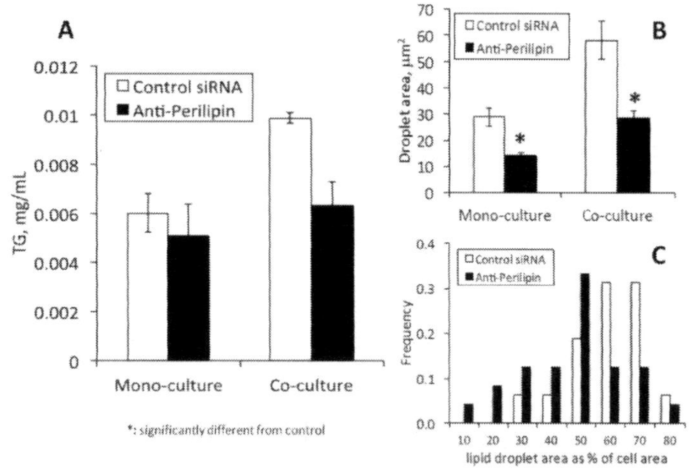

Fig 2 (A) Triglyceride accumulation in 3D mono- and coculture treated with non-targeting (control) or anti-perilipin siRNA. Error bars are SEM. (B) Average lipid droplet size in 3D mono- and coculture treated with non-targeting (control) or anti-perilipin siRNA. Error bars are SEM. *Significantly different from respective control (n~1,000 droplets from 8 different images), p=0.0005 and p=0.03, respectively. (C) Histogram of droplet size in 3D monoculture.

(176h) Metabolomics-Based Assessment of Aquatic Animal Health

Vermeersch, K. A., Georgia Institute of Technology
Smith, M., Georgia Institute of Technology
Dove, A.,
Cipriano, R.,
Styczynski, M. P., Georgia Institute of Technology
Originally presented on: 11/4/2013 17:21:00 - 17:39:00

Veterinary care and monitoring of aquatic animals faces many obvious limitations, from difficulties identifying symptoms to limits on knowledge of the molecular and biochemical indicators of disease. The most common diagnostic approaches include observable phenotypes and traditional clinical chemistry assays, with only the most basic measures in those assays well-characterized for any given species. Metabolomics (system-wide measurement of the levels of small-molecule biochemical intermediates) is a promising, yet underexploited, method to advance our understanding of a variety of species-specific maladies, as well as our ability to monitor and diagnose aquatic animal health. The systems-scale and potentially untargeted nature of metabolomics makes it ideal to identify biomarkers in the understudied area of aquatic animal health, enabling earlier intervention whether in the field or in captivity.

Here, we present our work in two different aquatic animal systems: the bottlenose dolphin (*Tursiops truncatus*) and the Atlantic salmon (*Salmo salar*). We have studied the impact of pathogenic infections on each species (lobomycosis and furunculosis, respectively), and of inappetance due to salmon spawning, on the plasma metabolite profiles of the animals using two-dimensional gas chromatography coupled to mass spectrometry. In each case, we were able to differentiate between treatment and control groups of animals. These studies have provided significant novel biological insights into each system. In lobomycosis-infected dolphins, the accumulation of fatty acids and depletion of specific vitamin metabolites has suggested the existence of a previously-unexpected catabolic state in such animals. Our investigations of the metabolic impacts of spawning-induced inappetance in salmon – including changes in specific classes of fatty acids – suggest a distinct, but not dysfunctional, metabolic state in these animals, consistent with their natural life cycle. Taken together, our results suggest that metabolomics is a powerful tool to monitor, diagnose, and learn about aquatic animal health.

(205a) Evolving a User-Defined, Targeted DNA Methyltransferase

Chaikind, B., Johns Hopkins University
Ostermeier, M., Johns Hopkins University
Originally presented on: 11/4/2013 15:15:00 - 15:33:00

CpG methylation of promoters is an important form of epigenetic control in eukaryotic cells, causing transcriptional repression. It is therefore involved in many important cellular processes, such as cellular differentiation, X-chromosomal inactivation, and chromatin remodeling. As such, aberrant methylation patterns have been implicated in many disease states such as cancer, psychological disorders and a range of diseases caused by the deregulation of epigenetic markers. The ability to target methylation toward a specific site within the genome would have broad applicability as a tool to study the effect of single CpG methylation events on transcription and as a potential therapeutic to target diseases resulting from the hypomethylation of particular promoters.

We have designed an *in vitro* directed evolution strategy to create methyltransferases that recognize non-palindromic, user-defined sequences. The strategy is both simple and rapid; it has enabled us to quickly optimize an engineered, heterodimeric methyltransferase previously described in our lab. This methyltransferase consists of two heterodimeric methyltransferase fragments fused to separate zinc fingers, whose recognition sequences flanks a targeted CpG site. Fragments are engineered to have reduced affinity for one another, and do not efficiently assemble away from the target DNA sequence. Our newly optimized enzyme methylates ~80% of a desired site with minimal methylation at non-targeted sites. Early initial experiments in eukaryotic cells will be discussed.

(205b) Production of the Trimeric Influenza Hemagglutinin Stem Domain for Potentially Broadly Protective Influenza Vaccines

Lu, Y., Stanford University
Swartz, J. R., Stanford University
Originally presented on: 11/4/2013 15:33:00 - 15:51:00

The rapid dissemination of the 2009 pandemic H1N1 influenza virus emphasizes the need for universal influenza vaccines that would broadly protect against multiple mutated strains. Recent efforts have focused on the highly conserved hemagglutinin (HA) stem domain. Even though producing isolated domains of non-modular, multimeric ectodomain proteins has proven difficult, we report a method to rapidly produce the properly folded HA stem domain protein from influenza virus A/California/05/2009 (H1N1) using *Escherichia coli*-based cell-free protein synthesis (CFPS) and a simple refolding protocol. The CFPS platform provides a rapid, scalable and cost-effective means to combat pandemic and epidemic threats when a rapid response with large amounts of proteins are required. Several residues in the HA stem domain were mutated or removed to facilitate folding and trimeric assembly. The T4 bacteriophage fibritin foldon placed at the C-terminus of the HA stem domain induces trimer formation. The protein refolding conditions were also optimized. pH and the use of the detergent emerged as the most important factors. Multiple preparations indicate high production yields and process consistency. The correct immunogenic conformation of the stabilized HA stem domain trimer was confirmed using a commercial antibody, C179, that blocks influenza infection by binding to the HA stem domain trimer. This neutralizing antibody recognizes the new stem domain trimer with approximately 30 fold higher affinity relative to the full-length HA protein, most likely because of improved steric availability. Based on the insights gained, we believe that similar procedures can be rapidly developed for the other influenza HA stem domains needed for universal vaccine candidates. We also feel that this overall protocol provides a new paradigm for the rational design and evaluation of stable properly folded domains of other multimeric proteins.

(205c) Logical Computation By Chimeric Antigen Receptors Increases Tumor-Targeting Capability of Engineered T Cells

Chen, Y., University of California, Los Angeles

Originally presented on: 11/4/2013 15:51:00 - 16:09:00

T cells expressing tumor-targeting chimeric antigen receptors (CARs) have shown exciting promise in clinical trials, particularly in the treatment of B-cell leukemia. However, important challenges remain in the use of CAR-modified T cells, including off-target toxicity toward normal cells and susceptibility to mutational escape by targeted tumors. Due to a general lack of truly tumor-exclusive antigens, CARs typically target tumor-associated antigens that are also present on a subset of healthy cells, leading to on-target, off-tumor toxicity such as B-cell aplasia observed in patients treated with T cells expressing anti-CD19 CARs. Furthermore, patients who attain complete remission with anti-CD19 CAR-expressing T cells remain susceptible to relapse from the outgrowth of CD19- blast cells—i.e., mutational escape by tumor cells. One strategy to prevent mutation escape is to engineer CARs that recognize multiple antigen inputs and execute killing upon finding any one of the targeted antigens. Alternatively, CARs can be programmed to require the correct combination of input signals before triggering T-cell effector functions, thereby increasing targeting specificity. Here, we present an "OR-Gate" CAR that, when expressed in primary human T cells, efficiently lyse target cells expressing either CD19 or CD20, both of which are common markers of leukemic B cells. We demonstrate this dual-antigen targeting capability through quantitative measurements of target cell lysis and cytokine production. We will also present our designs for a "NOT-Gate" CAR and discuss its implications on our understanding of CAR signaling in relation to natural T-cell receptor signaling. Ongoing work focuses on widening the target differentiation response for the NOT-gate CAR and expanding the technology to additional antigen combinations for improved tumor-targeting specificity in T-cell immunotherapy.

(205d) Engineered Mono-Specific Proteins As In Vivo Imaging Agents That Target Tumor Vasculature

Kapur, S.,
Silverman, A. P.,
Cherf, G. M., Stanford University
Ye, A.,
Cochran, J. R., Stanford University
Originally presented on: 11/4/2013 16:09:00 - 16:27:00

Angiogenesis, the formation of neovasculature, is a key step in tumor growth and metastasis. Engineered proteins that bind to tumor vasculature receptors have great potential as non-invasive *in vivo* molecular imaging agents to diagnose disease, to predict response to particular treatment regimens, and to monitor response to therapy. Vascular endothelial growth factor receptor 2 (VEGFR-2) is the principal mediator of angiogenesis and consequently agents based on its activating ligand (vascular endothelial growth factor, VEGF) have attracted attention as *in vivo* imaging agents. However, these VEGF-based agents are agonists and thus are unlikely to be suitable for disease diagnosis and repeated serial imaging necessary for disease management and staging.

To address this short-coming we utilized protein engineering to develop a VEGF-based antagonist that displayed high-affinity for VEGFR2. We started with a single chain variant of wild-type VEGF (VEGFwt) that was previously shown to retain VEGFR2 binding, but prevented receptor dimerization and activation (VEGFmut). This was achieved by mutating one face of the VEGF ligand that converted it from a bivalent to a monovalent binder. However, this resulted in a significant loss in apparent affinity due to the loss in avidity and rendering the molecule unsuitable for *in vivo* targeting. Using rational and combinatorial approaches we affinity matured the parent (VEGFmut) resulting in the isolation of monovalent variants (VEGFa-m) that exhibited VEGFR2 affinities comparable to the bivalent parent (VEGFwt) of ~400 pM but retained antagonistic activity. In addition, these agents exhibited a marked shift in the preference for VEGFR2 versus VEGFR1 (a VEGFR family member that is not functionally implicated in active angiogenesis and is responsible for high renal uptake) compared to VEGFwt based agents.

In this presentation biochemical and biophysical characterization of the engineered mono-specific variants will be described. The performance of these agents with respect to the currently utilized VEGFwt based probes in *in vivo* tumor models will also be discussed.

(205e) Advances in Engineering a Human Cystathionine-y-Lyase for Systemic Lâ€'Methionine Depletion Cancer Therapy

Paley, O. M., University of Texas Austin
Lu, W. C., University of Texas Austin
Hu, J.,
Cheung, N. K.,
Stone, E. M., University of Texas at Austin
Georgiou, G., The University of Texas at Austin
Originally presented on: 11/4/2013 16:27:00 - 16:45:00

It has been known for nearly a half century that many human tumors are much more sensitive than normal tissues to L-methionine (L-Met) starvation. More recently, systemic L-Met depletion by administration of Pseudomonas putida methionine-γ-lyase (MGL) could effectively inhibit human tumors xenografted in mice. However, bacterial-derived MGLs are unstable in serum ($t_{1/2}$ = 1.9 ± 0.2 h) and are highly immunogenic in primates. Since the human genome does not encode a human MGL enzyme, we created *de novo* a methionine degrading enzyme by reengineering the structurally homologous pyridoxal phosphate-dependent human enzyme cystathionine-γ-lyase (hCGL). hCGL degrades L-cystathionine but displays no promiscuous activity toward L-Met. Rational design and scanning saturation mutagenesis led to the generation of a variant containing three amino acid substitutions (hCGL-NLV) that degraded L-Met with a k_{cat}/K_M of $5.6 \times 10^2\ M^{-1}\ s^{-1}$ and displayed a serum deactivation $t_{1/2}$ = 30 ± 3 h (PEGylated). *In vitro*, the cytotoxicity of hCGL-NLV toward neuroblastoma cell lines was essentially indistinguishable from that of the *P. putida* MGL. Furthermore, intravenous administration hCGL-NLV was able to sufficiently deplete serum L-Met to retard growth of neuroblastoma xenografts in a mouse model. In order to minimize dose requirement, we further engineered hCGL-NLV to identify next generation variants with improved kinetics. The current best variant has a five-fold higher k_{cat}/K_M as compared to hCGL-NLV and an improved *in vitro* cytotoxicity profile, with no appreciable loss of stability. We are currently testing this new variant with a variety of cancer cell lines as well as determining its PK/PD and efficacy in a mouse model.

(205f) A Suite of Anti-Fibroblast Activation Protein-Alpha Antibodies for Targeting the Tumor Microenvironment

Van Deventer, J. A., Massachusetts Institute of Technology
Rajan, S.,
Sidhu, S., University of Toronto
Wittrup, K. D., Massachusetts Institute of Technology
Originally presented on: 11/4/2013 16:45:00 - 17:03:00

An expanding body of work suggests that the tumor stroma plays an integral role in the initiation and development of epithelial cancers. Within the stroma, cancer-associated fibroblasts (CAFs) are involved in several functions that form key aspects of tumor progression, including immune suppression, metabolism, and invasion and metastasis. Despite the demonstrated importance of CAFs within the tumor microenvironment, targeting CAFs in a therapeutic context remains a challenge. The work presented here details the development of a suite of antibodies targeting multiple epitopes of fibroblast activation protein-alpha (FAP), a membrane-bound protease selectively overexpressed on CAFs in numerous epithelial cancers. The combination of a minimalist antibody fragment library and yeast surface display allowed for the isolation of more than 50 unique antibodies recognizing the human or murine forms of FAP, with close to 30 exhibiting cross-reactivity. The cross-reactive subset of antibody fragments was secreted in an antibody-like scFv-Fc format from yeast for further characterization of binding function. These antibody-like structures, usable without purification, led to the identification of binding proteins that recognize the membrane-bound form of FAP and further categorization of antibodies into those recognizing two or more distinct epitopes of the protein. Several of these fragments have been affinity matured and converted to full-length IgGs, with the conventional antibody formats retaining the binding function and distinct epitope recognition observed with scFv-Fcs. These antibodies are expected to provide biological insights into the roles of CAFs and therapeutic approaches to cancer treatment. Furthermore, the general antibody isolation and characterization approaches used in this work are accessible to many single-investigator laboratories and should be applicable to the development of antibodies against numerous therapeutic targets.

(205g) Engineering Ligand Biophysics to Enhance Tumor Targeting

Case, B., University of Minnesota
Hackel, B. J., University of Minnesota
Originally presented on: 11/4/2013 17:03:00 - 17:21:00

Proteins and small molecules can be engineered for specific, high affinity binding for applications in molecular targeting for tumor therapy and diagnostics. Yet, target affinity is only one element of delivery; biophysical parameters also play a significant role in dictating tumor uptake and off-target background. Molecular charge and hydrophilicity impact vascular permeability, plasma clearance kinetics, tumor retention, and physiological retention in multiple organs (especially kidneys and liver). Yet, we lack both fundamental understanding and systematic empirical data needed to empower rational engineering of charge and hydrophilicity in delivery molecules for specific applications.

We have demonstrated select cases of charge reduction as a means of reducing renal retention. Yet, this approach is not uniformly robust. We have now produced a systematic series of protein ligands with a range of net charge, total charge, and charge distributions. Conjugation of these ligands with positron emitters (^{64}Cu) via metal chelators enables dynamic positron emission tomography and excised tissue gamma counting for both rapid kinetics upon injection as well as multi-day longitudinal monitoring of biodistribution with whole-body information at sub-organ (1 mm) resolution. We will present data on tumor targeting and off-target background, especially renal retention and plasma clearance kinetics. The results are nuanced but indicate benefit from charge removal. Yet, this removal is often detrimental to solubility, stability, and recombinant production. Thus, we are also studying the ability to efficiently engineer charge using data from structural studies, phylogenetic sequence analysis, and high throughput screening. We have performed yeast surface display stability studies and high throughput production assays. Ongoing studies suggest that minimal experimental efforts can yield significant predictive value of mutational tolerance.

(205h) Selective Inhibition of Weak-Affinity Epitope-IgE Interactions Via Heterobivalent Inhibitor Design Prevents Mast Cell Degranulation

Handlogten, M. W., University of Notre Dame
Kiziltepe, T., University of Notre Dame
Serezani, A.,
Kaplan, M.,
Bilgicer, B., University of Notre Dame
Originally presented on: 11/4/2013 17:21:00 - 17:39:00

SELECTIVE INHIBITION OF WEAK-AFFINITY EPITOPE-IGE INTERACTIONS VIA HETEROBIVALENT INHIBITOR DESIGN PREVENTS MAST CELL DEGRANULATION

An Allergic response results from the activation of the adaptive immune system directed against noninfectious environmental substances (allergens). Cross-linking of immunoglobulin E (IgE) bound to its high affinity receptor, FcεR1, on the surface of basophils and mast cells initiates a complex signaling cascade that results in degranulation which is the release of preformed chemical mediators associated with allergic symptoms. The majority of the current therapies used to treat allergies and related adverse immunologic responses, such as autoimmune diseases, involve non-specific suppression of significant parts of the immune system resulting in the increased risk of infection. The development and use of selectively targeted therapeutics would decrease these risks. Our overall goal in this project was to engineer hetero-bivalent peptidic ligands that would competitively and selectively inhibit allergen binding to IgEs on mast cells, thereby inhibiting IgE clustering and preventing an allergic response. This was accomplished via designing hetero-bivalent inhibitors (HBIs) that simultaneously targeted two nearby sites located on the Fab domain of an IgE, the antigen binding site, and the less renowned "unconventional nucleotide binding site". Simultaneous bivalent binding to both these sites provided the hetero-bivalent inhibitors with enhanced avidity and selectivity for the target IgE, and enabled competitive inhibition of allergen binding to IgE. We previously demonstrated the utility of the HBI design with the most commonly used allergy model – the rat basophilic leukemia cell (RBL)/IgE^{DNP} system. In this system, a saturating concentration of a single monoclonal IgE is used to prime the mast cells, then the allergen, typically BSA presenting multiple copies of DNP, is added to the system and the degranulation response is observed. This system has been very useful for studying allergy, but lacks much of the complexity present in the natural system. First, in humans, mast cells are primed with serum derived polyclonal IgE resulting in mast cells presenting IgE specific for a wide range of antigens, of which only 0.1 – 20% is specific for any given allergen. Second, natural allergens typically contain 2 – 9 distinct immunodominant epitopes that possess a range of affinities for the polyclonal IgEs. Third, the method of conjugating multiple copies of a single hapten to scaffolds such as BSA results in poorly defined allergens that contain a wide range in the number of haptens per allergen. A matter that is further complicated as not all the haptens in these synthetic allergens are capable of binding to mast cell bound IgE due to steric constraints.

The limitations in the current model led us to develop a new system to study allergy and determine the effectiveness of our HBIs. The system we developed was based on the design of a series of heterotetravalent allergens (HtTAs) that were used to stimulate the degranulation of mast cells that were primed with 3 monoclonal IgEs each with a different target antigen. The HtTAs contained two sets of two haptens that were specific for a different IgE antibody. The third IgE antibody was used to model the orthogonal IgE (IgE of other specificity) on mast cells. The use of the third IgE allowed us to test the HtTAs with physiological relevant amounts of allergen specific IgE on the mast cell surface. The HtTAs proved to be potent stimulators of degranulation only when the mast cells were primed with both of the IgEs specific to the allergen. This was expected as bivalent allergens are poor stimulators of degranulation.

With our new more advanced allergy model we were able to successfully and specifically inhibit allergen-mediated mast cell degranulation using our HBIs in both *in vitro* and *in vivo* allergy models. Specifically, we designed an HtTA possessing two sets of two haptens each specific for a different monoclonal IgE. The first IgE-hapten interaction had a high affinity (K_d = 55 nM) while the second IgE-hapten pair had a weak affinity (K_d = 41 μM). The HtTA stimulated degranulation, following a bell-shaped dose response curve, thus illustrating the importance of weak binding epitopes in allergy. An HBI was designed to inhibit the binding of only 1 of the hapten-IgE pairs by conjugating an antigen binding moiety to a nucleotide binding moiety via a short ethylene glycol linker. The increased avidity and specificity of simultaneous bivalent binding to the antigen binding region and the nucleotide binding site on a single Fv of an IgE, inhibited allergen binding to the IgE. When the HBI was co-incubated with the allergen, degranulation was inhibited in a dose dependent manor. The binding of the HBI to one of the allergen specific IgEs essentially reduced the HtTA to a bivalent allergen which are not effective stimulators of degranulation. This result demonstrated that it is possible to completely inhibit allergen-mediated degranulation through the specific targeting of only a fraction of the allergen specific IgE on the mast cell. The long-term goal of this project is to develop a broadly applicable technology that can selectively inhibit the allergic responses induced by an allergen via competitively inhibiting its binding to the IgE antibodies with engineered hetero-bivalent ligands.

Force to Death and New Life, Some with Programming Vitality

Liaison Functions
AIChE Annual Meeting

Regenerative medicine is the quest to provide biological or hybrid synthetic-biological materials that can enable or facilitate therapeutic interventions , from transfusions and gene therapy to organ replacement. A long-standing goal is to ability to produce human blood cells for transfusion medicine. Among blood cells , platelets , needed for blood coagulation and vascular repair , are an expensive "product" in limited supply. This is due to the collection and processing steps from donated blood , and the fact that platelets cannot be frozen , but also due to the possibility of bacterial or blood-borne pathogen contamination. Can we produce platelets in a "blood factory"? This is recognized as a grand challenge that remains largely elusive. Platelets are produced from the large , polyploidy megakaryocytes (Mks) in the bone marrow and the lung vasculature , under a spectrum of biomechanical forces. I will discuss these forces and show how important they are for producing functional proplatelets and other , small , anuclear particles , which we classify as Mk microparticles (MkMPs). I will show the extraordinary ability of these MkMPs in programming stem and progenitor cells , and discuss their potential as a means for targeted delivery of nucleic acids and proteins. I will also discuss our explorations aiming to understand the mechanisms by which MkMPs act , and will argue for producing and using them for regenerative-medicine applications.

(247a) On-Chip Automated Phenotyping of Synapses in C. Elegans for High-Throughput Fully Automated Genetic Screens

San Miguel, A., Georgia Institute of Technology
Crane, M., Georgia Institute of Technology
Kurshan, P., Stanford University
Shen, K., Stanford University
Lu, H., Georgia Institute of Technology
Originally presented on: 11/5/2013 8:30:00 - 8:48:00

C. elegans, a soil dwelling nematode, is a highly studied multi-cellular organism that offers many experimental advantages, including a 3-day generation time and ease of culture. Its full genome sequence is known, as well as its stereotypical neuronal wiring. Being transparent, this nematode is excellent to study neurological processes *in vivo* with fluorescent gene reporters. To understand gene function, genetic screens are performed where a *C. elegans* population is randomly mutagenized and its progeny is scored to isolate mutants with a phenotype of interest. In this manner the function of many genes have been elucidated. However, the isolation of animals with strikingly altered phenotypes has reached a practical limit and the focus has now shifted to isolating animals with very subtle phenotypic differences. Aside from the difficulty of typical screens that require manual handling and inspection of a very large number of animals, the current challenge lies on the identification of mutants with very subtle phenotypes difficult to identify by eye. Performing screens based on fluorescent reporters of micron-sized features, such as synapses, presents an exceptionally difficult scenario.

Here, we benefit from microfluidic chips to enable easy worm handling and imaging in a worm-sorting device. Additionally, machine learning algorithms allow the extraction of relevant information from each image and consequent decision-making for worm sorting. This integrated approach is fully automated by incorporating external hardware control. With this platform, we are able to perform automated high-throughput imaging and sorting of thousands of worms based on quantitative synapse-related features. This method enables imaging, phenotyping and sorting about 100 times faster than manual handling. Not only does this method allow genetic screens to be performed in a simple, automated and fast manner, it also provides a platform for discovery of very subtle mutants with differences in micron-sized puncta distribution, which would otherwise be overlooked in a typical manual screen.

(247b) Functional Interplay Between Cell Cycle and Cell Phenotypes

Chen, W. C., Johns Hopkins University
Wu, P. H., Johns Hopkins University
Phillip, J., Johns Hopkins University
Khatau, S., Johns Hopkins University
Choi, J. M.,
Dallas, M., Johns Hopkins University
Konstantopoulos, K., Johns Hopkins University
Sun, S. X., Johns Hopkins University
Lee, J.,
Hodzic, D.,
Wirtz, D., Johns Hopkins University
Originally presented on: 11/5/2013 8:48:00 - 9:06:00

Cell cycle distribution of adherent cells is typically assessed using flow cytometry following detachment of the cells, which presumes the measurements of cell properties (cell size and nuclear size) and cell cycle distribution are still the same in the original environment. Here we use a high-throughput microscopy system to simultaneously analyze the cell-cycle phase of thousands of individual cells and their associated cell properties. This assay demonstrates that cell properties strongly depend on cell-cycle phase. By perturbing cell cycle with synchronization or inhibition of cell cycle regulator or changing nuclear size and shape with knockdown structural proteins (Nesprins, Lamin A/C), cell cycle and cell phenotypes always interfere with each other. With a linear combination equation, we could calculate how much effect on phenotypes is due to cell cycle redistribution or intrinsic phenotypes. This study highlights the functional interplay between cell cycle and cell phenotypes by showing the simultaneous, *in situ* measurements of cell cycle and phenotypes at single-cell resolution, a method that could be used in a wide range of applications in biology and biomedical research.

(247c) Role of Lysine Deacetylases in Megakaryocyte Differentiation

DeLuca, T.,
Kuo, H. Y.,
Konieczna, I.,
Mrksich, M.,
Miller, W. M., Northwestern University
Originally presented on: 11/5/2013 9:06:00 - 9:24:00

Low platelet count, or thrombocytopenia, is an important clinical issue for cancer patients either as a direct symptom of leukemia or as a side effect of chemotherapy. It is desirable to produce platelets *in vitro* for transfusions rather than relying solely on platelet donations. However, for this process to be feasible as a clinical option, significant advances must be made to increase the yield of platelets collected per input stem cell. Our lab previously found that nicotinamide (NIC), a form of vitamin B3, enhances the platelet-producing potential of megakaryocytes (Mks) produced in culture[1]. One role of NIC is to inhibit sirtuins, an NAD^+-dependent class of lysine deacetylase enzymes (KDACs). Since select inhibitors of sirtuins elicit similar enhancements in Mk maturation, we hypothesized that NIC mediates its effects by inhibiting sirtuins.

In the last decade, post-translational acetylation of proteins has been recognized as a prevalent modification that regulates protein function in a variety of signaling contexts[2]. Acetylation, which is regulated by KDACs and acetyl transfereases, has been demonstrated to be important for stem cell maintenance and the differentiation of many cell types including oligodendrocytes, cardiomyocytes and mesenchymal stem cells[3]. Additionally, two KDAC isoforms, KDAC1 and KDAC2, have been shown to directly impact hematopoietic stem cell differentiation[4]. We have been investigating the molecular regulation of KDACs in megakaryopoiesis, with a focus on the roles of sirtuins throughout the course of Mk differentiation. A better mechanistic understanding could then be exploited *in vitro* for platelet production or *in vivo*to provide the basis for developing better treatments for patients with megakaryocytic leukemias.

Gaining an understanding of sirtuin (SIRT) activity, rather than protein expression levels alone, is important because the SIRT proteins can be post-translationally modified to alter their endogenous activity, and these alterations would not be revealed through protein levels alone. We have developed a high-throughput system utilizing mass spectrometry to profile endogenous deacetylase activity in megakaryocytic cell lines. We validated that we can discriminate between activity of the Zn^{2+}-dependent KDACs (class I and class II) and the SIRTs (class III) using class-specific inhibitors. Results using a small panel of peptide substrates suggested that in the CHRF-288-11 (CHRF) Mk cell line, SIRT activity decreased throughout the course of PMA-mediated Mk differentiation. This trend was retained upon expansion to a library of 361 acetylated peptide substrates. We are extending our analysis to primary Mk cultures where we can profile KDAC activity in various culture populations that include mature Mks, immature Mks, and non-Mks. In CHRF cell lysates, SIRT1, 2, and 7 protein levels decreased after PMA stimulation, while SIRT3 protein levels actually increased. This highlights the importance of analyzing enzyme activity in addition to protein levels, and also the need for SIRT isoform-specific peptide substrate fingerprints. We aim to identify SIRT isoform-specific peptide fingerprints using Mk cell lines with SIRT overexpression/knockdown. Development of this assay system will be widely useful as tool to understand the roles of SIRTs in a variety of contexts. Currently, there is evidence in the literature that, in addition to differentiation, SIRTs play a role in tumorigenesis, regulation of metabolism, cardiovascular disease, and several other processes, but there is a lack of reliable tools to examine activity of these enzymes, which limits discovery of molecular mechanisms and isoform-selective inhibitors.

Additionally, we have utilized mouse models with the genes for specific sirtuin proteins (SIRT3 and SIRT7) knocked out of their genome to help us to understand the importance of these sirtuin proteins for *in vivo* hematopoiesis. We have found that SIRT3 and SIRT7 seem to be dispensable for *in vivo*megakaryopoiesis and platelet production based on steady state platelet counts, reticulated platelets, bone marrow Mk ploidy, bone marrow histology, and recovery after induced thrombocytopenia. This suggests that these two KDAC isoforms are not promising targets for enhancing platelet production.

1 Giammona, L. M. *et al.* Mechanistic studies on the effects of nicotinamide on megakaryocytic polyploidization and the roles of NAD+ levels and SIRT inhibition. *Experimental Hematology* **37**, 1340-1352.e1343, doi:10.1016/j.exphem.2009.08.004 (2009).

2 Choudhary, C. *et al.* Lysine Acetylation Targets Protein Complexes and Co-Regulates Major Cellular Functions. *Science* **325**, 834-840, doi:10.1126/science.1175371 (2009).

3 Kretsovali, A., Hadjimichael, C. & Charmpilas, N. Histone deacetylase inhibitors in cell pluripotency, differentiation, and reprogramming. *Stem Cells Int* **184154**, 8 (2012).

4 Wada, T. *et al.* Expression Levels of Histone Deacetylases Determine the Cell Fate of Hematopoietic Progenitors. *Journal of Biological Chemistry* **284**, 30673-30683, doi:10.1074/jbc.M109.042242 (2009).

(247d) Chemo-Mechanically Tunable Antibiotics-Based Hydrogel for High Throughput Generation of 3D Tumor Microenvironments (3DTMs)

Grandhi, T. S. P., Arizona State University
Potta, T., Arizona State University
Faust, J., Arizona State University
Rege, K., Arizona State University
Originally presented on: 11/5/2013 9:24:00 - 9:42:00

Despite their popularity, two dimensional cell cultures fail to capture the inherent complexity within a 3D tumor. Three dimensional co-culture systems are effective *in vitro* models of the tumor microenvironment, since they can better represent cell-cell contacts and interactions, existence of metabolic and nutrient gradients, necrotic core regions, and cellular resistance to therapeutics. Specifically, co-culturing epithelial cancer cells with their stromal/stellate counterparts provides insights into tumor-stromal interactions, cell-ECM interactions, epithelial-mesenchymal transition, and their role in tumor proliferation and metastasis. Available techniques to generate high throughput 3DTMs include hanging drop technique, rotary suspension and non-adhesive substrates for liquid overlay including Agarose, and Matrigel. However, these methods require continuous refreshing of media, and result in spheroids with highly polydisperse sizes.

Here, we describe the discovery and extensive characterization of an antibiotic-based hydrogel whose formation kinetics and chemo-mechanical properties were tunable based on temperature and on the molar ratios of the monomers. These hydrogels extensively facilitated 3DTM formation across multiple cell lines including those of prostate, breast, and pancreatic cancer cells co-cultured with stromal or stellate cells. Tunable chemo-mechanical properties resulted in the formation of one 3DTM per well, thereby providing exquisite control over size and heterogeneity. Further, 3DTM size was easily controlled by simply plating different cell numbers. These 3DTMs were characterized for cell viability using fluorescence stains and H & E staining, morphology using scanning electron microscopy, cell cycle profiles using flow cytometry, and response to chemotherapeutic drugs and their combinations. Taken together, these novel antibiotic-based hydrogels facilitate high-throughput generation of custom designed tumor microenvironments that can be used as *in vitro* tumor models for screening drugs and delivery systems, and for fundamental studies on tumor biology.

(247e) A High Throughput Screening Platform for Antibiotic Drug Discovery

Griswold, K. E., Dartmouth
Scanlon, T. C., Dartmouth College
Originally presented on: 11/5/2013 9:42:00 - 10:00:00

We describe an ultra-high throughput screening platform enabling discovery and/or engineering of antibacterial biomolecules. Individual recombinant expression hosts and bacterial pathogens are co-encapsulated within hydrogel-in-oil emulsions, hosts secreting bactericidal agents kill adjacent bacterial targets, and the corresponding gel microdroplets are selectively labeled using a fluorogenic viability probe. Importantly, the hydrogel matrix serves to physically link active expression hosts with cognate killed pathogens, and the fluorescent dye provides an easily detected phenotype. We have successfully employed both bulk emulsification and microfluidics technology to generate picoliter-scale gel microdroplets that are size-compatible with conventional flow cytometry. Using fluorescence activated cell sorting, libraries can be screened at rates exceeding 3,000 microdroplets per second. We have easily achieved throughputs of more than 5 million clones per day, demonstrating the capacity to outstrip conventional screening methods by orders of magnitude. In proof-of-concept experiments using a 10,000-fold excess of negative control cells, we efficiently selected yeast engineered to secrete lysostaphin, a powerful anti-Staphylococcal enzyme. As a practical extension of this work, we constructed a small metagenomic library derived from the native plasmids of *Staphylococcus simulans, Staphylococcus arlettae,* and *Staphylococcus equorum*, each a natural environmental competitor of *Staphylococcus aureus*. In this case we utilized an *E. coli* expression host and screened the library against a methicillin resistant *S. aureus* strain. In a matter of just 4 days, we isolated the native lysostaphin gene, which is naturally encoded on the native pACK1 plasmid of *S. simulans*. Our results establish the practical utility of the screening platform, and we anticipate that the accessible nature of our methods will enable others seeking to identify and engineer the next generation of antibacterial biomolecules.

(247f) Automated Image Analysis of Intracellular Lipid Droplets: Toward High-Throughput Screening of Anti-Obesity Agents

Sims, J. K., Tufts University
Rohr, B.,
Miller, E. L.,
Lee, K., Tufts University
Originally presented on: 11/5/2013 10:00:00 - 10:18:00

Automated Image Analysis of Intracellular Lipid Droplets: Toward High-Throughput Screening of Anti-Obesity Agents

Introduction and Motivation - Cellular hypertrophy of body fat, or white adipose tissue (AT), underlies many of the proposed mechanisms for obesity-related illnesses. In hypertrophic adipocytes, the lipid volume accounts for >90% of cell volume. Thus, an increase in adipose cellular lipid is equivalent to an increase in cellular volume. This direct coupling between cellular morphology and metabolism can be exploited to screen for the effects of agents designed to target specific metabolic pathways by monitoring the dynamics of cellular lipid droplets. Ideally, the lipid droplets are monitored noninvasively to enable repeated observations over time, which can reveal important information regarding the dynamics of lipid droplet formation. Â An attractive approach for monitoring cellular lipid droplets in cultured adipocytes is to combine optical microscopy and image analysis. In principle, contrast agents such as lipophilic dyes or labeled antibodies for lipid droplet-specific proteins are not required, as there is sufficient contrast between a lipid droplet and the cytoplasm surrounding the droplet. Importantly, forgoing the use of dyes or labels would not only lessen the experimental effort, thus aiding high-throughput analysis, but also reduce the likelihood of introducing additional measurement artifacts into the analysis.

Manual analysis of microscopy images is both time consuming, error prone, and subject to biases. This motivates the development of an automated image analysis algorithm. The major challenges arise from the complexity of the images. In a typical adipocyte culture derived by inducing preadipocytes or other precursor cells, differentiation occurs unevenly, leading to a highly heterogeneous population of cells and lipid droplets. Â Due to this complexity, relying on a fixed contrast thresholds to identify lipid droplets can lead to inaccurate results (Fig 1B). Â While manual tuning of the threshold on an image-by-image basis can improve the detection of lipid droplets, the improvement is only marginal, and is obtained at the cost of throughput and potential for bias. In this abstract, we present an image analysis algorithm that is capable of processing a wide range of images with little to no user input, and thus well-suited to high-throughput analysis of lipid droplets in cultured cells.

Image Analysis Algorithm ? Our algorithm identifies lipid droplets using four different characteristics of typical adipocyte images: droplets are (1) relatively light, (2) circular, and (3) surrounded by a relatively dark boundary, (4) which is also circular. To take advantage of the first two characteristics, the original grayscale image is converted into black and white images twice: once using a low threshold, and once using a high threshold. Instead of choosing a fixed threshold, our algorithm uses a dynamic threshold value, which adapts to the overall brightness of the entire image. When the high threshold is applied, the light interior of the lipid droplets and some light background noise are converted to white, while the rest of the image becomes black (Fig 2A). When the low threshold is applied, the dark rings surrounding the lipid droplets and some dark noise are converted to black, while the rest of the image becomes white (Fig 2B). By inverting the low-threshold image and filling in the holes, an image results that is white where the dark droplet boundaries (rings) and dark background noise originally were, and black everywhere else. An intersection of the two images yields an image that is white only where the original image was both light and surrounded by a dark boundary (Fig 2C). In addition to accurately identifying the lipid droplets, these processing steps also reduce noise in the image, because dark and light noise cannot occupy the same location in an image. A circle finding algorithm is applied to the low threshold image (Fig 2D and 2E), and the resulting image is then combined with the intersected image to generate the final processed image (Fig 1C).

Image Analysis Results The image analysis processing method was validated against experimental data. We differentiated 3T3-L1 preadipocytes in 12-well plates for 16 days. Every 4 days, we captured 300 images per well, in 6 different wells, then lysed the cells for enzymatic analysis. As expected, accumulated triglyceride increased over time (Fig 3). The image analysis and enzymatic assay data correlated significantly across time point groups (Fig 3A) and individual wells (Fig 3B). However, the correlation was stronger for time point groups (R^2=0.99). The weaker correlation for individual wells is likely due to the heterogeneity of adipocyte differentiation. As the images captured represent only a fraction of the total culture surface of a well, it is possible that the captured images do not fully represent all of the cells in the well.

Conclusions We developed a new image processing method to support noninvasive analysis of lipid droplets in cultured adipocytes. This method does not require destruction of the cells for endpoint analysis of cell metabolism or introduction of exogenous dyes or labels. The image analysis results correlate strongly with enzymatic assay data, especially for condition group averages. Prospectively, automated image analysis processing could be use to monitor the dynamics of lipid droplet formation in cells, and establish high-throughput screens for agents designed to affect these dynamics.

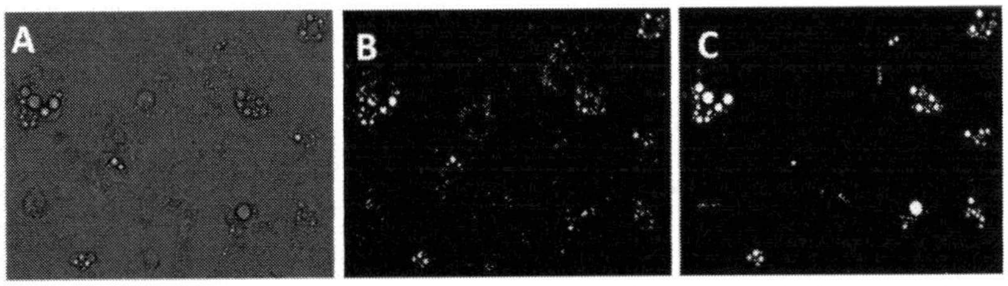

Figure 1: Comparison of currently available image analysis methods to our new algorithm. Original image of adipocytes in 3D culture (A), processed image using the currently available program (B), processed image using the new algorithm (C).

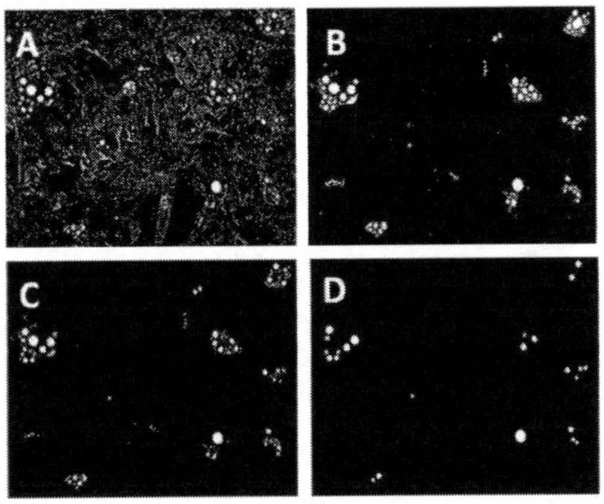

Figure 2: Image analysis algorithm. High threshold image (A), low threshold image (B), intersected image (C). Circle finding on the low threshold image (D).

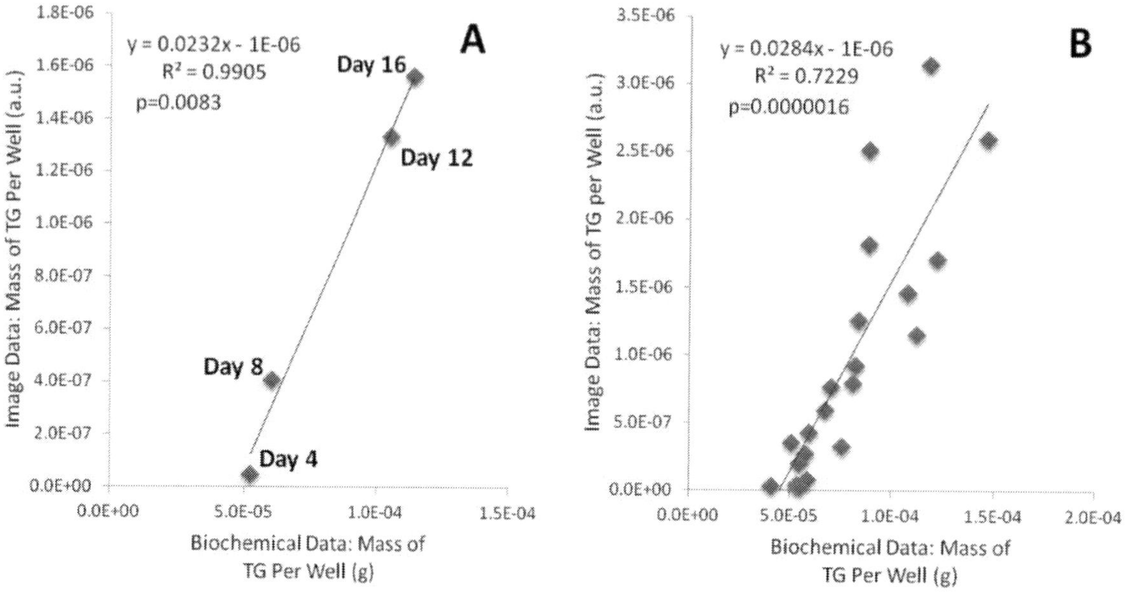

Figure 3: Comparison of triglyceride accumulation assessed

(247g) Quantitative Solution-Phase Profiling of Active Transcription Factors in Parallel

Bilgin, B., Michigan State University
Liu, L., Michigan State University
Chan, C., Michigan State University
Walton, S. P., Michigan State University
Originally presented on: 11/5/2013 10:18:00 - 10:36:00

Transcription factors (TFs) are key effectors of gene expression. TFs bind specific sites of chromosomal DNA and control the expression of thousands of genes to facilitate proper cell function in response to intracellular and extracellular stimuli. TF networks are highly interconnected so that one TF can be activated by many different stimuli, and, conversely, a single stimulus can activate many TFs. Thus, measuring the level of one TF in response to a stimulus potentially ignores other important events that contribute to the cellular response. Hence, parallel measurements of transcription factors are necessary to capture the whole picture of the instantaneous state of the cell. However, given that there are an estimated 1850 different human TFs, parallel TF profiling still remains an unsolved challenge.

We have sought to develop a new, scalable, flexible, and sensitive approach to analysis of transcription factor levels that complements existing parallel approaches. The goal for our method would be for it to be parallel (up to thousands of TFs simultaneously), quantitative, to have superior dynamic range to existing methods, to require small sample sizes (10^6 cells or fewer), and requiring no manipulation of the cells being studied prior to the assay. Using magnetic bead affinity separation, we have successfully measured the levels of human TFs, NF-kB and Ap1, in parallel with ~10-fold improved sensitivity over existing approaches. We have also quantified differences in the levels of NF-kB, Ap1 and TFIID in breast cancer cells following TNF-alpha stimulation and IKK inhibitor treatment relative to control cells. Here, we will describe current results along with further directions, including expansion to larger sets of TFs and analysis of complex biological processes.

(247h) Transcription Factor Activity Profiling Reveals New Insights Into Megakaryocyte and Erythrocyte Differentiation

Duncan, M. T., Northwestern University
Shin, S., Northwestern University
Mays, Z., Northwestern University
Shea, L. D., Northwestern University
Miller, W. M., Northwestern University
Originally presented on: 11/5/2013 10:36:00 - 10:54:00

Transcription factor activity profiling reveals new insights into megakaryocyte and erythrocyte differentiation

Mark T. Duncan,[1] Seungjin Shin,[1] Zachary Mays, Lonnie D. Shea,[2] William M. Miller[2]

Department of Chemical and Biological Engineering, Northwestern University, Evanston, IL, United States of America

[1]M.T.D. and S.S. contributed equally to this work. [2]Co-corresponding authors.

The differentiation of hematopoietic stem and progenitor cells (HSPCs) is a complex process that is regulated by a multitude of transcription factors (TFs). In particular, the lineage-choice at the bifurcation between the megakaryocyte (MK) and erythrocyte (E) lineages highly depends on the interactions within a network of TFs.[1] Post-lineage commitment, the maturation process of both red blood cells (RBCs) and MKs continues to depend on activation of the appropriate TFs at the appropriate time. We applied our recently developed dynamic TF activity array to study E/MK differentiation.[2,3] The assay is based on the lentiviral delivery of TF-specific binding elements that drive the expression of firefly luciferase, which is subsequently monitored by daily bioluminescence imaging of the same cells for 5-6 days. In the initial studies, we assessed the dynamic activity of a panel of 8 E/MK relevant TFs during hemin-mediated E and PMA-mediated MK differentiation of the bipotent cell line K562. We observed numerous trends in TF activities that were consistent with results from the literature. In particular, we observed that PMA strongly induced the activity of NF-E2 and FLI-1, both of which are required for terminal MK maturation, while PMA repressed the activity of the myeloid-promoting TFs PU.1 and c-myb. Consistent with its known degradation in response to PMA, GATA-1 activity was also rapidly reduced. Conversely, during hemin-mediated E differentiation, we found that NF-E2, GATA-2, and TAL1 activities were elevated. To confirm the specificity of the observed trends to the E/MK differentiation programs, we generated GATA-1 silenced K562 cells, which we confirmed were unable to differentiate towards either lineage.Â Profiling the TF activities in these cells after hemin or PMA treatment revealed that only activation of NF-E2 appeared to be independent of either differentiation program.Â This indicates that NF-E2 activation is not far downstream of the initial signaling events induced by the addition of PMA or hemin.

We further utilized our assay to explore terminal MK maturation and, in particular, polyploidization.Â For that purpose, we examined the effects of NIC, a potent enhancer of MK polyploidization,[4] on TF activities in the MK-committed cell line CHRF-288-11.Â Concomitant with increasing ploidy, NIC greatly enhanced the PMA-mediated increases in activities of NF-E2, FLI-1, and p53.Â Moreover, a preliminary experiment in primary MKs suggests that NIC does indeed significantly increase NF-E2 activity during normal MK differentiation.Â

Based on the results of these studies, we conclude that dynamic profiling of TF activity is a novel, useful tool that will complement more traditional, low-throughput assays (e.g. qPCR, microarray, western blot, ChIP) in understanding stem and progenitor cell differentiation. Â In particular, the dynamic, non-invasive measurements of TF activity provide a rapid means of identifying the critical TFs at different stages of the differentiation process.Â Additionally, the described technology can easily be extended to simultaneously monitor the activity of a large numbers of TFs, making it suitable for exploratory studies to identify TFs with previously unknown involvement in the differentiation process.

References

1 Dor , L. C. & Crispino, J. D. Transcription factor networks in erythroid cell and megakaryocyte development. *Blood* 118, 231-239, doi:10.1182/blood-2011-04-285981 (2011).

2 Bellis, A. D. *et al.* Cellular arrays for large-scale analysis of transcription factor activity. *Biotechnology and Bioengineering* 108, 395-403, doi:10.1002/bit.22916 (2011).

3 Weiss, M. S. *et al.* Dynamic, Large-Scale Profiling of Transcription Factor Activity from Live Cells in 3D Culture. *PLoS ONE* 5, e14026, doi:10.1371/journal.pone.0014026 (2010).

4 Giammona, L. M., Fuhrken, P. G., Papoutsakis, E. T. & Miller, W. M. Nicotinamide (vitamin B3) increases the polyploidisation and proplatelet formation of cultured primary human megakaryocytes. *British Journal of Haematology* 135, 554-566, doi:10.1111/j.1365-2141.2006.06341.x (2006).

(249a) Recent Challenges and Opportunities for Biotechnology: A Perspective From Commercial Manufacturing

Luo, J., Genentech, Inc
Mills, K., Genentech, Inc.
Bergevin, J.,
Originally presented on: 11/5/2013 8:30:00 - 8:50:00

It has been three decades since the first biopharmaceutical, recombinant human insulin, was licensed by the US Food and Drug Administration. Since then, biopharmaceutical industry has become the fastest growing, most dynamic and technology-intensive sector in the biomedical field. Mammalian cell culture remains the main source for antibody or therapeutic protein production and has demonstrated the capability to produce therapeutic recombinant proteins in large quantity.

For stable protein products like monoclonal antibodies, fed-batch processing has been widely accepted as a platform technology for large-scale manufacturing due to good process robustness, scalability and ease of operation. With the recent improvement in cell line engineering and cell culture media and process optimization to increase cell specific productivity, viable cell density, and culture longevity, titers have increased from below 1 g/L in the early 1990s to over 8-10 g/L more recently. Industry has moved to using platform technology, maximizing the speed to clinic and product launch. With this, the process optimization may be shifted from late stage process development to post launch (i.e. before the build-up of the newly-launched product's demand peak, or transfer the production to a second plant). This change in development philosophy increases the need to build a more experienced and scientifically-minded technical supporting group at commercial manufacturing site.

Several case studies will be discussed to demonstrate the recent challenges encountered in the commercial manufacturing, including the development of a barrier for adventitious agent, raw material variation, increased product mix, and implementation of quality by design. Overall, the focus of commercial manufacturing is on product quality and process consistency through improved process and product understanding.

(249b) Establishment of a Disposable Platform Process and Comparability Study to Stainless Steel Bioreactors

Bock, D.,

Originally presented on: 11/5/2013 8:50:00 - 9:10:00

Flexible platform processes and operation facilities are needed to manage a wide range of material demand and variability in unit operations. Boehringer Ingelheim has successfully established our global disposable platform process, which allows us to efficiently meet the demands of global trials and markets. A good understanding of the fundamental engineering aspects and comparability between the disposable and stainless steel bioreactors is important for process transfer between different bioreactor systems as well as process scale up and down.

This study investigated engineering parameters of disposable bioreactors such as volumetric mass-transfer coefficient, carbon dioxide removal and mixing time, and compared these parameters to stainless steel bioreactors at different scales. This study was performed at different agitation speeds and process transfer between systems was based on power input per volume and volumetric gas flow rates. Finally the comparability was evaluated with a mammalian cell cultivation platform process carried out in different systems.

(249c) Hydrodynamic Effects On Cell Aggregation and Cell Attachment of Suspension Cultures in Small-Scale Cell Culture Vessels

Zang, R., Genzyme
Hayes, T., Genzyme Corporation
Yin, J., Shire HGT
Originally presented on: 11/5/2013 9:10:00 - 9:30:00

Hydrodynamic Effects on Cell Aggregation and Cell Attachment of Suspension Cultures in Small-Scale Cell Culture Vessels

Ru Zang, Tom Hayes and Jin Yin

45 New York Ave, Genzyme, a Sanofi Company, Framingham, MA 01701

Abstract:

Small-scale vessels for suspension cultures such as shake flasks, spinners and wave bags have been widely used in cell culture processes. We have observed significant cell aggregation and cell attachment (CACA) issues when growing some cell lines in these vessels. In order to resolve CACA issues, different types of small scale cell culture vessels have been applied in this investigation and it was found that CACA was related to hydrodynamics in these vessels. Furthermore, CFD (Computational Fluid Dynamics) analysis of hydrodynamics was conducted to support the understanding of mechanism of CACA in vessels with different mixing patterns. This presentation will share some of our experience and understanding on CACA for suspension culture in different types of vessels. Instead of using anti-clumping agents, a potential approach to reduce CACA by controlling culture hydrodynamics will be discussed as well.

(249d) Stem Cell Expansion and Differentiation in Microcarrier-Based Bioreactors for Large Scale Production

Sart, S., Florida State University
Agathos, S., University of Louvain
Li, Y., FAMU-FSU College of Engineering
Originally presented on: 11/5/2013 9:30:00 - 9:50:00

Microcarriers have been widely used for various biotechnology applications because of their high scale-up potential, high reproducibility on regulating cellular behaviors, and their compliance with current Good Manufacturing Practices (cGMP). Recently, microcarriers have been investigated for stem cell expansion and differentiation, enabling potential scale-up of stem cell-derived products in large bioreactors. This presentation summarizes recent advances of using microcarriers for mesenchymal stem cell (MSC) and pluripotent stem cell (PSC) cultures in bioreactors. From the reported data, efficient expansion and differentiation of stem cells on microcarriers relies on their ability to modulate cell shape (round or spreading) and cell organization (i.e. aggregate size). Nonetheless, current screening of microcarriers remains empirical, and accurate understanding of how stem cells interact with microcarriers still remains unknown. Accurate characterization of biochemical and biomechanical properties of the microcarriers is required to fully exploit their potential in regulating stem cell fate decisions. Due to the evident variety of microcarriers, such detailed analyses should lead to the rational design of application-specific microcarriers, enabling reproducible effects on stem cell expansion and differentiation for large scale biological and biomedical applications in the pharmaceutical industry.

(249e) Sustainable Production of Omega-3 Fatty Acids By Metabolically Engineered Yarrowia Lipolytica Fermentation

Suwannakham, S., The Ohio State University
Zhu, Q., E.I. du Pont de Nemours and Company
Tyreus, B. D., E.I. du Pont de Nemours and Company
Boonyaratanakornkit, B., E.I. du Pont de Nemours and Company
Dellomonaco, C., Rice University
Fan, X., E.I. du Pont de Nemours and Company
Hong, R., E.I. du Pont de Nemours and Company
Sharpe, P. L., E.I. du Pont de Nemours and Company
Short, D. R., E.I. du Pont de Nemours and Company
Xie, D., DuPont Company
Xue, Z., E.I. du Pont de Nemours and Company
Yadav, N. S., E.I. du Pont de Nemours and Company
Odom, J. M., E.I. du Pont de Nemours and Company
Jackson, E. N., E.I. du Pont de Nemours and Company
Originally presented on: 11/5/2013 10:00:00 - 10:20:00

The omega-3 fatty acids, cis-5, 8,11,14,17-eicosapentaenoic acid (C20:5 or EPA) and cis-4,7,10,13,16,19-docosahexaenoic acid (C22:6 or DHA), have been shown to confer wide-ranging health benefits against a number of different diseases, e.g. cardiovascular disease. DuPont has developed an advanced platform technology of metabolic engineering of the GRAS oleaginous yeast, *Yarrowia lipolytica*, to produce high value oil and biomass with tailored omega-3 (EPA, DHA) or omega-6 (ARA, GLA) fatty acid compositions; providing a superior source for these essential molecules for applications in nutritional supplements, functional foods, infant foods, medical foods, pharmaceuticals, and animal feeds. We report here the development of clean and sustainable commercial products of EPA oil and EPA-rich biomass by fermentation. The developed yeast triacylglyceride oil has a unique fatty acid profile with less than 5% as saturated fatty acids and more than 55% as EPA (corresponding to the highest omega-3 fatty acid reported so far in the oil). The EPA oil (Newharvest™) has been used as human nutritional supplement; and the EPA-rich biomass has been used as salmon feed to raise brand salmon, Verlasso™ (www.verlasso.com). This EPA-rich biomass for feed reduced >75% fish oil consumption and produced premium salmon. Approaches in development of metabolically engineered strain candidates, strain evaluation and fermentation process for production of these two commercial products will be discussed in this talk.

(249f) Exometabolome Analysis Reveals Hypoxia At the Up-Scaling of a Saccharomyces Cerevisiae High-Cell Density Fed-Batch Biopharmaceutical Process

Aon, J.,

Originally presented on: 11/5/2013 10:20:00 - 10:40:00

Scale-up of industrial fermentation process occurs after the process is optimized in the small scale and is critical for a successful technology transfer and subsequent commercialization of the product of interest. Here we characterize process performance changes that were observed during the scaling-up of a *Saccharomyces cerevisiae* high-cell density fed-batch culture process to a commercial scale (10,000 L) bioreactor. The fermentation process was optimized and demonstrated to be robust and reproducible in small scale (10L) bioreactors. However, differences were observed at the large scale that affected process controllability including increases in wet cell weight, culture apparent viscosity and pure oxygen sparge supply. Mitigation of these differences to ensure process performance at large scale required changes in the dissolved oxygen and pressure set points as well as addition of diluents. To further understand the physiological changes between two scales, we performed a comprehensive quantitative and time-dependent analysis of the exometabolome of the culture at both scales. A metabolome profiling approach was used to identify groups of biochemicals associated to known metabolic pathways in *S. cerevisiae*. At both scales and different time points a diversity of intermediates was found from central carbon metabolism, and mevalonate/ergosterol synthesis pathway. The profiling revealed disturbances in key carbon intermediates and sterols at 10,000L scale, indicating hypoxia. The contribution of this work is the knowledge of how cells respond to such an environment at the large scale and how those responses are related to the process performance.

(249g) Smart Imaging Analysis for the On Line Observation of Bioprocesses At the Example of the Microalgae Crypteconinium Cohnii

MaaÃŸ, S., SOPAT GmbH
Emmerich, J.,
Junne, S.,
Neubauer, P.,
Originally presented on: 11/5/2013 10:40:00 - 11:00:00

Docosahexaenoic acid (DHA) is a polyunsatured fatty acid and belongs to the group of omega-3-fatty-acids. Due to its beneficial effect on human health DHA is a valuable natural product and is commercially manufactured from the microalgae *Crypthecodinium cohnii* [1]. The dinoflagellate grows in a certain culture medium heterotrophically to high cell concentrations.Under controlled and monitored conditions the microalgae produces a significant fraction of pure DHA [2]. The organism is very sensitive to various environmental conditions which make high yield production of DHA challenging. Detailed process understanding and exact process control is essential to accurately adjust process parameters to the requirements of the organism.Achieving certain cell densities and starting the DHA production phase, the process is switched from a nutrient-excess state to a nutrient-limiting state. Based on the morphologic and physiologic changes of the cells, the optimal modes how to switch between these two process stages can be defined. The integration of an optical detection method will be beneficial, when cell densities and different morphological states can be displayed and recognized automatically [3].

This study analyzes the influence of flash light and media contamination of an invasive photo-optical microscope through changes of optical density and DHA productivity. Cell concentration is measured automatically with the developed in-situ probe and compared to ex-situ cell counting methods. Additionally feature extraction with focus on changes in cell sizes, cell shapes and cell walls is performed. This information provides the basis for the development of automated feature recognition by a newly developed image analysis software.

[1] Wynn, J., P. Behrens, A. Sundararajan, J. Hansen and A. Kirk 2010. Production of Single Cell Oils by Dinoflagellates, Single Cell Oils. Z. Cohen and C. Ratledge.

[2] Mendes et al. 2008, Journal of Applied Phycology 21: 199-214

[3] Maaß et al. 2012, Computers and Chemical Engineering, DOI10.1016/j.compchemeng.2012.05.014

(254a) Effects of Allelic Regulation On Nanog Expression Heterogeneity in Embryonic Stem Cell Populations

Wu, J., State University of New York at Buffalo
Tzanakakis, E. S., State University of New York at Buffalo
Originally presented on: 11/5/2013 8:30:00 - 8:48:00

Nanog is a pluripotency transcription factor in embryonic stem cells (ESCs) and its expression pattern and function have been studied extensively in recent years. A distinct feature of ESC populations in vivo or in vitro is their heterogeneous expression of Nanog. A bimodal distribution of Nanog is observed with cells switching typically between states of $NANOG^{high}$ and $NANOG^{low}$ indicative of the robustness in Nanog expression heterogeneity. These observations have prompted the development of mathematical models featuring not only transcriptional fluctuations in Nanog but also feedback loops among Nanog and its partners, Oct4 and Sox2. However, Nanog was recently reported [1] to be allelically controlled in mouse ESCs (mESCs) adding another layer of regulation in the observed expression heterogeneity in addition to gene network activity. In this study, we developed a multiscale stochastic population balance equation (PBE) model for ESC ensembles to investigate the effects of allelic expression on Nanog heterogeneity.

A PBE system was constructed with each equation corresponding to a subpopulation of ESCs exhibiting a distinct allelic pattern of Nanog expression. The physiological state vector comprised cell size as indicator of cellular proliferation and the intracellular NANOG protein content. The model captures activities such as cell division and allelic regulation transpiring at longer time scales and fast occurring gene expression dynamics with transcriptional noise described by stochastic differential equations. Parameter calculation was based on experimental data carried out by our group [2] and others [1]. Numerical solutions to the PBE model were obtained via Monte Carlo simulations considering switching of the allelic pattern of Nanog expression as a Markov process.

Allelic regulation resulted in a trimodal distribution for Nanog. However, transcriptional noise caused the profile to become bimodal in agreement with findings from experiments by our group and others. The model also allows the calculation of the cell residence time in various modes of the distribution shedding light on the differential stability of the $NANOG^{low}$ and $NANOG^{high}$ states. Employing this framework, we showed that stem cell lines engineered with a reporter transgene knocked in an allelically regulated locus, exhibit nonmatching distributions of the endogenous and reporter proteins. This led us to investigate the performance of stem cell reporter systems with the transgene inserted in one or both alleles and with different or similar net production dynamics for the target and reporter proteins. The results show that the protein and reporter signatures are weakly correlated (Pearson product-moment correlation coefficient: 0.6) in typical stem cell lines with monoallelic insertion of the reporter gene. These findings point to pitfalls in the interpretation of the data from such reporter systems. Moreover, the PBE model presented here can be utilized to back-calculate the actual distribution of the native gene product from reporter signatures. Finally, the issue of whether stem cells with reduced Nanog expression can maintain their pluripotency under non-differentiating conditions is debatable. Aided by the PBE model, we analyzed data from stem cells with a single Nanog allele deletion. Our study shows that a subpopulation of these cells remains pluripotent while the deletion of a single Nanog allele does not simply reduce NANOG uniformly for all ESCs but modulates NANOG heterogeneity directly.

Multiscale models coupled to experimental studies will be essential in the design of efficient protocols for directing stem cell differentiation and for the reprogramming of cells [3]. Such models will also serve as a basis for the engineering and control of bioprocesses for the production of stem cell therapeutics.

1. Miyanari Y, Torres-Padilla ME (2012) Control of ground-state pluripotency by allelic regulation of Nanog. Nature 483: 470-473.

2. Wu J, Tzanakakis ES (2012) Contribution of Stochastic Partitioning at Human Embryonic Stem Cell Division to NANOG Heterogeneity. PLoS One 7: e50715.

3. Wu J, Rostami MR, Tzanakakis ES (2013) Stem cell modeling: from gene networks to cell populations. Current Opinion in Chemical Engineering 2: 17-25.

(254b) Enhancing the Traffic Flow of Enzymes Accelerates the Kinetics of Cellulose Decomposition

Chu, J. W., University of California, Berkeley
Shang, B., University of California - Berkeley
Originally presented on: 11/5/2013 8:48:00 - 9:06:00

Interfacial proteins operate under crowded conditions and exhibit a complex reaction machinery to facilitate biocatalysis. Important modern examples are cellobiohydrolases, which decompose crystalline cellulose microfibrils processively within the plant cell wall and represent key components in second-generation biofuel technologies. To identify rate-limiting mechanisms, we develop a systems-level simulation model of enzyme kinetics with a molecular-scale spatial resolution. The null-event kinetic Monte Carlo and stochastic quasi-static algorithms were used to overcome the stiffness of covering eight orders of magnitude of timescales. We found cellulose conversion is limited due to slow complexation and excessive jamming of enzyme molecules on the substrate. The ability of involving molecular details in three dimensions allows explicit representation of molecular crowding in propagating enzyme kinetics. The computational results suggest that optimization of complexation and decomplexation between enzyme and substrate is critical for enhancement of cellulose bioconversion rates.

(254c) A Systems Biology Approach for Investigation of the Influence of Environmental Nutrients On the Biofilm Formation of P. Aeruginosa

Xu, Z., Villanova University
Fang, X., Henry M. Jackson Foundation
Wood, T. K., Texas A&M University
Huang, Z., Villanova University
Originally presented on: 11/5/2013 9:06:00 - 9:24:00

A Systems Biology Approach for Investigation of the Influence of Environmental Nutrients on the Biofilm Formation of *P. aeruginosa*

Zhaobin Xu, Xin Fang, Thomas K. Wood, Zuyi (Jacky) Huang*

Abstract:

Biofilms have been frequently associated with human infections such as those of implanted devices. (Costerton et al., 2005) Antibiotic resistance of pathogens is strongly enhanced once they form biofilms. (Thien and O'toole, 2001) Investigation of the switch from planktonic to biofilm growth mode can facilitate the treatment of biofilm-associated pathogens. Ines Thiele from University of Iceland used the planktonic model to predict the growth of *Pseudomonas aeruginosa* in the biofilm environment and performed gene knockout to determine the gene target that can stop microbial growth in biofilms (Sigurdsson et al., 2012). However, the biofilm formation of the mutants has not been considered in this approach. Our group (Xu et al., 2013) presented the first systems biology approach to quantify the biofilm formation ability of single mutants

The availability of nutrients in the environment influences biofilm formation: The experimental results presented by Kristich et al. (Kristich et al., 2004) indicated that the nutrients that were more dilute tended to promote biofilm formation relative to nutrient-rich media. Some strains of *Escherichia coli* K-12 and *Vibrio cholerae* will not form biofilms in minimal medium unless supplemented with amino acids (Watnick et al., 1999). The availability of ions such as iron also plays an important role in biofilm formation (Banin et al., 2005). However, little research has been conducted to quantify the influence of nutrients such as amino acids, iron, phosphate, and sulfate on the microbial biofilm formation.

In this work, we develop a systems-level analysis approach to determine the influence of nutrients on the biofilm formation of *P. aeruginosa*. This work extends our previous approach to further quantify the biofilm formation ability of *P. aeruginosa* based on the availability of amino acids, iron, phosphate, and sulfate changes in the surrounding environment. Using flux balanced analysis(Edwards and Palsson, 2000), artificial hit and run sampling (Becker et al., 2007), we propose an approach to quantify the biofilm formation ability from the profile of flux change through those biofilm-associated reactions, and then further quantify the biofilm formation ability of *P. aeruginosa* in an environment with different availability of nutrients. Specifically, we predict the biofilm formation of *P. aeruginosa* upon the change of uptake rates of amino acids, iron, phosphate and sulfate. The change of uptake rates in the flux balance analysis mimic the change of availability of nutrients in the environment.

Our approach predict that *P. aeruginosa* switches from planktonic to biofilm growth upon the addition of most of amino acids into the minimal medium. This prediction is consistent with the experimental data presented in Bernier et al., 2011, which shows the amino acid effect on biofilm formation of *P. aeruginosa*. Our approach also predicted that the biofilm formation ability of planktonic *P. aeruginosa* is enhanced when limited ions such as iron, phosphate and sulfate are available in the surrounding environment. In particular, upon the limited availability of these ions, *P. aeruginosa* is predicted to slow down its growth rate but increase itsbiofilm formation. This may reveal the strategy used by *P. aeruginosa* to survive in the hostile environment. In addition, the developed platform can be used to predict the biofilm formation of *P. aeruginosa* in the environment conditions with the combination of different nutrient elements. This approach can provide directions for experimental work to quantify the planktonic-biofilm transition under a specific nutrient environment for other microorganisms based on cell metabolism.

References:

Banin E, Vasil ML, Greenberg EP.(2005) Iron and *Pseudomonas aeruginosa* biofilm formation. Proc Natl Acad Sci U S A. 102(31):11076-81.

Becker SA, Feist AM, Mo ML, Hannum G, Palsson BO, Herrgard MJ.(2007) Quantitative prediction of cellular metabolism with constraint-based models: the COBRA Toolbox. Nat Protoc 2:727-738.

Bernier SP, Ha DG, Khan W, Merritt JH, O'Toole GA. 2011. Modulation of *Pseudomonas aeruginosa* surface associated group behaviors by individual amino acids through c-di-GMP signaling. Res Microbiol. 162(7):680-8.

Costerton JW, Montanaro L, Arciola CR. (2005) Biofilm in implant infections: its production and regulation. Int J Artif Organs 28(11):1062-8.

Edwards JS, Palsson BO. (2000) Metabolic flux balance analysis and the in silico analysis of *Escherichia coli* K-12 gene deletions. BMC Bioinformatics 1:1-1.

Kristich CJ, Li YH, Cvitkovitch DG, Dunny GM. 2004. Esp
independent biofilm formation by *Enterococcus faecalis*. J Bacteriol.186(1):154-63.

Mah TF, O'Toole GA (2001) Mechanisms of biofilm resistance to antimicrobial agents. Trends Microbiol 9(1):34–39.

Sigurdsson G, Fleming RM, Heinken A, Thiele I, (2012). A systems biology approach to drug targets in *Pseudomonas aeruginosa* biofilm. PLoS One. 7(4):e34337.

Watnick PI, Fullner KJ, Kolter R. 1999. A role for the mannose-sensitive hemagglutinin in biofilm formation by *Vibrio cholerae* El Tor. J. Bacteriol. 181(11):3606–9.

Xu Z, Fang X, Wood TK, Huang ZJ (2013) A systems
level approach for investigating *Pseudomonas aeruginosa* biofilm formation. PLoS One.

(254d) Parameter Set Selection for Signal Transduction Models With Significant Uncertainties

Dai, W., Rensselaer Polytechnic Institute
Bansal, L., Texas A&M University
Word, D., Texas A&M University
Hahn, J., Rensselaer Polytechnic Institute
Originally presented on: 11/5/2013 9:24:00 - 9:42:00

Mathematical models composed of ODEs are widely used to describe the dynamic behavior of biological/biomedical systems. The accuracy of these models not only depends on the structure of the model which is determined by the physics, chemistry, and biology of the system, but also relies on adjustable parameters in the model, many of which are either taken from the literature or estimated using experimental data. A number of studies have investigated various aspects of parameter estimation of dynamic system [1-3]. However, before any parameter estimation is performed, it is important to determine if all parameters are numerically identifiable and, if not, then what subset of parameters can be accurately estimated [4]. Examples of particular interest for this work are models of signal transduction pathways as these models tend to contain a large number of parameters, many of which contain significant uncertainties and which cannot be estimated from available measurements.

Sensitivity analysis is a powerful tool to study how model parameter variations can qualitatively or quantitatively influence model behavior. As such, a variety of methods for parameter set selection based on sensitivity analysis have been proposed in the literatures. These methods include, but are not limited to, genetic algorithms [5], collinearity index methods [6], column pivoting methods [7], Gram-Schmidt orthogonalization methods [8], and clustering methods [9]. A systematic scheme for parameter set selection is based on optimality criteria computed from the Fisher information matrix which is closely related to the parameter-output sensitivity matrix [10].

It is important to note that all of the above mentioned methods for parameter set selection utilize local sensitivity analysis. The main drawback of local sensitivity analysis is that the sensitivity vectors are dependent on the parameter values which are not precisely known prior to parameter estimation. This may result in identification of parameter subsets that are suboptimal which can have significant impact on the model's prediction accuracy if the parameter uncertainty is large [11]. One alternative to local sensitivity analysis is global sensitivity analysis which simultaneously varies multiple parameters, often over a large range, so that the uncertainty description of parameters is incorporated into the sensitivity analysis procedure. Unfortunately, it is difficult to interpret the result from global sensitivity analysis, e.g. the Morris method [12], sampling-based method [13], and a variance-based method [14], for experiment design or parameter set selection [11].

This work addresses the challenges mentioned above for parameter set selection for dynamic systems under uncertainty. This is achieved by combing a hierarchical clustering method [9] and dynamic optimization techniques [15] allowing simultaneous variation of all parameters over a large range to quantify the effect of the uncertainty in the parameter space on the sensitivity vectors. Unlike the local approach where parameters are clustered according to their sensitivity vectors [9], the presented approach clusters the parameters based upon the sensitivity cones. Important challenges that are dealt with in this work arise for the computation of the sensitivity cones, not only because of the uncertainty description over all parameter, but also because not all cones have the same angle as the sensitivity vectors of some parameters are significantly more affected by uncertainty than those corresponding to other parameters.

One of the key contributions of this work is to determine the angle of each sensitivity cone and make use of this information to determine threshold cutoff values for grouping parameter via hierarchical clustering. The angel of the sensitivity cone is calculated by solving a dynamic optimization problem. The problem is formulated by discretizing the ODE system with the 3-point Radau collocation method in AMPL [16] and is solved with an interior-point nonlinear solver IPOPT [17] integrated with MA86 [18].

The proposed systematic scheme for parameter set selection for dynamic systems under uncertainty is applied to two examples, including a model of a signal transduction pathway. The examples are presented in detail and comparisons of the prediction accuracy for optimal and non-optimal parameter sets are discussed.

Reference

[1] Poyton A, Varziri MS, McAuley KB, McLellan P, & Ramsay JO. Parameter estimation in continuous-time dynamic models using principal differential analysis. *Computers & Chemical Engineering*2006;30:698-708.

[2] Ramsay JO, Hooker G, Campbell D, & Cao J. Parameter estimation for differential equations: a generalized smoothing approach. *Journal of the Royal Statistical Society: Series B (Statistical Methodology)* 2007;69:741-796.

[3] Lin Y, & Stadtherr MA. Deterministic global optimization for parameter estimation of dynamic systems. *Industrial & Engineering Chemistry Research* 2006;45:8438-8448.

[4] Chu Y, & Hahn J. Parameter set selection for estimation of nonlinear dynamic systems. *AIChE Journal*2007;53:2858-2870.

[5] Chu Y, & Hahn J. Integrating parameter selection with experimental design under uncertainty for nonlinear dynamic systems. *AIChE Journal*2008;54:2310-2320.

[6] Brun R, Reichert P, & Künsch HR. Practical identifiability analysis of large environmental simulation models. *Water Resources Research* 2001;37:1015-1030.

[7] Velez-Reyes M, & Verghese G. Subset selection in identification, and application to speed and parameter estimation for induction machines. Control Applications, 1995, *Proceedings of the 4th IEEE Conference on IEEE*. 1995. p. 991-997.

[8] Lund BF, & Foss BA. Parameter ranking by orthogonalization—Applied to nonlinear mechanistic models. *Automatica*2008;44:278-281.

[9] Chu Y, & Hahn J. Parameter set selection via clustering of parameters into pairwise indistinguishable

groups of parameters. *Industrial & Engineering Chemistry Research* 2008;48:6000-6009.

[10] Walter E, & Pronzato L. Qualitative and quantitative experiment design for phenomenological models —a survey. *Automatica* 1990;26:195-213.

[11] Chu Y, & Hahn J. Quantitative optimal experimental design using global sensitivity analysis via quasi-linearization. *Industrial & Engineering Chemistry Research* 2010;49:7782-7794.

[12] Morris MD. Factorial sampling plans for preliminary computational experiments. *Technometrics* 1991;33:161-174.

[13] Hornberger GM, & Spear R. Approach to the preliminary analysis of environmental systems. *Journal of Environmental Management* 1981;12.

[14] Atherton R, Schainker R, & Ducot E. On the statistical sensitivity analysis of models for chemical kinetics. *AIChE Journal* 1975;21:441-448.

[15] Srinivasan B, Palanki S, & Bonvin D. Dynamic optimization of batch processes: I. Characterization of the nominal solution. *Computers & Chemical Engineering* 2003;27:1-26.

[16] Fourer R, Gay D, & Kernighan B. AMPL: a modeling language for mathematical programming. 2002. *Duxbury Press*.

[17] Wächter A, & Biegler LT. On the implementation of an interior-point filter line-search algorithm for large-scale nonlinear programming. *Mathematical Programming*2006;106:25-57.

[18] HSL(2013). collection of Fortran codes for large-scale scientific computation. http://www.hslrl.ac.uk.

(254e) Analyzing High Fidelity Models in Systems Biology Using a Meta-Model Approach: Case Studies in Cellular Signaling and Acute Inflammation

Mathew, S., University of Pittsburgh
Banerjee, I., University of Pittsburgh
Originally presented on: 11/5/2013 9:42:00 - 10:00:00

Background:

'High fidelity' models have become a commonplace in systems biology [1]. These models encompass sufficient systems levels information on the interacting components enriched with details on regulatory modules like feedbacks and crosstalks. However, these models are associated with large number of parameters that are often *ad hoc*and only a few of them can be simultaneously inferred from experiments. Therefore, it is important to select and analyze the nonlinear contributions of these parameters to the model behavior to rank the parameters by their importance. While several techniques are available for such a ranking process, sensitivity analysis is a common approach. In biological systems, most often researchers use local estimates of sensitivity because of the computational cost associated with global analysis of high fidelity models. However, local analysis of such non-linearly interacting biological models can often be unreliable. Here we present a work-flow for global sensitivity analysis of high fidelity models using a meta-model approach that provides a sufficiently accurate estimate of the sensitivity indices without costly model evaluations.

Methods:

We based our global sensitivity analysis on random sampling high dimensional model representation (RS-HDMR) developed by Rabitz *et al* [2]. The method has been previously applied to high dimensional models in physical systems like atmospheric chemistry, combustion processes and genetic circuits [3]. Here, we are extending the method to biological systems with significant nonlinear interactions via feedbacks. The technique constructs a meta-model of the original model and by using orthogonal function decomposition techniques, evaluates the well-known Sobol' sensitivity indices of the input parameters to the model output. At the same time, the component functions of the resulting function decomposition provide a convenient way to analyze the input-output (IO) behavior of the model. The advantage of using RS-HDMR lies in the reduction in number of random samples required ($\sim 10^3$) to estimate the sensitivity indices as compared to computationally expensive brute-force Monte Carlo methods ($> 10^5$). For each case study selected, we did a complete analysis of the model IO behavior predicted by RS-HDMR and compared the accuracy of the Sobol' indices and the relative ranking of the parameters with brute force Monte Carlo methods.

Applications and results:

We selected two ordinary differential equation models as case studies for our current work. Model 1 describes insulin mediated activation of PI3K/AKT pathway with 20 state variables and 25 free parameters [4]. Model 2 describes an acute inflammation process mediated by endotoxin with 15 state variables and 51 free parameters [5]. We selected models that contain a wide range of reactions that are typically encountered in systems biology models. From our analysis, we found RS-HDMR to perform well for models of this complexity and there was considerable reduction in the computational time. Furthermore, we also gained insights into the model behavior in the high dimensional parameter space that was not possible by analyzing each parameter in isolation. In particular, we could efficiently discover regions of the parameter space with distinct behaviors that are experimentally or clinically relevant. RS-HDMR may therefore be a promising technique with sufficient mathematical foundation to be adapted to several problems of interest to systems biology community.

References:

1. Kiparissides A, Kucherenko SS, Mantalaris A, Pistikopoulos EN (2009) Global sensitivity analysis challenges in biological systems modeling. Industrial & Engineering Chemistry Research 48: 7168-7180.

2. Li G, Rabitz H (2012) General formulation of HDMR component functions with independent and correlated variables. Journal of Mathematical Chemistry 50: 99-130.

3. Li G, Rosenthal C, Rabitz H (2001) High dimensional model representations. The Journal of Physical Chemistry A 105: 7765-7777.

4. Sedaghat AR, Sherman A, Quon MJ (2002) A mathematical model of metabolic insulin signaling pathways. American Journal of Physiology-Endocrinology and Metabolism 283: E1084-E1101.

5. Chow CC, Clermont G, Kumar R, Lagoa C, Tawadrous Z, et al. (2005) The acute inflammatory response in diverse shock states. Shock 24: 74-84.

(254f) Stochastic Analysis of Chemical Reaction Networks Using Moment Closure Techniques

Smadbeck, P., University of Minnesota - Twin Cities
Kaznessis, Y., University of Minnesota
Originally presented on: 11/5/2013 10:00:00 - 10:18:00

Random noise has long been known to contribute significantly to the dynamics of chemical reaction networks in systems that contain species with low copy numbers (e.g. DNA in a bacterium). These systems are common in biology and are typically referred to as being far from the thermodynamic limit. The fact that random noise or fluctuations are vital to understanding the behavior of such systems necessitates the use of stochastic simulation algorithms. While it is getting progressively easier to run the large kinetic Monte Carlo (kMC) simulations necessary for the simulation of biological systems, many issues remain. When compared to deterministic models, for example, it is not easy to obtain steady-state results or perform prior analysis for stochastic models using kMC simulations.

Decades ago it was recognized that while individual kMC trajectories are non-deterministic, the underlying distribution and its statistics (or moments, the mean, variance, etc.) are indeed deterministic. The equations that govern the dynamics of these moments are rather simple to obtain as well. The major issue is that for any chemical network with 2nd order reactions the set of equations is *open*. The mean dynamics depend on the variance, the variance on the skewness, etc. In such cases a *moment-closure* technique is needed: a way to relate higher-order moments to lower-order moments so that the set of equations can be solved. Several moment-closure schemes have been developed over the years, both analytical (cumulant closure) and numerical (our Zero-Information Closure), but stochastic analysis using moment-closure, for the most part, has not been attempted.

Using several examples, most prominently the Brusselator oscillatory system, we show that moment-closure techniques may afford the ability to use deterministic analytical tools for analysis of stochastic simulations. In particular, steady-state distributions are obtained quickly, the resulting linearized Jacobian is determined, and an eigenproblem is solved to determine both eigenvalues and moment vectors. While the accuracy of such analysis is under investigation, the possibility of using moment-closure techniques to quickly identify interesting steady-state and dynamic behavior in stochastic simulation is tantalizing.

Many moment-closure techniques have been developed and explored in recent years, but relatively little work has been put into exploring the analytical consequences of such work. By changing the focus of stochastic simulation from the non-deterministic trajectories of kMC algorithms to the deterministic statistics of the ensemble distribution a whole panel of new analytical techniques becomes available. For small-scale networks in particular such analysis may be particularly useful in determining when interesting behavior occurs or model reduction is possible.

(254g) Optimally Controlled DNA Amplification

Marimuthu, K., Carnegie Mellon University
Chakrabarti, R., Carnegie Mellon University
Originally presented on: 11/5/2013 10:18:00 - 10:36:00

Due to the universal need to amplify DNA for biochemistry applications ranging from molecular cloning to DNA sequencing, DNA amplification or the Polymerase Chain Reactions (PCR) is the workhorse of nearly every modern molecular biology laboratory, as well as the burgeoning discipline of personalized medicine. Despite the apparent simplicity of the PCR reaction, the method is often fraught with difficulties that can decrease the cycle efficiency or result in competitive amplification of undesired side products. To date, nearly all approaches to PCR optimization have been based on either heuristic experimental sampling of reaction conditions or calculation of the effects of changes in reaction mixture composition on thermodynamic parameters. By contrast, the engineering discipline of control theory can automatically derive prescriptions for the optimal temperature cycling protocols of a PCR reaction, if a suitable kinetic model exists. We have developed the first sequence and temperature dependent kinetic model for PCR reactions, validating this model through comparison to experimental data. Using this kinetic model we have analyzed the controllability of PCR and developed a control strategy to maximize the DNA amplification efficiency. Further, we have formulated and solved a large scale constrained dynamic optimization problem and derive the optimal temperature profile that maximizes the amplification of a given target DNA sequence and reduces the overall reaction time.

(254h) Tailoring Treatment for Individual Patients: Bayesian Modeling and Control of Chemotherapeutics
Devaraj, J., Purdue University
LaÃnez, J. M., Purdue University
Ramkrishna, D., Purdue University
Originally presented on: 11/5/2013 10:36:00 - 10:54:00

Tailoring Treatment for Individual Patients: Bayesian Modeling and Control of Chemotherapeutics

Jayachandran Devaraj[a], *JosÃ© Miguel Laínez*[a], *Doraiswami Ramkrishna*[a,*]

[a] School of Chemical Engineering, Purdue University, 480 Stadium Mall Way, West Lafayette, IN 47907, USA

[*]Corresponding author at: School of Chemical Engineering, Purdue University, 480 Stadium Mall Way, West Lafayette, IN 47907, USA. Tel: +1 765 494 4066; fax: +1 765 494 0805; email: ramkrish@purdue.edu

Abstract

Inter-patient variation in drug response presents a great challenge in determining the optimum dose for an individual patient. In the present form of chemotherapy administration, patients are started with an average dose of drug and 'titrated' to a maximum tolerated dose through a trial and error approach. During this process, patients undergo several dose modifications and encounter severe episodes of acute side-effects. Many retrospective studies show that such endeavors are responsible for the long-term medical complication. A possible alternative would be a system theoretic framework adequately empowered by statistical and optimal control theory. Such a quantitative approach, if implemented carefully, will greatly enhance the decision making capabilities of the treating physicians.

In this work, we present a Bayesian modeling and robust model predictive control based approach to individualize the dosing of an important chemotherapeutic drug, 6-Mercaptopurine (6-MP). 6-MP is one of the key drugs in the treatment of many pediatric cancers, auto immune diseases and inflammatory bowel disease (IBD). 6-MP is a prodrug which is converted to 6-TGN, an active metabolite, through enzymatic conversion. An enzyme known as TPMT plays a key role in this metabolic conversion and determines the amount of 6-TGN produced. A genetic polymorphism observed in the TPMT enzyme produces a huge variation in drug response among patient population. The pharmacological model for 6-MP metabolism accounts for such variations in enzyme activities and predicts the 6-TGN concentration in humans. To circumvent the data scarcity in the clinical settings, a global sensitivity analysis based model reduction approach is used to minimize the parameter space. The complete and reduced models were compared in terms of goodness-of-fit and number of samples required in characterizing a new patient. Furthermore, we show how the availability of additional information on the biomolecular hierarchy can be integrated into Bayesian information and improve the prediction quality. With the final model, a model predictive control algorithm is used to optimize the dose scheduling with the objective of maintaining the 6-TGN concentration within its therapeutic window. The model and the control approach can be utilized in the clinical setting to individualize 6-MP dosing based on the patient's ability to metabolize the drug instead of the traditional standard-dose-for-all approach.

(270a) Engineering and Analysis of Antibody Fragments for Prion Disease Therapy

Doolan, K. M., University of Delaware
Colby, D. W., University of Delaware
Originally presented on: 11/5/2013 8:30:00 - 8:48:00

Prion diseases are propagated by the conversion of the native form of the prion protein (PrP^C) to an alternatively folded, disease causing form (PrP^{Sc}) catalyzed by the interaction of PrP^{Sc} with PrP^C. Due to its centrality in the pathogenesis of prion diseases, several antibodies have been developed for therapy that target PrP^C; however these antibodies have shown limited success in animal models. We have engineered a suite of six scFvs based on the most promising anti-prion antibody therapeutic, ICSM 18. The antibodies vary in affinity from 1 nM to 320 nM and have expression levels in yeast that cover two log orders, implying differential stability. Characterization of the antibodies in cell culture models of prion is ongoing. Additionally, to gain insight into the molecular level antibody-antigen interactions, we have developed a methodology to provide residue level epitope identification using yeast surface display and next generation DNA sequencing. We have validated this technology by comparing our findings for ICSM18 to the antibody-complex crystal structure, and applied the new method to identify critical residues that form the epitopes of three additional anti-Prion antibodies. Finally, we are gaining molecular level insight into the protein-protein interactions that mediate prion formation and propagation using the same technology. Such knowledge may lead to the identification of epitopes for more effect therapeutic antibodies for prion diseases. Combined, our work enables a quantitative analysis of the effects of affinity, stability, and epitope specificity on therapeutic efficacy and provides insight into the molecular basis of prion pathogenesis.

(270b) Design of Antibodies Specific for Misfolded Proteins

Perchiacca, J. M., Rensselaer Polytechnic Institute
Ranjan, S., Rensselaer Polytechnic Institute
Lee, C.,
Tiller, K., Rensselaer Polytechnic Institute
Julian, M.,
Meng, F.,
Tessier, P. M., Rensselaer Polytechnic Institute
Originally presented on: 11/5/2013 8:48:00 - 9:06:00

Antibodies that recognize toxic aggregated proteins associated with neurodegenerative disorders (e.g., Parkinson's disease) are valuable for therapeutic and diagnostic applications. We have developed a novel approach for designing domain antibodies specific for misfolded proteins that is inspired by the molecular interactions that mediate protein aggregation. We find that grafting small hydrophobic peptides from several amyloidogenic polypeptides (e.g., Alzheimer's and Parkinson's polypeptides) into the complementarity-determining regions of domain antibodies generates Grafted AMyloid-Motif AntiBODIES (gammabodies) that recognize aggregated proteins with nanomolar binding affinity (Perchiacca et al., *PNAS*, 2012; Ladiwala et al., *PNAS*, 2012). Gammabodies bind preferentially to aggregates containing the grafted peptides via homotypic interactions. We are currently optimizing the length, location and number of grafted amyloid peptides within single- and multidomain gammabodies to further enhance binding affinity and specificity.

(270c) De Novo Computational Design of Fully-Human Antibody Variable Domains

Maranas, C. D., The Pennsylvania State University
Pantazes, R. J., The Pennsylvania State University

Originally presented on: 11/5/2013 9:06:00 - 9:24:00

The *de novo* computational design of novel proteins remains a formidable, but essential, challenge with only rare successes to date. Antibodies are an excellent initial system to learn reliable *de novo* design principals, due to their many modular structural features and functions that are limited to binding and not catalysis. Previously we have developed and presented at AIChE the OptCDR method, for designing antibody complementarity determining regions, and a database of Modular Antibody Parts (MAPs) for reliably predicting antibody structures. We have built upon these successes to develop an Optimal Method of Antibody Variable region Engineering (OptMAVEn) to design fully human antibody variable domains from scratch to bind any specified antigen epitope.

The MAPs database, which can be used to predict antibody structures with an average all-atom RMSD of 1.900 angstroms, is an all-atom representation of portions of antibody structures. OptMAVEn begins by searching the MAPs database for the initial antibody that can best bind the specified antigen epitope. Once such an antibody is identified, a computational affinity maturation is carried out using our Iterative Protein Redesign & Optimization procedure, with restraints that ensure that all 9-mer sequences are fully human. If desired, a library of the most promising mutants can be identified once affinity maturation is completed. The utility of OptMAVEn is demonstrated through the design of broadly-neutralizing anti-influenza antibodies.

(270d) Computationally Driven Deimmunization of Therapeutic Proteins

Salvat, R. S., Dartmouth
Parker, A. S., Dartmouth
Bailey-Kellogg, C., Dartmouth
Griswold, K. E., Dartmouth
Originally presented on: 11/5/2013 9:24:00 - 9:42:00

Novel biotherapeutics have reshaped drug discovery and promise a continued revolution in disease therapy, yet therapeutic proteins present unique design challenges for the biomolecular engineer. One distinguishing risk factor is the prospect of eliciting an immune response upon administration to human patients. An anti-drug immune response can compromise therapeutic efficacy and even threaten patient safety. In the case of therapeutic antibodies, codified humanization technologies can mitigate much of the risk associated with immune reactions. These grafting-based deimmunization strategies are not, however, broadly applicable to other protein classes, such as therapeutic enzymes. Our capacity to effectively tap the full diversity of non-immunoglobulin therapeutic candidates will require more broadly applicable protein deimmunization technologies. To help meet this challenge, we have developed optimization algorithms that seamlessly integrate computational prediction of T cell epitopes and bioinformatics-based assessment of the structural and functional consequences of epitope-deleting mutations. Our methods address the fact that biotherapeutic deimmunization represents a dual-objective protein design space with tradeoffs between immunogenicity and stability or activity.

Here, we describe the experimental validation of our algorithms using the Enterobacter cloacae P99 β-lactamase, a component of Antibody Directed Enzyme Prodrug Therapies. We describe a high-throughput immunoassay for peptide-MHC II interactions, and we analyze the robustness of our algorithm in two complementary studies. In aggregate, we have constructed a large panel of engineered enzymes and analyzed their thermal stability, catalytic activity, constituent peptide affinity towards human MHC-II immune proteins, and, in one case, relative immunogenicity in humanized mice. The results demonstrate our algorithm's predictive power and in particular a robust capacity to generate highly active yet immunoevasive therapeutic candidates. These novel computational methods can be applied to virtually any protein target, and we anticipate that they could speed biotherapeutic development and mitigate the risk of immunogenicity-related complications.

(270e) Engineered Biosensor Proteins for Programming Novel Therapeutic Modalities

Dudek, R. M., Northwestern University
Daringer, N., Northwestern University
Leonard, J. N., Northwestern University
Originally presented on: 11/5/2013 9:42:00 - 10:00:00

Technologies enabling bioengineers to construct cell-based therapies that sense and respond to specific environmental cues may transform our ability to design and implement novel therapeutic strategies that overcome existing barriers to treatment. To date, we lack synthetic biology "parts" that couple exclusively extracellular cues, such as protein mediators of intracellular communication, to engineered circuits comprised of synthetic proteins, RNA, and/or DNA. To address this need, we are developing a suite of biosensor technologies for transducing extracellular stimuli into defined state changes within the cell. We have developed the first fully orthogonal cell surface biosensor platform, termed a modular extracellular sensor architecture (MESA), and have described the engineering of a generic dimerization-dependent signal induction mechanism, wherein receptor dimerization results in the release of an engineered transcription factor from the plasma membrane. Here, we present expansion of this technology to integrate both novel biosensor outputs and novel biosensor input modalities. First, we will describe MESA biosensors that reconstitute intracellular enzymatic activities in a dimerization-dependent fashion, providing novel opportunities for performing protein-based logic and coupling MESA output to transcription-independent circuits. We will also describe the development of biosensors that transduce extracellular enzymatic activity into intracellular state changes, which is of relevance for applications such as detection of proteases that typify disease states such as tumor invasion and autoimmune disease (e.g., matrix metalloproteinases). Like many synthetic biology technologies, these biosensor platforms provide powerful tools that may be applied to scientific inquiry and investigation of disease processes as well as novel therapeutic modalities, each of which will be discussed.

(270f) Evaluation of Three Vascular-Targeted Enzyme Prodrug Systems for Breast Cancer Therapy

Krais, J., University of Oklahoma
Van Rite, B., University of Oklahoma
Kurkjian, C., University of Oklahoma Health Sciences Center
Harrison, R., University of Oklahoma
Originally presented on: 11/5/2013 10:00:00 - 10:18:00

The use of targeted therapeutics is a promising approach for the development of new cancer treatments that seek to reduce the devastating side effects caused by the systemic administration of current drugs. This study evaluates three fusion proteins developed as enzyme prodrug therapies targeted to tumor vasculature. Cytotoxicity is localized to the site of the tumor using a protein fusion of annexin V and one of three enzymes: bacterial purine nucleoside phosphorylase (PNP), yeast cytosine deaminase (CD), or bacterial methionine-γ-lyase (MT). Annexin V acts as the tumor targeting component of the fusion protein as it has been shown to bind to phosphatidylserine expressed externally only on cancer cells and the endothelial cells of tumor vasculature. The enzymatic component of the fusion converts a non-toxic prodrug (fludarabine, 5-fluorocytosine, methylselenol) into a cytotoxic drug (2-fluoroadenine, 5-fluorouracil, selenomethionine). This conversion allows for increased impact at the site of the tumor, while decreasing side effects that would otherwise result from the high systemic concentrations necessary for non-targeted treatments.

Fusions of the three enzymes and human annexin V were produced and purified in *Escherichia coli*. Binding strength and stability studies were performed with human endothelial cells HAAE-1 and breast cancer cell lines, MCF-7 and MDA-MB-231. Results were qualitatively confirmed with fluorescence microscopy. The same cell lines were used for a cytotoxicity analysis of the enzyme prodrug treatment. *In vivo* studies were conducted using SCID mice and MDA-MB-231/GFP xenografts. A pharmacokinetic profile was obtained for each fusion protein. The PNP and MT studies examined possible synergism with a non-toxic level of docetaxel, a chemotherapeutic shown to increase externalization of phosphatidylserine *in vivo* for endothelial cells in the tumor vasculature.

Recombinant production and purification of the fusion proteins yield active proteins and promising *in vitro* results for all three enzyme prodrug systems. Our *in vitro* results show successful binding and killing of breast cancer and endothelial cells representative of tumor vasculature. The CD system yielded unsatisfactory *in vivo* results; however both the PNP and MT systems achieved tumor growth suppression for the duration of the treatment period with the strongest effect shown with MT. The results suggest that this treatment has a positive clinical therapeutic potential with minimal side effects.

(270g) Engineering Nanog Protein for Effective Protein Transduction: A Possible Alternative to Reverse the Effects of Organismal Aging On Mesenchymal Stem Cells

Liang, M. S., State University of New York at Buffalo
Son, S., SUNY Buffalo
Sinha, S., SUNY Buffalo
Andreadis, S. T., University at Buffalo, The State University of New York
Originally presented on: 11/5/2013 10:18:00 - 10:36:00

Previously, our group demonstrated that overexpression of a single transcription factor, Nanog, can reverse the effects of organismal aging on mesenchymal stem cells (MSCs) proliferation and myogenic differentiation [1]. We also found that short-term Nanog overexpression by using doxycycline-control system is sufficient to reverse the effects of aging, thereby suggesting the permanent genetic modification may not be necessary. These results prompted us to hypothesize that delivery of Nanog proteins into MSCs may overcome the effects of cellular senescence. However, it has been proven difficult to deliver proteins into cells, especially into MSCs. To address this problem, we first engineered Nanog proteins, which contain a cell penetrating peptide (nona-arginine) for efficient protein delivery. To increase the stability of Nanog protein in the cells, ubiquitination target sequence (PEST) was removed. These engineered Nanog proteins were expressed in *E. coli* (BL21) and purified by denaturation/renaturation process followed by preparative chromatography. Chloroquine was used to enhance endosomal escape of internalized Nanog proteins. Human hair follicle MSCs were used in this study. Our results have demonstrated that fusion of the CPP can efficiently increase the binding of Nanog proteins on the cells and help internalization. Also, removal of PEST sequence did not have detrimental effects on Nanog transcription factor activity and increased the stability, as demonstrated by its prolonged half-life in the cells. Chloroquine effectively promoted endosomal escape of Nanog proteins that might increase the chance of nuclear localization. Furthermore, we introduced the SV40 virus nuclear localization signal (NLS) to Nanog proteins and its effects on nuclear delivery after endocytosis are under investigation. This study provides a possible alternative to overcome the difficulty of Nanog protein transduction by rationally engineering Nanog protein. Our results imply that this strategy may be employed to reverse the effects of organismal aging or culture senescence on MSC without viral transduction and permanent genome modification.

[1] Han J, Mistriotis P, Lei P, Wang D, Liu S, Andreadis ST. Nanog reverses the effects of organismal aging on mesenchymal stem cell proliferation and myogenic differentiation potential. Stem Cells. 2012;30:2746-59.

(270h) Encapsulation and Delivery of Biomolecular Therapeutics Using the Bacteriophage MS2 Viral Capsid

Glasgow, J., University of California, Berkeley
Francis, M., University of California Berkeley
Tullman-Ercek, D., University of California
Originally presented on: 11/5/2013 10:36:00 - 10:54:00

The high specificity and vast diversity of proteins make them potential therapeutics for a wide variety of diseases. Unfortunately rapid clearance, immunogenicity, poor targeting, and endolysosomal degradation have rendered most proteins ineffective as drugs. Our labs have developed methods to encapsulate a wide variety of molecules inside the bacteriophage MS2 viral capsid. By disassembling the capsid into its constituent dimers and reassembling around negatively charged molecules, we have demonstrated the encapsulation of proteins, nanoparticles, polymers, and nucleic acids in a highly versatile, well-defined 27 nm protein shell. We are now applying this technique to the delivery of therapeutics to the cytoplasm of cancer cells using MS2 as a vehicle. The capsid has previously been shown to be an effective platform for targeted delivery of imaging agents, drugs, photodynamic therapy agents, and has even been shown to deliver biomolecules in cell culture. We are developing this system into a versatile device for delivery of any cargo molecule into the cytoplasm of cells via endocytosis and subsequent escape from the endosome. We are exploring this technique for use in triggering apoptosis in cancer cells by activating specific intracellular processes.

(276a) Genome Engineering of Escherichia Coli Enhances Site-Specific Incorporation of Unnatural Amino Acids in Cell-Free Protein Synthesis

Hong, S. H., Northwestern University
Kwon, Y. C., Northwestern University
Martin, R. W., Northwestern University
Marcum, R.,
Des Soye, B. J.,
Jewett, M. C., Northwestern University
Originally presented on: 11/5/2013 8:30:00 - 8:48:00

Site-specific incorporation of unnatural amino acids (uAAs) into proteins and biopolymers by amber suppression makes possible new chemical properties, new structures, and new functions. However, competition between release-factor 1 (RF1) and orthogonal tRNAs have limited the technology. Here, we describe the development of a high yielding cell-free protein synthesis (CFPS) platform from genomically recoded *Escherichia coli* strains lacking RF1. Specifically, occurrences of the amber stop codon TAG are reassigned to the synonymous TAA codon. Because the parent strain has not been previously optimized for CFPS, we exploited multiplex automated genome engineering (MAGE) to design and construct synthetic genomes that upon cell lysis lead to improved extract performance. To stabilize template DNA, messenger RNA, amino acids, and protein products in CFPS, we targeted the deletion of 17 potential negative effectors. More than 50 strains were generated and tested in CFPS, allowing us to catalogue the systems impact of making numerous gene deletions both individually and in combinations. The protein synthesis activities of our most productive cell extracts were more than five-fold greater as compared to the extract from the parent strain, achieving more than 1.2 mg/ml of superfolder green fluorescent protein (sfGFP). The yields of modified sfGFP containing *p*-acetylphenylalanine at single and multiple positions in our *in vitro* system were more than ten-fold higher than the reported *in vivo* results. Our work demonstrates improved CFPS activity using whole genome editing for extract strain development and promises CFPS systems as suitable platforms for generating unnatural biopolymers.

(276b) Impact of Recovery Time On Recombination Efficiency in Escherichia Coli

Reynolds, T. S., University of Colorado
Gill, R. T., University of Colorado

Originally presented on: 11/5/2013 8:48:00 - 9:06:00

With the rapid advancement of DNA synthesis technologies, the ability to efficiently recombineer user-defined DNA into organisms of interest has become increasingly important. Recombination efficiency in *Escherichia coli* has been the subject of many studies. Despite these efforts, there is still considerable variability in reported recombination efficiencies. We hypothesized that multiple chromosome copies within *E. coli* could partially explain reported variability. With more than one chromosomal target per *E. coli* cell, population phenotype expression levels could change as recombineered chromosomes are segregated to daughter cells. A simple predictive tool was developed to estimate cell population phenotype and genotype during outgrowth after recombination. We found that recombineered chromosomes were increasingly segregated within the population; leading to a dilution of the phenotype of interest but an increase in chromosomal homogeneity. These results were confirmed experimentally leading to the conclusion that recovery time should be accounted for when reporting recombineering efficiencies.

(276c) White-Boxing Genetic Circuit Modeling: Absolute TF Binding Free Energies From Fluorescence Measurements

Farasat, I., Pennsylvania State University
Salis, H., Pennsylvania State University
Originally presented on: 11/5/2013 9:06:00 - 9:24:00

Every genetic circuit model in existence today has been parameterized with proportional data. Fluorescence. Absorbance. Luminescence. These measurements are only proxies for the cell's internal protein concentrations, and yet we use them to record the cell's internal dynamics and to measure the biophysics of its genetic parts. A first-principles biophysical model that uses proportional data will always be left with an unknown proportionality constant, and this unknown factor will depend on many extrinsic and ever-changing variables. Beyond the creation of community-wide measurement standards, we investigated whether a set of properly designed experiments could measure the *instrinsic* biophysical properties of a genetic part inside the cell, using only proportional fluorescence as the measurement output. We then asked whether this intrinsic knowledge could be used to predict the behavior of a family of synthetic genetic circuits.

Specifically, we develop a high-throughput quantitative approach to measuring the absolute binding free energies of a transcription factor to its DNA operator sites, utilizing only protein reporters and fluorescence measurements. The expression output of the transcription factor's regulated promoter is recorded within a series of rationally designed genetic sandboxes that contain additional non-regulating TF-binding sites. Our first-principles quantitative model shows, and we experimentally demonstrate, that the ratio in fluorescence measurements between two genetic sandboxes yields a direct, absolute relationship to the TF's binding free energy, as measured in kcal/mol.

We validate this approach using three newly engineered repressors, measuring their TF binding free energies, and predicting the behaviors of entire families of genetic circuits. We validate these predictions experimentally and demonstrate that the resulting biophysical model of a TF-regulated promoter can predict its transcription rates across a wide range of TF expression levels. We use these models to identify optimal NOT-gate genetic circuits with maximal dynamic ranges and tunable input thresholds. The affirmative answers to our questions take us two steps closer to creating a genetic circuit engineering discipline that is free of context effects.

(276d) Toward Practical Applications of Synthetic Biology: Part Mining to Construction of Programmable Cells

Shopera, T.,
Immethun, C.,
Henson, W. R.,
Moon, T. S., University of California, San Francisco
Originally presented on: 11/5/2013 9:24:00 - 9:42:00

The past decade has witnessed the power of synthetic biology in the creation of synthetic devices and systems that help to understand complex biological systems. However, its potential for real-world applications has not been fully exploited. One of its promising applications is the construction of programmable cells that are able to integrate multiple environmental signals and to implement synthetic control over biological processes. Here, we harness a toolbox of genetic parts from bacteria to construct genetic circuits. To this end, genetic parts have been recruited from various organisms and assembled into genetic devices, which were characterized using fluorescent proteins as outputs. Our programmable cells contain environmental sensors to distinguish various conditions and multi-input logic programs with genetic memory to generate the output only when the user-defined conditions are met. We will present progress on the construction of such engineered cells toward practical applications, such as disease prevention and chemical production.

(276e) Rational Design of Synthetic Yeast Promoters Using Systems-Level Bioinformatics Models

Curran, K., The University of Texas at Austin
Crook, N., The University of Texas at Austin
Karim, A., The University of Texas at Austin
Gupta, A., The University of Texas at Austin
Alper, H., The University of Texas at Austin

Originally presented on: 11/5/2013 9:42:00 - 10:00:00

Promoters are a key component in metabolic engineering and synthetic biology. These elements enable expression of heterologous genes, provide regulation of genetic circuits, and establish the basis for wholly rewired cells capable of producing new biochemical products. As such, a large amount of effort has been expended to characterize and synthesize both endogenous and synthetic promoters. Traditionally, new promoters have been generated through random mutagenesis of native promoters, bioprospecting to find non-native promoters, and synthetic hybrid promoters. However, with the exception of the latter example, little progress has been made to rationally and systematically engineer promoters for a prescribed activity level. In this presentation, we describe a new approach—namely a model-driven promoter design tool. Specifically, we highlight recent progress in using models to rationally re-design endogenous yeast promoters. In each of these cases, the rationally designed promoters followed the expected transcriptional output. This work therefore represents a fundamentally new method for understanding promoter architecture with regards to expression level. Moreover, this work enables rational design of synthetic parts.

(276f) Functional Characterization of Saccharomyces Cerevisiae Promoters and Application in Pathway Engineering

Li, Z.,
Zhang, G.,
Li, C., Beijing Institute of Technology
Originally presented on: 11/5/2013 10:00:00 - 10:18:00

Synthetic biology is considered as an emerging research field that will bring new opportunities to biotechnology. Among the microbial platforms in synthetic biology, *Saccharomyces cerevisiae* is one of the most widely used organisms due to its well-characterized physiology and genetics, fast cell growth rate, and availability of abundant genetic tools.

Most metabolic engineering and synthetic biology studies have focused on selecting the appropriate enzymes, because enzyme homologs cloned from a variety of organisms could have different behavior in terms of expression and activity in a heterologous host. However, selection of proper promoters is also important especially for yield and productivity optimization. This is not only because the protein transcription level is determined by the promoter strength, but also because promoters could be regulated and behave differently under different growth conditions. The constitutive promoters have been reported mostly come from housekeeping genes, including glycolysis pathway genes and elongation factor. Here we report the cloning of eighteen constitutive promoters employing eGFP as a reporter from the *S. cerevisiae* genome and the comparison of their strengths. Five of them come from glycolysis pathway genes, the rest thirteen come from amino-acyl tRNA synthetase genes. To our knowledge, this is the first time report about the promoters that come from amino-acyl tRNA synthetase genes. By analyzing the sequence of the promoters, we found some transcription factors could affect the strengths of promoters. In the batch fermentation, the strengths of promoters that come from different pathways behave in different patterns. We chose the promoters with relatively high strength and cloned the xylan degradation pathway and xylose reductase genes under these promoters, and transformed into *S. cerevisiae*. The resulting strain synthesized xylitol directly from beechwood xylan, and therefore can be utilized as a consolidated bioprocessing (CBP) organism for biomass utilization and conversion. During the fermentation we found the promoter not only affected the expression of gene, but also the growth of the cell.

(276g) Sensing- and Communication-Dependent Control of Microbial Motility

Bower, A. G. W., Rensselaer Polytechnic Institute
Julius, A. A.,
Collins, C. H., Rensselaer Polytechnic Institute
Originally presented on: 11/5/2013 10:18:00 - 10:36:00

The long-term goal of this project is to enable the use of microorganisms for the controlled motion of micron-scale devices, known as microrobots. A key aspect of his design is the generation of strain of *E. coli* where motility is tightly regulated by a cell-cell communication, or quorum sensing (QS) signal. The second biological module will be a strain, or combination of strains, that recognize extracellular signal(s) and produce the QS signal recognized by the first strain. This approach is advantageous because the sensor strains can easily by swapped to recognize different signals or combined to recognize multiple signals, or perform more complex computation. The use of QS may also be useful for amplifying the signal when low concentrations of a target molecule are present. We have constructed a strain of *E. coli*, where its swimming motility is tightly controlled by the *esa* QS system. The esa system provides advantages over the well-known *lux* system because its regulator, EsaR, can act as a repressor, which leads to lower levels of background expression. This is expecially important when controlling genes where only a few copies can enable a change in phenotype. In this case, we have used EsaR to control the expression of a key motor protein. The ribosome binding site was also engineered to optimize the control of the expression of the motor protein and enable QS-dependent motility. In swim plate assays, we have shown that our new construct shows no swimming motility in the absence of QS signal. Interestingly, we have also observed directional swimming, where the swimming cells move up a gradient of QS signal generated by a second strain of cells producing the signal.

(276h) Highly Orthogonal Programmable Riboregulators for Complex Digital Logic in E. Coli

Green, A. A., Harvard University
Silver, P., Harvard Medical School
Collins, J. J., Howard Hughes Medical Institute, Boston University
Yin, P., Harvard Medical School
Originally presented on: 11/5/2013 10:36:00 - 10:54:00

A diverse array of synthetic genetic programs has been constructed inside living cells to carry out digital logic operations[1], record memories of biochemical events[2], and synthesize valuable chemicals[3,4]. These behaviors have been established largely using existing biological components rewired in novel ways to enable new capabilities. It remains a significant challenge, however, to engineer living systems with functionalities that extend beyond rudimentary logic operations. This challenge arises in part from the limited number of biological circuit components that operate independently from one another and the difficulty in creating robust approaches to combine these elements for predictable behavior in living cells. Here, we exploit synthetic networks of RNA self-assembly reactions to construct a versatile platform for complex digital logic in *Escherichia coli*. The building blocks of this computational platform are newly designed programmable riboregulators of gene expression consisting of pairs of cis-repressed mRNAs (crRNAs) and trans-activating RNAs (taRNAs), which hybridize with their cognate crRNAs to de-repress them. In contrast to previous work, these RNA-based regulatory systems exploit design concepts derived not from existing biological systems[5-7] but from hairpin-based strand displacement reactions developed in nucleic acid nanotechnology[8,9]. These design features greatly expand the RNA sequence space of potential riboregulators compared to earlier systems[7,10] and provide the riboregulators with a high degree of programmability, orthogonality, and modularity for carrying out digital logic. To demonstrate these capabilities, we have assembled a family of 48 programmable riboregulators that increase expression of a GFP reporter by a factor of 40 or more upon transactivation and have validated a set of 15 highly orthogonal riboregulators with cross-talk levels below 10% *in vivo*. We have used this expansive toolkit of synthetic biological components to construct a number of complex decision-making systems in living cells. These include a multi-input OR gate that activates protein translation only in response to one of six programmed taRNAs and an AND circuit that produces GFP only upon assembly of a two-strand RNA complex. These programmable riboregulators provide a robust platform for evaluating more complex computations in living cells, and also constitute a promising test bed for probing interactions between assemblies of multiple RNA molecules *in vivo*.

References

1. Moon, T. S., Lou, C., Tamsir, A., Stanton, B. C. & Voigt, C. A. Genetic programs constructed from layered logic gates in single cells. *Nature* **491**, 249-253 (2012).

2. Friedland, A. E., Lu, T. K., Wang, X., Shi, D., Church, G. & Collins, J. J., *Science* **324**, 1199-1202 (2009).

3. Steen, E. J., Kang, Y., Bokinsky, G., Hu, Z., Schirmer, A., McClure, A., del Cardayre, S. B. & Keasling, J. D., *Nature* **463**, 559-U182 (2010).

4. Delebecque, C. J., Lindner, A. B., Silver, P. A. & Aldaye, F. A., *Science* **333**, 470-474 (2011).

5. Isaacs, F. J., Dwyer, D. J., Ding, C. M., Pervouchine, D. D., Cantor, C. R. & Collins, J. J., *Nature Biotech.* **22**, 841-847 (2004).

6. Lucks, J. B., Qi, L., Mutalik, V. K., Wang, D. & Arkin, A. P., *Proc. Natl. Acad. Sci. USA* **108**, 8617-8622 (2011).

7. Mutalik, V. K., Qi, L., Guimaraes, J. C., Lucks, J. B. & Arkin, A. P., *Nature Chem. Biol.* **8**, 447-454 (2012).

8. Dirks, R. M. & Pierce, N. A., *Proc. Natl. Acad. Sci. USA* **101**, 15275-15278 (2004).

9. Yin, P., Choi, H. M. T., Calvert, C. R. & Pierce, N. A., *Nature* **451**, 318-322 (2008).

10. Callura, J. M., Cantor, C. R. & Collins, J. J., *Proc. Natl. Acad. Sci. USA* **109**, 5850-5855 (2012).

(294a) Designing Metabolite Biosensors for Bioprocess Monitoring Using Synthetic Biology

Polizzi, K. M., Imperial College London
Constantinou, A., Imperial College London
Goers, L., Imperial College London
Originally presented on: 11/5/2013 12:30:00 - 12:48:00

Synthetic biology is an emerging discipline which seeks to use engineering principles to design or redesign biological systems. One of the clear application areas for this technology is the design of new biomanufacturing systems for the production of useful compounds including specialty chemicals and biopharmaceuticals.

Sensors are key to any industrial manufacturing process, where they are used to monitor important variables that have been linked to product yield and quality and are often linked to process control systems that can take corrective actions if these variables are not within defined limits. Synthetic biology systems offer an opportunity to deploy biological versions of these sensors-- gene circuits and/or proteins which perform the same functions of monitoring and correction. These can be included as part of the system design and offer advantages in terms of ithe speed at which they can respond and the diversity of actions which can be selected.

Biosensors can be designed to exploit any part of biology as the recognition event including transcription, translation, or post-translational signal generation. Each of these has different advantages and utilities. In this talk, I will focus on two types of biosensors: transcription-based biosensors which utilise inducible promoters as the signal transduction element and post-translational biosensors based on Forster Resonance Energy Transfer (FRET) which use protein conformational change as a signal output. We have successfully applied both of these for monitoring the concentration of important small molecule metabolites in the production of biopharmaceuticals in mammalian cell culture. The targets of our biosensors include carbon and nitrogen sources and waste products generated by metabolism. I will cover all aspects of sensor design and validation from selecting the appropriate biological recogntion element through biosensors construction and testing. Finally, I will show how biosensor signal can be used predict more complex outcomes such as biomass or protein yield.

(294b) Development of Biosensors to Identify New Chemicals Against Dengue Fever Vectors

Shakalli Tang, M., Ohio State University
Wood, D. W., the Ohio State University
Originally presented on: 11/5/2013 12:48:00 - 13:06:00

The development of a bacterial biosensor containing a nuclear hormone receptor (NHR) has proven to be an effective alternative to current compound screening strategies that rely on hormone-mimicking properties. NHRs comprise one of the largest classes of protein targets because they can bind to a wide variety of endocrine disruptors in addition to their native hormones. Our allosteric biosensors can detect compounds that target NHRs and distinguish between their positive or negative effects on the receptor's activity. When this design is extended from human to insect NHRs, an opportunity arises to screen compounds for their potential role as insecticides against infectious disease vectors such as *Aedes aegypti*, the dengue fever mosquito. This project seeks to construct a nuclear hormone binding sensor in *E. coli* to favor the identification of those compounds that disrupt the endocrine system of this target insect. By incorporating the ligand-binding domain of *Aedes aegypti's* NHR 96 into our *E. coli* biosensing plasmid, we are able to detect its response to CITCO, a chemical compound known to be the NHR96 native ligand. After a thorough optimization of parameters such as incubation time, temperature, and dilution, we seek to attain a reproducible growth phenotype assay for this novel biosensing system. Furthermore, the specificity of a particular *A. aegypti* NHR to a given ligand will be harnessed to evaluate the compound's effectiveness as an insecticide.

(294c) New Molecular Probes for mRNA Detection

Blackstock, D.,
Chen, W., University of Delaware
Originally presented on: 11/5/2013 13:06:00 - 13:24:00

Numerous viral infections and life threatening diseases are characterized by a specific mRNA expression pattern, making it a reliable target for diagnostics and drug therapies. While focusing on the fast acting and highly specific characteristics necessary when creating worthwhile diagnostics, we have created a new generation of molecular beacons suitable for detecting specific mRNA molecules. The design tools include molecular beacon-type oligonucleotides in combination with zinc finger-fluorescent protein fusions for fluorescence signaling upon target binding. Ultimately the design allows for simple construction, easy operation, and modular mRNA targeting capabilities.

(294d) Aptamer Binding of Beta Amyloid for Early Detection of Alzheimer's Disease

Chan, H., Rensselaer Polytechnic Institute
Sorci, M., Rensselaer Polytechnic Institute
Agris, P., University of Albany, State University of New York
Belfort, M., University at Albany
Belfort, G., Rensselaer Polytechnic Institute
Originally presented on: 11/5/2013 13:24:00 - 13:42:00

Alzheimer's Disease (AD) is projected to be a major medical burden by 2050 and early detection will hopefully reduce this burden through prevention and, yet to be discovered, treatment measures. The leading candidates for the neurotoxic species causing AD are beta amyloid 1-42 ($A\beta_{1-42}$) oligomers. Detection in the cerebrospinal fluid of the variation in concentration of these structures and those of the protein tau could potentially be used for the diagnosis of AD. Here, we develop an affinity biosensor for early detection of AD based on specific RNA aptamer-$A\beta_{1-42}$ binding interaction. Aptamers were chemically immobilized and oriented onto a gold-coated crystal for measurement in a Quartz Crystal Microbalance with Dissipation (QCM-D). Both mass uptake (frequency change, Δf) and aptamer film rigidity (dissipation change, ΔD) were measured as a function of time to monitor aptamer immobilization and $A\beta_{1-42}$ binding. Sensitivity, selectivity and robustness of the $A\beta_{1-42}$ oligomer sensor will be presented.

(294e) Development of Anti-Capsular Polysaccharide Aptamers for High-Throughput Screening and Serotyping

Cress, B., Rensselaer Polytechnic Institute
Linhardt, R. J., Rensselaer Polytechnic Institute
Koffas, M. A. G., Rensselaer Polytechnic Institute
Originally presented on: 11/5/2013 13:42:00 - 14:00:00

Bacterial pathogens are often encapsulated by a viscous layer of polysaccharide known as the capsule, which serves as the first layer of protection against a diverse array of environmental pressures. The mucous-like layer formed by the capsular polysaccharide retains water to prevent desiccation and can even hinder opsonization and subsequent phagocytosis by host immune systems. A unique class of bacteria biosynthesize capsules that are identical to polysaccharides also produced in mammalian tissues and are thus non-immunogenic, further contributing to evasion of host immune response. Pathogens bearing these capsular polysaccharides are implicated in such human diseases as urinary tract infections, strep throat, and bacterial meningitis. Coincidentally these capsular polysaccharides are metabolic engineeering targets as they can also serve as precursors in the production of medically relevant glycosaminoglycans, or highly acidic linear polysaccharides such as heparin and chondroitin sulfate. The utility of a rapid and selective assay quantifying and differentiating between capsular polysaccharides is twofold. Engineered strains could be screened in a high-throughput manner and selected for increased capsular polysaccharide production levels. Furthermore, a molecular screen selective for specific polysaccharides could be utilized in disease diagnosis and in strain serotyping. As antibodies cannot be generated against these non-immunogenic polysaccharides due to their natural presence in host animals, we have instead opted to select DNA aptamers from an oligomer library against capsular polysaccharides purified from three distinct bacterial sources. Heparosan from *Escherichia coli* K5, chondroitin from *E. coli* K4, and hyaluronan from *Pasteurella multocida* were purified, as confirmed by NMR, from fermentation supernatant using anion exchange chromatography. Linkage of capsular polysaccharides to amine-functionalized magnetic microparticles by reductive amination with sodium cyanoborohydride was verified by using a method recently developed in our lab. Specifically, bead-immobilized capsular polysaccharides were digested into disaccharides with their cognate polysaccharide lyases, followed by 2-aminoacridone (AMAC) labeling and quantification in a single, highly sensitive reverse phase ultra-performance liquid chromatography-mass spectrometry (RP-UPLC-MS) experiment. Molecular beacons based on selected aptamers will be further developed for fluorometric high-throughput screens to evaluate the capsular polysaccharide production capacities of metabolically engineered microbes.

(294f) Use of Fluorescent Protein Fusions to Optimize Membrane Protein Expression in Anaerobic Photoheterotrophic Rhodobacter

Escotet, M. S., Pennsylvania State University
Erbakan, M., Pennsylvania State University
Kumar, M., The Pennsylvania State University
Curtis, W. R., The Pennsylvania State University
Originally presented on: 11/5/2013 14:00:00 - 14:18:00

We describe the use of fluorescent protein fusions as a method for optimizing and accurately quantifying membrane protein expression in Rhodobacter. Membrane proteins (MP) are a difficult class of proteins to study; however, their medical and biotechnical importance makes them compelling to address the associated challenges. Difficulties of expressing membrane proteins have been reported in number of other protein expression systems such as *Escherichia coli*, yeast and mammalian cell culture. The anaerobic photoheterotroph, *Rhodobacter sphaeroides*, has the desirable characteristic of large quantities of intracellular membranes (typically absent in prokaryotes) as a result of the need to assemble the photosynthetic apparatus which facilitates ATP formation in light, but without oxygen formation. This growth mode can be implemented in anaerobic photobioreactors where biomass yields of more than 9.2 grams per liter dry weight (gDW/L) have been achieved.

The production of the homologous water-transport membrane protein Aquaporin Z (rAqpZ) from *R. sphaeroides* was utilized as a model system to develop the screening approach. Expression levels were quantified by monitoring sample fluorescence and Western Blot densitometry (WBD). Initially, monomeric Banana YFP (mBanana) fusion was chosen as a fluorescent tag because antibodies are commercially available and there is no overlap between the fluorescence spectrum of the protein and that of *R. sphaeroides*. To demonstrate the utility of such correlation for optimizing protein expression, temperature effects were studied as a simple optimization parameter. Ambient and optimal growth temperatures (25°C and 32°C, respectively) for *Rhodobacter sphaeroides* were studied under anaerobic photoheterotrophic conditions, required to induce protein expression from the *puc*-promoter used to drive MP expression. Correlations between off-line sampled culture fluorescence and WBD were developed through the use of pure protein standards to demonstrate the feasibility of using fluorescence measurements as the screening method for quantifying membrane protein levels inside growing cells.

A substantially improved signal-to-noise ratio could be achieved by diluting cell suspension in bovine serum albumin (BSA) to provide for refractive index matching and adjust the relative magnitude of light scattering to absorption. Similar trends were observed in MP production per cell at both temperatures, although the productivity was much higher at 32°C due to faster growth. This work identified several opportunities for improving this screening method towards the development of an online monitoring method to track the expression of MPs *in vivo.*

Although the mBanana fluorescent fusion served as an effective method for correlating membrane protein concentration, these measurements could only be taken offline. The requirement for offline measurement results because variants of GFP (green fluorescent protein) such as mBanana are obligate aerobic fluorescent proteins and require oxygen exposure to fluoresce properly. This presents a significant experimental constraint due to the required anaerobic growth conditions. Maturation involves folding in the presence of oxygen, during which time the fluorescence level per protein increases. Measurements of mBanana-MP fusion maturation within Rhodobacter showed that mBanana required more than 40 hours to mature (20 times its literature value). This long maturation time further precludes the use of this protein for rapid and online monitoring. Ongoing research seeks to overcome this limitation by using a new generation of facultative anaerobic fluorescent proteins based on riboflavin-binding activation. We will present the effectiveness of using pGlow-Pp1 and pGlow-Bs1 derivatives commercially available from EvoCatal as fusion tags that will hopefully allow correlations with *in vivo* measurements as well as online productivity monitoring.

(294g) The Impact of Dissociation Constant On the Detection Sensitivity of Polymerization-Based Signal Amplification Reactions

Kaastrup, K., Massachusetts Institute of Technology
Chan, L.,
Sikes, H. D., Massachusetts Institute of Technology
Originally presented on: 11/5/2013 14:18:00 - 14:36:00

Many studies have demonstrated free-radical polymerization as a simple and inexpensive signal amplification technique for the detection of molecular recognition events indicative of disease states. Of the various chemistries that have been implemented, photopolymerization has shown promise as a point-of-care technology for its ability to generate micron-scale hydrogels in as little as 35 seconds. Photopolymerization-based amplification requires the localization of photoinitiators at a surface as a function of molecular recognition events. Much of the previous work developing this technique has explored the case in which the signal amplification rather than the binding kinetics of the molecular recognition is limiting. We provide the first systematic study of how dissociation constant impacts assay sensitivity using proteins with affinities typical of those encountered in molecular diagnostic applications involving protein-protein binding. We use experimental results to validate a mass-action kinetic model that may be used to predict assay performance by serving as an alternative/supplement to the currently employed empirical approach to polymerization-based amplification assay development. We show that this model, given the surface probe density and protein binding affinity, is able to predict whether or not the minimal surface initiator density required for polymerization will be exceeded at a particular protein concentration. In addition, the polymerization-based amplification assays were performed on novel activated agarose surfaces, an inexpensive alternative to the SAM modified glass surfaces typically used.

(294h) Creation of a Firefly Luciferase Reporter Cassette for Use in PCR-Mediated Gene Deletion and Fusion in Saccharomyces Cerevisiae

Ainsworth, W. B., Louisiana State University
Benton, M. G., Louisiana State University
Originally presented on: 11/5/2013 14:36:00 - 14:54:00

When exposed to environmental perturbations, organisms increase production of proteins implicated in survival. Assessing the transcriptional level of one or more of these responding proteins provides valuable information regarding the cellular surroundings. Reporter systems developed for quantification of promoter response typically rely on enzymatic reactions (β-galatosidase and luciferase) or fluorescent molecules (green fluorescent protein - GFP), however deficits in implementation (such as protein extraction for β-galactosidase and the necessity of expensive high-throughput machinery for GFP) severely limit biosensor applications. The development of *in vivo* luciferase reporter systems has led to inexpensive, sensitive and accurate promoter assays without the variability reported between *in vitro* samplings, but current luciferase reporter systems are largely inflexible to modifications to the promoter of interest. PCR-mediated gene deletion/complementation techniques that overcome promoter inflexibility have been developed for fluorophores-based reporter systems and show promise for extension to luciferase reporters. In our study, we report the creation of a novel vector system, which introduces a cytosol-localized *Photinus pyralis* luciferase [*LUC*(−SKL)*] capable of one-step, *in vivo* measurements into a promoter–reporter system via PCR-based gene deletion/complementation. Fluorescent microscopy is utilized to confirm cytosolic localization of the luciferase construct under *HUG1* promoter. Further, we demonstrate accurate promoter dose–response through comparisons of luciferase assays to that of a similar *HUG1*Δ::*yEGFP1* promoter–reporter system. Our data indicate that, alongside a significant decrease in experimental time, the luciferase reporter system is shown to be highly reproducible and correlate well to comparable reporter constructs. Though we have applied this construct directly to DNA damage biosensing, use extends to any application where promoter response levels are desired

(306a) Receptor Tyrosine Kinase Inhibitor Efficacy in Carcinoma Is Stiffness Dependent

Nguyen, T. V., University of Massachusetts Amherst
Herrick, W. G., University of Massachusetts Amherst
Sleiman, M.,
Moriarty, T.,
Peyton, S. R., MIT
Originally presented on: 11/5/2013 12:30:00 - 12:48:00

There is a critical need for model systems to support the efforts of drug development. In particular, several groups are creating tissue mimics that capture key features of the *in vivo* cell and tissue microenvironment to test true drug efficacy and latent toxicity in physiologically relevant microenvironments. Our lab has developed a high-throughput system to quantify carcinoma cell response to receptor tyrosine kinase (RTK) inhibitors during tumor growth and stroma stiffening.

This platform consists of a new class of hydrogels with an extremely wide range of mechanical properties that combine poly(ethylene glycol) (PEG), and phosphorylcholine (PC) zwitterions. The Young's modulus of these "PEG-PC" hydrogels can be tuned over four orders of magnitude, much greater than conventional PEG-diacrylate or PEG–dimethacrylate gels. We are synthesizing these PEG-PC hydrogels within a silane-treated glass 96-well plate, wherein we can control the modulus of the hydrogels from 900Pa to 10MPa, and couple full-length ECM proteins to the gel surface. These ECM proteins are made in combinations to capture the integrin-binding present pre-tumorigenesis (collagen IV, fibronectin, and laminin), and late in tumor progression (collagen I, III, and fibronectin). We have used this system to quantify carcinoma cell (liver: HEP3B, and breast: MDA-MB-231, SKBR-3, and BT549) response to sorafenib, an FDA-approved RTK inhibitor, as a function of tissue stiffness and integrin binding.

We have observed that tissue stiffening increases the resistance of carcinoma cells to sorafenib by nearly three-fold, and there is a synergistic effect on collagen-rich, stiff environments, compared with soft, pre-tumorigenic integrin-binding proteins. This stiffness-conferred chemo-resistance is dampened when sorafenib is co-administered with ROCK inhibitor, β1 integrin blocking antibodies, or an epidermal growth factor receptor (EGFR) inhibitor. Interestingly, these combinatorial treatments were more effective on triple negative breast cancer cells than Herceptin receptor overexpressing cells. Toward more rational identification of combinatorial treatments, we used a bead-based ELISA (MAGPix, Luminex) to identify JNK as a promising phospho-protein target, which provided survival signals during substrate stiffening. Dosing carcinoma cells with JNK inhibitor and sorafenib was the one combination, which eliminated stiffness sensitive resistance. Most excitingly, we have identified a sub-population of carcinoma cells, within these cell lines, which express qualities of cancer stem cells, and are enriched during substrate stiffening. We are currently investigating RTK inhibitor combinatorial treatments to eliminate this dangerous, drug resistant subpopulation of cells. We propose that this high-throughput biomaterial system may serve as a system that pharmaceutical companies can use to rule out false positives and potentially save billions of dollars in the drug development pipeline.

(306c) Modeling Liver Metabolism for Drug Development

German, C., Oklahoma State Unifersity
Madihally, S. V., Oklahoma State University
Originally presented on: 11/5/2013 13:06:00 - 13:24:00

Abstract

Successful evaluation of efficiency and safety of new medicines is a time consuming and expensive process. One way to improve the initial drug screening is to develop high-throughput 3D cultures that mimic liver metabolism. Liver tissue is targeted due to its pivotal role in metabolism of medicinal drugs, and hepatocytes are grown outside the body to test hepatotoxicity. Recent advances have used porous structures onto which cells are seeded and cultured in static cultures or in bioreactors. Media is introduced to the cells providing nutrients for cell survival and growth. Cells are subjected to initial drug concentrations to determine toxicity levels. However, lack of understanding of the culture conditions and limitations has hampered the process of developing high-throughput technologies. The objective of this study was to develop a liver metabolism model using computational fluid dynamics (CFD) for the purpose of creating an artificial environment capable of supporting three-dimensional liver tissue cultures.

We used a commercially available CFD software (Comsol Multiphysics 4.3a) to assess the effect of scaffold thickness, location within the culture, and porosity of the scaffold. A three dimensional static culture with a scaffold 14mm in diameter and of various thicknesses to be placed in a six-well plate was simulated. In static cultures, changes in the concentration of three components oxygen, estrogen and urea were evaluated for 48 h (typical media culture duration) using Michaelis–Menten rate constants from literature. Initial concentrations were either based on solubility or physiological level. Fluid properties were that of water, and effective diffusivities were based on Mackie-Meares relationship. Overall concentrations in the static culture were determined through multiple point analysis of Fick's Law and rate laws at nodes throughout the three dimensional structure. Model concentration distributions were tracked as scaffold thickness, elevation, and porosity were varied. A constant cell density was used and increased thickness lead to increased cell number.

Time-dependent concentration profiles were analyzed for 1mm, 0.75mm, 0.5mm, and 0.25mm thick scaffolds of 0.85 and 0.25 porosities in order to determine limitations due to scaffold thickness. These results showed that scaffolds thicker than 0.5mm scaffolds were not capable of supporting three-dimensional liver tissue cultures due to oxygen diffusion limitations. Scaffold elevations of 0mm, 0.5mm, 1mm, and 1.5mm from the bottom of the bioreactor, for each of the scaffold thicknesses, were examined to investigate limitations due to scaffold placement. Concentrations of all three molecules were affected by placement of the scaffold at 0mm as molecules could not diffuse into the scaffold through the bottom. Only oxygen concentration profiles were affected by a further increase in elevation as the static bioreactor is open to the atmosphere and oxygen can diffuse into the media from air.

Due to scaffold thickness limitations in the static culture, a flow-through bioreactor of identical dimensions was simulated. Scaffolds of 1mm and 2mm thicknesses were investigated at 0.25 and 0.85 porosities. The shape of the bioreactor was initially optimized for uniform nutrient distribution. Then concentration profiles were determined at steady-state for each molecule, using similar rate kinetic parameters in convective-diffusion equation. Scaffold elevation was set at 1mm from the bottom of the bioreactor. Fluid velocity was varied to obtain sufficient nutrient distribution for cell survival. Flow-through bioreactors were constructed in-house according to the simulated dimensions. Chitosan-gelatin scaffolds with 0.85 porosity were prepared by freeze drying. For static cultures, 6-well plates were used and experiments were performed for each media change using HepG-2 cells (from ATCC) and protocols recommended by the vendor. Simulation results were validated by measuring oxygen consumption and urea production. Using these conditions, one could develop multiple wells and bioreactors that could be used in rapid drug screening.

(306d) High Throughput Solvent Screening for Ethylcellulose Barrier Membranes in Pharmaceutical Applications

Rogers, T., The Dow Chemical Company
Schoener, C. A., The Dow Chemical Company
Curtis-Fisk, J.,

Originally presented on: 11/5/2013 13:24:00 - 13:42:00

High Throughput Solvent Screening for Ethylcellulose Barrier Membranes in Pharmaceutical Applications

Cody A. Schoener, Jaime Curtis-Fisk, True L. Rogers

Dow Pharma and Food Solutions, The Dow Chemical Company, Midland, MI, USA

Purpose: Solvent is a process variable that can influence barrier membranes composed of ethylcellulose for modified-release pharmaceutical applications. Due to a wide variety of solvents used in barrier membrane coating, high throughput research (HTR) can be utilized to quickly screen and identify successful or unsuccessful solvent systems. Therefore, HTR screening was used to explore a variety of solvent systems wherein a subset of those solvent systems were used for rheological, mechanical, and performance testing of ethylcellulose barrier membranes.

Methods: HTR robots were used to prepare 31 formulations in a Design of Experiments approach. The solvent systems were composed of ethanol, isopropanol, methanol, acetone, and/or water in which 10 wt% of ethylcellulose 10 cps (ETHOCEL™ Std 10 Premium, Dow Chemical) was dissolved by shaking overnight. Phase Identification and Characterization Analysis HTR instrumentation was used to image and analyze clarity of the solutions. A subse of these samples was selected that represented the range of clarity observed for Dynamic Light Scattering (DLS) analysis. From this set, samples that exhibited unique light scattering characteristics were subjected to rheological measurements, tensile tests, water permeation, and drug transport studies.

Results: Using HTR, the 31 formulations (solvent + ethylcellulose) were developed and imaged to determine their clarity. Samples ranged from nearly translucent to hazy, and in some cases the ethylcellulose remained un-dissolved (Figure 1). 10 samples ranging from translucent to hazy were further tested with DLS. DLS enabled characterization of monomodal and bimodal distributions of ethylcellulose conformation and reduced the DLS subset to 4 samples f further investigation (Figure 1). The pre-screening HTR methods resulted in formulations exhibiting unique properties, and allowed us to focus on four formulations for film tensile and permeation testing. Further results will be presented demonstrating how altering solvent combinations alter final modified-release performance.

Conclusion: This research will demonstrate relationships between solvent selection and modified release performance, and results can help pharmaceutical formulators identify harmful or advantageous formulation parameters.Â Furthermore, HTR reduces research time significantly, whether it is used to explore unique polymer/active pharmaceutical ingredient interactions or to optimize barrier membrane formulations.

FIGURE 1: Cross polarized images of ethylcellulose and DLS data of ethylcellulose (right) in a variety of solvent combinations (right).

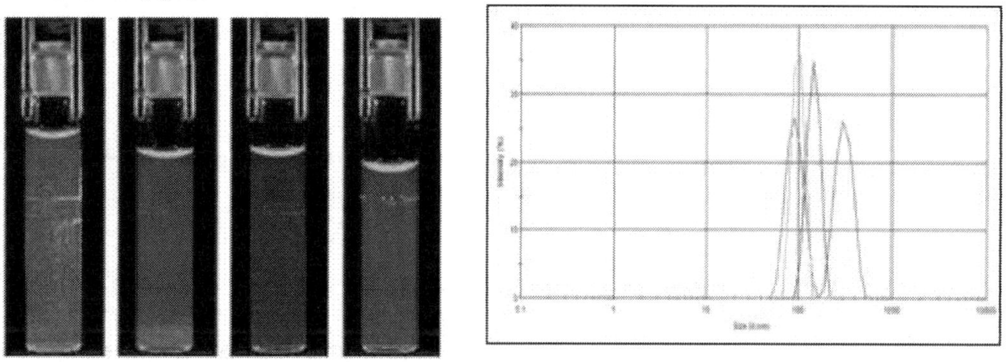

(306e) High Throughput Fluorescence Decay Measurements Using Time-Resolved Flow Cytometry

Houston, J. P., New Mexico State University
Cao, R., New Mexico State University
Vaziri, A.,
Houston, K.,
Jenkins, P. L., New Mexico State University
Originally presented on: 11/5/2013 13:42:00 - 14:00:00

Flow cytometry is a fluorescence-dependent measurement approach that enables the screening of single cells at a high-throughput level. Current flow cytometry assays involve tagging cells with several differently colored fluorophores so that through spectral analysis, cell surface features or intracellular targets can be discriminated and quantified. Yet owing to the severe spectral overlap that occurs between the emission of organic fluorophores, complex cytometry assays are becoming increasingly limited. In this contribution a high-throughput fluorescence lifetime-dependent flow cytometry system is characterized and evaluated in order to provide a new cytometric parameter that is not dependent on fluorescence intensity output. Here we present new time-resolved approaches for flow cytometry analysis that involve a multi-frequency methodology. By utilizing the idea that a laser excitation source is modulated as a square wave, where multiple odd harmonics are included in the signal, we can extract multiple phase shifts and correlate those to multiple fluorescence lifetimes that are present in each cell. Additionally, we test this novel approach on a range of applications including measurement of fluorescence lifetime changes of green fluorescent protein. Specifically, enhanced green fluorescence protein (EGFP) molecules are fused to a protein (human LC3 type B) that changes location within cells during the process of autophagy. The localization of the LC3 protein into punctae occurs during autophagy. Thus tagging EGFP to LC3 will indirectly cause the fused EGFP to group in to the punctae. Our fluorescence lifetimes are shown to be sensitive to the presence of diffuse EGFP-LC3 in the cytosol and to the organized EGFP-LC3 localization into punctate autophagosomes during autophagy induction. We verify this with low-throughput imaging and compare to less-quantitative standard fluorescence intensity-only flow cytometry. In general this work prefaces a new rapid compound-screening assay for effects on protein location. If autophagy or other biologically relevant phenomena can be quantitatively detected then time-resolved flow cytometry can become a powerful sorting and analysis tool.

(306f) Developing Droplet Microfluidic-Based High Throughput Screening System for Microbiology

Yuan, H.,
Tu, R., Tianjin Institute of Industrial Biotechnology, Chinese Academy of Sciences
Wang, Q., Key Laboratory of Systems Microbial Biotechnology, Tianjin Institute of Industrial Biotechnology, Chinese Academy of Sciences
Originally presented on: 11/5/2013 14:00:00 - 14:18:00

Microbial high-throughput screening system is derived from the traditional mutagenesis screening via the use of advanced automation and instrumental analysis technologies, with the characteristics of automation, standardization, and high throughput. This greatly exceeded the limitations of manual screening in the aspect of speed, efficiency and standardization. However, at present, various types of high-throughput screening systems only from some foreign companies are based on expensive and complicated large equipment. There is no mature, small, low-cost and easy-to-handle high-throughput screening system available so far. To address this issue, we developed a mchip-based fluorescence screening systems based on designing, and assembling the modular functions droplet microfluidic chip, constructing and debugging the accurate fluorescence detection device for dynamic target via, and gathering and processing the input and output signals with LABVIEW based software. Our research would contribute to successfully building the functional, miniature high-throughput system with the detection and sort throughput in the range of 100-1000 droplet per second for the application of screening effective microbes to produce industrial enzymes and bio-based chemicals.

(306g) Dynamic Viscoelasticity As a Rapid Single-Cell Biomarker

Sawetzki, T., Colorado School of Mines
Roth, K. B., Colorado School of Mines
Kasukurti, A., Colorado School of Mines
Eggleton, C., University of Maryland Baltimore County
Desai, S. A., National Institute of Allergy and Infectious Disease
Marr, D. W. M., Colorado School Of Mines
Originally presented on: 11/5/2013 14:18:00 - 14:36:00

Probing the mechanics of living cells has been successfully identified as a label-free biophysical property marking variations in the health condition of individual cells in a broad range of different biological systems. The exploitation of this promising biomarker in population-based studies, however, has been limited by the low measurement throughput of available methods. Besides the technical challenges of a cell-mechanics based cytometry, a fundamental hurdle is the signal response time of biological samples to external loading. Where the fluorescence signatures of common cell cytometers are virtually instantaneous, the cell stress relaxation occurs on the order of ~0.1 seconds, therefore greatly limiting the potential speed of current elastic property measurements.

To circumvent these limitations, we introduce a novel optical-based technique probing the cell response to dynamic optical force fields modulated at high frequencies, thus providing deeper insights into cell mechanics than common cell stretching approaches. Individual cell in microfluidic environment are repeatedly deformed by the anisotropic distribution of laser light, generating linear optical traps aligned with the streamlines of the flowing carrier liquid. Studying the viscous contribution to the cell relaxation mechanism provides an additional criterion that allows an extended classification of cellular systems at timescales where pure elasticity-based methods fail. As a well-explored model system, we investigate the viscoelastic properties of malaria-infected and uninfected red blood cells, demonstrating the destruction-free detectability of both populations at timescales at the order of 10 ms, far beyond the typical stress relaxation times of classic approaches. With our current configuration we realized throughputs of more than 20 cells/s and a total of ~750 cells/min, so far mainly limited by the rapidness of our data-intensive video-based detection.

(306h) Computational Recovery of Chromatin Conformations From Experimental Contact Probabilities

Meluzzi, D., University of California, San Diego
Arya, G., University of California, San Diego
Originally presented on: 11/5/2013 14:36:00 - 14:54:00

Chromatin is a nanoscale fiber resulting from the self-assembly of DNA and proteins within the nucleus of eukaryotic cells. The conformation of this fiber is believed to play a role in the regulation of essential cellular processes, such as DNA transcription, replication, recombination, and repair. However, the spatial organization of chromatin and its functional implications are not yet well understood. Here we present computational methods that can be used to determine chromatin conformations by analyzing DNA sequences obtained from high-throughput chromosome conformation capture (Hi-C) experiments. We use the DNA sequences to infer the probability of contact between different parts of chromatin in vivo. We then process these contact probabilities to recover possible ensembles of chromatin conformation, which can be further analyzed to correlate spatial properties with gene activity. We validate our computational methods against simulations of both Hi-C experiments and a coarse-grained polymer model representing the 30-nm chromatin fiber. We also compare computational predictions to results of fluorescence *in situ* hybridization (FISH) experiments. Our methods provide a means to deduce and compare the spatial organization of chromatin in different cell types, thus elucidating the interplay between chromatin organization and cellular processes.

(310a) Chimeric Receptor Engineering to Control Cell Fate and Its Applications

Nagamune, T., School of Engineering, The University of Tokyo
Kawahara, M.,
Originally presented on: 11/5/2013 12:30:00 - 12:55:00

When cells respond to changes in an extracellular milieu, membrane proteins expressed on the plasma membrane play a pivotal role as an interface between extracellular and intracellular milieus. If we could engineer membrane proteins, we could artificially regulate various cellular responses [1, 2]. Therefore, we have focused on cytokine receptors that are one of the membrane proteins involved in cell-fate signal transduction, and developed novel artificial receptors that enable signal transduction in response to a ligand quite different from original cytokines. In order to regulate various cellular responses with distinct ligands, an antigen-antibody system, having infinite combinations, was utilized as the ligand and its binding domain.

Until now, we have designed single-chain Fv (scFv)/cytokine receptor chimeras having different kinds of signaling domains, thereby regulating growth [3], migration [4] and death [5] of various mammalian cells including primary cells in response to a specific antigen. In addition, we also developed an antibody selection system based on growth induction of mammalian cells. In this system, a scFv library/growth-inducing receptor chimera is expressed in cells, and the cells expressing antigen-specific scFv can be easily selected based on the growth advantage in the presence of antigen. This panning-less, one-step selection system allowed us to select specific scFvs against not only extracellularly added antigen but also intracellularly expressed antigen. Our chimeric receptor approach, which provides cell-fate controlling and screening platforms, could be useful for regenerative medicine and antibody therapeutics.

References

[1] Kawahara, M., Ueda, H., and Nagamune, T. "Engineering cytokine receptors to control cellular functions." *Biochem. Eng. J.*, 2010, **48**, 283–294.

[2] Kawahara, M., and Nagamune, T. "Engineering of mammalian cell membrane proteins." *Curr. Opin. Chem. Eng.* 2012, **1**, 411-417.

[3] Kawahara, M., Chen, J., Sogo, T., Teng, J., Otsu, M., Onodera, M., Nakauchi, H., Ueda, H. and Nagamune, T. "Growth promotion of genetically modified hematopoietic progenitors using an antibody/c-Mpl chimera." *Cytokine*, 2011, **55**, 402-408.

[4] Kawahara, M., Shimo, Y., Sogo, T., Hitomi, A., Ueda, H. and Nagamune, T. "Antigen-mediated migration of murine pro-B Ba/F3 cells via an antibody/receptor chimera." *J. Biotechnol.*, 2008, **133**, 154-161.

[5] Tone, Y., Kawahara, M., Kawaguchi, D., Ueda, H., and Nagamune, T. "Death signalobody: Inducing conditional cell death in response to a specific antigen." *Hum. Gene Ther. Methods*, 2013, doi: 10.1089/hgtb.2012.147.

(310b) Nanobiocatalysis and Its Potential Applications

Kim, J., Korea University

Originally presented on: 11/5/2013 12:55:00 - 13:20:00

Enzyme reactions with their unique selectivity can be employed in various applications, but the short lifetimes of enzymes often inhibits their practical uses. Nanobiocatalysis, using nanostructured materials for enzyme immobilization, has gathered growing attention due to its recent successes in stabilizing the enzyme activity. This presentation will discuss recent developments of nanobiocatalysis to improve the enzyme stability by using crosslinked enzymes in mesoporous media and on nanofibers/nanotubes/nanoparticles. The approach of nanometer-scale enzyme reactors (NERs) stabilized enzymes in mesoporous media via a ship-in-a-bottle approach, which consists of enzyme adsorption and crosslinking. A similar approach resulted in the enzyme coating on the surface of electrospun nanofibers, which also stabilized the enzyme activity in a vivid way. Nanobiocatalytic systems with stabilized enzyme activity have been employed in various applications, including biofuel cells, antifouling, and carbon capture. Several successful examples of nanobiocatalytic applications will be introduced in details.

(310c) Development of Automated Cell Processing System for Clinical Application

Kino-oka, M., Osaka University

Originally presented on: 11/5/2013 13:20:00 - 13:45:00

In the last decade, cell and tissue therapies have encompassed a broad, rapidly growing field of medicine that involves the manipulation and administration of cells for the treatment of disease. Especially, the advances in tissue engineering have offered promising strategies for reconstructing and repairing defective tissues in vivo, enabling damaged tissue to be replaced with cultured tissues that meet the needs of the individual patients.

In the present study, the comparison of management between cell processing facilities with safety cabinet and with the isolator system reveals that isolator technology leads to reductions of the running cost as well as operational laboriousness in the small production, compared to the conventional clean room facility with safety cabinet. In addition, a novel design of manufacturing facility based on a flexible Modular Platform (fMP) has been proposed to realize that the individual modules can connect and disconnect flexibly with keeping the aseptic environment as well as biological containment in each module, leading to the compactness of aseptic processing area and quick change-over for multi-purposes and patients. This technology will make cost-saving with safety and security.

(310d) Olfactory and Taste Biosensor Using Human Receptors Combined With Nanodevices

Park, T. H., Seoul National University

Originally presented on: 11/5/2013 13:45:00 - 14:10:00

We have five senses, which are the sense of vision, sound, touch, smell, and taste. Among these five senses, the sense of smell and taste is the chemical sense, while the other three senses are physical sense. Regarding the sense of vision and sound, human developed the devices which can measure their activating stimuli, which are light and vibration. However, we do not have any device which can measure the smell and taste. We still depend on the sensory evaluation using our own nose and tongue. In the smelling process of the human olfactory system, the initiation step in the odor recognition is the binding of odorants to the olfactory receptor proteins. Similarly in the tasting process for the bitter, sweet and umami tastes, the initial recognition step is also the binding of tastants to the taste receptor protein. Therefore, human's olfactory and taste receptors can be used to develop the human-like bioelectronics nose and tongue. In this presentation, I am going to present a successful demonstration of olfactory and taste biosensor using human receptors combined with nanodevices.

(310e) Development of Microbial Cell Factories for Sustainable Biorefineries

Kondo, A., Graduate School of Engineering, Kobe University

Originally presented on: 11/5/2013 14:10:00 - 14:35:00

This presentation will provide an overview of recent development of microbial cell factories for sustainable biorefineries in Japan and worldwide. This presentation will provide an overview of recent development of microbial cell factories for sustainable biorefineries in Japan and worldwide. This presentation will provide an overview of recent development of microbial cell factories for sustainable biorefineries in Japan and worldwide.

(310f) Biomimetic Protein-Based Adhesive Biomaterials for Tissue and Biomedical Engineering Applications

Cha, H. J., Pohang University of Science and Technology
Originally presented on: 11/5/2013 14:35:00 - 15:00:00

Marine mussel adhesion is known to be mediated by mussel adhesive proteins (MAPs), which are secreted through the mussel byssus and have great potential as biologically and environmentally friendly adhesive biomaterials due to their biocompatibility and biodegradability. In addition, MAPs have strong adhesion ability even on wet surfaces due to unique amino acid arrangements and composition. However, researches using the natural amino acid composition of MAPs have been limited due to difficulties in obtaining sufficient quantities of MAPs for practical applications. We successfully and massively produced genetically redesigned new fusion MAP using a bacterial expression system, and this fusion protein showed significant adhesion ability. In addition, using the superior cell adhesion properties and fusion technology of the bioactive extracellular matrix (ECM) peptides, we discovered the great potential of MAPs as bio-functional smart biomaterials for tissue and biomedical engineering applications. We will also demonstrate the multi-functional roles of MAPs as coating materials for bone regeneration as well as components of nanofibrous scaffolds.

(326a) Modular Assembly of Designer Puf Proteins for Specific Post-Transcriptional Regulation of Endogenous RNA

Abil, Z., University of Illinois at Urbana-Champaign
Zhao, H., University of Illinois at Urbana-Champaign
Originally presented on: 11/5/2013 12:30:00 - 12:48:00

Due to their modular structure, RNA-binding proteins (RBP) are an attractive platform for the development of tools needed for the study of eukaryotic transcriptomes as well as the manipulation of RNA for therapeutic purposes. In this work, we utilize the Pumilio and FBF (PUF) homology domain as a scaffold for engineering RNA substrate specificity and functionalize it with a repressor domain for post-transcriptional down-regulation of endogenous mRNA.

PUF proteins are eukaryotic RBPs that are typically involved in translational regulation of developmental genes. The RNA-binding domain of PUF proteins ususlly contains 8 structural repeats, each containing 36 amino acids, two of which contribute to the specificity of the repeat to a corresponding nucleotide in the target RNA sequence. With the elucidation of the "code" for RNA recognition by PUF proteins, it was shown that tailored PUF proteins with predicted specificity could be engineered. In the past several years, engineered PUF proteins with designed specificities were successfully implemented for applications like enhancing or repressing translation, modulating splicing, and imaging of endogenous RNA.

One of the limitations to engineering PUF domains with novel specificities is the lack of a cloning platform capable of rapid and efficient introduction of multiple mutations in separate modules simultaneously. In this study, we report the implementation of "Golden Gate" cloning, a modular, type IIS restriction endonuclease-based cloning approach for engineering PUF-based RBPs. We have created a library in which all 8 structural repeats of human Pumilio 1 homology domain are inserted in separate plasmids, and each of them has 3 more variations for the recognition of a certain nucleotide, according to the published "code". Thus, our library of PUF repeat modules is potentially capable of generating a PUF protein with a specificity for any RNA sequence of 8nt, given that it can be solubly expressed.

In order to demonstrate the efficiency of this approach, we first used this assembly method for rapid construction of several mutant PUF domains with novel specificities and assayed their binding affinities. By replacing as many as 8/8 repeats using the assembly strategy, we were able to obtain a PUF mutant that binds specifically to its corresponding RNA with a K_D in a low nanomolar range. Second, we linked the RNA-binding activity of engineered PUF domains to a repressor activity of tristetraprolin (TTP), a key player in AU-rich element (ARE)-mediated mRNA decay, and used the protein fusion in a functional reporter system to assay the PUF domain activity in HeLa cells. Using a firefly luciferase reporter assay, we observed that TTP-PUF fusion represses gene expression in a sequence-specific and modular manner. Finally, we demonstrated the application of the engineered TTP-PUF fusion proteins in translational knock-down of human vascular endothelial growth factor A protein in HEK293 cell line.

This one-step modular assembly approach for PUF domain engineering would allow greater flexibility and speed at creating PUF proteins with user-defined specificities and thus facilitate the use of PUF-based designer RBPs as a tool in research and therapeutics.

(326b) Development of a Universal Protein Switch Platform Using Antibody Mimetic Proteins

Nicholes, N., Johns Hopkins University
Date, A., Johns Hopkins University
Kanwar, M., Johns Hopkins University
Beaujean, P., Johns Hopkins University
Ostermeier, M., Johns Hopkins University
Originally presented on: 11/5/2013 12:48:00 - 13:06:00

Ligand-activated enzymes (protein switches) can be constructed by the correct fusion of ligand-binding domains and enzymes. For example, maltose-activated beta-lactamases have been constructed by the insertion of beta-lactamase (BLA) into the maltose binding protein (MBP). In these protein switches, BLA catalytic activity is a function of maltose concentration. Creating new protein switches that are activated by different ligands requires the availability of the prerequisite ligand-binding domains (the input domain of the switch). The process of developing a new switch is also time consuming and labor intensive. To accelerate our ability to construct switches that are activated by new protein and small molecule targets, we have developed protein switches using monobody and DARPin antibody mimetics as input domains. First, MBP-activated beta-lactamases were created by a directed evolution approach involving the fusion of BLA with the MBP-binding DARPin off7 and the fusion of BLA with the MBP-binding monobody MBP-74. We identified switches that function both *in vitro* and in *E. coli* cells. Next, these MBP-activated switches were converted for activation by new protein targets simply by transplanting mutations known to cause binding to new targets in the antibody mimetic domain alone. These results outline a general strategy for the more rapid creation of protein switches.

(326c) Evolution of Novel Amine Dehydrogenase Properties

Bommarius, B., Georgia Institute of Technology
Au, S., University of Kentucky
Bommarius, A. S., Georgia Institute of Technology
Originally presented on: 11/5/2013 13:06:00 - 13:24:00

The reductive amination of ketones to produce chiral amines is an important transformation in the production of pharmaceutical intermediates. Recently, we obtained the first amine dehydrogenases (L-AmDH) from a leucine dehydrogenase [1] and from phenylalanine dehydrogenase (F-AmDH).[2] Here, we report on successful domain shuffling efforts to create further variants of amine DH with altered substrate specificity. As an example, we now can accommodate sterically very demanding side chains, such as adamantly groups, in the ketone substrate, which did not react at all beforehand. Next, we will elucidate the role of restricted libraries towards higher activity and broader specificity.

We also report on progress in processing AmDHs and in fusing them to cofactor regeneration enzymes. On large-scale, IMAC purification with his-tagged proteins is problematic. We have developed a fractionated precipitation, two-phase extraction purification protocol for the untagged AmDHs. In parallel, we have developed a fusion protein of AmDH with formate dehydrogenase (FDH) to regenerate NADH and report on the status.

References:

[1] M.J. Abrahamson, E. Vazquez-Figueroa, N.B. Woodall, J.C. Moore, A.S. Bommarius, Development of a Novel Amine Dehydrogenase for Synthesis of Chiral Amines", *Angew. Chem. Intl. Ed.* **2012**, *51*, 3969-3972

[2] M.J. Abrahamson, J.W. Wong, and A.S. Bommarius, "The rapid evolution of an amine dehydrogenase biocatalyst for the asymmetric production of chiral amines", *Adv. Synth. Catal.* **2013,** 000 (in press)

(326d) Stabilization of Enzymes in Ionic Liquids Via Modification of Enzyme Charge

Norwald, E.,
Kaar, J. L., University of Colorado Boulder
Originally presented on: 11/5/2013 13:24:00 - 13:42:00

Due to the propensity of ionic liquids to inactivate enzymes, the development of strategies to improve enzyme utility in these solvents is critical to fully exploit ionic liquids (ILs) for biocatalysis. We have developed a strategy to broadly improve enzyme utility in ILs based on elucidating the effect of charge modifications on the function of enzymes in IL environments. Results of stability studies in aqueous-IL mixtures indicated a clear connection between the ratio of enzyme-containing positive-to-negative sites and enzyme stability in ILs. Stability studies of the effect of [BMIM][Cl] and [EMIM][$EtSO_4$] on chymotrypsin specifically found an optimum ratio of positively-charged amine-to-negatively-charged acid groups (0.39). At this ratio, the half-life of chymotrypsin was increased 1.6 and 4.3 fold relative to wild-type chymotrypsin, in [BMIM][Cl] and [EMIM][$EtSO_4$], respectively. The half-lives of lipase and papain were similarly increased as much as 4.0 and 2.4 fold, respectively, in [BMIM][Cl] by modifying the ratio of positive-to-negative sites of each enzyme. More generally, the results of stability studies found that modifications that reduce the ratio of enzyme-containing positive-to-negative sites improve enzyme stability in ILs. Additionally, we have also investigated the underlying mechanism for the connection between stability and surface charge ratio by measuring IL binding as a function of the charge ratio. Understanding the impact of charge modification on enzyme stability in ILs may ultimately be exploited to rationally engineer enzymes for improved function in IL environments, including cellulases for improved biomass conversion in ILs.

(326e) Analysis of the Roles of Mutations in Nuclear Hormone Receptors in a Bacterial Biosensor System

Shi, C., Ohio State University
Shakalli Tang, M., The Ohio State University
Wood, D. W., the Ohio State University
Originally presented on: 11/5/2013 13:42:00 - 14:00:00

Mutations in nuclear hormone receptors (NHRs) usually lead to metabolic disorders, but are difficult to treat with existing drugs due to the aberrant behavior of the receptor. In our previous work, we created several hormone biosensors based on bacterial growth phenotypes, which allow hormone detection in both human and animal hormone receptors. In this study, we introduced six clinically relevant human mutations into our thyroid hormone receptor-β and peroxisome proliferator-activated receptor-γ sensor. For each group of mutated biosensors, several native and mimic ligands were tested. Compared to their respective wild-type native sensor, we found that these six mutations have varying effects on ligand binding. More specifically, some mutations lead to partial loss of ligand binding ability, while some fully disable ligand binding ability. These mutant biosensors can be thus used to efficiently screen for effective drug candidates to treat human patients that harbor these rare mutations.

(326f) A High-Throughput in Vitro Photosynthetic Hydrogen Production Screen

Shiigi, S., Stanford University
Koo, J., Stanford University
Swartz, J. R., Stanford University
Originally presented on: 11/5/2013 14:00:00 - 14:18:00

Coupling photosynthetic water-splitting with the production of hydrogen (H_2) is an attractive method to produce this important commodity chemical and potential, next-generation fuel. Photobiological H_2 production requires integration of hydrogen-producing enzymes, hydrogenases, into native photosynthetic electron transport chains. However, heterologous hydrogenases are unstable in the presence of oxygen, an unavoidable byproduct of photosynthesis, and inefficiently couple with the native photosynthetic machinery of organisms like cyanobacteria. Engineering hydrogenases for increased H_2 production is limited in part by a lack of robust, high-throughput screening methods. *In vitro* expression and screening of enzyme mutants would enable more rapid testing of larger library sizes than *in vivo* systems, in part due to challenges in genetic engineering of photosynthetic organisms.

Here, we will discuss the development and optimization of an *in vitro* photosynthetic H_2-production pathway using photosystems isolated from a cyanobacterium. Additionally we will describe the use of a high-throughput assay for assessment of H_2 production rates of hydrogenases coupled to the isolated photosystems in microtiter plates. This approach can be extended to the high-throughput screening of *in vitro* synthesized enzyme libraries for increased photobiological H_2 production rates. Additionally, this work will provide a powerful new assay to interrogate photosynthesis *in vitro* allowing fine-tuned control over the reaction environment by modulating enzyme and reagent concentrations.

(326g) Type III Secretion of Silk, Elastin and Resilin-Based Hybrid Biomaterials

Azam, A., University of California Berkeley
Tullman-Ercek, D., University of California, Berkeley

Originally presented on: 11/5/2013 14:18:00 - 14:36:00

Despite their integral role in the development of load-bearing materials for tissue culture and biomedical applications, recombinant expression of biopolymers remains challenging due to their inherent insolubility in cell cytoplasm. Monomer expression and purification is difficult because this class of proteins is characterized by repetitive, glycine-rich, lengthy hydrophobic domains that are prone to aggregation and proteolytic degradation *in vivo*. Engineering the type III bacterial secretion pathway enables a new strategy for customizable design, expression and export of pure biopolymer-forming proteins. Using this approach we are able to achieve high-titer recombinant production of pure silk, elastin and resilin monomers that can be cross-linked post-secretion for the fabrication of materials with a variety of tunable mechanical and chemical properties. We utilize a recursive cloning strategy to sequentially incorporate different functional modules into a genetic sequence encoding silk monomers, tropo-elastin and pro-resilin; we then transform the plasmid into *Salmonella typhimurium* and induce high-titer expression and extracellular secretion (10-100 mg/L) of the protein. The pure monomers are then cross-linked via chemical or enzymatic methods to produce insoluble, mechanically stable hybrid biomaterials. Due to genetic-level control over protein character, we can incorporate chemical moieties for conjugation with functionalized hydrogel-forming synthetic polymers and antibacterial peptide motifs for advanced functionality. With this talk, we will detail the synthesis and characterization of these biopolymers, and discuss the key advantages of our microbial production system.

(326h) Chemoenzymatic Labeling of Recombinant Proteins Enables Rapid In-Lysate Bioconjugation and Assay Development

Kulkarni, C., California Institute of Technology
Tirrell, D. A., California Institute of Technology
Kinzer-Ursem, T. L., California Institute of Technology
Originally presented on: 11/5/2013 14:36:00 - 14:54:00

The selective conjugation of proteins to reactive partners is central to a diverse array of engineering applications, including the development of biosensors, self-assembled materials, and therapeutics. Therefore, methods that enable controlled, site-specific incorporation of bioorthogonal reactive groups into proteins are highly useful. Bioconjugation often requires extensive effort to first purify a protein of interest, and purification protocols may expose the protein to conditions detrimental to its structure and function. Thus, the ability to perform in situ bioconjugation in a protein's native environment is advantageous. Here, we demonstrate co-translational protein labeling with an azide-bearing fatty acid by employing N-myristoyl transferase (NMT), a eukaryotic enzyme that appends myristic acid (Myr) to the N-terminus of its substrate proteins and that exhibits no activity toward bacterial proteins.

NMT-mediated protein labeling is quantitative and site-specific and can be engineered so that protein function is not inhibited by the addition of Myr-analogs. Here we report the labeling of proteins that are native NMT targets as well as proteins that are engineered to display NMT recognition sequences. In both instances, we find that addition of an azide-bearing fatty acid tag (12 azidododecanoic acid, 12-ADA) does not interfere with the proteins' native function. Furthermore, we demonstrate bioconjugation of 12-ADA-labeled proteins to surfaces directly from soluble cell extracts via strain-promoted azide-alkyne cycloaddition. We find that the cycloaddition reaction rapidly and specifically conjugates 12-ADA-labeled proteins to microarrays in a concentration-dependent manner. Additionally, we conjugate 12-ADA labeled proteins to surface plasmon resonance (SPR) chips and measure the kinetics of binding of partner proteins. Of particular significance is the demonstrated ability to immobilize tagged proteins directly from cell lysate without prior protein purification, thus enabling the measurement of protein activity and binding dynamics with fast, reproducible assay platforms and significantly improving the throughput and quantitative characterization of protein function.

(378a) Towards A Genetic Basis for Antibiotic Resistance in Chromobacterium Violaceum

Das, D.,
Raghunathan, A., National Chemical Laboratory
Originally presented on: 11/5/2013 15:15:00 - 15:33:00

The emergence of mutations in the genome is one of the major factors underlying evolution. Mutation as the cause of antibiotic resistance was studied in *Chromobacterium violaceum* for the antibiotics chloramphenicol and streptomycin. The wild-type *C. violaceum* was tested to be susceptible to low concentrations of both antibiotics (Chloramphenicol: MIC 8ug) (Streptomycin: MIC 10ug). Continuous exposure to the antibiotics at concentrations slightly lower than the MIC, resulted in resistance to the antibiotics. This adaptive evolutionary phenomenon allowed growth on agar plates at high antibiotic concentrations (MIC: 200ug). The growth rates of these evolved strains 2-30 fold higher at varying concentrations of antibiotic. All genetic changes were delineated during the evolution of antibiotic resistance using next generation sequencing (Ion Torrent^{TM}). This talk will focus on the results of NGS and the functional impact of the genetic variations detected including AcrAB drug efflux pumps and ribosomal proteins like rpsL.

(378b) The Small Rnome of Clostridium Acetobutylicum That Responds to Butanol and Butyrate Stress

Venkataramanan, K. P., University of Delaware
Jones, S., University of Delaware
McCormick, K. P., University of Delaware
Kunjeti, S. G., University of Delaware
Ralston, M., University of Delaware
Meyers, B. C., University of Delaware
Papoutsakis, E. T., University of Delaware
Originally presented on: 11/5/2013 15:33:00 - 15:51:00

In the genus *Clostridium*, which is of major importance to human and animal pathogenesis, human physiology, the carbon cycle, biorefinery and biotechnological application, very few regulatory non-coding small RNA (sRNA) have been so far identified. Significantly, no sRNAs have been reported as being involved in toxic metabolite stress or any stress. In *C. acetobutylicum*, we have previously predicted 113 sRNAs and have experimentally validated a subset of 31 of them. Here, using deep RNA sequencing (RNA-seq), we investigated the expression of the small RNome of *C. acetobutylicum* in response to two important toxic metabolites, butanol and butyrate. Butanol is the major metabolite of the solventogenic stationary phase while butyrate is the major product of the acidogenic growth phase. The goal of this study is to identify sRNAs that respond to butanol and/or butyrate stress, aiming to identify sRNAs that can be engaged to engineer strains tolerant to these toxic metabolites, but significantly also discover if and how sRNAs respond to metabolic stresses.

We collected a large set of temporal RNA-seq data from multiple experiments with biological replicates employing various levels of metabolite stress. We mined the RNA-seq data for robustly identifying the presence and differential expression of sRNAs by relying on the analysis, based on the RNA-seq data, of the previously predicted 113 and the validated subset of 31 sRNAS. This analysis enabled us to set metrics (read counts) for reliably identifying the presence of new sRNAs from RNA-seq data. Based on these metrics, we identified an additional set of 46 sRNAs not previously known or predicted for this organism. We then examined the differential expression under metabolite stress of all identified sRNAs and verified the presence of a select subset by Northern analysis. Based on the conservation of the putative Hfq binding module, we identified the subset of these sRNAs that are likely to be regulated by the well-conserved RNA chaperone Hfq. Finally, we examined the likely role of a few of these stress-related sRNAs by relating their expression pattern to identifiable physiological characteristics.

This is the first comprehensive study of the stress-related sRNome in a *Clostridium* organism. Elucidation of the role of sRNA in clostridial response to metabolite stress will be an essential component for understanding the complexity of the regulatory network that underlies these stress responses.

(378c) Discovery and Regulation of Biomass-Degrading Enzymes From Anaerobic Gut Fungi Using Next-Generation Sequencing

Solomon, K., University of California, Santa Barbara
Haitjema, C., University of California, Santa Barbara
Borges-Rivera, D., Broad Institute of Massachusetts Institute of Technology and Harvard
Thompson, D. A., Broad Institute of Massachusetts Institute of Technology and Harvard
Regev, A., Broad Institute of Massachusetts Institute of Technology and Harvard
O'Malley, M. A., University of California, Santa Barbara
Originally presented on: 11/5/2013 15:51:00 - 16:09:00

Lignocellulosic biomass is a renewable resource capable of supplying one third of the domestic liquid fuel demand. However, processes to extract fermentable sugars from this biomass involve expensive pretreatment and enzymatic digestion steps that rely on a limited number of characterized cellulases. Therefore, there is a critical need to identify new technologies for the economic conversion of crude biomass into sugars. Anaerobic fungi within the digestive tracts of large herbivores are powerful biomass degraders yet they remain a rich untapped source for new lignocellulolytic enzymes. Despite their lignocellulolytic behavior, the enzymes responsible are largely unknown due to limitations associated with the cultivation and molecular characterization of gut fungi. In this study, we sought to identify these enzymes through RNAseq, and capture their expression profiles to understand the hydrolytic behavior of gut fungi against lignocellulosic substrates.

We have isolated a novel gut fungus from the *Piromyces* genus, which we have employed as a model organism to study biomass degradation. The isolate exhibits strong cellulolytic activity against diverse cellulosic and lignocellulosic substrates, which is repressed in the presence of simple sugars (e.g. glucose). Using next generation strand specific RNAseq and the TRINITY bioinformatic platform, we have assembled the first reported *de novo* gut fungal transcriptome. Analysis of the transcriptome reveals transcripts encoding over a hundred novel lignocelluloytic enzymes with diverse catalytic activities, which are either freely secreted or associated with fungal cellulosome complexes. Furthermore, we have confirmed the biological presence of enzymes encoded by assembled transcripts, supporting the quality of the *de novo* assembly. Also embedded within the transcriptome are quantitative transcript abundance information and non-coding RNA such as antisense RNA, which we hypothesize to directly regulate the expression of these cellulases. In this talk, I will describe the subset of novel enzymes that we have identified, and discuss the global regulatory patterns observed under various growth conditions.

(378d) Microdevices and Materials for Genomic Analysis of Single Microbial Cells

Meagher, R. J., Sandia National Laboratories
Light, Y., Sandia National Laboratories
Yilmaz, S., Sandia National Laboratories
Rhee, M., University of Michigan
Singh, A. K., Sandia National Laboratories
Liu, P., Sandia National Laboratories
Originally presented on: 11/5/2013 16:09:00 - 16:27:00

Next-generation sequencing technologies have made whole-genome sequencing of microbes routine, as long as the microbe can be grown in pure culture to provide sufficient quantities of clonal DNA for library preparation. For microbes that are not cultured, which includes the vast majority of taxa both in the environment and within the human body, other techniques must be used to analyze microbial genomes. Metagenomics is useful for analyzing genes and pathways at the level of microbial communities, but assembly of complete microbial genomes from a mixed population is difficult except for the simplest communities. We have developed a suite of technologies that complement metagenomic techniques by analyzing the genomic content of single cells. We begin with complex microbial communities, isolate individual cells, and then amplify to prepare sequencing libraries. Key aspects of the technology include: (1) compartmentalization of single cells in nanoliter-volume droplets, as ideal "nanoreactors" for amplifying the genome of a single microbial cell, (2) multiple displacement amplification (MDA) using Phi29 DNA polymerase, with special consideration to purity requirements when amplifying single cells, and (3) labeling and sorting techniques based on fluorescence in situ hybridization (FISH), either before or after amplification, to identify interesting or novel microbes based on phylogeny (i.e. 16S rRNA sequence) or presence of functional genes. Our pipeline utilizes a combination of conventional instrumentation (FACS) and novel microfluidics, with the ultimate goal of producing sequencing libraries from single-amplified genomes (SAGs), with increased ability to target rare or novel microbes relative to "blind" sorting techniques.

(378e) High-throughput, Comprehensive Analysis of Eukaryotic Translation Initiation Motifs by FACS-Seq

Noderer, W., Stanford University
Wang, C., Stanford University
Originally presented on: 11/5/2013 16:27:00 - 16:45:00

The probability that a scanning ribosome will stop at a start codon and initiate translation is a function of the sequence surrounding the AUG. Unlike bacterial translation initiation, which uses well defined Watson-Crick base paring between the ribosome and the ribosome binding site, eukaryotic translation initiation is not well understood. Marilyn Kozak's consensus sequence provides helpful rules of thumb but is not quantitative. We have revisited this problem using a high-throughput, quantitative approach named FACS-seq which couples fluorescence activated cell sorting (FACS) with next-generation sequencing. Our mutational library consisted of random permutations of the 6 bases preceding and 2 bases following a start codon driving the expression of a fluorescence reporter. In agreement to Kozak's results, we find the -3 position to be the largest determinant of translation initiation. Contrary to Kozak's results, we find the +4 to be of little importance and instead find the two bases neighboring the -3 position to play a major role. A translation initiation prediction model incorporating the FACS-seq data was found to be in agreement with ribosomal footprinting data and was able to predict translation of upstream open reading frames. In principle, the FACS-seq method described here could be applied to any mutational library that can be paired with a fluorescence reporter.

(378f) Engineering Orthogonal Protein-Nucleic Acid Pairs for Cellular Circuits

Mansell, T. J., Johns Hopkins University
Lynch, S. A., University of Colorado at Boulder
Boyle, N. R., University of Colorado Boulder
Gill, R. T., University of Colorado Boulder
Originally presented on: 11/5/2013 16:45:00 - 17:03:00

Cellular circuits in synthetic biology often depend on inducible regulators like the *lac* and *tet* repressors. However, increased circuit complexity is currently limited by the availability of orthogonal regulators. We present a method for generation of directed libraries of regulators for orthogonal binding of DNA and RNA sequences in *E. coli*. Our unique selection-counterselection strategy allows for selection of repressors that do bind a target (i.e., novel) sequence but do not bind a pre-determined sequence of interest (e.g., wild type). We leverage the technologies of multiplex oligo synthesis and next-generation high-throughput sequencing to create libraries and glean sequence-activity relationship data from selections to iterate evolution in a rational, directed way.

(378g) Designing Synthetic Genes for Recombinant Expression in Synthetic Biotechnology

Chung, B., Bioprocessing Technology Institute (A*STAR)
Lee, D. Y., National University of Singapore

Originally presented on: 11/5/2013 17:03:00 - 17:21:00

Abstract

System-wide data, such as genomics, transcriptomics, proteomics and metabolomics, generated by high-throughput experimental techniques have fueled the rapid advancement in systems biology research over the past decade. This wealth of information generated in systems biology has equipped us with the knowledge to challenge the new frontier of synthetic biology. Hence, in this study, we demonstrate a methodology for employing omics data to computationally design synthetic genes with enhanced recombinant expression capabilities. In the biotechnology industry, recombinant DNA technology is commonly used to achieve high productivity of certain desired biochemical in host organisms such as *Escherichia coli*, *Saccharomyces cerevisiae*, *Pichia pastoris* and Chinese hamster ovary (CHO) cells. Accordingly, high heterologous expression levels of the recombinant genes in these host organisms are required. Nonetheless, these foreign genes are typically poorly expressed because they have not been evolved for efficient expression in the selected host organism. Consequently, molecular biology techniques have to be employed to either genetically engineer the host organism or modify the recombinant genes for enhanced heterologous protein expression. However, since the former approach is more challenging to implement and may adversely affect the host cells' physiology, the latter approach of redesigning the recombinant genes to generate synthetic genes is usually adopted. In this aspect, the issue of codon usage bias has been well-known to be a key factor affecting recombinant protein expressivity, more specifically at the step of translating mRNA to peptide chain (Gustafsson et al., 2004). Several codon optimization software tools, such as Gene Designer (Villalobos et al., 2006), JCat (Grote et al., 2005) and OPTIMIZER (Fuglsang, 2003), have been developed to generate synthetic genes with improved codon usage bias. Most of these earlier algorithms optimize the genetic sequence based on individual codon usage (ICU). Yet, the pairing of codons, sometimes referred to as codon context (CC), is also known to significantly affect mRNA translation efficiency (Hatfield and Roth, 2007; Coleman et al., 2008). Thus, it is relevant to incorporate both ICU and CC as design parameters under a multi-objective optimization framework to examine the relevant importance of these two parameters. Towards this end, we have developed a computational tool based on the Non-dominated Sorting Genetic Algorithm II (NSGA-II) to optimize the synthetic gene design with respect to both ICU and CC (Chung and Lee, 2012). This tool has been applied to optimize recombinant genes for heterologous expression in *P. pastoris* (Ahn et al., 2013) and CHO cells (Chung et al., 2013). The *in vivo*expression of the optimized synthetic genes in these two expression hosts has consistently revealed that CC-optimized genes were capable of higher expression levels than the ICU-optimized ones. This suggests that CC may be a more important design criterion than the widely considered ICU. Henceforth, the incorporation of the CC parameter into the codon optimization algorithm is expected to improve the efficacy of existing tools. In addition, the developed multi-objective codon optimization framework based on NSGA-II will be a good implementation strategy for future applications since it can be easily expanded to include more design parameters.

Acknowledgements

This work was supported by the National University of Singapore, Biomedical Research Council of A*STAR (Agency for Science, Technology and Research), Singapore, and a grant from the Next-Generation BioGreen 21 Program (SSAC, No. PJ009520), Rural Development Administration, Republic of Korea.

References

Ahn, J., Jang, M., Chung, B.K., Lee, D.Y., Lee, H., Jung, J.K., Choi, E.S., (2013) Comparison of relative importance of codon usage and codon-pair context as gene design parameters for improving protein expression in Pichia pastoris. *Submitted.*

Chung, B.K., Lee, D.Y., (2012) Computational codon optimization of synthetic gene for protein expression. *BMC Syst Biol*6, 134.

Chung, B.K., Yusufi, F.N.K., Mariati, Yang, Y., Lee, D.Y., (2013) Enhanced expression of codon optimized interferon gamma in CHO cells. *Submitted.*

Coleman, J.R., Papamichail, D., Skiena, S., Futcher, B., Wimmer, E., Mueller, S., (2008) Virus attenuation by genome-scale changes in codon pair bias. *Science*320, 1784-1787.

Fuglsang, A., (2003) Codon optimizer: a freeware tool for codon optimization. *Protein Expr Purif*31, 247-249.

Grote, A., Hiller, K., Scheer, M., Munch, R., Nortemann, B., Hempel, D.C., Jahn, D., (2005) JCat: a novel tool to adapt codon usage of a target gene to its potential expression host. *Nucleic Acids Res*33, W526-531.

Gustafsson, C., Govindarajan, S., Minshull, J., (2004) Codon bias and heterologous protein expression. *Trends Biotechnol*22, 346-353.

Hatfield, G.W., Roth, D.A., (2007) Optimizing scaleup yield for protein production: Computationally Optimized DNA Assembly (CODA) and Translation Engineering. *Biotechnol Annu Rev*13, 27-42.

Villalobos, A., Ness, J.E., Gustafsson, C., Minshull, J., Govindarajan, S., (2006) Gene Designer: a synthetic biology tool for constructing artificial DNA segments. *BMC Bioinformatics* 7, 285.

(378h) An Integrated Pipeline for Organism Engineering
Canton, B.,
Originally presented on: 11/5/2013 17:21:00 - 17:39:00

Ginkgo BioWorks is working to build a reliable and cost effective pipeline for engineering bacteria to produce fuels and high value chemicals. I will argue that there is no "silver bullet" technology to address this goal; instead, an integrated, balanced mix of technologies is the most plausible approach to making the engineering of biology more standard, more predictable, and hence more scalable. Our pipeline consists of tightly-integrated workflows for designing, building, and testing engineered organisms that each rely on a combination of software, hardware, and wetware. I will demonstrate how detailed data recording and analysis can enable the improvement and scaling of a novel DNA assembly scheme. Next, I will describe a high-throughput metabolomics workflow that captures a rich picture of the metabolic state of engineered cells. This metabolomics workflow depends on the latest hardware advances in mass spectrometry and custom software to process and analyze the large resultant datasets. Finally, I will show how we are deploying natural biological parts to control the behavior of engineered organisms in new ways.

(380a) Engineering of Microfluidic Systems for Rapid Detection of Food Pathogens

Ladisch, M., Purdue University
Ximenes, E., Purdue University
Bement, A., Purdue University
Originally presented on: 11/5/2013 15:15:00 - 15:40:00

Microbial contamination of fresh food causes severe illness, compromises food security and could disrupt intra- and inter-national trade in fresh foods. Methods that enable rapid detection of pathogens in both exported and imported foods are of importance for enhancing food safety. One of these methods is the concentration of microorganisms in food derived samples. This approach utilizes hollow fibers in an integrated microfluidic system that microfilters a large volume of food extract and concentrates bacterial cells present in the sample by 500 to 1000 fold in about an hour. The resulting small volume has a sufficient concentration of bacteria so that they may be effectively probed for presence of pathogens on a "first alert" basis. A positive result, obtained during the process of cell concentration, would signal the need for subsequent analysis by PCR, antibody based capture and identification, and/or plating coupled with light scattering identification. Rapid detection is important since it could help support both source and border interventions to guard against the distribution of pathogens in foods in a manner not previously possible. However, results from rapid methods, currently under development, must be coupled with statistical techniques that define the best way to selectively sample the huge quantities of vegetable produce that are shipped into and across the US. This paper presents the current status of microfluidic technology for rapid cell concentration and its validation in an automated prototype. Some initial thoughts on the possible impacts of rapid detection technology on national and international intervention alternatives and their consequences, both positive and negative, are briefly outlined.

(380c) Adventures in Bacterial Glycoengineering

DeLisa, M. P., Cornell University

Originally presented on: 11/5/2013 15:40:00 - 16:05:00

Carbohydrates add a level of diversity across all forms of life that is unparalleled by the information content of nucleic acids and proteins. The lack of a simple template to translate a glycan code into defined sugar structures contributes to this complexity and provides a challenge for efforts aimed at the production of biologically important glycans and glycoconjugates. With the discovery of glycoprotein synthesis in bacteria and functional transfer of glycosylation pathways between species, Escherichia coli cells have become a tractable host for understanding glycosylation and the underlying glycan code of living cells. Moreover, efforts to manipulate the pathways from sugar nucleotides to glycolipids to glycoproteins have transformed E. coli into a living factory for scalable, bottom-up production of complex glycoconjugates by design. Here, I will discuss our efforts to develop E. coli for the biosynthesis of a diverse array of glycan structures, which can be used to tailor the activity, stability, half-life, and immunogenicity of a given biopharmaceutical. I will also discuss our efforts to unify protein glycosylation in E. coli with the advanced tools of protein engineering such as surface display technologies. The result is a powerful new way to engineer the enzymes, pathways, end-products, and genomes of glyoengineered bacteria for creating the next generation of protein therapeutics and vaccines for a wide range of human diseases.

(408a) Engineering Novel Tandem Catalytic Reactions Using Organometallic Catalysts and Metalloenzymes

Denard, C. A., University of Illinois at Urbana Champaign
Huang, H., University of Illinois Urbana-Champaign
Lu, L.,
Bartlett, M. J.,
Hartwig, J. F., University of California, Berkeley
Zhao, H., University of Illinois at Urbana-Champaign
Originally presented on: 11/6/2013 8:30:00 - 8:48:00

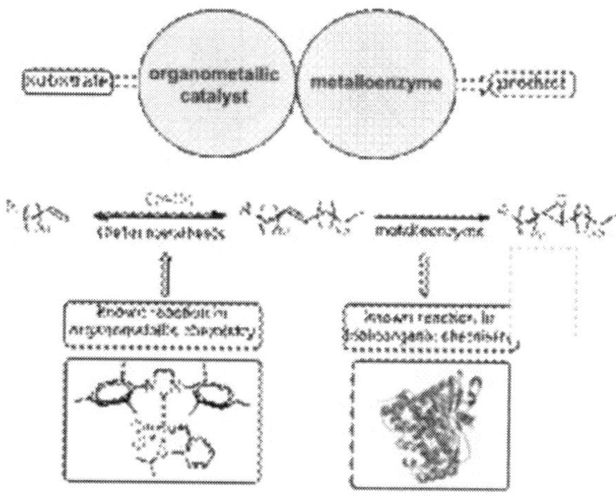

Figure. Cooperative tandem catalysis by two different classes of transition metal catalysts

Organometallic chemistry and bioinorganic chemistry are two prominent disciplines of chemistry. Organometallic chemistry lies at the intersection of organic and inorganic chemistry, while bioinorganic chemistry lies at the intersection of biological and inorganic chemistry. Transition metal complexes lie at the core of both disciplines, but synthetic chemistry that exploits the complementary reactivity of the two systems in a cooperative fashion has rarely been developed. Often, the two classes of catalysts effect the same transformation with different rates and scope, but more often, they catalyze completely different types of reactions that are conducted in different vessels if they are used in the same synthetic sequence. Recently, synthetic catalysts and enzymes are beginning to be combined in new ways to work in tandem; however, there have been no reports in which organometallic catalysts and metalloenzymes have been combined and shown to work cooperatively. Such reactions are challenging to achieve, in part because the conditions in which these catalysts operate are typically different, and mutual inactivation of the catalysts typically occur when they are combined in the same medium. In our work, we use a biphasic reaction and show that 90% yield of a single epoxide can be obtained from a mixture of alkenes, a ruthenium metathesis catalyst and a P450 enzyme by a combination of a dynamic equilibration of alkenes and a selective epoxidation of the cross-metathesis products (Figure). We have expanded the scope of this tandem reaction to include unnatural P450 substrates, such as alkenyloxybenzoic acids, and P450 mutants with different regioselectivities. These results show the potential of combining the two classes of catalysts as synthetic catalysts become more tolerant of functional groups and enzymes become more tolerant of organic media. We are also exploring and evolving new metalloenzymes with different selectivities on internal alkenes for the selective epoxidation of dynamic *cis-trans* mixtures of alkenes established by transition metal catalysts. Lastly, we are creating hybrid transition-metal-metalloenzyme catalysts to create a P450 enzyme capable of performing both oxidation and metathesis reactions.

(408b) Towards Understanding Lasso Peptide Biosynthesis and Regulation Through Genome-Mining

Maksimov, M. O., Princeton University
Link, A. J., Princeton University
Originally presented on: 11/6/2013 8:48:00 - 9:06:00

Lasso peptides are a growing class of ribosomally-synthesized and posttranslationally modified peptides (RiPPs) that are folded into a structure resembling a threaded lasso. This class of molecules shares a conserved gene cluster architecture that includes an exporter protein in addition to the maturation enzymes that tailor the gene-encoded polypeptide precursor into the lasso form. Using conserved elements in the precursor peptide and the maturation enzymes, we developed a genome mining approach that identified 79 putative lasso peptide biosynthesis clusters in a broad range of bacterial phyla (1). We have experimentally validated two of these clusters by expressing and characterizing three novel lasso peptides encoded by them and solved the structures of two of them by NMR. We present these results and our findings regarding the heterologous expression and purification of these molecules. Additionally, the gene architectures of these clusters deviate from the typical lasso peptide architecture by having a predicted protease instead of a transporter protein. We have experimentally characterized this enzyme and applied a bioinformatics approach to re-examine lasso peptide gene cluster architectures in light of this discovery.

1. Maksimov MO, Pelczer I, & Link AJ (2012) Precursor-centric genome-mining approach for lasso peptide discovery. *Proceedings of the National Academy of Sciences* 109(38):15223-15228.

(408c) Combinatorial Engineering of Isoprenoid Pathway for Improved Amorpha-4,11-Diene Production in Budding Yeast

Yuan, J.,
Ching, C. B., Singapore Institute of Technology
Originally presented on: 11/6/2013 9:06:00 - 9:24:00

Synthetic biology has been emerging to produce value-added compounds with numerous applications in the food, agriculture, chemical, and pharmaceutical industries. In the present study, our group attempted to optimize the metabolic flux of isoprenoid pathway in budding yeast for improved production of terpene derivatives. Specifically, combinatorial engineering of isoprenoid pathway through shuffling pathway gene copy number into the genome was carried out. By exploiting carotenoid biosynthesis as screening module, we successfully created a library through current combinatorial engineering approach, with yeast colonies appeared to have various intensities of orange color. Eight colonies from the library with intense orange color were picked out and characterized by PCR for the validation of genome integration events. As proof of concept that our carotenoid overproducers could serve to boost the titer of other terpenoid production, the engineered strains were subsequently transformed with pRS425-ADS for testing the production of amorpha-4,11-dine. From our preliminary result, the carotenoid overproducers from the library showed significantly high level of amorpha-4,11-diene over the parental strain harboring pRS425ADS. Among them, the best producer obtained at this moment yielded more than 6-fold of final product over the parental strain. Our result clearly suggested that combinatorial engineering approach through tuning the gene copy number into genome for stable expression can be very promising to balance the metabolic flux of isoprenoid pathway and it should be straightforward to exploit the current platform for engineering other metabolic pathways as well. Moreover, anther beauty of our approach is that multiple rounds of combinatorial engineering might be performed to further balance the metabolic flux after marker removal and/or simply through increasing selection pressure to facilitate more genome integration events.

(408d) Engineering Glucose Oxidase to Improve Electron Transport in Enzymatic Fuel Cells

Kim, H., University of Delaware
Chen, W., University of Delaware
Originally presented on: 11/6/2013 9:24:00 - 9:42:00

Enzymatic fuel cells, specifically their electrode compartments, lose much of their effectiveness due to inefficient electron transfer between the active center of the redox enzymes and the electrode. It is often shown that the power loss can be mitigated using immobilization techniques, such as chemical modifications targeting random surface residues, encapsulation, or entrapment methods. However, such techniques often negatively affect enzymatic activity, making biological methods that properly expose the active center more desirable. In this presentation, glucose oxidase (GOx) was displayed on the yeast surface and fused to a gold binding peptide to bring the GOx in close proximity to the electrode for enhanced electron transport between the active center and the electrode. The increased potential in current production due to the enhanced attachment of GOx to the electrode surface and the effect of tethering mediators in between GOx and the electrode to further aid electron transport will be discussed.

(408e) Engineering Enzymes for Hydrogen Production From Biomass

Liong, S., Stanford University
Smith, P. R., Stanford University
Swartz, J. R., Stanford University
Originally presented on: 11/6/2013 9:42:00 - 10:00:00

Hydrogen is a potential energy carrier to replace hydrocarbon fuels and reduce emissions from their consumption. Hydrogen production from renewable cellulosic biomass is favored for its nearly carbon-neutral cycle, placing high interest on its cost-effective technology development. We have previously demonstrated the feasibility of using an unpurified cell extract engineered for hydrogen production from glucose. The endogenous pentose phosphate pathway is coupled with a synthetic electron transfer pathway composed of three key proteins: ferredoxin-$NADP^+$ reductase (FNR), ferredoxin, and a clostridial hydrogenase. However, the major challenge lies in increasing the electron flux from NADPH to hydrogen. In particular, low turnover numbers have been observed for the electron transfer from FNR to ferredoxin in the pathway. This work is aimed to investigate the application of FNR-hydrogenase fusion proteins for improved turnover number in hydrogen production. The fusion protein consists of FNR from *Anabaena sp.* and [Fe-Fe] hydrogenase from *C. pasteurianum*, physically linked by a polypeptide chain and expressed in *E. coli* strain. We also varied the linker length and its amino acid composition for increased electron transfer. Compared to un-fused enzymes added at equimolar concentrations, the fusion FNR-hydrogenase provides up to 5-fold improvement in hydrogen production rates.

(408f) Evaluating the Effect of Attachment Orientation On the Activity and Stability of Immobilized Enzymes

Wu, J. C., Brigham Young University
Smith, M. T., Brigham Young University
Werner, C., Brigham Young University
Bundy, B. C., Brigham Young University
Originally presented on: 11/6/2013 10:00:00 - 10:18:00

Due to the overwhelming demand for environmentally friendly manufacturing processes, the multibillion dollar biocatalysis industry is rapidly expanding. Biocatalytic systems enable stereo-, chemo-, and regio-specificity in chemical manufacturing, reducing the production of wasteful byproducts. This specificity is particularly valuable in industries such as the pharmaceutical industry, where removal of chemically similar but physiologically harmful waste products is essential. Traditional biocatalytic processes suffer from limitations in enzyme stability, leaching, recoverability, and reusability. Given the high cost of enzyme production, these limitations significantly hinder the cost-effectiveness of biocatalysis for industrial applications. These traditional processes of enzyme immobilization such as adsorption, cross-linking, and entrapment provide some improvements to stability, recoverability, and reusability but suffer from enzyme leaching, complicated or toxic conjugation procedures, and a lack of attachment orientation control. The limited number of canonical amino acids presents an additional obstacle to site-specific attachment orientation control as it is difficult to target a single attachment location. Here, we build upon recent advancements in unnatural amino acid incorporation to demonstrate a biocompatible, bioorthogonal, and covalent enzyme immobilization process that improves protein stability and enables attachment orientation control. This system, which we refer to as the Protein Residue-Explicit Covalent Immobilization for Stability Enhancement or PRECISE system, permits the covalent attachment of enzymes at potentially any specific residue onto a surface. Using this process, we report the effect of attachment orientation on enzyme activity, durability, and stability in harsh and mild environments.

(408g) Transcriptomics- and Metabolomics-Driven Discovery of a Novel Indolic Secondary Metabolite Pathway in Arabidopsis Thaliana

Rajniak, J., Stanford University
Sattely, E., Stanford University
Originally presented on: 11/6/2013 10:18:00 - 10:36:00

Plant secondary metabolism is an exceptionally rich source of both small molecules of therapeutic importance and enzyme catalysts with unique functions. However, owing largely to the complexities of their genomic and cellular organization, plants remain underutilized as a source of novel chemistries. Here, we describe the transcriptomics- and metabolomics-driven discovery of a novel secondary metabolite pathway in the model plant *Arabidopsis thaliana*. Starting with a single target gene in the cytochrome P450 family, we have systematically uncovered the biosynthesis of the novel metabolite indole-3-carbonyl nitrile (ICN), containing a unique alpha-keto nitrile moiety not previously seen in any plant or microbial species, from tryptophan. Conjugation of ICN with cysteine-containing peptides *in vivo* represents a new biosynthetic route towards thiazole and thiazoline ring-containing derivatives, similar to those produced by a number of bacterial nonribosomal peptide synthetases. Correlation analyses of publicly available transcriptomics data followed by metabolomic analysis of relevant homozygous knockout lines and *in vitro* biochemical assays enabled us to directly demonstrate the function of two new cytochrome P450 enzymes and an FADH-containing reductase, and thereby reconstitute essentially the entire ICN pathway. Our approach shows the power of an integrated "omics" strategy for the discovery of potentially useful biocatalysts and novel small molecular scaffolds.

(408h) Proteomic Analysis of Pseudomonas Putida Reveals An Organic Solvent Tolerance-Related Gene Mmsb

Ni, Y., Virginia Commonwealth University
Qian, X., Jiangnan University
Song, L., Jiangnan University
Originally presented on: 11/6/2013 10:36:00 - 10:54:00

Organic solvents are toxic to most microorganisms. However, some organic-solvent-tolerant (OST) bacteria tolerate the destructive effects of organic solvent through various accommodative mechanisms.

In this work, we developed an OST adapted strain *Pseudomonas putida* JUCT1 that could grow in the presence of 60% (v/v) cyclohexane. Two-dimensional gel electrophoresis was used to compare and analyze the total cellular protein of *P. putida* JUCT1 growing with or without 60% (v/v) cyclohexane. Under different solvent conditions, 5 high-abundance protein spots were identified by MALDI-TOF/TOF spectra. The corresponding genes of these five proteins, *arcA*, *PSEEN1080*, *mmsB*, *tsf*, and *PSEEN0851*, were separately expressed in *E. coli* to evaluate their effect on OST properties of the host strain. In the presence of 4% (v/v) cyclohexane, *E. coli* transformants harboring *mmsB* (3-hydroxyisobutyrate dehydrogenase, 3-HIBDH) could grow to 1.70 OD_{660}, whereas cell growth of *E. coli* JM109 (the control) was completely inhibited by 2% (v/v) cyclohexane. Other transformants also showed an increased resistance to cyclohexane as well as other organic solvents. Of these 5 genes, *mmsB* exhibited the most prominent effect on increasing OST of *E. coli*. Less oxidation product of cyclohexane was detected because *mmsB* transformants might help keep a lower intracellular cyclohexane level.

The OST-related function of *mms* and its homologous gene in *E. coli*, *zwf* (glucose-6-phosphate dehydrogenase, G6PDH), was also investigated. It was noted that the OST of *E. coli* host was severely impaired by *zwf* knockout, and its complementation resulted in significantly enhanced OST in 4% (v/v) cyclohexane than parent *E. coli*. Furthermore, recombinant *E. coli* strains carrying *mmsB* showed better OST capacity than that harboring *zwf*. To achieve a stable gene expression, *mmsB* was integrated into the genome of *E. coli* JM109(DE3) by red homologous recombination. The result strain *E. coli* JM109(DE3) (Δ*endA*::*mmsB*) was successfully applied as an OST host of whole-cell biocatalyst in an aqueous/butyl acetate biphasic system with 82.8% higher product yield.

This study demonstrates a feasible approach for elucidating OST mechanisms of microorganisms, and provides molecular basis to construct organic-solvent-tolerant strains for industrial applications.

(446a) Role of Protein Confinement On Enzyme Function and Stability

Kunkel, J.,
Asuri, P., Santa Clara University
Originally presented on: 11/6/2013 8:30:00 - 8:48:00

Research over the past few decades have attempted to answer how proteins behave in molecularly confined or crowded environments when compared to dilute buffer solutions. This information is vital to understanding *in vivo* protein behavior as the cell environment is very crowded with the average macromolecular space much smaller than the macromolecules themselves. In our study, we attempt to address this question using a model enzyme – horseradish peroxidase (HRP) encapsulated in agarose hydrogels of different porosities. Our studies reveal that the activity of the agarose encapsulated HRP is lower than that of the solution phase enzyme. However, the encapsulated enzyme retains a higher percentage of its activity in the presence of denaturants. Moreover, we observe that the degree of confinement (achieved using the different percentages of agarose – 0.5%, 2%) has an effect on the stabilization effect; HRP encapsulated in 2% agarose is significantly more stable than HRP encapsulated in 0.5% agarose. Furthermore we observed similar results through structural measurements of HRP denaturation using 8-anilinonaphthalene-1-sulfonic acid assay. Our work demonstrates the utility of hydrogels to study protein behavior in highly crowded environments; furthermore, these observations are in accord with previous studies of macromolecular crowding on protein structure and function.

(446b) Modeling the Oxidation of Methionine Residues in Proteins

Chennamsetty, N., Bristol-Myers Squibb Company
Quan, Y., Bristol
Nashine, V., Bristol-Myers Squibb Company
Sadineni, V., Bristol-Meyers Squibb Company
Lyngberg, O. K., Bristol Myers Squibb
Qiu, D.,
Li, B.,
Krystek, S.,
Originally presented on: 11/6/2013 8:48:00 - 9:06:00

We report the use of molecular modeling to predict the oxidation propensity of methionine residues in proteins in aqueous solutions by hydrogen peroxide. Oxidation of methionine to the sulfoxide form is one of the major degradation pathways for therapeutic proteins. Oxidation can occur during processing or storage of pharmaceuticals and it often reduces or eliminates biological activity. Identifying the methionines most liable for oxidation may help to overcome the oxidation either through direct mutation or formulation. We use a molecular model based on atomistic simulations called 2-shell water coordination number (2SWCN) to predict the oxidation rates for several model proteins and therapeutic proteins. We compare the 2SWCN modeling results against experimental values and with other widely used models that are based on static and simulation average of the solvent accessible area (SAA) for either the side chain or the sulfur atom in the methionine residue. We find that the 2SWCN model and simulation averaged SAA are highly accurate in predicting the oxidation propensity of methionine residues for the proteins tested. Thus, the models have sufficient predictive power to enable further protein engineering or to guide formulation approaches to stabilize the unstable methionine residues.

(446c) Towards Understanding the Structure and Function of Alcohol Acetyltransferase Lipid Droplet Localization in S. Cerevisiae

Lin, J. L., UC Riverside
Wheeldon, I., UC Riverside
Originally presented on: 11/6/2013 9:06:00 - 9:24:00

Across eukaryotes metabolic activity of neutral lipid synthesis is localized to intracellular lipid droplet. Through specific protein domains, essential enzymes in the biosynthesis of triacylglycerols are associated with the outer membrane of the endoplasmic reticulum (ER) and ultimately to lipid droplets that emanate from the ER. Within these pathways, there are clear links between protein structure and function in terms of both lipid droplet association and enzymatic activity. In *Saccharomyces cerevisiae*, alcohol acetyltransferase 1 (ATF1) is also localized to lipid droplets. This enzyme catalyzes the reaction of acetyl-CoA and short chain alcohols to produce the corresponding ethyl esters, including the industrially useful solvent ethyl acetate and the flavor compound isoamyl ester (banana-like flavor and aroma). Compared to the triacylglycerol synthesis pathways there is little known about the structure of ATF1. The crystal structure has not been solved and there is no acceptable template from which an accurate homology model can be predicted. Here, we compare the sequence structure of ATF1 to proteins that are known to localize on lipid droplets and identify the domains responsible for localization. Truncation and localization studies reveal that both N- and C-terminal domains (amino acids 1-24 and 502-525) are necessary, but not sufficient, for ATF1 lipid droplet targeting. Similar to perilipin A, a structural protein localized to lipid droplets of adipocytes, internal hydrophobic domains of ATF1 are necessary for targeting. At present, the biological impetus for ATF1 localization is unknown. In this work we use our insights into the structure of ATF1 to explore the hypothesis that localization enhances ester synthesis pathway activity. We explore this hypothesis through comparative studies of ATF1 and cytosolic ATFs from *Solanum lycopersicum*, and the yeasts, *Pichia anomala* and *Kluyveromyces lactis*.

(446d) Biochemical Insight Into Fungal Cellulosome Architecture and Regulation
Haitjema, C., University of California, Santa Barbara
Solomon, K., University of California, Santa Barbara
O'Malley, M. A., University of California, Santa Barbara
Originally presented on: 11/6/2013 9:24:00 - 9:42:00

Anaerobic gut fungi are the primary colonizers and decomposers of plant biomass in monogastric herbivores and ruminants. Their ability to hydrolyze complex lignocellulosic materials is of great promise to cellulosic bioenergy development, since enzymatic breakdown of biomass is a stringent bottleneck. It is hypothesized that cellulolytic activity within gut fungi is mediated by large cellulolytic complexes called fungal cellulosomes. While bacterial cellulosomes are well-characterized, the molecular architecture of fungal cellulosomes is quite poorly understood. We characterized the composition and structure of fungal cellulosomes from the gut fungal isolate *Piromyces sp. finn.* A practical approach was developed to isolate cellulosome components from fungal cultures under a variety of substrate growth conditions. Following the purification of fungal cellulosomes, we combined tandem mass spectroscopy with strand-specific transcriptomic sequence information to identify the key components of these cellulosomes. To our knowledge, these studies are the first to reveal the entire enzymatic ensemble that comprise fungal cellulosomes. In addition, we will provide evidence that most dockerin-fused enzymes bind to high molecular weight proteins as putative scaffolding molecules, which have long remained elusive for fungal cellulosomes. We will connect these results to the regulation patterns observed for fungal cellulosomes in the presence of cellulosic and non-cellulosic substrates.

(446e) Novel Screening Method for Improving Monoclonal Antibody Selection and Formulation

Sule, S. V., Rensselaer Polytechnic Institute
Wu, J.,
Geng, S.,
Tessier, P. M., Rensselaer Polytechnic Institute
Originally presented on: 11/6/2013 9:42:00 - 10:00:00

A key challenge in developing therapeutic monoclonal antibodies (mAbs) is their complex self-association behavior at high concentrations (>50 mg/mL). Unfortunately, the low purities, small quantities and large numbers of mAb candidates available during antibody discovery are incompatible with conventional methods of measuring mAb self-association. We report a method (affinity-capture self-interaction nanoparticle spectroscopy, AC-SINS) for identifying mAbs with low self-association propensity that is robust at extremely low antibody concentrations (<50 ug/mL) and in the presence of contaminants. Gold nanoparticles are conjugated with polyclonal antibodies specific for human mAbs, and these conjugates are used to capture mAbs from dilute solutions. We find that the interparticle distance-dependent plasmon wavelength of the gold conjugates redshifts in a manner that is well correlated with light scattering measurements conducted at three orders of magnitude higher antibody concentrations. We are currently using AC-SINS for identifying mAb candidates with high solubility during antibody discovery as well as for improving mAb formulation during antibody development.

(446f) Direct Observation of the Formation of Dynamic Clusters in Concentrated Monoclonal Antibody Solutions

Wagner, N. J., University of Delaware
Liu, Y., NIST
Yearley, E. J., University of Delaware
Godfrin, P. D., Univesity of Delaware
Perevozchikova, T.,
Zarraga, I., Genentech Inc.
Porcar, L., Institut Laue-Langevin
Originally presented on: 11/6/2013 10:00:00 - 10:18:00

The development of human therapeutics based on monoclonal antibody (mAb) and related products have evolved rapidly with almost exponential growth in market value. They have found wide application in treating cancer, chronic inflammatory diseases, transplantation, infectious diseases and cardiovascular disease because drugs based on antibodies generally have very high selectivity, low side effects, good reproducibility, and a high success ratio. The importance of mAb based biopharmaceuticals motivates fundamental research into problems related to their manufacture and clinic applications. One industrially important challenge is to control the viscosity of concentrated mAb protein solutions. A high viscosity hinders the large-scale production, purification, storage and delivery of these drugs. In particular, a viscosity exceeding ~50 mPa•S -s prevents delivering drugs via subcutaneous (SC) injection. This can be easily exceed at the high, protein concentrations, in the range 100~200 mg/ml, that are required for effective syringe injection.

This anomalous viscosity effect is hypothesized to arise from the formation of dynamic clusters. Here we report the direct observation of these dynamic clusters. mAb cluster size and morphology are estimated by combining neutron spin echo, small angle neutron/x-ray scattering and computer simulations. The dynamic cluster formation is shown to be controllable by the added salt concentration. Cluster formation leads to a dramatic increase in the solution viscosity, conclusively proving the hypothesis. Our results can aid in the development of mAb formulations with desired physical properties.

(446g) Gauging Colloidal and Thermal Stability in Human IgG1 - Sugar Solutions Through Diffusivity Measurements

Rubin, J., Georgia Institute of Technology
Sharma, A., Georgia Institute of Technology
Bommarius, A. S., Georgia Institute of Technology
Behrens, S. H., Georgia Institute of Technology
Originally presented on: 11/6/2013 10:18:00 - 10:36:00

Monoclonal antibodies are the fastest growing class of biotherapeutics. Ensuring their colloidal stability in liquid dispersions is crucial for maintaining therapeutic efficacy and economic viability. Sugars are often added to increase the colloidal and thermal stability of protein; yet, determining which sugar is the most stabilizing requires time and sample-consuming stability tests. Here we show that the extent of stabilization by different sugars can be gauged by analyzing the proteins' diffusive virial coefficient k_D. This protein interaction parameter is measured conveniently in a non-invasive, high-throughput manner using dynamic light scattering. It is found to correlate closely with experimental dimerization rate constants and melting temperatures for antibodies in different sugar solutions. The proposed analysis thus provides a rapid test of the subtle differences between inherently similar sugar-protein interactions; it should greatly facilitate the formulation of protein therapeutics. Among the tested sugars we observe a trend of disaccharides providing more efficient stabilization than monosaccharides. Circular dichroism spectroscopy further suggests that the sugar induced thermal stabilization involves a reversibly formed intermediate protein conformation between the native and fully denatured state.

(446h) Local Mobility and Degree of Retention of Native Secondary Structure Contribute to the Stability of Recombinant Human Growth Hormone in Glassy Lyophilized Formulations

Randolph, T. W., University of Colorado
Xu, Y.,
Carpenter, J. F., University of Colorado
Cicerone, M. T., National Institute of Standards and Technology
Originally presented on: 11/6/2013 10:36:00 - 10:54:00

Using recombinant human growth hormone (rhGH) as a model, we created a variety of glassy, lyophilized formulations and studied how degradation rates depended on the the degree of retention of protein secondary structure and on glassy-state mobilities. Secondary structures were detrmined from solid-state FTIR analysis of amideI bands; local and global mobilities were determined by neutron scattering and calorimetry, respectively. Rate constants for degradation (aggregation, deamidation, and oxidation) strongly correlated with both protein structure and local mobility in the glass. Degradation rate constants were inversely correlated with global mobility, as quantified by proximity to glass transition temperatures and by entalpy relaxation measurements .

(447a) Low Abundance of Respiratory Nitrate Reductase Is Essential for Escherichia Coli in Resistance to Aminoglycoside and Cephalosporin

Wang, S., Xiamen University
Ma, Y., Sun Yat-sen University
Guo, C., Sun Yat-sen University
Li, H.,
Peng, X.,
Originally presented on: 11/6/2013 8:30:00 - 8:48:00

The increasing incidence of antibiotic-resistant bacteria has posed a serious threat to human health and aquaculture today and thus has become a scientific issue worldwide. A line of evidence has indicated that four mechanisms contribute to the bacterial resistance, including modification or hydrolysis of enzymes, modification of drug targets, activation of efflux pump systems, and reduce of outer membrane permeability. Despite these progresses over past sixty years, our understanding of the resistant mechanisms is still incomplete, which should be one of the major causes resulting in antibiotic-resistant bacteria out of control. Thus, further investigation of bacterial resistant strategies to antibiotics may be necessary for combating antibiotic-resistant bacteria.

The recently developed technologies are expected to facilitate the study of bacterial antibiotic-resistant mechanisms. For example, protein-protein interactions, being important for the majority of biological functions including antibiotic resistance, may be potentiated using highly throughput screening methods of 2-D native PAGE (N-PAGE) and 2-D blue native PAGE (BN-PAGE) based proteomics. The two methods have been proven especially useful for investigation of native protein complexes in the context of proteomics. Comparatively, N-PAGE based on proteomics may offer more information on labile protein complexes containing metabolic enzymes than BN-PAGE based proteomics. Using the two approaches, researchers have investigated membrane and outer membrane protein complexes in *Escherichia coli* and identified a series of homogeneous and heterogeneous complexes. The identified membrane-related protein complexes may contain cytoplasmic and periplasmic proteins since approximately 30 - 40% *E. coli* proteins may function in the membrane of the cell envelope, in which there are soluble proteins tethered to the inner and outer membranes through hydrophobic patches, lipid moieties, or charge interactions or in membrane protein complexes. On the other hand, recent reports have indicated that metabolic enzymes are related to antibiotic resistance, and the regulation of metabolism elevates susceptibility of bacterial persisters to antibiotics. We have also shown that Na(+) NQR complex is essential for *Vibrio alginolyticus* in resistance to balofloxacin using N-PAGE based proteomics coupled with functional validation. We hypothesized that the comparative N-PAGE approach might provide metabolic-related protein complexes that contributed to antibiotic resistance for further functional investigation since many of metabolic enzymes are detected in membrane protein fraction. Here, N-PAGE based proteomics is used to identify a metabolic protein complex in membrane fraction of *Escherichia coli* K12 BW25113 and then the role of the complex in bacterial antibiotic resistance is further investigated.

Separation of membrane proteins and detection of Nar complex in *E. coli* K12 BW25113 membrane protein fraction using N-PAGE based proteomics. N-PAGE was used to separate the membrane proteins of *E. coli* K12 BW25113. Approximately a hundred of clear protein spots were observed on each of the 2-DE gels stained with cCBB R-250, most of which were located in the range of 25-120 kDa with minimal streaks. The streaks could be resulted from the diffusion of the same proteins during the first dimensional native-PAGE as reported previously. These well visualized protein spots were excised from 2-DE gels and analyzed by MALDI/TOF-TOF analysis. A hundred of proteins were successfully identified. They represented ninety-four unique proteins since five proteins had more than one spot. We interestingly found that two components, NarG and NarH, of Nar complex were lined, which did not show in our previous N-PAGE to detection of *E. coli* DH5α membrane protein fraction. Nar complex includes four polypeptides NarG, NarH, NarJ and NarI of molecular weight 138.7, 57.7, 26.5 and 25.5 kDa, respectively. NarJ and NarI did not appear in the gel probably due to the low molecular weight. NarG and NarH were further validated in 2-DE Western blotting. The complex is capable of reducing nitrate using normal physiological substrates and is clearly the major respiratory Nar in *E. coli* since it accounts for 98% of the Nar activity when fully induced. The result provides a chance to investigate a role of the complex contributed into bacterial energy metabolism in bacterial antibiotic resistance. In addition, sixteen out of the 94 proteins were validated using Western blotting. We identified the subunits A and B of the BAM complex in two distinct protein complexes. The BAM complex consists of one outer membrane protein, BamA (formerly YaeT), and four lipoproteins, BamB, BamC, BamD and BamE (formerly YfgL, NlpB, YfiO and SmpA, respectively). The other three were absence, which might be related to protein abundance, molecular weight and binding force. Further pathway analysis was carried out and obtained 90 annotation out of the 119 proteins identified. Among the 90 annotated proteins, 76 were enriched in 16 pathways. They are mainly involved in carbon metabolism, nitrogen metabolism, oxidative phosphorylation. Furthermore, we investigated the relationships between protein complexes containing altered abundance of protein(s) and antibiotic resistance.

Validation of Nar complex in *E. coli* K12 BW25113 membrane proteins. Then, the Nar complex was validated using two assays. First, far-Western blotting was used to investigate the interaction of NarH with NarG. Our results indicated that the bait protein was NarH when NarG was used as the prey protein. Then, the complex was further validated by N-PAGE analysis of *narH* and *narG* deletion mutants. This was based on the hypothesis that when one protein within a protein complex is absent, the others may disappear from N-PAGE gel. Our results showed that loss of *narG* caused disappearance of NarH, while deletion of *narH* resulted in loss of NarG in the same line, meaning that NarG-NarH complex existed in the line. These results indicate that Nar complex detected by the N-PAGE based proteomics is reliable, and thus may be used for investigation of antibiotic resistance.

Characterization of Nar complex in five antibiotic-resistant *E. coli* strains. In order to investigate whether abundance of Nar complex components was altered to resist antibiotics, the N-PAGE based proteomics approach was used to detect NarG and NarH levels in SM-R, GEN-R, CAZ-R, TET-R and NA-R. Significant changes were detected, showing that NarG and NarH disappeared in the SM-R, CAZ-R, TET-R and decreased in GEN-R and NA-R. These results not only indicate that Nar complex may play an important role in antibiotic resistance, but also show that the role is related the classes of antibiotics.

Functional characterization of *narG* and *narH* deletion mutants in response to five antibiotics. We next performed functional characterization of the Nar complex proteins using *narG* or *narH* deletion mutants. The performance was carried out using two antimicrobial susceptibility assays: MIC assay and survival capability assay. Deletion of *narG* or *narH* resulted in distinct elevation of MICs in medium with SM or GEN, and two-fold decrease of MICs in medium with TET, and no significant changes in medium with CAZ or NA. The survival capability assay was also carried out to further investigate functional characterization of the mutants due the assay is more sensitive than MIC assay. These bacteria including *ΔnarI* and *ΔnarJ* were cultured in tubes with series of two-fold dilution of SM, GEN, CAZ, TET and NA concentrations from 0.625 to 5 μg/mL, 0.1953-1.1719 μg/mL, 0.0781-0.3125 μg/mL, 0.78125-3.25 μg/mL, 1.25-7.5 μg/mL, respectively. The resulting cultures were assayed by measurement of optical density at 600 nm and the survival capability was determined by division of OD values of the cultures with an antibiotic over those without the drug. The survival rates of the mutants cultured in medium with SM, GEN or CAZ were significantly increased ($P<0.01$) in a dose-dependent manner. However, lower survival capabilities were detected in the mutants cultured in medium with TET and NA. Generally, the results obtained from survival capability assay were in consistence with those obtained from MIC assay. In summary, the results obtained from functional characterization of *ΔnarG* and *ΔnarH* were consistent with the changes of NarG and NarH at the protein abundance which was detected by 2-DE analysis of these antibiotic-resistant bacterial strains.

In summary, N-PAGE based proteomics approach was used to investigate *E. coli* K12 BW25113 membrane protein complexes and resulted in the detection of NarG and NarH, two main components of Nar complex. The decreased abundance of the two proteins was characterized in SM-R, GEN-R, CAZ-R, TET-R and NA-R. Functional tests using *narG*- and *narH*-deleted mutants demonstrated distinctly elevated resistance to aminoglycosides and CAZ. It is documented that low abundance of Nar complex is required for *E. coli* resistance to aminoglycosides and CAZ, which is the first report on that *E. coli* Nar metabolism alters bacterial resistance to antibiotics. This finding can serve to propose novel strategies for treating infections caused by aminoglycoside- and CAZ-resistant *E. coli* through activation of Nar activity.

ACKNOWLODGEMENTS

This work was sponsored by grants from NSFC projects (30972279, 40976080), Guangdong Provincial Science and technology projects (2012A031100004). Doctoral Fund of Ministry of Education of China (100171110029)

(447b) Cell Type-Specific Proteomic Profiling in Caenorhabditis Elegans

Yuet, K. P., MIT
Doma, M. K.,
Ngo, J. T.,
Sternberg, P. W., California Institute of Technology
Tirrell, D. A., California Institute of Technology
Originally presented on: 11/6/2013 8:48:00 - 9:06:00

In a complex eukaryote like *C. elegans*, cell and tissue heterogeneity restricts the usefulness of large-scale, mass spectrometry-based analysis of an organism's proteome. As worm cell lines are not available and enriching for specific cells or tissues is challenging, researchers cannot systematically identify low abundance proteins expressed in specific cells or tissues from whole-worm lysates. To isolate protein from specific cells, we have engineered *C. elegans* phenylalanyl-tRNA synthetases capable of appending azide- or alkynyl-bearing analogs of phenylalanine to $tRNA^{Phe}$. As these analogs are not substrates for any of the wild-type aminoacyl-tRNA synthetases in *C. elegans*, we achieved cell type selectivity by spatially restricting the expression of a mutant synthetase using cell type-specific promoters. Because proteins from those cells that express the mutant synthetase -- and only those cells -- contain the analogs, they can be enriched by treatment with alkyne- or azide-functionalized biotin reagents and subsequent affinity chromatography for identification by LC/MS-MS.

(447c) Phospho-Proteomics Reveals Oncogenic Phospho-Tyrosine Signaling Networks in Cancers Lacking Mutated Or Amplified Tyrosine Kinases

Graham, N. A., California Institute of Technology
Drake, J. M., University of California, Los Angeles
Tahmasian, M., University of California, Los Angeles
Pienta, K. J., University of Michigan, Ann Arbor
Witte, O. N., University of California, Los Angeles
Graeber, T. G., University of California, Los Angeles
Originally presented on: 11/6/2013 9:06:00 - 9:24:00

Introduction: Many human tumors exhibit aberrant activation of tyrosine kinase signaling due to mutation or DNA amplification of tyrosine kinases (eg, EGFR). In contrast, prostate cancer and some subtypes of glioblastoma (GBM) lack mutated or amplified tyrosine kinases. To investigate the regulation of phospho-tyrosine signaling in the absence of tyrosine kinase mutations or amplifications, we applied quantitative, label-free mass spectrometry to a GBM cell line, a mouse model of prostate cancer and human metastatic castration resistant prostate cancer (CRPC) biopsies.

Materials and Methods: Human GBM cells (U87) were treated with glucose starvation, epidermal growth factor (EGF), hydrogen peroxide (H_2O_2) or the phosphatase inhibitor vanadate. Mouse models of prostate cancer were generated by expressing combinations of non-tyrosine kinase oncogenes commonly found in prostate cancer (constitutively active AKT, androgen receptor (AR), ERG and K-Ras G12V) in the prostate *in vivo* regeneration model system (Lawson et al, 2009). Human treatment naÃ¯ve and metastatic castration resistant prostate cancer (CRPC) samples were obtained from the UCLA Translational Pathology Core Laboratory and the University of Michigan Rapid Autopsy Program. Quantitative, label-free mass spectrometry was performed on peptides immunoprecipitated using a pan-specific anti-phospho-tyrosine antibody (Drake et al, 2012; Graham et al, 2012).

Results and Discussion: U87 cells, which lack mutated or amplified tyrosine kinases, were treated with glucose starvation, EGF, H_2O_2 or vanadate. Each stimulus induced a unique signature of phospho-tyrosine signaling (Fig. 1a, n = 46 peptides). Unexpectedly, glucose starvation induced phosphorylation of proteins associated with focal adhesions, which suggested additional experiments that revealed a positive feedback loop between tyrosine kinase signaling and NADPH oxidase activity within focal adhesions (Graham et al, 2012). Next, analysis of mouse models of prostate cancer expressing non-tyrosine kinase oncogenes (eg, activated Akt and overexpressed ERG) (Drake et al, 2012) revealed oncogene-specific signatures of phospho-tyrosine signaling (Fig. 1b, n = 147 peptides), including an enrichment of EGFR substrates in Akt-ERG tumors. Finally, analysis of human prostate cancer patient samples demonstrated differential activation of tyrosine kinase signaling in treatment naÃ¯ve and metastatic CRPC tissues (Fig. 1c, n = 126 peptides), including the presence of Src kinase substrates in metastatic tumors.

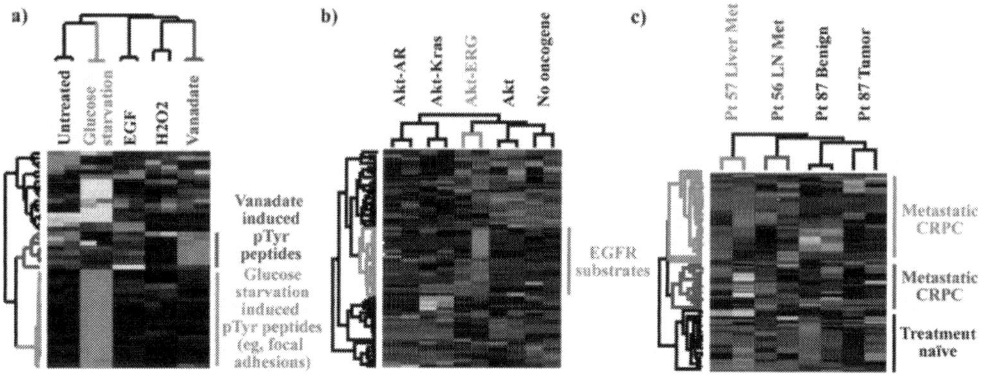

Figure 1. Hierarchical clustering of phospho-tyrosine peptides in a) U87 cells, b) mouse models of prostate cancer c) human CRPC patient (Pt) samples (LN, lymph node; Met, metastasis). Rows represent phospho-tyrosine peptides and columns represent samples run in duplicate. Red and green denote high and low phosphorylation, respectively.

Conclusions: We applied quantitative, label-free mass spectrometry to investigate regulation of phospho-tyrosine signaling in systems without mutated or amplified tyrosine kinases. This analysis revealed that a) metabolic perturbation (ie, glucose starvation) induces a positive feedback loop leading to focal adhesion-related tyrosine kinase signaling in U87 cells, b) the non-tyrosine kinase oncogenes Akt and ERG activate EGFR signaling in a mouse model of prostate cancer, and c) metastatic human CRPC biopsies exhibit elevated Src signaling compared to treatment naÃ¯ve biopsies. Together, these data demonstrate that cell lines and tumors lacking tyrosine kinase mutations or DNA amplifications can exhibit elevated tyrosine kinase signaling, suggesting that tyrosine kinase inhibitors may be effective therapeutics even in these cancers.

Acknowledgements: N.A.G. is supported by the UCLA Scholars in Oncologic Molecular Imaging (SOMI) program, NIH grant R25T CA098010. J.M.D. is supported by the Department of Defense Prostate Cancer Research Program W81XWH-11-1-0504.

References: Drake, J.M., *PNAS* (2012), 109(5), 1643-1648; Graham, N.A., et al, *Mol Syst Biol* (2012), 8:589; Lawson, D.A., et al, *PNAS* (2009).

(447d) Ovarian Cancer Metabolomic Profiles Differ in Ovarian Cancer Initiating Cells

Vermeersch, K. A., Georgia Institute of Technology
Styczynski, M. P., Georgia Institute of Technology

Originally presented on: 11/6/2013 9:24:00 - 9:42:00

A key emerging area in cancer research is the study of stem-like "cancer initiating cells" (CICs) in tumor populations. CICs are a major concern for effective cancer treatment, as they can self-renew, can differentiate into cancer cells, and exhibit chemoresistance; they are thus suspected as a primary cause of cancer recurrence. Very little is known about the molecular workings of these cells, and we must start to understand them in order to effectively treat CICs. Metabolism is widely known to be dysfunctional in cancer, most famously via the Warburg effect, but there is limited knowledge about CIC metabolism and about cancer metabolism as a whole. Metabolomics is the study of cellular metabolism as a whole by attempting to measure the levels of as many metabolites as possible. Metabolite levels integrate the effects of gene regulation, environmental perturbations, and numerous other levels of regulation.
By profiling the metabolism over time, we can obtain a systems-level understanding of the metabolic dynamics of CICs, which could profoundly affect our understanding, or even treatment, of cancer and CICs.

Here we present the results of an *in vitro* glucose deprivation experiment to model *in vivo* tumor conditions in ovarian cancer cells (OCCs) and ovarian CICs (OCICs). Extracellular and intracellular metabolite samples were taken over a period of two days and were analyzed using two dimensional gas chromatography – mass spectrometry (GCxGC-MS). To our knowledge, this is the first metabolic profiling of OCIC metabolism. Distinct metabolic profiles were found for the two cell types and between the control and glucose-deprived cells. Interestingly, independent of the differences imposed on the cells by glucose deprivation, there is a characteristic set of intracellular metabolites that only appear in OCCs, but not OCICs. These metabolites are spread across multiple pathways, suggesting system-wide metabolic alteration of OCICs from OCCs and potentially serving as indicators of CIC transition or even as therapeutic targets.

(447e) Integrated Energetics and Flux Analysis Reveals Differential Metabolic Reprogramming in Cancer Cells

Achreja, A., Indian Institute of Technology Madras
Yang, L.,
Nagrath, D., Rice University
Originally presented on: 11/6/2013 9:42:00 - 10:00:00

Cancer cell metabolism was investigated using systems biology tools such as metabolic flux analysis (MFA) to reveal intracellular cancer cell metabolic flux distribution. To obtain a complete role of nutrient-induced deprivation in metabolic reprogramming of cancer cells, we need to use metabolic flux analysis to determine unknown intracellular fluxes. Here, we have developed an integrated energy balance and flux balance analysis framework for analyzing cancer cell metabolism. Our framework allows estimation of intracellular fluxes and reveals specific adaptations and dysregulation of metabolic pathways in ovarian cancer cells as a result of the nutrient deprivation conditions. Our findings indicate that NAD+/NADH ratio regulates glutamine's direct effect on lactate production irrespective of the bioenergetic substrate requirement for different kinds of cancer cells. Our findings, using quantification of metabolites pertaining to central carbon and nitrogen metabolism provided clues as to which pathways relevant to glycolysis, the pentose phosphate pathway, the TCA, urea cycles and lipid cycles, amino acid and nucleotide metabolisms may be altered by nutrients in cancer cells. This resulted in identification of dominant factors which control cancer metabolism. We believe that development of our oncosecretomics based metabolic strategy provides an unprecedented opportunity for the development of nutritional supplementation that can deregulate the adaptations and alterations induced by various nutrients in cancer metabolism.

(447f) Temporal Metabolic Profiling of Plasma in Response to Endotoxemia in Humans

Kamisoglu, K., Rutgers, The State University of New Jersey
Sleight, K., Rutgers, The State University of New Jersey
Calvano, S. E., Robert Wood Johnson Medical School
Corbett, S. A., Robert Wood Johnson Medical School
Androulakis, I. P., Rutgers University
Originally presented on: 11/6/2013 10:00:00 - 10:18:00

Elective administration of bacterial endotoxin (LPS) to healthy human subjects has been used as a reproducible experimental procedure providing mechanistic insights into how cells, tissues and organs respond to systemic inflammation. Low doses of LPS transiently alter many physiologic and metabolic processes in a qualitatively similar manner to those observed after acute injury and systemic inflammation (1, 2); thus allow the analysis of body's responses to infectious stress at many physiological levels for building better understanding of the relevant pathology. Analysis of the metabolic response, in this regard, is of special interest since metabolic composition of a tissue is uniquely altered in response to stimuli due to collective effects of the regulations at transcriptional and translational levels. Therefore, concentrations of metabolites of a sample at a given time, i.e. the ?metabolome? (3), can be thought of as the metabolic fingerprint representative of the state of the body at that time and give information about the dominant regulatory mechanisms. Although the changes in the major metabolites, such as lipids, amino acids and glucose, has been previously documented for human endotoxemia (4); current study constitutes the first attempt of a complete metabonomic analysis that describes the coherent temporal patterns of the metabolic landscape in plasma following exposure to LPS.

Metabonomic analysis is comprised of global biochemical profiles determined in human plasma samples from healthy subjects. Nineteen subjects were administered either placebo (saline, n=4) or LPS (n=15) by a bolus injection and blood samples were collected at 5 time points within a 24 hour post-treatment period. Global biochemical profiles obtained by GC/MS and LC/MS/MS platforms represented information on 366 biochemicals including amino acids, short peptides, carbohydrates, lipids, nucleotides, cofactors and vitamins, xenobiotics and intermediate products of major energy production pathways. Within this dataset, we first identified those metabolites which have differential temporal profiles between control and LPS groups. The most significant difference between the two groups was at 6hr time point, which was also an inflection point separating development and recovery phases of the LPS induced metabolic changes. Then, through consensus clustering (5), we identified subsets of the metabolites with common coherent profiles. This analysis yielded two clusters with opposing directionality as shown in Figure 1. The first cluster (16 metabolites) was up-regulated within the first 6hr and down-regulated by the 24th hr and was mostly composed of metabolites from pathways related to lipid metabolism. The second cluster (21 metabolites), in contrast, was down-regulated within the first 6hr post-LPS, and then up-regulated by the 24th hr. Strikingly

15 out of 21 metabolites in this cluster were amino acids or derivatives and an additional 2 were dipeptides.

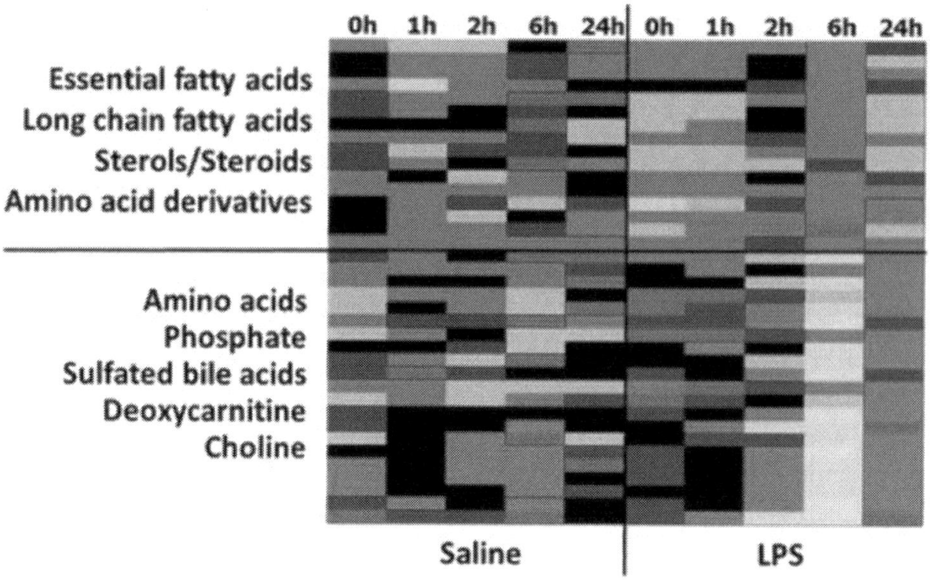

Figure 1: Differential patterns of metabolic response to LPS. Two clusters of plasma metabolites reflect two distinct patterns with opposing temporal directionality.

Preferential early clearance of amino acids can be attributed to hepatic uptake for the synthesis of acute phase proteins, while build-up of fatty acids in the plasma can be due to extensive lipolysis as a result of cytokine-mediated changes in lipid metabolism. Also, early build-up of some precursor molecules in sterol/steroid related pathways may indicate the propensity for hormone biosynthesis to suppress inflammation and promote recovery. These results highlight the changes in lipid and amino acid metabolism in response to infectious stress while also reflecting the most-closely associated plasma metabolites pointing out to co-regulatory schemes governing that response.

References

1. Lowry SF. Human endotoxemia: a model for mechanistic insight and therapeutic targeting. Shock. 2005;24 Suppl 1:94-100.

2. Calvano SE, Coyle SM. Experimental human endotoxemia: a model of the systemic inflammatory response syndrome? Surg Infect (Larchmt). 2012;13(5):293-9. PMCID: 3503465.

3.Nicholson JK, Lindon JC. Systems biology: Metabonomics. Nature. 2008;455(7216):1054-6.

4. Fong YM, Marano MA, Moldawer LL, Wei H, Calvano SE, Kenney JS, et al. The acute splanchnic and peripheral tissue metabolic response to endotoxin in humans. J Clin Invest. 1990;85(6):1896-904.

5. Nguyen TT, Nowakowski RS, Androulakis IP. Unsupervised selection of highly coexpressed and noncoexpressed genes using a consensus clustering approach. OMICS A Journal of Integrative Biology. 2009;13(3):219-37.

(447g) Metabolic Regulation for Double Increasing Balofloxacin Concentrations in Vibrio Alginolyticus

Peng, X.,
Yang, M., Tibtet vocational technical college
Zhao, X.,
Wang, S.,
Tian, Y.,
Guo, C.,
Liu, X.,
Li, H.,

Originally presented on: 11/6/2013 10:18:00 - 10:36:00

Metabolome refers to the complete set of small-molecule metabolites to be found within a biological sample. Metabolomics approach provides the global qualitative and quantitative analysis of the small-molecule metabolites and shows great promise as a mean to identify biomarkers of drug efficacy and understand metabolic mechanisms. In the present study, metabolomics carrying cellular metabolites of *Vibrio alginolyticus* has been developed to analyze dynamic changes in metabolites induced upon stress with double increasing concentrations of balofloxacin (BLFX). The objective was twofold. First, we were interested in identifying metabolites that, being specifically regulated in response to BLFX exposure, might be involved in BLFX-stressed biomarkers. Second, we wanted to exploit the metabolomics data to identify, among the BLFX-stressed metabolites, compounds capable of being reasonable proofs describing the characteristic feature of *V. alginolyticus* in response to the double increasing concentrations of BLFX. We hypothesize that the achievement of the two objectives will tell us a new story on bacterial metabolic response to gradient increasing concentrations of antibiotic.

Effect of double increasing concentrations of BLFX on bacterial growth. *alginolyticus* cells were cultured in LB medium separately with 0, 1/64, 1/32, 1/16, 1/8, 1/4 and 1/2 MIC of BLFX. These cultures were harvested when the bacterial cell density in medium without the drug arrived at approximately 1.0 OD. The resulting OD was negatively related to the double increasing concentrations of BLFX. Compared with the BLFX-con0, bacterial survival rates were 98.5 ± 1.34, 97.2 ± 0.78, 95.0 ±0.39, 92.0±0.61, 87.8 ± 0.54, 83.4 ± 0.48 in BLFX-con64, BLFX-con32, BLFX-con16, BLFX-con8, BLFX-con4 and BLFX-2, respectively. Growth ability of the bacterium was a negative correlation to the double increasing concentrations of the drug in medium. Significant difference was found at $P < 0.01$ between them. These results indicate an association of the inhibitory role in bacterial growth with BLFX concentrations.

Characteristics of dynamically metabolic response to double increasing concentrations of BLFX. To investigate a differential metabolome of *V. alginolyticus* in response to BLFX, GC/MS based metabolomics approach was used to identify metabolic responses to the double increasing concentrations of BLFX. Quantitative data throughout the entire time course were obtained for 89 compounds from 128 peaks after deletion of nine peaks that were below the limit of quantitation or too unstable to yield reliable information. OPLS-DA approach distributed the seven sample groups into different levels with the double increasing concentrations of BLFX in 3-D scatter plot, in which BLFX-con0 and BLFX-con64 were located in the bottom face and in the interface between the bottom and lateral faces, respectively. The others were clearly separated from BLFX-con0 and the separation was trended with the double increasing BLFX concentrations in the lateral face. Dynamics of the 89 compounds was shown in a clustered heat-map format. The format showed that BLFX-con0, BLFX-con64 and BLFX-con32, BLFX-con16, BLFX-con8 and BLFX-con4, and single BLFX-con2 were clustered together. In the first two clusters, there was significant distance between BLFX-con0 and the other two, and between BLFX-con4 and the other two, indicating the quantitative/qualitative pattern of metabolic changes in response to the double increasing BLFX concentrations. These results show that the metabolic profiles have significantly adjusted as a result of the increased antibiotic stress.

Identification of dynamically differential metabolites in response to double increasing concentrations of BLFX. For determining the characteristic compounds in BLFX, the multi-class unpaired method of SAM was used to identify the metabolites whose concentration differed significantly among all samples collected across the double increasing concentrations of BLFX. Despite these groups being only double-fold elevation of BLFX concentrations, they could be differentiated from SAM analyses. SAM identified 47, 64, 67, 56, 59 and 78 peaks representing 35, 48, 47, 48, 46 and 53 metabolites in BLFX-con64, BLFX-con32, BLFX-con16, BLFX-con8, BLFX-con4 and BLFX-2, respectively. The 44 positively significant metabolites and 9 negatively significant metabolites from 78 peaks detected in BLFX-2, containing the all changes occurred in the other samples, in order of decreasing significance palmitelaidic acid, phosphoric acid, tetradecanoic acid, adenosine, morphinan, oleic acid as the first six metabolites. Interestingly, ratios of metabolite areas with BLFX doses were not exactly proportional in every concentration for majority of the metabolites with differential levels, although changing tendency of the differential metabolites was detected. These results indicate that BLFX-stressed *V. alginolyticus* showed a quantitative/qualitative change in the response.

Then, PCA was separately used to identify significant di erences in the data set. The first six metabolites from factor weight were similar in the seven comparisons in order of decreasing significance L-proline, butanedioic acid, silanamine, phosphoric acid, L-alanine, glycine.

Potential markers of interest were further extracted from S-plots constructed following analysis with OPLS-DA. Fourteen biomarkers were identified from the loading S-plots for metabolic pathway analysis since they showed statistically significant ($P < 0.05$) difference in abundance. Out of them, the first eight existed only in comparison cross the seven groups (BLFX-con all), in which negative regulation of phosphoric acid, palmitelaidic acid, glutamine, glycine, 5'-adenylic acid and 9H-purin-6-amine and positive regulation of butanedioic acid, L-proline, adenosine were characterized. VIP analysis confirmed the reliability of the first four biomarkers phosphoric acid, butanedioic acid, L-proline and palmitelaidic acid in the total comparison. The other four were validated in comparison between two groups. However, the regulation by each of these compounds did not run through whole responsive process to the double increasing BLFX concentrations. They showed an alternating renewal process of positive/negative regulation. For example, negative and positive regulations were detected in BLFX-con64 vs -con0 (-0.18), BLFX-con16 vs –con32 (-0.19) and BLFX-con32 vs –con64 (0.20), BLFX-con8 vs –con16 (0.53), BLFX-con4 vs –con8 (0.66), BLFX-con2 vs –con4 (0.86) for phosphoric acid, respectively. Generally, an alternating renewal positive/negative regulation was detected in first four comparisons, BLFX-con64 vs -con0, BLFX-con32 vs – con64, BLFX-con16 vs –con32 and BLFX-con8 vs –con16, and then changing tendency kept until the last comparison BLFX-con2 vs –con4 during the stress processes for each of the key metabolites. There were two change points, BLFX-con64 vs -con0, BLFX-con16 vs –con32, in the response process. These results obtained from S-plot were consistent with the finding obtained from the clustering. These results not only indicate the importance in the structure of advance by degrees for details, but also show the quantitative/qualitative pattern of metabolic changes in response to the double increasing concentrations of BLFX.

Identification of differential metabolic pathways in response to double increasing concentrations of BLFX. Furthermore, ingenuity network analysis was used to perform more detailed analysis of pathways and networks influenced by the double increasing concentrations of BLFX. Out of the significantly altered 77 metabolites, 66 were used for the analysis in deletion of 11 unclear pathway metabolites. To identify possible pathways that were affected by BLFX, metabolites contributing to the separation of control and groups with the double increasing BLFX concentrations were analyzed in structure of advance by degrees using MetPA. The analysis revealed metabolites, which were identified together were important for the bacterial response to the drug, were responsible for 27-39 pathways. When the impact-value threshold was set to 0.10, eight potential targets pathways were filtered out. They were inositol phosphate metabolism, glycine, serine and threonine metabolism, alanine, aspartate and glutamate metabolism, pantothenate and CoA biosynthesis, arginine and proline metabolism, glyoxylate and dicarboxylate metabolism, tryptophan metabolism, methane metabolism. However, the Raw p, the original p value calculated from the enrichment analysis, analysis showed an alternating renewal process of significant/insignificant changes in the double increasing concentrations of BLFX. For example, significant difference was detected in BLFX-con64 (0.001), BLFX-con32 (0.03) and BLFX-con8 (0.02), but not in BLFX-con16 (0.09), BLFX-con4 (0.32), BLFX-con2 (0.16) for glycine, serine and threonine metabolism (Table 3), suggesting an efficient strategy for the bacterium in response to BLFX through alternating renewal modification of pathways.

Identification of promoting constructive metabolism in response to double increasing BLFX concentrations. We further detected bacterial dry weight and protein content for understanding whether the bacterium response to BLFX mainly through promoting constructive metabolism or catabolism. First, bacterial dry weight was measured. An increasing tendency with the double increasing BLFX concentrations was detected with correlation coefficient ($P < 0.05$). Approximately 10 – 20 % dry weight was increased in medium with 1/8 – 1/2 MIC of BLFX. Then, significantly elevated total protein content was determined at $P < 0.01$ with a positive correlation between the double increasing BLFX concentrations and protein content. These results indicate that *V. alginolyticus* enhances protein constructive metabolism in response to BLFX.

In summary, our findings indicate that *V. alginolyticus* mounts an alternating renewal positive/negative regulation of metabolites and discontinuity modification of pathways for the increasing BLFX concentrations, forming characteristic changes of the quantitative/qualitative pattern. These findings have significant implication in controlling BLFX-resistant *V. alginolyticus* by adjusting metabolites and metabolic pathways.

This work was sponsored by grants from "973" project (2012CB114406) and NSFC projects (30972279, 40976080). Doctoral Fund of Ministry of Education of China (100171110029).

(447h) Metabolic Analysis of Saccharomyces Cerevisiae to Evaluate the Relationship Between Nitrogen Utilization Efficiency and Ethanol Tolerance During Alcoholic Fermentation

Lozada Contreras, M., University of California, Davis
Block, D. E., University of California Davis
Originally presented on: 11/6/2013 10:36:00 - 10:54:00

During alcoholic fermentation, the sugar utilization activity of yeast cells decreases in the presence of alcohol, followed eventually by a loss in cell viability. Understanding ethanol tolerance should allow its modulation in yeast strains with other desirable characteristics. Commercial yeast strains, given an identical initial nitrogen concentration and composition, will form different concentrations of biomass. Theoretically, strains that use nitrogen more efficiently to create more biomass will be able to tolerate higher levels of ethanol produced. The goal of this project is to understand the relationship between nutrient environment, nitrogen metabolism, and specific growth rate in *Saccharomyces cerevisiae* on a fundamental level, in order to develop a rational approach to strain modification for increasing ethanol tolerance. This will be achieved by determining the extracellular and intracellular concentrations of key metabolites in yeast collected during the exponential and stationary phases. Thirty-two commercial yeast strains were compiled with a wide range of growth and ethanol tolerance attributes. Nitrogen utilization efficiency and fermentation kinetic parameters were studied in each strain. Maximum optical densities for these yeast ranged from 2.68 to 6.17. A complete metabolic analysis will be performed to investigate the differences in carbon and nitrogen metabolism using supernatant and cell samples saved from fermentations using GC-MS.

(448a) Using An Agent-Based Model of Mycobacterium Tuberculosis Infection to Understand the Roles of Interleukin-10 During Disease Progression

Cilfone, N. A., University of Michigan
Kirschner, D. E., University of Michigan Medical School
Linderman, J. J., University of Michigan
Originally presented on: 11/6/2013 8:30:00 - 8:48:00

Introduction

Mycobacterium tuberculosis (Mtb) is the causative agent of the infectious disease tuberculosis (TB). Over 2 billion people are infected with *Mtb*, causing 2-3 million deaths each year. Although the root cause of the disease has been known for over a century, we still lack a fundamental understanding of many of the processes occurring during infection. The granuloma is a structure that forms in the lungs as a result of the immune response to inhaled *Mtb*. Granulomas consist of *Mtb* and infected macrophages in the center, an inner cuff of mainly resting and activated macrophages, and an outer cuff of predominantly T and B lymphocytes. Formation of a granuloma relies on many coordinated immunological processes directed by the molecular mediators known as cytokines[1]. Interleukin-10 (IL-10), an anti-inflammatory cytokine, is thought to be a critical regulator of many processes during granuloma formation and function, but understanding its importance has been difficult due to the complexity and heterogeneity of a granuloma and the myriad of cellular and signaling processes acting across multiple spatial and temporal scales. Experiments attempting to elucidate these phenomena are either difficult or currently infeasible, thus we have developed a complementary *in silico* approach to study the immune response to *Mtb*and its physiological consequences.

Methods

We developed a computational model of *Mtb* infection, validated the model against experimental observations in non-human primates (NHPs), and subsequently used the model to determine how IL-10 contributes to granuolma pathology and progression. We built a hybrid agent-based model (ABM) of *Mtb* infection that describes immune processes over three different biologoical scales: tissue, cellular, and molecular[2]. At the molecular scale we modeled single-cell level receptor-ligand trafficking events, diffusion, and degradation of IL-10 and tumor necrosis factor-α (TNF-α), a pro-inflammatory cytokine, using ordinary and partial differential equations and linked these models to a tissue and cellular scale ABM[2]. The ABM includes both macrophages and T cells, each with multiple states. We model three bacterial sub-populations: intracellular *Mtb*, extracellular *Mtb*, and non-replicating *Mtb*. The ABM represents multiple immune mechanisms using a set of rules and interactions, such as chemokine-induced cellular recruitment, growth of *Mtb*, macrophage phagocytosis of *Mtb*, cellular apoptosis, and T cell killing. We validated our ABM against observations in the NHP model of *Mtb* infection for wild type, *Tnf-α*$^{-/-}$, and *Ifn-γ*$^{-/-}$ scenarios[2]. We utilize uncertainty and sensitivity analysis to identify model parameters that have significant effects on model output and to create cell-specific IL-10 deletions to ascertain the roles of IL-10 production from different immune cells.

Results

Our *in silico* model predicts that multiple groups of IL-10 parameters, representing processes relevant to cytokine synthesis, signaling, and spatial distribution, control concentrations of TNF-α and IL-10 in a granuloma environment and eventually determine granuloma outcome, at the tissue scale, over the long-term. We demonstrate that each group of parameters is balancing a trade-off between host-induced tissue damage and bactericidal processes through various IL-10 mechanisms. We determine, for the first time, that the balance of TNF-α and IL-10 concentrations is an essential mediator of *Mtb* infection control and prevention of host-induced tissue damage. Our results predict that granulomas having higher average concentrations of IL-10 than TNF-α promote containment of bacteria and prevention of host tissue damage instead of bacterial clearance with high levels of healthy tissue damage.

We create models of cell-specific IL-10 deletions, including infected macrophage *Il10*$^{-/-}$, activated macrophage *Il10*$^{-/-}$, and regulatory T cell *Il10*$^{-/-}$, which suggest different roles for IL-10 produced by infected macrophages and activated macrophages. We show that production of IL-10 by activated macrophages is used as a host-protective mechanism to regulate tissue damage and control macrophage activation, while production of IL-10 by infected macrophages is used as an immune evasion mechanism to promote persistent infection. We then explore possible immunomodulation intervention strategies, such as the addition of anti-IL-10R antibodies, in order to limit *Mtb*-induced immune evasion mechanisms.

Conclusions

Our modeling approach represents a critical step towards fully understanding the role of IL-10 and its effects on *Mtb* infection outcome in the long-term. In addition, the hybrid ABM platform we developed will allow us to rapidly explore new intervention strategies, such as immunomodulation, to modify the immune response to *Mtb*, narrowing the large design space for future non-human primate experiments.

Acknowledgements

We acknowledge funding from NIH (R01 EB012579 and R01 HL110811).

References

1. Flynn, J. L. & Chan, J. Immunology of tuberculosis. *Annual review of immunology* **19**, 93–129 (2001).

2. Cilfone, N. A., Perry, C. R., Kirschner, D. E. & Linderman, J. J. Multi-Scale Modeling Predicts a Balance of Tumor Necrosis Factor-α and Interleukin-10 Controls the Granuloma Environment During Mycobacterium tuberculosis Infection. *PloS one* (2013) - Accepted.

(448b) Vessel Wall Transport: Can Increasing the Endothelial Cell Expression of the Membrane Protein Aquaporin-1 Slow Down Pre-Lesion Atherosclerosis?

Joshi, S. D., City College and Graduate Center of City University of New York
Raval, C. B., Gradute Center at the City University of New York
Jan, K. M., College of Physician and Surgeons, Columbia University
Rumschitzki, D. S., The City College of The City University of New York
Originally presented on: 11/6/2013 8:48:00 - 9:06:00

Macromolecular (e.g., low-density lipoprotein (LDL) cholesterol) transport across the arterial wall appears to be the earliest pre-atherosclerotic triggering event. This transport occurs mainly by advection: Transmural (i.e., across the vessel wall) pressure differences (i.e., blood pressure) drive plasma (water and small solutes) across the monolayer of endothelial cells (EC) that lines the inside of the artery wall. One EC every few thousand has junctions that are temporarily widened, some because they are either dying, dividing, or have stigmata, and the transmural flow can advect large tracers through them into the wall. A wall parameter, the hydraulic conductivity (L_p), defined as the ratio of the transmural water flux to hydrostatic pressure difference, is central to the understanding of this transmural water transport. On one hand, it advects macromolecules like LDL into the arterial subendothelial intima (SI) through these rare, widened EC leaks and spreads it in the SI, thereby giving it the chance to bind to extracellular matrix in the SI layer and possibly to trigger the start of atherosclerotic lesion formation. On the other hand, the overall water flow across the normal (non-leaky) endothelium dilutes the local SI LDL concentration, thereby likely slowing binding kinetics, and washing not-yet-bound lipid further into the wall. Our group's discovery of the presence of the ubiquitous water channel membrane protein Aquaporin-1 (AQP) in rat aortic endothelial cells suggests a new possibility of water transport across the normal (non-leaky) endothelium, alongside the generally accepted paracellular route. Interestingly, we found that chemically blocking AQPs or knocking them down with siRNA changes rat aortic wall Lp in a strongly pressure dependent manner. We then proposed a new theory that agrees well with this otherwise perplexing experimentally-observed pressure-dependence. One upshot of this theory is that it suggests that AQPs contributes about 30% to the phenomenological endothelial Lp at low transmural pressures.

In the present work we investigate the effect of AQPs on the overall transport of various species, like Horseradish Peroxidase (HRP) and LDL, across the vessel endothelium and its further spread in the vessel wall using the advection-diffusion model developed by Huang et al. (*Journal of Biomechanical Engineering*, vol 116, 1994, pp. 430-445) and later improved by Zeng et al. (*American Journal of Physiology*, Heart Circ Physiol, 302, 2012, pp. 1683-1699). We extend this model by incorporating the trancellular water flow through AQPs and use some important measurable phenomenological parameters like the hydraulic conductivity of the endothelium plus intima (Lp_{e+i}). We also include the intima thicknesses corresponding to a given $Lp_{e+i.}$ Our local filtration model mentioned above allows us to calculate how these important

(448c) Platelet Count Control in Immune Thrombocytopenic Purpura Patient: Optimum Romiplostim Dose Profile

Tsai, C. H., University of Delaware
Sandler, S. I., University of Delaware
Ogunnaike, B. A., University of Delaware
Originally presented on: 11/6/2013 9:06:00 - 9:24:00

Patients with immune thrombocytopenic purpura (ITP), a disease characterized by abnormally low platelet count, are susceptible to excessive bleeding as a direct consequence. While the problem of low platelet count can be addressed fundamentally either by slowing down the rate of platelet destruction or by increasing platelet production, or both, one of the more effective means of treating ITP patients is to increase platelet production with romiplostim, a relatively new drug that has recently been approved for treating patients who have not responded to other treatments. However, current romiplostim treatment strategies tend to result in undesirable responses where platelet counts oscillate between dangerously low values followed extremely high peaks, as a direct consequence of the complex nonlinear dynamics associated with platelet count regulation. The objective of this study is to employ a model-based approach to determine the optimum dose profile of romiplostim required to maintain a steady platelet count of 70×10^9/L for a specific ITP patient.

Using clinical platelet count data obtained from the patient in response to a series of subcutaneously applied doses of romiplostim, a standard pharmacokinetics and pharmacodynamics (PKPD) model was modified, validated, and analyzed first to obtain insight into the patient's physiological characteristics. The customized model was then subsequently used to investigate, via simulation, the performance of three control strategies ("fixed dose" open-loop control, "variable dose" discrete PI feedback control, and "variable dose" open-loop optimal control) implemented for both weekly and biweekly treatment regimens. Our results indicate that with both treatment frequencies, the "fixed dose" open-loop control strategy resulted in unacceptable sustained oscillating platelet counts while the PI feedback control and optimal open-loop control led to stable platelet count profiles after about 50 days but only for weekly injections. Additional details concerning the controller designs and performance under various conditions will be discussed along with other notable observations, for example: that the time span required to achieve a stable platelet count is very sensitive to the values of initial doses administered at the very beginning of the treatment; and that, for the specific patient in question, regardless of control strategy, it is nearly impossible to maintain stable, non-oscillatory platelet count with biweekly injections.

(448d) Design of a Drug Infusion Controller for Reducing Post-Operative High Blood Pressure

Mailhiot, S., Rensselaer Polytechnic Institute
Newman, M., Rensselaer Polytechnic Institute
DiNardi, N., Rensselaer Polytechnic Institute
Maziarski, K., Rensselaer Polytechnic Institute
Baral, E., Rensselaer Polytechnic Institute
Hasekioglu, T., Rensselaer Polytechnic Institute
Hahn, J., Rensselaer Polytechnic Institute
Originally presented on: 11/6/2013 9:24:00 - 9:42:00

Fast acting intravenous vasodilators are often given to postoperative hypertensive patients in the hospital. Due to varied patient response to different concentrations of a drug, vasodilators must be titrated during infusion. Currently, vasodilator titration is adjusted manually over time: a clinician periodically checks patient blood pressure and adjusts the infusion rate of vasodilator from a smart infusion pump based on hospital protocols written from package inserts provided by the drug manufacturer. This method of titration is time consuming for the clinician and can be inaccurate and imprecise, which clearly presents an opportunity for automating the drug infusion rate adjustment.

This is a problem that has been investigated by many different researchers in the past. Sodium nitroprusside was used as a drug for decades for this type of problem and it has been well characterized. A model for the patient's response to the administration of the drug has been developed by Slate *et al.*(Slate, & Sheppard, 1982; Slate, Sheppard, Rideout, & Blackstone, 1980) and has been used by many researchers for controller design (Isaka, & Sebald, 1993). Close to one hundred publications of different control strategies using sodium nitroprusside had been reported by 1993 (Isaka, & Sebald) and papers are still published on this topic today. However, sodium nitroprusside is becoming less clinically relevant and is being replaced by Cleviprex (Grossman & Messerli, 2011) as the drug of choice for treating postoperative hypertension. The effect that Cleviprex has on the dynamic behavior of the mean arterial blood pressure of a patient differs significantly from what is known for sodium nitroprusside, and existing control strategies may not be appropriate when Cleviprex is used.

This work develops a dynamic model of the mean arterial blood pressure of patients who are given Cleviprex and then uses this model to develop a controller. It should be noted that this work is the first that develops a model for Cleviprex for treating postoperative hypertension. Data for model development is taken from Bailey (2002) and Singla (2008). Cleviprex is a fast acting drug and the response of a patient to infusion of the drug can be appropriately captured by a first order transfer function. However, there is significant patient-to-patient variability and the gain of the model can vary by an order of magnitude. This is a point that needs to be addressed for controller design: the patient is first given a continuous low dose of the drug and the response is analyzed after 3 minutes. The patient is then categorized as sensitive, nominal, or insensitive in his/her response to the drug and a separate PI controller has been designed for each case. It is important to note that this model identification step is possible due to the fast acting nature of Cleviprex, whereas this was impractical for sodium nitroprusside. In a next step, the chosen PI controller is used within a closed-loop control framework to lower the blood pressure to a desired level. Specific attention has been paid to include clinically-relevant design parameters such as upper/lower bounds for desirable blood pressure, upper limits on the drug dosage, keeping the infusion rate constant once the blood pressure has stabilized, and most importantly, alarming the clinician to unexpected behavior.

The controller is tested with the developed patient model using Monte Carlo simulations. Patient sensitivity, weight, and target blood pressure range are varied during testing. The controller proves effective for reducing the blood pressure for 98% of simulated patients while not exhibiting adverse effects, e.g., extensive overshoot, in any of the investigated cases.

References

Bailey, J. M., Lu, W., Levy, J. H., Ramsay, J. G., Shore-Lesserson, L., *et al.,* "Clevidipine in Adult Cardiac Surgical Patients," *Anesthesiology,*96 (5), pp. 1086-1094 (2002).

Grossman, E. and Messerli F. H., "Hypertensive Emergencies," Chapter 4.6, in "Oxford Desk Reference: Cardiology," H. F. Tse, G. Y. Lip, and A. J. S. Coats, Eds., University Press, Oxford, NY, pp. 137-138 (2011).

Isaka, S., and Sebald, A. V., "Control Strategies for Arterial Blood Pressure Regulation," *IEEE Transactions on Biomedical Engineering,*40 (4), pp. 353-363 (1993).

Singla N., Warltier, D. C., Gandhi, S. D., Lumb, P. D., Sladen R. N., *et al.,* "Treatment of Acute Postoperative Hypertension in Cardiac Surgery Patients: An Efficacy Study of Clevidipine Assessing Its Postoperative Antihypertensive Effect in Cardiac Surgery-2 (ESACPE-2), a Randomized, Double-Blind, Placebo-Controlled Trial," *Anesthesia & Analgesia,*107 (1), pp. 59-67 (2008).

Slate, J. B., and Sheppard, L. C., "Automatic Control of Blood Pressure by Drug Infusion," *IEE Proceedings-A-Science Measurement and Technology,*129 (9), pp. 639-645 (1982).

Slate, J. B., *et al.,* "Closed-loop Nitroprusside Infusion: Modeling and Control Theory for Clinical Application," *IEEE International Symposium on Circuits and Systems 1980,* pp. 482-488 (1980).

(448e) Model-Based Identification of Therapeutic Intervention Targets in Calcium-Related Pathologies

Christie, C. R., Virginia Polytechnic Institute and State University
Achenie, L. E. K., Virginia Polytechnic Institute and State University
Ogunnaike, B. A., University of Delaware
Originally presented on: 11/6/2013 9:42:00 - 10:00:00

Calcium (Ca) homeostasis is the maintenance of a stable plasma Ca concentration in the human body in the presence of Ca variability in the physiological environment (e.g. by ingestion and/or excretion). For normal physiological function, the total plasma Ca concentration must be maintained within a very narrow range (2.2-2.4mM) [1]. Meeting such stringent requirements is the task of a regulatory system that employs parathyroid hormone (PTH) and calcitriol (CTL) to regulate Ca flux between the plasma and the kidneys, intestines and bones. Chronic imbalances in plasma Ca levels are associated with disorders of the regulatory organs, which cause abnormal hormonal secretion and activity. These changes in hormonal activity may lead to long-term problems, such as, osteoporosis (increased loss of bone mineral density), which arises from primary hyperparathyroidism (PHPT) – hyper secretion of PTH [2].

We previously proposed a representation of the physiological process of Ca regulation in the form of a control system block diagram comprising: sensor, controller, actuators and process [3]. Mathematical models were generated for each component sub-process by applying standard conservation of mass principles augmented with available mechanistic information. The resulting model is a system of ordinary differential equations (ODE) that describe the dynamics of: (i) PTH in the controller; (ii) calcitriol from the kidneys in the actuator; (iii) bone cells (osteoblasts and osteoclasts) from the bone in the actuator; and (iv) plasma calcium (the "process"). The overall model was subsequently validated with published clinical data. Finally known pathological conditions were accurately simulated through induced irregularities in the control system components which represent defects in the Ca homeostatic system. The irregularities include defects of the sensor, controller and/or actuator.

In this presentation, we focus on our approach to identifying potential targets for therapeutic intervention in PHPT and chronic kidney disease (CKD). We apply sensitivity analysis to models of the pathological conditions corresponding to PHPT and CKD to identify the intrinsic parameters that have the most influence on the key hormones/ions (PTH, CTL, Ca and phosphate). Additionally, we explore the extent to which each of these parameters affects the different hormones/ions and explain the physiological significance of these findings in the two pathologies. Furthermore, we show that our analysis is consistent with the clinical results of calcimimetic therapy in PHPT and CKD. Calcimimetics constitutes a class of drugs used to reduce PTH secretion through targeting of the calcium-sensing receptors on the parathyroid gland. We anticipate using this approach in hypothesizing therapeutic intervention targets for other Ca-related pathologies.

1. Kovacs, W.J. and S.R. Ojeda, *Textbook of Endocrine Physiology*. 6th ed2012, Oxford: Oxford University Press. xii, 462 p.

2. Fraser, W.D., *Hyperparathyroidism*. Lancet, 2009. **374**(9684): p. 145-58.

3. Christie, C.R., L.E.K. Achenie, and B.A. Ogunnaike, *A Control Engineering Perspective to Modeling Calcium Regulation and Related Pathologies*, in *Annual Meeting October 28- November 2,*2012.

(448f) Cardiac Alternans Annihilation By Optimal and Model Predictive Control

Yapari, F.,
Dubljevic, S., University of Alberta
Originally presented on: 11/6/2013 10:00:00 - 10:18:00

Cardiac alternans are defined as the beat-to-beat oscillations of the action potential duration (APD) in paced cardiac cells. This phenomena has been identified as a possible precursor to ventricular arrhythmia and even sudden cardiac death. The annihilation of these alternans is therefore a promising anti-arrhythmic strategy that requires more exploration within the realm of cardiac implantable devices.

In this work, optimal and model predictive controllers are implemented on the small amplitude of alternans parabolic partial differential equation (PDE). This PDE was developed by [1], and describes the dynamics of the amplitude of alternans along a cable of cardiac cells with a finite length. In our proposed control strategy, both electrical and mechanical stimuli are utilized to annihilate the alternans along the length of the cable. The electrical stimulus represents a pacer which is applied at the cable's boundary, while the mechanical stimuli are spatially distributed along the cable and affect the APD by altering the intracellular calcium dynamics.

In previous work, it has been demonstrated that the successful annihilation of cardiac alternans can be achieved using a low dimensional optimal controller [2]. However, optimal controllers might violate the system's naturally present physiological constraints. This provides the motivation to develop a model predictive controller for the cardiac system in which the constraints are explicitly accounted for. We demonstrate that by using the MPC formulation of [3], stabilization of the alternans along the cable is achieved. Furthermore, satisfaction of the input constraints arising from the actuator limitations (pacing limitations) as well as state constraints which are naturally present in the cardiac system is also achieved through this MPC formulation.

Finally, we address the issue of uncertainties in the PDE parameters by studying the robust stability of the closed loop system. These uncertainties arise from the fact that the PDE parameters are obtained by applying a pacing algorithm on a single cell ionic model, and therefore will vary depending on the choice of ionic model.

References

[1] B. Echebarria and A. Karma Instability and spatiotemporal dynamics of alternans in paced cardiac tissue. Physical Review Letters, vol. 88 (2002), no. 20, pp. 208101.1-208101.

[2] S. Dubljevic and P.D. Christofides Optimal mechano-electric stabilization of cardiac alternans. Chemical Engineering Science, vol. 63 (2008), pp. 5425-5433

[3] K.R. Muske and J.B. Rawlings Linear model predictive control of unstable processes. J. Proc. Cont., vol. 3 (1993), no. 2, pp. 85 - 96

(448g) Autocatalytic Release of Heme in Sickle and Normal Erythrocytes

Aich, A.,
Freundlich, M., cole Nationale Sup rieure de Chimie de Paris
Petrova, E. V.,
Li, Y.,
Dinu, B.,
Airewele, G.,
Vekilov, P. G., University of Houston
Originally presented on: 11/6/2013 10:18:00 - 10:36:00

Autocatalytic release of heme in sickle and normal erythrocytes

Anupam Aich[1,[1]], Melissa Freundlich[1,2, [1]], Elena V. Petrova[1], Ye Li[1], Bogdan Dinu,[3] Gladstone Airewele,[3] and Peter G. Vekilov[1,4,*]

[1] *Department of Chemical and Biomolecular Engineering, University of Houston, Houston, TX 77204, USA*

[2] *École Nationale Sup rieure de Chimie de Paris, 11, rue Pierre et Marie Curie, 75231 PARIS Cedex 05 FRANCE*

[3] *Texas Children's Hospital, Baylor College of Medicine, 6621 Fannin St. Houston, TX 77030, USA*

[4] *Department of Chemistry, University of Houston, Houston, TX 77204, USA* Abstract

The mutation responsible for sickle cell anemia has two main consequences: the mutant hemoglobin, HbS, polymerizes, and it partially releases its heme. Here we determine the concentration of free heme in sickle, sickle trait, and normal erythrocytes. We develop a novel assay for heme quantification, which is robust, sensitive to 100 pM, and accurate within 10%. We separate the free heme present in the red cells from the hemoglobin by dialysis. We use the heme in the dialysate to reconstitute horse radish peroxidase in excess of the apoenzyme, and measure the rate of enzymatic decay of peroxide. We find that the average concentration of free heme in sickle cell patients is 41±12 mM, in sickle trait individuals—35±4 mM, and in normal adults—22±9 mM. We find that the release of heme in both sickle cell patients and normal adults is autocatalytic, i.e., the presence of free heme induces stronger heme release. The found unexpectedly high concentration of free heme suggests that it may be an important factor in the pathology of sickle cell anemia. The autocatalytic release may contribute to the known variability and unpredictability of the disease.

[1] These two authors contributed equally to this work.

(448h) Biofilm Mitigation Via Magnetic Hyperthermia

O'Toole, A., University of Iowa
Nuxoll, E., University of Iowa
Originally presented on: 11/6/2013 10:36:00 - 10:54:00

Bacterial biofilm infection is a common (~ 2 to 4%) complication for recipients of surgically implanted medical devices. Due to the tremendous increase in antibiotic resistance when these bacteria enter the biofilm phenotype, present treatment requires explantation and replacement of the device, often with multiple surgeries and always with much longer patient recovery time. The specific objective of this study is to determine the feasibility of using heat to deactivate these biofilms *in situ*. While extreme heat is routinely used to sterilize (e.g. autoclaves), such temperatures have a severe cost within the body. Despite extensive studies on thermal deactivation of bacteria in the planktonic phenotype over a wide range of temperatures (e.g., pasteurization protocols), surprisingly little is known about the thermal deactivation of biofilms except under extreme (>150 °C) conditions. Here, we report the deactivation of *Psuedomonas aeruginosa* biofilms cultured at 37°C, 5% CO_2 for 24 hours in a drip-flow reactor and subjected to heat shocks on the range of 50°C to 80°C for durations of 1 to 30 minutes. Heat shocks were delivered by immersion in thermostatted media for the prescribed time and the resulting concentration of colony forming units (CFU/mL) were quantified using direct enumeration. Up to 9 orders of magnitude reduction in CFU concentration was observed, indicating that thermal deactivation is a reasonable approach to biofilm mitigation. By integrating this approach with a magnetic nanoparticle implant coating, we aim to treat implant infections *in situ* without explantation or device replacement.

(453a) Scalable Production of Dopaminergic Neuron Progenitors for Treating Parkinson's Disease

Lei, Y., University of California, Berkeley
Schaffer, D., University of California, Berkeley
Originally presented on: 11/6/2013 8:30:00 - 8:48:00

Introduction: Cell replacement therapies (CRTs) for Parkinson's disease (PD) based on human pluripotent stem cells (hPSCs), including human embryonic stem cells (hESCs) and induced pluripotent stem cells (iPSCs), are promising. Prior studies with human fetal tissues containing dopaminergic neuron progenitors (DA) have shown this promise, but they have problems with scale and ethics, which hPSCs can solve. hPSCs have the capacities for indefinite in vitro expansion and, due to recent advances, can be efficiently differentiated into DA progenitors or neurons. However, a GMP compliant and cost effective culture system is required to produce these cells at various scales before they can be broadly utilized in the clinic. It is becoming clear that current 2D-based cell culture system cannot generate the massive numbers of cells needed for such therapeutic application. As an illustrative calculation, in several clinical trials for PD, patient benefit was associated with the survival of ~200,000 transplanted dopamine neurons, and the survival of hPSC-derived DA neurons in animal models is often ~5% or less. At a typical density of ~5,000 DA neurons/cm^2 in 2D culture system, and with over a million Americans with PD, culturing sufficient numbers of DA neurons would require ~0.5 km^2 of cell culture surface (equivalent to ~100 football fields), not to mention the area required to expand the parent hPSCs. In this presentation, we will introduce a simple, defined, scalable and GMP compliant 3D culture system for producing DA progenitors. We will also present data on using these cells for treating PD with rodent model.

Materials and Methods: hPSCs were first expanded to a target number before differentiation into DA progenitors, both in the 3D system. PSCs were dissociated into single cells, and cultured in the 3D system with a fully defined medium for 5 days before the next passage. Cells were continuously expanded until the target number was reached. Following expansion, a DMEM/F12 medium supplied with a small molecule blend was used to convert the PSCs into DA progenitors, which were then implanted into the striatum of rat to evaluate their safety, maturation, and integration. The whole process was thus done in a fully defined system with only recombinant growth factors and no matrix proteins

Results and Discussion: With the 3D culture system, hPSCs could be efficiently expanded with a high replication rate (20-fold over 5-day per passage), yield (~2.0x10^7 cells/ml), and purity (~95% Oct4+) through single cell inoculation. We developed a small molecule cocktail to convert these hPSCs into DA progenitors with efficiency close to a published protein-based protocol that has been used in 2D. This approach successfully yielded DA progenitors in the new 3D culture system, resulting in a much higher yield (~8x10^7 cells/ml) than the conventional 2D culture system. The ability of these cells to survive, mature into TH+ DA neuron and function in vivo will also be presented.

Conclusions: In summary, we developed an efficient GMP compliant system along with a small molecule based protocol for producing DA progenitors. This system has the capacity to produce sufficient DA progenitors at scales necessary for clinical translation and application.

(453b) Alginate Encapsulation of Human Embryonic Stem Cells Enhances Its Directed Differentiation to Pancreatic Islet-Like Cells

Richardson, T., University of Pittsburgh
Kumta, P. N., University of Pittsburgh
Banerjee, I., University of Pittsburgh
Originally presented on: 11/6/2013 8:48:00 - 9:06:00

Introduction: Type 1 diabetes (T1D) is an autoimmune disease affecting myriadpeople worldwide whereinthe beta cells ofthe pancreasare destroyed resulting in insulin dependence. The Edmonton Protocol of islet transplantation affords the feasible option for treatment of diabetes. However shortage of donor islets combined with immune rejection limits islet transplantation from becoming a viable therapy. We propose alginate encapsulation of insulin producing cells derived from human embryonic stem cells (hESC) is a potentially viable therapy for type 1 diabetes, due to the unlimited supply of hESCs and the immunoisolation capability of alginate capsules.

Materials and Methods: We developed a stage wise directed differentiation protocol to derive islet-like cells from hESCs. Definitive endoderm was induced with ActivinA and Wnt3A for 4 days followed by pancreatic progenitor induction with Cyclopamine and Cyclopamine with Retinoic Acid for 2 days each; respectively. Maturation was induced by Nicotinamide for 2 days and Nicotinamide with DAPT for 7 days. A single cell suspension of ROCK inhibitor treated hESCs weresuspended in 1.1% alginate and 0.2% gelatin followed by subsequent encapsulation by drop wise addition into a 100mM $CaCl_2$ bath with 10mM HEPES. hESCs were encapsulated at 3 different stages: after maturation to islet-like cells, the definitive endoderm (DE) stage or as undifferentiated (UD) hESCs. The hESCs were then continued with remaining differentiation protocol based on the encapsulation stage. Islet maturation was analyzed by gene expression of insulin, *MAFA*, glucagon, and *PDX1*. Further characterization was done by immunostaining, flow cytometry, and protein content of islet-specific markers. Additionally, the gene expression of extracellular matrix (ECM) and cell adhesion molecules (CAM) was analyzed by qRT-PCR array. Finally, key molecules of the TGFβ and WNT pathway were analyzed at the DE stage to further investigate the effect of alginate encapsulation on hESC differentiation.

Results: Our results to date show that encapsulation of fully matured cells result in low viability, and significantly lower gene expression of mature markers compared to tissue culture plastic controls (TCP), prior to encapsulation. However, when hESC derived DE cells were encapsulated and further induced to islet like cells; the cells exhibited strong maturation markers significantly higher than TCP controls. The viability of the encapsulated cells, however, still remained low. Finally, we evaluated the encapsulation of UD hESCs and allowed them to progress through all stages of differentiation under encapsulation. The encapsulated UD hESCs remained viable and grew into large colonies towards the end of the differentiation protocol. Very encouragingly, many of the differentiation markers analyzed were even stronger when cells were encapsulated in contrast to TCP controls. To further analyze the maturation of hESCs fully differentiated under encapsulation, we measured the intracellular protein content of the islet-specific hormone c-peptide. The encapsulated cells showed significantly higher c-peptide (0.28pg c-pep /ug total-protein) than TCP controls (0.085pgc-pep/ug total-protein). Upon glucose challenge, the hESCs fully matured under encapsulation secreting approximately 19 pg/ml of c-peptide into the media. The ECM and CAM array was analyzed at the DE and PP stage for encapsulated cells, normalized to cells differentiated on TCP. While most genes were down regulated, NCAM1 and VCAM1 were upregulated at the DE stage and MMP7, 15 were upregulated at the PP stage. Investigation of the TGFβ pathway showed an increased ratio of pSMAD3/pAKT and pSMAD2/pAKT under encapsulation, compared to the TCP control.

Discussion: Our results indicate that islet-specific maturation of hESCs is not only a feasible option in the 3D encapsulated alginate system but is alsoa configuration that appears to significantly enhance maturation of hESCs. This is enhanced maturation is likely due to higher levels of pSMAD3/pAKT under encapsulation. C-peptide secretion upon glucose stimulation shows that the islet-like cells are able to fully cleave proinsulin into mature c-peptide and insulin for secretion. Further studies are currently underway to adapt this protocol to microencapsulated cells to further mature and test the in-vivo functionality of these encapsulatedcells in diabetic mice.

(453c) Engineering Cardiomyocytes From Diverse Human Pluripotent Stem Cell Lines For Cardiac Cell Therapy With Optimized Cell Yield and Efficiency

Parikh, A., State University of New York at Buffalo
Tzanakakis, E. S., State University of New York at Buffalo
Originally presented on: 11/6/2013 9:06:00 - 9:24:00

Human embryonic and induced pluripotent stem cells (known as pluripotent stem cells or hPSCs) can be a renewable source of cardiomyocytes for treating heart diseases which are leading causes of morbidity and mortality. Recent advances in the generation of patient-specific induced PSCs greatly improve the chances of clinical success. Yet, robust methods are still under development for producing cardiomyocytes across different hPSC lines having varying lineage-specific differentiation propensities. Therapeutic realization also hinges on the transplantation of sufficient numbers of cells (approximately 1.5×10^9 per damaged myocardium (1)) pointing to the need for pertinent large-scale bioprocesses. To resolve these issues, we have assessed particular signaling pathways related to heart tissue formation such as those induced by bone morphogenetic proteins (BMPs), Wnt, Nodal/Activin, and FGF ligands for efficient cardiac differentiation, and investigated methods for improving the yield of cardiomyocytes across a variety of hPSC lines.

A novel serum-free, defined medium formulation was developed by accounting for the signaling mechanisms linked to cardiogenesis aiming at increasing the reproducibility of in vitro commitment of hPSCs to heart cells. Human PSCs were efficiently directed in 48 hours to mesoderm-oriented primitive streak using BMP and Wnt ligands. Treated cells were characterized by high expression levels of Brachyury *(T)*, *MIXL1*, *EOMES* and *MESP1* assessed by qPCR and immunofluorescence. The total fraction of T^+ cells determined by flow cytometry was more than 95%. Subsequently, manipulation of physiologically-relevant signaling pathways that interact and participate in mesoderm patterning such as Wnt, Nodal/activin, BMP and FGF resulted in different mesoderm derivatives. Therefore, a specific combination of these pathways was utilized to coax these cells further along cardiovascular lineages expressing *KDR, cKIT, Nkx2.5, GATA4* and *MEF2C*. Beating cardiomyocytes emerged within a week and the appearance of beating clusters increased till 12 days of differentiation displaying several beating foci (>10 per cm^2) that corresponded to beating of more than 80% of the cells in each dish. During this period, cardiac markers such as *β-MHC, TBX20, MLC2a* and *ANF* were strongly upregulated. Expression was higher than that seen in cells treated with conventional protocols for cardiogenic differentiation. Upon further maturation, corresponding upregulation was observed by western blot analysis and immunofluorescence of cardiac proteins such as NKX2.5, GATA4, MEF2C, α-actinin, cardiac troponins and the cardiac gap junction protein connexin-43. When subjected to electrophysiological measurements, cells displayed cardiac action potentials indicating approximately 60% ventricular cells and a mix of atrial and sinoatrial node (pacemaker) cells. Connections among neighboring contractile foci in each culture were observed leading to integrated beating activity. Single cells were also capable of displaying contractions and their coupling with neighboring cells led the emergence of binucleated cell clusters typical of cardiomyocytes. Their contractile activity was modulated in a dose-dependent, organotypic fashion by incubation with diltiazem, isoproterenol and phosphodiesterase inhibitors. Up to 90% of committed cells expressed particular cardiac markers as assessed by flow cytometry yielding a significant number of cardiac cells per stem cell initially seeded.

An optimized protocol for cardiogenic differentiation of hPSCs was developed based on the rational manipulation of the core signaling pathways involved in cardiac cell commitment. Application of the method improved the efficiency and yield across different hPSC lines. Ongoing work concentrates on translating our findings to the generation of functional cardiomyocytes from hPSCs in scalable cultivation systems (2). The outcome of our studies is expected to facilitate the development of stem cell-based therapies for heart diseases.

Acknowledgements: Funding support has been provided by the National Institutes of Health (NHLBI, R01HL103709) and the New York Stem Cell Science Trust (NYSTEM, contract C024355) to EST. AP is a recipient of the Mark Diamond Research Fund award.

References:

1. Jing D, Parikh A, Canty JM, Tzanakakis ES. Stem Cells for Heart Cell Therapies. Tissue Eng Part B Rev.14:393-406. 2008.

2. Jing D, Parikh A, Tzanakakis ES. Cardiac cell generation from encapsulated embryonic stem cells in static and scalable culture systems. Cell Transplant.19:1397-412. 2010.

(453d) A Multiscale Approach to Stem Cell-Based Chondrogenesis for Cartilage Repair

Chou, C. L., Case Western Reserve University
Rivera, A. L., Case Western Reserve University
Sakai, T., Cleveland Clinic Lerner Research Institute
Drazba, J., Cleveland Clinic Fourndation
Caplan, A. I., Case Western Reserve University
Goldberg, V. M., Case Western Reserve University
Welter, J. F., Case Western Reserve University
Baskaran, H., Case Western Reserve University
Originally presented on: 11/6/2013 9:24:00 - 9:42:00

Tissue engineering is a promising method for long-term repair of cartilage lesions; however, current tissue-engineered cartilage constructs have inferior mechanical properties compared to native cartilage. Other researchers attribute this problem to a lack of an oriented structure in the constructs at the microscale that is present in the native tissue. Â In this study, we utilized contact guidance to develop constructs with microscale architecture for improved chondrogenesis and function. In this system, we used microscale channels on the scaffold surface to guide mesenchymal stem cells (MSCs) undergoing chondrogenesis to produce oriented extracellular matrix (ECM). Â Stable channels of varying microscale dimensions were formed in collagen-based and polydimethylsiloxane membranes via a combination of microfabrication and soft-lithography. Human MSCs were selectively seeded in these channels. The chondrogenic potential of MSCs seeded in these channels was investigated by culturing them for 3 weeks under differentiating conditions with transforming growth factor β1 and then evaluating the subsequent synthesized tissue for mechanical function and by type II collagen immunohistochemistry. We demonstrated selective seeding of viable MSCs within the channels (**Fig. 1**). MSC aligned and produced mature collagen fibrils along the length of the channel in smaller linear channels of widths 25-100 Âμm compared to larger linear channels of widths 500-1000 Âμm. Further, substrates with microchannels that led to cell alignment also led to superior mechanical properties compared to constructs with randomly seeded cells or selectively seeded cells in larger channels. The ultimate stress and modulus of elasticity of constructs with cells seeded in smaller channels increased by as much as four folds (**Fig. 2**). We demonstrated that microscale guidance is useful to produce 2-dimensional oriented cartilage structures with improved mechanical properties. Furthermore, we extended the 2-dimensional construct findings to a 3-dimensional construct for preclinical experiments. We successfully created 3-dimensional constructs for in vivo applications by rolling up 2-dimensional MSC-seeded collagen based scaffolds and formed 3-dimensional rolled-up large scale (3.5 mm in diameter Ã— 18 mm in length) cartilage constructs with microscale architecture to guide the differentiating MSCs to produce oriented ECM. Histology and immunohistochemistry indicated extensive glycosaminoglycan and collagen type II production in the 3-dimensional constructs, which are both indicative of chondrogenesis (**Fig. 3**). Our results show that the microscale guidance channels incorporated within the 3-dimensional cartilage constructs led to the production of aligned cell-produced collagenous matrix and enhanced mechanical function. The tissue modulus of elasticity of the 3-dimensional cartilage constructs containing guidance channels (1.16-2.07 MPa) increased by as much as six times compared to constructs without channels, which are higher than most current tissue-engineered constructs (**Fig. 4**). Overall, these findings offer new insight into how microscale guidance channels can regulate matrix deposition and long term construct development. Furthermore, our results provide new parameters which help optimize the design of functional MSC-based tissue-engineered cartilage using collagen-based substrates and can be used to fabricate large clinically useful MSC-based cartilage constructs with superior mechanical properties.

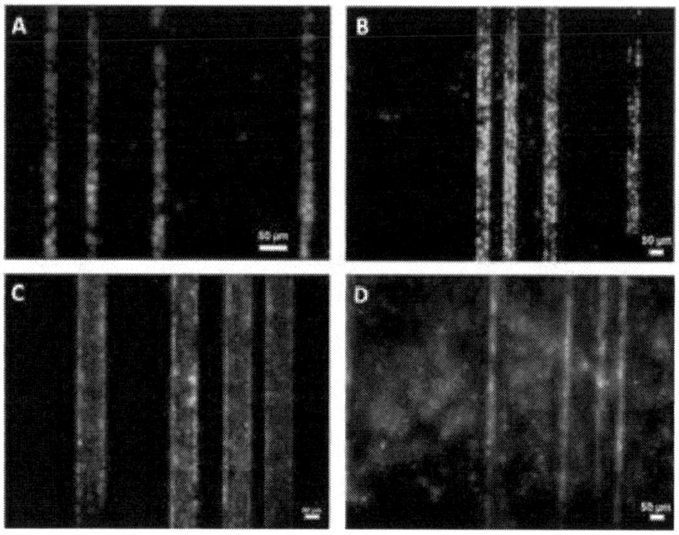

Fig. 1 Fluorescent images of cells in linear channels of widths 25 μm (**A**), 50 μm (**B**), and 100 μm (**C**) show selective seeding compared to untreated channel surfaces (**D**). Scale bars: 50 μm

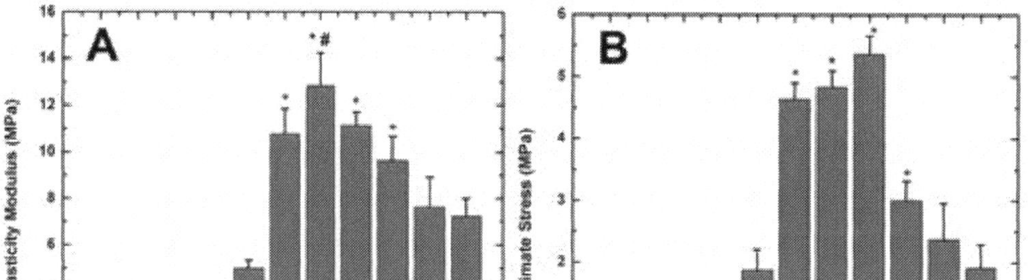

Fig. 2 Effect of microscale guidance on mechanical properties. Collagen membranes were subject to tensile testing until failure. Mean values of elasticity modulus (**A**), and ultimate stress (**B**) are shown. **BK** and **EDC** represent respectively uncrosslinked and EDC crosslinked collagen membranes without microchannels or MSCs. **C25** and **C100** represent collagen membranes consisting of 25 μm and 100 μm linear channels only, without MSCs. All other channels were seeded with MSCs. **Ran:** EDC crosslinked collagen membranes without channels; **25**, **50**, **100**, **250**, **500** and **1000** represent widths (μm) of linear channels in the EDC crosslinked collagen membranes (**A**) * indicates statistically significant difference ($p < 0.05$) compared to data with no channels (**RAN**). # indicates statistically significant difference ($p < 0.05$) compared to data with 1000 μm channels (**1000**). (B) * indicates statistically significant difference ($p < 0.05$) compared to data with no channels (**RAN**), 250 (**250**), 500 (**500**) and 1000 μm (**1000**) -wide channels.

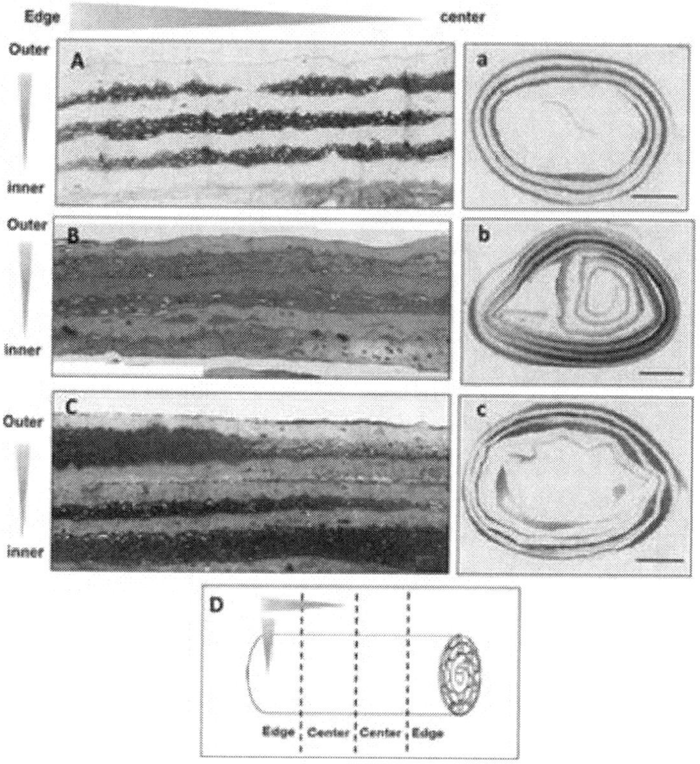

Fig. 3 Cartilage tissue constructs formed by the role up process. Three different microchannel designs were used. A-C indicate longitudinal sections and a-c indicate cross sections. Scale Bar=100 μm (A-C), 1 mm (a-c)

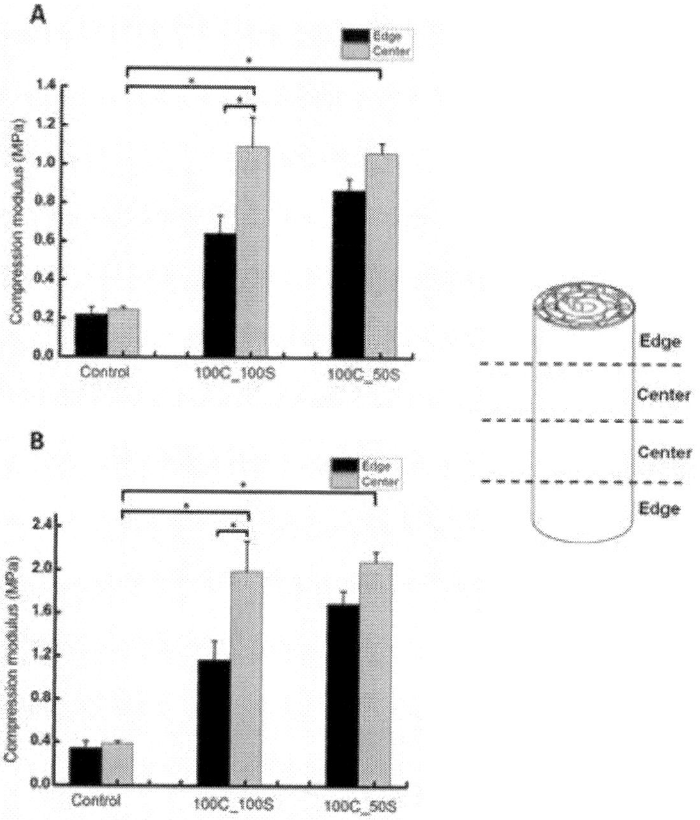

Fig. 4 Mechanical properties of cartilage constructs with microchannel guided ECM. Compressive modulus of the whole construct (A) and the chondrogenic tissue (B) are shown.

(453e) Directed Evolution of Adeno-Associated Virus for Enhanced Gene Delivery to Pluripotent and Adult Neural Stem Cells

Kotterman, M. A., University of California at Berkeley
Asuri, P., University of California at Berkeley
Jang, J. H., University of California, Berkeley
Vazin, T., University of California at Berkeley
Schaffer, D. V., University of California at Berkeley
Originally presented on: 11/6/2013 9:42:00 - 10:00:00

Stem cells are characterized by their ability to self-renew and to generate one or more mature cell types. Both pluripotent stem cells (PSCs) and neural stem cells (NSCs) present attractive targets in gene therapy and/or cell replacement therapies to treat nervous system injury or disease. Unfortunately, most gene delivery vectors are not capable of targeted gene delivery to stem cells. Adeno-associated virus (AAV) is a promising gene delivery vector that has been used in nearly 100 clinical trials to date. We have applied directed evolution, a high-throughput molecular engineering approach, to create AAV variants capable of efficient and specific transduction of human pluripotent stem cells or adult neural stem cells. Here, we report the *in vitro* and *in vivo* infection properties of our novel AAV variants and demonstrate the variants' utility for genetic engineering of stem cells.

Following a directed evolution approach consisting of one round of evolution and three selection steps on human embryonic stem cells, a variant of AAV2 was isolated and found to mediate increased transduction of several human embryonic stem cell (hESC) and induced pluripotent stem cell (iPSC) lines. In addition, the variant was capable of gene-targeting frequencies of up to 0.1%, approximately 10-fold higher than the targeting frequencies in pluripotent stem cells obtained using naturally occurring AAV serotypes. Furthermore, using zinc finger nucleases in conjunction with AAV-mediated gene targeting, this variant achieved gene-targeting frequencies of 1.2%, a >10-fold increase in the already efficient AAV-mediated gene targeting in hESCs. This capability can enable the use of hPSCs for *in vitro* studies of disease mechanisms, developmental processes, and drug discovery and toxicity.

Furthermore, following a directed evolution approach consisting of three rounds of evolution and ten selection steps on adult rat NSCs, a variant containing a peptide insertion was isolated and found to mediate a 50-fold increase in the transduction of rat NSCs. The variant also exhibited increased transduction of murine NSCs, human fetal NSCs, and human embryonic stem cell-derived neural progenitor cells *in vitro*. Moreover, the variant was capable of gene-targeting frequencies of up to 0.16%, approximately 5-fold higher than the targeting frequencies in rat NSCs obtained using naturally occurring AAV serotypes. To build upon the *in vitro* success of this variant, the *in vivo* infection properties of this novel AAV variant were analyzed. It was capable of efficient and selective transduction of rat and mouse neural stem cells in the adult hippocampus compared to several natural AAV serotypes. Delivery of a constitutively-active β-catenin transgene demonstrated that variant r3.45 can be used to study mechanisms through which key regulatory genes are acting in NSCs and that r3.45 can be used to deliver transgenes that can induce changes in the NSC population in the hippocampus.

(453f) Stem Cell Pluripotent State Transition: Going From Embryonic Stem Cells to An Extraembryonic Precursor State and Back

Cho, D. S., University of Minnesota
Sharma, S., University of Minnesota
Verfaillie, C., Katholieke Universiteit Leuven
Hu, W. S., University of Minnesota
Originally presented on: 11/6/2013 10:00:00 - 10:18:00

Stem cells, in particular pluripotent stem cells, hold great potential for regenerative medicines. Their practical use in clinical applications will require a large and steady supply of cells. These pluripotent stem cells in culture are prone to spontaneous deviation from a high degree of homogeneity. Furthermore, direct transplant of pluripotent stem cells bears the risk of tumorigenicity. Because of these shortcomings, highly multipotent stem cells derived from pluripotent stem cells that still possess the differentiation capacity to the desired lineage may serve as a means of expanding cell mass for clinical applications. In many cases, the maintenance and expansion of those multipotent stem cells is more robust and amenable to scalable processes.

Multipotent adult progenitor cells (MAPCs) derived from rat bone marrow are one of the multipotent stem cells. Their differentiation state appears to be equivalent to multipotent extraembryonic endoderm (ExEn) precursor (XEN-P) cells which can be derived from rodent blastocysts. Both cell types express pluripotency-associated transcription factor, Oct4, at high levels very close to that of rat embryonic stem cells (rESCs); however, unlike rESCs, they also express extraembryonic endoderm-specific genes including Gata6, Gata4, Pdgfra, and Sox7 at high levels, but lack Nanog and Sox2.

We uncovered trajectories that allow rESCs to transit to XEN-P state, and, with a restricted window of time, to revert back to ES state. When grown in the presence of PDGF, EGF, and LIF, rESCs undergo morphological and gene expression changes to become like MAPCs and XEN-P cells within 10 days. Interestingly, the ExEn precursor-like cells revert to ESCs when transferred to mouse embryonic fibroblast (MEF) feeder cells with supplements of two inhibitors (2i), MEK and GSK inhibitors, although only with a low frequency. The differentiated cells from rESCs express extraembyonic endoderm-specific markers, Gata6, Gata4, Pdgfra, Sox7, and Sox17 as highly as XEN-P cells and MAPCs without loss of Oct4 expression. Immunostaining results suggest that the expression of Nanog and Gata6 in the ExEn precursor-like cells is not mutually exclusive, even though their simultaneous expression in a cell is rarely reported and is generally thought to be mutually repressive of each other's expression. We have constructed a mathematical model describing the network consisted of Oct4, Sox2, Nanog, and Gata6 that contribute to the regulation of pluripotency. The experimental ESC-XEN-P transition system allows the dynamics of this gene network to be studied through treatments with signaling molecules and genetic manipulations. This model system of pluripotent state transition is a valuable tool for better understanding of the underlying regulation of pluripotency and differentiation.

(453g) Lyophilization of Human Pluripotent Stem Cells for Long-Term Storage Based On Intracellular Loading of Trehalose By Cell Cenetrating Ceptides

Zhang, L.,
Wei, Y.,
Xu, Y.,
Li, C., Institute of Process Engineering, Chinese Academy of Sciences
Xu, X., Institute of Process Engineering, Chinese Academy of Sciences
Originally presented on: 11/6/2013 10:18:00 - 10:36:00

Human pluripotent stem cells (hPSCs) including human embryonic stem cells (hESCs) and human induced pluripotent stem cells (hiPSCs) hold great potential for cell therapy and regenerative medicine. Efficient preservation of hPSCs without affecting their survival and functions is essential for cell-based clinical applications. There is growing interest in lyophilization of cells since stabilization of mammalian cells in desiccated state can significantly simplify the storage and transportation and provide enormous benefit for clinical applications. At present, freeze-drying method has been successfully applied in the preservation of diverse kind of cells, including sperm and platelet. However, the possibility of long-term preservation of hPSCs using this method has not yet been studied. Due to a higher glass transition temperature and dihydrate crystalline structure, trehalose can provide a better stability to the system than other carbohydrate. However, the major challenge in using trehalose for freeze-drying of cells is to overcome the impermeability of the plasma membranes to introduce them at high intracellular concentrations. In this study, a novel cell penetrating peptide (CPP) was designed to deliver trehslose intracellularly based on the molecular simulations. A sequence rich in basic amino acid combined with a sequence having the ability of coupling with trehalose through non-covalent bonding was developed. We evaluated the cytotoxicity of the designed CPP to hPSCs and further examined the efficiency of the designed CPP entering into hPSCs. The capacity of intracellular trehalose loading using the designed CPP was also determined. The results indicated that CPPs has high loading rate of trehalose and no significant toxic effects. In addition, freeze-dried hPSCs loaded with trehalose after rehydration showed significantly higher cell viability than the control. The establishment of freeze-drying method of hPSCs loaded with trehalose based on CPPs may prove a powerful tool for long-term and stable storage of hPSCs for various clinical and research applications.

(453h) Shear, Microparticles and Platelet-Like Particles in the Interplay Between Platelet Biogenesis and Megakaryopoiesis

Jiang, J., University of Delaware
Papoutsakis, E. T., University of Delaware
Originally presented on: 11/6/2013 10:36:00 - 10:54:00

Shear stress has been shown to be an important differentiation cue for platelet biogenesis (Baruch et al., *Blood* **114**:1975 (2009)). Yet, the mechanisms by which shear flow enables platelet biogenesis and megakaryopoietic (Mk) maturation remains largely unexplored. Separately, there has been an enormous activity in exploring the role of microparticles (MPs; also referred to as microvesicles) in the biology and differentiation of a variety of tissues and organs. Yet, little is known on the biogenesis and role of MPs from megakaryocytes or platelets. Here we report our studies on the role of shear stress on the generation of platelet-like particles (PLPs), and Mk-derived MPs (MKMPs) from mature Mks *in vitro*, and their role in Mk maturation.

Human Mks were differentiated from primary stem and progenitor cells (SPCs; $CD34^+$ cells) and seeded into Von Willebrand Factor-coated parallel flow slides. The continuous laminar flow was applied to Mks at defined shear stress level for different duration times. Cell fragments and particles were harvested after the flow application. When medium flow at shear stress level of 1 dyn/cm^2 was applied, 5.5-fold higher numbers of PLPs (diameter of 2-3 μm) were generated after 2 hr compared to static conditions. 30 min exposure to 4 dyn/cm^2 (the upper limit of physiological shear range) increased the generation of PLPs by 180% compared to1 dyn/cm^2 for 2 hrs. The quality and functionality of these PLPs were also studied based two standard platelet functionality assays: P-selectin exposure and fibrinogen binding assays. After activation, the % P-selectin$^+$ PLPs from flow conditions was 2.35-fold higher than in PLPs from static conditions. Similar results were found regarding to fibrinogen binding. These results show that PLPs generated from flow condition have better quality than PLPs generated under static condition.

Mature Mks can also generate $CD41^+$ MKMPs with diameter < 1 μm (Flaumenhaft et al., *Blood* **113**:1112 (2009)), but a biological role for these MKMPs remains to be discovered. We found that shear stress enhances MKMP generation at day 10 of culture. When Mks were exposed to shear flow at 2.5 dyn/cm^2, 30-50 times more MKMPs were generated after 2 hrs of flow application. To explore the physiological function of MKMPs that is unknown, MKMPs were cultured together with hematopoietic progenitor cells (HPCs) under cytokine-free condition. Our results show that there were more 2N, 4N and 8N or higher polyploidy Mks in cultures with MKMPs than in control cultures. These data suggest that MKMPs can communicate with HPCs to promote their differentiation to Mks in the absence of Mk-specific cytokines. Data regarding the mechanisms by which this communication takes place will be presented and discussed in the context of the *in vivo* milieu.

(463a) High Cell Density Fed-Batch Fermentation of Bacillus Megaterium for Enhanced Polyhydroxybutyrate Production From Sugarcane Molasses

Kanjanachumpol, P.,
Tolieng, V.,
Kulpreecha, S.,
Thongchul, N., Chulalongkorn University
Originally presented on: 11/6/2013 12:33:00 - 12:51:00

Polyhydroxybutyrate (PHB) which belongs to the group of polyhydroxyalkanoates (PHAs) is one of the most well known compostable biopolymers. It is recognized as the completely biosynthetic polymer with the remarkable biocompatibility properties. PHB as the thermoplastics can be used in various applications including agriculture, marine, and biomedical fields. Due to its renewable property and similar characteristics to the petroleum based polymer (polypropylene), PHB has recently become more attractive.

PHB is commonly synthesized and accumulated by several microorganisms as the intracellular carbon and energy sources. Although PHB has the wide range application, the commercial market is still limited. This is because of the relatively high production cost compared to that of the commodity petroleum based plastics. Many efforts have been developed to reduce the production cost in order that PHB can be more competitive to the petroleum based plastics. Those include 3 major attempts: I. using the low cost feedstocks such as the renewable biomass locally available nearby the production site; II. screening and modifying the robust microorganisms that could rapidly synthesize and accumulate a large amount of PHB inside their inclusion bodies, and III. enhancing the fermentation and recovery processes to obtain high PHB recovery yield.

Bacillus megaterium BA-019 previously screened and identified in our lab is the promising PHB producer because it can produce and accumulate a large amount of PHB inside the cell. This strain has low nutrient requirements and it is also capable of utilizing the low cost substrates such as sugarcane molasses and urea. In this work, enhanced PHB production from low cost sugarcane molasses medium was observed in high cell density cultivation via fed-batch fermentation using the appropriate feeding method. The improved PHB production via high cell density cultivation of *B. megaterium* BA-019 with balanced initial total sugar concentration and carbon to nitrogen (C/N) weight ratio was observed. In the 10 L stirred fermentor operated at 30 °C, pH 7.0, 600 rpm, and 1.0 vvm air, with the initial total sugar concentration of 60 g/L and urea at the C/N weight ratio of 10:1, 32.48 g/L cell biomass with the corresponding PHB weight content of 26.94% and volumetric productivity of 0.73 g/L×h were obtained from batch cultivation. Continuing cultivation by intermittent feeding of the sugarcane molasses along with urea at the C/N weight ratio of 12.5:1 gave much improved biomass and PHB production (90.71 g/L biomass with 45.84% PHB content and 1.73 g/L×h PHB productivity). Similar biomass and PHB yields were obtained in the 90 L stirred fermentor when using the impeller tip speed as the scale up criterion. It is believed that with the results obtained from this study, further development of commercial PHB production will become more promising.

(463b) Industrial Platform Design for Large Scale Production of Pleuran: An Immunomodulator/Anticancer Polysaccharide From Pleurotus Ostreatus

El Enshasy, H. A., Institute Bioproduct Development, Universiti Teknologi Malaysia
Originally presented on: 11/6/2013 12:51:00 - 13:09:00

For centuries, macrofungi have been used as food and medicine in different parts of the world. This is mainly attributed to their high nutritional value as potential source of carbohydrates, proteins, amino acids and minerals. In addition, they are also rich in many bioactive metabolites of high medicinal values such as lectins, polysaccharides, phenolics and polyphenolics, terpenoids, ergosterols and volatile organic compounds. Thus, mushrooms played crucial role as important component in folk medicine especially in Africa, Middle East, China and Japan since long time. During the last 30 years of intensive research, mushroom extracts showed many potential medicinal applications as immumomodulator, antitumor/anticancer, antibacterial and antiviral, antioxidant, antihypoglycemic and as active medicines in the prevention of cardiovascular diseases through their action as anti-atherosclerosis agents. Pleuran (CAS # 159940-37-1) is one of the most important glucose polymers and belongs to (β-(1,3/1,6)-D-glucan group. Like other hexose polymers, it has the general chemical formula of $(C_6H_{10}O_5)_n$., and alkali soluble polysaccharide. This bioactive polysaccharide is produced by the third most famous mushroom in the world *Pleurotus ostreatus* (widely known as oyster mushrooms). This polysaccharide can significantly stimulate human immune system and also has also anticancer, antiviral and antibacterial activity. The present demonstrates the design of large scale production platform for pleuran production as exopolysaccharide during submerged culture cultivation of *P. ostreatus*. In addition, the presented platform could be also applied as universal platform for submerged cultivation of mushrooms for the production of bioactive metabolites.

(463c) Detection of Low-Concentrated Bacteria On a Simple and Rapid Electrophoretic Platform

Azadi, G., Brown University
Park, J.,
Shapley, N. C., Rutgers, the State University of New Jersey
Tripathi, A., Brown University
Originally presented on: 11/6/2013 13:09:00 - 13:27:00

Detection of pathogenic microorganisms in low concentration is essential in many areas of research and technology, especially in food and pharmaceutical industry. In recent years, capillary electrophoresis has gained increasing attention as a fast and sensitive technique for detection of bacteria. However, the conventional techniques involve the integration of pre-concentration steps for detection of low concentrated bacteria. Here we propose a rapid electrophoretic method for detection of *E.*coli, independent of pre-concentration steps, using a cationic surfactant (CTAB) on simple electrophoretic platform, consisting of two cuvettes as reservoirs and a plastic microcapillary. Our technique is capable of detecting 100 cells/ml, in order of minutes (10 min).

(463d) Modeling the Food During Mastication

Gray-Stuart, E.,
Jones, J. R., University of Massey
Bronlund, J. E., Massey University
Paterson, T., Massey University
Originally presented on: 11/6/2013 13:27:00 - 13:45:00

Mastication breaks down food so that it can be swallowed safely and so that the particles are the right size for digestive processing. To avoid some of the complexities of modelling the changes in food and bulk properties during mastication, this model identifies the important properties for 'safe swallowing' using a HAZOP on the passageway from the mouth to the stomach. These are used to define physical property thresholds which must be met, whether for partial or final swallowing. Thresholds can be defined for temperature, volume, deformation, liquid expression, adhesion and particle size. The model then interrogates the mechanical testing regime within the mouth to identify the relationships between measurement and the properties. Using these, the model tracks the development of the identified threshold properties as a function of the rate processes that occur during mastication: absorption, size reduction, dissolution, melting and mixing. A discretized approach is used. In this way, the model is focussed on predicting the bolus properties necessary for a safe swallow.

(463e) Tackling Obesity Without Sugar Or Sugar Substitutes: The Production of a Taste-Bud Modifying Protein in Edible Plants

Minocha, S., University of New Hampshire
Matthews, A.,
Originally presented on: 11/6/2013 13:45:00 - 14:03:00

Due to the obesity epidemic in the United States and the rapid increase in the incidence of type-2 diabetes, there is major emphasis on the production of calorie-free and sugar-free foods, preferably without using synthetic sugar substitutes. Since sweetness is as much a matter of the presence of a sweet substance as it is the matter of its perception by taste buds, modifying taste buds could be used as an alternative to changing one's perception of sweetness rather than meeting its needs. Proteins like Miraculin possess such taste-bud modifying properties that make them of particular interest for this application. This protein modifies the taste buds by making acidic foods taste sweet. Due to the low bioavailability and short shelf life of the plant that produces Miraculin, *Richadella dulcifica,* the large-scale use of the protein has proven to be very cost-prohibitive. In order to make the protein more accessible and using it without purification, the gene that encodes for Miraculin has been cloned and the protein has been produced in transgenic plants. Our research is aimed at optimizing the production of Miraculin in routinely consumed plants (e.g. cucumber or tomato) by targeting production of the protein to the edible portions of the plant (e.g. fruits) using a tissue/organ specific promoter sequences and to test *in vitro* models for evaluating the functionality of this protein. The idea would be to eat something like a cucumber slice before enjoying a drink or eat the dessert made from limejuice.

(463f) Zero-Waste Processing of Protein Rich Side-Streams

Kangas, P., VTT Technical Research Centre of Finland
Nappa, M.,
Lantto, R.,
Slizyte, R., SINTEF Fisheries and Aquaculture Ltd.
Navickas, K., Aleksandras Stulginskis University
Originally presented on: 11/6/2013 14:03:00 - 14:21:00

Background

This study evaluates possibilities of complete exploitation of selected high protein and high oil containing industrial co-streams by applying environmentally benign biomechanical processing technologies. Value components for food and skin care products are main focus. Remaining residue is simultaneously exploited for soil improvement. Rapeseed and fish processing co-streams have been chosen as raw materials, as they have high enough quality for food applications.

The quantities of these side-streams are significant. For example, if the volume of the co-stream from Norwegian salmon aquaculture is 275 000 t annually, there are 55 000 t of lipids and 41 000 t of proteins available. Likewise if 270 000 t of rapeseed press cake is produced by SME cold pressers in Germany, there are 27 000 t of lipids and 95 000 t of protein available for refining.

This study presents three developed process concepts for zero-waste processing of food side-streams: (i) thermo-enzymatic process for recovery fish oil and protein hydrolysate from fish side-stream, (ii) enzymatic process for fractionating proteins and peptides from rapeseed press-cake, and (iii) co-generation of heat and fertilisers out of residue from previous processes, namely fish bones, hulls or solids from hydrolysis.

The main emphasis of this study is to evaluate techno-economic feasibility of proposed process concepts. Mass and energy balances of concepts are estimated, raw material demand, utilities used and intermediates or products produced are calculated and variable production costs are evaluated. In addition, rough estimates for the fixed operating costs and investment costs are given. Finally, a profitability analysis is conducted based on the variable costs and fixed costs.

As the degree of uncertainty is high at early stages of process design, Monte Carlo simulation are applied when estimating variable and fixed costs in order to cope with this uncertainty. By applying this methodology, ranges and distribution can be used for the main process parameters, prices and also for investment. As a result of calculations, a probability distribution of outputs, e.g. profit, is obtained.

Processing rapeseed press-cake

Rape-seed press cake is milled and enzymatically treated. Mixture is flash heated and possibility of CO2 extraction to remove oil fraction is evaluated. Second enzymatic hydrolysis is conducted. To reduce water consumption in the process, it is proposed that consistency of enzyme treatments and extractions can be increased to at least 15% or dry methods to be used. Solid residue and hydrolysate is obtained after phase separation. Hydrolysate might be concentrated or precipitated. Solid residue is dried and can be used as raw material for co-generation of heat and fertilisers.

Processing fish filleting co-stream

Proposed concepts for fish-filleting co-stream process combine the gentle thermal treatment and the enzymatic hydrolysis. This thermo-enzymatic process begins with mechanical grinding of the fresh fish filleting co-stream. Moist raw material is then heated up for separation of the premium oil. Phase separation is utilised in order to recover the first oil fraction. The remaining raw material is further processed by enzymatic means. The flow of raw material into the enzymatic stage decreases due to the preliminary heat treatment resulting in a decreased amount of enzymes needed. After enzymatic treatment, the raw material is heated up in order to terminate the enzymatic activity. Phase separation conducted in a tri-canter is used to recovery second oil fraction. The solid fraction containing insoluble proteins, lipids and bones can be used as material for extraction valuable ingredients like phospholipids or can be dried and used as feed. Fish protein hydrolysate can be further concentrated by utilising an evaporator or dried depending on the specifications of the final product.

Co-generation of energy and fertilisers out of residue

Due to the seasonal nature of the need for fertilisers, a process for improving the preservability of residue, especially from fish side-stream processing, is needed. A process with anaerobic digestion of residue and cattle manure on a regional scale is proposed. The function of a biogas plant, regardless of the end use of the biogas generated and organic fertiliser obtained, is to provide treatment capacity for the residues. Process includes storage capacity for residue at ambient temperatures. Liquid manure and residue are mixed together before feeding to the anaerobic digester. The anaerobic digestion process was assumed to be a wet, single stage, continuously fed process operating at mesophilic temperatures and a hydraulic retention time of 40 days. Operation of the biogas plant requires heat and electricity which is generated from an on-site CHP plant using biogas as a fuel. Remaining substrates are used as fertilisers on fields.

Summary

This study presents three process concepts for zero-waste processing of food-industry side-stream. Raw materials are fish filleting side-stream and rapeseed press-cake. End-products are food and skin-care products as well as oil, feed, biogas and fertilisers. This study focuses mainly evaluates the techno-economic feasibility of proposed processes and provide insight of the opportunity to utilise these food processing side-streams.

Acknowledgements

The APROPOS project is part funded by the European Commission's SeventhFramework Programme for research and technological development over the period 2007 to 2013.

(463g) Production of Fructooligosaccharide Prebiotics: Effect of Cultivation Parameters On Fructosyltransferase Production

Ahmed Elsayed, E., King Saud University, Faculty of Science
Noor El-Deen, A.,
A. Farid, M.,
A. Wadaan, M., King Saud University, Faculty of Science
Originally presented on: 11/6/2013 14:21:00 - 14:39:00

Recently, a great attention has been paid to the use of dietary carbohydrates as prebiotic functional foods. Among the new commercially available products, fructooligosaccharides (FOS), which are microbially produced from sucrose, have attracted special interest due to their valuable properties and, thus, have a great economic potential for the sugar industrial branch. They are non-cariogenic sweeteners of low caloric value, as they are not hydrolyzed by the gastro-intestinal enzymes, promoting selectively the growth of the bifidobacteria in the colon, helping to eliminate the harmful microbial species to human and animal health and preventing colon cancer. FOSs have been also found to reduce cholesterol, phospholipids and triglyceride levels in blood. FOS have been mainly produced by microbial fructosyltransferase (FTase) enzymes.

The present work outlines bioprocess optimization for different cultivation parameters affecting the production of FTase by *Penicillium aurantiogriseum* AUMC 5605. The optimization involves both traditional as well as fractional factorial design approaches. Additionally, the production process will be compared under batch and fed-batch conditions. Finally, the optimized process conditions will be applied to 5-L stirred tank bioreactor cultivations.

(463h) Kinetic Study of Galacto-Oligosacharides Production From Lactose and Whey Using Immobilized b-Galactosidase

Guio, F.,
Sanchez, O.,
Serrato, J. C., Universidad Nacional de Colombia
Originally presented on: 11/6/2013 14:39:00 - 14:57:00

Kinetic study of Galacto-Oligosacharides Production from Lactose and Whey using immobilized b-Galactosidase

Guio, Felipe [1], Serrato, Juan [1], Sánchez, Óscar [1]*

1. *Department of Chemical Engineering, School of Engineering, Universidad Nacional de Colombia, Bogota, Colombia.*

** Corresponding author*

Galacto-oligosaccharides (GOS) are prebiotics used in functional food formulations. GOS are produced during lactose hydrolysis by b-galactosidase. In this study GOS production was evaluated using lactose and dry whey with Lactozym 6500 L free and immobilized. The synthesis was performed at 30, 40 and 50°C, pH 5.5, 6.0 and 6.5, lactose concentration 300, 400 and 500 g L^{-1} and enzyme concentration 1.5, 3 and 4.5 U ml^{-1}. B-galactoside was immobilized by entrapment in calcium alginate and covalent adsorption in silica gel 60. Both immobilized enzymes were characterized with ONPG before GOS production to determine activity retention. The obtained GOS yield for free enzyme was 32.7% and 24.5% using lactose and dry whey, respectively. Immobilization on silica gel 60 showed enzyme activity retention of 41.8%, compared with entrapment technique with only 18.4%. GOS yield with immobilized enzyme on silica gel 60 was 22.9% using lactose and 11.2% with dry whey as substrate. Two kinetic models for GOS production were proposed to experimental data adjustment for free and immobilized enzyme on silica gel 60. They were based on a reaction mechanism described by elementary equations (engineering approach) and Michaelis-Menten kinetic equations (classical approach). The model based on an engineering approach adjusted experimental profiles with an overall average percentage error of 13.74 and 16.85, using lactose and whey in GOS production with immobilized enzyme, respectively. This immobilization technique and kinetic study could be studied in a continuous GOS production system using the same enzyme several times.

(469a) Analysis and Engineering of Modular Polyketide Synthase Systems

Lowry, B., Stanford University
Khosla, C., Stanford University
Originally presented on: 11/6/2013 12:30:00 - 12:48:00

Polyketide synthases (PKSs) are enzyme complexes that are incredible cellular machines involved in synthesis of molecules known as polyketides. Polyketides are complex, organic molecules that have a variety of therapeutic applications, serving as antibacterial, anti-viral, and anti-cancer drugs. A specific polyketide synthase, known as a modular PKS, assembles these compounds using an assembly line of modules (repeating protein subunits within the enzyme complex) that perform a step-by-step synthesis of the final product. Since these enzymes are modular by nature, PKSs appear to be very attractive candidates for protein engineering. Researchers have envisioned that by creating hybrid PKSs (assembly lines whose components come from PKSs originating in different organisms), novel polyketides could be synthesized, possessing new therapeutic properties. Over the past two decades, attempts have been made to engineer new PKS complexes, but very few attempts have been met with success. Furthermore, the body of knowledge relating to how full, intact PKS assembly lines function is very limited.

With regards to improving engineering strategies for building hydrid PKSs, our lab has recently developed a model that explains the protein-protein interactions which are critical in the processing of polyketide intermediates. Using this model as a guide, we have engineered new and improved non-natural PKS interfaces. These studies have brought a deeper understanding of how these enzymes perform their catalysis and will hopefully pave the way for construction of novel PKS assembly lines. In addition, we have recently demonstrated the full, *in-vitro* reconstitution of the 6-deoxyerythronolide B synthase (DEBS) which produces the macrocyclic precursor to the erythromycin family of antibiotics. This reconstituted system has shed light on the kinetics and mechanism of full assembly-lines. The convergence of both of these areas of research will hopefully make the engineering of novel PKS systems for biosynthesis of new polyketide therapeutics a reality.

(469b) Production of Plant Alkaloids in Saccharomyces Cerevisiae

Siddiqui, M. S., Stanford University
Smolke, C. D., Stanford University
Originally presented on: 11/6/2013 12:48:00 - 13:06:00

Natural products or compounds derived from natural products comprise most successful drug molecules. However, their discovery, synthesis, and scalable manufacture remain ad hoc, tedious, and biased. As a result, many proposed drug candidates cannot be seriously explored or tested, and the number of successful new drugs that further improve human health has stagnated. For example, the benzylisoquinoline alkaloids (BIAs) are a class of plant secondary metabolites that exhibit diverse pharmacological activities, including antiviral, antiprotozoan, and analgesic activities. However, these molecules cannot be economically produced by traditional chemical synthesis strategies; current commercial production relies on plant extraction isolation of select BIAs. As such, a microbial platform for the production of BIAs has the potential to reduce the cost and increase the scale of current production strategies, as well as expand the diversity of molecules accessible beyond those that accumulate in plants.

We are developing engineered strains of *Saccharomyces cerevisiae* for the production of diverse BIA molecules. In particular, I will discuss the development of yeast strains for efficient production of BIAs *de novo* from simple carbon and nitrogen sources. The ultimate precursor of all BIA molecules is the amino acid tyrosine and we have developed strains that can produce early BIA precursors from fed tyrosine. Under native regulation, allosteric feedback inhibition of a key enzyme in aromatic amino acid biosynthesis limits carbon flux towards tyrosine. The introduction of previously identified mutations to this enzyme alleviates this allosteric inhibition in our engineered yeast strains, thereby increasing carbon flux through aromatic amino acid biosynthesis and ultimately allowing us to engineer strains that do not require fed tyrosine for BIA production. Additional modifications to the regulation of native yeast enzymes, both in central metabolism and in aromatic amino acid biosynthesis, further improved BIA production in engineered strains, but only when made in combination with the key mutations conferring allosteric feedback resistance. These yeast strains optimized for early BIA production will provide a platform for the introduction of plant enzymes for the synthesis of more complex and diverse natural products.

(469c) Unraveling the H-cluster Biosynthetic Pathway: Investigations of Radical SAM Chemistry Catalyzed by the Hydg Maturase

Kuchenreuther, J. M., UC Davis
Myers, W. K., UC Davis
George, S. J., UC Davis
Shiigi, S. A., Stanford University
NejatyJahromy, Y.,
Cramer, S. P., UC Davis
Swartz, J. R., Stanford University
Britt, R. D.,
Originally presented on: 11/6/2013 13:06:00 - 13:24:00

[FeFe] hydrogenases rapidly produce H_2, which has stimulated interest in their application for renewable energy technologies. In addition, their catalytic iron center termed the H-cluster has inspired the design of many synthetic catalysts, albeit few with ability to evolve H_2, and a better understanding of how the H-cluster is made could further the development of H_2-based technologies. HydE, HydF, and HydG are three enzymes that assemble the H-cluster and activate [FeFe] hydrogenases, and recent work has shed light on H-cluster biosynthesis such as the formation of the cofactor's CO and CN ligands from free tyrosine. Nonetheless, challenges with experimentally probing the reaction pathway have left several aspects unresolved due in part to the O_2 sensitive nature of both the maturation proteins as well as the H-cluster. In this work, we combine multiple techniques to study the HydG maturase and its radical SAM chemistry. We use rapid freeze quench methods along with EPR spectroscopy in order to examine the radical SAM reaction and determine the kinetics of its products. In doing so, we identify a novel benzylomethyl-4-olate (BM4O) radical species that derives from free tyrosine radical, which presumably forms during the generation of CO and CN^- molecules. We also use stopped-flow infrared (SF-FTIR) spectroscopy to examine the HydG chemistry *in situ*. Specifically, we observe the concurrent formation of CO and CN^- species during the radical SAM reaction, which we attribute to a HydG-specific iron compound coordinated by at least two CO and one CN^-. We then utilize our cell-free hydrogenase maturation platform in conjunction with ENDOR spectroscopy to trace ^{57}Fe nuclei from HydG to the [FeFe] hydrogenase, providing conclusive evidence that the HydG maturase synthesizes an iron compound precursor of the H-cluster. From this work, and in contrast to the current paradigm, we present a new reaction sequence for the assembly of the H-cluster and the maturation of [FeFe] hydrogenases.

(469d) Characterization of the Biosynthetic Pathway of Physostigmine

Liu, J., University of California, Berkeley
Ng, T., University of California, Berkeley
Ad, O.,
Zhang, W., University of California, Berkeley
Originally presented on: 11/6/2013 13:24:00 - 13:42:00

Physostigmine is a tryptophan-derived natural product with a rare pyrroloindole skeleton. Because of its ability to reversibly inhibit acetylcholinesterase, physostigmine has been used to treat a variety of neurological disorders, including glaucoma and Alzheimer's disease. However, despite its medicinal uses, little is known about the biosynthesis of physostigmine and pyrroloindole alkaloids in general. In this work, we identified the biosynthetic gene cluster for physostigmine in the producer, *Streptomyces griseofuscus*, and confirmed its function by heterologous expression. Gene deletion mutants and *in vitro* assays with purified enzymes further enabled us to identify the set of enzymes essential for physostigmine biosynthesis and propose an enzymatic pathway for the transformation of tryptophan into physostigmine. We find that the formation of the pyrroloindole skeleton proceeds by an unexpected N-acetylated intermediate that is necessary for methylation of C-3 of the indole ring although the subsequent ring closure occurs with the hydrolysis of the acetyl group. This characterization of the physostigmine biosynthetic pathway provides the basis for future work in engineering the biosynthesis of physostigmine intermediates and analogues.

(469e) Designing a Synthetic Golgi Reactor

Royle, K., Imperial College London
Jimenez del Val, I., Imperial College London
Miller, A.,
Polizzi, K. M., Imperial College London
Kontoravdi, C., Imperial College London
Originally presented on: 11/6/2013 13:42:00 - 14:00:00

Recombinant gene expression has revolutionised the pharmaceutical and healthcare industry; in 2011, 16 of the top 50 selling drugs worldwide were derived from recombinant technology [1] with sales of $53.4 billion in the US [2].

Glycosylation, the post-translational covalent addition of oligosaccharide chains to the protein structure, is a critical quality attribute of biopharmaceuticals. Indeed, glycosylation affects protein folding, stability, immunogenicity and functionality. Of the 212 currently approved biotherapeutics, 78 are glycosylated [3]. There are two prominent forms of glycosylation, N-linked and O-linked glycosylation, of which the former is a paramount concern in biotherapeutics. In essence, it stems from the addition of a core glycan $GlcNAc_2Man_9Glc_3$ onto a sequence of Asn-X-Ser/Thr residues in the polypeptide. Subsequently, the core structure goes through a variety of trimming and addition reactions to produce the particular glycan structure [4].

As with all biological processes, heterogeneity occurs in the structures produced. Variations occur in the addition and trimming processing reactions, resulting in moderately different glycan structures, affecting the pharmacokinetic behaviour of the product. The latter is extremely important, such that the consistent glycosylation of biotherapeutics is a real challenge to industry.

To target this, we are designing and developing a synthetic Golgi reactor which will disengage the trimming and addition reactions from the cell. First and foremost, this allows protein production in a microbial host, greatly enhancing yield. Secondly, extracting the process from the cell allows greater control thereby reducing heterogeneity. Finally, it provides flexibility, such that therapeutics with optimised glycosylation profiles for enhanced functionality and stability can be designed.

The initial glycan structure will be transferred *in vivo*, by a partially 'humanised' strain of *Pichia pastoris*. The early glycosylation reactions are common between yeast and mammalian systems, up to the production of a $Man_8GlcNAc_2$ structure. There the two pathways diverge: yeast glycoproteins undergo substantial mannosylation while the human profile requires removal of mannose and addition of N-acetylglucosamine residues. In the 'humanised' strain of *P. pastoris*, the initial enzyme of the mannosylation cascade (α-1-6-

mannosyltransferase) has been knocked out. In its place, a mouse α1,2-mannosidase IA has been inserted. Consequently, recombinant proteins produced with this system contain N-linked glycans with the core, human $Man_5GlcNAc_2$ structure.

To complement the *in vivo* core glycosylation, we are designing an *in vitro* synthetic Golgi reactor which links upstream protein production with downstream processing. The reactor was initially conceived of as a series of modular columns, each containing a glycosylation enzyme immobilised in a peptide hydrogel. The three enzymes selected were N-Acetylglucosaminyltransferase-1 (GnTI), Alpha Mannosidase 2 (MsnII) and N-Acetylglucosaminyltransferase-2 (GnTII) as these are the first three enzymes in the human glycosylation profile.

The dimensions and process parameters presented an optimisation problem. For instance, longer columns provide a higher residence time, increasing the percentage of oligosaccharide conversion. Long columns, however, encounter problems with pressure drop. Similarly, low flow rates can also increase oligosaccharide conversion but such a strategy would negatively impact yield and some reactions are affected by co-product inhibition. In contrast to a trial and error experimental optimisation of parameters, we have implemented the kinetics from a previously validated model of glycosylation in the Golgi apparatus published by this group [6] in the reactor scenario. This allowed the optimisation of flow rate, column length and enzyme concentration.

The *in silico* model provided a first approximation of the reactor design; to start to develop the reactor in practice, we have expressed the three glycosylation enzymes. These enzymes have been cloned into a commercial *P. pastoris* expression vector with an N-terminal truncation removing the transmembrane region. In its place, we have incorporated the specific peptide sequence for incorporation into a peptide hydrogel [7]. Preliminary characterisation of the enzymes has been carried out, using a combination of fluorogenic substrates assays, HPLC and MALDI TOF/TOF MS confirmation of product structure. Indeed, as the kinetics of the glycosylation enzymes are derived from in solution studies, it was important to analyse their behaviour when immobilised to hydrogels. For proof-of-concept, we are using the K3 Kringle fragment as a substrate 'therapeutic' for the tethered glycosylation enzymes.

References

[1] EvaluatePharma (2012) World Preview 2018: Embracing the Patent Cliff.

[2] Aggarwal, S. (2012) *Nat Biotechnol*, **30**, 1191-1197.

[3] Kyriakopoulos, S. & Kontoravdi, C. (2013) *Eur J Pharm Sci*, **48**, 428-441.

[4] Kornfeld, R. & Kornfeld, S. (1985) *Annu Rev Biochem*, **54**, 631-664.

[5] Christen, M. T., Frank, P., Schaller, J. & Llinas, M. (2010) *Biochemistry*, **49**, 7131-7150.

[6] Jimenez del Val, I., Nagy, J. M. & Kontoravdi, C. (2011) *Biotechnol Prog*, **27**, 1730-1743.

[7] Roberts, D., Rochas, C., Saiani, A. & Miller, A. F. (2012) *Langmuir*, **28**, 16196-16206.

(469f) Combinatorial Biosynthesis of Pharmaceutically Important Simple Coumarins in Bacteria

Lin, Y., the University of Georgia
Sun, X., Beijing University of Chemical Technology
Yuan, Q., Beijing University of Chemical Technology
Yan, Y., the University of Georgia
Originally presented on: 11/6/2013 14:00:00 - 14:18:00

Coumarins are a group of benzopyrone-type natural products that can be classified into four categories: simple coumarins, pyrone-substituted coumarins, furanocoumarins, and pyranocoumarins. Simple coumarins consist of the simplest coumarin skeleton (1,2-benzopyrone) and its hydroxylated, alkylated and glycosylated derivatives. Coumarins and their derivatives have demonstrated a vast array of therapeutical effects, such as antibacterial, anti-inflammatory and anti-coagulant activities. For instance, the well-known synthetic 4-hydroxycoumarin derivatives warfarin, phenprocoumon and acenocoumaroyl are among the most widely prescribed oral anticoagulants over the world. Moreover, extensive research into their pharmacological properties has revealed their therapeutic roles in the treatment of cancer and AIDS. Despite the pharmaceutical importance of coumarins, relatively limited information is available regarding their biosynthesis in nature. Couamrins are usually considered to be derived from the plant phenylpronanoid pathway; while the formation of 4-hydroxycoumarin (the synthetic precursor of warfarin) was prove to be the result of the infection of melilotoside-containing plant materials by molds. Here we present the development of novel biosynthetic mechanisms to produce several pharmaceutically important simple coumarins in *Escherchia coli*, including umbelliferone, scopoletin and 4-hydroxycoumarin. In this work, artificial biosynthetic pathways were designed, validated and optimized via combinatorial biosynthesis and metabolic engineering approaches. This work not only demonstrates promising application potentials, but also advances understanding of the chemistry in coumarin biosynthesis and lays the foundation for extending the pathways to produce more complicated coumarin derivatives for their biomedical and clinical property study.

(469g) Enzymatic Synthesis of Prebiotic Human Milk Oligosaccharide Analogs Using Five - Galactosidases From Bifidobacterium Longum Subsp. Infantis ATCC 15697

Jimenez Espinoza, R.,
Mills, D. A., University of California Davis
Block, D. E., University of California Davis
Originally presented on: 11/6/2013 14:18:00 - 14:36:00

Human milk is established as the ideal nutrition for term infants. Specifically, it contains a large variety of prebiotic human milk oligosaccharides (HMO) with the potential to provide selection for beneficial probiotic organisms in the gut. Synthesis of complex oligosaccharides resembling those found in human milk would therefore be desirable as a nutritional additive to infant formula. We investigated the production of galactooligosaccharides (GOS) using five β-galactosidases cloned from *Bifidobacterium longum* subsp. *infantis* ATCC 15697. One ml reactions containing 30% w/v initial lactose concentration were followed over 46 hours and samples analyzed by HPLC. A Central Composite Design was used to establish optimum pH and temperature for transgalactosylation activity (i.e., maximum initial rate of reaction). Kinetic parameters (i.e., V_{max} and K_m) were subsequently calculated using lactose as the sole reactant. All enzymes showed transgalactosylation activity, with two of them producing GOS with a degree of polymerization higher than 3. Optimum pH varied from 5.9 to 7.4 and optimum temperature from 46 to 64 °C. V_{max} varied more than 10x fold. While formation of GOS is important, HMO are typically decorated with fucose and sialic acid residues that improve the prebiotic efficacy. Therefore, production of decorated GOS was also investigated by using different combinations of lactose, sialyllactose and fucosyllactose as building block donors. Products were analyzed with LCMS.

(469h) Antimicrobial Mechanism of Resveratrol-Trans-Dihydrodimer Produced From the Enzymatic Oligomerization of Resveratrol Mediated By Soybean Peroxidase

Mora-Pale, M., Rensselear Polytechnic Institute
Bhan, N., Rensselear Polytechnic Institute
Masuko, S., Rensselaer Polytechnic Institute
Linhardt, R. J., Rensselaer Polytechnic Institute
Koffas, M. A. G., Rensselaer Polytechnic Institute
Dordick, J. S., Rensselaer Polytechnic Institute
Originally presented on: 11/6/2013 14:36:00 - 14:54:00

Many natural polyphenols possess low antimicrobial activities which can be improved by using them as scaffolds to generate libraries of their oligomers. We have performed the *in vitro* diversification of resveratrol mediated by soybean peroxidase, and isolated two dimers (resveratrol-*trans*-dihydrodimer, and pallidol) and tested them as antimicrobials. In particular, resveratrol-*trans*-dihydrodimer was found to be an effective antimicrobial against Gram positive strains *B. cereus* (MIC = 15.0 μM), *Listeria* (125.0 μM), and *S. aureus* (MIC = 62.0 μM) as well as against Gram negative *E. coli* (MIC = 122.5 μM; in the presence of efflux pump inhibitor Phe-Arg-β-naphthylamide), while pallidol did not show significant antimicrobial activity. Further we deciphered the mechanism of action of the *trans*-dihydrodimer by transcriptomic, biochemical and flow cytometry analysis. Gene expression microarray and qRT-PCR data from *B. cereus* indicated down regulation of ABC transporters (ATP binding proteins). Flow cytometry analysis showed a strong depolarization of cell membrane upon treatment with resveratrol-*trans*-dihydrodimer, while ATP measurement indicated an accumulation of overall ATP (>2 fold); down regulation of ABC transporters that break down ATP is one of the individual events that contributes for ATP accumulation.

Alternately, a decrease in nucleic DNA content of *B. cereus* and *E. coli* upon incubation with the resveratrol-*trans*-dihydrodimer was also observed by flow cytometry. An *in vitro* activity assay of DNA gyrase from *E. coli* showed a strong inhibitory effect caused by the resveratrol-*trans*-dihydrodimer, which is consistent with previous reports regarding the inhibitory effect of polyphenols on DNA gyrase activity. This enzyme requires ATP as a source of energy for supercoiling DNA and its inhibition stops the breakdown of ATP. This observation suggests that resveratrol-*trans*-dihydrodimer could be responsible for inhibiting other enzymes that require ATP resulting in the increase of overall ATP. Finally, we have demonstrated that the enzymatic oligomerization of resveratrol is an attractive strategy for generating effective antimicrobials with potential therapeutic application especially due the low toxicity of resveratrol derivatives against mammalian cells.

(508a) A Study of Mutations That Cooperatively Increase Leakage in the Cytochrome P450 BM3 Reaction Cycle

Lim, J. B., MIT
Barker, K., MIT
Jiang, L.,
Molina, V.,
Saifee, J.,
Sikes, H. D., Massachusetts Institute of Technology
Originally presented on: 11/6/2013 12:30:00 - 12:50:00

Many redox enzymes of biotechnological utility consist of oxidase and reductase domains encoded on a single polypeptide chain. Coupling between the oxidase and reductase domains is important for efficient use of cofactors and for maximal enzyme lifetime in cases where uncoupling leads to undesirable reactive species, and yet the physical basis for coupling remains poorly understood. We report a series of experiments that explores the relationship between mutations and coupling in a member of the cytochrome P450 family of enzymes. Three different mutants of cytochrome P450 BM3 have been previously reported in the literature to enhance the leak rate, that is, oxidation of nicotinamide adenine phosphate (NADPH) in the absence of any other substrate besides oxygen. An increased leak rate corresponds to a lesser degree of coupling between the oxidase and reductase domains in this enzyme. We have combined the mutations in all possible combinations, resulting in four additional mutants with even higher NADPH oxidation activity in the absence of native fatty acid substrates. These new mutants also had decreased thermostabilities, as determined by T_{50} measurements, relative to their parents. For each mutant, we used horseradish peroxidase and 2,2'-azino-bis(3-ethylbenzothiazoline-6-sulphonic acid) to measure the hydrogen peroxide content of the reaction product. Finally, we measured the total turnover number of each mutant, that is, the total amount of NADPH consumed by a mutant within the enzyme's lifetime.

(508b) Towards a Fully Defined In Vitro Maturation of [FeFe] Hydrogenases

De La Paz, L., Stanford University
Swartz, J. R., Stanford University
Originally presented on: 11/6/2013 12:50:00 - 13:10:00

The sustainable production of hydrogen for renewable energy technologies positions it as a promising alternative to current fossil fuel based technologies. By replacing fossil fuel based H_2 production with a biological process, we can exploit nature's biological catalysts, hydrogenases. The [FeFe] hydrogenases are specific and productive in hydrogen evolution, but application of these complex enzymes is limited by inefficient and expensive production methods. The development of a fully defined *in vitro* hydrogenase maturation system will elucidate the assembly requirements for producing [FeFe] hydrogenases so they can be used commercially and enable technologies for carbon neutral hydrogen production. These enzymes are characterized by complex active sites, and knowledge of how the complex "H-cluster" active site is assembled and installed into the polypeptide remains unclear. The goal of this project is to identify the substrates and overall chemical mechanisms required for active site assembly and installation.

In particular, the *in vitro* [FeFe] hydrogenase maturation system seeks to activate purified *Clostridium pasteurianum* CpI apoprotein using *E. coli* cell extract containing recombinant *Shewanella oneidensis* HydE, HydF, and HydG maturases and exogenous small molecules. CpI is an [FeFe] hydrogenase containing a complex active site, the "H-cluster," and four accessory Fe-S clusters involved in the delivery of electrons to the active site. Our group has previously developed an *in vitro* activation for CpI. However, activation can only be achieved in the presence of a crude *E. coli* cell extract, and its composition is undefined. Because previous results have demonstrated the installation of Fe-S centers in limiting the maturation of hydrogenases, the proteins involved in Fe-S assembly were explored in this context. The *isc* operon is a well-characterized biosynthetic system involved in the assembly of cellular Fe-S proteins, and is believed to maintain a general housekeeping biosynthetic function. To determine if these were the unknown components in the extract responsible for incurring full hydrogenase maturation, the *isc* proteins in *E. coli* were cloned, expressed, and purified with N- and C- terminal StrepII tags in order to use a streptavidin/biotin purification procedure. The activity of each purified proteins was confirmed with an assay specific to its purported function, and their role in the apoCpI activation was evaluated.

Because the *isc* proteins only demonstrated the ability to stimulate 1% activation versus full activation with a crude lysate, work is in progress to fractionate the cell extract using chromatography to allow further evaluation and identification the substrates required to fully define an *in vitro* [FeFe] hydrogenase activation. These findings will provide a deeper understanding of the mechanism for assembly and installation of the active site. This knowledge will also enable *E. coli* genetic engineering for high-level hydrogenase production. Finally, a fully functional *in vitro* hydrogenase maturation system can be used as a platform for high throughput evolution projects.

(508c) Pathobiology of Crohn's Disease Risk Nod2 Mutation

Mann, J., University at Buffalo
Park, S., University of Pennsylvania
Originally presented on: 11/6/2013 13:10:00 - 13:30:00

Crohn's disease (CD) is an autoimmune disease in the gastrointestinal tract that affects 400,000 ? 600,000 individuals in the US. A number of genes have known association with the disease, including Nod2, which is an intracellular protein that recognizes bacterial cell wall component, muramyl dipeptide, and activates the NF-kB pathway. Three Nod2 mutations, R702W, G908R, and 1007fs, increase the odds of developing CD in a dose dependent manner. Although the genetic link between Nod2 mutations and CD is well established, the mechanism by which these mutations contribute to the disease phenotype is not known. Based on the location of the CD risk mutations, it is expected that the mutations affect ligand binding and/or regulation of Nod2 activation. Elucidating the mechanism of Nod2 activation is important so that the pathobiology of the mutations can be correctly modeled. To understand Nod2 signaling at a molecular level, we modeled the mechanism of Nod2 activation using biochemical data collected on Nod2 as well as structural data on other members of the NB-ARC family, including Apaf-1 and CED-4, in both active and inactive conformations. The resulting three step activation mechanism for Nod2 was confirmed using FRET based assays. The study further led to a hypothesis regarding the molecular basis of R702W pathobiology. We present data from functional studies that suggest loss of ligand sensitivity in the R702W mutant, which can be reversed using a designed peptide inhibitor. The study points to a possible therapeutic strategy to treat a subclass of CD patients carrying the R702W mutation.

(508d) Microfluidic Platforms for Time Resolved Laue Crystallography

Perry, S. L., University of Illinois at Urbana-Champaign
Pawate, A., University of Illinois at Urbana-Champaign
Schieferstein, J. M., University of Illinois at Urbana-Champaign
Guha, S., University of Illinois at Urbana-Champaign
Srajer, V., Argonne National Laboratory
Ren, Z., Argonne National Laboratory
Kenis, P. J. A., University of Illinois at Urbana Champaign
Originally presented on: 11/6/2013 13:30:00 - 13:50:00

Microfluidic chips for combinatorial analyses such as protein crystallization have advanced tremendously in the past decade. These platforms take advantage of parallel processing and small volumes, while creating an environment free of inertial or convective effects and providing exquisite control over local conditions and gradients. The ability to create highly reproducible microenvironments on chip has proven to be a tremendous asset for improving the consistency and quality of protein crystals grown on-chip. Coupling these benefits with advanced *in situ* protein crystallography techniques, microfluidic technology is now poised to facilitate a wide range of structure-function studies than have previously been accessible.

Standard X-ray diffraction methods are performed at cryogenic conditions that freeze out enzyme motion and collect frames of data on a time-scale of seconds. However, polychromatic Laue crystallography enables ultra-fast data collection at biologically relevant temperatures, matching the timescale of analysis to the chemical and structural changes in the protein. Two challenges associated with time-resolved Laue data collection are (i) the severe radiation damage associated with room temperature data collection using an intense polychromatic X-ray beam and (ii) the need to simultaneously trigger all of the protein molecules in a crystal for diffraction analysis.

We use high throughput X-ray transparent microfluidic platforms for protein crystallization to address these challenges. The fine control over composition on-chip facilitates the easy growth of a large number of isomorphous crystals. X-ray diffraction analysis of these crystals can then be performed in a serial fashion, collecting only a small amount of data from each individual crystal so as to avoid radiation damage. Here we report the *in situ* time-resolved structural analysis of photoactive yellow protein from protein crystals grown and analyzed on-chip. The reaction is laser-initiated and diffraction images are collected at time delays ranging from nanoseconds to milliseconds. Electron density difference maps generated from merged data can be used to clearly track the expected conformational changes of the chromophore with time, validating our approach. Future efforts will be focused on extending these capabilities to enable chemical triggering of enzymatic reactions (i.e. ligand addition) through the development of crystallization arrays containing flow-cell geometries.

(508e) Identification of CDR H3-Like Loop Conformations in Non-Immunoglobulin Proteins

Weitzner, B. D.,
Dunbrack, R. L. Jr., Fox Chase Cancer Center, Institute for Cancer Research
Gray, J. J., Johns Hopkins University
Originally presented on: 11/6/2013 13:50:00 - 14:10:00

The third complementarity-determining region loop on the immunoglobulin heavy chain (CDR H3) often provides a large portion of the antigen-binding contacts, providing motivation for antibody structure prediction algorithms to model this loop accurately. Despite significant effort, H3 structure prediction has remained elusive. One common strategy for loop modeling in non-antibody proteins is to find template loops with similar anchor positions, identical sequence length, and similar amino acid sequences. Because of the genetic mechanism that gives rise to the H3 loop, it is unclear whether their conformations are unique to antibodies. Although the other five CDR loops (L1, L2, L3, H1 and H2) have been shown to adopt canonical structures that can be reliably predicted by loop length and chain sequence, H3 conformations have proven to be unclassifiable, save for the fact that the majority contain an unusual "kink" or "bulge" at the C-terminal end of the loop. We set out to determine whether this feature is observed in other proteins, and if so, whether it confers structural properties similar to those observed in CDR H3. To that end, we searched a set of high-quality non-antibody structures for regions that match the geometry of the residues immediately surrounding the loop. By incorporating the C-terminal "kink"/"bulge" constraint into our search, we were able to identify a set of over 2,500 non-antibody loops that share a number of structural properties with H3 loops, namely: (1) position and prevalence of local structural features such as beta turns; (2) existence of non-antibody loops with RMSDs within 2.0 Å of nearly all H3 loops of length ≤ 15; (3) greater than expected occurrence of a salt bridge across the base of the loop. These results suggest that CDR H3 is similar to non-antibody loops, thus providing a set of templates that can be used to find starting structures for H3 loop structure prediction and design.

(508f) Novel Framework for Beta-Sheet Topology Prediction Using Low-Homology Template-Based Constraints

Kieslich, C. A., Princeton University
Smadbeck, J., Princeton University
Khoury, G. A., Pennsylvania State University-University Park
Floudas, C. A., Princeton University
Originally presented on: 11/6/2013 14:10:00 - 14:30:00

Accurate prediction of beta-sheet topology is a major unresolved challenge in the area of protein structure prediction. Current state-of-art approaches utilize sequence alignments, secondary structure assignments, and pairwise potentials to derive rank ordered lists of solutions [1-3]. Due to the large combinatorial complexity that arises for even a small number of beta strands, Mixed Integer Linear optimization (MILP) models have been proposed to identify the optimum topology [1-2]. Involvement of non-local tertiary contacts can make the prediction of beta-sheets based on sequence information alone very difficult. Therefore, we propose a novel framework for beta-sheet topology prediction, which utilizes structural templates of low-homology to derive likely beta-strand pairs that serve as constraints for a MILP model.

Given a query sequence, structural templates are first identified using a modified version of SPARKS-X [4]. Distance constraints are extracted from each template and are used as input for CYANA [5] to generate structural models. The initial set of structural templates is reduced using hierarchical clustering based on pairwise GDT, a measure of protein structure similarity. The beta-sheet topology of each template structure is extracted based on sequence alignments to the query and secondary structure assignment. A final set of structural templates is selected based on clustering of the template beta-sheet topologies, and a set of observed strand pairs is obtained.

The presented MILP model utilizes pairwise potentials calculated by BetaPro [6], as well as template-based constraints derived as described above. Additional constraints, initially proposed by Subramani and Floudas [2], are also imposed to ensure that only biologically relevant topologies are generated. Ultimately, a rank-ordered list of likely beta-sheet topologies is produced. We present results for the application of the proposed framework to all beta and mixed alpha-beta proteins of the PDBSelect25 data set, as well as to the most difficult targets from recent CASP competitions.

1. Klepeis, J. L.; Floudas, C. A. Prediction of Beta-Sheet Topology and Disulfide Bridges in Polypeptides. Journal of Computational Chemistry 2003, 24, 191-208.

2. Subramani, A.; Floudas, C. A. β-sheet Topology Prediction with High Precision and Recall for β and Mixed α/β Proteins. PLoS ONE 2012, 7 (3), e32461.

3. Ho, H.K.; Zhang, L.; Ramamohanarao, K.; Martin, S. A survey of machine learning methods for secondary and supersecondary protein structure prediction. Methods Molecular Biology 2013, 932, 87-106.

4. Yang, Y.; Faraggi, E.; Zhao, H.; Zhou, Y. Improving protein fold recognition and template-based modeling by employing probabilistic-based matching between predicted one-dimensional structural properties of query and corresponding native properties of templates. Bioinformatics 2011, 27(15), 2076-82.

5. López-MÃ©ndez, B.; Güntert, P. Automated protein structure determination from NMR spectra. Journal of American Chemical Society 2006, 128, 13112-13122.

6. Cheng, J.; Baldi, P. Three-stage prediction of protein beta-sheets by neural networks, alignments and graph algorithms. Bioinformatics 2005, 21, 75-84.

(508g) The Role of Primary Sequence and Secondary Structure in the Hydrolysis of Peptide Bonds in Proteins

Lauer, T. M., Massachusetts Institute of Technology
Wood, G. P. F., Massachusetts Institute of Technology
Farkas, D., MedImmune
Bishop, S., MedImmune
Sathish, H., MedImmune
Samra, H., MedImmune
Trout, B. L., Massachusetts Institute of Technology
Originally presented on: 11/6/2013 14:30:00 - 14:50:00

Proteins stored in liquid formulations are susceptible to a wide range of degradation pathways. Due to the requirement for high purity (>95%) throughout the shelf-life (1-2 years) of a pharmaceutical product, even these relatively slow reactions can drastically impact the feasibility of turning a drug candidate into a successful drug product. A significant source of fragmentation in proteins is the non-enzymatic hydrolysis of the peptide backbone. This reaction decreases the amount of active proteins in solution and produces new species which can enhance the rate of other degradation pathways such as aggregation. The primary sequence, the secondary structure and the pH have all been identified as important factors in determining the rate of hydrolysis. However, a molecular level understanding of the role that these factors play in the hydrolysis mechanism is still lacking. In the present study we employ high-level *ab initio* quantum chemistry calculations and machine learning algorithms to probe this complex mechanism

Experimental work indicates that a number of residues may promote hydrolysis. However, aspartic acid is typically the most susceptible and the combination of Asp–Pro is particularly sensitive. Therefore we have initially focused our attention on the mechanistic aspects of Asp–Pro hydrolysis. Through the use of *ab initio* quantum chemistry methods, we investigated the hydrolysis of this amino acid sequence in tripeptides systems. Our results show that the side chain of the aspartic acid interacts strongly with a neighboring water molecule during the rate-limiting step. This interaction promotes the disassociation of the water molecule, which then adds to the protein backbone. Our results concerning the role of proline in promoting hydrolysis show that it does not lower the enthalpy of the reaction. Our current hypothesis is that it lowers the activation energy for the hydrolysis reaction through inductive effects. Specifically, because proline is the only naturally occurring amino acid containing a secondary amine, it may promote the stabilization of the transition structure through charge delocalization.

These results shed light on the role of primary sequence, but because of the model systems are relatively small the effect of secondary and tertiary structure is not adequately accounted for. In order to investigate these, we gain insight by performing classical molecular dynamics simulation of whole proteins. Ensembles of protein conformations are then correlated to known aspects of the mechanism. Initial results from the simulation of an immunotoxin suggest that secondary structure elements affect rates of hydrolysis through two primary means – steric interactions and hydrogen bonding.

(508h) Applying Mechanical Analysis Tools to the Study of Allosteric Regulation Pathways in Proteins

Ribeiro, A. A. S. T.,
Ortiz, V., Columbia University
Originally presented on: 11/6/2013 14:50:00 - 15:10:00

Allostery is a phenomenon in which the function of a protein is regulated by the propagation of a signal via small conformational changes. Allostery plays a key role in many mechanistic pathways in biology, and for this reason, knowledge on how to analyze allosteric proteins and predict pathways of allostery has a wide range of potential applications, including drug design and protein engineering. Because many of the conformational changes involved in protein allostery include small deformations at a local level, we hypothesize that knowledge of the modulus of elasticity at each atom position or within specific regions in a protein can provide valuable information about the local mobility and stability under conformational deformations. Here, we expand to the analysis of allosteric proteins, a method previously developed to calculate local elastic moduli at different positions in a material (J. Chem. Phys. 128, 224504 (2008)), in order to build a 3D map of local mechanical properties that can be used to identify pathways of allostery in proteins and protein networks. We apply this method to the analysis of a preliminary set of allosteric proteins, and compare our results to previous ones reported from both computational and experimental studies.

(509a) Interstitial Fluid Pressure Modulates Gene Expression in Engineered Models of Breast Tumors

Piotrowski, A., Princeton University
Tien, J.,
Nelson, C. M., Princeton University
Originally presented on: 11/6/2013 12:30:00 - 12:48:00

Tumor development alters the normal cellular processes that maintain tissue integrity and homeostasis, and introduces changes to the tissue surrounding the tumor as well as the tumor cells themselves. Moreover, tumorigenesis and invasion are regulated by both the physical and chemical properties of the interstitial microenvironment. Here, we examined the effects of interstitial fluid pressure (IFP) on the expression of both epithelial and mesenchymal genes in engineered three-dimensional (3D) aggregates of MDA-MB-231 human breast cancer cells. We found using experimental analysis and finite element computational models that the pressure-induced convection of tumor-conditioned media from the base to the tip of the cell aggregate controlled the invasive phenotype at the tip. This suggests that IFP affects the migratory and invasive phenotype of cancer cells. It is therefore likely that IFP affects the expression of genes related to the invasive phenotype. Mesenchymal gene expression in particular is expected to be altered by IFP.

(509b) Engineered Tissues to Quantify the Biology of Tumor Spread

Barney, L. E., University of Massachusetts, Amherst
Dandley, E. C., University of Massachusetts, Amherst
Peyton, S. R., MIT
Originally presented on: 11/6/2013 12:48:00 - 13:06:00

Metastasis is the leading cause of fatality for women diagnosed with breast cancer. The most common anatomical sites of distant tumor growth include the brain, lung, liver, and bone, and it is well known that this metastatic spread in breast cancer is not random. Rather, different clinical subtypes of breast cancer exhibit unique patterns of metastatic site preference, called tissue tropism. As two examples, luminal A breast cancer spreads most aggressively to bone, whereas triple negative breast cancer shows the most homogeneous metastatic spread. Given the physical and chemical diversity of these secondary tissue sites, we hypothesize that there is a relationship between the biophysical and biochemical properties of the tissue, and the ability of cells within a particular subtype of breast cancer to adhere, migrate, and grow at these secondary sites.

To quantify this relationship, we have created Metastatic Engineered Tissues (METs), which capture the integrin-binding properties of the secondary site tissues often recipient of breast cancer spread (brain, lung, and bone), and quantified the behavior of a range of cell lines that represent a subset of the known clinical subtypes of breast cancer, including their adhesion, polarization, motility, and proliferation. The spreading rates of MDA-MB-231s, BT549s, MCF-7s, and SkBr3s is heterogeneous, and cell polarization is the *in vitro* metric that best reflects the known metastatic aggression for each subtype. To evaluate MET-specific responses in the presence of microenvironment-specific chemical perturbations, we have dosed cells with a range of integrin antibodies ($\beta 1$, $\alpha 2$, and $\alpha 6$) as well as stimulated with EGF. Cell response to EGF and integrin antibody is tissue specific, and, excitingly, this "MET sensitivity" reflects the known clinical site-specificity in human patients documented by Kennecke et al. When the full data set is clustered, and regression is performed on migration speed and spreading rate data, strong evidence emerges that an antibody for $\beta 1$ is the most promising clinical therapeutic to block metastatic spread to bone, brain, and lung tissue.

In parallel, we have generously received cells from the Massague lab that specifically home to brain, lung, and bone tissue. These cells have been serially passaged in mice, and are a subpopulation of the MDA-MB-231 parental cells. Interestingly, the MET-specificity of these tissue-specific cell lines only emerges in the presence of EGF. By regressing against the tissue-specific populations, MET-specific measurements, such as cell polarization on brain and spreading rate on lung, have emerged as putative predictors of tissue-specific spread. We are currently extending these studies to cells directly from patients at the UMass Medical School, and, connected with accompanied pathology reports and simple multivariable statistical modeling, our goal is to use this tropic fingerprinting to predict metastatic spread in patients, leading to patient-specific therapy, and improved patient surveillance.

(509c) Identifying Determinants of EGFR-Targeted Therapeutic Efficacy Using Computational Modeling

Monast, C. S., University of Pennsylvania
Lazzara, M. J., University of Pennsylvania
Originally presented on: 11/6/2013 13:06:00 - 13:24:00

For decades, clinicians have attempted to antagonize aberrant growth factor receptor-mediated signaling in cancer using antibodies and small-molecule inhibitors designed to interfere with specific mechanistic steps leading to receptor activation and phosphorylation. Significant dose-response variability is observed for these drugs among cells and tumors even in the absence of receptor mutations, but the basis for this is not fully understood. Here, using the epidermal growth factor receptor (EGFR) as a model system, we implement a computational model of EGFR phosphorylation dynamics to identify the factors that determine the efficacy of EGFR-targeted kinase inhibitors and antibodies that compete with EGFR ligands. Our results: 1. identify different kinetic processes as preferentially controlling the efficacy of antibodies versus inhibitors; 2. suggest that the efficacy of inhibitors and antibodies may be favored by the expression of certain naturally expressed EGFR ligands versus others; and 3. suggest new therapeutic design principles. As an example, the model predicts that the efficacy of kinase inhibitors is more sensitive than that of antibodies to perturbations in the activity of protein tyrosine phosphatases (PTPs) regulating EGFR because the time scale for drug-bound receptors to eventually become phosphorylated after drug dissociation is closer to the time scale of receptor dephosphorylation for inhibitors than for antibodies. This result suggests that variation in PTP expression or activity among cells is more likely to explain differential cellular response to kinase inhibitors than to antibodies. Conversely, antibody efficacy is more sensitive to perturbations in ligand binding kinetics, suggesting that local variation in the expression of EGFR ligands is more likely to explain differential cellular response to antibodies than to kinase inhibitors. Our results also specifically highlight that kinetic considerations beyond those reflected by equilibrium binding affinities determine therapeutic efficacy. For example, for a constant kinase inhibitor binding affinity, decreasing rates of inhibitor binding and unbinding is predicted to promote inhibitor efficacy by increasing the time scale with which the receptor must remain unphosphorylated. Conversely, for constant antibody affinity, decreasing rates of antibody binding and unbinding impair antibody efficacy as the antibody becomes increasingly less able to compete with ligand binding. Finally, by integrating these considerations for kinase inhibitors and antibodies, the model identifies minimum concentrations of the two drugs that can be combined to maximally reduce receptor phosphorylation.

(509d) An Algorithm for Patient-Specific Cancer Chemotherapy Scheduling

Ho, T., University of Arkansas
Parker, R. S., University of Pittsburgh
Originally presented on: 11/6/2013 13:24:00 - 13:42:00

The annual U.S. cancer care expenditure is projected to reach $147 billion in 2013 [1]. An improvement in cancer understanding and treatment is needed to ease the economic burden and improve patient outcomes. Advances have been made in developing new cancer treatments such as targeted therapies and immunotherapy. However, a systemic treatment is needed when tumors are not surgically resectable or when tumors reach a detectable size (small mestatasis to remote sites is assumed) [2]. Chemotherapy remains the most common method used for systemic cancer treatment [3]. Chemotherapy affects both healthy and diseased tissues. A balance between successful killing of cancer cells and minimizing the toxicity to the host remains a challenge for clinicians in deploying chemotherapy treatments.

The balance between efficacy and toxicity creates a challenging and interesting problem for chemotherapy scheduling optimization. The objective is to minimize the tumor volume within pharmacokinetic, toxicity, and logistical constraints related to the chemotherapy regimen for a particular drug/tumor pair. Here we present a clinically-relevant formulation of the multiple input chemotherapy scheduling problem using the billion-dollar (sales) per year agent docetaxel ($1.2 billion in 2012) coupled with granulocyte-colony stimulating factor (G-CSF), which mitigates the primary toxicity associated with docetaxel administration.

Docetaxel is a widely used chemotherapeutic for the treatment of breast cancer, non-small cell lung cancer, prostate cancer, and head and neck cancer. However, the primary adverse effect of docetaxel is myleosuppression, or neutropenia, a low absolute neutrophil count (ANC) in the plasma (Grade 1: ANC $>1.5\text{x}10^6$ cells/ml; Grade 2: $1.0\text{x}10^6$ cells/ml $\leq$ ANC $<$ 1.5x10 cells/ml; Grade 3: 0.5x10 cells/ml $\leq$ ANC $<10^6$ cells/ml; Grade 4: ANC $< 0.5\text{x}10^6$ cells/ml) [4]. Depending on the administration schedule, a patient treated with docetaxel may suffer severe neutropenia, especially, at higher doses of docetaxel (*e.g.*, 100 mg/m^2 infusion every three weeks) [5, 6]. Other studies have shown decreased neutropenia at a dose level of 35 mg/m^{2}delivered weekly for 6 weeks (followed by two weeks off). Interestingly, anti-tumor efficacy was unchanged [7,8].

Granulocyte-colony stimulating factor (G-CSF) is a protein that induces the production of neutrophil progenitors, leading to an increase in white blood cell counts. In patients with low neutrophil counts, G-CSF has been shown to reduce neutropenia when injected into patients from day 2-12 after a docetaxel injection on day 1 [9]. G-CSF injection therefore provides an additional actuator to blunt the severe neutropenic response induced by some chemotherapies.

The chemotherapy scheduling problem that we pose and solve will be model-based; hence we require models of drug pharmacokinetics (PK), pharmacodynamic (PD) effect, and toxicity. We use a rigorously-derived reduced-order form of our previously developed physiologically-based model of docetaxel PK [10] to drive a novel reduced-order version of our docetaxel neutropenia model [11], which includes G-CSF PK and PD. We subsequently synthesize the tumor growth and antitumor efficacy model specific to docetaxel and prostate cancer. The problem objective is to minimize tumor volume at the end of every 6 week cycle with manageable toxicity.

The optimization problem is subject to dynamics constraints via mathematical models of docetaxel PK, PD efficacy, and toxicity. Clinically-relevant logistical constraints include: (i) the number of permissible drug doses per week (5); (ii) maximum docetaxel dose (240mg/m^2 per 6 week cycle); (iii) maximum infusion duration (4 hours); (iv) and G-CSF usage only in the cases of most severe neutropenia (Grade 4 toxicity). Model-predicted week to week ANC above Grade 3 neutropenia is also considered in the docetaxel scheduling problem. The nonlinearities in the toxicity model, the tumor growth model, and drug efficacy model in combination with integer variables associated with toxicity and logistical constraints result in the docetaxel scheduling problem formulating as a mixed-integer nonlinear programming (MINLP). Using variable transformation, linearization, and piecewise continuous linear approximation, the MINLP will be reformulated with reduced complexity and solved using GAMS/CPLEX or Pyomo. Building on the results of [12], we will compare the single-cycle results with a receding horizon formulation where multiple cycles are synthesized simultaneously, with recalculation when data (tumor measurements, ANC, etc.) becomes available. The overall objective is to provide clinicians with a model-based tool to assist in improving treatment efficiency and ultimately patient treatment outcome and survival.

[1] L. Zhu, L.W. Pickle, K. Ghosh, D. Naishadham, K. Portier, H.S. Chen, H.J. Kim, Z. Zou,J. Cucinelli, B. Kohler, et al. Predicting us-and state-level cancer counts for the currentcalendar year. *Cancer*, 2012.

[2] E. Chu and V. T. DeVita. Physicians' cancer chemotherapy drug manual 2006. Jones & Bartlett Pub, 2006.

[3] S. E. Clare, F. Nakhlis, and J. C. Panetta. Molecular biology of breast metastasis: The useof mathematical models to determine relapse and to predict response to chemotherapy inbreast cancer. *Breast Cancer Research*, 2(6):430, 2000.

[4] L. E. Friberg, A. Henningsson, H. Maas, L. Nguyen, and M. O. Karlsson. Model ofchemotherapy-induced myelosuppression with parameter consistency across drugs. Journal of Clinical Oncology, 20(24):4713{4721, 2002.

[5] S. D. Baker, M. Zhao, C. K. K. Lee, J. Verweij, J. R. Zabelina, Y.and Brahmer, A. C.

Wolf, A. Sparreboom, and M. A. Carducci. Comparative pharmacokinetics of weekly and every-three-weeks docetaxel. Clinical Cancer Research, 10(6):1976-1983, 2004.

[6] S. Bernard, J. Blair, and M.C. Mackey. Oscillations in cyclical neutropenia: new evidence based on mathematical modeling. *Journal of Theoretical Biology*,223(3):283 - 298, 2003.

[7] G. de Bock, H. Putter, J. Bonnema, J. van der Hage, H. Bartelink, and C. van de Velde. The impact of loco-regional recurrences on metastatic progression in early-stage breast cancer: a multistate model.

Breast Cancer Research and Treatment, 117:401-408, 2009.

[8] T. Yokoi, Tamaki, T., Shimizu, T., & Nomura, S. A pilot study of a metronomic chemotherapy regimen with weekly low-dose docetaxel for previously treated non-small cell lung cancer. *Lung Cancer*, *3*, 15-20, 2012.

[9] A. Bamias, G. Aravantinos, C. Deliveliotis, D. Bafaloukos, C. Kalofonos, N. Xiros, A. Zervas, D. Mitropoulos, E. Samantas, D. Pectasides, et al. Docetaxel and cisplatin with granulocyte colony-stimulating factor (g-csf) versus mvac with g-csf in advanced urothelial carcinoma: a multicenter, randomized, phase iii study from the hellenic cooperative oncology group. *Journal of Clinical Oncology*, 22(2):220-228, 2004.

[10] J.A. Florian. Modeling and dose schedule design for cyclic-specific chemotherapeutics. Dissertation, University of Pittsburgh, 2008.

[11] T. Ho, Clermont, G., & Parker, R. S. A model of neutrophil dynamics in response to inflammatory and cancer chemotherapy challenges. *Computers & Chemical Engineering*, 51, 187-196, 2013

[12] J.M. Harrold and R.S. Parker. Clinically relevant cancer chemotherapy dose scheduling via mixed-integer optimization. *Computers & Chemical Engineering*, 33(12):2042-2054, 2009.

(509e) Tumor Derived Exosomes As Local Modulators of Immunogenic Field Effects in Melanoma

Wu, Y., West Virginia University
Chambers, E., West Virginia University
Klinke, D. J., West Virginia University
Originally presented on: 11/6/2013 13:42:00 - 14:00:00

Therapeutic modulation of immune checkpoints is one of the most recent clinical advances in the treatment of melanoma. Conceptually, immune checkpoints limit the clonal expansion of T cells. Antibodies against these immune checkpoints, such as anti-CTLA4 mAb, enable increased clonal expansion and enhance immune cell infiltration into the tumor microenvironment. Clinical results using anti-CTLA4 mAb demonstrate a durable clinical response but only in a subset of patients. Identifying the mechanisms of resistance is central to broadening the clinical benefit. Pre-clinical models indicate that the efficacy of immune checkpoint modulation depends inversely on tumor load [1]. We have previously identified that malignant melanocytes secrete a number of soluble factors that have the potential to alter the efficacy of immune cells that enter the tumor microenvironment [2]. The objective of this study was to characterize the biological influence of tumor-derived exosomes on immune cell function. In particular, we analyzed exosomes derived from three different cell models of melanoma: B16F0, a cell model of metastatic malignant melanoma; Cloudman S91, a cell model of immunogenic melanoma; and Melan-A, an immortalized melanocyte cell line. We used SEM and TEM to characterize the morphology of exosomes isolated from cell-conditioned media. The exosomes contained receptors derived from the parent cell as demonstrated by identification of interleukin 12 receptor beta 2 (IL-12Rβ2) on B16F0 exosomes by flow cytometry, western blot and immune-gold TEM. Furthermore, the profile of messenger RNA contained within B16F0-derived exosomes was compared against the parental cell using microarray gene expression analysis. Functional annotation and pathway enrichment methods were used to infer putative functional roles of exosomes from coding RNA. We also found that the coding RNA was functionally active as exosomes from B16F0 cells, but not Cloudman S91 and Melan-A exosomes, induced the synthesis of SHP2 in targeted immune cells. This is interesting as SHP2 is the negative regulator of T cell receptor signaling that is activated by the immune checkpoint CTLA4. Anti-CTLA4 mAb is intended to inhibit SHP2 activation. The SHP2 results also help explain the dependence of immune checkpoint modulation on tumor load observed in pre-clinical animal models. Collectively, the results suggest that exosomes help establish a density-dependent field effect by altering the activity of immune cells that enter the tumor microenvironment.

References

1. van Elsas A, Hurwitz AA, Allison JP. "Combination immunotherapy of B16 melanoma using anti-cytotoxic T lymphocyte-associated antigen 4 (CTLA-4) and granulocyte/macrophage colony-stimulating factor (GM-CSF)-producing vaccines induces rejection of subcutaneous and metastatic tumors accompanied by autoimmune depigmentation." J Exp Med. 1999 Aug 2;190(3):355-66.

2. Kulkarni YM, et al. "A quantitative systems approach to identify paracrine mechanisms that locally suppress immune response to Interleukin-12 in the B16 melanoma model." Integr Biol (Camb). 2012 Aug;4(8):925-36.

(509f) An Automated System for the Design of Optimal Personalized Chemotherapy Protocols for the Treatment of Leukemia

Pefani, E., Imperial College of London
Panoskaltsis, N., Northwick Park Hospital
Velliou, E., Imperial College London
Fuentes, M.,
Mantalaris, A., Imperial College
Georgiadis, M. C., Imperial College London
Pistikopoulos, S. N., Imperial College London
Originally presented on: 11/6/2013 14:00:00 - 14:18:00

An automated system for the design of optimal personalized chemotherapy protocols for the treatment of Leukemia.

E. Pefani[a], N. Panoskaltsis[b], E. Velliou[a], M. Fuentes[a], A. Mantalaris[a], M.C. Georgiadis[a], E.N. Pistikopoulos[a]

[a]Centre for Process Systems Engineering, Department of Chemical Engineering, Imperial College London, South Kensington Campus, London SW7 2AZ, UK

[b] Department of Hematology, Imperial College London, Northwick Park & St. Mark's Campus, London, HA1 3UJ, UK

Keywords: Mathematical modeling, chemotherapy optimization, cell cycle models, pharmacokinetics, pharmacodynamics

In the UK (Cancer Research UK data, 2008), it is estimated that more than 1 in 3 people will be afflicted with cancer in their lifetime. For one such cancer, leukemia - a neoplasm of the blood and bone marrow (BM) - 1 in 71 men and 1 in 105 women will be affected, with incidence sharply rising in adults over the age of 50. Approximately 40% of those affected with leukemia will have Acute Myeloid Leukemia (AML), a cancer of the BM and blood wherein blood cells are unable to develop or function normally, are overproduced at an immature stage of development and overtake any normal elements remaining in the BM and blood. This uncontrolled growth compounds the morbidity and mortality due to the disease by inhibiting development of healthy blood and immune cells through multiple mechanisms (Panoskaltsis 2003; Panoskaltsis, 2005).

The most common treatment for most types of leukemia is intensive chemotherapy. It is well-known that this therapy can be life-threatening with only relatively few patient-specific and leukaemia-specific factors are considered in current protocols; choice of chemotherapy, intensity and duration often depends on either the availability of a clinical trial, the treating physician's experience or the collective experience of the treating centre, with significant international protocol variability.

There is therefore a need to optimize current treatment schedules for cancers such as AML to limit toxicities and to improve clinical trial pathways for new drugs to enable personalized healthcare. Towards this end, mathematical modeling is undoubtedly a useful tool for the automation of treatment design due its advantages in exploring extensive datasets in combination with scientific principles in order to capture a systems dynamics and subsequently provide better insight for process enhancement (Dua et. al, 2008; Pefani et. al, 2013).

In this work, the prospective advances to the clinical treatment design through a mathematical model are discussed. A mathematical model and model analysis is presented for induction chemotherapy treatment for AML using Daunorubicin (DNR) and Cytarabine (Ara-C) anti-leukemic agents, a standard intensive treatment protocol for AML. The proposed model combines critical targets of drug actions on the cell cycle, together with pharmacokinetic (PK) and pharmacodynamic (PD) aspects providing a complete description of drug diffusion and action after administration. The current developed model has been created in the gPROMs environment (PSE, 2003) and consists of a model simulator and an optimizer comprised of a closed-loop system for the design of optimal personalized chemotherapy protocols (Figure 1).

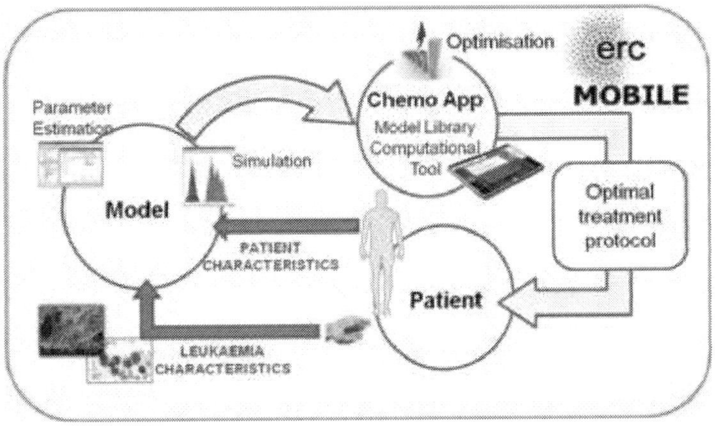

Figure SEQ Figure * ARABIC 1: Schematic representation of the proposed closed loop system for the design of optimal personalized chemotherapy protocols.

The required inputs for both the simulator and the optimizer consist of patient, disease and drug information. Patient information includes gender, age, weight and height, whereas, disease information is that acquired as standard practice from the BM diagnostic test ? leukemic blast percentage in the BM aspirate and the BM cellularity from the trephine biopsy. Extra information to that which is currently used in clinical practice will be the cell cycle characteristics of the S-phase duration and the total cell cycle duration. Moreover, PK information of drug elimination rate in liver and kidneys is required and is readily available in the product specification provided by the pharmaceutical manufacturer. PD information is also required that consists of drug dependent parameters characterising the effect of the anticancer agent on the cell population.

Results of sensitivity analysis indicate that the cell cycle times are the critical model parameters affecting treatment outcome. Clinical data of 6 patients are used in order to estimate the cell cycle time distribution and characterise the intra- and inter- patient variability on cycling times. Moreover, the derived mathematical model is used as an application for an optimization problem. The benefits of optimization are presented and an optimization problem is developed for chemotherapy treatment schedule. The objective of chemotherapy treatment in AML is the reduction of cancer cells to eventual eradication and to ensure that more normal cells remain in the BM. Â Moreover, the number of normal cells is constrained to have a maximum total reduction of 3-log during one chemotherapy cycle.Â An optimization algorithm is presented in the current work for the scheduling of the optimal treatment design based on the control drug use, dose load and dose duration. This optimization problem is solved for one of the patient case studies (P016). P016 received two chemotherapy cycles of treatment and by treatment completion residual disease was detected. For this reason optimization is applied and the optimal treatment schedule for this specific case study is derived. At completion of the optimized treatment protocol, the leukemic cell population is further minimized with a difference of 3Â·10^9 cells when compared to that of the standard treatment protocol. Furthermore, at treatment completion the normal cell population is also lower than that prior to treatment, but it is still higher than the leukemic cell count with a difference of 5.3Â·10^9 cells, as is the desired treatment outcome.

Model analysis reveals the utility of mathematical modeling in gaining better insight into the disease and into normal tissue dynamics during treatment with chemotherapy. There is future potential for the amelioration ofÂ treatment design that will be defined on a case-by-case basis and would be dependent on disease characteristics (tumor-specific characteristics, tumor burden, cell cycle times, normal cell population) as well as patient-specific characteristics (gender, age, weight and height). This design would provide the opportunity to personalize treatment protocols through the use of optimization methods.

Acknowledgment

This work is supported by European Research Council (MOBILE, ERC Advanced Grant, No: 226462), the Richard Thomas Leukaemia Fund and CPSE Industrial Consortium.

References

Dua, P., Dua, V., Pistikopoulos, E., 2008. Optimal delivery of chemotherapeutic agents in cancer. Computers & Chemical Engineering, 32(1-2), 99-107.

gPROMS, Introductory user's guide, release 2.2, Process Systems Enterprise Limited, London, U.K; 2003.

Panoskaltsis N, Reid CDL, Knight SC. Quantification and cytokine production of circulating lymphoid and myeloid cells in acute myelogenous leukemia (AML). Leukemia 2003; 17:716-730.

Panoskaltsis N. Dendritic cells in MDS and AML ? cause, effect or solution to the immune pathogenesis of disease? Leukemia 2005; 19:354-357.

Pefani E., Panoskaltsis N., Mantalaris A., Georgiadis M. C., Pistikopoulos E. N.. Design of optimal patient-specific chemotherapy protocols for the treatment of Acute Myeloid Leukaemia (AML). Computers and Chemical Engineering Journal, In Press, Available Online 1 March 2013.

(509g) The Role of Reductive Carboxylation in Warburg-Like Renal Cancer Cells

Gameiro, P. A., Massachusetts Institute of Technology
Metelo, A.,
Carro, R. P.,
Larrubia, P. L.,
Iliopoulos, O., Massachusetts General Hospital
Stephanopoulos, G. N., Massachusetts Institute of Technology
Originally presented on: 11/6/2013 14:18:00 - 14:36:00

Mammalian cells must consume glucose and glutamine to fuel biosynthetic processes necessary for growth. Acetyl coenzyme A (AcCoA) is a major precursor for lipid synthesis

being primarily generated from glucose-derived pyruvate through the citrate shyttle and ATP citrate lyase (ACLY) in the cytosol. We have shown that hypoxic

and "pseudo-hypoxic" renal cancer cells switch to reductive carboxylation (RC) of a-ketoglutarate as the primary route of AcCoA and de novo fatty acids synthesis. To study the

mechanism(s) of RC induction in hypoxia, we investigated the role of hypoxia-inducible factors (HIF) in renal cell carcinoma (RCC) cells deficient in the

von Hippel-Lindau (VHL) tumor suppressor, which express HIF consititutively even under normoxia conditions (Warburg-like cells). To this end, we took advantage of a panel

of disease associated VHL mutants that differ in their ability to bind and degrade HIF. Using 13C glucose and glutamine tracers, we show that HIF expression is necessary and

sufficient to promote RC in RCC cells under normoxia. Using [1–13C] glutamine, we monitored for the first time the *in vivo* RC activity in VHL-deficient tumors

growing as xenografts in mice. Furthermore, HIF expression renders VHL-deficient cells sensitive to glutamine deprivation *in vitro*, and systemic administration of glutaminase inhibitors

suppresses the growth of RCC cells as mice xenografts. These observations may lead to viable therapeutic strategies targeting the hypoxic compartment of RCC tumors.

(509h) Design Considerations for a Dual-Interaction Drug: Drug Clearance By Feedback Control

Figueroa, I., Merck Research Laboratories
Carballo-Jane, E., Merck Research Laboratories
Camacho, R., Merck Research Laboratories
Lin, S., Merck Research Laboratories
Meehl, M., Merck Research Laboratories
Song, Y., Merck Research Laboratories
Shinsky-Bjorde, N., Merck Research Laboratories
Beaumont, M., Merck Research Laboratories
Tomazela, D., Merck Research Laboratories
Chen, H. S., Merck Research Laboratories
Antonenko, S., Merck Research Laboratories
Ma, H., Merck Research Laboratories
Mengesha, W.,
Zhang, Y. W., Merck Research Laboratories
Judo, M., Merck Research Laboratories
Vives, F., Merck Research Laboratories
Ayanoglu, G., Merck Research Laboratories
Seghezzi, W., Merck Research Laboratories
Cho, C., Merck Research Laboratories
Fayadat-Dilman, L., Merck Research Laboratories
Raoufi, F., Merck Research Laboratories
Bowman, E., Merck Research Laboratories
Marian, M., Merck Research Laboratories
Tabrizi, M., Merck Research Laboratories
Originally presented on: 11/6/2013 14:36:00 - 14:54:00

Chemical or physical actors with the ability to interact with two or more different agents have traditionally been used in various areas of process engineering—from detergents and surfactants to newly application of janus particles in electronics and biomedical sciences. In this work, we explore the design considerations of a 'dual interaction drug'. Unlike traditional drugs that mainly interact with one kind of receptors (effect receptors, ER), this drug is designed to additionally compete with other physiologically relevant players for a second type of receptors (clearance receptors, CR). These receptors act as scavengers and provide a biomarker-regulated clearance pathway. When the biomarker concentration is reaching the lower bound of the desired range, increased CR-mediated clearance should restrict the dual-interaction drug's pharmacological activity due to a decrease in its pharmacokinetic (PK) resident time. The fine balance between the ER's affinity and density and the CR's affinity and density is a critical determinant for this drug's activity. Understanding this balance across species is a key consideration to guide the design of this novel drug. Using a systems pharmacology approach, we are determining the ability of a dual interaction drug and its biomarker to compete for the clearance pathway via a combination of novel experimental and PK-PD modeling approaches.

(512a) Submillisecond Pulses of Fluid Shear Stress Suppress Chemoattractant-Induced Neutrophil Activation

Mitchell, M., Cornell University
King, M. R., Cornell University
Originally presented on: 11/6/2013 12:30:00 - 12:48:00

Submillisecond Pulses of Fluid Shear Stress Suppress Chemoattractant-Induced Neutrophil Activation

Michael J. Mitchell[1], Michael R. King[1]

[1]Department of Biomedical Engineering, Cornell University, Ithaca, New York USA

Introduction: Fluid shear stress (FSS) is a ubiquitous component of the microenvironment for leukocytes, as they carry out the physiological processes of initial adhesion to the endothelium, the formation of pseudopods, and migration into tissues (1). Our lab has recently shown that leukocytes exposed to low venular fluid shear stresses (0.1-4.0 dyn/cm^2) develop a resistance to chemoattractant-induced activation (2). However, leukocytes can transiently encounter FSS values >1000 dyn/cm^2 in the heart, near the walls of large blood vessels, and at vessel bifurcations. Herein, we evaluated the effect of submillisecond FSS pulses on the neutrophil response to chemoattractants.

Materials and Methods: Neutrophils isolated from peripheral human blood following informed consent were exposed to 1-10 submillsecond FSS pulses via perfusion through microscale conduits using a high-pressure syringe pump, followed by exposure to the chemoattractant formyl-methionyl-leucyl-phenylalanine (fMLP). Expression and spatial distribution of L-selectin, activated $\alpha_M\beta_2$ integrins, and formyl peptide receptors (FPR) were analyzed using flow cytometry and confocal microscopy. Neutrophil morphology and the appearance of pseudopods were imaged using brightfield and phase contrast microscopy, with images used to measure neutrophil shape factor using an edge detection function. Expression of phosphorylated proteins including p38 MAPK and c-Abl were determined using a Bio-Plex phosphorylated protein detection system.

Results and Discussion: In the absence of fMLP stimulation (Fig. 1A), neutrophils exhibited minimal activation when exposed to either static conditions or 5 submillisecond FSS pulses (wall shear stress (WSS) = 6400 dynes/cm^2, exposure time = 0.89 ms). Neutrophil samples exposed to static conditions followed by

10 nM fMLP stimulation resulted in a 36% loss of L-selectin and 34% $\alpha_M\beta_2$ integrin activation (Fig. 1A). However, neutrophil exposure to 5 submillisecond FSS pulses followed by stimulation with 10 nM fMLP reduced L-selectin loss to 16%, and $\alpha_M\beta_2$ integrin activation to 9% (Fig. 1A). No significant difference in L-selectin shedding and $\alpha_M\beta_2$ integrin activation was found between sheared and nonsheared neutrophils, whereas a significant reduction in fMLP-induced L-selectin shedding and $\alpha_M\beta_2$ integrin activation was found in neutrophils exposed to FSS pulses (Fig. 1B). Neutrophil resistance to fMLP-induced activation was FSS pulse-dependent, as 5 pulses of FSS were required to induce neutrophil resistance to activation (Fig. 1C).

Conclusions: We have shown that brief exposure to submillisecond pulses of FSS can induce a mechanosensitive response in human neutrophils. Exposure to high FSS pulses reduce neutrophil activation in the presence of bacterial peptides, and likely serve as a physiological regulator of neutrophil activation and invasion into tissues. Insight into the molecular basis of the neutrophil response can reveal important drug targets for inflammation and hemostasis.

Acknowledgements: This study was supported by the National Institutes of Health, grant No. HL018208.

Figure 1: Submillisecond FSS pulses suppress fMLP-induced neutrophil activation. (A) Neutrophils exposed to static conditions or 5 submillisecond pulses of FSS (WSS= 6400 dyn/cm2, pulse time = 0.89 ms) followed by 10 nM fMLP treatment. As indicators of activation, neutrophils were labeled for L-selectin expression and $\alpha_M\beta_2$ activation and analyzed using flow cytometry. (B) Neutrophil L-selectin shedding and αMβ2 activation from n = 4 donors, when exposed to FSS and/or fMLP. (C) Neutrophil fMLP-induced L-selectin shedding and αMβ2 activation after exposure to 1-10 submillisecond FSS pulses.

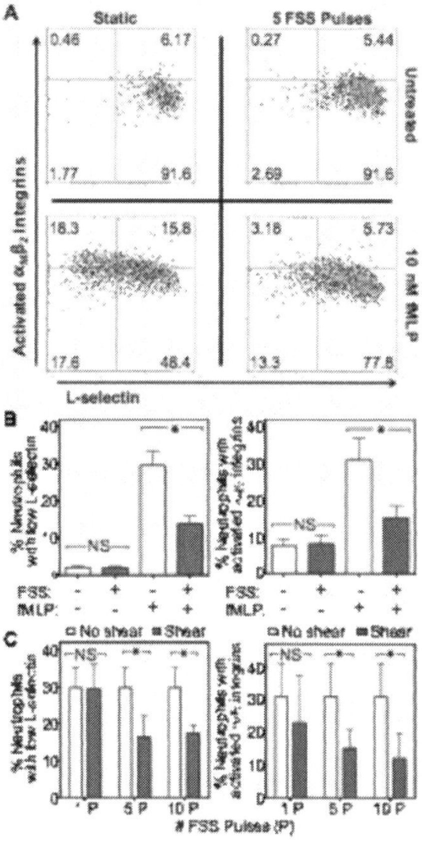

References:

1. Springer TA. Traffic signals for lymphocyte recirculation and leukocyte emigration: The multistep paradigm. *Cell*. 1994 Jan;76(2):301–14.

2. Mitchell MJ, King MR. Shear-Induced Resistance to Neutrophil Activation via the Formyl Peptide Receptor. *Biophysical Journal*. 2012 Apr 18;102:1804–14.

(512b) Influence of Size and Shape of Vascular-Targeting Carriers On Neutrophil Adhesion to An Inflamed Endothelial Monolayer in Blood Flow

Onyskiw, P., University of Michigan
Eniola-Adefeso, L., University of Michigan
Originally presented on: 11/6/2013 12:48:00 - 13:06:00

Ligand-based targeting of diseased vasculature is attractive for improving diagnostics and therapeutic treatment of cardiovascular diseases. Many cardiovascular events, including plaque development and reperfusion-injury, originate at sites of vascular inflammation. Receptors such as E-selectin are often targeted for imaging and drug delivery because of its specific upregulation by endothelial cells, the monolayer of cells lining the internal wall of blood vessels, during inflammation. Researchers have investigated the effects of size and shape on the adhesion of E-selectin targeting carriers to inflamed endothelial cells; however, little work has been done in the presence of blood cells - specifically neutrophils, which interact with the vessel wall through E-selectin during an inflammatory response. Little is known about particle-neutrophil dynamics in blood flow and whether targeted carriers affect neutrophil adhesion to the inflamed vessel wall. Here we investigate the effect of carrier size and shape on neutrophil adhesion to an inflamed endothelial monolayer using E-selectin targeting polystyrene microspheres (0.5, 2 and 5 μm) and ellipsoidal rods in human blood flow at physiological shear conditions using a parallel plate flow chamber. Our results show that E-selectin targeting carriers influence neutrophil adhesion and the response is dependent on the carrier's size, shape, and hemodynamics.

(512c) Elucidating Design Principles Underlying Attractive and Repulsive Gradient Sensing in Eukaryotic Chemotaxis

Krishnan, J., Imperial College London
Alam-Nazki, A., Imperial College London
Originally presented on: 11/6/2013 13:06:00 - 13:24:00

Many cells respond to gradients of concentrations of chemicals in their environment by exhibiting directional migration, a process known as chemotaxis. This widely occuring process plays very important biological and physiological roles. In many cases, the movement is up-gradient, and the process is referred to as chemoattraction, and in some other cases, the movement is down-gradient and the process is referred to as chemorepulsion. It has been experimentally found that different eukaryotic cells are capable of both chemoattraction and chemorepulsion, either to different chemicals, or in some cases to the same chemical but under different conditions. It is thus important to examine what features in the wiring of the signalling networks and their organization allow the cell to exhibit both these features, and how this affects the signal transduction in each case.

We do this by systematically analyzing two cases, motivated by their postulation and discussion in the experimental literature. The first is that of a "polarity switch" and the second is the role of competing signalling effects at the gradient sensing. For each of these cases, we will discuss how these basic network wiring possibilities may allow for the cell to exhibit attractive and repulsive response, and then examine how this works upstream of different qualitative signalling behaviour such as adaptation and spontaneous polarization. This is achieved by "connecting" the relevant modules in-silico. This systematic approach reveals the behaviour of each design feature, the capabilities and constraints involved. This approach provides a "systems skeleton" for understanding how cell signalling is organized to exhibit multiple migratory responses, and how cell may either conform to or bypass these constraints

We will then discuss the implication of our various results in the context of the known signalling networks of different concrete eukaryotic cell types. Overall this provides a design framework for examining and elucidating design principles of attractive/repulsive gradient sensing in multiple systems and a basis for controlling cell migration based on this feature.

(512d) Role of a Tandem Hormone-Binding Element in Insulin Recognition By the Insulin Receptor

Vashisth, H., University of Michigan
Abrams, C. F., Drexel University
Originally presented on: 11/6/2013 13:24:00 - 13:42:00

Insulin regulates blood glucose levels in higher organisms by binding to and activating insulin receptor (IR), a constitutively homodimeric glycoprotein of the receptor tyrosine kinase (RTK) superfamily. Therapeutic efforts in treating diabetes have been significantly impeded by the absence of structural information on the activated form of the insulin/IR complex. Many biophysical, biochemical, and structural studies strongly suggest the possibility of plasticity in hormone and the receptor structure. Specifically, insulin is proposed undergo a conformational change in the C-terminus of its B-chain, while receptor flexibility may in part be due to rearrangement of the so-called CT-peptide, a tandem hormone-binding element of IR. Using various atomistic simulation approaches, we have constructed all-atom structural models of hormone/receptor complexes in the presence of CT in its crystallographic position and a thermodynamically favorable displaced position. The displacement of CT is confirmed by recent structural studies on minimal receptor constructs that can bind insulin. When aligned into a full ectodomain structure of receptor, the new insulin-bound structures suffer many severe steric clashes with these domains. Therefore, the likely structural rearrangements in these domains on insulin binding yet remain unclear. Starting with our displaced-CT structural model of insulin/IR complex and by creating a simulated target density map from the crystallographic data, we have generated a structural model of insulin/IR complex without any structural overlap of hormone with receptor domains. This refined structural model provides significant insights into a new hormone/receptor contact surface other than the classical insulin recognition surface on the receptor. The results presented in this work have further implications for the design of receptor-specific agonists/antagonists.

(512e) Development of Partial Agonists Targeted At Receptor Tyrosine Kinases for Cancer Drug Discovery

Cullum, R. L., Auburn University
Riese, D. J. II, Auburn University
Gupta, R. B., Auburn University
Originally presented on: 11/6/2013 13:42:00 - 14:00:00

Development of Partial Agonists Targeted at Receptor Tyrosine Kinases for Cancer Drug Discovery

Richard L. Cullum*, David J. Riese, II**, and Ram B. Gupta*

*Department of Chemical Engineering, Auburn University, Auburn, AL 36849

**Department of Pharmacal Sciences, Auburn University, Auburn, AL 36849

According to the American Cancer Society, it is estimated that about 580,350 Americans will die of cancer in 2013. This is almost 1,600 people per day. Figures as large as this indicate that there needs to be a push in the improvement of chemotherapeutics. In recent years, receptor tyrosine kinases (RTKs) have been shown to be promising targets for cancer drug discovery. RTKs are cell surface receptors that regulate normal cell growth and development, but also function in the growth of cancer tumors. Targeting RTKs with ligands that function as partial agonists could potentially prohibit the proliferation of tumor cells. The partial agonists would compete with the full agonists for receptor binding and therefore preventing the full agonists from stimulating coupling to cell proliferation.

ErbB4 (HER4) is a member of the ErbB family of receptor tyrosine kinases, a family that also includes the epidermal growth factor receptor (EGFR/ErbB1), ErbB2 (HER2/Neu), and ErbB3 (HER3). Conflicting reports in the literature suggest that ErbB4 may function either as an oncogene or as a tumor suppressor in human malignancies.[1,2] Thus, our long-term goal is to develop ErbB4 ligands that can be used to elucidate ErbB4 function and ErbB4 antagonists that could be used to treat ErbB4-dependent tumors. Our strategy is based on the observation that the Q43L mutant of the naturally-occurring ErbB4 agonist Neuregulin 2beta (NRG2β) functions as a partial agonist at ErbB4; NRG2β/Q43L stimulates ErbB4 tyrosine phosphorylation, but does not stimulate ErbB4 coupling to cell proliferation and competitively antagonizes agonist stimulation of ErbB4 coupling to cell proliferation. Therefore, we seek to develop medium- and high-throughput assays for stimulation of ErbB4 tyrosine phosphorylation and for stimulation of ErbB4 coupling to cell proliferation. Molecules that stimulate ErbB4 tyrosine phosphorylation but do not stimulate ErbB4 coupling to cell proliferation are highly likely to function as ErbB4 antagonists. Here we describe our efforts to develop these assays.

The goal of the first assay will be to screen for molecules that stimulate ErbB4 tyrosine phosphorylation. After seeding CEM/ErbB4 cells in 96-well dishes, they will be stimulated with a library of potential ErbB4 ligands, NRG2β/Q43L, or PBS. The Q43L mutant will act as the positive control for stimulation of tyrosine phosphorylation and the PBS will act as the negative control. After stimulation, the CEM/ErbB4 cells will be lysed and the lysates will be added to 96-well dishes that have been coated with a capture antibody that recognizes ErbB4. An anti-phosphotyrosine antibody, an HRP-conjugated secondary antibody, and a colorimetric assay will then be used to quantify ErbB4 tyrosine phosphorylation. Several parameters will be varied to optimize detection and quantification of ErbB4 tyrosine phosphorylation. These parameters include the amount of ErbB4 capture antibody, the amount of CEM/ErbB4 lysate, the concentration of NRG2β/Q43L, the amount of anti-phosphotyrosine antibody, and the amount of the HRP-conjugated secondary antibody.

The second assay will be designed to detect molecules that do or do not stimulate ErbB4 coupling to cell proliferation. To accomplish this, BaF3/ErbB2+ErbB4 cells that have been seeded in a 96-well dish will be stimulated with NRG2β, NRG2β/Q43L, the library of potential ErbB4 ligands, and PBS. Cells treated with NRG2β will serve as the positive controls. Cells treated with NRG2β/Q43L will serve as the negative controls. Cells treated with PBS will serve as the diluent controls. After stimulation, the cells will be incubated for 48-144 hours before a tetrazolium salt and a colorimetric assay are used to analyze the ErbB4 coupling to cell proliferation. In an effort to maximize detection of ErbB4 coupling to cell proliferation the following parameters will be varied: number of BaF3/ErbB2+ErbB4 cells, NRG2β and NRG2β/Q43L concentrations, incubation time, and tetrazolium salt dye incubation time.

The development of these assays will be of great importance in the search for ligands that function as partial agonists for ErbB4. Once all experimental parameters have been optimized, libraries of potential ErbB4 ligands such as a series of monoclonal antibodies could be effectively scanned in hope of discovering ligands that competitively bind with full agonists for the ErbB4 receptor. Partial agonists at ErbB4 would be promising candidates for the establishment of a new paradigm in cancer treatment.

(1) Mill, C. P.; Gettinger, K. L.; Riese II, D. J. *Experimental Cell Research* **2011**, *317*, 392–404.

(2) Koutras, A. K.; Fountzilas, G.; Kalogeras, K. T.; Starakis, I.; Iconomou, G.; Kalofonos, H. P. *Critical Reviews in Oncology/Hematology* **2010**, *74*, 73–78.

(512f) Aortic EC Aquaporin-1 Expression and Function in Whole Vessels and Culture

Rumschitzki, D., Department of Chemical Engineering, City College of City University of New York
Raval, C. B., Gradute Center at the City University of New York
Tarbell, J. M., Gradute Center at the City University of New York
Jan, K. M., College of Physician and Surgeons, Columbia University
Sedes, O., The City College of New York
Originally presented on: 11/6/2013 14:00:00 - 14:18:00

ABSTRACT:

Transmural fluid transport in conduit arteries is important for processes depending on convective transport such as paracrine signaling, gas and nutrient supply and catabolite removal. Arterial endothelial cell (EC) aquaporin-1 (AQP-1) expression and function may be regulated in tandem with these events. The activation of cultured endothelial membrane adenyl cyclase production either by arginine vasopressin (AVP) or forskolin (FSK) influences both the formation and cell location of AQP-1. AVP acting via receptors significantly increases the content of AQP-1 content in confluent bovine aortic EC (BAEC) cultures with most of it perinuclear. FSK activates plasma membrane adenyl cyclase increasing cAMP with no increase in cell AQP-1 content but a redistribution of much of the protein to the cell periphery. The cultured EC hydraulic conductivity (Lpe) was much greater after FSK exposure despite no change in AQP-1 synthesis. FSK significantly decreased paracellular permeability indicating that the increased Lpe was by a transcellular mechanism probably secondary to the AQP-1 shuttling from a perinuclear site to the plasma membrane. In contrast to FSK, which produced a much greater Lpe per AQP, the Lpe increment after AVP was equivalent to the increment in AQP-1 synthesis. ECs on intact rat aorta exposed to physiologic physical and humoral stimuli had a much different response to FSK. In contrast to the BAEC the rat AEC AQP-1 was virtually all in the surface membrane, mostly on the abluminal side. FSK exposure in this system, in contrast to the cultured cells, provoked a significant increase in AQP-1 accumulation, mostly abluminal, and a shift in Lpe equivalent to this increment. These data demonstrate differences in the response of EC AQP-1 to cAMP depending both on the stimulus i.e. AVP vs FSK and on whether the cell has recently proliferated and had not been exposed to physiologic physical and humoral stimuli or had been quiescent and programmed by these stimuli.

(512g) Investigating Inter-Cellular Signaling in Engineered Liver Mimetic Tissues

Vu, L., Virginia Tech
Rajagopalan, P., Virginia Tech
Originally presented on: 11/6/2013 14:18:00 - 14:36:00

In vivo, cytokine signaling can generate both pro- and anti-inflammatory effects such as cell death, generation of reactive oxygen species, expression of cytochrome P450 enzymes, progression towards carcinomas, as well as liver regeneration. Interleukin 6 (IL-6), a cytokine released by Kupffer cells (KCs), is commonly associated with liver injury and regeneration in hepatocytes. *In vitro* studies on the effects of cytokines have been conducted primarily on hepatocyte monolayers (HMs) or collagen sandwich (CS) cultures. However, neither culture is physiologically relevant. Due to dedifferentiation of *in vitro* primary hepatic cultures, studies on cytokine signaling are often conducted within 24 h in culture, and therefore provide limited information on longer term effects. We have assembled engineered multi-cellular organotypic hepatic models that closely recapitulate the *in vivo* architecture and cellular composition. Since these tissues maintain hepatic phenotype for extended time periods, they are ideal to conduct detailed investigations into the effects of cytokine secretion on multiple liver cell types.

The present study is focused upon the perturbations in liver function and signaling in the presence of exogenous and endogenous IL-6. Organotypic liver models were assembled using primary hepatic cells obtained through the surgical excision of rat livers. Cellular models that closely mimic the architecture and *in vivo* composition were assembled using detachable polyelectrolyte multilayers to mimic the Space of Disse [1]. These liver models are typically assembled with primary hepatocytes, KCs, and liver sinusoidal endothelial cells (LSECs). HM and CS cultures served as controls. All hepatic cultures were maintained for up to 12 days and hepatic functions were monitored.

The introduction of exogenous IL-6 into HM, CS and liver model cultures elicited different responses. HM and CS cultures both exhibited a reduction in albumin production upon addition of IL-6. Interestingly, CS cultures recovered and secreted basal albumin levels within 5 days. Another very interesting trend was the level of IL-6 measured in spent culture medium 24 h subsequent to the administration of this cytokine. It was observed that liver models (comprised of LSECs and hepatocytes) had approximately two-fold higher levels of IL-6 in spent culture medium in comparison to HM and CS samples. These trends were observed at two separate time points (days 4 and 8) suggesting that LSECs in the liver models could be contributing through an endogenous release of this cytokine. Our ongoing efforts are to investigate how exogenous administration of IL-6 activates these pathways using a combined proteomics and transcriptional approach as compared to endogenous signaling.

Cytokines are important signaling molecules that often elicit multiple effects. We have demonstrated that the use of an organotypic hepatic model facilitates the study of cytokines at later time points. In addition, we hypothesize that hepatic non-parenchymal cells, such as LSECs, are also playing an important role.

References:

1. Larkin A. L. and coworkers, Tissue Engineering Part C: Methods, In Press, 2013.

(512h) Novel Interactions Between the NF-kB and BMP Signaling Pathways in the D. Melanogaster Embryo

Carrell, S., North Carolina State University
Reeves, G. T., North Carolina State University
Originally presented on: 11/6/2013 14:36:00 - 14:54:00

Receptor-mediated signals play a crucial role in tissue patterning of developing organisms, stem cell maintenance, and adult homeostasis. Many diseases, most notably cancer, occur when cells misinterpret or ignore signals regarding proliferation, migration, and/or apoptosis, implying that these signaling pathways must be tightly regulated. Since these same signals are highly conserved between species, we study the regulation of signaling pathways in the context of patterning the dorsal-ventral (DV) axis in the early *Drosophila melanogaster* embryo.

When the embryo is roughly 2 hours old, two conserved signaling pathways act to pattern the DV axis: the Dorsal pathway and the Dpp pathway. On the ventral side of the embryo, the transcription factor Dorsal, homologous to NF-κB, triggers expression of the genes that initiate the DV pattern in the *Drosophila* embryo. Dorsal is present in a nuclear gradient, with the highest concentration at the ventral midline and a steady decay to about 40% of the embryo's circumference. Signaling through the Toll receptor initializes the Dorsal gradient in the developing embryo; on the ventral side of the embryo, the activated ligand SpÃ¤tzle binds to Toll, which then phosphorylates the inhibitor protein Cactus (homologous to I-κB), marking it for degradation. In the absence of Cactus, Dorsal can enter the nuclei and activate expression of target genes in a concentration-dependent manner. Dorsal also acts as a transcriptional repressor, limiting some genes to expression on the dorsal half of the embryo.Â One of those dorsally-expressed genes is *decapentaplegic* (*dpp*), which encodes Dpp, a bone morphogenic protein (BMP) ligand. Dpp is present in a graded fashion, establishing DV gene patterns beyond Dorsal's reach. Dpp signals through the receptor Thickveins (Tkv), phosphorylating the receptor Smad protein MAD; phospho-MAD (pMAD) can be used as a readout of BMP signaling. See (A) for a schematic of the two signaling pathways. Note: dorsal is up, ventral is down. Together, the nuclear gradient of Dorsal and BMP signaling work together to pattern the DV axis of developing *Drosophila* embryos.Â

Because robustness of tissue patterning is essential for proper embryonic development, regulatory loops must exist to ensure correct placement of target genes in the face of perturbed conditions. As an example, embryos with a half dose of Dorsal protein survive to adulthood. We have analyzed the Dorsal gradient in these heterozygotes by modeling the nuclear gradient of Dorsal as a Guassian, using the equation:

C(x) = Aexp[-x^2/2σ^2] + B

where C is the concentration of nuclear Dorsal, x is the normalized DV coordinate (0 being the ventral midline), A represents the amplitude, σ, the width, and B is the basal level of Dorsal present in the nuclei at the dorsal midline. As shown in (B), we found that the Dorsal gradients in these embryos are not only shorter in amplitude as compared to wild type, as expected, but statistically wider and fundamentally different in shape, having a broader domain of peak nuclear concentration. These differences point to the existence of Dorsal gradient regulation in circumstances in which protein levels are compromised. We will present evidence that the Dpp signaling network can alter the Dorsal gradient. When levels of MAD (the BMP signal transducer) were overexpressed, we found that the width of the gradient (as measured by σ) expanded significantly. Therefore, feedback through the Dpp signaling network is a prime candidate for enhancing the robustness of patterning of DV gene expression, as shown in (C). These interactions between the NF-κB and BMP signaling pathways may be necessary to ensure robust gene expression in the developing *Drosophila* embryo.

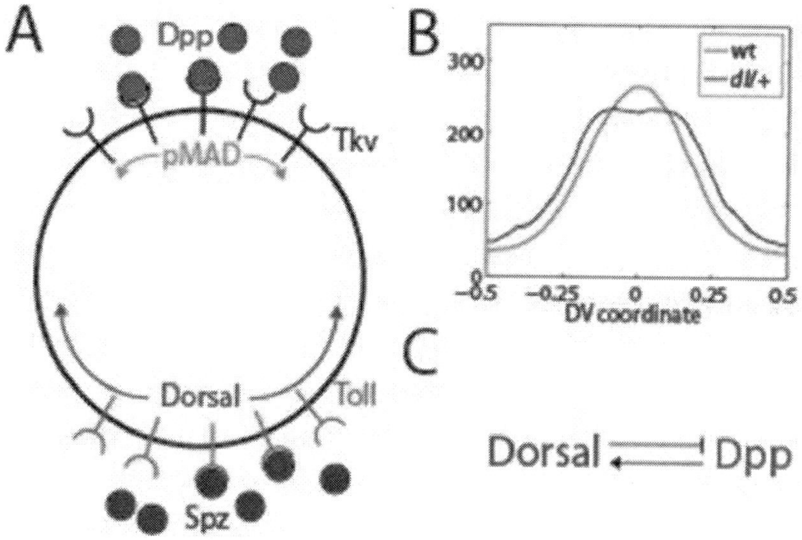

(516a) PDGF and TGF- 1 Crosstalk Is Essential for Mesenchymal Stem Cell Stiffening

Ghosh, D., Georgia Institute of Technology
McGrail, D. J., Georgia Institute of Technology
Dawson, M. R., Georgia Institute of Technology
Originally presented on: 11/6/2013 12:30:00 - 12:48:00

Mesenchymal stem cells (MSCs) are multipotent progenitor cells which can be isolated from bone marrow (BM), expanded rapidly and exhibit multi-lineage differentiation potential [1]. MSCs spontaneously home to tumor and wound healing sites, which secrete pro-inflammatory molecules that mediate MSC migration. In tissue microenvironments, MSCs secrete cytokines and chemokines for recruitment of other cell types (e.g. fibroblasts and endothelial cells), deposit and remodel local extracellular matrix, and differentiate into stromal cells. The role of MSCs in the tissues is dependent on niche-specific mechanical and chemical stimuli. For example, the tumor stroma is often more rigid than surrounding tissues with 13-fold higher rigidities of high-grade invasive ductal carcinoma compared to normal fibroglandular tissue [2]. Cells can feel this differential local tissue rigidity and respond by differentiating into specific cell types (e.g. myofibroblasts). In vivo microenvironments also include a multitude of growth factors that stimulate cell survival, migration, proliferation and differentiation using both autocrine and paracrine signaling mechanisms. Transforming growth factor-β1 (TGF-β1) [3, 4] and platelet derived growth factor-BB (PDGF-BB) [4-5] are two such widely studied growth factors that have been recognized to play key roles in tumors and regenerating tissues [7, 8].We systematically characterized the microscopic mechanical properties of MSCs, including their cytoskeletal rheology, morphology, cellular adhesion, and cell elongation, in response to TGF-β1 and PDGF-BB (referred as PDGF), individually and in combination. Small-molecule signaling inhibitors SB-505124 (1μM) and JNJ-10198409 (50nM) were used to block TGF-βRI and PDGFR-β dependent signaling, respectively. Affymetrix 430.2 mouse genome arrays were used to analyze genome wide expression of treated and untreated MSCs.

Although PDGF alone did not alter the morphology of MSCs, TGF-β1 individually and in combination with PDGF resulted in drastic elongation of MSCs and condensed both actin and microtubule organization. The viscoelasticity of MSCs, measured with multiple particle tracking microrheology (MPTM), was also severely affected by TGF-β1, alone and in combination with PDGF. The elastic modulus (G') of the cytosol was increased 9-fold in TGF-β1 treated cells; whereas, the combination of PDGF and TGF-β1 treatment resulted in 100-fold change in the G' compared to control cells after 24 hours. Blocking either TGF-β1 or PDGF signaling with inhibitors reversed the soluble factor induced cellular stiffening to control levels. Flow Cytometry results indicated significant increase in cell surface expression of β1 (CD29) and α_v(CD51) integrins for these conditions. Finally, a genome-wide microarray analysis revealed TGF-β1 dependent regulation of genes involving cytoskeletal actin-binding proteins (ABPs). In conclusion, MSCs undergo dramatic elongation and stiffening under the influence of TGF-β1 (individually and combination with PDGF). However, our studies reveal that PDGF signaling is essential in activating ABPs required for TGF-

$\beta 1$dependent stiffening.

Acknowledgements: Funding provided by the Georgia Tech and Emory Center for Regenerative Medicine and NSF.

1. Pittenger, M.F., et al., *Multilineage potential of adult human mesenchymal stem cells.* Science, 1999. **284**(5411): p. 143-7.

2. Samani, A., J. Zubovits, and D. Plewes, *Elastic moduli of normal and pathological human breast tissues: an inversion-technique-based investigation of 169 samples.* Physics in medicine and biology, 2007. **52**(6): p. 1565.

3. Frantz, S., et al., *Transforming growth factor beta inhibition increases mortality and left ventricular dilatation after myocardial infarction.* Basic research in cardiology, 2008. **103**(5): p. 485-492.

4. Pierce, G., et al., *Platelet-derived growth factor (BB homodimer), transforming growth factor-beta 1, and basic fibroblast growth factor in dermal wound healing. Neovessel and matrix formation and cessation of repair.* The American journal of pathology, 1992. **140**(6): p. 1375.

5. Chen, L., et al., *Paracrine factors of mesenchymal stem cells recruit macrophages and endothelial lineage cells and enhance wound healing.* PloS one, 2008. **3**(4): p. e1886.

6. Crowe, M.J., T. Doetschman, and D.G. Greenhalgh, *Delayed wound healing in immunodeficient TGF-beta 1 knockout mice.* J Invest Dermatol, 2000. **115**(1): p. 3-11.

7. Leight, J.L., et al., *Matrix rigidity regulates a switch between TGF-beta1-induced apoptosis and epithelial-mesenchymal transition.* Mol Biol Cell, 2012. **23**(5): p. 781-91.

(516b) Surface Immobilization of P-Selectin Glycoprotein Ligand-1 On Mesenchymal Stem Cells Enables Selectin Mediated Cell Tethering and Rolling

Lo, C., State University of New York at Buffalo
Antonopoulos, A., Imperial College
Dell, A., Imperial College London
Haslam, S., Imperial College London
Lee, T., State University of New York at Buffalo
Neelamegham, S., State University of New York
Originally presented on: 11/6/2013 12:48:00 - 13:06:00

Mesenchymal stem/stromal cells (MSCs) are an important candidate for cell-based therapy since they can be easily isolated and expanded, secrete beneficial paracrine factors, and differentiate into multiple lineages. Since the endothelium at sites of injury and inflammation often express adhesion molecules belonging to the selectin family, methods to endow MSCs with selectin-ligands can enhance the efficacy of cell delivery and tissue engraftment. Here, we describe a construct 19Fc[FUT7$^+$], where the first 19 amino acids of the pan-selectin ligand PSGL-1 (P-selectin glycoprotein ligand-1) was fused to a human IgG1. When expressed in HEK293T cells over-expressing the α(1,3)fucosyltransferase FUT7, 19Fc[FUT7$^+$] is decorated by a core-2 sialyl Lewis-X sialofucosylated O-glycan. The non-covalent coupling of this protein onto MSC surface using palmitated protein G (PPG) enhanced cell binding to E- and P-selectin under hydrodynamic shear, without altering MSC multipotency. MSCs functionalized with 19Fc[FUT7$^+$] were captured/tethered onto stimulated endothelial cell monolayers at wall shear stresses up to 4 dyn/cm^2. Once captured, the cells rolled robustly up to the highest shear stress tested, 10 dyn/cm^2. Unlike previous work where MSCs could only be captured onto selectin-bearing substrates at low or no-flow conditions, the current work presents a robust strategy to enable leukocyte-like capture and rolling.

(516c) Ectopic Expression of Nanog Up-Regulates Srf and Reverses the Loss of Myogenic Differentiation Capacity of Human Mesenchymal Stem Cells Due to Senescence

Mistriotis, P., University at Buffalo, State University of New York
Liang, M. S., State University of New York at Buffalo
Andreadis, S. T., University at Buffalo, The State University of New York
Originally presented on: 11/6/2013 13:06:00 - 13:24:00

Human Mesenchymal Stem Cells (hMSCs) show great promise as an autologous smooth muscle cell source for vascular tissue engineering and regenerative medicine. However culture senescence and organismal aging reduce the proliferative and myogenic differentiation capacity of MSCs. To address this issue, we recently reported that ectopic expression of Nanog can restore the lost contractile phenotype due to aging and enhance the proliferative potential of ovine MSCs (Han et al., 2012, Stem Cells). In the current study we aim in identifying the molecular mechanism that mediates the rejuvenating effects of Nanog in MSCs. MSCs isolated from hair follicle or bone marrow were transduced with a tetracycline regulatable vector that carries the Nanog gene. This system enables Nanog expression upon Doxycycline (Dox) treatment. After serial passaging, MSCs were induced to senescence and Dox was added to the culture medium. Subsequently, the effect of Nanog on proliferative and myogenic capacity of MSCs were evaluated and compared with early passage (EP) MSCs. Ectopic expression of Nanog in late passage (LP) MSCs significantly increases the proliferation and decreases the percentage of senescent cells as indicated by SA-β-GAL staining. Furthermore, we show that Nanog controls the expression of key regulators of senescence such as cyclin inhibitors and DNA methyltransferases. Next we studied the myogenic differentiation upon Nanog expression. SMC-specific markers such as ACTA2 and Calponin were highly upregulated in Nanog^{+}LP MSCs resembling EP MSCs. In agreement, the contractility in response to vasoactive agonists was also enhanced. Interestingly, we also report that Nanog reverses the loss of the expression of key myogenic transcription such as SRF, MRTF-b, SMAD2 and MYOCD due to senescence. Notably, we identified that Nanog co-immunoprecipitatets with SRF and enhances the transcriptionaly activity of the SRF DNA binding site (CArG box). At this framework, SRF knockdown abolished the effect of Nanog on myogenic markers and contractility. On the other hand, over-expression of SRF in LP MSCs enhances the myogenic phenotype resembling the effect of Nanog. Taken together these data suggest that Nanog reverses the effects of senescence on MSC proliferation and myogenic differentiation capacity, thereby increasing the potential of MSC from aged donors for cellular therapy and tissue regeneration.

(516d) Tissue Development of Pluripotent Stem Cell-Derived Neural Progenitor Aggregates Regulated By Human Mesenchymal Stem Cell Secretome

Sart, S., Florida State University
Liu, Y., Florida State University
Ma, T., Florida State University
Li, Y., FAMU-FSU College of Engineering
Originally presented on: 11/6/2013 13:24:00 - 13:42:00

Neural progenitor cells (NPCs) derived from pluripotent stem cells (PSCs), including embryonic stem cells (ESCs) and induced PSCs, emerge as a promising tool for the potential treatment of various neurological diseases. Regulation of PSC-derived NPC microtissues is critical for the cell survival, proliferation, and differentiation both in vitro and in vivo. It is hypothesized that human mesenchymal stem cell (hMSC) secretomes promote PSC-derived NPC microtissue attachment, survival, proliferation, and migration through the modulation of PSC-NPC microenvironment. To test this hypothesis, PSC-derived NPC aggregates were replated in the presence of hMSC secretomes generated under hypoxia and normoxia. The adherent cell number, cells in S phase of cell cycle, and cell viability were significantly promoted in hMSC secretomes which were produced under normaxia or hypoxia. The migration length, neurite extension, and the yield of neural and glial differentiated cells were also increased. The hMSC secretomes were found to stimulate the endogenous secretion of extracellular matrices from PSC-derived NPC aggregates. Moreover, molecules regulating the signaling from fibroblast growth factor (FGF)-2, transforming growth factor (TGF)-β1, and brain-derived neurotrophic factor (BDNF) were identified to contribute to the cellular response through the pathway inhibition with PD173074, a FGFR1/FGFR3 inhibitor, SB431542, a SMAD2 inhibitor, and anti-proBDNF respectively. PSC-derived NPC aggregates in hypoxic hMSC secretome may represent a suitable combination to promote engraftment and neurogenesis *in vivo*.

(516e) Influence of Polymeric Substrate Topography On the Orientation, Proliferation and Transdifferentiation of Mesenchymal Stem Cells for Neuroregeneration Strategies

Sharma, A. D., Iowa State University
Zbarska, S., Iowa State University
Petersen, E. M., Iowa State University
Marti, M. E., Iowa State University
Sakaguchi, D. S., Iowa State University
Mallapragada, S. K., Iowa State University
Originally presented on: 11/6/2013 13:42:00 - 14:00:00

Nerve regeneration is a complex biological process. Currently, autologous nerve grafts are considered the "gold standard" for treatment of severe peripheral nerve injuries (PNI). Autologous grafting involves transplanting part of a healthy nerve from the same individual to the site of injury. Consequently, ethical issues and probability of tissue rejection are minimized. However, donor site morbidity and requirement of multiple surgeries are some of the serious concerns in utilizing this technique. As such, research into alternative routes is essential for development of experimental strategies to facilitate nerve regeneration.

A promising alternative, which is the main objective of this work, is to enhance neuronal regeneration following PNI by incorporating interdisciplinary approaches at the intersection of neuroscience, biomaterials, tissue engineering and stem cell therapies. Our previous work has shown that micropatterned biodegradable polymeric conduits preseeded with Schwann cells and sutured to the ends of transected sciatic nerves promote nerve regeneration and functional recovery in rats. Biocompatible and biodegradable polymeric materials were used for engineering conduits to provide mechanical support and orientation to the regenerating axons, and to facilitate localized delivery of neurotrophic factors at the site of injury. Micropatterns on the inner lumen of the conduits provide guidance to the regenerating axons. However, there is no readily available and relatively non-invasive source of Schwann cells.

Therefore, our current focus is on the transdifferentiation of bone marrow derived mesenchymal stem cells (MSCs) into a Schwann cell-like phenotype, as a substitute for myelinating Schwann cells for promoting regeneration in a PNI model. Bone marrow-derived MSCs can be extracted easily using less invasive surgical procedures in comparison to autologous nerve grafts. Also, MSCs have been suggested as an attractive cell type for neural regeneration purposes owing to their high proliferation rate, self-renewing ability and differentiation capabilities into several lineages.

MSCs isolated from Brown Norway rat were characterized for multi-potency and rat MSC markers. These isolated MSCs were grown on micropatterned Polystyrene (PS) and Poly(lactic acid) (PLA) films. A comparative study of morphological and molecular differences was carried out by growing the MSCs on half micropatterned-half smooth (non-patterned) polymeric films.More than 90% of cells were found to be oriented (0°-20°) in the direction of micropatterns on both PS and PLA patterned substrates. No specific orientation was observed for cells on the smooth side of the substrate. Elongation of cells on the patterned polymeric substrates was found to be approximately 3 times that of the cells grown on smooth substrates. MSCs were treated with bromodeoxyuridine (BrDU) for 12 hours to check their proliferation. Immunolabeling of these cells with an anti-BrDU antibody revealed that 30-40% of the cells proliferated on both PS and PLA substrates. Proliferation of cells on PLA substrates was observed to be higher than PS substrates. No significant effect due to micropatterned or smooth substrates was observed, although, the smooth PLA substrates resulted in maximum proliferation.

The impact of micropatterned polymeric films on transdifferentiation of MSCs towards SC-like phenotypes was also tested. A comparison of transdifferentiated MSCs (tMSCs) and undifferentiated MSCs (uMSCs) using molecular markers such as S100β and $p75^{NTR}$ was carried out. Significantly higher numbers of tMSCs expressed S100β and $p75^{NTR}$ as compared to uMSCs, and the percentage of cells expressing these markers were quantified on all four substrate types. No appreciable variation was found in the degree of transdifferentiation for any substrate type. However, PLA substrates had slightly higher number of transdifferentiated cells as compared to PS, and non-patterned PLA caused maximum transdifferentiation compared to other substrate types. Results revealed that micropatterns strongly influence orientation and morphology of MSCs but have negligible effect on growth and transdifferentiation.

Fabrication and transplantation of tMSCs seeded on micropatterned PLA conduits in a rat peripheral nerve injury model will be conducted for investigating the combined ability of micropatterns and tMSCs in promoting nerve regeneration.

(516f) An Engineered Liver Graft With Improved Blood Compatibility and Prolonged Survival

Kim, Y., Virginia Polytechnic Institute and State University
Berendsen, T., MGH/Harvard Medical School
Ozer, S., Center for Engineering in Medicine (CEM) at Massachusetts General Hospital â€" Harvard Medical School - Shriners Hospital for Children
Yarmush, M. L., Massachusetts General Hospital, Harvard Medical School
Uygun, B., Center for Engineering in Medicine, Massachusetts General Hospital, Harvard Medical School, Shriners Hospital for Children
Originally presented on: 11/6/2013 14:00:00 - 14:18:00

Introduction: The only definitive treatment for end-stage liver disease is orthotopic liver transplantation and it is limited by the shortage of available donor organs. To address the issue, we developed a novel method to prepare transplantable liver grafts using perfusion decellularized liver matrix (DLM). While we successfully showed that the engineered liver grafts were functional in vitro and could be transplanted in rats, the post-transplantation survival time of the recipient animals was significantly short (~8 h) due to complications related to the high thrombogenicity of the grafts. An ideal engineered liver graft should enable undisrupted blood circulation through the graft in order to allow for its therapeutic testing in a liver failure model. Therefore, we hypothesized that the deposition of an anti-coagulant, heparin, on the DLM scaffold surface will render the DLM blood compatible, and consequently will prevent blood clotting when exposed to blood upon transplantation and prolong the post transplantation survival time of the recipient animal.

Materials and Methods: Heparin was immobilized on the interior surface of DLM via layer-by-layer (LbL) deposition technique. Heparin and poly(diallydimethly ammoniumchloride) were used as negatively charged and positively charged polyelectrolytes, respectively. Four different conditions of heparin LbL films were deposited at two different bilayers of heparin (4 and 8 BL) and at two different concentrations of heparin (2 g/L and 4 g/L). The heparin LbL film was verified via toluidine blue staining and was quantified by hexosamine assay. The ability of the heparinized DLM (hDLM) to prevent blood coagulation was tested via 2-h *ex vivo* diluted blood perfusion. The hDLM scaffolds were recellularized with primary hepatocytes and the hepatic functions of recellularized hDLM were observed over 5-day culture period. The recellularized hDLM scaffolds were perfused with diluted blood for 24-h *ex vivo* to evaluate the extent of blood coagulation prior to transplantation. The engineered liver grafts that showed the maintenance of hepatic functions and the reduced blood coagulation *in vitro*were transplanted via inferior vena cava anastomosis with the use of cuff technique. Animal survival time was monitored over 3 weeks and the transplanted liver grafts were harvested 1d post-transplantation for histological assessment.

Results and Discussion: Four different types of hDLM scaffolds were prepared to verify heparin immobilization via toluidine blue staining. The amount of heparin deposited ranged 20-80 mg depending on the number of heparin bilayers and the concentration of heparin used. The 2-h *ex vivo* blood perfusion showed that the time and extent of blood clot formation varied depending on the level of heparin in the scaffolds. The hDLM deposited with 8 bilayers of heparin at 4 g/L showed the most significant reduction in the blood clot formation. Albumin and urea production of the recellularized hDLM scaffolds were comparable to that observed in unheparinized DLM over 5d of *in vitro* perfusion culture, indicating heparin deposition has no effect on the maintenance of hepatic functions over the culture time. As a result of 24-h *ex vivo*diluted blood perfusion, the heparinized liver grafts exhibited reduced blood coagulation as observed by reduced flow resistance when compared to unheparinized liver graft. Animals that received the engineered liver grafts as auxiliary grafts survived for longer than 3 weeks. Histological analyses showed the significant prevention of blood coagulation in the heparinized liver grafts post-transplantation.

Conclusions: The engineered liver grafts with heparin layers exhibited enhanced survival by reducing blood clot formation post-transplantation. The generation of transplantable and functional engineered liver grafts that can serve as a reliable alternative for liver transplantation will offer a great potential for developing treatment of liver diseases.

(516g) Maturation of Implantable Vascular Grafts in An Ovine Model Using Small Intestinal Sub-Mucosa: Do We Need Pre-Seeding of Smooth Muscle Cells?

Andreadis, S., State University of New York at Buffalo
Swartz, D. D., University at Buffalo, The State University of New York
Row, S., State University of New York at Buffalo
Peng, H. F., SUNY at Buffalo
Schlaich, E. M., SUNY at Buffalo
Originally presented on: 11/6/2013 14:18:00 - 14:36:00

The purpose of this study is to engineer vascular grafts which have the ability to grow, remodel and function normally within the host. Tissue engineered arterial substitutes with cell-seeded biomaterial scaffolds have been proven to perform better at the blood-material-interface as compared to synthetic alternatives. In this study, we aim to study the time-dependent process of host cell migration, maturation and remodeling of graft tissue with fully functional and implantable arterial grafts made from hair follicle derived smooth muscle cells (HF-SMC), ovine pulmonary artery endothelial cells on a scaffold of Small Intestinal Submucosa (SIS). Early events of immune responses and fibrotic tissue formation around the implanted vessels as well as cell seeding effect on long-term patency and functionality of SIS based vessels are reported.

SIS sheets were rolled into tubular constructs with fibrin glue using a previously optimized technique (*Peng 2011, Cell, Tissue and Organ*). Using a customized bioreactor, arterial shear stress was achieved within our constructs using optimized parameters sustaining patency of our grafts with the use of an arterio-venous shunt bypass model that we developed in our laboratory. SIS grafts, with or without initial seeding of HF-SMC were evaluated when implanted as an inter-positional carotid graft in an ovine model at 1, 4 and 12 weeks.

The implanted tissues were monitored biweekly using Doppler ultrasound to document patency and measure blood flow rates through the grafts. At 1, 4 and 12 weeks we also performed angiography, which confirmed uniform bilateral blood flow rates through the carotids. We observed early cellularization of graft by host (before the 1 month time point starting at 1 week), followed by differentiation and maturation of permeated host cells, which remodel the TEVs in-vivo. Our data directly correlate with functionality and mechanical properties of explanted grafts. Over time in the host, SIS grafts developed architecture similar to native artery and vascular functionality due to infiltration of host cells.

The patency of the grafts and their capacity for remodeling and development of vascular function suggest that this approach is very promising and may be of clinical significance. Our grafts exhibited structural and functional properties similar to native artery and provided appropriate chemo-attractant signals for maturation in-vivo in a time-dependent manner. Biomaterial vascular grafts thus made provide an implantable arterial replacement for vascular regeneration, with potential for off-the-shelf design.

(516h) Regenerating Articular Cartilages By Genetically Engineered Ascs: Prolonged TGF-b3/BMP-6 Expression Improved Chondrogenesis and Restored Zonal Structure

Hu, Y. C., National Tsing Hua University

Originally presented on: 11/6/2013 14:36:00 - 14:54:00

Adipose-derived stem cells (ASCs) hold promise for cartilage regeneration but their chondrogenesis potential is inferior. Here we employed a baculovirus system that exploited FLPo/Frt-mediated transgene recombination and episomal minicircle formation to genetically engineer rabbit ASCs (rASCs). The baculovirus system conferred prolonged and robust TGF-β3/BMP-6 expression in rASCs cultured in porous scaffolds, which critically augmented rASCs chondrogenesis and suppressed osteogenesis/hypertrophy, leading to the formation of cartilaginous constructs with improved maturity and mechanical properties in 2-week culture. Twelve weeks after implantation into full-thickness articular cartilage defects in rabbits, these engineered constructs regenerated neocartilages that resembled native hyaline cartilages in cell morphology, matrix composition and mechanical properties. The neocartilages also displayed cartilage-specific zonal structures without signs of hypertrophy, ossification and degeneration, and eventually integrated with host cartilages. In conclusion, the baculovirus-engineered rASCs that persistently express TGF-b3/BMP-6 improved the chondrogenesis, *in vitro* cartilaginous constructs production and *in vivo* hyaline cartilage regeneration, thus representing a remarkable advance in cartilage engineering.

(528a) Mechanism and Specificity of An Acyltransferase Domain From a Modular Polyketide Synthase

Dunn, B., Stanford University
Khosla, C., Stanford University
Originally presented on: 11/6/2013 15:15:00 - 15:33:00

Natural products have historically been an invaluable resource in the discovery and development of therapeutically active compounds. The polyketide class of natural products has seen huge success in the commercial drug arena, acting as antibiotics, anticancer agents, immunosuppressants, and other widely-used therapeutics. This biological diversity stems largely from the structural diversity of these small molecules, many of which are produced in an assembly-line fashion by large enzymatic complexes called modular polyketide synthases (PKSs). Each module of a PKS is comprised of several domains, and each domain plays a distinct role in extending and chemically modifying the growing polyketide chain. The acyltransferase domain (AT) is responsible for selection and incorporation of simple Coenzyme A- (CoA-) linked building blocks, and therefore acts as a gatekeeper to polyketide structural diversity. Engineering of the AT domain for incorporation of novel building blocks is an attractive approach in the production of new therapeutic polyketides.

This work presents the characterization of an AT domain from the 6-deoxyerythronolide B synthase (DEBS), the PKS responsible for producing the aglycone precursor to the antibiotic erythromycin. DEBS is arguably the most highly studied PKS and often acts as a model system for understanding modular polyketide biosynthesis. We have utilized a continuous, coupled enzymatic assay to examine the steady state kinetics of AT-catalyzed transacylation. The transacylation reaction includes the formation of an acyl-AT intermediate and subsequent attack by a nucleophilic phosphopantetheine prosthetic group on the adjacent acyl carrier protein (ACP). We discuss the substrate (α-carboxyacyl-CoA) and ACP specificity of the transacylation reaction, focusing on the mechanistic determinants of this specificity. This knowledge is necessary for successful engineering of the specificity of AT domains.

Previous attempts to engineer AT domains for novel extender unit incorporation have focused primarily on the swapping of an entire domain for one with alternate substrate specificity. These attempts often come at the expense of the appropriate protein-protein interactions necessary for successful polyketide chain extension and therefore lead to drastically decreased product titers. A more conservative and increasingly popular approach for engineering AT domain specificity involves the targeted mutagenesis of the active site or adjacent residues. We present kinetic data on the "best" AT site-directed mutants characterized in the literature to date, and reveal that these mutants have drastically attenuated activity *in vitro*. This decreased activity (rather than enhanced specificity) is likely what allows these mutants to incorporate a nonnative substrate *in vivo*. This reveals extensive room for improvement in the engineering of AT specificity by site-directed mutagenesis. We present briefly our future plans for tackling this engineering challenge, with the concomitant goal of generating novel polyketides by incorporation of diverse building blocks.

(528b) Mapping the Protein Interaction Between a Clostridial [FeFe] Hydrogenase and Its Redox Partner, Ferredoxin

Shiigi, S., Stanford University
Swartz, J. R., Stanford University
Originally presented on: 11/6/2013 15:33:00 - 15:51:00

Electron transport is a fundamental process underlying energy transfer in central metabolism. As electron transport is largely mediated by protein-protein interactions, understanding and manipulating the interactions between electron-carrying proteins can be used to direct electron flow in metabolic pathways engineered to produce reduced compounds including next-generation fuels like hydrogen. [FeFe] hydrogenases from *Clostridia* catalyze the reaction combining protons and electrons to produce hydrogen and are of significant interest for the development of hydrogen-producing metabolic pathways. The hydrogenase binds to and receives electrons from an electron carrying protein, ferredoxin. However, the interaction between these hydrogenases and ferredoxins is not well understood.

A protein-protein docking model (ClusPro) was used to generate a structural model of the interaction between the ferredoxin and hydrogenase. This model was then used to predict surface mutations designed to disrupt specific types of interactions that might be occurring at a protein-protein interface. Here, we will discuss the construction of ferredoxin and hydrogenase mutant libraries and our assessment of the effects of these mutations on the interaction between these proteins. This study will provide new, molecular-level insight into the protein interaction that facilitates electron transfer between the *Clostridial* hydrogenase and its native redox partner. These studies can be further extended to enhance the interaction between these proteins, thereby enabling the construction of more efficient hydrogen production pathways.

(528c) An Engineered Peptide for Non-Invasive Optical Imaging of Brain Tumors

Moore, S. J., Stanford University
Hayden Gephart, M. G., Stanford University
Bergen, J. M., University of Washington
Su, Y. S., Howard Hughes Medical Institute
Rayburn, H., Howard Hughes Medical Institute
Scott, M. P., Howard Hughes Medical Institute
Cochran, J. R., Stanford University

Originally presented on: 11/6/2013 15:51:00 - 16:09:00

The majority of central nervous system (CNS) tumors carry grave clinical prognoses due to the limited effectiveness of surgical resection, radiation, and chemotherapy. Thus, improved strategies are critically needed for brain tumor visualization and targeted treatment. We demonstrate that mouse cerebellar medulloblastoma (MB) can be targeted and illuminated with a fluorescent, engineered cystine-knot (knottin) peptide that binds with high affinity to $\alpha_v\beta_3$, $\alpha_v\beta_5$, and $\alpha_5\beta_1$ integrin receptors. This integrin-binding knottin, denoted EETI 2.5F, was evaluated as a molecular imaging probe in both orthotopic and genetic models of MB. Following tail vein injection, fluorescence arising from dye-conjugated EETI 2.5F was localized to the tumor compared to the normal surrounding brain tissue, as measured by optical imaging. The intensity of the imaging signal correlated with tumor volume. Due to its unique ability to bind to $\alpha_5\beta_1$ integrin, EETI 2.5F showed superior in vivo and ex vivo brain tumor imaging contrast compared to other engineered integrin-binding knottin peptides and to c(RGDfK), a well-studied integrin-binding peptidomimetic. Next, EETI 2.5F was fused to an antibody Fc domain (EETI 2.5F-Fc) to determine if a larger integrin-binding protein could also access intracranial brain tumors. EETI 2.5F-Fc, conjugated to a fluorescent dye, illuminated MB following intravenous injection and was able to distribute throughout the tumor parenchyma. In contrast, brain tumor imaging signals were not detected in mice injected with EETI 2.5F proteins containing a scrambled integrin-binding sequence, demonstrating the importance of target specificity. These results highlight the potential of using EETI 2.5F and EETI 2.5-Fc as targeted molecular probes for brain tumor imaging.

(528d) Performance and Applications of a Yeast Surface Display-Based Biocatalyst

Eby, J.,
Peretti, S. W., North Carolina State University
Originally presented on: 11/6/2013 16:09:00 - 16:27:00

A whole-cell biocatalyst has been developed using surface display (SD) on recombinant *Saccharomyces cerevisiae*. Enzymes can be produced by the yeasts and attached to the outer cell surface covalently or non-covalently. Two enzymes have been displayed: lipase B from *Candida antartica* (CALB) and lipase from *Photobacterium lipolyticum* M37 (M37L). To characterize the activity and protein concentration on the cell surface, a flow cytometry method has been developed. Active enzymes are labeled with a reactive probe that is linked to a detectable dye via a "click" reaction. All proteins produced as part of the SD construct are then labeled via a C-terminal epitope tag. The amount of active and total protein can be measured in respective flow cytometry fluorescence channels.

The performance of the whole-cell biocatalyst was evaluated in several ways. The cells were lyophilized and assays for hydrolytic and synthetic activity were performed. The assays are based on the release of *para*-nitrophenol and the formation of propyl laurate, respectively. The catalysts were also used in the synthesis of several commercial products, including glycerol carbonate (a value-added derivative of glycerol).

(528e) Solvent-Enhanced Biotransformation of Chemicals By Beauveria Bassiana As Biocatalyst

Gonz lez, R., University of Iowa
Peeples, T. L., University of Iowa
Nicolau, F., University of Iowa
Originally presented on: 11/6/2013 16:27:00 - 16:45:00

This research evaluates the oxidative capacity of n-alkane-induced fungal *Beauveria bassiana,* a versatile whole cell biocatalyst used in the biotransformation of steroids and in the production of valuable organic compounds. n-Alkane solvents such as dodecane, and hexadecane serve as carbon sources during growth of *B. bassiana. These hydrocarbons* induce the expression of oxidative enzymes and enhance the growth rate of B. bassiana in glucose and glycerol culture media. This research aims to optimize yield and selectivity in the hydroxylation of unfunctionalized carbons on DHEA (dehydroepiandrosterone, 3β-hydroxyandrost-5-en-17-one) to 11α-hydroxy-DHEA, as result of virulence enhancement conditions. DHEA is an essential endogenous steroid male-hormone and provides the opportunity to study the activation of unfunctionalized carbons; an attractive reaction that produce valuable intermediates for chemical synthesis in pharmaceuticals. Results include growth curve analysis and biomass yield of *B. bassiana* in glucose and glycerol media. Biotransformation was successful in glycerol medium and no reaction inhibition effect was found in comparison to biotransformation on glucose-based medium. 1H-NMR was used to characterize metabolites. An analysis of the n-alkane effect on biocatalyst performance, including impact on reaction selectivity as well as correlations between biomass and reaction yields will be presented.

(528f) Protein Engineering for the Generation of Resveratrol Analogues

Bhan, N., Rensselear Polytechnic Institute
Koffas, M. A. G., Rensselaer Polytechnic Institute
Originally presented on: 11/6/2013 16:45:00 - 17:03:00

Type III polyketide synthases have been studied extensively over the past few decades, but most of these structural-functional studies have been carried out on chalcone synthase (CHS) like enzymes, which carry out a C6-C2 Claisen condensation of their polyketide intermediates. Stilbene synthases share 75-90% sequence similarity with CHS and form the same tetraketide intermediate, but cyclize the product via a C1-C7 aldol condensation. This cyclization requires a thioesterase-like hydrolysis step made possible by the hydrogen bonding network (HBN) present only in STSes, involving residues S338, T132 and E192 and a water molecule. Here we studied the effect of mutating functionally important residues in STS. Similar mutations have been carried out in CHS but we predicted that new products would be formed because of the different cyclization of intermediates by STS. Interestingly a single mutation, S338V, converted the STS to CHS possibly due to the disruption of the HBN. Mutant G256L completely lost its ability to handle the bulky p-coumaroyl-CoA substrate while a T197G mutant was still able to produce resveratrol while forming at the same time an increased amount of a more hydrophobic compound which is yet to be identified. We plan to further feed 15 other mutants created with different substrates and attempt to identify products formed.

(528g) Optimization and Characterization of Candida Cylindracea Lipase Produced From Palm Kernel Cake By Solid-State Bioconversion

Alam, M. Z., International Islamic University Malaysia
Elgharbawy, A. A.,
Salleh, H. M., International Islamic University Malaysia
Originally presented on: 11/6/2013 17:03:00 - 17:21:00

Lipases, (triacylglycerol acylhydrolase, EC.3.1.1.3) are one of the most important classes of industrial enzymes. The role of lipases is to hydrolyse triglycerides into diglycerides; monoglycerides, glycerol and fatty acids. Lipases from microbial sources are currently receiving particular attention due to of their potential applications in the industry, food technology and medicine. The high cost of enzyme production is still representing a challenge in large-scale industrial processes. To overcome this, the use of microorganisms along with agro-industrial residues such as palm kernel cake (PKC), which found in abundant amounts in Malaysia, as substrates to produce the enzyme could reduce the cost and contribute in adding value to low-cost material in the market.

This study was focused on the optimization of two factors and the characteristics of *Candida cylindracea* lipase for potential of applications through solid state bioconversion of palm kernel cake (PKC), a by-product of oil palm industry. A statistical design, face centred central composite design (FCCCD) was used to identify thirteen experiments to optimize two significant factors: yeast extract and inoculum. Fermentation was conducted in shake flasks employing solid-state fermentation for 72hrs at 30°C and moisture content of 70.0%. As a result, a maximum lipase activity of 400.0±5.0 U/g dry PKC was achieved on PKC as basal medium supplemented with 2.5 (% w/v) of yeast extract, 2.0 (% v/w) Tween-80, 0.5 (%v/w) olive oil and inoculum concentration of 7.0 (%v/w) with average cell count of 2.0 Í10^7 cells/ ml. Analysis of variance (ANOVA) proved the significance of the model with R^2 value of 0.9893 at $p< 0.05$ which indicated a successful and a reliable design. The characteristics of lipase showed that the enzyme was stable for 24hrs at pH 7.0 and 8.0. On the other hand, the lipase produced retained its full activity at temperature ranges of 25-45°C while 73% and 68% of the activity was retained at 50°C and 55°C respectively after 90 min of incubation. Development of this process should be conducted in the future for better lipase production and recovery at large scale.

(528h) Selective Esterification of Tuna Free Fatty Acids With Immobilized Pseudomonas Cepacia and Thermomyces Lanuginosus Lipases to Produce Esters Rich in Docosahexaenoic Acid

Chaurasia, S. P., Malaviya National Institute of Technology
Sharma, A., Banasthali Vidyapith, Banasthali
Dalai, A. K., University of Saskatchewan
Originally presented on: 11/6/2013 17:21:00 - 17:39:00

In the present work, selective esterification of DHA from Tuna fish oil free fatty acids (TFFAs), using immobilized lipases from *Pseudomonas cepacia* (PCL) and *Thermomyces lanuginosus* (TLL) have been studied. The suitable reaction conditions have been determined as lauryl alcohol to FFAs ratio, 2:1 (wt/wt), pH 7.0, buffer to FFAs ratio,1:1 (wt/wt), temperature 50 °C and agitation speed 800 rpm. At the suitable reaction conditions, 47% and 41% esterification have been obtained in 6 h for immobilized PCL and TLL, respectively. Reusability of both the immobilized lipases for esterification has also been carried out and the activity retention was stabilized around 29% and 21% after 5 cycles of repeated use of immobilized PCL and TLL, respectively. Kinetic study of esterification has also been carried out using four different substrate concentrations and conducting the reaction for 56 h. Kinetic parameters using Oliveira model has been obtained by fitting in the MATLAB program, such as K_{cat} = 488.2 (h^{-1}), K_B = 105 (μmoles of DHA esters/ml); $K_{m(B)}$ = 20.5 (μmoles of DHA esters/ml); $K_{m(A)}$ = 6.1 (μmoles of DHA esters/ml)for immobilized PCL and K_{cat} = 267.17 (h^{-1}), K_B = 777.3 (μmoles of DHA esters/ml); $K_{m(B)}$ = 211.1 (μmoles of DHA esters/ml); $K_{m(A)}$ = 118.75 (μmoles of DHA esters/ml)for immobilized TLL. Regression coefficient (R^2) such as 0.99 and 0.98have been obtained for immobilized PCL and TLL respectively, indicating a good fitting of the modelfor both the lipases.

(544a) Peptoid-Based Klvff Peptide Mimics to Inhibit Amyloid Beta Aggregation

Turner, J. P. Jr., University of Arkansas
Servoss, S., University of Arkansas
Originally presented on: 11/6/2013 15:15:00 - 15:33:00

Alzheimer's disease (AD) is a neurological disorder that slowly develops, leading to dementia, rising healthcare expenses, and eventually death. AD is characterized by the build-up of the amyloid beta (Aβ) protein. As the Aβ proteins begin to aggregate, plaques form in the spaces between nerve cells, resulting in cellular death. Currently, there is no cure for AD; however, research has found that the peptide KLVFF is a key binding sequence for Aβ folding. By utilizing the peptide KLVFF sequence, researchers have shown to inhibit and speed up aggregation, as well as change the morphological state of Aβ. Unfortunately, peptide KLVFF has a low bioavailability. To overcome this, we have designed peptoid-based mimics of KLVFF to interact with Aβ. Structurally similar to peptides, peptoid side chain groups are amended to the amide group rather than the α-carbon. As a result, peptoids are immune to proteases and their small size is unlikely to induce an immune response, making peptoids ideal candidates for drug design.

Peptoid mimics were designed to mimic peptide KLVFF so peptoids could effectively bind to Aβ. Helical length, charge, and secondary structure were then investigated to determine peptoid-Aβ activity. Preliminary work has shown Aβ aggregation can be inhibited by 82-98% in 5-fold molar excess (peptoid inhibitor to Aβ monomer). Furthermore, we believe we can alter current peptoid sequences to develop a novel sequence that can inhibit Aβ aggregation in a 1:1 ratio or less of peptoid inhibitor to Aβ monomer and disaggregate fibrils.

(544b) The Effect of Peptoids On Aß Aggregation and NF-KB Activation in Alzheimer's Disease

Wolf, L. M., University of South Carolina
Moore, K. A., University of South Car olina
Turner, J. P. Jr., University of Arkansas
Moss, M., University of South Carolina
Servoss, S. L., University of Arkansas
Originally presented on: 11/6/2013 15:33:00 - 15:51:00

Introduction:

Alzheimer's disease (AD) is a costly and devastating illness, affecting millions within the geriatric population. It is characterized by progressive neurological degeneration, manifesting in memory and cognitive impairment until eventual death. Although the exact mechanism for AD is currently unknown, it is hypothesized that the disease etiology lies within the aggregation and accumulation of the protein amyloid-β (Aβ), whose oligomeric and soluble fibril aggregates are toxic to neuronal cells. Thus, inhibition of Aβ aggregation is one therapeutic strategy. Peptoids are peptide derivatives that resist proteolytic degradation through a small structural modification in the backbone, leaving the side chain chemistry to largely resemble the original structure. This study examines the therapeutic potential of three peptoids, designed to mimic the KLVFF hydrophobic core of Aβ and incorporate a neutral, positive, or negative charge. These peptoids are assessed for their abilities to abate Aβ aggregation and reduce the NF-κB activation secondary to Aβ exposure.

Materials and Methods:

To examine the effect of peptoids on formation of large aggregates, SEC-purified $A\beta_{1-40}$ monomer (20 μM) was incubated alone or with peptoid, aggregation was stimulated via agitation, and aggregate formation monitored via dot blot. To examine the effect of peptoids on the formation of small, oligomeric aggregates, Ab_{1-42} was solubilized in DMSO with or without peptoid and diluted into PBS to initiate oligomer formation. Oligomer size was assessed via Western blot and SDS-PAGE. In addition, oligomer hydrophobicity was examined using ANS fluorescence. Finally, the effect of peptoids on Ab physiological activity was assessed using activation of NF-kB in SHSY5Y human neuroblastoma cells.

Results and Discussion:

Ab aggregation in the presence of peptoids let to a reduction in the formation of fibrillar aggregates, with the positively charged peptoid having the most pronounced effect, significantly reducing aggregate formation by 57.4 ± 23.6%. (Fig 1) Ab oligomer formation was also affected by the presence of peptoids. All three peptoids reduced the quantity and size of oligomers formed, with the positively charged peptoid again possessing the greatest inhibitory capabilities. In contrast, only the neutral and positive peptoids also altered oligomer conformation, evidenced by the increased exposure of hydrophobic patches that elevated ANS fluorescence. These changes, however, had no significant impact on NF-kB activation.

Conclusions:

The three peptoids, which varied in charge, all exhibited inhibitory capabilities toward Ab aggregation; however the magnitude of that inhibitory capability and the manner of aggregate structure modification was dependent upon the peptoid charge. These results indicate that each of the peptoids interact differently with Ab. The peptoid-induced alterations in Ab aggregation, however, failed to decrease NF-kB activation. Thus, the peptoids examined here are ineffective as a treatment option for AD. However, the ability of peptoids to interact with Ab and alter aggregation suggests that other peptoid structures may qualify as a valid potential treatment for AD.

(544c) Effects of Osmolytes On Amyloid Protein Surface and Membrane Activity

Chi, E., University of New Mexico
Anaya, J.,
Hernandez, M.,
Originally presented on: 11/6/2013 15:51:00 - 16:09:00

The misfolding and aggregation of largely intrinsically disordered amyloid proteins into beta-sheet rich fibrillar deposits are implicated in the pathogenesis of over 20 neurodegenerative diseases, including Alzheimer's and Parkinson's diseases. There are currently no cures for these disorders due to, at least in part, a lack of understanding of the mechanisms by which amyloid proteins aggregate. We have shown that interfaces, such as the air/water and lipid membrane interfaces, can catalyze the formation of partially folded or structurally compacted aggregation competent intermediates that subsequently irreversibly assemble into higher order fibrillar aggregates. In this study, we examine the effects of osmolytes, sucrose and urea, on the surface and membrane activities of the amyloid-beta protein in an effort to better understand the effects of the cellular environment on interface-induced amyloid protein aggregation. To evaluate the surface activity of amyloid-beta, adsorption isotherms (surface pressures vs. time) of amyloid-beta to the air/subphase interface was measured in a Langmuir trough on subphases containing varying concentrations of sucrose (0 – 1 M) and urea (0 – 6 M). Membrane activity was measured by carrying out constant pressure insertion assays into anionic DMPG and zwitterionic DPPC lipid monolayers in a Langmuir trough. The preferentially excluded sucrose is found to significantly decrease the lag time and increase the rate of amyloid-beta absorption to the air/water interface. Additionally, finally surface pressure reached by the adsorption of the amyloid-beta protein also increased with increasing sucrose concentration in the subphase. Our results thus indicate that sucrose increases surface and membrane activities of amyloid-beta. Urea, a chaotrope that preferentially bind to protein backbone and increases protein solubility in solution, surprisingly, also increased amyloid-beta surface activity. Our data thus far indicate that the crowded cellular environment can significantly enhance driving forces for interface-induced amyloid-protein aggregation. Investigations are currently underway to evaluate the effect of urea on amyloid-beta membrane activity and the effects of osmolytes on interface-induced amyloid-beta aggregation.

(544d) Transthyretin As Both Sensor and Scavenger of Toxic Abeta Oligomers

Murphy, R. M., University of Wisconsin-Madison
Yang, D. T., University of Wisconsin-Madison
Cho, P. C., University of Wisconsin-Madison

Originally presented on: 11/6/2013 16:09:00 - 16:27:00

Transthyretin (TTR) is a homotetrameric transport protein, assembled from monomers that each contains two four-stranded beta-sheets and a short alpha-helix and loop. In the tetramer, the 'inner' beta-sheet forms a hydrophobic pocket while the helix and loop are solvent-exposed. Beta-amyloid (Abeta) aggregates bind to TTR, and the binding is significantly reduced in mutants L82A (on the loop) and L110A (on the inner beta-sheet). Wild-type TTR, but not L82A or L110A, protected against Abeta toxicity, providing a direct link between TTR-Abeta binding, and TTR-mediated cytoprotection. Protection is afforded at substoichiometric (1:100) TTR:Abeta molar ratios, and binding of Abeta to TTR is highest for partially aggregated materials and decreased for freshly-prepared or heavily aggregated Abeta, suggesting that TTR binds selectively to soluble toxic Abeta oligomers. A novel technique, nanoparticle tracking, is used to show that TTR arrests Abeta aggregation by both preventing formation of new aggregates and inhibiting growth of existing aggregates. TTR tetramers are normally quite stable. However, the quaternary structure of wild-type and L110A, but not L82A, was destabilized when co-incubated with Abeta, suggesting that Abeta binding to L82 triggers tetramer dissociation. Thus, partially *folded* (and aggregated) Abeta triggers partial *unfolding* of TTR. Interestingly, TTR monomers bind more Abeta than do tetramers, presumably because the hydrophobic 'inner' sheet is solvent-exposed upon tetramer disassembly. Therefore, partial loss of TTR quaternary structure is advantageous, by enhancing further binding of Abeta aggregates. Taken together, these results suggest a novel mechanism of action for TTR: the EF helix/loop 'senses' the presence of soluble toxic Abeta oligomers, triggering destabilization of TTR tetramers and exposure of the hydrophobic inner sheet, which then 'scavenges' these toxic oligomers and prevents them from causing cell death.

(544e) Anthocyanidins Modulate Amyloid-ß Aggregation and Attenuate Reactive Oxygen Species Associated With Alzheimer's Disease Pathogenesis

Tseng, J. H., University of South Carolina
Wang, Y., University of South Carolina
Moss, M. A., University of South Carolina
Originally presented on: 11/6/2013 16:27:00 - 16:45:00

Alzheimer's disease (AD), one of the most widespread neurodegenerative diseases, is characterized by confusion, language skill breakdown, long-term memory loss and other cognitive impairments. Existing therapies for AD only provide symptomatic benefits but do not stop or slow AD pathogenesis. Increasing evidence revealing the multi-factorial nature of AD suggests that a multi-target treatment may be an effective therapeutic strategy.

The generation of excess reactive oxygen species (ROS) has been detected in AD brain, and epidemiological studies have demonstrated a correlation between increased intake of antioxidant capable polyphenols and reduced incidence of AD. On the other hand, polyphenols such as tyrosol, resveratrol, and epicatechin have been reported for their capability to alter the formation of amyloid-β protein (Aβ) aggregates, which deposit to form amyloid plaques that are a hallmark of AD pathology. Anthocyanidins (ACs) are a group of polyphenols commonly found in daily diets, such as blueberries and purple cabbages. In this study, a family of ACs was studied for their ability to serve as multi-target agents for AD.

Conformational-specific antibodies were utilized to detect the appearance of fibrillar Aβ species during monomer aggregation. ACs all showed some ability to reduce the formation of fibrillar Aβ species. Moreover, inhibitory capability correlates with the number of hydroxyl groups on the 6-membered aromatic ring. ACs also extended the lag time to aggregate formation. Thus, their ability to inhibit Aβ oligomerization was also examined and confirmed for ACs that have a larger number of hydroxyl groups. An oxygen radical absorbance capacity (ORAC) assay was further introduced to evaluate the anti-oxidant activity of selected ACs. All ACs showed significant anti-oxidant activity. In summary, ACs are promising compounds that not only reduce Aβ aggregate formation, but also attenuate ROS to serve as multi-target drugs for AD.

(544f) Detection and Structural Analysis of Misfolded Tau Protein in Neurodegenerative Disease

Morozova, O. A., University of Delaware
March, Z. M., University of Delaware
Colby, D. W., University of Delaware

Originally presented on: 11/6/2013 16:45:00 - 17:03:00

Paired helical filaments (PHFs) of tau protein are a pathological hallmark of several neurodegenerative disorders including Alzheimer's disease (AD). We have developed a technique for sensitive detection of PHFs and faithful amplification of their distinct conformation. We show that when recombinant tau protein is incubated with PHFs isolated from humans with AD, the nucleation-associated lag phase is shortened. We find that as little as 100 pg of misfolded tau protein accelerates the conversion of recombinant tau into fibrils, and that misfolded tau may be detected by this method in both AD brain samples (N=3/3 positive, 0/3 negative) and AD cerebrospinal fluid samples (N=13/14 positive, 0/8 negative). Additionally, we found that the recombinant tau amyloid formed in the presence of AD PHFs shares many of the structural features of the AD brain-derived seed. In comparison, tau amyloids formed with heparin as an inducing agent—a common biochemical model of tau misfolding—are structurally distinct from AD PHFs. Ultrastructural analysis by electron microscopy, circular dichroism, and GdnHCl denaturation showed that AD PHF seeded recombinant tau fibrils are similar in structure and stability to tau fibrils isolated from AD brain tissue, while heparin-induced recombinant tau fibrils have a significantly different morphology, secondary structure, and stability. Our results show that AD brain PHF seeds act as a conformational template for the formation of recombinant tau fibrils, while heparin induces a distinct recombinant tau fiber conformation. Such an *in vitro* biochemical model may be of improved utility for structural studies specific to different tauopathies and for drug screening approaches that use recombinant tau. The combined detection and structural analysis of misfolded tau by amyloid seeding also suggests possible new specific biomarkers for diagnosing and monitoring different tauopathies.

(544g) Elucidating the Role of N-Terminal Huntingtin Fragment in Polyglutamine Aggregation

Shukla, D., Stanford University
Pande, V., Stanford University
Originally presented on: 11/6/2013 17:03:00 - 17:21:00

Huntington's disease is cause by the aggregation of poly-glutamine (35-40 residues) repeats forming toxic fibrils in the neuronal cells leading to the cell death. These repeats are attached at the N-terminal of the Huntingtin protein. It has been shown that the in the presence of the 17 residues N-terminal fragment, N17 of the Huntingtin protein, the poly-glutamine repeats aggregate faster. Several mechanisms have been proposed to explain the role of N17 fragment including poly-glutamine induced unfolding of N17 (Williamson et. al., J. Mol. Bio., 2010) or a conformational change in N17 to expose hydrophobic residues (Dlugosz et. al., J. Phys. Chem. B, 2010) leading to the N17 aggregation. Aggregation of N17 fragment brings the poly-glutamine repeats closer and induces the fibril formation. Furthermore, phosphorylation of serine residues in N17 leads to changes in conformation and toxicity (Atwal et. al., Nat. Chem. Biol., 2011) In this talk, we report results from an extensive molecular dynamics simulations of both the phosphorylated and unphosphorylated N17 fragment of Huntingtin protein along with pathogenic (35 residues) and non-pathogenic poly-glutamine repeats (15 residues) to understand the influence of polyglutamine repeats on the conformational preferences of the N17 fragment and viceversa.

(544h) The Effects of Salt On the Free Energy Surface for Protein Folding

Du, H., University of Arkansas
Qian, X., University of Arkansas
Originally presented on: 11/6/2013 17:21:00 - 17:39:00

Protein structure and conformation determines its physical property and biological function. Many human pathologies including Alzheimer's, Parkinson's diseases and type 2-diabetes are due to protein misfolding and aggregation associated with the formation of insoluble β-sheet amyloid fibrils. However, earlier efforts to understand protein folding and aggregation often fail to take into account the presence of salt ions in solution, which is ubiquitous under physiological conditions. Salt ions play a critical role in many biological processes including protein solubility, stability, denaturation and aggregation, as well as regulate neuron activities. Here classical molecular dynamics (MD) simulations were conducted to elucidate the effects of salt ions on the mechanisms, dynamics and the free energy surfaces during protein folding, unfolding and aggregation. Here the folding and unfolding of a protein proxy poly (N-isopropylacrylamide) as well as lysozyme in the presence and absence of salt ions were investigated. The effects of salt ions on protein folding/unfolding dynamics and associated free energy are rather complex and will be discussed in this presentation.

(560a) Polymer-Based Multilayer Coatings for the Rapid Release of DNA From Surfaces: A Weak Polyelectrolyte Approach

Yu, Y., University of Wisconsin-Madison
Si, Y., UW-Madison
Bechler, S. L., University of Wisconsin - Madison
Liu, B., UW-Madison
Lynn, D. M., University of Wisconsin-Madison
Originally presented on: 11/6/2013 15:15:00 - 15:33:00

Thin films and coatings that promote the release of DNA from surfaces play important roles in the development of new gene delivery systems. These materials have utility as tools for basic biomedical research and they are useful as platforms for the local delivery of DNA in therapeutic contexts. Degradable surface coatings that contain DNA provide straightforward approaches to achieving spatial control over DNA delivery (e.g., release of DNA from the surface of an object results in delivery of DNA to cells in the immediate vicinity of that object). One challenge confronting the development of such coatings, however, is to design materials that degrade or erode on time scales that are appropriate in the context of a broad range of different applications (e.g., to achieve sustained release, rapid release, or to exert tunable temporal control). Another substantial challenge lies in fabricating thin films that do not simply release DNA, but that release it in a form that promotes internalization and fruitful processing by cells. This talk will focus on the design of ultrathin polymer-based multilayer coatings designed to release DNA rapidly in physiological environments.

Our group has used methods for layer-by-layer assembly to fabricate thin polymer multilayers (~100 nm thick) consisting of plasmid DNA and cationic poly(beta-amino ester)s (PBAEs). These films erode gradually and release DNA and DNA/polymer aggregates via hydrolysis of the ester groups in the backbones of the degradable polymer layers. Films having this basic design can be used to coat the surfaces of indwelling devices (e.g., intravascular stents), and can be used to promote localized transgene expression *in vitro* and *in vivo*. Although this "degradable polymer/degradable film" approach is useful, it is generally limited to the design of coatings that erode and release DNA for extended periods of time (e.g., ranging from several days to several weeks or months). These coatings are thus not particularly well suited for use in contexts that require rapid or short-term transfer of DNA (e.g., over periods ranging from several seconds to minutes or a few hours). Approaches to achieve accelerated or contact-mediated transfer of DNA will be critical in applied contexts that are inherently time-limited (e.g., balloon-mediated vascular interventions) and in other applications (e.g., transdermal delivery of DNA using microneedles) where short-duration transfer is also desirable.

We recently reported the basis of a new approach to accelerate the release of DNA from polymer multilayers. This approach is based on incorporation of poly(acrylic acid) (PAA; a biocompatible polyanion) to design coatings that disassemble rapidly at physiological pH (through a mechanism that involves pH-dependent ionization of PAA and subsequent disruption of ionic interactions, rather than through a mechanism that involves the hydrolysis of a cationic PBAE). These coatings release DNA up to 24 times faster than films designed using PBAEs and DNA alone (e.g., over 3 hrs vs. 3 days). Here, we demonstrate that this 'weak polyelectrolyte' approach can be used to destabilize coatings formed from DNA and non-degradable linear polyethyleneimine (LPEI, a commonly used gene delivery polymer) that are otherwise stable for prolonged periods. Strategic incorporation of nano-layers of PAA during the assembly of LPEI/DNA films yields coatings that release DNA into solution in less than 3 hours (with ~80% released in the first 5 minutes) upon incubation in physiologically relevant media. The results of *in vivo* experiments also demonstrate (i) that inflatable balloon catheters coated with these LPEI/DNA/PAA films can be used as quick-release coatings for the contact-transfer of DNA to injured vascular tissue in rats (e.g., over periods of 20 minutes), and (ii) that this approach can be used to promote robust and localized expression of reporter genes (e.g., beta-galactosidase) in injured rat arterial tissue *in vivo*.

In addition to promoting substantially faster release, this weak polyelectrolyte approach is potentially enabling for two other reasons. First, exploiting changes in the ionization of a weak polyelectrolyte as a means to drive film disruption removes the need to include layers of cationic polymers that are hydrolytically degradable. Second, the removal of this constraint provides opportunities to integrate a much broader range of functional polymers – including conventional cationic gene delivery polymers such as LPEI, but also other anionic weak polyelectrolytes – into quick-release coatings. When combined, these attributes suggest a basis for the design of thin, conformal coatings that could promote more rapid release and more effective DNA delivery and transgene expression than other existing PBAE-based multilayer platforms. Results of physicochemical characterization and additional *in vitro* assays will be presented, and opportunities for the application of this approach to the contact-mediated vascular delivery of therapeutic DNA constructs using film-coated inflatable catheter balloons will be discussed.

(560b) Synergistic Combination of Coated Microneedles and Biochemical Approaches for Oligonucleotide Delivery to Cells in 3D Tissues

Luo, Z., University of California-Davis
Ye, T., University of California-Davis
Ma, Y., Texas Tech University
Gill, H. S., Texas Tech University
Nitin, N., University of California-Davis
Originally presented on: 11/6/2013 15:33:00 - 15:51:00

Oligonucleotides (ONs) have the potential to provide significant clinical impact by enabling novel therapeutic and disease diagnosis strategies. For example, molecular beacons could be used for disease detection, while anti-RNA ON probes could be used to knockdown protein expression in cells to provide a therapeutic action. However, intracellular delivery of ONs, especially in 3D tissues is challenging. Two major barriers limit successful intracellular delivery of ONs in an intact 3D tissue. The first challenge is related to the inability to achieve wide-spread uniform delivery of ONs in the tissue such that all cells in the tissue can be exposed to a sufficient therapeutic dose of ONs. The second challenge is related to the poor intracellular delivery of ONs. The objective of this study was to synergistically combine (i) microneedles as devices to precisely deliver ONs into tissues to achieve extensive intra-tissue distribution of ON probes, with (ii) biochemical approaches including use of streptolysin O (SLO) and cholesterol-conjugation to enhance penetration of these widely dispersed ONs into cells of the 3D tissue.

Microneedles were uniformly coated with ONs. Co-insertion of microneedles coated with ONs and SLO into 3D tissue models resulted in delivery of ONs into both the cytoplasm and nucleus of cells. Within a short incubation time (35 min), ONs were observed both laterally and along the depth of a 3D tissue up to a distance of 500 μm from the microneedle insertion point. Similar wide-spread intra-tissue distribution of ONs was achieved upon delivery of ON-cholesterol conjugates. We also demonstrate that delivery of a cholesterol-conjugated anti-green fluorescent protein (GFP)-ON using coated microneedles resulted in reduction of GFP expression in HeLa cells in 3D culture. In conclusion, the results of this study provide a novel approach for efficient intracellular delivery of ONs in tissues.

(560c) Cellular and Molecular Factors Contribute to An in Vitro Refractory Period Between Sequential DNA Complex Deliveries

Kelly, A. M., University of Nebraska - Lincoln
Martin, T. M., University of Nebraska-Lincoln
Plautz, S. A., University of Nebraska-Lincoln
Pannier, A. K., University of Nebraska-Lincoln
Originally presented on: 11/6/2013 15:51:00 - 16:09:00

Repeated delivery of DNA complexes to cells, tissues, or organisms is required for multiple biological applications such as the maintenance of a therapeutic effect in gene therapy, the triggering of specific cellular responses to alter transfection, and the study of genetics in general. Previous studies report on a putative *in vivo* refractory period, where cells are unlikely to be transfected a second time after previously being transfected. To further understand this effect, *in vitro* studies were conducted to determine the molecular factors attributed to the refractory event.

Microarray analysis, RT-PCR, and flow cytometry studies were conducted to identify the genes and cellular mechanisms that might contribute to the refractory period. For the microarray studies, DNA complexes were delivered to HEK 293T cells, and gene expression data were collected 2 to 24 hours post complex delivery. At these same time points, post complex delivery, total RNA was extracted from cells and the amount was quantified via RT-PCR. Finally, flow cytometry analysis was conducted on HEK 293T cells that were first treated with complexes containing a plasmid encoding for a fluorescent reporter gene, then, 2 to 24 hours afterwards, treated with complexes containing a secondary fluorescent reporter gene. Flow cytometry was conducted 24 hours after the second plasmid was delivered.

Microarray data revealed altered gene expression profiles in cells recently treated with complexes, as compared to cells treated at later time points. These data indicate different molecular mechanisms guiding initial cell responses to complex delivery as opposed to those mechanisms promoting transfection. In addition, results from RT-PCR demonstrates a decrease in total RNA at earlier time points, indicating a possible shut-down of mRNA production immediately following DNA complex delivery. Furthermore, flow cytometry analysis shows decreased transfection efficiency of a secondary plasmid, as compared to the efficiency of single plasmid delivery. Together, these data demonstrate that cellular mechanisms are initiated after DNA complex delivery that may be responsible for the refractory period during which cells are less likely to respond to additional DNA complex treatments. Elucidation of the cellular and molecular factors underlying the refractory period will allow for the exploitation of these molecules by cell priming or modification of DNA delivery systems to overcome the effects of the refractory event and facilitate more efficient transfection of repeat DNA deliveries.

(560d) Acid-Degradable Polyketal Nanoparticles Carrying MnSOD siRNA to Restore Intrinsic Apoptosis in Tamoxifen-Resistant Breast Cancer Cells

Cho, S. K.,
Kemp, J. A., University of California, Irvine
Pedram, A.,
Levin, E. R.,
Kwon, Y. J., University of California, Irvine
Originally presented on: 11/6/2013 16:09:00 - 16:27:00

Drug-resistance in cancer cells is a major cause of cancer relapse and eventually leads to failure of chemotherapy. RNA interference (RNAi) has been suggested to be a promising treatment option for supporting chemotherapy. RNAi can restore cancer cells' sensitivity towards chemotherapeutic drugs by silencing the proteins responsible for increased resistance. Specifically, manganese superoxide dismutase (MnSOD), an enzyme which converts DNA-damaging reactive oxygen species (ROS) to less harmful ROS, has been known to be up-regulated in tamoxifen-resistant breast cancer cells. By silencing MnSOD, intrinsic apoptosis pathways can be recovered and re-sensitize cells to tamoxifen.

Despite immense potential, RNAi therapy is limited due to delivery issues especially for invivo applications. In order to develop safe and efficient siRNA carriers for invivo applications, an acid-degradable polyketal (PK) layer was introduced to poly(amidoamine) (PAMAM)/siRNA dendriplexes via surface-initiated photo-polymerization. The PK layer is hypothesized to provide the protection of siRNA as well as avoid detrimental aggregation in serum. An acid-degradable PK layer can also promote efficient escape from endosomal compartments. Photo-polymerization enables simple tuning of the shell thickness, thereby offering control over release kinetics.

siRNA-PAMAM/PK core-shell nanoparticles (NPs) were prepared with MnSOD-specific siRNAs. Resulting NPs showed narrow size distribution and avoided aggregation in serum contrary to dendriplex cores. Core-shell structure provided protection of siRNA from biological macromolecules and stimuli-responsive release. Intracellular trafficking by confocal laser scanning microscopy showed efficient release from endosomal compartments as hypothesized. Invitro transfection in tamoxifen-resistant breast cancer cell line MCF7-BK-TR demonstrated that delivery of MnSOD-siRNA resulted in knockdown of MnSOD levels and reversed tamoxifen-resistance. Moreover, inclusion of NPs in the transplanted tamoxifen-resistant tumor model significantly suppressed the growth of the tumor. Recovery of apoptotic activity in tumors treated with NPs was confirmed by Annexin-V staining, TUNEL staining, and cleaved Caspase-7 immuno-blotting.

In conclusion, acid-degradable polyketal nanoparticles were prepared to complex and protect siRNA from biological macromolecules and promote efficient intracellular trafficking. Our novel in vivo model with tamoxifen-resistant breast cancer demonstrates the reversal of drug resistance by MnSOD knockdown, proving the pre-clinical significance of this study.

(560e) Dendrimer-siRNA Complexes and Their Aerosol Formulation for Lung Delivery Applications

Conti, D. S., Wayne State University
Brewer, D., Wayne State University
Grashik, J., Wayne State University
Avasarala, S., Wayne State University
da Rocha, S. R. P., Wayne State University
Originally presented on: 11/6/2013 16:27:00 - 16:45:00

RNA interference (RNAi) promotes silencing of gene expression in a post-transcriptional manner. RNAi-is a potential therapeutic in the treatment of several medically relevant pulmonary diseases, such as lung cancer, cystic fibrosis, asthma, chronic obstructive pulmonary disease, respiratory syncytial virus, and severe acute respiratory syndrome viral infection, and there are several ongoing clinical trials related to lung and siRNA. However, progress in this area has been hindered largely by the lack of efficient carriers capable of overcoming the lung architecture, extra and intracellular barriers present in the lung tissue to the efficient delivery of siRNA to the cell cytosol, and also due to formulation challenges.

In this work we report the preparation and characterization of dendriplexes between siRNA and poly(amidoamine) (PAMAM) dendrimers, and their *in vitro* gene knockdown efficiency in a model lung alveolar epithelium cell line stably expressing eGFP. We also show strategies to formulate and deliver siRNA and their dendriplexes via oral inhalation administration using pressurized metered-dose inhalers (pMDIs). We study the gene knock-down efficiency of the nanocarriers as a function of the N/P ratio and also examine the effect of the propellant environment on siRNA activity. We observed that the knockdown efficiency was not affected when the siRNA was kept in contact with the propellant during long periods of time. Finally, we show that the proposed particle engineering strategies are effective in producing aerosols conducive to the delivery of siRNA in the form of dendriplexes to the deep lungs. The respirable fraction of the proposed particle engineering strategies was determined to be at least 64%, with mass median aerodynamic diameter close to the 2 mm target. This work is relevant as it demonstrates the potential of formulating and delivering siRNA-nanocarrier complexes to the lungs using the inexpensive and widely employed portable inhalers, the pMDIs.

(560f) Modification of Fiberless Adenovirus With a Polymer/Peptide Conjugate for Enhanced Transduction and Reduced Immune Response

Nigatu, A., Oklahoma State University
Kupgan, G., Oklahoma State University
Ramsey, J. D., Oklahoma State University
Originally presented on: 11/6/2013 16:45:00 - 17:03:00

Recombinant adenovirus vectors are relatively safe and have been used widely as gene delivery vectors in gene therapy trials. Adenovirus capsid and fiber proteins, however, mediate an inflammatory response and initiate clearance of the virus in patients with preexisting immunity. In addition, adenovirus has a tropism toward cells that have abundant coxsackie-adenovirus receptor (CAR) and the inability to efficiently infect cells that lack this receptor. The adenovirus fiber protein plays a role in both of these drawbacks. Clearance of the vector, or immune inactivation, is the result of a reaction of neutralizing antibodies against epitopes on the surface of the virus (e.g. hexon and fiber proteins), and the virus tropism is based on the interaction of the fiber proteins with CAR. Adenovirus vectors could be significantly improved if the fiber protein could be replaced with an alternate material that would function in place of the protein while reducing the associated drawbacks.

In our study we removed the fiber protein and developed a vector composed of fiberless adenovirus and a cell-penetrating peptide/polyethylene glycol conjugate (CPP-PEG) with the aim of diminishing the drawback associated with the virus. The CPP (Tat and Penetratin) aided in transporting the virus into cells in a CAR-independent manner while PEG acted to reduce the innate and adaptive immune responses to the virus particle. The type of CPP, PEG molecular weight, and degree of PEGylation were varied to maximize the transduction efficiencies and minimize the immune response. The resulting vector showed 30-fold enhancement in gene transduction on cells lacking the CAR receptor while retaining 75 % of its transduction efficiency on cells with CAR. Modifying the vector with CPP-PEG lowered the level of proinflammatory cytokine IL-6 produced by mouse macrophage cells by 80 % compared to unmodified adenovirus. Furthermore, the vector was able to avoid immune inactivation by neutralizing antibodies. The results of this study demonstrated that the function of the fiber protein was effectively replaced with a CPP-PEG conjugate, which in turn reduced the tendency of the adenovirus to be neutralized by inactivating antibodies and introduced the ability to efficiently transform cells with and without the native receptor.

(560g) In Vivo Evaluation of AAV-Mediated Gene Transfer in Airways of Mice With Acute Or Chronic Respiratory Infection

Myint, M., University of Pennsylvania
Limberis, M. P., University of Pennsylvania School of Medicine
Diamond, S. L., University of Pennsylvania
Originally presented on: 11/6/2013 17:03:00 - 17:21:00

Individuals affected by cystic fibrosis are atypically susceptible to bacteria that can colonize the lung indefinitely. These infections are not initially fatal, but their chronic nature results in prolonged inflammatory responses that can lead to serious complications, permanent damage, and eventual death. Gene therapies for cystic fibrosis are a promising treatment for the disease, yet little research has been done to evaluate gene transfer in the context of a chronic inflammatory state of the airway. In this study, two models were developed to evaluate AAV-mediated gene transfer in such environments. The bacterium *Bordetella bronchiseptica* RB50 was used to establish a chronic, but non-lethal, respiratory infection in C57BL/6 mice. An inoculum of ~10^5 CFUs allowed this bacterium to persist in the upper and lower respiratory tracts for at least 16 days and decreased gene expression by almost 3-fold when compared to uninfected controls (average nasal cavity luminescence 1.98×10^8 vs. 7.09×10^7 photons/sec, p-value < 0.01). In another model, *Pseudomonas aeruginosa* PAO1 was used to infect surfactant protein D (SP-D) knockout mice. With an inoculum of 10^5-10^6 CFUs, the bacteria persisted for only two days in the nasal cavity but caused a similar reduction in gene expression of approximately 2.5-fold (average nasal cavity luminescence 1.98×10^8 vs. 7.78×10^7 photons/sec, p-value < 0.05) at day 7 post-infection when compared to uninfected controls. These results confirm that on-going or nearly resolved infections in the respiratory tract decrease efficiencies of AAV-mediated gene transfer in the airway.

(560h) Directed Evolution of Adeno-Associated Virus for Enhanced Evasion of Human Neutralizing Antibodies

Kotterman, M. A., University of California at Berkeley
Hwang, B. Y., University of California, Berkeley
Stone, D., University of California, Berkeley
Koerber, J. T., University of California, Berkeley
Couto, L., The Children's Hospital of Philadelphia
Mingozzi, F., The Children's Hospital of Philadelphia
High, K., Howard Hughes Medical Institute
Schaffer, D. V., University of California at Berkeley
Originally presented on: 11/6/2013 17:21:00 - 17:39:00

Gene delivery vectors based on adeno-associated viruses (AAV) have yielded positive results in numerous preclinical disease models and recently in human clinical trials for several disease targets, leading to the first market approval of a clinical gene therapy. However, the high prevalence of anti-capsid neutralizing antibodies, due to widespread exposure to numerous AAV variants and serotypes within the human population, decrease the efficacy of AAV gene therapy. This pre-existing immunity, as well as the subsequent development of immunity due to vector administration, can impede the broader implementation of AAV gene therapy and should be addressed to build upon successful AAV results in immune privileged sites. Directed evolution has proven to be a powerful approach to generate AAV vectors with novel capabilities, and our results demonstrate that AAV can be engineered to significantly overcome neutralization by anti-AAV antibodies, both *in vitro* and *in vivo*.

Starting with an AAV2 *cap* gene containing point mutations isolated from previous selections using individual human serum samples, the *cap* gene was then subjected to saturation mutagenesis at additional immunogenic amino acid positions determined by ourselves and others. In addition to AAV2 mutagenesis, a shuffled AAV library was selected for resistance to IVIG neutralization. Of the twelve variants chosen for individual analysis from the saturation mutagenesis and shuffled libraries after nine rounds of screening against human IVIG, all twelve could resist higher antibody concentrations than wild-type AAV2. Variant Shuffle 100-3, which withstood a 35-fold higher *in vitro* IVIG concentration for neutralization than wild-type AAV2, was still capable of transducing approximately 10% of cells in the presence of 1 mg/mL IVIG. In addition, variant SM 10-2 required a 6-fold higher *in vitro* IVIG concentration for neutralization than wild-type AAV2. Furthermore, variants Shuffle 100-3 and SM 10-2 showed enhanced transduction in the presence of sera samples from individual patients excluded from a hemophilia B clinical trial. For *in vivo* analysis, recombinant Shuffle 100-3 and SM 10-2 encoding luciferase were administered via a tail vein injection in naïve mice. SM 10-2 displayed similar *in vivo* tropism to AAV2, except for 7-fold higher transduction of the heart, 5-fold higher transduction of the lungs, and 4.5-fold lower transduction of the liver. By comparison, Shuffle 100-3 exhibited 4-fold higher transduction of the brain, 3-fold higher transduction of the lungs, and 27-fold higher transduction of muscle than AAV2. The ability of SM 10-2 and Shuffle 100-3 to evade antibody neutralization *in vivo* was examined by passively immunizing mice with human IVIG prior to AAV injection. SM 10-2 had significantly higher heart, liver, and muscle transduction than AAV2 in the presence of IVIG, as measured by luciferase enzyme activity. Additionally, Shuffle 100-3 had significantly higher heart and muscle transduction compared to AAV2. Analysis of the post-AAV serum from these mice showed that the variants required equal or higher *in vitro* serum concentrations for neutralization than wild-type AAV. These findings represent the most broadly evasive variants described to date. The isolation of such novel clones resistant to anti-AAV antibodies may enable the future treatment of patients with high antibody titers that are currently ineligible for AAV gene therapy.

(582dl) Effect of Different Ratio of Air-CO2 Mixing Feed On the Growth of Nannochloropsis Oculata in Batch Photobioreactors

Razzak, S., King Fahd University of Petroleum and Minerals
Ilyas, M.,
Hossain, M. M., King Fahd University of Petroleum & Minerals

Microalgae have recently received growing attention given its prospects as a source of renewable energy and its potential for CO_2 capture. Microalgae culture is of increasing value given that: i) algae can be cultivated on non-agricultural land using wastewater, ii) algae can provide a high yield on a per unit of light irradiated area, iii) algae growth requires CO_2 and nutrients that can be obtained from wastewater and fossil fuel combustion and iv) algae contains high oil and starch making possible the production of high quality biodiesel. Thus, algae culture can contribute to CO_2 fixation, wastewater treatment and can be a source of bio-energy. As in the Middle East countries especially in Saudi Arabia region huge amount of CO_2 emitted from industrial and automobile sources (power plant, desalination, cement). Due to the scarcity of sweet water, reuse of wastewater also a necessary with minimum energy and costs. With targeting the CO_2 mitigation and wastewater treatment *Nannochloropsis Oculata* were cultured in batch photo-bioreactors. The six different CO_2 concentrations of 2%, 4%, 6%, 8%, 10% and 12% in Air are used to study the effect of growth kinetics, cell density and dry biomass production rate. Results of the finding used for optimize operating parameters for culture developing of *Nannochloropsis Oculata* with maximum CO_2utilization.

Acknowledgement: *The author would like to acknowledge the support provided by King Abdulaziz City for Science and Technology (KACST) through the Science & Technology Unit at King Fahd University of Petroleum & Minerals (KFUPM) for funding this work through project No. NSTIP # 11- ENV1644-04, as part of the National Science, Technology and Innovation Plan.*

(582dy) Engineering of An Endogenous Phytoene Desaturase Gene As a Dominant Selectable Marker for Chlamydomonas Reinhardtii Transformation and Enhanced Biosynthesis of Carotenoids

Liu, J.,

There have been several markers used for *Chlamydomonas reinhardtii* transformation, consisting mainly of bacterial antibiotic-resistant genes such as the phleomycin resistance gene and an aminoglycoside 3'-phosphotransferase type VIII gene. These exogenous genes undergo likely epigenetic silencing and proper endogenous promoters and/or enhancers (e.g., intron) are commonly required for stable expression in algae. Engineering of an endogenous gene as the selectable marker may be ideal. *C. reinhardtii* is highly sensitive to the herbicide norflurazon that specifically targets the phytoene desaturase (PDS), a carotenogenic enzyme which synthesizes the first colored carotenoid ζ-carotene. In the present study, a *PDS* gene from *Chlamydomonas reinhardtii*was cloned and characterized. Functional complementation analysis revealed *C. reinhardtii* PDS (CrPDS) catalyzes the conversion of phytoene to the colored carotenoid ζ-carotene. A single amino acid substitution, L505F, enhanced its desaturation activity by 29%, as indicated by an *in vitro* enzymatic assay. In addition, CrPDS-L505F exhibited 27.7-fold higher resistance to the herbicide norflurazon. Glass bead-mediated delivery disclosed a high transformation efficiency of *C. reinhardtii* with *CrPDS-L505F*, demonstrating clearly that the engineered endogenous *CrPDS* is a dominant selectable marker for *C. reinhardtii* and also possibly for other green algae. Furthermore, the expression of *PDS* could enhance the intracellular carotenoid accumulation of transformants, opening up the possibility of engineering the carotenogenic pathway for improved carotenoid production in microalgae.

(582bt) Isobutyraldehyde Production From Escherichia Coli By Removing Aldehyde Reductase Activity

Rodriguez, G., UC Davis
Atsumi, S., UC Davis

Increasing global demand and reliance on petroleum-derived chemicals will necessitate alternative sources for chemical feedstocks. Currently, 99% of chemical feedstocks are derived from petroleum and natural gas. Renewable methods for producing important chemical feedstocks largely remain unaddressed. Synthetic biology enables the renewable production of various chemicals from microorganisms by constructing unique metabolic pathways. Here, we engineer *Escherichia coli* for the production of isobutyraldehyde, which can be readily converted to various hydrocarbons currently derived from petroleum such as isobutyric acid, acetal, oxime and imine using existing chemical catalysis. Isobutyraldehyde can be readily stripped from cultures during production, which reduces toxic effects of isobutyraldehyde. We adopted the isobutanol pathway previously constructed in *E. coli*, neglecting the last step in the pathway where isobutyraldehyde is converted to isobutanol. However, this strain still overwhelmingly produced isobutanol (1.5 g/L/OD_{600}(isobutanol) vs 0.14 g/L/OD_{600} (isobutyraldehyde)). Next, we deleted *yqhD* which encodes a broad-substrate range aldehyde reductase known to be active toward isobutyraldehyde. This strain produced isobutanol and isobutyraldehyde at a near 1:1 ratio, indicating further native isobutyraldehyde reductase (IBR) activity in *E. coli*. To further eliminate isobutanol formation, we set out to identify and remove the remaining *IBR*s from the *E. coli* genome. We identified 7 annotated genes coding for IBRs that could be active toward isobutyraldehyde: *adhP*, *eutG*, *yiaY*, *yjgB*, *betA*, *fucO*, *eutE*. Individual deletions of the genes yielded only marginal improvements. Therefore, we sequentially deleted all seven of the genes and assessed production. The combined deletions greatly increased isobutyraldehyde production (1.5 g/L/OD_{600}) and decreased isobutanol production (0.4 g/L/OD_{600}). By assessing production by overexpression of each candidate *IBR*, we reveal that AdhP, EutG, YjgB, and FucO are active toward isobutyraldehyde. Finally, we assessed long-term isobutyraldehyde production of our best strain containing a total of 15 gene deletions using a gas stripping system with *in situ* product removal, resulting in a final titer of 35 g/L after 5 days.

(582ax) Development of Novel Feed Additives Through the Production of Recombinant Proteins in Microalgal Chloroplasts

Potvin, G.,
Zhang, Z., University of Ottawa

Microalgae offer many advantages as systems for recombinant protein production, as they combine the high growth rate and ease of cultivation of microorganisms with the ability to perform post-transcriptional and post-translational modifications of plant systems. Given that many microalgal species are deemed safe for consumption, and are in fact of considerable nutritional value, they present an interesting potential platform for the production of orally delivered therapeutics and vaccines, or as animal feed additives, while avoiding extensive purification costs associated with conventional recombinant microorganism fermentation-based production systems.

Despite advances over the last decade, however, high-level (i.e. economically viable) expression of recombinant products in microalgae remains anecdotal, and a comprehensive understanding of the factors and systems involved in product yield regulation remains elusive. Several proteins of industrial and therapeutic interest have successfully been expressed at sufficient levels in microalgal chloroplasts, but expression of other genes in otherwise identical expression cassettes is not met with the same success. Despite these drawbacks, microalgae are increasingly regarded as versatile and advantageous recombinant protein expression platforms.

Our research focuses on the development and cultivation of novel recombinant microalgal strains which express and store enzymes that improve nutrient utilization and degradation of non-digestible matter by animals. When fed to crop animals, such microalgal additives not only provide the nutritional benefits of the enzymes, already recognized and exploited in animal feeds through purified additives, but also provide the value-added nutrition of microalgal biomass, while conferring environmental benefits through decreased nutrient waste. The expression of different gene constructs, under the control of different 5'UTR/promoter sequences, and designed using different codon optimization algorithms, is being investigated, in conjunction with the enzymatic activities of cell culture lysates. The production of products destined for other applications, notably enzymes of therapeutic or medical interest, is also being pursued.

(582cd) Analysis of Photomixotrophic Metabolism in Synechocystis Sp. PCC 6803 Via Inverse Tracer-Metabolic Flux Analysis

You, L., Washington University in Saint Louis
He, L., Washington University in St. Louis
Berla, B. M., Washington University in St. Louis
Pakrasi, H., Washington University
Tang, Y. J., Washington University

Analysis of photomixotrophic metabolism in *Synechocystis* sp. PCC 6803 via Inverse Tracer-Metabolic Flux Analysis

Cyanobacteria have gained considerable attention in recent years for their potential applications in biotechnology. For nearly half a century, extensive research has been undertaken to characterize photosynthetic metabolism in cyanobacteria. However, questions still remain about the presence of a complete TCA cycle or the glyoxylate shunt. We present here a novel method termed Inverse Tracer-Metabolic Flux Analysis to identify and quantify photosynthetic metabolism in *Synechocystis sp.* PCC 6803. The labeling information of metabolites under the use of ^{12}C-tracers (such as glutamate and glyoxylate) in a ^{13}C-based background ($NaH^{13}CO_3$ and U-$^{13}C_6$ glucose), were measured by GC-MS to reveal several metabolic insights. First of all, we found that TCA pathway, instead of being a cyclic or an incomplete pathway appeared to be a unique branched structure in *Synechocystis*, which terminated in the production of succinate. Secondly, the *in vivo* activity of isocitrate lyase, which involved in the glyoxylate shunt, was not detected under our experimental conditions. The flux analysis revealed that glyoxylate can be synthesized through photo-respiration and can serve as a glycine precursor. Finally, the oxidative pentose phosphate pathway (OPP pathway) showed a measurable flux under glucose-based mixotrophic conditions. This study described an economical labeling approach to precisely characterize functional metabolisms.

(582cu) Engineering Enzymes With Nanostructures

Ge, J., Tsinghua University
Liu, Z., Tsinghua University

Enzymatic catalysis has the advantage over chemical synthesis of high enantio- and regioselectivity, which provides enormous opportunities to use enzyme-catalyzed reactions to synthesize molecules, particularly those with complex structures such as fine chemicals and pharmaceutical intermediates. We focus on the strategy of utilizing nanostructured materials to improve the performance of enzymes for biocatalysis applications.[1,2] Three examples of nanostructured enzyme catalysts will be introduced in this talk with their applications in enzymatic synthesis of therapeutic molecules: 1) A general method for creating hybrid enzyme-inorganic nanoflowers in which the enzyme has markedly enhanced activity and stability, making such enzyme nanoflowers suitable for catalyzing important chemical transformations;[3,4] 2) A type of soluble and recyclable enzyme-polymer nanoconjugates used in organic media, which is highly solubilized in organic solvents and facilitates substrate transport, allowing highly efficient enzymatic synthesis of therapeutic compounds; 3) A substrate-'imprinted' enzyme-polymer nanogel, which has high affinity to the substrate and enhanced activity for the synthesis of therapeutic molecules.[5]

Keywords: nanobiocatalysis, non-aqueous enzymatic catalysis, immobilized enzyme, nanostructures

References

(1) Ge, J.; Yang, C.; Zhu, J.; Lu D.; Liu, Z.* ***Topics in Catalysis*** 2012, **55**, 1070-1080.

(2) Wang, R.; Zhang, Y.; Lu, D.; Ge, J.*; Liu, Z.*; Zare, R. N.* ***Wiley Interdisciplinary Reviews: Nanomedicine and Nanobiotechnology*** 2013, doi: 10.1002/wnan.1210.

(3) Ge, J.; Lei, J.; Zare, R. N.* ***Nature Nanotechnology*** 2012, **7**, 428-432.

(4) Zhu, L.; Gong, L.; Zhang, Y.; Wang, R.; Ge, J.*; Liu, Z.*; Zare, R. N.* ***Chemistry-An Asian Journal*** 2013, doi: 10.1002/asia.201300020.

(5) Wang, R.; Zhang, Y.; Huang, J.; Lu, D.; Ge, J.*; Liu, Z.* ***Green Chemistry*** 2013, doi: 10.1039/c3gc40465j.

(582ar) Impact of Sursfactants On Asymmetric Bioreduction of Aromatic Ketones Catalyzed By Yeast Cells

Yang, F.,
Liang, M.,
Liu, X., Jiangnan University

Chiral aromatic alcohols are important intermediates of pharmaceuticals, pesticides and other industrial chemicals. These chiral alcohols can be obtained from the asymmetric reduction of prochiral aromatic ketones catalyzed by yeast cells. Yeast cells are widely used due to their cheap and good reduction properties. For these asymmetric bioreduction, besides major effect factors, some additives can also change the reaction rate and selectivity, eventually affecting the substrate conversions and enantiomeric excesses. Surfactants are a kind of interesting additives because they can affect cell growth, metabolism, cell membrane permeability or intracellular enzyme activity. In this paper, the asymmetric reduction of aromatic ketones catalyzed by yeast cells were studied in presence of surfactants and the corresponding (*S*)-1-aryl-1-ethanols were obtained. The effects of surfactants on the conversion and enantiomeric excesses were investigated. The results indicated that the conversions could be remarkably affected and the enantiomeric excesses were less affected by the surfactants. The conversions of aromatic ketones (40 mmol/L) reached 95.7~99.0% in the system containing 0.5 g/L NP-4 (nonylphenol polyoxyethylene ether-4), exceeding 20% than the surfactant-free system. The enantiomeric excesses of product (*S*)-1-aryl-1-ethanols only changed 1~2 % in the reaction system with NP-4. The other surfactants were also applied for additives of asymmetric reduction of aromatic ketones. The results stated clearly that the nonionic surfactants were a kind of better additives than cationic surfactants or anionic surfactants in the asymmetric bioreduction of aromatic ketones catalyzed by yeast cells.

(582ck) Global Alterations in the Secretome of B. Subtilis and S. Cerevisiae

Frayne, E. G., University of Phoenix

Thio-phosphate is a useful tool for manipulating nucleic acids in vivo as it results in nuclease resistant phosphorothioate linkages and as such can also be used to modulate protein synthesis. Two dimensional SDS polyacrylamide gel electrophoresis was successfully used to reveal changes in the pattern of protein expression in B. subtilis and S. cerevisiae. Both organisms secrete proteins and global alterations were readily detected in the secretome of these organisms using thio-phosphate. In both organisms, a shift in the distribution of proteins occurred favoring less abundant and/or heterologous proteins at the expense of naturally more abundant proteins. This shift appeared to result from the incorporation of thio-phosphate into the mRNA backbone, preferentially stabilizing less stable mRNAs at the expense of more stable mRNAs. The results revealed a clear correlation between the % total mass of a protein and the fold change in B. subtilis. Furthermore, in S. cerevisiae total protein secretion was enhanced several fold. Thio-phosphate is a potentially useful tool for manipulating protein synthesis and related outputs in micro-organisms. It also helpful for the detection of less prevalent proteins in the secretome. In B subtilis 853 protein spots were detected compared with 200 proteins previously detected in 2D gels.

(582ao) Enhanced and Sustained Transgene Expression Mediated By a Binary Baculoviral System

Sung, L. Y., National Tsing Hua University
Hu, Y. C., National Tsing Hua University

Baculovirus (BV) mediates transient expression (< 7 days) partly due to its large genome. To prolong the expression, we developed a binary system whereby the transgene (flanked by attP/B, loxP or Frt sites) in the substrate BV was excised by the recombinase (ΦC31o, Cre or FLPo) expressed by a second BV and recombined into smaller minicircles. The recombination efficiency conferred by ΦC31o was lower (≈40~75%), but approached ≈90~95% by Cre and FLPo in various cell lines (HEK293, HeLa, BHK, etc.) and stem cells (e.g. human adipose-derived stem cells, hASCs). Compared with FLPo, Cre exerted higher expression level and cell viability (≈95~100%), thus the Cre-based BV was engineered to incorporate additional *cis-acting* element to help the replication of the minicircles, inculuding: *oriP*/EBNA1, S/MAR or human origin of replication (*ori*). In HEK293 cells, only *oriP*/EBNA1 prolonged the transgene expression and maintained the episomal minicircles for 30 days without inadvertent integration, while the BV genome was degraded in 10 days. Upon delivering *bmp2* or *vegf* genes, the efficient recombination/minicircle formation substantially prolonged and enhanced the growth factor expression in hASCs, and the prolonged BMP2 expression concomitantly ameliorated the osteogenesis of hASCs, a stem cell with poor osteogenesis potential. Altogether, this binary BV system exploiting Cre-mediated recombination and *oriP*/EBNA1 effectively enhanced and sustained the transgene expression in dividing and non-dividing cells, thereby broadening the applications of BV.

(582dc) A Novel Protein Structure Refinement Method: Blind Predictions During CASP10, and An Optimized Protocol to Maximize Performance

Khoury, G. A., Pennsylvania State University-University Park
Tamamis, P., Princeton University
Pinnaduwage, N., Princeton University
Smadbeck, J., Princeton University
Kieslich, C. A., Princeton University
Floudas, C. A., Princeton University

The protein structure prediction problem remains unsolved. Stated succinctly, given a primary amino acid sequence, determine accurately the full three-dimensional structure of the protein. Protein structure refinement addresses a different problem; given a three-dimensional structure, perform a set of operations that will consistently improve the accuracy of that structure towards its experimentally defined structure, and improve its structure quality removing unphysical features such as clashes.

The refinement problem is plagued by many challenges, with most methods consistently degrading the structure, rather than improving it, as demonstrated by the blind predictions during the two most recent Critical Assessment of Techniques for Protein Structure Prediction (CASP) experiments [1, 2]. This is a result of many generated models being significantly similar to one another and the inability of current forcefields to distinguish between such better or worse models. This may not be surprising given predicted structures have already been optimized to a conformation that is in a deep local minima, and the probability of making a movement that will improve the structure is much smaller than one's chance of degrading it.

In this presentation, we introduce a novel optimized framework that consistently refines structures, both in terms of improved structure accuracy (GDT_TS and RMSD) and quality (# of clashes and Ramachandran violations). The initial version of our refinement method was ranked 5th place in the international blind structure prediction experiment CASP10 among all groups. The method employed biological filters based on the high correlation observed between sequence length and the number of hydrogen bonds, as well as SASA as a function of a protein's molecular weight. The method achieved the peak-performance for CASP target TR722 indicating the maximum refinement in GDT_TS among all groups and methods. During post-CASP assessment, the method was automated and optimized to include a support vector machine classification model to select refined vs. degraded structures, as well as enrich the probability of selecting a refined structure from a set of decoys, and enhanced with a molecular dynamics refinement step. Each step is tuned to maximize its performance and the probability of refining the structure. The optimized method that will be presented has been benchmarked on CASP10, 9, 8, and 7 refinement targets achieving between a 66% to 80% refinement success rate in terms of GDT_TS in 1 model. An online webtool has been made to disseminate the method used during CASP10 to the broader community and is freely available for academic use at http://atlas.princeton.edu/refinement. The optimized method will be available online for use in the near future.

References

1. Center, P.S.P. *10th Community Wide Experiment on the Critical Assessment of Techniques for Protein Structure Prediction* 2013 [cited 2013 4/15/2013]; Available from: www.predictioncenter.org/casp10/.

2. MacCallum, J.L., et al., *Assessment of protein structure refinement in CASP9.* Proteins: Structure, Function, and Bioinformatics, 2011. **79**(S10): p. 74-90.

(582bb) Efficient Production and Molecular Design of Microbial Surfactants for Next-Generation Biochemicals

Konishi, M., Kitami Institute of Technology
Ikarashi, M.,
Hatada, Y.,
Nagahama, T.,
Horiuchi, J. I., Kitami Institute of Technology

Biosurfactants (BSs) are amphiphilic compounds produced from carbon neutral materials by microorganisms. The compounds have excellent surface active properties and a great potential for applications in cosmetics, pharmaceutical and food industries. We have isolated several types of BS producing microorganisms to investigate the diversity of BS producing microorganisms the variety of molecular structures of BSs. Mannosylerythritol lipids (MELs) produced by several species in the genera *Pseudozyma* in basidiomycetus yeast have a variety of the major products. *Pseudozyma antarctica, Ustilago maydis,* and related species produces 4-*O*-[4', 6'-*O*-diacetyl-2',3'di-*O*-alka(e)noyl-β-D-mannopyranosyl]-D-erythritol, named MEL-A, and *Pseudozyma hubeiensis* produces 4-*O*-[4'-*O*-acetyl-2',3'di-*O*-alka(e)noyl-β-D-mannopyranosyl]-D-erythritol, named MEL-C. We have isolated *Pseudozyma hubeiensis* SY62 from deep-sea. SY62 produce 129 g/L of MELs from glucose and olive oil after optimization of the culture conditions. The productivity of MELs by SY62 is the highest record in published papers in the best of our knowledge. SY62 is one of candidates for mass production of MEL in industrial process.

In order to investigate the difference of molecular structure of major MELs secreted, draft genome sequencing of a good MEL producing strain, SY-62, in *P. hubeiensis* were performed by Illumina HiSeq system in this study. The sequence reads from pair-end library (400 bp) were initially assembled into 74 contigs using Augstus v1.2.08. The contigs total genome size of 18,442,938 bp. The G+C content were 56.5%. The cording regions and their functions were predicted using MetaGeneAnnotator 1.0 and NCBI BLAST 2.2.18. The rRNA and tRNA genes were found using RNAmmer and tRNAscan, respectively. The draft genome is composed of 7,523 putative-coding genes or open reading frames (ORFs), 26 rRNA and 121 tRNA genes. The genes of the cluster, *emt1*, *mac1*, *mac2*, *mmf1*, and *mat1* were observed in the genome of *P. hubeiensis* SY62. These translated amino acids sequences shows identities of 79.7, 71.6, 62.1, 82.2, and 57.8% to the corresponding genes in *U. maydis*. The low identity of Mat1 between *P. hubeiensis* and *U. maydis* seemed to associate the differences of the specificity of acetyltransferase and the resulting difference of acetylation of major products.

We are developing deleted mutants producing the variety of MEL intermediates in the synthetic pathway. The intermediates will enhance the diversity of MEL derivatives. We aim to develop a designable biosurfactant production of SY62 using draft sequencing and deleted mutation. We discuss the designable biosurfactant production concept in this presentation.

(582db) The Uptake of Cu in Tyrosinase Affects the Monophenolase/Diphenolase Activity Ratio

Fishman, A., Technion - Israel Institute of Technology

Tyrosinase belongs to the type-3 copper enzyme family, containing a di-nuclear Cu center, CuA and CuB. It is mainly responsible for melanin production in a wide range of organisms. Although Cu ions are essential for the activity of tyrosinase, the mechanism of Cu uptake is still unclear. We have recently determined the crystal structure of tyrosinase from *Bacillus megaterium* (TyrBm) and revealed that this enzyme has tighter binding of CuA in comparison to CuB. Investigating Cu accumulation in TyrBm, it was found that Cu presence has a more significant effect on the diphenolase activity. Therefore, by decreasing the concentration of Cu, the diphenolase/molophenolase ratio was increased by 2-fold. Using a rational design approach, we identified five variants having an impact on Cu uptake. We have found that a major role of the highly conserved Asn205 residue is to stabilize the orientation of the His204 imidazole ring in the binding site, thereby promoting the correct coordination of CuB. Further investigation of these variants revealed that Phe197, Met61 and Met184, which are located at the entrance to the binding site, not only play a role in Cu uptake but are also important for enhancing the diphenolase activity. In this study we propose a mechanism of Cu accumulation as well as an approach to changing the selectivity of TyrBm towards L-dopa production.

(582di) Directed Evolution Of Nitrobenzene Dioxygenase For The Production Of The Antioxidant Hydroxytyrosol

Fishman, A., Technion - Israel Institute of Technology

Nitrobenzene dioxygenase (NBDO) is known to add both atoms of molecular oxygen to the aromatic ring of nitrobenzene to form catechol. It is assembled of four subunits of which the alpha subunit is responsible for catalysis. As an oxidizing enzyme it has potential use in the detoxification of industrial waste and the synthesis of pharmaceuticals and food ingredients, however, not much work had been done studying its structure-function correlations. We used several protein engineering approaches (neutral drift libraries, random libraries, two types of focused libraries and family shuffling) to engineer NBDO for the production of the highly potent antioxidant, hydroxytyrosol (HTyr), from the substrate 3-nitrophenyl alcohol (3NPA). We obtained a triple mutant, F222C/F251L/G253D, which is able to oxidize 3NPA 375-fold better than wild-type with a very high regioselectivity. In total, we identified four positions which are important for acquisition of new specificities, of which only one is well-known and studied. Based on homology modeling it is suggested that these mutations increase activity by vacating extra space within the active site for the larger substrate and also by hydrogen bonding to the substrate. The best variant had acquired a stabilizing mutation which was beneficial only in this mutant. Thus we have achieved two goals, the first is the enzymatic production of HTyr, and the second is valuable information regarding the structure-function correlations of NBDO.

(582ak) Encapsulation of Heterologous Pathways Within the Pdu Microcompartment of Salmonella Enterica

Slininger, M. F., UC Berkeley
Tullman-Ercek, D., University of California

Bacterial microcompartments are subcellular protein structures that encapsulate certain pathways in bacterial cells. They provide some of the same function as the organelles of eukaryotes.Â A microcompartment is made up of many hexameric and pentameric shell proteins that fit together to form a polyhedral shape about 100 nm across. Â These protein shells may be used to enhance kinetics and product yield of a series of enzymatic reactions by imparting several advantages: increased proximity of enzymes, isolation of toxic intermediates, insulation from inhibitors in the cytosol, and concentration of substrates. These effects have been exhibited by native microcompartments such as the carboxysome in chemoautotrophs and the propanediol utilization (Pdu) microcompartment in *Salmonella.* It is of interest to learn whether these benefits are also observed for heterologous pathways localized to the microcompartment. To that end, three pathways were expressed in *Salmonella* and localized to the Pdu microcompartment. Each was chosen to examine a putitative benefit of encapsulation. Here, we discuss the functionality of each encapsulated pathway as compared to the same pathway in the cytosol.Â Moreover, we present a strategy for optimizing the Pdu shell proteins for compatibility with non-native metabolites, a potentially critical barrier to successful heterologous pathway encapsulation.

(582bc) Exopolysaccharides and Thermostable Proteins From Thermophile

Wang, J., South Dakota School of Mines and Technology
Salem, D. R., South Dakota School of Mines and Technology
Sani, R. K., South Dakota School of Mines & Technology

Exopolysaccharides and thermostable proteins from thermophile

Jia Wang[1], David R. Salem[1, 2], and Rajesh K. Sani[1]

[1]Department of Chemical and Biological Engineering, SDSM&T, Rapid City, SD 57701
[2]Department of Materials and Metallurgical Engineering, SDSM&T, Rapid City, SD 57701

Recently thermophilic microorganisms have been recognized as potential producers of exopolysaccharides (EPSs) having novel and unusual characteristics and functional activities. Thermophiles have several advantages for EPSs production compared to mesophiles e.g., short fermentation period, decrease in viscosity at high temperatures, and non-pathogenic products. Thermophilic EPSs have unique structures and chemical-physical characteristics which could be exploited for several applications in the near future. In addition, thermostable proteins synthesized by thermophilic bacteria play important roles in survival of these thermophiles under harsh environmental conditions. The knowledge on the structural and functional properties of EPSs and thermostable proteins from thermophiles is very limited. We therefore isolated several thermophilic *Bacillus* and *Geobacillus* from the deep biosphere of the Homestake Gold Mine and local compost facilities.

One of the isolate, *Geobacillus* sp. (WSUCF1) produced unique type of EPSs. WSUCF1 genome revealed the presence of 25 ORFs of glycosyltransferases that are responsible for the biosynthesis of polysaccharides. Among the various carbon sources, cellobiose was found to be the most appropriate carbon source for EPSs production (60.47 mg EPS/L) when grown in a batch reactor at 60°C. The EPSs were isolated using ethanol precipitation and purified. The chemical-physical analyses of the EPSs revealed 94.31% carbohydrate and 5.68% protein and trace amount of nucleic acids. Fourier Transform Infrared and Nuclear Magnetic Resonance spectroscopy results showed several unique features of the EPS isolated from WSUCF1 including the presence of large number of methyl ketones. Further characterizations of EPSs from WSUCF1 in-terms of determination of molecular weight, identification of sugar components, optical rotation value, differential scanning calorimetry (DSC) and thermogravimetrical analysis (TGA) are continuing. Additionally, thermostable proteins produced by WSUCF1 are currently being characterized. The EPSs and thermostable proteins of WSUCF1 strain have given an evidence for their potential applications in industrial processes. Future investigations of the EPSs and thermostable proteins from WSUCF1 will likely reveal more functional properties in regard to new opportunities in various industrial fields including bio-based polymeric materials.

(582au) Production of Bifunctional Molecules By Clostridium Sp. BOH3

Rajagopalan, G., National University of Singapore (NUS)
He, J., National University of Singapore (NUS)
Kun-Lin, Y., National University of Singapore (NUS)

Title

Production of bifunctional molecules by *Clostridium* sp. BOH3

Abstract

Clostridium sp. BO3 produces high yield of butanol and volatile fatty acids (predominantly acetic and butyric acids) from simple sugars and polysaccharides. When sp. BOH3 is cultured in reinforced clostridial medium with glycerol and sugar alcohols including mannitol, galactitol, sorbitol and xylitol, it produces bifunctional molecules including succinic acid and 1,3 propanediol (PDO). These bifunctional molecules are important feedstocks for various applications. For instance, succinic acid is as a precursor for the production adipic acid, 1,4 butanediol, terahydrofuran and g-butyrolactone. Similarly, PDO is used for the production of polymers, cosmetics and lubricants. The strain BOH3 produces 6.8Â±0.2 g/l of PDO and 1.3 g/l of succinic acid as major fermentation products from 30 g/l of glycerol. Meanwhile, it produces high level of succinic acid (>5g/l) and low concentration of (< 2 g/l) of PDO from sugar alcohols. Among the selected sugar alcohols, mannitol (30 g/l) supports 7.2Â±0.2 g/l of succinic acid and 1.3Â±0.2 g/l of PDO after 5 days of fermentation. In batch fermentation, a higher initial concentration (> 30 g/l) of glycerol and mannitol did not inhibit the growth of sp. BOH3, and did not support the production of succinic acid and PDO. While employing fed-batch process, addition of 30 g/l of glycerol and mannitol after 3 days of fermentation is able to enhance the production of PDO and succinic acid to 12.6Â±0.2 and 10.6Â±0.3 g/l respectively. Furthermore, addition of Na_2S at lower concentration (0.2 g/l) supports the PDO production (15.6Â±0.2 g/l) from glycerol. Meanwhile, $CaCO_3$, and $NaHCO_3$ (2.5-5 g/l) enhance the production of succinic acid to 16.9Â±0.5 g/l from mannitol. Overall, the fed-batch production process result the yield of PDO and succinic acid as 0.26Â±0.005 g /g of glycerol and 0.28Â±0.008 g / g of mannitol, respectively. To best of our knowledge, this is first report deals the production of succinic acid by solventogenic *Clostridium* strain.

(582o) Soft Immobilization of Redox Mediator for Electroenzymatic Synthesis

Lee, H.,
Lee, H. J.,
Lee, S. H.,
Park, C. B., Korea Advanced Institute of Science and Technology
Won, K., Dongguk University-Seoul

Enzyme-catalyzed reactions have emerged as an alternative strategy to conventional chemical synthesis for the production of pharmaceuticals, food, cosmetics, and even bulk chemicals. Particularly, oxidoreductases, one of the major classes of enzymes, are very promising in that they catalyze reactions involving oxygen insertion, hydrogen extraction, or electron transfer. In spite of their efficiency, however, the application of redox enzymes to chemical industries has been very limited; a major hindrance to their applications is that nicotinamide coenzymes, which most of them require as reduction or oxidation equivalents, are too expensive to be supplied in stoichiometric quantities. Therefore, efficient regeneration of the coenzymes is essential for successful applications of redox enzymes to industrial synthetic processes.

The reduced form of nicotinamide cofactor (nicotinamide adenine dinucleotide), NADH, is employed for more diverse enzymatic reactions than its counterpart (NAD^+): reduction catalyzed by dehydrogenases; hydroxylations and epoxidations by oxygenases. To regenerate NADH in vitro, chemical, electrochemical, or enzymatic methods have been employed. Electrochemical regeneration, in which NAD^+ is reduced by cathodic potential, is often favored owing to no requirement of impurities such as reducing agents, secondary enzyme, and co-substrate when compared with the other methods. Moreover, electrons are one of the cheapest redox equivalents available.

The NAD^+/NADH redox pair has a formal potential of −0.32 V vs. the standard hydrogen electrode (pH 7), but direct electrochemical reduction of NAD^+ requires a large overpotential. In addition, the direct reduction on unmodified electrodes mainly leads to the formation of an inactive dimer and isomers. In order to overcome these limitations of the direct reduction, redox mediators capable of transferring electrons from electrodes to NAD^+ were introduced. Among them, pentamethylcyclopentadienyl-2,2'-bipyridine-chloro-rhodium(III) $[Cp^*Rh(bpy)Cl]^+$, which is hydrolyzed in the form of $[Cp^*Rh(bpy)H_2O]^{2+}$ in aqueous medium, has been the most widely used, showing high regioselectivity (>99%) over a broad pH and temperature range.

For long-term operation of electroenzymatic processes, not only enzymes but also mediators should be immobilized. However, much effort has not been made to immobilize the organometallic mediator in an electrocatalytically-active form onto electrodes. It was found that substituents bearing nucleophilic moieties such as S- or N-containing groups on the bipyridine ligand interfered with the formation of the active Rh hydride complex, resulting in inactivation of the mediator. This means that immobilization methods based on covalent binding are not suitable for this redox mediator.

In this work, soft immobilization of the Rh complex mediator is carried out. The organometallic mediator is immobilized on solid electrode surfaces using non-covalent interactions. The electrochemical reactions were performed in a conventional three-electrode cell with a potentiostat (Gamry G750, USA): the working, counter, and reference electrodes were mediator-immobilized electrodes, a platinum wire, and Ag/AgCl (saturated KCl), respectively. NAD^+-reducing activity and repeated usability of the immobilized mediator are examined using cyclic voltammetry and spectrophotometry. Cyclic voltammograms and amperometric responses are recorded with mediator-immobilized electrodes in the absence and in the presence of NAD^+. The immobilized mediator is repeatedly used and its catalytic activity is measured after each use. Finally, the indirect electrochemical NADH regeneration is coupled with synthesis reactions catalyzed by dehydrogenases.

References

[1] Lee, H. J.; Lee, S. H.; Park, C. B.; Won, K. *Chem. Commun.* **2011**, *47*, 12538-12540.

[2] Lee, S. H.; Won, K.; Song, H.-K.; Park, C. B. *Small* **2009**, *5*, 2162-2166.

[3] Lee, J. S.; Lee, S. H.; Kim, J.; Park, C. B. *J. Mater. Chem. A*, **2013**, *1*, 1040-1044.

[4] Song, H.-K.; Lee, S. H.; Won, K.; Park, J. H.; Kim, J. K.; Lee, H.; Moon, S.-J.; Kim, D. K.; Park, C. B. *Angew. Chem. Int. Ed.* **2008**, *47,* 1749-1752.

[5] Walcarius, A.; Nasraoui, R.; Wang, Z.; Qu, F.; Urbanova, V.; Etienne, M.; Göllü, M.; Demir, A. S.; Gajdzik, J.; Hempelmann, R. *Bioelectrochemistry* **2011**, *82*, 46-54.

(582cl) Metabolic Profiling of Terpenoid Indole Alkaloids in Transgenic Catharanthus Roseus Hairy Roots

Zhao, L., Iowa State University
Shanks, J. V., Iowa State University

Plant-derived natural products play an important role in the pharmaceutical industry. The terpenoid indole alkaloids (TIAs) are prime examples and some of them have been developed as leading antineoplastic pharmaceuticals, such as vincristine and vinblastine. The natural yield of those valuable drugs is very low and chemical synthesis of those compounds is not very successful because of complex structure, which makes them extremely expensive. Despite significant progress in recent years, much remains unknown in these pathways, including uncharacterized enzymes or transcriptional regulation. Therefore, manipulation of the TIA pathways and analysis of transcription and metabolite profiles in transgenic *Catharanthus roseus*hairy root lines will improve the understanding of the network fluxes and regulation in TIAs pathways.

The bottleneck in developing the metabolic profiling methods of TIA pathways is that there are few commercialized available metabolite standards to be used to identify and quantify the metabolites in hairy root lines, which leads to an incomplete metabolic profile and a metabolic flux map that cannot reflect the real situation in the hairy root lines. Therefore, it is essential to get more identified metabolite standards to complete the metabolic profiling. In metabolomics study, LC-tandem MS technique has been applied in separation, detection, potential identification and quantification of low-concentrated metabolites with the advantages of high sensitivity and selectivity. Along with LC-tandem MS, NMR is also a leading analytical tool to investigate metabolism.

In this study, several metabolites were extracted and purified from either *C. roseus* hairy roots or an engineered yeast strain, NMR and LC-tandem MS were applied to identify the structure and MS fragment patterns of each metabolite. The newly identified metabolites were used as standards to develop the metabolic profiling of TIAs in hairy root lines.

(582p) Simulation of Turbulence Effects On Red Blood Cell Trauma

Ozturk, M., University of Oklahoma
O'Rear, E. A., University of Oklahoma
Papavassiliou, D. V., The University of Oklahoma

Blood flow is often laminar, but in some cases like the coronary artery, the ascending aorta and in arteriovenous grafts, turbulent blood flow can occur in vivo. In near prosthetic heart devices, such as ventricular assist devices, artificial hearts, blood pumps, and artificial valves, turbulence occurs much more often [1] . Locally high shear stresses and significant pressure fluctuations are some of the negative effects of turbulence in blood. In addition, physical contact within these devices produces mechanical forces. Such flow conditions can often cause hemolysis that can be experimentally observed through hemoglobin release from erythrocytes or red blood cells, RBC, because of destruction or trauma of the RBC. Design of such devices, therefore, depends on understanding and predicting the effect of turbulent shear stresses on RBC. However, the structure and the effects of turbulence on the cells remains unclear, and application of computational fluid dynamics in conjunction with better correlations to predict blood damage could significantly improve the development of blood contacting devices.

Blood trauma is in general considered to be a function of time exposure to high stresses and of the magnitude of the shear stresses. Furthermore, recent work out of our laboratory has indicated that extensional stresses can also be detrimental to RBC [2]. Computational techniques that can predict the detailed structure of turbulence can be used to investigate the effect of turbulent eddy structures on cell damage. In this work, we use Reynolds-averaged Navier-Stokes models of turbulence (kappa-epsilon) to simulate classic Couette flow hemolysis experiments [3, 4]. The goal is to relate the hemolysis seen experimentally to the characteristics of turbulence in the flow. Reynolds stresses and Kolmogorov scales are calculated and correlated to the observed hemolysis. After discussing the validation of our simulations, we will discuss details of the flow that are not easily available experimentally, such as turbulence kinetic energy dissipation rates, but are important to RBC trauma. An understanding of the scales and the structure of turbulence that is mostly responsible for blood trauma would lead to rapid design of heart assist devices that minimize blood damage.

References

1. Antiga, L. and D.A. Steinman, *Rethinking Turbulence in Blood.* Biorheology, 2009. **46**: p. 77-81.

2. Down, L.A., D.V. Papavassiliou, and E.A. O'Rear, *Significance of Extensional Stresses to Red Blood Cell Lysis in a Shearing Flow.* Annals of Biomedical Engineering, 2011. **39**(6): p. 1632-1642.

3. Kameneva, M.V., et al., *Effects of Turbulent Stresses upon Mechanical Hemolysis: Experimental and Computational Analysis.* ASAIO Journal, 2004. **50**: p. 418-423.

4. Sutera, S.P. and M.H. Mehrjardi, *Deformation and Fragmentation of Human Red Blood Cells in Turbulent Shear Flow.* Biophysical journal, 1975. **15**.

(582bu) Microbial Production of 2,3 Butanediol and L-Lactate From Seaweed Hydrolysate Using Metabolically Engineered Escherichia Coli

Mazumdar, S., Korea University

A variety of biofuel and biorefinery products have been produced from engineered *Escherichia coli* till date. Most of these products had been derived from simple sugars in its pure form, rather than deriving it from alternative, renewable and carbon neutral sources, such as marine alga biomass. Engineering *E. coli* to use algal hydrolysate can make these an attractive carbon source for the industrial production of value added fuels and chemicals. In two separate studies we have reported the engineering of *E. coli* by a combination of gene deletion and synthetic pathway incorporation, for the efficient utilization of algal hydrolysate under microaerobic condition to produce BA (2,3 butanediol + acetoin) and L- Lactate. In one study engineered strain produced ~19g/L of BA at a yield of 0.43g/g, while second study demonstrated the homofermentative route to produce 37.7g/L of high optical purity L-lactate. Our studies had shown that algal biomass would be competitive with lignocellulosic biomass and that brown seaweed can be used as a feed stock for industrial production of other chemicals.

(582q) Potential of Thermophilic Bioprocessing of Lignocellulosic Biomass for Generation of Biofuels

Kainth, A. S., South Dakota School of Mines and Technology
Bhalla, A., South Dakota School of Mines & Technology
Bibra, M., SDSMT
Kumar, S., South Dakota School of Mines & Technology
Sani, R. K., South Dakota School of Mines & Technology

Thermophiles and their enzymes will likely play important roles in many kinds of bioprocessing (e.g., in conversion of non-food biomass into biofuels including bioethanol, biodiesel, biobutanol, and biogas, bioremediation of contaminated aquifers, and production of pharmaceuticals). Thermophilic bioprocessing of lignocellulosic biomass to biofuels, through the use of elevated temperatures, offers several advantages including improved hydrolysis of cellulosic substrates, higher mass-transfer rates leading to better substrate solubility, lowered risk of contamination, and increased flexibility with respect to process design thus improving the overall economics of the process. Using soil/biofilm samples of deep biosphere (1.34 km) of the Homestake Gold Mine, Lead, SD, compost facility, and Hot Springs State Park in Thermopolis WY, we have isolated several thermophilic cellulose- and xylan-degrading pure cultures belonging to the genera *Brevibacillus*, *Paenibacillus*, *Bacillus*, and *Geobacillus*. Unique characteristics of cellulases and xylanases produced by mine and compost thermophiles include optimum temperatures of >70°C, pH ranges from 4 - 8, and high thermostability (e.g., at 60°C, 50% of cellulases and xylanases activities were retained in 35 and 23 days of incubation, respectively). Currently, we are working on heterologous expression of thermo-stable enzymes, without reducing enzymatic activity and thermo-stability, in suitable hosts so as to have higher production of these industrially important enzymes. Along with this, we are trying to elucidate the structure-function relationship of these thermostable enzymes and the structural factors that leads to their thermo-stability. The significant and immediate impact of ongoing research will be not only on biofuel research, but also on the national economy through the development of an efficient, economical, and sustainable biofuel production process.

(582bq) Toward Microbial Factories: Mechanism and Kinetics of Enzyme Encapsulation in Bacterial Microcompartments

Jakobson, C., UC Berkeley
Tullman-Ercek, D., University of California

Bacteria such as *Salmonella enterica* lack organelles to spatially organize their metabolism, but have been found to express proteinaceous microcompartments (MCPs) which can perform this function. These compartments allow the optimization of enzymatic reactions by concentrating enzymes and reaction intermediates, sequestering toxic intermediates, and insulating the compartmentalized pathways from other cellular processes. *S. enterica*, in particular, employs a MCP for 1,2-propanediol utilization (Pdu). Two different 18-amino acid N-terminal peptides from native Pdu enzymes are sufficient to direct the compartmentalization of heterologous proteins in the Pdu MCP, allowing the encapsulation of heterologous metabolic pathways. The Pdu MCP can serve as a model system to elucidate the mechanism of enzyme encapsulation and the kinetic effects on encapsulated pathways.

The Pdu MCP is thought to sequester a toxic intermediate, propionaldehyde, of the Pdu pathway. The mechanism of this sequestration is unknown. We aim to elucidate the contributions from differential MCP pore permeability, enhanced enzyme kinetics due to increased intermediate concentrations, and increased concentrations of enzymatic cofactors such as coenzyme A. We investigate the Pdu MCP by an *in vitro* system, whereby Pdu MCPs with specified contents can be purified and the behavior of the compartmentalized enzymes measured. These results will aid the design of custom MCPs for heterologous pathways with enhanced enzyme kinetics and improved toxin sequestration.

(582af) Rational Approach for the Improvement of Cell Culture Media Based On the Life Cycle Management of An Investigational Product

McDermott, S., AbbVie Bioresearch Center
Racicot, C., AbbVie Bioresearch Center
Hossler, P., Abbott Bioresearch Center
Fann, J. C. H., Abbott Bioresearch Center

Early stage upstream process development for mammalian cell culture is focused on generating material at sufficient levels on the most efficient timeline to enable clinical studies. During this stage, a top-down level development approach may be pursued to improve platform cell culture media through enrichment strategies. The use of select surfactants enabled highly enriched chemically-defined feed media that led to improved titer performance in a CHO based expression system. As the product matures through its lifecycle, the focus shifts to building a more complete understanding of the media impact on the process. At this stage, a bottom-up level approach was pursued to uncover the key media components that resulted in the improved titer performance from the enriched feed. These studies utilized robotic liquid handling to generate thousands of media variants. Utilizing a DOE-based approach, these media were tested in a high-throughput fashion in 24 well plates to examine how the CHO cell line responded to changes in media composition. This strategy not only can lead to an improved feed media addition strategy but can also shed light on process robustness. Both approaches increased antibody titers in CHO cell lines at g/L levels compared to the original platform feed strategy and enabled a better understanding of product quality control.

(582aw) Engineering Rhodobacter Sphaeroides for Direct Electron Transfer By Heterogenous Expression of Redox-Active Components From Electrogenic Species

Cheng, D., The Hong Kong University of Science and Technology
Wang, W., The Hong Kong University of Science and Technology
Chow, K. L., The Hong Kong University of Science and Technology
Hsing, I. M., The Hong Kong University of Science and Technology

Engineering *Rhodobacter*Â *Sphaeroides* for direct electron transfer by heterogenous expression of redox-active components from electrogenic species

Danhui Cheng[1,2], Wanyang Wang[1,2], King L Chow[1,2,4], and I-Ming Hsing[1,2,3]*

Bioengineering Graduate Program[1], Division of Biomedical Engineering[2], Department of Chemical and Biomolecular Engineering[3], Division of Life Science[4], the Hong Kong University of Science and Technology, Clear Water Bay, Kowloon, Hong Kong, SAR China

(* Corresponding Author)Â Fax: 852-2358-0054, E-mail : kehsing@ust.hk

Direct electron transfer between photosynthetic bacteria and electrode surface has been pursued in the study of bioelectricity generation through a solar powered microbial fuel cell, with the purpose to eliminate the usage of redox mediators and enhance the efficiency of power generation. In nature, a few species, including *Shewanella oneidensis* and *Geobacter sulfurreducens* are able to ?respire' on insoluble iron oxide through direct electron flow from cytoplasmic oxidative reactions to metal oxide/electrode surface. This feature attributes to special electron transfer conduit composed of cytochromes and structural proteins. A previous study has shown that expressing key components in the electron conduit of *Shewanella* can convert *E.coli* into microorganism vehicle capable of direct electron transfer. [1]

In this presentation, we aim to construct a direct electron transfer pathway in a photosynthetic purple non-sulfur bacterial species, *Rhodobacter sphaeroides* via heterologous expression of MtrCAB complex from *Shewanella oneidensis.* MtrCAB consists of periplasmic decaheme cytochrome MtrA, outer membrane β-barrel protein MtrB, and outer membrane decaheme cytochrome MtrC. Through the interaction of MtrA with cytoplasmic membrane components, electron flow can be channeled from cytoplasmic quinol pool to the periplasmic region. MtrA can then interact with MtrC through MtrB and direct the electron to outer membrane. The electron can be ultimately delivered to an external metal oxide/electrode surface via the direct contact between the surface and MtrC.

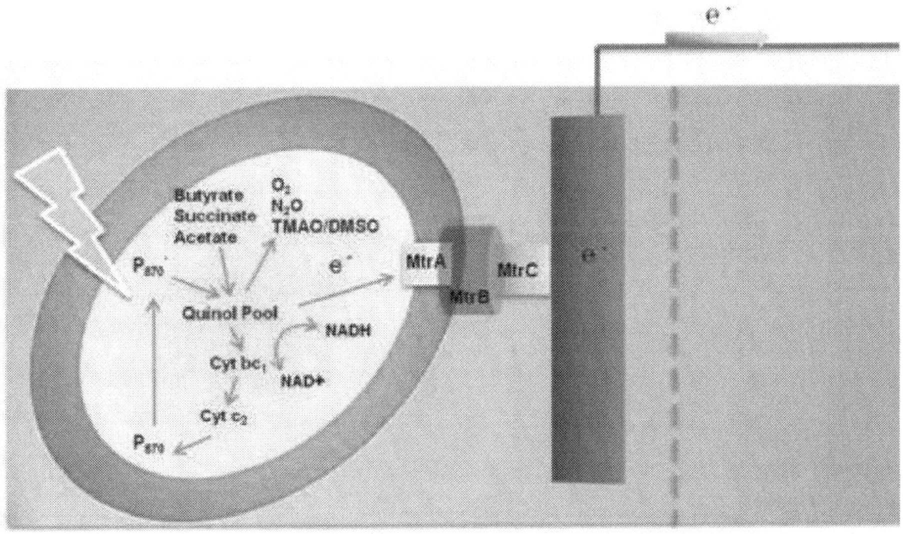

Figure. Engineering *Rhodobacter* for direct electron transfer by expression of MtrCAB complex for application in a solar powered microbial fuel cell.

In our preliminary study, decaheme cytochrome MtrA was first expressed in *Rhodobacter*. Using the native signal peptide from *Rhodobacter* cytochrome c_2, we successfully localize MtrA in the periplasmic region. Both TMB-based heme staining and absorption spectrometry have shown that the heme group is correctly incorporated in MtrA and confers its redox activity. The MtrA elevates the reduction rate of soluble ferric iron in *Rhodobacter* when a highly reduced substrate, butyrate, is provided (i.e., when the cytoplasmic quinol pool of *Rhodobacter* is over-reduced). The preliminary results implicate feasibility of redirecting electrons from cytoplasmic oxidative reactions to the extracellular electron pathway in *Rhodobacter* through this synthetic electron conduit.

[1] Jensen, H. M., Albers, A. E., Malley, K. R., Londer, Y. Y., Cohen, B. E., Helms, B. a, Weigele, P., et al. Engineering of a synthetic electron conduit in living cells. Proceedings of the National Academy of Sciences of the United States of America, 107(45), 19213?8(2010).

(582ay) HEK 293 Platform Bioprocess Development for Second Generation Programs Leveraging Knowledge From First Generation Processes and An Existing In-House CHO Platform Process

Dobrowsky, T. M., Biogen Idec
McElearney, K., Biogen Idec
Kshirsagar, R., Biogen Idec
Ryll, T., Biogen Idec

A proprietary, chemically defined platform bioprocess was developed for an HEK 293 expression system through the adaptation of existing CHO based platform bioprocesses. Expedited development timelines and partner programs originally demanded the use of vendor medium dependent processes for two glycoprotein producing HEK cell lines. This later led to difficulty identifying causes of variability in metabolite concentration, process control, and raw material quality. Implementation of proprietary, chemically defined medium and feeds during the second generation process development allowed for easier process optimization. The effect of an existing platform for CHO cell expression systems was initially tested on a single HEK cell line. The process was then adjusted and tested on a second HEK cell line to determine applicability and subsequently adjusted further. This development work revealed issues including metabolic differences from existing CHO programs as well as cellular aggregation, shear sensitivity, sensitivity to dissolved gas concentrations, and metal ion dependency.

(582r) Harnessing Flux Balance Analysis for Simulating Biological Evolution

Winkler, J. D., Texas A&M University
Kao, K., Texas A&M University

Current evolution simulators rely on artificial fitness landscapes that are meant to represent a collection of genotypes that may be generated through random mutation during population growth. Despite the many successes of these tools and other theoretical approaches that make similar approximations, a representative genome that allows for direct linkage between metabolism *in silico* and *in vivo* as the basis for simulated evolution would provide experimenters with a powerful new method to predict mutations of interest prior to experimental evolution. To address this issue, we have developed a genetic algorithm that represents individual cells using metabolic flux models, providing a direct linkage between genotype and metabolic behavior of all mutants during *in silico* evolution. After demonstrating the functionality of the simulator with an example model, gene expression data for *Escherichia coli* under n-butanol and isobutanol stress is utilized to identify possible compensatory, non-trivial metabolic changes that may occur during adaptive evolution to increase biofuel tolerance. Previously collected experimental data were found to agree with the simulator predictions, demonstrating the effectiveness of this approach for strain engineering purposes.

(582br) Engineering Pentose Sugar Fermentation in Zymomonas Mobilis for Efficient Fuel Ethanol Production From Lignocellulosic Feedstocks

Dunn, K., University of Illinois at Urbana-Champaign
Rao, C. V., University of Illinois, Urbana Champaign

The bacterium *Zymomonas mobilis* is of interest for the commercial production of fuel ethanol from lignocellulosic feedstocks due to its high ethanol productivity. *Z. mobilis* has been engineered to ferment all of the sugars present in cellulosic biomass, but the digestion of five carbon sugars by the strain is still slow and inefficient. The inefficiencies are present because *Z. mobilis* displays preferential sugar digestion—only after all six carbon sugars in the culture medium are depleted will five carbon sugars be consumed. We hypothesize that this is not due to an internal regulatory mechanism, but is instead due to the fact that *Z. mobilis* cannot efficiently transport pentose sugars into the cell. Therefore, in our work, we have investigated the effect on ethanol production of introducing dedicated pentose transporters in recombinant *Z. mobilis*. Our data reveal that the strains expressing the transporters can produce up to twice as much ethanol as the strains without them, indicating that this is an effective approach toward engineering cellulosic ethanol production by the strain. In addition, we have adapted pentose-fermenting *Z. mobilis* to growth on these sugars as sole carbon source. We have isolated strains with improved metabolic characteristics and are working to elucidate the genetic origins of these traits.

(582s) Effect of Sodium Chloride On Properties of Thermoplastic Starch/Poly(butylene adipate-co-terephthalate) Blend

Punchangcharoen, C.,
Sane, A., Kasetsart University

Poly(butylene adipate-co-terephthalate) (PBAT) is a biodegradable aliphatic-aromatic copolyester with high strength and water resistance. However, the use of PBAT in packaging industry is still limited due to its high cost and poor oxygen barrier property. Alternatively, melt blending PBAT with biodegradable thermoplastic starch (TPS) could be a solution because TPS is less expensive and has higher oxygen barrier property than PBAT. However, our preliminary work showed that the PBAT/TPS blend comprising high contents (i.e., 50 wt%) of TPS was partially miscible. In general, the differences in interfacial tension and viscosity among polymers play an important role in the blend morphology. Either adding a compatibilizer or reducing viscosity difference between polymers can enhance the miscibility of immiscible blends. Because melt viscosity of TPS was higher than that of PBAT, sodium chloride (NaCl) was incorporated to decrease the TPS viscosity. In addition, none has reported the effect of NaCl on properties of PBAT/TPS blend. Thus, the objective of this work was to investigate the effect of NaCl on rheological, morphological, mechanical, and barrier properties of PBAT/TPS (50:50 wt/wt) blend. It was found that the presence of NaCl (0.7-1.4 wt%) resulted in increasing TPS melt flow index and improved the miscibility as well as tensile properties of PBAT/TPS blend. Additionally, the compatibility and tensile modulus of PBAT/TPS blend increased with NaCl concentration. Adding 0.35 wt% NaCl led to increased tensile strength and elongation at break of PBAT/TPS blend from 5.9 to 9.1 MPa and from 380 to 630 %, respectively. However, further increasing NaCl content to 0.7 wt% resulted in decreased tensile strength and elongation at break by ~20 %. The presence of NaCl (0.35-0.7 wt%) considerably decreased oxygen permeability of the PBAT/TPS blend from ~18,000 to ~4,300-5,600 cc.mil/m^2.day.atm. Furthermore, increasing NaCl content increased oxygen barrier property while slightly decreased moisture resistance of PBAT/TPS blend.

(582cs) Sensitive, Rapid, Amplifi cation-Free Immunoassay Using Labeled Bacteriophage, Magnetic Capture Beads and Image Processing Techniques

Kim, J., University of Houston
Adhikari, M., University of Houston
Dhamane, S., University of Houston
Litvinov, J., University of Houston
E. V. Hagstr m, A., University of Houston
Kourentzi, K., University of Houston
Strych, U., University of Houston
Willson, R. C., University of Houston
Conrad, J. C., University of Houston

An amplification-free, imaging immunoassay is designed using labeled bacteriophage M13 as highly detectable "phage tags" and 1 μm magnetic beads as the immunosensing surface. The immunoassay sandwich is formed by a capture antibody immobilized on the surface of magnetic beads, the analyte and then a second anti-analyte antibody coupled to the phage tag. The phage tags are stained specifically with FITC-labeled anti-M13 antibodies for fluorescence detection at the level of a single label. Magnetophoresis is used to transport magnetic beads toward the coverslip surface in two minutes, which reduces the duration of the assay and improves its sensitivity. The magnetic beads and the fluorescently-labeled complexes are imaged at the same focus area using brightfield and fluorescence microscopy, respectively. By comparing the locations of beads and fluorescent complexes, the beads with attached phage tags are selectively identified using automated high-throughput image analysis. The target concentration is directly estimated from the fraction of beads bearing phage tags without the need for experimental standard curves. This assay can be easily modified to detect a wide range of proteins and pathogens.

(582dj) Protein Polymerization Technique Using Horseradish Peroxidase Catalyzed Tyorsine Coupling Reaction

Minamihata, K., The university of Tokyo
Nagamune, T., The University of Tokyo
Goto, M., Kyushu University
Kamiya, N., Kyushu University

Cross-linking reaction of proteins offers an enhancement effect of a distinct function of proteins due to the multivalent effect. In addition to that, conjugation of different kinds of proteins creates hybrid biomaterials possessing multiple functions of the building block proteins and such protein conjugates often show great performance caused by the synergistic effect. In this context, we have focused on developing an alternative protein cross-linking technique that enables to generate protein conjugates composed of multiple numbers of proteins. The horseradish peroxidase (HRP) mediated tyrosyl radical coupling reaction was utilized to achieve such protein cross-linking reaction. HRP catalyzes the oxidation reaction of various phenolic compounds by decomposing hydrogen peroxide and generates phenoxy radicals. These radicals then react with each other to form polymers. One of the 20 natural amino acids, tyrosine (Tyr or Y), has phenolic moiety in its side chain. Therefore Tyr can be recognized by HRP and polymerize upon HRP catalysis as well. Our strategy involves the genetic engineering of target proteins to introduce a highly flexible peptide tag containing tyrosine residues (hereafter called Y-tag). We anticipated that HRP recognizes the tyrosine residues in the Y-tag due to steric hindrance and electrostatic interaction between the target proteins and HRP. As model proteins, we selected *E.coli* alkaline phosphatase and streptavidin. The wild-type of these proteins showed almost no reactivity against HRP treatment, while the Y-tagged proteins showed high reactivity and resulted highly cross-linked protein conjugates. By using this methodology, we have demonstrated protein homoconjugation and protein heteroconjugation and also we very recently reported the control of the Y-tag and HRP mediated proteins conjugation reaction by using electrostatic interactions. In this report, we would like to summarize the recent progress of Y-tag and HRP mediated protein cross-linking techniques.

(582dm) Tyrosinase-Inhibitory and Antioxidant Activities of Oligopeptides From Yeast Protein Hydrolysate

Wu, Y., Jiangnan University
Tang, W., Jiangnan University

Yeast(Saccharomyces cerevisiae) proteins were separately hydrolysed using four different types of proteases to study antioxidant and tyrosinase-inhibitory of their peptides. The antioxidant and tyrosinase inhibitory activities of yeast peptide were mainly influenced by hydrolysis time and enzyme type. The result revealed that papain hydrolysate obtained after 30min hydrolysis (PYPH) displayed a higher DPPH radical scavenging activity and the highest mushroom tyrosinase-inhibitory activity. PYPH also showed good superoxide anion and hydroxyl radical scavenging, effective lipid peroxidation activities, and copper-chelating inhibition. Then PYPH was purified sequentially by ultrafiltration, gel filtration, and data showed the peptide with molecular weight below 5 KDa (PYPH-III) had the strongest tyrosinase–inhibitory and antioxidant activities. Further fraction separation and purification are being carried out on RP-HPLC and by LC-MS/MS to identify the amino acids sequences. It is expected that PYPH and/or its isolated peptides may be useful as cosmetics and food ingredients, especially in cosmetics as whitening agent. Because compared with ordinary whitening agent, it is not only good tyrosinase inhibitory performance, but also can provide nutrients for the skin.

(582t) Droplet-Based Microfluidic Device for Small Animal Drug Screening

Rahman, M., Texas tech University
Khan, Z. S., Texas Tech University
Bussel, F. V., Texas Tech University
Szewczyk, N.,
Blawzdziewicz, J., Texas Tech University
Vanapalli, S. A., Texas Tech University

The millimeter-scale nematode *Caenorhabditis elegans (C. elegans)* is a model organism for research in biology and medicine. These nematodes have a simple nervous system, short lifespan and are genetically tractable. These attributes have made *C. elegans* an attractive animal model for drug discovery and testing. Multiwell plates aided by automated pipetting represent the prevailing paradigm for performing drug-screening assays with *C. elegans*. In addition to requiring large volumes of reagents, automated image-based detection of the response of the worm to drugs in well plates is difficult due to the presence of multiple animals per well, non-uniform illumination conditions particularly at the well edges where the worms tend to reside, and three-dimensional swimming motion of worms in the unconfined environment of the well.

We report the development of a microfludic device that captures *C. elegans* in arrays of static droplets surrounded by an oil phase. The device consists of an array of trap chambers (3000x150 μm), optimized to trap mostly single animals. The shallow depth of the trap chambers and their small size maintains the nematode in the field of view and the device transparency reduces non-uniform illumination. An automated microscope stage acquires movies of each trap continuously or after a set time interval. In this work, assays were performed in the microfluidic device using the drug levamisol HCl that causes paralysis due to prolonged activation of nicotinic acetylcholine (nACh) receptors on body wall muscle. We also assess the effect of surrounding oil phase on the efficacy of drug action on *C. elegans*.

Thus, the developed microfluidic device solves many of the issues commonly encountered with drug screening in well plates. Our device is simple, as worms can be trapped using a single pressure source or hand pipetting. We will also discuss the potential of the device to perform on-chip dilution and obtain the full kinetics of drug action at single animal resolution with sophisticated image-based readouts.

(582u) Small Reaction Volume Trend to Produce Monodisperse Gold Nanoparticles in the Biosynthesis By Bacteria

Lv, H., Tianjin University
Sun, J., Tianjin University
Dai, L., Tianjin University
Shi, M., Tianjin University

With the development in the field of biotechnology, syntheses of nanoparticles with biomass or biomolecules become more and more popular. In the past several years, many success in the biosyntheses have been reported and various of nanomaterials were produced in this way. Besides the nanomaterials in different morphology, both metal and nonmetal, some achievement in theories are also suggested to give an explaination to the mechanism of the biosynthesis process.

Compared with syntheses methods, e.g. traditional chemical methods and physical methods, biosyntheses show significant eco-friendly advantage and most biomass or biomolecules only need a mild condition to reduce the metal ions to nanoparticles. According to our knowledge, metal nanoparticles obtained by microbes always show a well monodispersity, which attracts quite a lot attentions. However, some obstruction still restrict the development on a large scale of the biological method, e.g. it is difficult to prepare a large amount of microbes for a commercial process, and separating the nanoparticles from the biomass is also inefficient. As a result, the researchers turn to the imitation of the biological process, and different biomass have been used as reducing agents, in order to find out the key components which play a important role in the morphology control and metal ions reducing. Some amino acids and proteins are reported as the key components and related explainations are also stated,while several experiment are carried on with those, which finally show an unsatisfactory result that only with the key components, the amino acids and proteins, product will have a wider particle size distribution compared with nanoparticles synthesized by microbes.

What leads to the better monodispersity in microbes□Besides the research focused on the chemical components, we try to get a new explaination which considered the reaction volume as well as the reducing agents. In our experiment, the structure of microbes is ignored because existing knowledge is still weak on the mechanism of the transportation and reducing in the cell of microbes which let it very difficult to prefabricate a man-made microbe. We except to figure out some ideas to improve the monodispersity of existing biological methods, and with a simple analysis to the factor leading to the polydispersity, we finally contribute the polydispersity of the nanoparticles synthesized by simple biomass compared with those synthesized with complex microbes to the unbalance of the reaction rate, diffusion rate and randomness of Nucleation, leading to a relative microcosmic concentration distribution in the close zone of every nucleus or nanoparticles, which make the particle size distribution wider.

(582bs) Frontier Manufacturing: Scaling Up Synthetic Biology

Polizzi, K. M., Imperial College London
Hallett, J. P., Imperial College
Kontoravdi, C., Imperial College London
Shah, N., Imperial College London

Synthetic biology has the potential to revolutionise the way we make a host of consumer products from materials and energy to food and medicine. In order for this impact to be realised, we must find the best way to translate laboratory discoveries into operating industrial production processes. The challenge here is to transition from existing factories into the factories of the future.

Today many consumer products are made from fossil resources using synthetic chemistry techniques. In the future we will need to reduce our dependence on petroleum products and move to renewable resources. At the same time, the advent of synthetic biology techniques for rapidly tailoring biological systems for manufacturing purposes will allow us to transition away from synthetic chemistry and into more environmentally friendly production mechanisms using cells. We will tackle the question of how to undergo this transition smoothly by working with our industrial partners on real-world applications in two consumer areas (therapeutics and chemicals manufacturing).

Developing these future biofactories will require the invention of some new generalised technologies to underpin the new manufacturing processes. This will draw on our world-class expertise in synthetic biology, biocatalysis, sustainable manufacturing, novel solvents (ionic liquids), and process systems engineering.

We will need new biologically based sensors in order to be able to monitor the production processes as they occur to ensure the product quality (and to allow us to intervene if necessary). We will also need new, more robust production cells that can tolerate the high levels of compounds they make and new microreactors and/or compartmentalisation strategies for using enzymes when whole cells are not required. Because the transition will not happen overnight, we will need to develop intermediate production methods that combine biological and chemical catalysts. This will require solvents that are less toxic to proteins and cells and proteins that are engineered to be more robust in the presence of chemicals. In order to develop processes that are economical and efficient (minimal energy and water usage), we will create computer models to compare alternatives. The most promising processes will be implemented in the factories of our industrial partners.

We have chosen two challenge areas in which to test our new technologies. The first is healthcare, specifically the manufacture of medicinal compounds and therapeutic proteins. These are already largely

(582dh) Enhancing the Activity of Glutamate Decarboxylase From Lactobacillus Brevis By Directed Evolution

Mei, L., Zhejiang University
Lin, L.,
Hu, S.,
Huang, J.,
Yao, S. J., Zhejiang University
Lei, Y.,
Hu, G.,
Yu, K.,

Glutamate decarboxylase (GAD, EC4.1.1.15) can catalyze the decarboxylation of L-glutamate to γ-aminobutyrate (GABA). Due to the physiological functions of GABA, such as the regulation of neurological disorders, and the induction of hypotensive, diuretic effects, development of foods and pharmaceuticals containing highly concentrated GABA is in demand. In our previous study, *gad*, the gene coding one glutamate decarboxylase (GAD) from *Lactobacillus brevis* CGMCC 1306, was cloned and its soluble expression was accomplished. In order to improve its activity, error-prone PCR, followed by a screening, was conducted. A mutant Q51H with much higher activity towards L-MSG (55.4 mM/(min×$mg_{Protein}$), 120% higher than that of the wild type at pH4.8) was screened from the mutant library. In order to investigate the potential role of this site in the regulation of enzymatic activity, site-directed saturation mutagenesis at site 51 was carried out and three specific mutants, N-terminal truncated GAD, Q51P, and Q51L, were identified. Kinetic parameters of the three mutants and Q51H were characterized and revealed that aspartic acid at site 88 and N-terminal domain were essential to the activity as well as correct folding of GAD. This study not only improved the activity of GAD and enlarged its optimal pH range, but also shed new light on the structure-function relationship of GAD.

(582v) Electrospun Clay: Fabrication of Moldable 3-Dimensional Fibrous Scaffolds

Lee, S., Yonsei University
Cho, S. H., Yonsei University
Jeong, U., Yonsei University
Jang, J. H., Yonsei University

Electrospinning has been extensively utilized in a variety of biomedical applications due to its capabilities to induce high surface/weight ratio, versatile substructures and morphology, and especially biomimicity. Nevertheless electrospun fibres still have critical drawbacks to be applied in many biomedical trials: two-dimensional (2D) sheet-like constructs instead of three-dimensional (3D) structures. Recently, Xu et al. [1] presented a novel fabrication method of fluffy 3D electrospun matrices using polystyrene (PS) without manipulation of the apparatus. However, utilization of PS, which is not a biodegradable polymer, ultimately limits its utilization in *in vivo* applications. In this study, 3D scaffolds composed of biodegradable polymers were fabricated using a modified protocol. By employing the specific leaching technique, biodegradable and clay-like electrospun matrices that can be shaped freely into various shapes with a mold was fabricated. Furthermore, the combinatorial approach using electrospun clay with viral vector delivery [2] was also investigated. Adeno-associated viral vectors were adsorbed onto the scaffold for subsequent delivery to adherent cells. Mechanical, chemical and biological properties of electrospun clay-based scaffolds and, finally, the efficacy of viral delivery via 3D electrospun clay for potential tissue engineering scaffolds will be discussed. Reference: [1] B Sun et al. Nanoscale, 2012, [2] S Lee et al. Acta Biomater, 2011

(582al) Supercooling: A Viable Non-Freezing Preservation Method of Hepatocytes

Usta, O. B., Massachusetts General Hospital/Harvard Medical School/Shriners Hospital for Children
Kim, Y., Center for Engineering in Medicine (CEM) at Massachusetts General Hospital Harvard Medical School - Shriners Hospital for Children
Ozer, S., Center for Engineering in Medicine (CEM) at Massachusetts General Hospital Harvard Medical School - Shriners Hospital for Children
Bruinsma, B., Center for Engineering in Medicine, Massachusetts General Hospital, Harvard Medical School, Shriners Hospital for Children
Lee, J., University of Michigan, Ann Arbor
Demir, E.,
Berendsen, T., MGH/Harvard Medical School
Puts, C.,
Izamis, M. L., Center for Engineering in Medicine, Massachusetts General Hospital, Harvard Medical School, Shriners Hospital for Children
Uygun, K., Center for Engineering in Medicine, Massachusetts General Hospital, Harvard Medical School, Shriners Hospital for Children
Uygun, B., Center for Engineering in Medicine, Massachusetts General Hospital, Harvard Medical School, Shriners Hospital for Children
Yarmush, M. L., Center for Engineering in Medicine (CEM) at Massachusetts General Hospital Harvard Medical School - Shriners Hospital for Children

Supercooling preservation holds the potential to drastically extend the preservation time of organs, tissues and engineered tissue products, and fragile cell types that do not lend themselves well to cryopreservation or vitrification. We investigate the effects of supercooling preservation (SCP at -4 ^{o}C) on primary rat hepatocytes stored in cryovials and compare its success to that of static cold storage (CS at +4 ^{o}C) and cryopreservation. We consider two preservation solutions a) Hypothermosol (HTS-FRS) and b) University of Wisconsin Solution (UW) and a range of preservation temperatures (-4 to -10 ^{o}C). We find that there exists an optimum temperature (-4 ^{o}C) for SCP of rat hepatocytes which yields the highest viability; at this temperature HTS-FRS significantly outperforms UW solution in terms of viability and functional characteristics (secretions and enzymatic activity in suspension and plate culture). With the HTS-FRS solution we show that the cells can be stored for up to a week with high viability (~56 %); moreover we also show that the preservation can be performed in large batches (50 million cells) with equal or better viability and no loss of functionality as compared to smaller batches (1.5 million cells) performed in cryovials.

made using biological systems, but the existing processes are expensive and complicated. Also, in the future, it would be more efficient to make these medicines as and when they are needed (point-of-care manufacture). Our goals are to make simpler, more cost effective, point-of-care manufacturing systems using a combination of the above mentioned platform technologies: enzyme microreactors, specialised cells, and biosensors. Our second target is to produce bulk chemicals without the need for petroleum inputs. This will require us to adjust our manufacturing techniques for renewable inputs (such as biomass) and to develop new processes that use biology and/or environmentally friendly chemistry to do the conversions. Synthetic biology has never been attempted on such a large scale. Our challenge will be to adapt our parts, devices, and systems to operate at this level.

We are currently recruiting a large number of postdoctoral researchers to join our team. If you are interested to hear more about the project, please come by the poster!

(582bw) Application of 13C Metabolic Flux Analysis to Understanding the Physiology of Engineered Geobacillus Thermoglucosidasius

Ward, C. E., Imperial College London
Leak, D. J., University of Bath

Application of ^{13}C metabolic flux analysis to understanding the physiology of engineered *Geobacillus thermoglucosidasius*

C.E. Ward* and D.J. Leak

*c.ward10@imperial.ac.uk

* Department of Life Sciences, Imperial College London, South Kensington, London SW7 2AZ.

Geobacillus thermoglucosidasius is a facultative anaerobe capable of mixed acid fermentation, with optimal growth conditions between 45 °C and 70 °C. It has emerged as a promising platform for next generation biofuel production from lignocellulosic biomass due to its ability to effectively metabolise pentose and hexose sugars and longer chain oligomers as well as the increased efficiency that can be achieved in the industrial process through use of a thermophile.

Metabolic engineering of the fermentation pathways of the organism has resulted in a 4-fold increase in ethanol yield from glucose, this has been achieved by redirecting metabolic flux from the major fermentation products of lactate and formate through to ethanol. The high ethanol yielding strain, named DL66, has disruption of both the lactate dehydrogenase and pyruvate formate lyase genes and upregulation of the pyruvate dehydrogenase complex under anaerobic conditions. A further, intermediary, strain termed DL44 has been developed, which carries just the lactate dehydrogenase knock out.

^{13}C metabolic flux analysis (^{13}C-MFA) is a powerful technique that aims to determine the flow of metabolites through the metabolic pathways of the cell using a combination of experimental and computational techniques. Biomass cultured with an isotopically labelled substrate is analysed for the carbon-13 content of stable metabolites and specifically for the distribution of the isotope isomers – or isotopomers, which make up the entire metabolite pool. Proteinogenic amino acids provide an abundant and stable source of isotopomer pools and as such are routinely employed in ^{13}C-MFA; the isotope distribution found in amino acids is indicative of the metabolic pathways that have led to their synthesis and thus can be used to trace the flow of carbon in metabolism.

Measured rates of external fluxes from the cell, such as substrate uptake and fermentation product efflux, as well as biomass precursor requirements are used to determine the total flow of carbon into and out of the cell. These are combined with the isotopomer distributions and a metabolic network model detailing the carbon metabolism of interest, with all carbon atom transitions between metabolites, in computational analysis which provides the metabolic flux of the cell.

We have sought to apply ^{13}C-MFA to our strains of *G. thermoglucosidasius* in order to a) develop an intricate understanding of the metabolic flux distribution in our wild type strain, DL33, and the two engineered strains DL44 (Δ*ldh*) and DL66 (Δ*ldh*Δ*pfl* ↑*pdh*) and b) to identify targets for rational metabolic engineering that will contribute to the iterative strain development process, with the intention of increasing the yield or efficiency of bioethanol production in the industrial process.

Experimental data has been obtained by carrying out small-scale continuous cultivations of the three strains under fermentative, microaerobic and aerobic conditions, in order to determine the differences that may exist not only between strains but also through comparison of different states of metabolism. Cells were grown on a mixture of 20% U-^{13}C glucose, 80 % naturally labelled glucose and an amino acid mixture consisting of threonine, serine and glutamic acid, required by the organism for growth under oxygen limiting conditions.

Using biomass that has been cultivated to isotopic steady state, the relative abundance and distribution of carbon-13 atoms in the proteinogenic amino acids of the cell were measured using GCMS analysis. Extracted amino acid labelling patterns were corrected for naturally occurring isotopes. External fluxes of substrate uptake and efflux of metabolites and fermentation products were measured by HPLC analysis, to enable absolute values of flux to be calculated.

Metabolic network models of *G. thermoglucosidasius* have been created reflecting the genetic background of the wild type and engineered strains under both oxygen limited and aerobic growth. Metabolic models have been constrained based on literature-derived biomass precursor requirements of the organism and physiological and thermodynamic considerations, in order to create working models of metabolism of *G. thermoglucosidasius* for ^{13}C-MFA. The 13CFLUX2 suite of programs is being employed to calculate metabolic flux using the metabolic network models.

Preliminary analysis of the isotopomer distribution of the amino acids has revealed previously unobserved characteristics of the strains. Strain DL44 (Δ*ldh*) grown under microaerobic and fermentative conditions has revealed a labelling pattern which is similar to that seen under less oxygen limited (but not under oxygen-limited) conditions in other strains, indicating potentially altered flux in this strain compared to the wild type and other engineered strain. Additionally, under both aerobic and fermentative metabolism flux is observed from central metabolism to the supplemented amino acids glutamic acid, threonine and serine, indicating that these anabolic pathways are at least partially active, even under oxygen limiting conditions.

Ongoing metabolic flux analysis of these data and measured external fluxes, using the metabolic network models and 13CFLUX2 software show that these observed differences in isotopomer distribution reflect significant differences in metabolic flux between wild type and engineered strains, which will be presented and discussed.

(582bx) The Effects of NADPH Reducing Power On Isopropanol Production in Clostridium Tyrobutyricum

Jiang, W., The Ohio State University
Yang, S. T., Ohio State University

The effects of NADPH reducing power on isopropanol production in *Clostridium tyrobutyricum*

Wenyan Jiang and Shang-Tian Yang

Isopropanol is one of the most widely used solvents in the world. It is also one of the secondary alcohols that can be used as a direct or partial replacement for gasoline. Since the metabolic pathways and regulation of natural isopropanol producing strains are not well known, the final isopropanol titer can not satisfy rapidly increasing industrial demand. In our previous studies, we genetically engineered *Clostridium tyrobutyricum* to produce isopropanol by overexpressing *C. acetobutylicum ctf* (acetoacetyl-CoA transferase), *adc* (acetoacetate decarboxylase) and *C. beijerinckii adh* (secondary alcohol dehydrogenase) genes. However, since one extra NADPH is required to convert acetone into isopropanol, which disrupts overall redox balance within cells, the isopropanol titers of the mutants consuming glucose cannot surpass 1.2 g/L. Therefore, in this study, two different NAD kinase genes and one transhydrogenase gene were cloned into engineered *C. tyrobutyricum* separately to boost NADPH formation. The enzyme activity assay was performed to confirm that the genes were successfully overexpressed and the enzymes were functionally activated. The fermentation kinetics of the mutants shown increases in isopropanol titers, yields and productivities.

(582bi) High-Throughput Optimization of Genetically Refactored Nitrogenase Gene Clusters

Woodruff, L. B. A., Massachusetts Institute of Technology
Smanski, M. J., Massachusetts Institute of Technology
Gordon, D. B., Massachusetts Institute of Technology
Nicol, R., Broad Institute of MIT and Harvard
Voigt, C. A., University of California, San Francisco

We constructed, sequence-verified, and screened a large library (~3 Mb) comprising diverse permutations of refactored nitrogenase clusters to enable extraction of design rules for tuning refactored gene clusters. Genetic refactoring allows for accessing, transferring, and optimizing highly complex and tightly regulated biological processes, such as gene clusters. By removing known or potentially unidentified regulatory components and constructing a modular genetic system built from parts, genetically refactored systems enable fine-tuning of these large multicomponent processes. Previous work in our group demonstrated refactoring of the nitrogenase cluster from *Klebsiella oxytoca* and developed a hierarchical cloning strategy to construct refactored clusters from scarlessly assembled transcriptional units. Building upon this work, here we present the construction and screening of 120 whole refactored nitrogenase clusters in an *E. coli* host. This library was designed to investigate the operon structure, the number of transcriptional units comprised (4-11), and the orientation of genes and operons in the refactored nitrogenase cluster. We verified the sequence of the constructed ~23 kb refactored nitrogenase clusters using next-generation sequencing and sequence assembly tools Velvet and Pilon. Additionally, we are investigating the integration of next-generation sequencing technologies to enable increased throughput for building and optimizing large multipart DNA constructs.

(582bj) Human Fc Receptor Fragment Production for Antibody Drugs Purification By Escherichia Coli

Iwatani, K., Kyoto Institute of Technology
Oe, S.,
Kumada, Y., Kyoto Institute of Technology
Kishimoto, M., Kyoto Institute of Technology

Antibody drugs are recently expected for use in molecularly-targeted therapy. However, their production costs are still high, and therefore innovation of producing technique is necessary. One of the factors that arises the market price of antibody drugs is in purification process, and thus we must develop a low-cost affinity purification method instead of protein A affinity chromatography. Human Fc receptor protein can be used for purification of Antibody drugs, and its partial fragment can be produced by recombinant bacteria, yeast, or mammalian cells. In this study, we use recombinant *Escherichia coli*, which can be expected to grow faster than yeast or mammalian cells. Moreover, we can use inexpensive medium in *E.coli* culture compared with that in mammalian cell culture. Furthermore, the scale-up of culture system of *E.coli* is much easier than that of mammalian cells. The heterologous gene can be introduced into *E.coli* cells with plasmid vectors. The recombinant *E.coli* is often used for expression of the heterologous protein.

In this study, we tried to produce Fc receptor fragment in large quantities in order to reduce the cost of purification process in the antibody drugs production. After transformation of *E.coli* cells with the expression vector for Fc receptor fragment, we cultivate the recombinant *E.coli* using flask or Jar-fermenter. In the flask culture, *E.coli* cells grew slowly and produced small amount of affinity ligand because the nutrients in the medium was not enough for long-term cultivation. In the case of the culture using Jar-fermenter, we controlled temperature of culture broth, agitation speed, pH, dissolved oxygen (DO), and feeding rates of nutrients. At the first stage of cultivation for 16h, we carried out a batch culture. After depletion of glucose in the culture medium, we started a fed-batch culture by DO-stat (controlling feeding rate of nutrients based on the value of dissolved oxygen for maintaining optimum culture condition).When *E.coli* cells efficiently grew in exponential growth phase, the expression of target protein was started by addition of IPTG into the medium. During cultivation, we measured cell growth, glucose concentration, and production amount of Fc receptor fragment.

Consequently, using the Jar-fermenter system, we obtained 123mg/L of target protein, which is 400 times larger than that of the flask culture. We are going to improve and optimize the culture condition for increase in productivity of Fc receptor fragment.

(582w) Copper Nanocluster Based Fluorescent Sensors for Sensitive and Selective Detection of Kojic Acid in Food Stuff

Gao, Z., Tianjin University
Qi, W., Tianjin University
Wang, M., Tianjin University
Su, R., Tianjin University
He, Z., Tianjin University

Kojic acid (KA), a fungal metabolic product, has been widely used for its preservative actions against both chemical and microbial degradation [Bentley, 2006]. However, safety assessment for KA remains controversial and it is restricted for use by the US Food and Drug Administration for its potential carcinogenicity, embryotoxicity and teratogenicity [Burdock et al., 2001]. Hence, it is highly desirable to develop highly sensitive and selective sensing systems for KA.

In this work, we utilized BSA-capped copper nanoclusters (CuNCs) to develop a new fluorometric method for the determination of kojic acid with high sensitivity and selectivity. The as-proposed biosensor was sensitive for the detection of KA.The linear range for KA was 0.2Â¨C50 μM. The detection limit can be as low as 0.07 μM. Several real foodstuff samples spiked with KA, including sauce and vinegar are analyzed using the sensing system, and experimental results show that this fluorescent sensor exhibits excellent recoveries. The concept of our fluorometric sensing is illustrated in Scheme 1.

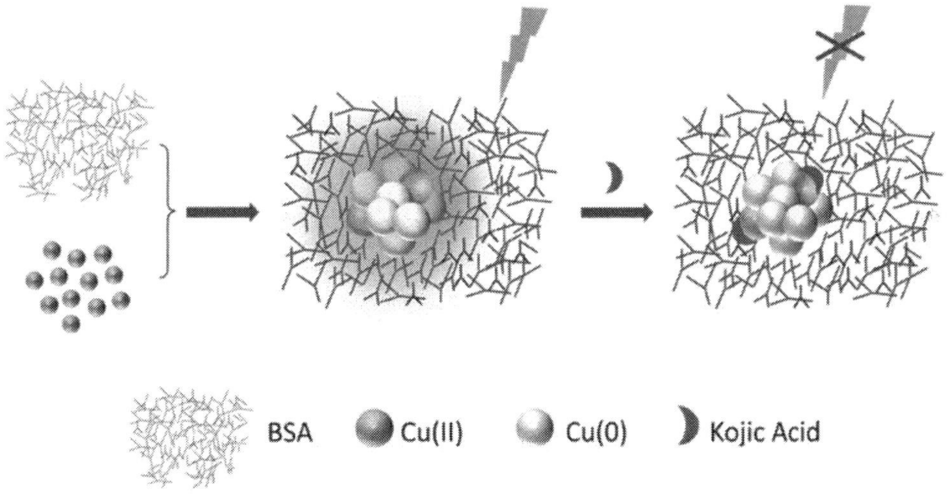

Scheme 1. Schematic representation of the CuNCs-based sensor for kojic acid.

(1) Firstly, BSA capped CuNCs were prepared and purified.The aqueous solution of the CuNCs exhibited violet fluorescence under UV light. After addition of KA, the original fluorescence of the CuNCs became weak. The corresponding fluorescence spectrum showed that the maximum emission peak at 407 nm disappeared (Figure 1A). The fluorescence intensity at 407 nm was plotted as a function of KA concentration and the value decreased gradually with increasing amount of KA until a plateau was reached (Figure 1B). A good linear relationship between the fluorescence intensity and KA concentration over the range of 0.2Â¨C50 μM (Figure 1B, inset) was obtained, with a detection limit of 0.07 μM. For an excellent sensor, high selectivity is a matter of necessity. To investigate whether our system is specific for KA, interference of various coexisting substance in the determination of kojic acid was investigated (Figure 1C).

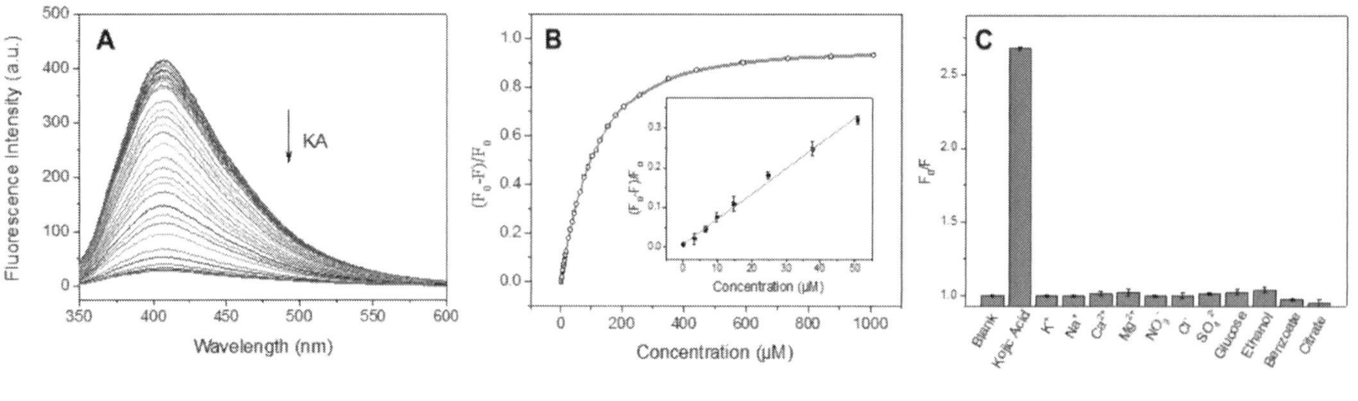

Figure 1. (A) Fluorescence emission spectra of the CuNCs in the presence of increasing concentrations of KA (0Â¨C1000 μM). (B) Plots of the fluorescence quenching rate at 407 nm as a function of the KA concentration (0Â¨C1000 μM). Inset is the linear range for KA. (C) The responses of interfering substance to CuNCs.

(2) In this system, several factors that may affect the quenching efficiency of KA were optimized, such as the reaction time, pH value, buffer system, and temperature. To understand the response rate of the fluorescence signal of the CuNCs to KA, the time-dependent fluorescence intensity upon addition of 150 μM KA was monitored. When KA was added to the CuNCs solution, the fluorescence intensity at 407 nm decreased rapidly within 3 min and changed little over time. Although the fluorescence intensity of CuNCs was dependent on temperature, the quenching rate of KA was not influenced. This process was notable as it involved green chemistry, simple, stable, fast and highly selective.

(3) Furthermore, the mechanism of the sensitive fluorescence response of BSA-CuNCs toward KA has also been investigated. It is observed that the surface of the clusters is stabilized by a small amount of Cu^{2+}, which should have strong and specific interactions with KA. The selective binding of KA to copper ions promotes the formation of copper kojate precipitation [Barham, 1939] on the surface of CuNCs, and thus to statically quench the fluorescence of CuNCs as given in scheme 1. A new peak at 314 nm appeared on the absorption spectra of CuNCs in the presence of increasing concentrations of KA, which proves the formation of a new complex, namely copper kojate in this case. Additionally, the excited state lifetime of the BSA-CuNCs in the absence and presence of KA was compared (2.454 ns and 2.416 ns, respectively). The result showed that the lifetime of the CuNCs hardly changed with the addition of KA, further implying static quenching. Based on these results, we concluded that the mechanism for the fluorescence quenching behavior was a static quenching process.

This work was supported by the Natural Science Foundation of China (Nos 21276192, 51173128 and 31071509), the Ministry of Science and Technology of China (Nos 2012YQ090194, 2012BAD29B05 and 2012AA06A303), and the Ministry of Education (No. NCET-11-0372).

References

1. Barham, H. Industrial & Engineering Chemistry Analytical Edition, 11, 31-33,1939

2. Bentley, R. Natural Product Reports, 23, 1046-1062,2006

3. Burdock, G.A., Soni, M.G., Carabin, I.G. Regulatory Toxicology and Pharmacology, 33, 80-101,2001

(582dq) Refolding and Immobilization of Peptide-Tag-Fused VHH Antibodies and Its Application to the Sensitive Immunoassays

Yamakawa, K., Kyoto Institute of Technology
Kumada, Y., Kyoto Institute of Technology
Kishimoto, M., Kyoto Institute of Technology

Antibodies can specifically recognize a target molecule with strong binding affinity. Therefore, they have been utilized as diagnostic reagents as well as drugs. The conventional antibodies have been considered to have some problems for use in such fields, for example, high production cost and low binding activity after immobilization due to lower density and unfavorable orientation. On the other hand, recombinant antibody fragments such as Fab, scFv, and VHH have been developed to overcome these problems.

Here, we focus on VH domain of heavy-chain antibodies (VHH antibodies) from camel for use as a ligand antibody in immunoassay. VHH antibodies are functional recombinant antibodies that consist only of the VH domain of heavy-chain. Therefore they are expected to be the smallest size for antigen recognition. Its molecular weight is less than one-tenth of whole antibody. Therefore, immobilization density must be higher than those of whole antibody and scFv. Furthermore, VHH antibodies may easily refold because they have only single intramolecular disulfide bond. For above reasons, VHH antibodies are expected to be useful ligand among the recombinant antibody fragments.

In this study, we investigated to produce VHH antibodies by E. coli cells and immobilize them onto the various plastic plates for use in immunoassay. We recently developed affinity peptide-tags, which can strongly recognize the surfaces of polystyrene (PS-tag:RIIIRRIRR) and poly(methylmethacrylate) (PMMA-tag:DVEGIGDVDLVNYFEVGATYTFNK).

Consequently, we successfully produced VHH antibodies genetically-fused with both PS-tag and PMMA-tag, and they were efficiently refolded by dialysis method. Both VHH antibodies refolded showed strong binding activities against the model antigen, hCG. Both PS-tag-fused VHH and PMMA-tag-fused VHH were successfully immobilized onto the surface of hydrophilic polystyrene plates and the VHH-immobilized plates detect the antigen with much high sensitivity. Furthermore, PMMA-tag-fused VHH could be used as a ligand for maxisorp and PMMA plates, indicating that PMMA-tag possesses relatively strong affinity against a variety of plastic plates. Thus, peptide-tag-fused VHH as a ligand will be significantly useful for sensitive immunodiagnosis.

(582ah) Effect of Agitation Rate in Hybridoma Cell Culture On Monoclonal Antibody Production

Nakajima, Y., Kyoto Institute University
Kumada, Y., Kyoto Institute of Technology
Kishimoto, M., Kyoto Institute of Technology

In the medical field, the use of monoclonal antibody has attracted much attention in late years. It can be used not only for the diagnosis and treatment of cancer and other deceases but also for the purification processes of drug medicine which needs high purity. In cultivation of animal cells using large tanks, understanding about effect of agitation on cell growth and antibody production is needed for improvement of production efficiency. In the cultivation using an industrial scale tank, mixing by stirring is necessary for homogenizing cell population, oxygen, nutrient and temperature in culture broth. However, animal cells are vulnerable to shear stress by agitation. In this study, we used a 10(AAT)CloneA as a model hybridoma cell. 10(AAT)CloneA was cultured in Jar-fermenter (total volume 0.5 L) . Here we investigated the effect of agitation rate on the efficiency of antibody production.

10(AAT)CloneA (ECACC) was adapted to serum free medium(invitrogen) and batch culture in jar-fermenter. In each culture, temperature and pH were controlled at 37℃,7.4 ,respectively. We set the seven patterns of agitation rate, and examined the effect of agitation rate on the cells growth and antibody production. As the reference data, static cultures were carried out ,using a polystyrene petri dish with a diameter of 60mm(IWAKI), at 37 ℃,5%CO_2. We took a sample during cultivation once a day .We measured viable cell density and total cell density. Antibody concentration was also measured by Sandwich ELISA. We investigated the change of the protein expression inside cells with the variation of agitation rate, by using 2-dimensional electrophoresis and image analysis.

The viable cell density increased until 4 days and then decreased in cultivations at any agitation rate. This indicated cell proliferation was not severely damaged up to 360 rpm agitation rate. On the other hand, the antibody production in which agitation rate is more than 60 rpm, was only 60~70 % production of petri dish culture. The antibody production at 30 rpm agitation rate was more than antibody production in petri dish culture. Oxygen volumetric mass transfer coefficient ($k_L a$) are 6.6(h^{-1}) in the jar-fermenter at 30 rpm agitation rate and 2.9(h^{-1}) in the petri dish culture, respectively. These results suggested that oxygen was supplied sufficiently and cells growth and antibody production are not severely damaged in culture broth at 30 rpm of agitation rate. To investigate the change of the protein expression inside cells caused by the difference in agitation rate, cells samples were obtained from cultures at 30 rpm and at 300 rpm, and petri dish cultivation, and their proteins were analyzed by protein analysis based on 2-dimensional electrophoresis and image analysis. From these results, it is considered that the amount of antibody production at 300 rpm of agitation was lower than production at 30 rpm because the nutrients needed for antibody production might be exhausted by the expression of proteins of major metabolic pathway. Many of proteins expressed inside cells in culture of petri dish were not expressed inside cells in culture of Jar-fermenter, which suggested that the metabolism of cells in petri dish is much different from the one in the jar-fermenter.

(582dr) A Facile Resolution of (R, S)-Mandelic Acid Via Lipase-Catalyzed Esterification

Tong, L.,
Wang, M., Tianjin University
Qi, W., Tianjin University
Su, R., Tianjin University
He, Z., Tianjin University

The resolution of racemic mixtures to prepare enantiomerically-pure pharmaceuticals and synthetic intermediates is of great commercial relevance, and different strategies are employed including use of lipases. Lipases are now widely used as synthetic tools in organic chemistry because of their broad substrate specificity and high stereoselectivity, even toward unnatural substrates. In particular, besides hydrolysis, lipases can also catalyze several related reactions - esterifications, transesterifications and ammoniolysis.

Enantiomers of mandelic acid and their derivatives are valuable chemicals widely utilized for resolution and synthetic purposes. Typically, the enzymatic resolution of (R, S)-mandelic acid (MA) is achieved via lipase-catalyzed hydrolysis or ammoniolysis of mandelic acid derivatives [Wang et al, 2007], which could be prepared using chemical reactions. Apparently, extra costs, solvents and time are needed in the preparation of mandelic acid derivatives. In order to solve the problem, we proposed a new approach, that is, one-step lipase-catalyzed esterification to resolve MA (Scheme 1).

In our study, Novozym 435 and CALB (lipase from *Candida antarctic B*) were used to catalyze the esterification of MA and 1-pentanol. Novozym 435 showed higher enantioselectivity toward (R)-MA than free CALB. After an investigation into possible factors, we attributed this phenomenon to changes in the inherent selectivity of the enzyme derived from the immobilization on hydrophobic interface.

Subsequently, measurements of enzyme activity and enantioselectivity at different initial substrates concentrations were carried out. With increasing initial MA concentrations ranging from 0 to 100 mM, both of enzyme activity and enantioselectivity enhanced. According to experimental data, we obtained Michaelis constants $K_m(R)$, $K_m(S)$ for (R)-, (S)-MA and the corresponding kinetic parameters $k_{cat}(R)$, $k_{cat}(S)$, respectively. In contrast to the improvement of enzyme activity due to increased MA concentration, the enzyme activity decreased and enantioselectivity had no change when the initial 1-pentanol concentration gradually increased. The phenomenon revealed that 1-pentanol presented a significant inhibition toward the enzyme. Since no changes in enantioselectivity appeared, we could draw a conclusion that the inhibition of 1-pentanol toward lipase was not selective inhibition.

To decrease the inhibition of 1-pentanol, *n*-hexane was utilized as the solvent due to its low *logP*. Compared with the reaction in non-solvent system, the conversion (x) and the substrate enantiomeric excess (*ee(s)*) values both improved with increasing volume of *n*-hexane. Furthermore, when the volume ratio of 1-pentanol to *n*-hexane was 1/7, the reaction time decreased from 72 h to 48 h. Through the investigation into solvent *logP* varying with the ratio of 1-pentanol to *n*-hexane, we found that the increased enzyme activity not only derived from the decreased inhibition of 1-pentanol, but also resulted from the changes in solvent *logP*. An *ee(s)* of 98.7%, which achieved the requirement of the industry, was obtained with 96.7% conversion after 48 h of reaction.

In addition, effects of triethylamine as an additive on the enzyme activity and enantioselectivity were investigated [Tsai et al. 2006]. The experimental results showed that the enzyme activity was inhibited by triethylamine, which differed from previous studies [Fritz Theil, 2000]. The lower enzyme activity induced by addition of triethylamine can be probably due to the change of surface properties of lipase or the formation of acid-base pair, which hindered the formation of acyl-enzyme intermediate.

In conclusion, we employed one-step lipase-catalyzed esterification to resolve MA and demonstrated the feasibility of the new approach. Moreover, effects of initial substrate concentrations, *n*-hexane amount and triethylamine as an additive on the enzyme activity and enantioselectivity were investigated.

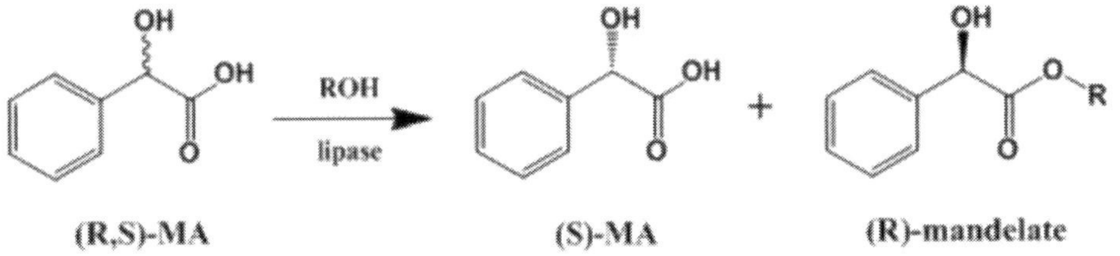

Scheme 1. Resolution of (R, S)-MA via lipase-catalyzed esterification.

This research was supported by the NSF of China (51173128, 31071509, 21206113), the Ministry of Science and Technology of China (Nos. 2012YQ090194, 2013AA102204, 2012BAD29B05), the Program for New Century Excellent Talents in Chinese University (NCET-08-0386), and Beiyang Young Scholar Program (2012).

References

1. Wang P. Y., Chen T. L., Tsai S. W. and Kroutil W. Hydrolytic resolution of (R,S)-2-hydroxycarboxylic acid esters in biphasic media: implication for rate-limiting formation or breakdown of tetrahedral intermediates in acylation step, *Biotechnology and Bioengineering*, 98, 30-38, 2007

2. Chen C. C., Chen T. L. and Tsai S. W. Altering lipase activity and enantioselectivity in organic media using organo-soluble bases: implication for rate-limiting proton transfer in acylation step, *Biotechnology and bioengineering*, 94, 201-208, 2006

3. Fritz T.Enhancement of selectivity and reactivity of lipases by additives, *Tetrahedron*, 56, 2905-2919, 2000

(582ed) Adsorption Characteristics of Whole Antibody and Scfv Onto Various Plastic Surfaces for Improvement of Sensitivity in ELISA

Miyamoto, R.,
Kumada, Y., Kyoto Institute of Technology
Kishimoto, M., Kyoto Institute of Technology

Immunoassay method utilizing antigen-antibody interaction has been widely used for biochemical research as well as clinical diagnosis. Among the immunoassay method, Enzyme-linked immunosorbent assay(ELISA), in which antibody-immobilized polystyrene(PS) plate is used as a solid support, is the most popular in the world. However, the conventional ELISA methods have several disadvantages as follows.

In Sandwich ELISA, whole antibody which is generally produced by hybridoma has been adopted as a ligand antibody. However, production cost is much expensive and thus, diagnosis price became also expensive. Furthermore, whole antibody is greatly huge protein and therefore it is hardly immobilized with high density and suitable orientation on a surface of plastic plate. Most of plastic plates for ELISA are made of PS. PS has been selected of low manufacturing cost as well as easy fabrication. However, PS may not be certainly optimum plastic for antibody immobilization.

In order to improve these problems, we fabricated new microtiter plates that are made of various materials including PS, poly(methyl methacrylate)(PMMA), and the others, and investigated to characterize antibody adsorption phenomena. On the other hand, we also investigated to immobilize a recombinant antibody fragment, especially scFv which consist of V_H and V_L domains connected by flexible linker $(G_4S)_3$. ScFv rapidly produced by cultivating recombinant *E. coli*cells and easily purified using affinity chromatography. The total production cost for scFv is approximately one-tenth for whole antibody. Although scFv is obtained in insoluble fraction called inclusion body, it can be refolded by dialysis of rapid dilution, and then it can recover original antigen-binding activity. Consequently, it can be immobilized with much higher density onto plastic than conventional whole antibody because its molecular weight is about 30 kDa. In recent years, our research group developed new scFvs that are genetically fused with material binding peptides such as PS-tag and PMMA-tag. Both of them can be successfully immobilized onto the surfaces of plastics, and consequently, they could maintain much high antigen-binding activity with suitable orientation.

Particularly, it was found that PMMA-tag fused scFv(scFv-PM) was efficiently refolded by dialysis and this binding peptide has a potential to recognize not only PMMA but also some other plastic surfaces.

In this study, we selected whole antibody and scFv-PM of anti-Thyroid Stimulating Hormone(TSH) antibody as models, and aimed adsorption characteristics of these antibodies to different plastic surfaces to immobilized then with high density and remaining binding activities. Also, we evaluate adsorption amount as well as signal intensity detected by ELISA for improvement of sensitivity.

(582by) Microbial Production of 2,3-Butanediol From Sugarcane Molasses Using Engineered Enterobacter Aerogenes

Jung, M. Y., Korea university
Oh, M. K., Korea University

Development of renewable energy or bio-based chemicals is now receiving lots of attention. 2,3-Butanediol (2,3-BD) is an example of bulk chemicals produced by biotechnological routes. Through chemical reaction such as dehydration and esterification, 2,3-BD can be converted into various chemical compounds including 1,3-butadiene, PUMAs (Polyurethane-melamides), and methyl ethyl ketone (MEK; butan-2-one). *Enterobacter aerogenes* is a facultative anaerobe, which can produce significant amounts of 2,3-BD from a wide range of carbon sources and has high growth rate. Furthermore, the *E. aerogenes*genome sequence has been revealed recently and the method of gene disruption was developed successfully.

One of the biggest hindrances to be overcome in the biorefinery process is the price of the carbon source that accounts for a large portion of the cost of microbial 2,3-BD production. Sugarcane molasses is considered to be a suitable fermentative carbon source due to its low-price and high sugar content. However, sugarcane molasses was not efficiently utilized by *E. aerogenes*. To improve sucrose utilization in *E. aerogenes*, a transcriptional repressor (ScrR) of sucrose utilization operon in genomic DNA was disrupted by using the lambda red recombination method. In addition, we have conducted lactate dehydrogenase (*ldhA*) gene knockout because lactate is mostly produced as byproduct. The mutant strain showed significantly enhanced the sucrose utilization when sucrose or sugarcane molasses was used as a carbon source. In batch fermentation with sugarcane molasses, 2,3-BD production of mutant strain (Δ*ldhA* Δ*scrR*) increased 60% compared to that of the wild type strain. Finally, 2,3-BD production reached 98.69 g/L at 36 h of cultivation in fed-batch fermentation. This result indicates that the metabolically engineered strain would be a promising 2,3-BD producer, employing an economic carbon source, sugarcane molasses.

(582bk) Characterization and Engineering of Acyltransferase Domains in the Pursuit of Novel Polyketide Therapeutics

Dunn, B., Stanford University

Natural products have historically been an invaluable resource in the discovery and development of therapeutically active compounds. The polyketide class of natural products has seen huge success in the commercial drug arena, acting as antibiotics, anticancer agents, immunosuppressants, and other widely-used therapeutics. This biological diversity stems largely from the structural diversity of these small molecules, many of which are produced in an assembly-line fashion by large enzymatic complexes called modular polyketide synthases (PKSs). Each module of a PKS is comprised of several domains, and each domain plays a distinct role in extending and chemically modifying the growing polyketide chain. The acyltransferase domain (AT) is responsible for selection and incorporation of simple Coenzyme A- (CoA-) linked building blocks, and therefore acts as a gatekeeper to polyketide structural diversity. Engineering of the AT domain for incorporation of novel building blocks is an attractive approach in the production of new therapeutic polyketides.

This work presents the characterization of an AT domain from the 6-deoxyerythronolide B synthase (DEBS), the PKS responsible for producing the aglycone precursor to the antibiotic erythromycin. DEBS is arguably the most highly studied PKS and often acts as a model system for understanding modular polyketide biosynthesis. We have utilized a continuous, coupled enzymatic assay to examine the steady state kinetics of AT-catalyzed transacylation. The transacylation reaction includes the formation of an acyl-AT intermediate and subsequent attack by a nucleophilic phosphopantetheine prosthetic group on the adjacent acyl carrier protein (ACP). We discuss the substrate (α-carboxyacyl-CoA) and ACP specificity of the transacylation reaction, focusing on the mechanistic determinants of this specificity. This knowledge is necessary for successful engineering of the specificity of AT domains.

Previous attempts to engineer AT domains for novel extender unit incorporation have focused primarily on the swapping of an entire domain for one with alternate substrate specificity. These attempts often come at the expense of the appropriate protein-protein interactions necessary for successful polyketide chain extension and therefore lead to drastically decreased product titers. A more conservative and increasingly popular approach for engineering AT domain specificity involves the targeted mutagenesis of the active site or adjacent residues. We present kinetic data on the "best" AT site-directed mutants characterized in the literature to date, and reveal that these mutants have drastically attenuated activity *in vitro*. This decreased activity (rather than enhanced specificity) is likely what allows these mutants to incorporate a nonnative substrate *in vivo*. This reveals extensive room for improvement in the engineering of AT specificity by site-directed mutagenesis. We present briefly our future plans for tackling this engineering challenge, with the concomitant goal of generating novel polyketides by incorporation of diverse building blocks.

(582ds) Porous Cleas of Î²-Mannanase Prepared By Macromolecular Cross-Linker

Wang, M., Tianjin University
Zhen, Q.,
Qi, W., Tianjin University
Su, R., Tianjin University
He, Z., Tianjin University

Cross-linked enzyme aggregates (CLEAs), a carrier-free enzyme immobilization approach, prepared by precipitation and subsequent cross-linking using glutaraldehyde, have attracted increasing attentions in recent years because of its advantages of high stability and enzyme activity due to the exclusion of carrier and the excellent property of repeated utilization.[1, 2] However, mass transfer limitation was considered as the main shortcoming for the application of CLEAs because of its conglomeration structure formed by glutaraldehyde.

Therefore, in this paper, the multi-aldehyde macromolecule of linear dextran polyaldehyde (2000 kD) as well as branched dialdehyde starch (350-750 kD), **Fig. A**, were introduced as cross-linker to construct the porous structure and enhance the mass transfer property of beta-mannanase CLEAs.

Scanning electron microscope revealed the inner structure of CLEAs prepared by different cross-linker. For the CLEAs cross-linked by glutaraldehyde (**Fig. B-a)**, the SEM image exhibited a smooth surface with few pores. It was estimated that this structure had the seriously steric hindrance effect on macro-substrates, which limited the substrates to enter into the inner CLEAs. Thus, beta-mannanase CLEAs showed the low activity (4.68%). When using the linearly macromolecular dextran polyaldehyde (2000 kD) as a cross-linker, beta-mannanases were bonded along its long chain, which not only prevented the lump of enzyme molecules but also formed a porous structure of CLEAs for macro-substrates to get into the inner CLEAs (**Fig. B-b**). These pores decreased the steric hindrance and obviously increased the activity of CLEAs (32.01%). Unlike the linear macromolecule, dialdehyde starch (350-750 kD) is dendritic. It bonded enzyme molecules along its interlaced branch, which also led to the congestion of enzyme and the steric hindrance of CLEAs (**Fig. B-c**). Even then, the activity of CLEAs prepared with dialdehyde starch (20.98%) was still higher than that cross-linked by glutaraldehyde due to the flexibility of enzyme.

This research was supported by the NSF of China (20976125, 31071509, 51173128, 21206113) and Tianjin (10JCYBJC05100), the Ministry of Science and Technology of China (Nos. 2012YQ090194, 2013AA102204, 2012BAD29B05), the Program for New Century Excellent Talents in Chinese University (NCET-08-0386), and Beiyang Young Scholar Program (2012).

References

[1] U. Roessl, J. Nahalka and B. Nidetzky, *Biotechnol. Lett.*, 2010, **32**, 341-350.

[2] L. Cao, F. van Rantwijk and R. A. Sheldon, *Organic Letters*, 2000, **2**, 1361-1364.

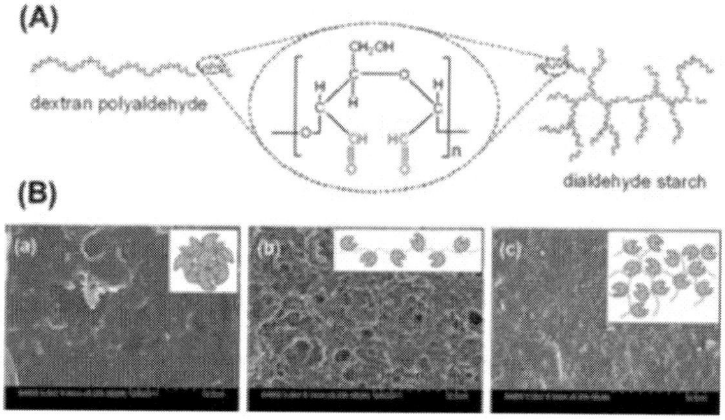

Figure (A) The molecular structure of dialdehyde starch and dextran polyaldehyde; (B) The morphology of beta-mannanase CLEAs cross-linked by (a) glutaraldehyde, (b) dextran (2000 kDa) and (c) dialdehyde starch (magnified 5 k times).

(582cm) A Novel Ailift Photobioreactor With Flashing Light Effect for Microalgae Cultivation

Zhang, Q., Institute of Process Engineering, Chinese Academy of Sciences
Xue, S., Institute of Process Engineering, Chinese Academy of Sciences
Wu, X.,
Yan, C.,
Li, Y.,
Cong, W.,

Microalgae can be used as a food for both human and animals, pharmaceuticals and health product, it also can be used to fix atmospheric CO_2 and produce biofuel. Among the photobioreactors used for microalgae cultivation, the airlift photobioreactors were widely reported. However, there is still much space to enhance the biomass production in these photobioreactors. It is well known that the flashing light effect (FLE) could have a considerable influence on the efficiency of light utilization and photosynthesis. In this work, a novel airlift photobioreactor with flashing light effect was developed. When the cells move in the airlift photobioreactor, they will pass through light region and dark region alternatively, so the FLE of microalgae is produced. The effect of light intensity, light fraction and the gas rate on biomass production of *Scenedesmus* were investigated. The results showed that the biomass production of *Scenedesmus* increased with an increase of the light intensity. And the higher the light intensity, the lower of the light fraction is needed to get a great enhancement of the biomass production. With increasing the gas rate, the biomass production of *Scenedesmus* firstly increased and then decreased.

(582cw) Immobilization of Cellulase On a Silica Gel Substrate Modified Using a 3-Aptes Self-Assembled Monolayer

Zhang, D.,
Hegab, H., Louisiana Tech University
Lvov, Y.,
Palmer, J. D., Louisiana Tech University

Cellulase was immobilized onto silica gel surfaces which was pretreated with (3-Aminopropyl) triethoxy-silane (3-APTES). Glutaraldehyde was used as a cross-linker between the cellulase layer and the 3-APTES layer. The hydrolysis of the carboxymethylcellulose sodium salt (CMC) solution made in pH=6 acetate buffer was investigated using the free and immobilized cellulase at different temperatures. Storage and pH stability were also investigated. The amount of protein in the enzyme solution was determined by the fluorescamine protein assay. As a result, cellulase was successfully immobilized on the modified silica gel surface, and little enzyme desorption was observed during the hydrolysis of the CMC solution. The highest activities of both immobilized and free cellulase appear at 50°C. The specific activity of the immobilized cellulase is 7% compared to the similar amount of free cellulase. Significant activity over multiple reuses was observed. The seventh batch achieved 82% activity of the initial batch, and the fifteenth batch retained 31% (with little decrease through 26 cycles of reuse). It was observed that the immobilized cellulase retained 48% of its initial activity after 4 days, and 22% even after 14 days.

(582at) Transgenic Expression of Thermomyces Lanuginosus and Candida Antarctica Lipases in Plants for the Enzymatic Production of Biodiesel

Deng, Y.,
Gagnon, M., University of New Hampshire
Minocha, S., University of New Hampshire
Vasudevan, P., University of New Hampshire

Enzymatic transesterification with lipase as the catalyst eliminates soap formation. Unlike alkali-based reactions, the products can easily be collected and separated. Moreover, enzymes require much less alcohol to perform the reaction, and can be reused despite some loss in activity at the end of each cycle. We have genetically engineered plants to constitutively express a lipase for biodiesel production from spent oils. We have cloned the gene of a lipase with known transesterification activity from *Thermomyces lanuginosus*, and *Candida antarctica*. Cloning of TL enzyme involved isolation of total RNA, reverse transcription of the mRNA into cDNA and PCR amplification of the lipase gene using specific primers. The gene was first inserted into a cloning vector (pCR8/GW/TOPO) and sequenced to confirm its identity. The gene has been inserted into a plant destination vector (pGWB408 and pMDC83) via LR clonase reaction. *Nicotiana tabacum* (tobacco) leaf was transformed with the lipase gene using *Agrobacterium tumefaciens* (strain GV3101) and transferred onto selection media plates. We have also grown *Arabidopsis thaliana* from seeds that were transformed using the floral dip transformation method. The recombinant enzyme was collected from the genetically engineered plants, purified, and tested for both hydrolytic and synthetic activity. The activity results will be compared with enzyme catalysts from commercial *Thermomyces lanuginosus and Candida antarctica,* and the effect of solvent addition will also be presented.

(582am) Constraints-Based Flux Balance Analysis In Cancer Cell Growth and Metabolism

Navare, C., National Chemical Laboratory
Uthup, R., National Chemical Laboratory
Prasad, M., National Chemical Laboratory
Sadhukhan, P., National Chemical Laboratory
Raghunathan, A., National Chemical Laboratory

Several types of cancer cells/tissues adapt and evolve their metabolism as a strategy for increased cell growth and proliferation leading to malignancy. The genome-scale human metabolic model is thus an excellent platform to study a complex multi-hit, multi phenotype disease like cancer. Constraints-based flux balance analysis allows for integration of heterogeneous data-types as well as analysis to compute phenotype. The focus of this talk is the development and analysis of tissue specific models for cancer. Global microarray gene expression data is used to develop cell-line and patient models for lung cancer using from legacy data using the recent genome scale model of human metabolism, Recon2 as the basis. The algorithm for integration of the gene expression data defines a scaling function to convert expression levels to fix the maximum capacity of every reaction in the network. The scaling function allows for introducing parameters for not only post-transcriptional and translation regulatory effects that modulate protein concentration but also the microenvironment of the cells, that dictate activity of the proteins. Linear optimization results in a computed value for the optimal growth phenotype of the cell lines comparable to experimental values. The cell line models are further validated using in-house growth and substrate uptake rate data and phenotypic micro-arrays Biolog^{TM}. The predictions of clinical models using tissue data and blood data will be compared to the classical cell line models. The clinical models are interrogated for growth rates of patient always higher than the normal cohort. Integration of results of routine blood work, specialized clinical tests like PET (Positron Emission Tomography) scans allow for better prognosis but also fundamental understanding of the disease. The challenge for predictive differentiation of progressive disease states and their ability to predict outcomes of therapy will also be discussed.

(582dt) Immobilization of Amylase On Metal Ceramic Powder: Application for Efficient Starch Hydrolysis

Wu, Z.,
Wang, M., Tianjin University
Qi, W., Tianjin University
Su, R., Tianjin University
He, Z., Tianjin University

Amylase, a kind of hydrolase which can convert starch into reduced-sugars, syrups and dextrin, is an important biocatalyst in food industry. Usually, amylase immobilized on the solid carriers which have suitable hardness, density and porosity, seems to be the most appropriate way for industrial application (Singh et al., 2013); while physical adsorption between carrier and enzyme would just supply weak affinity and covalent cross-linking may lead to the loss of enzyme activity. In recent years, due to the high immobilization efficiency, simple immobilization method and improved stability, enzyme immobilized through metal affinity has attracted increasing attention (Ho et al., 2004). The metal ceramic, with significant mechanical strength, electrical conductivity and abrasive resistance, would be a suitable carrier for enzyme immobilizaton. While the applications of metal ceramic on enzyme immobilization have been little reported.

In this work, the ceramic powder with high metal content (MCP) were prepared as a new kind of carrier for amylase immobilization. Cu-MCP showed the best affinity to alpha-amylase and exhibited maximal activity (Fig.1a). The effects of metal content, chelation time and amylase concentration on the immobilized amylase (Cu-MCP-amylase) activity were investigated. Under the optimal preparation conditions, the activity of Cu-MCP-amylase could reach 31.5 U/g (54.5 % of activity yield). Next, the characterizations confirmed that the Cu-MCP was a superior carrier with significant mechanical strength, considerable surface area, stable environment, homogeneous surface and low cost. Meanwhile, the properties of Cu-MCP-amylase were determined. It had optimal pH 7.0 and optimal temperature 70°C with improved thermal stability. Finally, Cu-MCP-amylase packed in a column reactor (Fig.1b) and showed significant catalytic capacity and stability, which could retain 80% of initial catalytic capacity after 40 recycle times. Thus, MCP as a novel immobilization carrier, would enhance mass transfer and supply stable environment, which could provide an efficient and practical method for enzyme immobilization in industrial application.

This research was supported by the NSF of China (20976125, 31071509, 51173128, 21206113) and Tianjin (10JCYBJC05100), the Ministry of Science and Technology of China (Nos. 2012YQ090194, 2013AA102204, 2012BAD29B05), the Program for New Century Excellent Talents in Chinese University

(NCET-08-0386), and Beiyang Young Scholar Program (2012).

References

[1] Ho, L.F., Li, S.Y., Lin, S.C., Hsu, W.H. 2004. Integrated enzyme purification and immobilization processes with immobilized metal affinity adsorbents. *Process Biochemistry*, **39**(11), 1573-1581.

[2] Singh, R.K., Tiwari, M.K., Singh, R., Lee, J.-K. 2013. From Protein Engineering to Immobilization: Promising Strategies for the Upgrade of Industrial Enzymes. *International journal of molecular sciences*, **14**(1), 1232-1277.

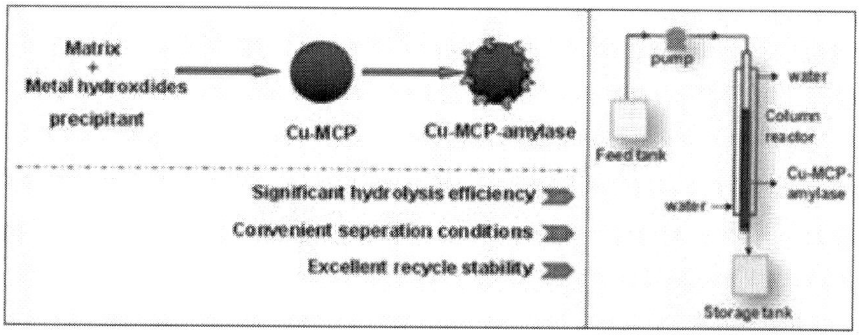

Fig.1 Amylase immobilization and application for starch hydrolysis

a) Preparation of Cu-MCP-amylase. b) Application of Cu-MCP-amylase for starch hydrolysis

(582ec) Engineering a Proteolytically Stable and Biologically Active Fibronectin

Karuri, N. W., Illinois Institute of Technology
Zhang, C., Illinois Institute of Technology
Desai, R.,

Fibronectin is an essential an essential component of the extracellular matrix. It is proteolytically degraded in cellular responses such as migration and proliferation. However, excessive proteolysis of fibronectin is associated with tumor invasion and chronic wound healing. A study is presented on stabilizing fibronectin by conjugating it to polyethylene glycol through lysine residues. Proteolytic stability of the conjugates was determined by adding protease and quantifying the number and amount of proteolytic fragments with time. The activity of fibronectin conjugated to polyethylene glycol was measured by its ability to engage in intermolecular interactions with cells, fibronectin fragments and other extracellular matrix components. This study demonstrates that conjugating polyethylene glycol to fibronectin stabilizes it against proteolytic degradation. The study also shows that proteolytic stability and activity of the fibronectin conjugates is dependent on the length of polyethylene glycol and the extent of polyethylene glycol conjugation. The study provides new insight into potential mechanisms of bioengineering fibronectin for therapeutic uses.

(582bd) Rational Engineering of the Dxp Pathway for Overproduction of Isoprenoids in Bacteria

Zhou, K., Singapore MIT Alliance
Zou, R., Singapore MIT Alliance
Zhang, C., Singapore MIT Alliance
Stephanopoulos, G., Massachusetts Institute of Technology
Too, H. P., National University of Singapore

Isoprenoids are a family of natural products (>55,000 examples identified), including many high-value compounds that can be used as pharmaceuticals, nutraceuticals and fragrances. Isoprenoids are naturally synthesized via two pathways, mevalonate pathway and deoxyxylulose phosphate (DXP) pathway. We are interested in exploiting the DXP pathway for overproduction of isoprenoids, because this pathway is more redox balanced and more efficient in terms of pathway yield. To rationally engineer the DXP pathway for isoprenoid overproduction in *Escherichia coli* and *Bacillus subtilis*, we have in this study developed a series of methods, which can reliably quantify gene transcription, enzyme solubility and metabolic intermediates in the pathway. For the first time, we have found that one phosphorylated intermediate of the DXP pathway (MEC) could be actively effluxed by the cell, diverting a large portion of carbon flux away from microbial isoprenoid synthesis. The methods developed and the rate-limiting step identified in this study, should be very useful in future practice of using the DXP pathway for high level production of valuable isoprenoids in bacteria.

(582dn) A Potent Multi-Cellulase Conjugate System Using a DNA Scaffold to Enhance Cellulose Degradation

Kamiya, N., Kyushu University
Mori, Y.,
Tanaka, T.,
Ichinose, H.,

Cellulosic biomass consisting of plant matter is the most abundant carbon-neutral renewable resource for the production of bioenergy and biomaterials. Research into the use of biorefineries, which convert cellulosic biomass to valuable fuels and products, has attracted much attention. To enhance degradation of plant cell wall cellulose, several anaerobic microorganisms produce a multi-cellulase complex called a cellulosome. In this study, to mimic a natural cellulosome, we have designed a novel artificial cellulosome as a form of multi-cellulase conjugate assembled onto double-stranded DNA. We anticipated the preparation of this new conjugate in a controllable manner because the length and the amount of the cross-linking site on a DNA scaffold can be tuned by enzymatic protein modification techniques. We chose endoglucanase (EG) as the conjugated cellulase and multiple EGs were site-specifically labeled on DNA scaffolds using a protein cross-linking enzyme, microbial transglutaminase. The catalytic performance of the resultant DNA-$(EG)_n$ conjugates was evaluated by using a crystalline cellulosic substrate (Avicel). We found that the multi-EG aligning strategy worked to enhance enzymatic saccharification compared to the free EGs without the DNA scaffold. The present results may lead to further development of DNA-based designer cellulosomes.

(582b) Computational Design of New Peptide Inhibitors for Amyloid Beta (A) Aggregation in Alzheimer's Disease: Application of a Novel Methodology

Gur, M., School of Medicine, University of Pittsburgh
Eskici, G.,

Alzheimer's disease is the most common form of dementia. It is a neurodegenerative and incurable disease that is associated with the tight packing of amyloid fibrils. This packing is facilitated by the compatibility of the ridges and grooves on the amyloid surface. The GxMxG motif is the major factor creating the compatibility between two amyloid surfaces, making it an important target for the design of amyloid aggregation inhibitors. In this study, a peptide, experimentally proven to bind Aβ40 fibrils at the GxMxG motif, was mutated by a novel methodology that systematically replaces amino acids with residues that share similar chemical characteristics and subsequently assesses the energetic favorability of these mutations by docking. Successive mutations are combined and reassessed via docking to a desired level of refinement. This methodology is both fast and efficient in providing potential inhibitors. Its efficiency lies in the fact that it does not perform all possible combinations of mutations, therefore decreasing the computational time drastically. The binding free energies of the experimentally studied reference peptide and its three top scoring derivatives were evaluated as a final assessment/valuation. The potential of mean forces (PMFs) were calculated by applying the Jarzynski`s equality to results of steered molecular dynamics simulations. For all of the top scoring derivatives, the PMFs showed higher binding free energies than the reference peptide substantiating the usage of the introduced methodology to drug design.

(582dk) Engineering 3-Dimensional Protein Scaffolds for Biocatalysts Assembly

Chen, Q., University of Delaware
Sun, Q., University of Delaware
Wang, S. W., University of California
Chen, W., University of Delaware

In nature, enzymes are often assembled into complex structures that are more effective than individual enzymes in converting raw materials into useful products. Many biologically-derived scaffolds have been used for single and multi-enzyme assembly and have shown increased overall reaction rates in one- or two-dimensional configurations. Analogously, we hypothesize that the use of three-dimensional scaffolds could further enhance the performance of biocatalysts. The E2 core of the pyruvate dehydrogenase complex from *Bacillus stearothermophilus* is a genetically-modifiable 60-mer protein nanocage that is thermally stable at temperatures as high as 70°C. In this talk, we will discuss different approaches in pairing the E2 nanocage with a thermophilic endoglucanase, CelA from *Clostridium thermocellum*, resulting an enzyme assembly that benefit from both the synergy gained from clustering and stability at more enzymatically favorable high reaction temperatures. This E2-based approach offers a flexible platform upon which multi-enzyme cascade reactions can be assembled and studied at the nanoscale.

(582as) Multi-Level Optimization Framework for the Identification of Multi-Objective Design Principles in Metabolic Networks

Vaskan, P., University Rovira i Virgilii
Guill n-Gos lbez, G., Universitat Polit cnica de Catalunya
Jimenez, L., University Rovira i Virgili

While in optimization scientists establish the objective and search the way to accomplish it, in natural systems the emergence of new designs results from the evolution driven by natural selection. While one can argue that natural systems are optimized by natural selection, it is unclear which is the objective function and constraints (i.e., optimization problem) being solved. Determining these design principles is a central topic in systems biology.

The identification of design principles can be posed as a reverse optimization problem, that is, we know the actual solution (the actual system) and we aim to determine which criteria (if any) are being optimized. Previous attempts to solve this problem have relied on single-objective models that assume that there is a single universal objective function that drives an organism's evolution. In a recent work, however, Sauer and co-workers have shown that there might be more than one fitness function driving the metabolic machinery (Schuetz et al., 2012), and that the optimization of a given criterion might vary according to the environmental conditions.

Bearing this in mind, in this work we propose a new optimization tool for indentifying the set of objective functions that describes in the best manner the operation of metabolic networks. Our approach is inspired by the bi-level linear programming framework proposed by Burgard and Maranas (2002), which is adequately extended in order to deal with several objective functions simultaneously. Particularly, given a set of experimental observations, the goal of the analysis is to infer the form of the multi-objective optimization problem that shapes the adaption of microorganisms to the environment. To this end, we have developed a multi-level linear programming formulation based on flux balance analysis that includes an outer problem and a set of inner models. The outer problem seeks to optimize the least square difference between the experimental observations and the optimal solution predicted *in silico*, while the inner problems, which are defined for every experimental condition, optimize a linear combination of objectives subject to the weights imposed by the outer problem. The overall multi-level model seeks those combinations of weights that make each experimental condition optimal. A nonzero weight implies that the objective is biologically meaningful, while a zero weight implies the converse. Binary variables are added in order to limit the number of plausible fitness functions, making it possible to evaluate the effect of considering an increasing number of objectives.

We reformulate this multi-level model into a single mixed-integer linear program (MILP) by replacing the inner problems by the corresponding Karush–Kuhn–Tucker conditions. This requires definition of auxiliary binary variables for each inequality present in the inner problems. Such binaries take a value of one if the constraint is active and zero otherwise. The reformulated MILP, which can be solved by standard branch and cut methods, provides, for each experimental condition being analyzed, the set of meaningful objective functions and associated weights. Hence, our approach allows testing a large number of objectives functions from which the best ones according to the calculated error are finally identified.

The capabilities of our methodology are illustrated through its application to the central metabolism of Escherichia coli using flux data derived from isotopomer analysis. Particularly, we consider as surrogates for cellular fitness functions a set of reaction fluxes involving production/consumption of ATP and redox potential, as well as biomass growth rate, to predict 13C-determined in vivo fluxes. One of the main advantages of our approach is that it allows studying several objectives under various conditions, as oppose to the approach by Burgard and Maranas (2002), which was developed under the assumption that there is a single universal objective function that is valid under all possible environmental conditions. Numerical results show that experimental observations can be well explained by a reduced number of objectives (i.e., 3), and that there are different combinations of objectives leading to similar errors. Furthermore, biomass formation is always among the objectives found, which indicates its importance as a fitness function in evolution.

References

- Schuetz, R., Zamboni, N., Zampieri, M., Heinemann, M., Sauer, U. (2012). Multidimensional optimality of microbial metabolism, Science, 601:336.
- Burgard, C. Maranas. (2002). Optimization-Based Framework for Inferring and Testing Hypothesized Metabolic Objective Functions. Biotechology and bioengineering, 82(6): 670-677.

Food, Pharmaceutical & Bioengineering Division

(582ce) Determination of the Metabolic Pathways for Linear Hydrocarbon Biosynthesis in the Green Algae Botryococcus Braunii Strain UTEX572

Brown, M. T., Iowa State University

The green algae *Botryococcus braunii* has drawn considerable interest as a potential source of drop-in biofuels due to its unique ability to accumulate high levels of hydrocarbons. Several biochemical mechanisms have been proposed for non-isoprenoid hydrocarbon biosynthesis in this and a number of other organisms, most notably the head-to-head condensation and elongation-decarboxylation pathways from fatty acid precursors, but definitive characterization of these and other possible mechanisms have largely remained elusive. In this study, structural characterization of the monoene, diene, and triene hydrocarbon classes and their fatty acid precursors in the UTEX 572 strain of *B. braunii*, as well as NMR and GC-MS analyses of stable isotope fed cultures, point to a likely elongation-decarboxylation biosynthetic pathway that is responsible for hydrocarbon biosynthesis. Additionally, a comparative transcriptomic analysis of hyper and hypo-hydrocarbon accumulating conditions in *B. braunii* has provided a potential genetic basis for the synthesis of hydrocarbons from fatty acids in this algae.

(582cv) Functional Role of the Tryptophan Residue On the Transmembrane Domain of the IRE1 Protein

Cho, H., Michigan State University
Stanzione, F., Colorado School of Mines
Sum, A. K., Colorado School of Mines
Chan, C., Michigan State University

The inositol-requiring 1α protein (IRE1α) is a type I transmembrane protein important in altering gene expression in response to endoplasmic reticulum (ER) stress. IRE1α contains an N-terminal luminal domain that is an ER stress sensor and a C-terminal cytosolic domain that carries protein Ser/Thr kinase and endoribonuclease activities. Under ER stress conditions, the Bip protein dissociates from the luminal domain of IRE1α, promoting self-association of the IRE1α protein. The face-to-face dimerization of the kinase domain facilitates trans-autophosphorylation, subsequently activating the RNase domain. The active form of IRE1α participates in reducing ER stress by upregulating genes involved in enhancing ER protein-folding capacity and degrading unfolded or misfolded proteins. We recently found that the transmembrane (TM) domain of the IRE1α protein dimerizes/oligomerizes in a membrane-mimic environment, and maintains its helical structure. However, functional information on the TM domain of the IRE1α protein is currently unavailable. We performed molecular dynamics (MD) simulations using two model TM peptides inserted in a POPC bilayer. The simulation results revealed that four potential residues (S450, L453, L454, and W457) are located on the dimerization interface of the TM model peptide. TOXCAT and SDS-PAGE analyses showed that alanine mutation of W457 residue dramatically reduces the oligomerization of the TM domain. In addition, mutational studies with MD simulations suggest that W457 serves as a driving force for the TM dimerization process and both the aromatic interaction and hydrogen bond formed by the W457 residue are crucial for stabilizing the dimerization of the TM domain. We further confirmed that the W457A mutation reduces both the kinase and RNase activities of the IRE1α protein in cells. Through the combination of computational and experimental studies, we gained insight into how the TM domain impacts the functions of IRE1α.

(582x) Conversion of Racemic Profens to (S)-Profens

Mortazavi, S., University of Texas at El Paso
Salvador, J.,

Conversion of Racemic Profens to (*S*)-Profens

Saideh Mortazavi, Chemistry Doctoral Student

Abstract

Some non-steroidal anti-inflammatory drugs (NSAIDs) such as profens (ibuprofen, naproxen, fenoprofen, ketoprofen, or flurbiprofen) can exist and are sold worldwide as a mixture of mirror image compounds called enantiomers (a racemic mixture) where one mirror image has the desired pain, fever, and inflammation reducing effects (referred to as (*S*)-enantiomer) and the other undesired mirror image (referred to as (*R*)-enantiomer) does not or may even have harmful side effects. Thus the efficacy of profen drugs can be improved and side effects reduced by isolating and administering only the desired (*S*)-profen as opposed to consuming its racemic mixture, (*R)* and *(S*)-profens. In addition, the manufacture of pure single enantiomer pharmaceuticals is a multibillion dollar a year industry and is now required by the Food and Drug Administration (FDA).

Whereas previous methods wasted the undesired profen enantiomer, (*R*)-profens, we propose to convert several examples to the desired mirror image drug, (*S*)-profens, using a combination of relatively inexpensive chemical and enzymatic methods.

The *Candida rugosa* lipase catalyzed Dynamic Kinetic Resolution of racemic ibuprofen methyl ester was optimized to produce (*S*)-ibuprofen in 3 days. The best concentration of various buffers for these reactions was found to be 0.5 M at optimal pH.Â The commercial lipase mixture was found to be acidic. Dimethylformamide was determined to be a better co-solvent to maintain the reaction pH than dimethylsulfoxide, with evidence of the latter functioning as an oxidizing agent. Lower concentrations of ibuprofen methyl ester and higher stirring rates both led to faster conversions. The minimal amount of lipase needed was determined to be 20 mg/ml buffer. Reaction of (*R*)-ibuprofen methyl ester under the optimized conditions excluding the lipase led to no racemization indicating that conversion of (*R*)-ibuprofen methyl ester to (*S*)-ibuprofen is not catalyzed by the base.

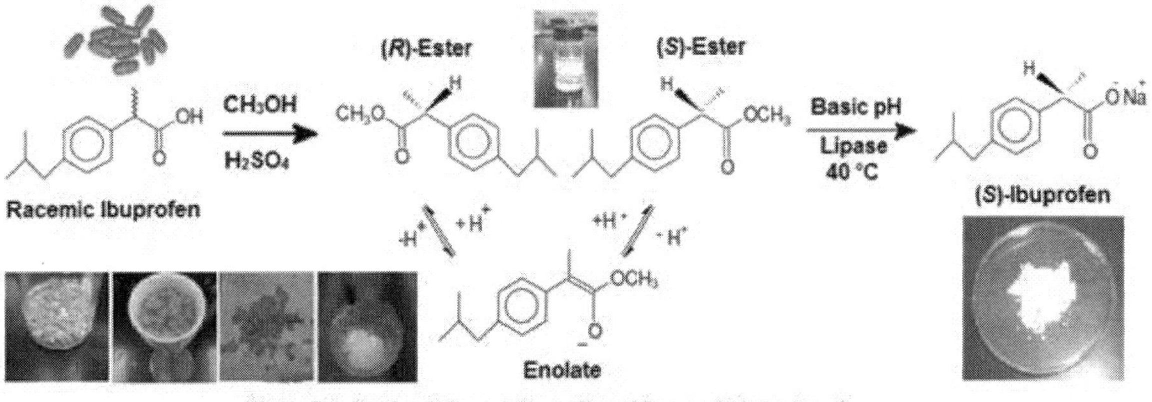

Fischer Esterification of Racemic Ibuprofen and Proposed Mechanism of Dynamic Kinetic Resolution of Racemic Ibuprofen Methyl Ester to (*S*)-Ibuprofen

(582cj) In Silico Metabolic Model of Gordonia Alkanivorans to Study Its Desulfurization Characteristics

Aggarwal, S., National University of Singapore
Karimi, I. A., National University of Singapore
Ivan, G.,

In silico Metabolic Model of *Gordonia alkanivorans* to Study its Desulfurization Characteristics

Shilpi Aggarwal[1], I A Karimi[1], Gregorius Reinaldi Ivan[1]

[1]Dept of Chemical & Biomolecular Engineering, National University of Singapore, Singapore 117576

In absence of any competitive and economical renewable fuel, fossil fuels continue to be the major source of energy worldwide. However, fossil fuels contain various heteroatom based compounds that on combustion lead to the release of various pollutants such as oxides of carbon, nitrogen, and sulfur. Of these, oxides of sulfur (SO_x) have attracted major attention owing to the harmful effects that they exert on both environment and human health. As such, the governments have laid stringent regulations to limit the sulfur content in fuels [1] thereby, making desulfurization an important step in their pre-processing. However, the prevalent method for desulfurization, namely hydrodesulfurization is energy-intensive, expensive, and requires extreme conditions of temperature and pressure to desulfurize certain recalcitrant sulfur compounds such as benzothiophene (BT), dibenzothiophene (DBT), and their derivatives present in fossil fuels. Clearly there is a need to improve the existing desulfurization methods and develop more efficient and economical ones. The researchers have identified biodesulfurization as a potential alternative.

Biodesulfurization is a process in which either microbial whole cells or enzymes catalyze the removal of sulfur atom from the compounds present in fossil fuels. Compared to hydrodesulfurization, biodesulfurization is less energy intensive, more specific in action, and is relatively economical [1]. Several bacterial strains belonging to *Rhodococcus*, *Gordonia*, *Mycobacterium*, etc. have been isolated for their ability to desulfurize BT, DBT, and their derivatives. Of these *G. alkanivorans* is of high interest as it exhibits higher desulfurization activity for several recalcitrant compounds [2]. However, the desulfurization rates obtained with the wild type *G. alkanivorans* are too low for any commercial application [3]. Even the genetically engineered strains are incapable of giving the desired level of activity. This is mainly because most genetic manipulations normally target the elements of desulfurization pathway, while the other host functions affecting the desulfurization activity in *G. alkanivorans* are unknown. A cellular phenotype such as desulfurization activity depends on the complex interactions among the various metabolic pathways and biochemical reactions [4] occurring within the cell. Therefore, it is critical to study desulfurization by *G. alkanivorans* as an integral part of its metabolism in a holistic manner.

In this work, we reconstruct a genome scale metabolic model of *G. alkanivorans* for a holistic study of its metabolism. The model consists of main metabolic pathways such as central, amino acids, nucleotide, nitrogen, and sulfur metabolism. It can aid in understanding the metabolic architecture of *G. alkanivorans*, and its host functions related to desulfurization. The model depicts the dependence of the desulfurization pathway on the various intracellular activities. We validate the model using the experimental data available in literature. The model is found to predict the biomass growth closely using the experimental DBT uptake rates [5] as inputs. Also, it shows that *G. alkanivorans* prefers BT over DBT as reported in literature [2]. The analysis of our model suggests that ethanol is the best carbon source for obtaining higher desulfurization activity and growth rates with *G. alkanivorans*. In addition, *in silico* experiments using our model shows the effect of various amino acids and vitamins on the desulfurization activity of *G. alkanivorans*. The flux variability and flux sum analyses with the model show the importance of various pathways and metabolites in determining the extent of desulfurization and growth of *G. alkanivorans*. Finally, the model provides several useful insights into the interactions between the desulfurization activity and the other parts of the metabolism in *G. alkanivorans*.

References

1. Soleimani, M., A. Bassi, and A. Margaritis, Biodesulfurization of refractory organic sulfur compounds in fossil fuels. Biotechnology Advances, 2007. 25(6): p. 570-596.

2. Alves, L., et al., Desulfurization of dibenzothiophene, benzothiophene, and other thiophene analogs by a newly isolated bacterium, Gordonia alkanivorans strain 1B. Applied Biochemistry and Biotechnology - Part A Enzyme Engineering and Biotechnology, 2005. 120(3): p. 199-208.

3. Kilbane Ii, J.J., Microbial biocatalyst developments to upgrade fossil fuels. Current Opinion in Biotechnology, 2006. 17(3): p. 305-314.

4. Raman, K. and N. Chandra, Flux balance analysis of biological systems: Applications and challenges. Briefings in Bioinformatics, 2009. 10(4): p. 435-449.

5. Rhee, S.K., et al., Desulfurization of dibenzothiophene and diesel oils by a newly isolated Gordona strain, CYKS1. Applied and Environmental Microbiology, 1998. 64(6): p. 2327-2331.

(582dv) Production of Flavor Esters By Candida Antarctica Lipase Type B Immobilized On Magnetic Nanoparticles

Santos, K. P.,
Farias, M. Y.,
Souza, M. C. M. D., Universidade Federal do CearÃ¡
Freire, R.,
Barreto, A. C. H.,
Fechine, P. B.,
Gon alves, L. R. B., Universidade Federal do CearÃ¡

The esters of short-chain fatty acids are an important group of flavor and fragrance compounds widely used in food, beverage, cosmetic and pharmaceutical industries. These aromas are typically extracted from natural sources, but some problems arise with respect to availability due to environmental problems. Thus natural catalysts are required and products obtained from the enzyme mediated reactions are considered natural. Lipases produced by *Candida sp*. are well-established enzymes for biocatalysis purposes, especially *Candida antarctica* lipase type B (CALB). Therefore, in this work, *Candida antarctica* lipase B (CALB) immobilized onto iron magnetic nanoparticles was evaluated as biocatalyst for the synthesis of flavor esters. Methyl and ethyl butyrate were synthesized by esterification of butyric acid with methanol and ethanol in different solvents. The nanoparticles were produced by the co-precipitation method. Modifications were carried out on the nanoparticles' surfaces with aminopropyltriethoxysilane and glutaraldehyde. The immobilization was performed in the presence of 100mM bicarbonate buffer, pH 10, at 25 °C. The process parameters (solvents, temperature, substrate concentrations, molar ratio, stirring speed and reaction time) were studied. The optimum conditions for both esters were heptano as solvent, 25 °C, 0,5 mol/L and 0,4 mol/L of substrates, molar ratio of 1:1, amount of biocatalyst: 10 mg, 150 rpm and 8 h of reaction. Maximum conversion of methyl butyrate (93,9 \%) and ethyl butyrate (96,8 \%) were achieved after 8 h of reaction at 25 °C for CALB immobilized onto magnetic nanoparticles. The synthesis of flavor esters was also evaluated by using novozyme 435, a commercial biocatalyst, for comparison. Operational stability (consecutive cycles of synthesis of esters) were performed and CALB immobilized onto magnetic nanoparticles retained 74 % of its initial activity after 12 cycles, while novozym 435 retained 81 %.

(582dw) Enzyme Active Site Design and Combinatorial Mutagenesis for Catalytic Modification in Bacterial Beta-Lactamase

Shi, Z.,

Beta-lactamase is naturally produced in some bacteria to hydrolyze beta-lactam compounds. As a result, these bacteria acquire antibiotic-resistance and become major public health hazards in hospitals and other medical facilities. Extensive research on this enzyme family and related proteins has led to much improvement in defense against bacterial infection. In this study, class C beta-lactamase from *Enterobacter cloacae* P99 is used as a template for enzyme structure and activity characterization. Based on crystallographic analysis of the P99 enzyme, a group of amino acid residues were identified in the substrate binding pocket (1). They were implicated in contacting the beta-lactam core structure and side groups (2, 3). To further examine the roles of these positions in catalytic turnover, substitutions which permit viable enzyme functions were generated in silico by applying computational modeling methods. To assay biological activities of these customized mutant enzymes, a variant library was created by combinatorial mutagenesis to incorporate the predicted substitutions into target residues in the wild-type template. To do so, DNA sequences encoding the wild-type and mutant enzymes were generated by synthetic gene assembly. These gene sequences were then cloned into expression vector to allow efficient protein production in E. coli BL21 derived host cells. Enzymes were purified by metal affinity column chromatography utilizing the C-terminal 6X histidine tag. Enzymatic activities were quantified for a variety of beta-lactam substrates by in vivo growth assay to determine minimal inhibitory concentrations and by in vitro kinetic analysis (3). All mutants were found active towards hydrolyzing multiple beta-lactam antibiotics in the cephem and penam families including cephalothin and ampicillin. Some mutant enzymes exhibited activities comparable to or higher than the wild-type enzyme, whereas activities decreased in many others. Interestingly, a subset of mutants were found to hydrolyze extended-spectrum cephalosporins more efficiently than the wild-type enzyme. Among them, mutant enzymes containing substitutions in Tyr221 in the omega loop were more active in hydrolyzing cefotaxime and ceftazidime, while their activities towards cephalothin were much reduced. These observations are consistent with previous reports for this family of enzymes, suggesting that conformational changes in the substrate binding pocket play important role in beta-lactam specificity (2, 4). Further characterization of the mutants is underway to quantify their kinetic activities towards cephem and penam antibiotics in relation to structural modification in the active site. Moreover, serine beta-lactamases are known to evolve from penicillin-binding proteins involved in bacterial cell wall synthesis. Some PBPs display transpeptidase activities and share very similar scaffolds as the class C beta-lactamases (5). Thus, the mutants are also being assayed for DD-peptidase activities to assess whether by modifying the active site, these beta-lactamases may acquire peptidase functionality. In the long term, these studies have implications in protein and inhibitor drug design such as with antimicrobial applications.

REFERENCES

1. Lobkovsky, E., Moews, P.C., Liu, H., Zhao, H., Frere, J.-M. and Knox, J.R. (1993) Evolution of an enzyme activity: crystallographic structure at 2-A resolution of cephalosporinase from the ampC gene of Enterobacter cloacae P99 and comparison with a class A penicillinase. Proceedings of National Academy of Sciences 90, 11257–11261.

2. Zhang, Z., Yu, Y., Musser, J.M. and Palzkill, T. (2001) Amino acid sequence determinants of extended spectrum cephalosporin hydrolysis by the class C P99 beta-lactamase. Journal of Biological Chemistry 276, 46568-46574.

3. Goldberg, S. Iannuccilli, W., Nguyen, T., Ju, J. and Cornish, V.W. (2003) Identification of residues critical for catalysis in a class C beta-lactamase. Protein Science 12, 1633-1645.

4. Carter, B.T., Lin, H., Goldberg, S.D., Althoff, E.A., Raushel, J. and Cornish, V.W. (2005) Investigation of the mechanism of resistance to third-generation cephalosporins by class C beta-lactamases by using chemical complementation. ChemBioChem 6, 2055-2067.

5. Massova, I. and Mobashery, S. (1998) Kinship and diversification of bacterial penicillin-binding proteins and beta-lactamases. Antimicrobial Agents and Chemotherapy 42, 1-17.

(582av) A Probabilistic Model to Describe the Dual Phenomena of Biochemical Pathway Damage and Biochemical Pathway Repair

Sadhukhan, P., National Chemical Laboratory
Banerji, A.,

Biochemical pathways emerge from a series of Brownian collisions between various types of biological macromolecules within separate cellular compartments and in highly viscous cytosol. Functioning of biochemical networks suggests that such serendipitous collisions, as a whole, result into a perfect synchronous order. Nonetheless, owing to the very nature of Brownian collisions, a small yet non-trivial probability can always be associated with the events when such synchronizations fail to emerge consistently; which account for a damage of a biochemical pathway. The repair mechanism of the system then attempts to minimize the damage, in the pursuit to bring restore the appropriate level of synchronization between rectant concentrations. Present work presents a predictive probabilistic model that describes the various facets of this complicated and coupled process(damaging and repairing). By describing the cytosolic reality of Brownian collisions with Chapman-Kolmogorov equations, the model presents analytical answers to the questions, with what probability a fragment of any pathway may suffer damage within an arbitrary interval of time? and with what probability the damage to a pathway can be repaired within any arbitrary interval of time?

(582du) Evaluation of the Influence of pH and Temperature On Enzymatic Activity of Alfa-Acetolatato Decarboxylase

Lemos, C. M., Universidade Federal do CearÃ¡
Fran a, W. L. Sr., Universidade Federal do CearÃ¡
Souza, T. C.,
Gon alves, L. R. B., Universidade Federal do CearÃ¡

During beer production process, the main fermentation products are ethanol and carbon dioxide, but also other products are obtained at a lower amount. An important carbonyl compound produced in the fermentation and present in the beer is the diacetyl, since it causes an undesirable organoleptic characteristic to the product, which is the flavor of butter. Efficient control of diacetyl concentration can be achieved with strategies for the prevention of the formation of the precursor (α-acetolactate) or by increasing the rate of chemical decarboxylation of it. The use of the enzyme a - acetolactate decarboxylase has been proposed to overcome the formation of diacetyl converting a - acetolactate directly into acetoin. The use of a - acetolactate decarboxylase makes it possible to shorten primary fermentation of beer and even decrease the maturation time with respect to the amounts of diacetyl. The objective of this work is to analyze the effects of pH and temperature on the enzymatic activity of a - acetolactate decarboxylase from *Bacillus subtilis*. The reaction medium used for evaluation of thermal and pH stability was composed of MES buffer at a concentration of 0.05 mM. The temperatures tested were 10 °C, 20 °C, 30 °C, 40 °C and 50 °C and the pHs were 4.5, 5.5, 6.5, 7.5 and 8.5.The enzyme (soluble or immobilized) was retained at each temperature and pH condition for 60 minutes. The temperature and pH values were chosen based on studies of the enzyme a - acetolactate decarboxylase derived from other microorganisms.

(582cn) Optimization of Sugar Yield and Input Energy By Step-Wise Reduction of Agitate Rate During Lignocellulose Hydrolysis

Guo, H.,
Su, R., Tianjin University
Qi, W., Tianjin University
He, Z., Tianjin University

Reaching the high sugar concentration and inputting the low energy during enzymatic hydrolysis of lignocellulosic biomass is crucial for the production of bioethanol and other chemicals. To achieve this goal, the initial high solid loadings and total low mixing energy consume are required in the hydrolysis step. The less mass transfer resistance resulted from high mixing power would be beneficial to obtain high sugar concentration in the short time [1], but the shear deactivation of cellulase seems to reduce the hydrolysis efficiency, especially at the extra high agitate rate [2]. Therefore, it is a key task to find the optimal agitate rate at different hydrolysis time scale at a given solid loadings for producing the high sugar concentration in the low cost.

In this paper, we investigated the effect of agitate rates on the hydrolysis of lignocellulose and found a new mixing strategy to optimize the hydrolysis process. Firstly, the shear deactivation of cellulase was studied under different agitate conditions. Then the time-course of glucose concentration was examined under different agitate rates at various solid loadings to find the optimal agitate condition for the hydrolysis process. Moreover, for the whole hydrolysis process, the agitate condition was suggested to be a step-wise reduction according to the changes of glucose concentration. The main ideas and conclusions are summarized as follows.

(1) A significant cellulase deactivation for pure enzyme solution occurred after 24 h at the agitate rate of 150RPM, leading to the decreases in glucose concentration in the latter hydrolysis step. The deactivation extents of a total cellulase mixture were related with the incubating time and mixing intensity [3]. At low solid loadings, a little increased glucose yield were found after 24h hydrolysis under agitate rate of 150 rpm than that under static condition; whereas a significant increase on the glucose yield could be seen at high solid loadings (10%) after the agitate rate increased from 0 to 150 rpm.

(2) The optimal agitate rate at the given solid loading was examined to increase hydrolysis efficiency. At low solid loadings, there was only a little increase in the glucose concentration at 150 rpm than that at the static state after 72h hydrolysis, whereas a significant decreased glucose concentration was found at 200 rpm due to the serious inactivation of enzymes. At high solid loadings, there was a greater increase in glucose yields under agitate condition (150 rpm and 100 rpm) than those without mechanical agitation. The results

revealed that extra high agitate rate have a serious negative effect on the cellulose hydrolysis due to the cellulase deactivation. However, the appropriate agitate rate would lead to higher glucose concentration, especially at high solid loadings.

(3) The hydrolysis of corncob were further optimized by using the step-wise reduction of the agitate rate. Adequate mixing is required to ensure sufficient contact between the substrate and enzymes and to promote heat and mass transfer within the reaction vessel. But it was not necessary to agitate continuously at the high rate for achieve the high sugar concentration [4]. The optimized mixing form could not only benefit the enzymatic hydrolysis process by reducing the energy consumption and limiting the cellulase deactivation, but also provide a new mixing form for the hydrolysis of lignocellulose at high solid loadings.

The authors acknowledge the financial supports received from the National Natural Science Foundation of China (No. 21276192 and 20976125), Open Funding Project of the State Key Laboratory of Bioreactor Engineering, and the Ministry of Education (Grant Nos. NCET-11-0372, 20110032130004 and B06006).

References

1. Palmqvist, B., M. Wiman, and G. LidÂ¨Â¦n, *Effect of mixing on enzymatic hydrolysis of steam-pretreated spruce: a quantitative analysis of conversion and power consumption.* Biotechnology for Biofuels, 2011. **4**(1): p. 10.

2. Gunjikar, T.P., S.B. Sawant, and J.B. Joshi, *Shear deactivation of cellulase, exoglucanase, endoglucanase, and β-glucosidase in a mechanically agitated reactor.* Biotechnology Progress, 2001. **17**(6): p. 1166-1168.

3. Ye, Z., K.M. Hatfield, and R. Eric Berson, *Deactivation of individual cellulase components.* Bioresource Technology, 2012. **106**: p. 133-137.

4. Ingesson, H., et al., *The effect of shaking regime on the rate and extent of enzymatic hydrolysis of cellulose.* Journal of Biotechnology, 2001. **88**(2): p. 177-182.

(582cf) A Flow-Cytometry Method for Optimizing Transformation Conditions in Bacteria

Woolston, B., The Pennsylvania State University
Rismani-Yazdi, H.,
Stephanopoulos, G., MIT

The application of metabolic engineering to non-traditional host organisms is often limited by the lack of transformation protocols for these organisms. Developing such protocols requires optimization of the method used for DNA delivery. However, there are few well-developed assays specifically for analyzing the efficiency of DNA uptake, rather than the transformation process as a whole. The objective of this study was to develop an assay to track and optimize the delivery of plasmid DNA to bacteria cells. Plasmid DNA covalently labeled with the Cy5 fluorophore was used for electroporation and sonoporation of competent cells, and the proportion of cells that had taken up plasmid determined by flow cytometry (FC). To address the issue of cell viability following transformation, cells were treated with propidium iodide (PI) just prior to FC. Fluorescence microscopy was used to confirm the presence of labeled plasmid within the cytoplasm of the cells. The assay was first validated in *E. coli*, then used to establish optimal transformation conditions for the gram-positive anaerobe *Moorella thermoacetica.* The optimized protocol allowed DNA delivery to up to 3% of the cells viable after electroporation. The assay is applicable to a variety of organisms, and will thus be useful in expanding the range of hosts amenable to metabolic engineering.

(582ap) Use of the Osteopontin Promoter for Expression-Targeted Cancer Gene Therapy

Chen, X., Tulane University
Godbey, W. T., Tulane University

Compared to chemotherapy, expression-targeted cancer gene therapy is a promising treatment that could deliver therapeutic levels of tumor cell death without bystander effects. To aid in reducing such side effects, it is critical to find DNA regulatory elements that are associated with strong specificity and high efficacy in cancer cells.

Osteopontin (OPN) plays an important role in cancer invasion and metastasis and is over-expressed in various types of malignant cancers. The regulatory elements of the *opn* gene are therefore attractive candidates for expression-targeted gene therapy. The ^{P}opn promoter (^{P}opn) was extracted from the human genome via the polymerase chain reaction, and deletion analysis was used to assess important control regions within the promoter. The engineered plasmids were used to demonstrate that ^{P}opn yields excellent selectivity for the cancer cells tested. The maximum ^{P}opn driven green fluorescent protein expression intensity in the cancer cells tested was as strong as in transfections where the cytomegalovirus promoter (^{P}cmv) was used: the average value was 102.69% of the average value obtained using ^{P}cmv. However, the maximum intensity of ^{P}opn-driven expression in normal cells was only 11.57% of the average value obtained using ^{P}cmv to drive transcription. This combination of strength and selectivity suggests that ^{P}opn has excellent therapeutic potential in expression-targeted gene therapy. For instance, the promoter could be used to safely drive the expression of pro-apoptotic genes in cancer cells.

The correlation between endogenous *opn* mRNA levels and ^{P}opn-driven transgene expression was also investigated. As has been noted in the literature with several other promoters used for expression-targeting, there was no apparent correlation between endogenous *opn* mRNA levels and ^{P}opn-driven transgene expression. However, when different isoforms of the OPN protein were taken into account, an obvious inverse relation was obviated with OPN-A (Pearson correlation coefficient $r=-1$, $p=0.0003$, $n=4$). A possible mechanism to explain the inverse relation involves mRNA splicing in generating the various endogenous isoforms of osteopontin, where OPN-A levels may be decreased for the cell to produce the other two isoforms (OPN-B and OPN-C). When transgenes under the control of a single exon are delivered into cells, the splicing of *opn* mRNA is not a factor for transgene expression. The data observed herein suggest that, in (cancer) cells strongly utilizing ^{P}opn, OPN-A levels will be reduced but ^{P}opn-driven transgene expression will be increased.

Single-nucleotide polymorphisms (SNPs) within ^{P}opn were also examined. SNPs are often overlooked in expression-targeted gene delivery, but may hold an important key to customized medicine. We examined six different SNP combinations and found that different cell types respond differently to distinct SNP constructs. The implications of this are that, for expression-targeted gene therapy to be successful in the clinic, plasmids should be tailored for individual patients. A one-size-fits-all approach should no longer be expected to yield a successful genetic treatment for cancer.

(582be) A Systems Approach for Metabolic Engineering of Tryptophan in Escherichia Coli

Immanuel, S. C.,
Gupta, A., National Chemical Laboratory
Raghunathan, A., National Chemical Laboratory

Biosynthesis of L-tryptophan is strictly controlled at several steps in *Escherichia coli*. The economic importance of this amino acid and its role as a precursor in many downstream pathways for secondary metabolites, makes it imperative to metabolically engineer a host strain for its over production. Prospective design analysis using *in silico* metabolic network models is essential to the success of identifying target genes for optimization of desired metabolite production. More over, combination of prospective design, high throughput data generation and evolutionary engineering hold great promise. This talk focuses on the development and analysis of a tryptophan over producing strain of *E. coli* using a gene deletion strategy to channel flux through the aromatic amino acid production branch to tryptophan. LC-MS analysis showed about ten-fold increase in tryptophan biosynthesis in a multiple knockout ΔTrpRΔpgi *E. coli* strain. LC-MS has been further used to characterize the intracellular profiles of amino acid production to study non-intuitive coupling between biosynthesis of amino acids. Further, in order to make the strain function efficiently in a hemi-cellulosic environment, directed evolution under suitable selection pressures activated a cryptic operon for native biosynthesis of a putative β xylosidase in the *E. coli* strain.

(582cx) Introducing Nitroreductase Functionality in Ene Reductases

Park, J. T., Georgia Institute of Technology
G mez Ramos, L. M.,
Bommarius, A. S., Georgia Institute of Technology

Nitroreductases are enzymes that reduce nitro compounds to corresponding nitroso, and hydroxylamino compounds. In the past 20 years, they have gathered interest in several applications such as degradation of xenobiotic nitroaromatics, and activation of prodrugs in medical applications. However, the use of nitroreductases in these fields is limited due to the lack of thermal and kinetic stability, which is shown by their low melting temperatures and high deactivation constants.

To overcome this limitation, we created a nitroreductase from a stable ene reductase scaffold through protein engineering methods such as rational design, and site-saturation mutagenesis. In this study, we characterize nitroreductases from *Salmonella typhimurium*, *Enterobacter cloacae*, and *Mycobacterium smegmatis*, to identify residues significant for catalysis and substrate binding. With this knowledge, mutations are performed to engineer nitro reduction functionality from ene reductases, namely xenobiotic reductase A (XenA) from *Pseudomonas putida*, YersER from *Yersinia bercoviei* and KYE1 from *Kluyveromyces lactis*. The novel enzymes are characterized with steady-state kinetics and conversion studies.

(582d) Selective Recovery From the Interstitial Spaces of Leaves

Kingsbury, N., University of California, Davis
McDonald, K. A., University of California, Davis

In planta production of apoplast-targeted cellulose degrading enzymes may be a valuable approach for efficient production of concentrated enzyme preparations that can be used for saccharification of cellulose in pretreated feedstocks. A vacuum infiltration-centrifugation method has been developed for recovery of transiently-produced cellulase enzymes in the apoplast wash fluid (AWF) of tobacco leaves. A single AWF recovery at the time of peak production following transient agroinfiltration of harvested tobacco leaves removes up to 67% of the Acidothermus cellulolyticus thermostable endo-1,4-β-glucanasecatalytic domain (E1cd) activity. However, recovering AWF daily from agroinfiltrated leaves more than tripled the amount of the enzyme activity recovered in the AWF and doubled the amount of active E1cd produced overall. Although the mechanism for the yield increase is not yet understood, rinsing the apoplast of agroinfiltrated leaves during the production/incubation period has potential advantages of improved protein stability, secretion kinetics, and leaf health. The AWF collected also had higher purity and enzyme concentration, up to 124- and 19-fold improvements respectively compared to recovery from homogenate extracts, facilitating its use in the saccharification process.

(582bf) Effect of Heterogeneous Cultivation Conditions On Mammalian Cell Productivity and Product Quality

Neunstoecklin, B., ETH Zurich
Lularevic, M., Institute for Chemical and Bioengineering
Morbidelli, M., ETH Zurich
Soos, M., ETH Zurich

The demand of therapeutic proteins like monoclonal antibodies (mAb) to treat cancer or chronicle diseases is increasing from year to year. Due to the required post translational modifications like glycosylation, mammalian cells are used exclusively to produce all currently approved mAbs. To fulfill the demand of the market, large-scale cell cultures up to 20000 L became industrial standard. Up scaling to those volumes becomes therefore a fundamental necessity to guarantee process performance, product amount and product quality. Apart from regular process parameters like temperature, pH or dissolved oxygen, it is of outmost importance to be aware of scale dependent parameters influencing the process. Reactor geometry, shear stress distribution, dissolved O_2 and CO_2distribution and gas mass transfer rates can be very different in between scales, resulting in non-desired variations of process performance when changing scales.

To study these parameters an experimental two zone model, capable of simulating such scale dependent parameters individually or in combination, was developed. The device consists of two fully controlled interconnected bioreactors capable of simulating shear heterogeneity, dissolved oxygen and carbon dioxid oscillations as well as pH perturbations. Parameters can be controlled individually with frequency, magnitude and exposure time adjusted to the condition of the desired large scale reactor. To demonstrate the capability of the system several case studies will be presented including the effect of i) different aeration strategies, applying different bubble sizes, oxygen fractions and gas flow rates, ii) the combined effect of DO oscillation and elevated hydrodynamic stress and iii) the heterogeneity of pH via acid and base addition. Culture performance, the amount of produced mAb and its quality were monitored using state of the art analytical methods.

The obtained results indicate limitations of currently used lab scale bioreactors for process development and the influence of scale dependent parameters on productivity and product quality. In contrast the proposed system can be used to test the operating window of various process parameters under quasi production scale conditions and therefore provide better understanding on the effect of these parameters on the mammalian cell cultivation.

(582cz) Multiscale Investigation of Biomolecular Systems Dynamics

Gur, M., School of Medicine, University of Pittsburgh
Zomot, E., School of Medicine, University of Pittsburgh
Madura, J. D., Duquesne University
Bahar, I., School of Medicine,University of Pittsburgh

Global transitions of proteins, which usually occur on the time scale of microseconds or more, are beyond the capacity of full-atomic conventional molecular dynamics (MD) simulations. Due to this limitations elastic network models, such as the Anisotropic Network Model (ANM), have been widely used to simulate such transitions. However, a direct comparison between the global dynamics obtained by MD simulation long enough to potentially provide hints about global changes is still missing. Here we perform such comparison by analyzing the MD simulations of two systems, one longer than a millisecond on the equilibrium dynamics of bovine pancreatic trypsin inhibitor (BPTI) and the other, of several microseconds on the gating mechanism of an archaeal aspartate transporter, GltPh, that serves as a model for human excitatory amino acid transporters.[1] It is shown that, despite depending only on the contact topology, the ANM successfully captures the global dynamics of these extensive MD simulations and defines the pathways that underlie the passage between substates. In order to take advantage of such features while simulating at all-atom level, we propose a new methodology[2] which deforms the structure collectively along the modes predicted by the ANM, while evaluating the interactions and energetics via a full-atomic MD simulations. Application to adenylate kinase, an allosteric enzyme composed of three domains, CORE, LID and NMP, shows that an energy-barrier crossing occurs during the NMP movements and originates from a switch between interdomain salt bridges. In addition, we explore the functional transition mechanisms of the dopamine transporter (DAT), an integral membrane protein that terminates chemical neurotransmission in the central nervous system by removing dopamine from the synapse. An intermediate substate of DAT occluded to both the extra- and intracellular regions is identified along the transition between the inward- and outward-facing substates.

1. Gur M, Zomot E, Bahar I. Global Motions Exhibited by Proteins in Micro- to Milliseconds Simulations Concur with Anisotropic Network Model Predictions. Journal of Chemical Physics, 2013, submitted.

2. Gur M, Madura J, Bahar I. Global Transitions of Proteins Explored by a Multiscale Hybrid Methodology: Application to Adenylate Kinase. Biophysical Journal, 2013, submitted.

(582an) Stabilization of a Fiberless Adenovirus Gene Vector

Kupgan, G., Oklahoma State University
Choudhari, S., Oklahoma State University
Flynn, N., Oklahoma State University
Nigatu, A., Oklahoma State University
Vupputuri, S., Oklahoma State University
Picking, W., Oklahoma State University
Picking, W., Oklahoma State University
Ramsey, J. D., Oklahoma State University

The ubiquitous nature of the coxsackie adenovirus receptor (CAR) and undesirable interactions between cell surface proteins and the adenovirus fiber and capsid proteins lead to a vector that has both limited and promiscuous tropism. Fiberless adenovirus, however, has the potential to reduce the promiscuous tropism commonly associated with adenoviral-based gene vectors but removal of the fiber protein has been shown to negatively affect the virus stability. To identify the most stable conditions for formulation and storage, we explored the effects of pH and temperature on stability of the virus particle. Our results indicated that the stability of fiberless adenovirus was significantly increased when the virus was kept in mildly acidic conditions, around pH 6. Further, the stability of fiberless adenovirus could be improved by including excipients. The results from screening a library of GRAS excipients showed that non-ionic surfactants and amino acids tended to increase the thermal transition temperature of the virus particle, and more importantly, improved biological activity after prolonged thermal stress. In conclusion, our data demonstrated that instability of fiberless adenovirus could be ameliorated by carefully controlling the pH and stabilizing excipients, thereby leading the way to development of gene vectors or vaccines based on fiberless adenovirus.

(582bl) Secretory Expression of Laccase From Geobacillus

Castagnaro, T.,
Wu, K., University of Illinois

Laccases are multi-copper oxidases that can oxidize a disparate range of organic substrates using oxygen and producing water as a byproduct without requirement for additional reactive compounds, making them excellent candidates for industrial oxidants. Compared to fungal laccases which are difficult to produce in active forms in heterologous hosts due to incorrect glycosylation, bacterial laccases are much easier to express heterologously and are more active and stable at high temperatures, pH and salt concentrations. However, most bacterial laccases studied are intracellular enzymes or periplasmic proteins. Purification of these proteins compared to extracellular enzymes can considerably increase the downstream processing costs.

Geobacillus is a genus of Gram-positive thermophilic bacteria, many of which have been found to naturally secrete proteins at high levels (Unpublished data). Genome analysis of several Geobacillus species revealed the presence of laccase with a signal peptide for secretion. In this work, we screened a collection of 170 Geobacillus species for their functional secretion of laccase and heterologously overexpressed these secretory laccases as well as intracellular laccase from *Bacillus subtilis* in *Geobacillus thermoglucosidasius*. Signal peptides of these laccases were engineered to improve the secretion efficiency. We also investigated the secretion pathways and tuned the expression of copper chaperons of the host to enhance the functional expression and secretion of laccase. These results will aid in the understanding of the mechanisms of copper binding, laccase folding and secretion. In addition, using laccase as an example, we demonstrated functional secretion of otherwise intracellular enzymes by signal peptide engineering.

(582co) Wastewater Treatment in Cellulosic Biorefineries Via Bioelectrochemical Systems

Borole, A. P., Oak Ridge National Laboratory

Wastewater in a biochemical refinery generating alcohol fuels from biomass contain acetate and phenolic compounds which are inhibitory to fermentation organisms. In order to realize water recycle, complete removal of these substrates is important. Typical wastewater treatment processes such as anaerobic digestion cannot handle such streams easily due o their high salt content and presence of inhibitory levels of acetate, furfural, etc. Electroactive biofilms which breakdown organics to generate electrons and protons can be employed to generate hydrogen via use of electroactive hydrogen production catalysts at the cathode.

Experimental demonstration of hydrogen production from biorefinery wastewater using a microbial electrolysis cell is presented. Coulombic efficiencies over 60% were obtained. A currrent density of 7 A/m2 was also achieved using a laboratory scale device. Commercial development of such a process has significant environmental and process benefits due to the potential of generating renewable hydrogen. Preliminary results showing proof of principle will be presented.

(582cc) Tuning Beauveria Bassiana Xenobiotic Metabolism

Nicolau, F., University of Iowa

Beauveria bassiana is a filamentous fungus widely used in organic chemistry to perform regio- and steroselective reactions due to its simplicity to handle and catalytic robustness. Researchers in the area of green chemistry including our group and others have applied *B. bassiana* and similar fungal systems in the synthesis of pharmaceuticals and other specialty chemicals through selective oxidations. The major industrial application of *B. bassiana* has been as an environmentally friendly insecticide. As a result, most work in understanding the complex regulation of cellular functions has been center towards virulence enhancement for pests control. This work gives insight into the catalytic machinery involved in oxidation reactions which have not been fully explored. Our approach is to combine the improvements in pest management practices and biopesticide efficacy to improve catalysis.

B. bassiana can enhance the yield and selectivity of oxidative biotransformation through the expression of key oxidoreductases under different growth conditions, starvation, or environmental stress. In presence of n-alkanes, nC_{12} to nC_{30}, it is possible to enhance the oxidative potential of *B. bassiana* by the induction of CYP450. CYP450s are oxidative enzymes involve in the first phase of xenobiotic metabolism. They are expressed in low levels and in several isoforms. *B.bassiana's genome* encodes for 83 cytochrome P450s, whose expression and/or induction varies by different alkane lengths and xenobiotic. In addition, the use of botanicals and insecticides, which act as xenobiotics, has shown a synergistic effect in controlling major pest. We explore these approaches by evaluating the changes in yield and productivity of the sulfoxidation of 2-[(diphenylmethyl)thio]acetic acid. We also evaluate the amount of oxidases present in the membrane-bound protein fraction. The success of these efforts will provide new insight on *B. bassiana's* oxidative performance.

(582ab) Development and Production of Novel VLP Vaccine Platform for B Cell Lymphoma

Chan, W., Stanford University
Swartz, J. R., Stanford University

B cell lymphoma occurs when a clonal population of white blood cells or lymphocytes becomes cancerous. One feature of B cell lymphoma is that the variable region of the surface antibody (known as the idiotype) is the same across all cancerous cells. This makes the idiotype an ideal target for the treatment of B cell lymphoma. Our treatment strategy is to conjugate the idiotype antibody variable region fragment (scFv) to the surface of a virus-like particle (VLP), as well as immunostimulants, such as granular monocyte colony stimulating factor (GMCSF) and CpG DNA. VLPs are made up of coat protein monomers that rapidly self-assemble into a nanoparticle. By incorporating an azide functional group into the coat protein monomer and an alkyne functional group into a biomolecule of interest, we can use azide-alkyne Huisgen cycloaddition to covalently link the biomolecule to the surface of the VLP. We are able to make a nanoparticle vaccine that elicits a strong immune response specifically against the B lymphoma cells. In mouse trials, mice dosed with the nanoparticle vaccine had a humoral antibody response that was four times stronger than mice dosed with the scFv alone. We are currently working on an improved nanoparticle vaccine by increasing the stability of the VLP and by conjugating additional immunostimulatory molecules.

(582cp) Design and Technoeconomic Analysis of a Biomanufacturing Facility for Production of a Biodefense Agent

McDonald, K. A., University of California, Davis
Burkle, R.,
Chew, C., UC
Li, C.,
O'Brien, K.,
Paridon, K.,
Tu, T.,
Wang, J.,

Organophosphate (OP) nerve agents are easily weaponized and present a clear danger to soldiers in combat zones as well as civilians. Butyrylcholinesterase (BuChE), a glycoprotein naturally found in human blood at low levels, has the ability to inactivate OP agents before they can cause harm. Currently BuChE is purified from donated blood, however the high production cost (~$10,000/dose) and low supply limit its application for broad protection of our warfighters. In this project, a detailed conceptual design and economic analysis of a novel large scale biomanufacturing facility for production of recombinant BuChE is presented. BuChE is produced transiently in wild type *Nicotiana benthamiana*, a relative of tobacco, using vacuum agroinfiltration. The plants are then homogenized and the recombinant protein is recovered and purified. However, *N. benthamiana* lacks the enzymes needed to add certain sugars to the glycoprotein; these sugars are important to maintain an appropriate circulatory half-life of the anti-nerve agent. An *in vitro* enzymatic process is designed to add these sugars to BuChE, with recombinant enzymes produced in *Escherichia coli*. SuperPro Designer®, a computer-based modeling tool, is used to design and simulate the unit operations, perform material and energy balances, schedule batch operations, determine the capital investment and production costs, and quantify the environmental impact of the facility.

(582cy) Controllable Depolymerization of Heparin By the Combinations of Heparinases I/II/III for the Production of Low Molecular Weight Heparins

Wu, J.,
Zhang, C.,
Li, Y.,
Mei, X.,
Xing, X.,

Low molecular weight heparins (LMWHs) produced by the degradation of heparin has been widely used as anticoagulant. Enzymatic depolymerization of heparin by heparinases is a promising way for LMWHs production due to their mild reaction conditions, high selectivities and controllabilities. Here, different heparinases and their combinations were used to depolymerize heparin. The results demonstrated that heparinase I and heparinase II could depolymerize heparin efficiently but the anticoagulant activities of depolymerization products were destroyed significantly. Furthermore, the anticoagulant activity ratio (anti-factor Xa activity/anti-factor IIa activity) of depolymerization products increased significantly during the depolymerization process, which implied that the action patterns of heparinase I and heparinase II were not exolytic and processive depolymerizing mechanism from the reducing end of heparin, were reported for the first time. However, heparinase III couldn't efficiently depolymerize heparin but could protect the anticoagulant activities of depolymerization products to the largest extent compared with heparinase I and heparinase II. The combination of heparinase III and heparinase I or heparinase II could exert the advantage of heparinase III to protect the anticoagulant activities to the largest extent during the depolymerization process and that of heparinase I or heparinase II to efficiently depolymerize heparin, and produce LMWHs with higher anticoagulant activities than that produced by Heparinase I and Heparinase II individually. The combination of heparinase III and heparinase I, exhaustive depolymerization by heparinase III followed by the controlled depolymerization by heparinase I, was the best combination for the production of LMWHs with high anticoagulant activities.

(582cq) The Effect of Bark Inclusion On Bioethanol Production in a Spruce-to-Ethanol Process

Frank , B., Lund University
Wallberg, O.,
Galbe, M., Lund University

In order to face up to climate change, one of our greatest environmental challenge, greenhouse gas emissions have to be reduced. One way to fulfill this intention is to replace fossil fuels in the transport factor, with biofuels, such as bioethanol. The production of fuel ethanol from softwood is well documented; and spruce is considered a promising abundant biomass source for bioethanol production in Sweden. Low-cost forest residues will contain large amounts of bark but the inclusion of bark in the feedstock have not be investigated to a large extent. The composition of spruce bark is very complex. The bark could have an adverse effect on the wood-to-ethanol process in many ways. For instance, the high lignin content of spruce bark can render the material more recalcitrant to hydrolysis, it could also contain several compounds such as extractives and phenolics that could be inhibitory to the yeast. Therefore, the inclusion of spruce bark in the production process was investigated.

The influence of bark has been studied by increasing the amount of bark from 0% to 100% (based on dry weight). The mixed materials were impregnated with SO_2 (2.5% w/w moisture) prior steam pretreatment at 190°C for 5 min. Pretreated materials have been evaluated by analysis of solid and liquid phases. Enzymatic hydrolysis, fermentation of the liquid phase (fermentability) and simultaneous saccharification and fermentation (SSF) were performed to determine the sugar and ethanol yield. Enzymatic hydrolysis have been performed at 5% water-insoluble solid (WIS), while SSF have been performed at 10% WIS content in order to avoid the impact of high WIS content. It has been planned to continue the study with performing SSF at high WIS content in batch, fed-batch mode and also with pre-hydrolysis in order to become aware of the best possible process configuration that can be applied to achieve high ethanol concentration in the broth and also high ethanol yield, thus to reduce production cost of the process. In mixtures, where bark inclusion had a negative effect, compared to just spruce chips, methods, such as running low pressure steam through the material before pretreatment, treatment with boiling water before or after pretreatment and leaching with hot dilute acid, can also be the target of future research to eliminate the negative effects of bark on the process.

(582bm) Generation of Resveratrol Analogues By Enzymatic Oligomerization and Protein Engineering

Bhan, N., Rensselear Polytechnic Institute
Mora-Pale, M., Rensselear Polytechnic Institute
Koffas, M. A. G., Rensselaer Polytechnic Institute
Linhardt, R. J., Rensselaer Polytechnic Institute
Dordick, J. S., Rensselaer Polytechnic Institute
Masuko, S., Rensselaer Polytechnic Institute

Most natural polyphenols have low antimicrobial activity. In this work, we tried to generate analogues of resveratrol via either enzymatic oligomerization or site-directed mutagenesis of stilbene synthase (STS). Enzymatic oligomerization produced two oligomers, resveratrol-*trans*-dihydrodimer and Pallidol. The *trans*-dihydrodimer showed commercially relevant antimicrobial activity against gram-positive bacteria, *B. cereus* (MIC = 15.0 μM), *Listeria* (125.0 μM), and *S. aureus* (MIC = 62.0 μM). Further, we identified it's mechanism of action to be via altering membrane structure and polarization, and inhibiting DNA synthesis. This was deciphered by transcriptomic, biochemical and flow cytometric analysis. Gene expression microarray and qRT-PCR data from *B. cereus* indicated down regulation of ABC transporters (ATP binding proteins). Flow cytometry analysis showed depolarization of cell membrane upon treatment with resveratrol-*trans*-dihydrodimer. A decrease in nucleic DNA content of *B. cereus* and *E. coli* upon incubation with the *trans*-dihydrodimer was also observed by flow cytometry. Further, *in vitro* activity of DNA gyrase was strongly inhibited by the resveratrol-*trans*-dihydrodimer.

Next we created 15 mutants of stilbene synthase and tried to create resveratrol analogues by feeding different substrates. Type III polyketide synthases have been studied extensively over the past few decades, but most of these structural-functional studies have been carried out on chalcone synthase (CHS) like enzymes, which carry out a C6-C2 Claisen condensation of their polyketide intermediates. Stilbene synthases share 75-90% sequence similarity with CHS and form the same tetraketide intermediate, but cyclize the product via a C1-C7 aldol condensation. This cyclization requires a thioesterase-like hydrolysis step made possible by the hydrogen bonding network (HBN) present only in STSes, involving residues S338, T132 and E192 and a water molecule. Here we studied the effect of mutating functionally important residues in STS. Similar mutations have been carried out in CHS but we predicted that new products would be formed because of the different cyclization of intermediates by STS. Mutant G256L completely lost its ability to handle the bulky p-coumaroyl-CoA substrate while a T197G mutant was still able to produce resveratrol while forming at the same time an increased amount of a more hydrophobic compound which is yet to be identified.

Finally, we have demonstrated that the enzymatic oligomerization of resveratrol is an attractive strategy for generating effective antimicrobials with potential therapeutic application. And further we plan to study the antimicrobial activity of resveratrol analogues created by mutating the functionally important residues of STS. The mechanism of action of which can be identified by carrying out studies similar to those carried out for the resveratrol-*trans*-dihydrodimer.

(582e) Expression of Recombinant Human Butyrylcholinesterase in Nicotiana Benthamiana and Its Postproduction In-vitro Glycosylation Modification

Alkanaimsh, S., UC Davis
Nandi, S., University of California, Davis
Hwang, M. S., University of California, Davis
Guerrero, A.,
Li, Y.,
Arzola, L., UC Davis
Hashimoto, B.,
Tu, A., University of California, Davis
Phu, M.,
Dandekar, A. M., University of California, Davis
Falk, B. W., University of California, Davis
Chen, X.,
Lebrilla, C.,
Rodriguez, R.,
McDonald, K. A., University of California, Davis

Human butyrylcholinesterase (hBChE EC 3.1.1.8) is a 574 amino acid cholinesterase-hydrolyzing enzyme. Organophosphates (OP) are highly toxic inhibitors of the acetylcholine-hydrolyzing enzymes like hBChE. The resulting accumulation of acetylcholine can lead to respiratory collapse and death. Current therapies are based on elevating the serum levels of OP bioscavengers like BChE. The major limitation of this therapy is high cost, with plasma-derived hBChE costing more than $10,000/treatment. Limitations like cost and availability necessitate an alternative expression platform capable of large scale, low-cost production of a fully active and efficacious recombinant hBuChE (rhBChE). The development of an effective rhBChE is a pressing national security concern in terms of protecting the nation's warfighters and civilian population from the threat of attack with OP agents.

We describe the use of viral amplicon-based gene expression systems based on either *Tobacco mosaic virus* (TMV) and *Cucumber mosaic virus* (CMV) to express functional rhBChE in *Nicotiana benthamiana* using transient agroinfiltration. Because hBChE is a glycoprotein with nine potential N-glycosylation sites, the glycan structures of the plant-made rhBChE were also characterized. For each expression system, two constructs were made to target the protein to the apoplastic compartment (Apo) of the plant cell or to retain the protein at the endoplasmic reticulum (ER). As expected the variation in subcellular localization resulted in different glycosylation patterns of the recovered butyrylcholinesterase. Retaining the protein at the ER yields a high mannose N-glycans structure, while targeting the protein to the apoplastic compartment yields a complex N-glycan structure. In each of the four different constructs (Apo-TMV, ER-TMV, Apo-CMV and

ER-CMV), a 3X FLAG tag is used at N-terminal of the protein for protein identification and purification. Here, we report the expression level of the four different expression systems and the N-glycan structure of the ER retained and apoplast targeted proteins.

Six days post infiltration, the expression level of functional rhBChE using the ER-TRBO, Apo-TRBO, ER-CMVar and Apo-CMVar expression systems are 14.6, 6.6, 0.74 and 2.9 mg/kg FW respectively. The glycan structure of purified ER-retained butyrylcholinesterase reveals almost all the N-glycan having a high mannose structure with insignificant amount of paucimannosidic-type N-glycan. While, the glycan structure of Apo targeted protein reveals a mixture of N-glycans consisting mainly of complex N-glycans (80%), high mannose structure (10%), and paucimannosidic-type N-glycan (10%). We found all nine sites occupied by typical plant glycans.

It is also known that plants are incapable of sialylating glycoproteins naturally and that sialylation is essential for the normal serum half-life of proteins. To increase the number of sialic acid residues per rhBChE molecule, we systematically added GlcNAC, galactose and sialic acid to branch-termini of plant N-glycans using multistep enzymatic reactions (i.e., *in vitro* sialylation). While preliminary results are encouraging, more characterization of ER-retained and apoplast-targeted rBuChE is needed. Because 30% of all commercial biopharmaceuticals are glycoproteins, these results could make plant-made pharmaceuticals a viable alternative to mammalian or insect expression systems.

(582da) A Multi-Compartment Skin Penetration Model Coupled to a Physiology-Based Biokinetic Model for Chemical Exposure Assessment

Sarigiannis, D., Aristotle University of Thessaloniki
Handakas, E.,
Karakitsios, S., Aristotle University of Thessaloniki

Reliable predictions of chemical transport through skin are important for topical drug, transdermal drug and cosmetics design, as well as chemical risk and safety assessment. The skin is a complex organ and a living membrane. Mathematical models of skin have been developed in order to predict and measure the transportation of chemicals, drugs and cosmetics through the skin. These models have provided an essential contribution to the evaluation of dermal exposure as well as toxicity assessment. Moreover, they play a significant role in analyzing experimental data and in reducing the number of necessary toxicological experiments. In this study a multi-compartmental Physiology Based ToxicoKinetic (PBTK) model of skin has been developed that describes explicitly penetration, absorption and potential metabolism of chemical compounds under different realistic exposure scenarios. The model was applied in the case of BPA and to model its contact with human skin. The generic character of the model is supported by calculating input diffusion and partition coefficients from correlations with the octanol/water partition coefficient. The detailed skin absorption model was coupled to a generic human PBTK model, giving us the capability to capture the contribution of dermal route of exposure to the overall bioavailability of a toxic compound, under complex aggregate exposure scenarios.

The skin has been modeled with a two layer structure: Stratum corneum (SC) and viable epidermis (VE). The stratum corneum has been described as a "bricks and mortar" structure. Current understanding of solute binding to keratin and other corneocyte constituents of the SC encompasses mostly equilibrium binding, and macro-scale parameterizations of transient binding for a few solutes. Our work leverages the latest and emerging results on homogenization theory and rates of binding to produce a broad mechanistic SC model that quantifies transient solute binding in terms of coexisting free and bound concentration fields, and is parameterized at the microscopic scale. This is critical to realistically describe actual chemical exposure, in which most of the penetration often occurs before reaching steady state.

The detailed skin model was coupled to a generic whole-body PBTK model, giving us the capability to capture the contribution of the dermal route of exposure to the overall bioavailability of toxic compounds, capturing complex aggregate exposure situations. The modeling framework was developed in the dynamic modelling environment acslX and applied to bisphenol A. Bisphenol A (BPA) is one of the highest industrial volume chemicals produced worldwide. The major volume of BPA is used for the production of polycarbonate plastic as well as a basic component in production of the epoxy resin. Various common consumer products contain or are made by polycarbonate plastic such as household electronics and baby bottles. Epoxy resin is used in the majority of food and beverage cans. Moreover, BPA is commonly used in paper industry and particularly as color developer in thermal and copy paper. Hence, BPA has been found in thermal paper of sale receipts and money at a mean concentration of 13.3 g/kg. BPA is characterized as an estrogen with endocrine disrupting properties mediated via multiple molecular mechanisms, which depend on its metabolism and, consequently, the possible exposure routes. Additionally, recent studies have examined the neurotoxicity of BPA, highlighting that even low maternal exposure to BPA is associated to neurodevelopmental defects. Parameterization of the skin model for BPA was based on recent in vivo studies relevant to BPA skin metabolism and permeation. The results showed that BPA was extensively metabolized in viable epidermis, and the major BPA metabolites produced by the skin were BPA mono-glucuronide and BPA monosulfate, accounting together for 27% of the daily administered dose. Evaporation has also been calculated based on the REACH technical guidance. Hence, the effect of skin exposure scenarios compared to oral and inhalation exposure scenarios was investigated.

The major question related to skin absorption of BPA is the potential bioavailability differences that might arise compared to oral exposure, which is considered as the dominant exposure route for the majority of exposure scenarios considered by the competent regulatory authorities in the EU. This concern arises from the fact that oral intake of BPA is accompanied by an extensive and very rapid first-pass metabolism (phase II glucuronidation), resulting in very low bioavailability. On the contrary, our modelling analysis has indicated that inhalation exposure to BPA results in 6-fold higher bioavailability compared to oral exposure. To elucidate potential differences, we investigated the difference in steady state plasma concentration for a given dose, using the EFSA Tolerable Daily Intake (TDI) of 50 μg/kg_bw/d as the reference point. We also simulated the concentration time profile of an oral dose of 50 g/kg/day (which corresponds to the TDI) in blood and liver. In blood, the free plasma BPA concentration after oral dose of 50 μg/kg_bw/d is equal to 0.16 μg/L, while the corresponding free plasma concentration after the administration of the same dose dermally is up to 0.3 μg/L. In contrast, peak concentration in the liver is 14 fold higher after oral as compared to dermal administration of the same dose, due to the well-known first pass metabolism of BPA.

In terms of actual exposure scenarios, the case of thermal paper containing BPA deserves special attention, since it can be heavily loaded with BPA. Unlike in plastics and epoxy resins, BPA is present in this type of paper primarily as a free monomer, and is therefore directly available for skin uptake. In a relevant study of 13 samples analysed, they were found to contain between 8–17 g/kg (0.8–1.7%) BPA. In the same study, the average level of receipt and receipt-like papers amounted to 1.4% and 1.6%. Of these thermal papers, car park tickets and bus tickets were notable for levels as high as 3.2% and 2.3%, respectively. These results show that BPA is not only present in thermal printing papers but also that BPA is taken up on the surface of the fingers when receipts are handled by the cashier and the customer. Accordingly, exposure to BPA of cashiers, was estimated up to 71 μg/day. From an extensive literature review, this was found to be the worst case scenario regarding dermal exposure to BPA. Based on this assumption, we evaluated the intra-day time course of BPA between an exposed individual from the general population (assuming an oral daily intake of 1.5 μg/kg_bw/d) and a cashier receiving an additional dose of 71 μg/day during a 10-hour shift. When BPA is administered also dermally the corresponding quasi steady-state concentration is close to the oral exposure scenario, achieved in about three to four days. This is due to the long absorption half-life during which BPA is delivered from the skin into systemic circulation. In the case of dermal exposure, the lack of first pass metabolism is less important due to the slow rate of absorption and the respective metabolism occurring in skin. In this case, the diurnal peak concentrations are higher only by 60% than the exposure levels corresponding to the typical EFSA exposure scenario. However, the contribution of this exposure route to the area under the curve (AUC), i.e. the integrated quantity of bioavailable BPA, is more significant, since the slow rate of absorption contributes to continuous presence of BPA in plasma within the day.

Additional investigation of the potential bioavailability differences that might exist because of dermal administration showed that BPA does not pose concern, since the lack of first-pass metabolism is decompensated by the slow rate of absorption, the extent of skin metabolism and more significantly to the low exposure. The above lead to the conclusion that dermal consumer exposure scenarios related to BPA arising from paper recycling (except in the case of thermal paper for cashiers), might be considered as negligible.

(582dg) Plantseed: Utilizing Functional Annotation of Plant Protein Families to Generate Metabolic Models Capable of Phenotypic Predictions

Seaver, S., Argonne National Laboratory
Henry, C. S., Argonne National Laboratory

Over the past two decades, there has been progress towards the development of *in silico* metabolic reconstructions of microbial organisms that enable researchers to explore the complex relationship between genotype and phenotype. The progress has advanced along three paths: the accelerated annotation of enzymes and reactions, the refinement of algorithms and techniques for both propagation of annotation and metabolic analysis, and the growth in the number of sequenced genomes. These paths have come together through the use of subsystems technology in the form of ModelSEED, a framework for the automated reconstruction of metabolic models that has been applied to construction over 13,000 genome-scale models.

To capitalize on the growth of next-generation sequencing and the interest in exploring metabolic pathways in plants, we present PlantSEED, a unique collaboration between teams at Argonne and the University of Florida to create a specialized niche within the ModelSEED ecosystem geared towards the development of plant metabolic reconstructions. The teams are working towards a tight integration between the current annotation of biochemistry with highly homologous and iso-functional plant-specific protein families (FIGfams) to build an ensemble of subsystems that can be utilized to automatically generate metabolic models that can be used to generate predictions of metabolic phenotypes in plant tissue.

(582f) Dynamic Adhesion Assays for Understanding the Significance of CD151 in the Metastatic Cascade

Fischer, J., University of Kentucky
Shearer, A., University of Kentucky
Yang, X., University of Kentucky
Trinkle, C., University of Kentucky
Eitel, R. E., University of Kentucky
Anderson, K. W., University of Kentucky

Cancer is the second leading cause of death in the United States, with metastatic lesions resulting from circulating tumor cells (CTCs) most often causing the mortality. The basic steps of the metastatic cascade, from initial tumor formation to growth at a secondary site, are a linear sequence of steps composed of a network of mechanisms that are not well understood. One of these steps, the adhesion of cancer cells to endothelial cells during dissemination, is the focus of this research. Specifically, the influence of CD151, a gene that encodes a cell surface glycoprotein that is known to complex with integrins and enhance cancer cell motility, invasion and metastasis was studied. To study the influence of CD151, a prostate cancer cell line (PC3) and a breast cancer cell line (MCF-10A) were employed in dynamic adhesion assays in which the expression of CD151 was knocked down on the cancer cells using shRNA gene silencing. Both the knocked down and unmodified lines were utilized to quantify adhesive strength and frequency of adherence to flow-adapted cell monolayers composed of Human Umbilical Vein Endothelial Cells (HUVECs). A parallel plate flow chamber system was utilized to investigate these properties under flow at various shear stresses ranging from 0.25 dynes/cm^2 to 15 dynes/cm^2 to cover the range of most physiologically relevant shears. Brightfield and fluorescent images were analyzed to determine cancer cell adherence over time as well as retention rates after a light wash and a period of subjection to a given shear stress. Preliminary results indicated that the presence of CD151 leads to enhanced initial adhesion of prostate cancer cells to statically-grown HUVECs in adhesion trials, but retention rates appeared to be unaffected by CD151 expression once those cells became firmly adhered. Results from PC3 and MCF-10A adhesion to flow adapted endothelial monolayers will be presented.

(582bn) Fermentation of Agrobacterium Tumefaciens for Large-Scale Transient Expression of Recombinant Proteins In Plants

Leth, I.,
McDonald, K. A., University of California, Davis

Production of proteins through *in planta* transient expression offers an alternative to conventional microbial and mammalian cell culture systems. This platform is particularly appealing because of its rapid and relatively low-cost implementation and its ease of scale-up. Transient expression is induced by introducing a gene construct to the nucleus plant cell, where it is expressed for a period of time without being stably integrated into the plant genome. A common method for inducing transient expression is to use the gene transfer capability of *Agrobacterium tumefaciens*. *Agrobacterium*-mediated transient expression can be induced in whole plants or in pre-harvested plant tissue by "agroinfiltration," a process that introduces an *Agrobacterium* suspension to the interior of a plant leaf, allowing the bacteria to transfer the target gene to the plant cells using its natural virulence machinery. Large-scale transient expression of recombinant proteins in plants is a relatively new area, and studies are underway to optimize the stages of the process in order to make it economically competitive. One area that has not been examined is the fermentation of *Agrobacterium* for agroinfiltration at a large-scale. We are investigating the effects of growth conditions including temperature, pH, and media on *Agrobacterium* growth kinetics and gene transfer capability, with the goal of identifying optimal processing conditions for growing *Agrobacterium* at large scale in a 100L working volume bioreactor. Specific growth rates and biomass yield parameters for a range of laboratory *Agrobacterium* strains have been identified, and the effect of growth temperature and media composition on recombinant protein expression level in plants have been examined. Maximum specific growth rates were found to vary from 0.13 to 0.38 h^{-1}, and biomass yield was found to be in the range of 0.66-0.76 g biomass/g sucrose, depending on growth conditions and strain. It was determined that growth temperature and bacterial harvest time did not significantly affect transgenic protein expression level after agroinfiltration, allowing greater flexibility in large scale operation design. A defined media for Agrobacterium fermentation has been developed and direct agroinfiltration with the bacterial growth media has been demonstrated to be effective, which may allow the elimination of a costly processing step when preparing Agrobacterium for large scale vacuum infiltration.

(582ca) Real-Time Monitoring Of Bacterial Growth and Biofilm Formation With Surface Plasmon Resonance Imaging

N. Abadian, P., northeastern University
Goluch, E. D., Northeastern University

In this study, for the first time we used a Surface Plasmon Resonance imaging (SPRi) system to monitor bacteria in real time on the sensor surface as they move, grow and form biofilms. SPRi provides label-free and sensitive detection of changes in refractive index that take place within approximately 200 nm of the gold sensor surface. Changes in the refractive index are caused by biomolecules displacing water at the sensor surface. Difference images are generated over time, which show the changes that occur at the surface during the experiment when compared to a reference image chosen at the beginning of the experiment. Combining microfluidics with our SPRi system provides added control while studying bacterial behavior. By placing microchannels on the surface of the gold surface we limit the bacteria to two dimensions and keep them close to the surface, which allows us to study bacterial growth and movement inside the channels.

Two different strains of *Pseudomonas aeruginosa* PA14 were used for this experiment: Wild Type, which produce biofilm, and PelA mutant, which grow without producing biofilm. The two bacteria were injected into separate microchannels and placed into the SPRi device. The instrument obtained a difference image of the sensor surface every three seconds. We continuously monitored the channels for up to 24 hours as bacteria grew and formed biofilms. We were able to differentiate between the wild type bacteria and the mutant strain. The wild type bacteria attached to the surface and quickly formed large bright spots as they produced extracellular matrix around them, while the mutant bacteria attached to the surface and grew as small discrete bright spots. For the first time, without functionalizing the surface we continuously monitored two different strains of the same bacteria in a single experiment over 1-cm-square area of the sensor surface. This study provides valuable information about bacterial growth and biofilm formation. This new technique can be used to study the effect of different chemicals and surface chemistries on bacterial growth behavior.

(582g) Proteome Analysis of Secretome From Thermophilic Lignocellulose Degrading Microbe Geobacillus Sp. WSUCF1

Bhalla, A., South Dakota School of Mines & Technology
Uppugundla, N., Michigan State University
Dale, B., Michigan State University
Balan, V., Michigan State University
Sani, R. K., South Dakota School of Mines & Technology

Second-generation feedstock, especially inedible lignocellulosic biomass is a potential source for sustainable biofuels production. Several microorganisms such as bacteria and fungi have been reported to produce cellulases and xylanases for effective breakdown of lignocellulosic biomass. It is important to identify novel enzyme from diverse microbes that can effectively hydrolyze biomass, improve sugar yields and reduce the processing cost. In this regard, *Geobacillus* sp. WSUCFI was cultivated on different substrates, namely xylan, avicel, untreated corn stover, Ammonia Fiber Expansion (AFEX^{TM}) pretreated corn stover and Extractive Ammonia (EA) pretreated corn stover. Spectral counting method was used to determine the composition of secretome and relative abundance of cellulolytic and hemicellulolytic enzymes produced by the microbe. Our results showed that protein profiles in the secretome vary with different substrates and pretreatment conditions due to complexity of the structures and composition. Secretomes with high yields of enzymes were observed when Avicel and AFEX corn stover were used as substrates. These secreted enzymes were then selected to carry out enzymatic hydrolysis of lignocellulosic biomass. Different combinations of crude secreted enzymes were used and compared with the conversions obtained from commercial fungal enzymes mixtures. Purified commercial enzymes such as CBH I and CBH II were also doped to study the enzyme synergy. Hydrolysis experiments were carried out using enzyme loadings of 15 and 60 mg/g glucan at 55°C for 24 hours. Promising results were obtained for glucan and xylan conversion using enzyme obtained from *Geobacillus* Sp. WSUCFI and found to be comparable to sugar conversions when commercial enzymes were used.

(582h) Production of Biosurfactant By a Microorganism Isolated From a Brazilian Mangrove: Influence of Oxygen Transfer On the Bioreactor

Franca, I. W., Universidade Federal do Cear
Lemos, J. A., Universidade Federal do Cear
Oliveira, D. Sr., Universidade Federal do Cear
Souza, T.,
Lemos, C. M., Universidade Federal do Cear
Melo, V. M. M., Universidade Federal do Cear
Sant'ana, H. B., Universidade Federal do Cear
Gon alves, L. R. B., Universidade Federal do Cear

The aim of this work is to evaluate the agitation and oxygen rate, on order of optimize the biosurfactant production for a scale-up operation. Biosurfactants are surface-active agents which are amphiphilic biological compounds produced extracellularly or as part of the cell membrane by a variety of yeast, bacteria and filamentous fungi. Advantageously they can be produced from renewable resource substrates or industrial waste products. One of the most important properties of biosurfactants is the reduction of surface tension. These materials show low critical micelle concentration (CMC) values, surface tension reduction of the fermentation broth to less than 30 mN/m, and interfacial tension below 1 mN/m. In addition, several biosurfactants have been reported to have manifold biological activities covering antibiotics, fungicides, insecticides, antiviral and antitumoral agents, immunomodulators or specific toxins or enzyme inhibitors, which make them also interesting for pharmaceutical applications. In comparison to their chemically synthesized equivalents they have many advantages. They are environmentally friendly, biodegradable, less toxic and non-hazardous. They have better foaming properties and higher selectivity. They are active at extreme temperatures, pH and salinity as well, and can be produced from industrial wastes and from by-products. This last feature makes cheap production of biosurfactants possible and allows utilizing waste substrates and reducing their polluting effect at the same time. Nevertheless, one of the major obstacles in the biosurfactant commercial production is high production costs. However, the cost can be reduced by strain improvement, medium composition optimization by statistical methods, using alternative inexpensive substrates, or through the bioreactor operation and design. There is a lack of fundamental knowledge about biosurfactant production scaling up. The most utilized criterion in scaling up of aerobic fermentations is maintaining similarity between the volumetric oxygen transfer coefficient (KLa) at both analyzed systems. In order to develop suitable technology of commercialization, carrying out tests in bioreactors is essential. Therefore, a reactor with an integrated foam collector was designed for biosurfactant production using a microorganism (ICA 56) isolated from a brazilian mangrove. Previous works have showed that the aeration rate and agitation increases the oxygen transfer on the bioreactor, favoring biosurfactant production by a *Bacillus* strain. At 200 rpm and 1 $L.min^{-1}$ O_2, the biosurfactant production was around 200 mg/L, and the surface tension of the foam collected was about 30 $dyna.cm^{-1}$.

(582az) Antimicrobial Peptides Targeting Gram-Negative Pathogens, Produced and Delivered By Lactic Acid Bacteria

Volzing, K. G., University of Minnesota
Borrero, J., University of Minnesota
Sadowsky, M., University of Minnesota
Kaznessis, Y. N., University of Minnesota

We present recombinant *Lactococcus lactis* that produce and secrete heterologous antimicrobial peptides with activity against Gram-negative pathogenic *Escherichia coli* and *Salmonella*. In an initial screening, the activities of numerous candidate antimicrobial peptides, made by solid state synthesis, were assessed against several indicator pathogenic *E. coli* and *Salmonella* strains. Peptides A3APO and Alyteserin were selected as top performers based on high antimicrobial activity against the pathogens tested and on significantly lower antimicrobial activity against *L. lactis*. Expression cassettes containing the signal peptide of the protein Usp45 fused to the codon optimized sequence of mature A3APO and Alyteserin were cloned under the control of a nisin-inducible promoter nisA and transformed into *L. lactis* IL1403. The resulting recombinant strains were induced to express and secrete both peptides. A3APO- and Alyteserin-containing supernatants from these recombinant *L. lactis* inhibited the growth of pathogenic *E. coli* and *Salmonella* by up to 20-fold, while maintaining the host's viability. This system may serve as a model for the production of antimicrobial peptides by lactic acid bacteria to target Gram-negative pathogenic bacteria populations.

(582aq) TAL Effectors As Artificial Transcription Factors For Synthetic Biology Applications

Politz, M., University of Wisconsin-Madison

Synthetic biology promises to deliver advances in the production of commodity chemicals and fuels as well as novel genetic circuits and biosensors. In order to accomplish these feats, artificial transcription factors that could be engineered to target any DNA sequence of interest would be highly beneficial. Previous efforts have either used zinc finger technology, antisense RNA, or site-specific DNA binding proteins (native or heterologous) to accomplish design goals, but these strategies require screening or the use of a particular DNA binding site. Transcription activator-like effectors (TALEs), which may be designed to bind any sequence of interest, have recently emerged as an alternative for the regulation of gene expression at the transcriptional level. These plant virulence factors from *Xanthomonas* species have already been used for genome editing and as transcription factors in eukaryotes, including in human cell lines, and hold the potential to adjust the expression levels of genes involved in pathways of interest in prokaryotic organisms.

The potential clinical applications of TALEs, which have been the major focus to date, do not necessitate a quantitative understanding of TALE DNA binding activity. On the other hand, metabolic engineering applications, such as the repression of genes in central metabolism to increase the yield of a desired product, would benefit greatly from a quantitative understanding of the relationship between TALE primary structure and activity in order to finely tune the expression of genes and associated metabolite concentrations. To assess the validity of this approach, a TALE was designed to target the *lacO1* operator. This designer transcription factor was able to repress expression of a red fluorescent reporter gene downstream of an IPTG-inducible promoter in *Escherichia coli*.[1] Further, the expression of this TALE was shown to alter the ratio of expression of two genes in an artificial operon harboring the *lacO1* operator in the intergenic space.[1] While these proof-of-concept experiments illustrate that TALEs retain their functionality in prokaryotes, efforts are currently underway to gain a quantitative understanding of how TALE structure affects activity *in vivo*. In addition, this technology will be applied to improve the yields of free fatty acids in an overproducing strain of *E. coli*.

1. Politz, M. C.; Copeland, M. F.; Pfleger, B. F., *Chemical Communications* **2013**.

(582ee) Structural and Activity Characterization of Irradiated Antimicrobial Peptide (WLBU2) for Use in Blood Processing and Biomedical Applications

Cheung, D., Oregon State University
Ryder, M., Oregon State University
Schilke, K., Oregon State University

Introduction: Sepsis afflicts approximately 750,000 individuals and kills 140,000 in North America annually, costing an estimated $17 billion/year to treat sepsis.[1] Current treatment involves administration of IV fluids, antibiotics, vasopressors, and other medications, which are suboptimal at best, resulting in prolonged hospital stays. Passing blood through a sorbent device (hemoperfusion) to specifically remove targets such as endotoxin and bacterial cells holds promise for rapid treatment of acute sepsis. Clinical use of hemoperfusion in this context is based on immobilized polymyxin B (PMB),[2,3] but remains limited owing to ineffective endotoxin removal, and serious complications (e.g., nephrotoxicity, neurotoxicity, monocyte stimulation, substantial protein loss) associated with PMB and PMB-based devices.[2,4-7] The synthetic, cationic amphiphilic peptide (CAP) WLBU2 has greater antimicrobial activity than PMB, works against a much broader spectrum of Gram-positive and Gram-negative bacteria, and shows higher selectivity for pathogenic entities over host cells.[8,9] The long term goal of work conducted in our laboratory is the covalent tethering of WLBU2 to pendant polyethylene oxide chains for capture of LPS. One strategy for covalently attaching these chains to surfaces involves the use of γ-irradiation.[10,11]Â In this work the effect of γ-irradiation on the structure and function of WLBU2 is investigated.

Materials and Methods:

WLBU2 Preparation

Lyophilized WLBU2 was purchased from Genscript (Piscataway, NJ). HPLC-grade water and deuterated water (D_2O) were used to make stock WLBU2 solutions at 5 mg/mL. Solutions were aliquoted and diluted to 0.3 and 1.0 mg/mL and irradiated to 3 kGy by a ^{60}Co source (OSU Radiation Center). Further dilutions and sample preparations were performed using the irradiated samples with HPLC or deuterated water.

Proton NMR Spectroscopy

Proton nuclear magnetic resonance (^1H-NMR) spectra were obtained using a 400 MHz Robinson NMR instrument with TopSpin 2.1 software at 25 °C using 1 mg/mL WLBU2 in D_2O with trimethylsilyl propanoic acid (TMSP) added as a spectrum reference. Each sample was measured using 64 scans.

UV-Vis Spectroscopy

Ultraviolet-visible (UV-vis) absorbance measurement scans of peptide solutions were obtained between 200 and 400 nm at 1 nm intervals, using a Genesys 6 UV-vis spectrophotometer. Measurements were performed at 20 °C in a quartz cuvette.

CD Spectroscopy

Circulary dichroism (CD) measurements were taken with a Jasco J-815 CD Spectrometer at room temperature between 180 and 280 nm with samples in water or $HClO_4$ using 5 scans per experimental trial. Aliquots of WLBU2 and irradiated WLBU2 were diluted to 0.3 mg/mL in water and 0.2 mg/mL in 0.2 M or 0.5 M $HClO_4$ to induce α-helix conformation. DichroWeb was used to deconvolute and determine the percent helicity of the samples by using the CONTIN or CDSSTR preset methods.

Radial Diffusion Bacterial Inhibition Assay

The assay was performed on *E. coli* (DH5α) and *P. pentosaceus*. Luria broth (LB) and Lactobacilli MRS broth were used to grow *E. coli* and *P. pentosaceus*, respectively, in suspended media at 37 °C overnight to at least 4 x 10^8 CFU/mL. Bacteria were added to the under layer for a final concentration of 4 x 10^7 CFU/mL and poured into plastic petri dishes. Small wells were punched in the cooled bacteria-under layer gel using a truncated P-1000 pipette tip. Peptide samples were added into the wells and allowed to diffused for 2 hours. An over layer containing nutrients was poured on top of the under layer and cooled to harden. Plates were incubated in overnight at 37 °C. Image analysis was performed using ImageJ software. Trials were performed in triplicates.

Results and Discussion: The deviation from standard WLBU2 spectra indicated a structural change upon irradiation (Figure 1a). The NMR and UV-vis spectra strongly suggest that irradiation oxidized or opened the tryptophan ring. Although structure was altered, γ-WLBU2 activity against Gram-positive and Gram-negative bacteria was not negatively affected (Figure 1b).

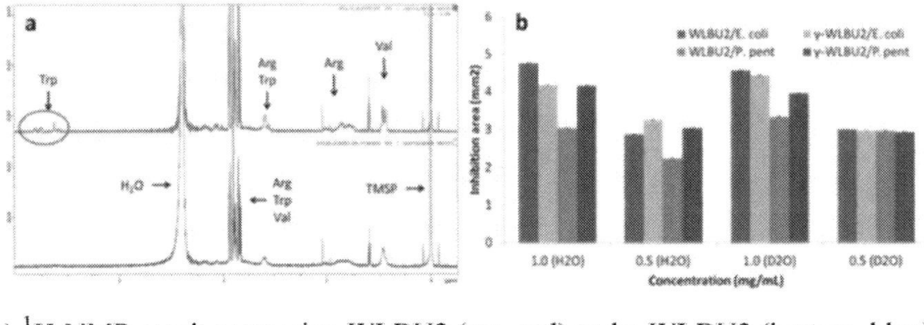

Figure 1: (a) ^{1}H-NMR result comparing WLBU2 (top, red) and γ-WLBU2 (bottom, blue). Large signal peaks at 0 and 4.8 represent the reference and H_2O, respectively. Arginine and valine signals were approximately the same from 0 to 6 ppm. Loss of the signal associated with tryptophan (7 to 7.6 ppm) indicates a chemical change. (b) Mean diameter of WLBU2 kill zones against *E. coli* and *P. pentosaceus*. A small effect of concentration was observed. Deuterated water (D_2O) had no effect on bacterial inhibition, with or without WLBU2 peptide.

Conclusions: The data presented indicate that while WLBU2 undergoes a chemical and structural change upon γ-irradiation, this change does not negatively impact the activity of WLBU2 against Gram-positive or Gram-negative bacteria. Further investigation will be conducted to elucidate the exact structural changes of WLBU2 as well as the extent of change in capture affinity of WLBU2 once tethered to pendant PEO chains.

References

1. Hancock RE, Sahl H-G. Antimicrobial and host-defense peptides as new anti-

infective therapeutic strategies. Nature biotechnology 2006;24(12):1551-1557.

2. Davies B, Cohen J. Endotoxin removal devices for the treatment of sepsis and septic shock. The Lancet Infectious Diseases 2011;11(1):65-71.

3. Shoji H. Extracorporeal endotoxin removal for the treatment of sepsis: endotoxin adsorption cartridge (Toraymyxin). Therapeutic Apheresis and Dialysis 2003;7(1):108-114.

4. Matsuno N, Ikeda T, Ikeda K, Hama K, Iwamoto H, Uchiyama M, Kozaki K, Narumi Y, Kikuchi K, Degawa H. Changes of cytokines in direct endotoxin adsorption treatment on postoperative multiple organ failure. Therapeutic Apheresis 2001;5(1):36-39.

5. Vincent J-L, Laterre P-F, Cohen J, Burchardi H, Bruining H, Lerma FA, Wittebole X, De Backer D, Brett S, Marzo D. A pilot-controlled study of a polymyxin B-immobilized hemoperfusion cartridge in patients with severe sepsis secondary to intra-abdominal infection. Shock 2005;23(5):400-405.

6. Ikeda T. Hemoadsorption in critical care. Therapeutic Apheresis 2002;6(3):189-192.

7. Ueno T, Sugino M, Nemoto H, Shoji H, Kakita A, Watanabe M. Effect over time of endotoxin adsorption therapy in sepsis. Therapeutic Apheresis and Dialysis 2005;9(2):128-136.

8. Gough M, Hancock RE, Kelly NM. Antiendotoxin activity of cationic peptide antimicrobial agents. Infection and immunity 1996;64(12):4922-4927.

9. Scott MG, Yan H, Hancock REW. Biological properties of structurally related α-helical cationic antimicrobial peptides. Infection and immunity 1999;67(4):2005-2009.

10. Ryder MP, Schilke KF, Auxier JA, McGuire J, Neff JA. Nisin adsorption to polyethylene oxide layers and its resistance to elution in the presence of fibrinogen. Journal of colloid and interface science 2010;350(1):194-199.

11. Schilke KF, McGuire J. Detection of nisin and fibrinogen adsorption on poly (ethylene oxide) coated polyurethane surfaces by time-of-flight secondary ion mass spectrometry (TOF-SIMS). Journal of colloid and interface science 2011;358(1):14-24.

(582eb) E. Coli Derived Substrate Quantifies ADAMTS13 Protease Activity in Blood Plasma Using Flow Cytometry

Gogia, S., State University of New York at Buffalo
Lo, C., State University of New York at Buffalo
Neelamegham, S., State University of New York

The metalloprotease ADAMTS13 regulates the size of the multimeric blood protein von Willebrand factor (VWF). Reduction or inhibition of ADAMTS13 activity in humans can lead to a bleeding disorder called Thrombotic Thrombocytopenic Purpura (TTP). In this paper, we describe a flow cytometry based assay to quantify ADAMTS13 activity in human clinical samples using an *E. coli* derived 77-amino acid peptide substrate that is based on a mutated form of the A2-domain of VWF. This substrate was tagged using NHS (N-hydroxysuccinimide)-fluorescein at the N-terminus and Maleimide-PEG(Polyethylene glycol)$_n$-biotin at the C-terminus. PEG repeat unit length was 2 in the case of 'A2-77p-short' and it was 77 for 'A2-77p-long'. Biotin-streptavidin bonding was used to immobilize the substrates onto streptavidin coated microspheres. In flow cytometry assays, fluorescein signal associated with A2-77p-short/A2-77p-long decreased inversely with recombinant ADAMTS13 concentration. A2-77p-long also detected plasma ADAMTS13 activity levels down to 5% of normal activity levels within 30 min. While previous substrates were limited in their ability to detect ADAMTS13 activity when clinically elevated levels of bilirubin are present, the current substrate immobilization strategy along with the specific protein design overcame this problem. As shown, A2-77p-long cleavage could be measured in the presence of bilirubin up to 400 μM. Further, enzyme catalytic activity could be tuned by varying buffer calcium concentration with lower divalent ion concentrations enhancing cleavage rate. Overall, the new peptide substrates allow rapid quantitation of ADAMTS13 activity in blood plasma and highlight a role of calcium in regulating VWF proteolysis activity.

(582dd) Molecular Determinants for Enzyme Stabilization By Insertional Fusion to a Thermophilic Maltose Binding Protein

Pierre, B., Polytechnic Institute of New York University
Kim, J. R., Polytechnic Institute of New York University

Insufficient stability of enzymes limits their practical applications. While various strategies that improve enzyme stability have been reported, they frequently suffer from a compromise of other intrinsic properties, such as enzymatic activity. Moreover, conventional methods for enzyme stabilization have a limited generality - a method working for one enzyme does not necessarily stabilize other enzymes. Recently, our lab has proposed insertional fusion of an enzyme into a thermophilic maltose binding protein as a potentially general method of stabilizing various enzymes without compromising other intrinsic properties. In the presented study, through construction and characterization of a set of insertional fusion complexes, we examined the molecular determinants of stabilization of beta-lactamase (BLA) upon insertion into the thermophilic maltose binding protein from Pyrococcus furiosus (PfMBP). The fusion complexes tested differed in spatial orientation and interdomain distance of the two fused domains. We found that insertional fusion to PfMBP affected the relative energetic states of folded, unfolded and intermediate forms of BLA. Insertional fusion to PfMBP also significantly decreased aggregation of the BLA domain. Collectively, our thermodynamic and kinetic analyses using comprehensive biochemical/biophysical tools have shed light into an understanding of how insertion into PfMBP stabilized BLA.

(582bo) Engineering Lipid Droplet Targeting Domains in S.Cerevisiae

Ho, P.,
Lin, J. L., UC Riverside
Wheeldon, I., UC Riverside

In eukaryotes, lipid droplets are dynamic membrane-bound lipid stores that expand and regress to regulate neutral lipid storage and metabolism. As the active compartment involved in fatty acid storage and mobilization, the lipid droplet membrane presents a novel intracellular target for enzyme localization. We envision the targeting of enzymes and enzyme pathways from a variety of industrially interesting pathways including the biosynthesis and downstream modification of triacylgylcerols (TAGs) and sterol esters (SE). Here, we investigate and characterize a series of putative lipid droplet-targeting domains with the goal of engineering a set of genetic fusions that will target a protein or enzyme of interest to lipid droplets in *Saccharomyces cerevisiae*. Among the N-terminal lipid droplet tags identified in previous mammalian cell studies, murine Viperin (1-42), murine ALD1 (1-27), and human AAM-B (1-28) appear to retain lipid droplet targeting in *S. cerevisiae*. Fluorescence microscopy and western blot analysis of lipid droplet fractions are used to confirm targeting and quantify the targeting specificity of each domain. Additional variable repeats of basic amino acid residues are incorporated in the juxtamembrane and C-terminal regions, and may further facilitate sorting from the endoplasmic reticulum to lipid droplets. We anticipate that the targeting domains developed here will be useful in the metabolic engineering of *S. cerevisiae,* the engineering of reaction pathways in other yeast species such as *Yarrowia lipolytica,* and in the study of eukaryotic lipid droplet biology.

(582ac) Application of Multivariate Analysis of Metabolic Models for Predictability of Cell Culture Performance and Quality Attributes; An Industry Prospective

Migliore, N., Janssen Pharmaceutical Companies of Johnson and Johnson
Tomlin-Verni, K., Janssen Pharmaceutical Companies of Johnson and Johnson
Cunningham, J., PDMS
Mehrman, S., PDMS
Bruwer, M. J., ProSensus Inc.
Schaefer, E., Janssen Pharmaceutical Companies of Johnson and Johnson
Abad, M.,

Cell culture processes are highly multivariate in nature yet to date much of the analysis and optimization of these processes has been univariate. Traditional multivariate techniques can be applied to analyze and optimize cell culture processes as well as increase the predictability of product quality attributes by correlating the outcome of a process to multiple inputs and process parameters as they vary over the course of the batch. PLS models can be built using primary input and output parameters or as combinations of primary parameters and parameters derived from unique fundamental principles such as engineering correlations or metabolic relationships. In this presentation the principles of PLS modeling have been applied to multiple comprehensive datasets with the following objectives:

1. Combine a comprehensive dataset for fed-batch CHO processes including in-line measurements of pH, dissolved oxygen, and temperature averaged at hourly rates; at-line measurements of cell density and extra-cellular metabolites performed daily; off-line measurements of complex intra and extra cellular metabolites performed at critical points of inflection as defined by the process; off-line quality analysis of the mAb expressed including glycosylation, charge variants, and purity at completion of the process.
2. Look at variations to this dataset which would include derived parameters from both engineering calculations and metabolic signatures of the culture. The goal of developing metabolic signatures is to be able to use a minimum number of measured metabolites which can identify different cell metabolic states as they relate to the PLS model.
3. Determine a predictable set of latent variables from in-process data that correlate with variation in the process and predict variability in product quality attributes namely glycosylation, charge variants, and purity.
4. Identify the key individual measurements driving the model in predicting the quality attributes named above.
5. Implement in-line or at-line cell culture measurement techniques, to collect more frequent data points for the key individual measurements correlating to predictability of quality attributes and implement at-line cell culture measurement techniques, such as amino acid analysis and TCA cycle intermediates, where in-line solutions are not practical

This presentation will discuss how to determine key cell culture measurements required to build a predictable model for quality attributes. This presentation will also discuss which approaches may be robust for routine process development and which assays are most critical to be run for routine process development, scale up and troubleshooting.

(582bg) Engineering Human Paraoxases for Quorum Sensing Quenching

Li, C. X.,
Ge, X., University of California Riverside

Bacterial biofilms are responsible for persistent infections such as cystic fibrosis, chronic wounds, etc. Mounting evidence implicated that formation of biofilms and expression of virulence factors by the pathogenic bacteria are controlled through intercellular communication mediated by quorum sensing (QS). QS signal molecules, *e.g.* acyl homoserine lactones (AHLs), have been therefore recognized as attractive pharmaceutical targets for fighting antibiotics resistance. Unlike antimicrobials that inhibit cell growth, enzymatic hydrolysis/degradation of QS molecules cause less selection pressure for bacterial resistance. Toward the development of QS-quenching enzymes, bacterial lactonases have been extensively studied but present a daunting challenge caused by their immunogenicity. Here we apply both rational and combinatorial approaches to engineer human paraoxase 2 (huPON2) as a novel agent to fight bacterial infections. Particularly, wild type huPONs are extremely difficult for expression in their soluble format and exhibit low activities. By replacing the extruded hydrophobic helixes with short linker peptide libraries, huPON2 clones able to be solubly expressed were successfully isolated. In addition, these clones clearly demonstrated their lactonase activities toward 3-oxo-C_{12}-HSL. To further engineer huPON2s as potent therapeutics, directed evolution has been applied to improve their enzymatic specific activities toward targeted QS lactones found in *Pseudomonas aeruginosa*, *Enterobacter cloacae*, etc clinically significant pathogens.

(582ba) Understanding the Functional Oligomerization State of Peridinin-Chlorophyll a-Proteins in the Dinoflagellate Symbiodinium

Jiang, J., Washington University
Zhang, H., Washington University in St. Louis
Blankenship, R., Washington University in St. Louis
Lo, C. S., Washington University in St. Louis

Symbiodinium is the most commonly found group of endosymbiotic dinoflagellates in corals. Peridinin-chlorophyll *a*-proteins (PCP) are one of the major light harvesting complexes in photosynthetic dinoflagellates. The X-ray crystal structure of MFPCP (main form PCP) of *Amphidinium carterae* has revealed a trimer [1], while HSPCP (high-salt form PCP) from the same strain and recombinant PCP crystallized as monomers [2, 3]. It is still unclear, whether the trimer of *Amphidinium carterae* MFPCP is an exception, or trimerization is common among PCPs. Our study utilized native mass spectrometry, gel filtration chromatography, chemical cross-linking and small-angle neutron scattering (SANS) to reveal the functional oligomerization state(s) of PCP in the dinoflagellate *Symbiodnium*. We found that most PCP in aqueous buffer is monomeric, with the presence of a small amount of trimeric PCP. We conclude that PCP monomer is able to fulfill its function, e.g. light-harvesting, but the oligomerization states *in vivo* remains to be understood.

[1] E. Hofmann, P.M. Wrench, F.P. Sharples, R.G. Hiller, W. Welte, K. Diederichs, Structural basis of light harvesting by carotenoids: peridinin-chlorophyll-protein from *Amphidinium carterae*, Science, 272 (1996) 1788-1791.

[2] T. Schulte, F.P. Sharples, R.G. Hiller, E. Hofmann, X-ray structure of the high-salt form of the peridinin-chlorophyll *a*-protein from the dinoflagellate *Amphidinium carterae*: modulation of the spectral properties of pigments by the protein environment, Biochemistry, 48 (2009) 4466-4475.

[3] T. Schulte, D.M. Niedzwiedzki, R.R. Birge, R.G. Hiller, T. Polivka, E. Hofmann, H.A. Frank, Identification of a single peridinin sensing Chl-*a* excitation in reconstituted PCP by crystallography and spectroscopy, Proc Natl Acad Sci U S A, 106 (2009) 20764-20769.

(582i) Polymer-Conjugated Enzyme Biohybrid Carriers for Detoxification Processes-Internal Diffusion and Kinetics

Compton, R.,
Eggert, M.,
Byrne, M. E., Auburn University
Chambers, R. P., Auburn University

Polymer-Conjugated Enzyme Biohybrid Carriers for Detoxification Processes-Internal Diffusion and Kinetics

Robert J. Compton(1), Matthew W. Eggert(1,2), Mark E. Byrne(1,2), Robert P. Chambers(1)

1. Department of Chemical Engineering, Auburn University

2.Biomimetic & Biohybrid Materials, Biomedical Devices, and Drug Delivery Laboratories, Department of Chemical Engineering, Auburn University

Biohybrid materials, or materials utilizing biological materials in their rational design, are prime candidates to act as therapeutic carriers for detoxification processes. Our group has previously introduced a biomimetic hybrid structure capable of effective ethanol reaction catalyzed by alcohol dehydrogenase covalently bound to a suitable hydrogel through multipoint attachment. This biohybid material is yeast alcohol dehydrogenase covalently attached to a poly(MAA-coPEG200MA-coPEG200DMA) pH sensitive hydrogel network via UV free-radical polymerization following functionalization of the enzyme amino groups with acryloyl chloride. The acryloyl chloride enables multipoint attachment of the polymer to the enzyme by acylating the enzyme at multiple amino groups.

The biohybrid material, alcohol dehydrogenase covalently attached to the hydrogel polymer through multipoint attachment, was studied extensively to determine its intrinsic rate expression and kinetic parameters, including enzyme activity, ethanol, acetaldehyde, [NAD^+] and NADH kinetic constants, for comparison to those for the free enzyme. The effect of internal diffusion on the performance of this biohybrid material was investigated experimentally by using hydrogels of varying characteristic lengths and varying enzyme loading. Internal diffusion effects were also studied through computer simulations using the intrinsic rate expression together with literature diffusivities of ethanol, acetaldehyde, [NAD^+] and NADH to develop appropriate Thiele moduli to properly interpret the above internal diffusion experimental effectiveness factor kinetic data.

Biohybrid materials, although quite complex, show tremendous potential as multi-enzyme therapeutic carriers for detoxification processes.

(582dz) Positive Inotropy - a Likely Mechanism for Lipid Reversal of Local Anesthetic Toxicity

Akpa, B. S., University of Illinois at Chicago

In small animal experiments and a growing number of clinical cases, lipid therapy – the intravenous infusion of an oil-in-water emulsion – has been observed to rapidly reverse the cardiotoxic effects of lipophilic drugs. Patients experiencing otherwise fatal poisonings have recovered a spontaneous cardiac function in as little as 60 seconds following prolonged cardiac arrest. The mechanism by which toxicity reversal by lipid infusion occurs is poorly understood. The variety of toxins and doses to which patients are exposed (as well as the obvious ethical impossibility of inducing cardiac failure) limits the opportunity to conduct clinical trials for lipid resuscitation. *In vitro* observations support the often-cited lipid sink theory of therapeutic action – oil droplets are thought sequester lipophilic molecules from the aqueous blood plasma. However, competing theories exist. In small animal experiments, IV lipid has been observed to act as a positive inotrope – directly improving cardiac function. It is proposed that this effect may be primarily responsible for the dramatic recoveries observed in lipid reversal of anesthetic-induced cardiotoxicity. In an effort to quantitatively probe the merits of lipid sequestering vs. inotropy as a dominating mechanism of resuscitation, a physiologically based pharmacokinetic-pharmacodynamic model (PBPK-PD) has been developed.

The PBPK-PD model was developed in-house. For both the human model and rat model, the simulation represents both bupivacaine and lipid pharmacokinetics and accounts for plasma binding of bupivacaine by 2 classes of plasma proteins, red blood cells, and plasma lipid. In all cases, bupivacaine binding is modeled as concentration-dependent and saturable. The pharmacodynamic model of cardiac function allows for the negative chronotropic effect of bupivacaine, the positive inotropic effect of lipid, and the flow promoting influence of IV fluid infusion. Observables include heart rate, mean arterial pressure, and cardiac output. The pharmacokinetic model includes no fitting parameters. The PD model includes a negative feedback control loop whose parameters are estimated from *in vivo* small animal data. The rat PBPK-PD model was used to simulate a low-dose scheme known to cause toxicity that resolves without CPR intervention. Lipid therapy was simulated as operating via (i) a sequestering mechanism, (ii) a positive inotropic mechanism, and (iii) a combination of the two.

The human bupivacaine PK model yielded quantitative agreement with available pharmacokinetic data. The lipid sink mechanism was predicted to yield (1) an increase in total plasma concentration, (2) a decrease in unbound concentration, and (3) a decrease in tissue content of bupivacaine. Bupivacaine concentration falls rapidly in the heart, but by less than 15%. In the rat model, the inotropic effect resulted in a decrease in tissue concentration, but yielded a more rapid recovery of cardiac function. The combined effect of the sink and inotropic mechanisms was similar to that of the inotropic mechanism alone. This study suggests that, in isolation, the lipid sink effect may not be sufficient to guarantee a reversal of local anesthetic systemic toxicity. The positive inotropic effect of lipid is, however, predicted to promote rapid recovery of cardiac function.

(582cr) Robust Sample Generation and Data Analysis for the Application of Metabolomics in Bioprocessing

Janakiraman, V., Case Western Reserve University
Zhang, A., Biogen Idec
Kshirsagar, R., Biogen Idec
Ryll, T., Biogen Idec

In recent years there has been a rise in –omics approaches to understand cellular phenotype with applications in bioprocessing. In particular, metabolomics involves the identification of small molecule metabolites which offer a more accurate snapshot of the cellular metabolism as a result of process and environmental parameters. Few instances in the literature have looked at quench and extraction methods to better preserve the cell metabolism. This work will emphasize the importance of a well hypothesized study design as a first step for the successful application of global metabolite profiling. Further, the development of a robust method for sample collection will be reviewed by discussing the impact of factors such as cell pellet handling, quenching of metabolism, time of exposure and cell pellet resuspension. The amount of data generated from such a study is typically large and insights will be provided on data analysis from sample studies.

(582de) Stabilization of Vaccines in Silk

Zhang, J., Auburn University
Panilaitis, B., Tufts University
Kaplan, D. L., Tufts University

There currently exists a need for new vaccine delivery systems able to improve storage stability and vaccine efficacy. Sensitive biological compounds, such as vaccines, traditionally require a time-dependent 'cold-chain' to maximize therapeutic activity. All vaccines require constant cold storage and special processing to ensure the necessary potency is maintained. The WHO estimates nearly 50% of all vaccines are wasted each year due to improper storage and delivery. To address these limitations, we demonstrate a novel silk-based vaccine delivery system able to entrap and stabilize the live measles, mumps and rubella (MMR) vaccine at ambient conditions and higher, eliminating the need for specialized storage needs. Using silk, a mechanically robust biomaterial with a unique block copolymer structure that provides a durable matrix, we demonstrate enhanced stabilization of these labile compounds and also provide mechanistic insight.

MMR® II (Merck & Co., Inc., USA), a commercial lyophilized powder vaccine, was used to demonstrate the stabilization effects of silk matrices. Standard MMR-silk films were prepared by casting an aliquot of MMR-silk solution on Teflon-coated molds and air-dried. Lyophilized MMR-silk films were prepared by freeze-drying aliquot of the same solution. Vaccine potency assays were conducted via inoculation of Vero cells (African green monkey kidney cells). The viral RNA from infected cells was extracted using TRIzol/chloroform, converted to cDNA and quantitated using qPCR. The structural characteristics of protein stability were measured using solid-state differential scanning calorimetry (DSC) and liquid-state nano-DSC. Aggregation of viral particles as a function of temperature was examined by dynamic light scattering (DLS).

Encapsulation in silk film (air-dried or lyophilized) enhanced measles, mumps, and rubella viral particle stability when stored at 25°C, 37°C and 45°C and exhibited improved residual potency over the MMR vaccine alone. For the measles component of the vaccine, after 6 months stored at 25°C, the MMR-silk films retained 83.9% potency compared to 74.5% for the MMR vaccine powder. At 37°C, silk films retained 56.5% potency compared to 9.9% from the powder. At 45°C, the vaccine powder lost all potency after 20 weeks in storage while the silk films retained 53.4% activity after 24 weeks. The mumps and rubella components displayed similar trends in potency retention. Lyophilized MMR-silk films reduced the residual moisture of the complex and after 6 months in storage at 37°C and 45°C, the lyophilized films retained $\geq$ 85% residual potency for all viral components of the vaccine.

The DSC thermogram of the MMR powder showed a glass transition (T_g) at 68.9°C, while the lyophilized MMR-silk films showed a T_g at 89.2°C. The nano-DSC thermogram of the purified viral particles in silk solution showed a transition point (T_p) at 68.3°C, compared to that of the viral particles in water at 16.8°C. The increase in T_g and T_p values are indicative of silk stabilizing the conformations of the protein components of the vaccine and the pure viral proteins, respectively. Conformational changes, such as protein unfolding, may affect the stability of the viral proteins by inducing aggregation and thus reducing vaccine potency. DLS results showed the purified virus solution exhibited an increase in mean effective diameter around 16°C while viral particles in silk solution did not show signs of aggregation until 70°C, indicating silk provided structural stability that prevented thermal-induced aggregation of the viral proteins. Silk is composed of large hydrophobic domains interspersed with small hydrophilic regions that form ordered crystalline domains (β-sheets) that stabilize via physical crosslinks. This assembly forms structural pockets that can immobilize bioactive molecules and improve their stability by minimizing water content and reducing protein chain mobility, thus preventing a transition from the native to denatured state. The unique chemistry, structure and assembly of silk make this protein polymer an attractive candidate for the stabilization of bioactive molecules at elevated temperatures.

Silk reduced the temperature-induced protein unfolding and subsequent aggregation by providing structural stability to the vaccine and elevated the temperature at which the viral proteins denature. These results show the potential of encapsulation of labile therapeutic molecules in silk to transform the approach to storage of these molecules and minimize the need for adherence to the 'cold-chain'.

(582bz) Rewiring Yeast Sugar Transporter Preference Through Modifying a Conserved Protein Motif

Young, E. M., University of Texas at Austin
Tong, A.,
Alper, H., The University of Texas at Austin

Metabolic engineering is a powerful platform for controlling cellular production of biomolecules. However, the field is still limited by few effective methods to alter the carbon source preference of an organism. For the case of lignocellulosic biomass conversion, recombinant strains of *Saccharomyces cerevisiae* are limited in their catabolism of xylose. Specifically, both sugar transport and assimilation steps limit pentose utilization in these strains. While much progress has been made in altering the pentose catabolic steps, transporter engineering has only recently been shown to be effective. In this talk, we will discuss efforts in the identification, directed evolution, and rational modification of major facilitator superfamily transporters for improved xylose uptake in *S. cerevisiae*. In particular, through this work, we have identified a conserved motif that is responsible for sugar transport efficiency and selectivity. By modifying this region, we have rewired the function of a transporter from a generalist capable of transporting both glucose and xylose into a specialist that is only able to support the transport of xylose. This engineering of the first yeast xylose transporter capable of supporting robust cell growth is of strong interest for biofuels applications. Additionally, this work has increased the overall understanding of yeast sugar transporters and represents another step forward for molecular transporter engineering.

(582ad) Using Design of Experiment (DOE) Principles Toward Titer Optimization in Perfusion Cell Culture

Zustiak, M. P., Johns Hopkins University
Caple, M. V., SAFC Biosciences (Member of Sigma-A

Over the past several decades tremendous improvements have been made in biotherapeutic protein production through both cell line engineering and media development. Titers of over one gram per liter are expected even for a non-optimized 14-day fed-batch process producing a monoclonal antibody. However, not all potentially therapeutic molecules are as stable as an IgG, and a continuous perfusion process is required. While the risks and economics of a perfusion process can be debated, perfusion cell culture allows for a unique system to test the effects of specific parameters or media components on culture performance by taking advantage of the steady state nature of perfusion. Unfortunately for process developers, there is no high-throughput scale down model available for perfusion cell culture. In the current study, we work toward developing a valid high-throughput scale down model for specific parameters in perfusion cell culture by evaluating several significant factors.

Initially, batch experiments were conducted using a factorial design for media development. One particular media component, component X, showed a significant improvement in batch titer over a relatively small concentration range. A dose-response experiment was then conducted to determine the saturation and potential inhibitory or toxic effects of this component. While toxic effects were not seen for concentrations over three logs, saturation was seen at approximately one log increase in starting component concentration, which corresponded to a tripling in product titer. While an optimum was not determined, a point well into the saturation curve was chosen for further study and use in a perfusion process. Unlike the batch study, increases in perfusion titer were highly variable. These variations were later tied to Lot-to-Lot variability in the basal media, which is a major risk factor in perfusion processes. Not only were the basal levels of component X found to be highly variable (50-150%), it was also determined that the small variability in incoming osmolality and pH correlated well with the variability in titer. To further explore this correlation, a response surface design was implemented in a batch shake flask model, looking at the impact of starting pH, osmolality and the concentration of component X. Our results indicate a significant impact from both osmolality and the concentration of component X, but not from the starting pH. To investigate potential mechanisms of increased titer, additional studies were conducted around the mode of osmolality enhancement as well as mock perfusion to better control pH. Previous studies have shown cell specific productivity to be highly pH sensitive and thus, developing a representative scale down model for perfusion that captures this sensitivity is highly desired. To that end, a full perfusion study was conducted around pH, osmolality, and the concentration of component X which was then compared with the results from the scale down model.

With this study we have explored the use design of experiment (DOE) techniques to move from component and parameter screening to a scale down model for perfusion processes development.

(582cg) Rhamnolipid Biosurfactant Production and Engineering By Evolution in Escherichia Coli Towards Enhanced Oil Recovery

Han, L., Tianjin Institute of Industrial Biotechnology, Chinese Academy of Sciences
Liu, P., College of Pharmacy, State Key Laboratory of Medicinal Chemical Biology, Nankai University
Lin, J.,
Wang, Q., Key Laboratory of Systems Microbial Biotechnology, Tianjin Institute of Industrial Biotechnology, Chinese Academy of Sciences

Rhamnolipids have the potential to play the important role in enhanced oil recovery (EOR) as the substitutes of chemical surfactants, and mainly produced from *Pseudomonas aeruginosa* that is pathogenic strain and limited or not suitable for future industrial development. Here, *Escherichia coli* strains were designed to produce either mono-rhamnolipids or di-rhamnolipids by introducing different gene combinations related to rhamnolipid biosynthesis from *P. aeruginosa*. Rhamnolipids from engineered strains showed different performance in EOR due to the different chemical structures. The directed evolution and subsequent site-directed mutagenesis of rhamnosyltransferase 1 resulted in the production of new types of rhamnolipids with the increased ratio of C10-C10 congener and different performance in EOR. Computational modeling of mutations elucidates the function and specificity for the production of rhamnolipids. Creation of these mutations fundamentally provides evidence for new biocatalyst design and the corresponding production of rhamnolipids, by which expanded the possibility to improve oil recovery.

(582j) Fitselect: A Synthetic Gene Circuit to Guide Combinatorial Metabolic Engineering Experiments

Freedman, B. G., Virginia Tech
Senger, R. S., Virginia Tech

The synthetic gene circuit "Feedback Inhibition of Transcription for Substrain Selection (FITSelect)" was developed to guide enrichment of a genomic over-expression library in *Escherichia coli* BL21 for enhanced production of L-arginine. The genomic library was derived from degenerate oligonucleotide priming of environmental samples using a novel procedure that allows for the use of error-prone PCR while producing 1-5Kb reads. This procedure resulted in significantly enhanced L-arginine concentrations upon library enrichment. The FITSelect circuit can contain both native and orthogonal parts including the *argR* promoter, which is responsive to L-arginine concentrations, and a growth inhibitory gene. Two inhibitory genes were investigated in this research: (i) the Control of Cell Death (*ccdb*) gene and (ii) a gene encoding a microfiber-forming peptide. Intracellular levels of CCDB are inversely related to the cell growth rate, and microfiber formation has been shown to significantly impact cell physiology and density. In continuously subcultured mixed cultures, substrains with the highest growth rate will ultimately dominate, allowing for selective isolation of strains with lower CCDB levels and higher L-arginine concentrations. Likewise, cells producing and incorporating microfibers are easily separated due to their larger size and density. This research compares these two competing approaches in deriving a culture with superior production of L-arginine.

(582k) Aggrecan As a Chondroinductive Signal in Interpenetrating Network Hydrogels for Cartilage Tissue Engineering

Rennerfeldt, D., University of Kansas
Gehrke, S. H., University of Kansas
Detamore, M., University of Kansas

We recently introduced interpenetrating network (IPN) hydrogels composed of agarose and poly(ethylene glycol) diacrylate (PEGDA) as a promising material for cartilage tissue engineering. We have shown that swelling and mechanical properties of such IPNs can be tuned by selecting appropriate monomer concentrations and PEGDA molecular weights while maintaining encapsulated chondrocyte viability. In the current study, we investigated a higher molecular weight than we have previously and found that IPNs consisting of 15% w/v PEGDA (molecular weight = 10 kDa) and 2% w/v agarose exhibited nearly a four-fold increase in compressive modulus over single-network PEGDA gels (197 kPa vs. 56.8 kPa) with only a slight decrease in swelling ratio (11.8, vs. 16.8 in pure PEGDA gels).

In addition, we have recently shown that the incorporation of aggrecan, a proteoglycan largely responsible for the stiffness and swelling properties of cartilage, improved encapsulated chondrocyte viability and matrix synthesis in agarose/PEGDA IPNs. In our presentation, we will report the findings of a study in progress with a different, more clinically relevant cell source: bone marrow stromal cells (BMSCs). The aggrecan has not adversely affected the swelling ratio or the mechanical performance of the IPNs, as expected. Therefore, the results of viability and gene expression assays from our ongoing studies will allow us to test our hypothesis that we may be able to control the cell response independent of the already superior mechanical performance of the IPNs.

(582l) Mechanical Phenotyping of Tumor Cells Using a Multiplexed Microfluidic Cell Squeezer Device

Kamyabi, N.,
Khan, Z. S., Texas Tech University
Vanapalli, S. A., Texas Tech University

Studies have indicated that highly aggressive cancer cells have distinct mechanical properties compared with benign and less aggressive cells. We are investigating the potential of cell mechanics as a biophysical marker for cancer diagnosis. To establish the significance of mechanical properties for cancer diagnosis, a high throughput method is desired. Although techniques such as atomic force microscopy are very precise, they are limited in throughput for cellular mechanical property measurements.

To develop a device for high throughput mechanical characterization of tumor cells, we have fabricated a microfludic cell squeezer device that contains multiple multiplexed micrometer-scale pores. Fluid flow is used to drive cells into these pores mimicking the flow-induced passage of circulating tumor cells through the vasculature. By integrating high resolution imaging, the device allows for the characterization of cell sizes and passage times through the squeezer. We have tested highly and lowly invasive prostate cancer cell lines, and have found that the passage times are capable of differentiating between cell lines with differing invasiveness. We will additionally report on the capability of our device to perform rheological measurements by testing simple model systems including viscous and viscoelastic drops and soft elastic particles.

(582m) The Complimentary Use of Sophisticated (SIMS) and Crude (Nile red) Chemical Screening to Improve Algae As a Biofuels Production Platform

Yoo, J., The Pennsylvania State University
Jackson, L., The Pennsylvania State University
Johnson, R., The Pennsylvania State University
Geveke, B., The Pennsylvania State University
Rajangam, A., The Pennsylvania State University
Mohandass, R.,
Winograd, N., The Pennsylvania State University
Curtis, W. R., The Pennsylvania State University

We present several lines of research focused on improving algae oil production. Secondary Ion Mass Spectrometry (SIMS) is being used to better understand the extracellular matrix of the hydrocarbon-producing alga *Botryococcus braunii* race B which consists of triterpene oils as well as a surrounding polysaccharide network. Nile red provides a crude means of correlating oil content for screening of different algae lines as well as different growth conditions for oil formation. These two methods therefore represent the opposite ends of the spectrum in terms of in depth, data-intensive chemical profiling and rapid, crude production screening.

SIMS CHEMICAL IMAGING: *Botryococcus braunii* is a hydrocarbon-rich microalga that has a poorly understood and complex extracellular matrix within which the algae grow as a colony of cells. Additionally, a polysaccharide network surrounding *B. braunii* colonies increases the difficulty of studying and cultivating these valuable algae. SIMS provides an unprecedented depth of spatially-resolved chemical information in 2 to 3 dimensions of *B. braunii*. In this manner, one can develop a chemical image of several chemical species within the algae. This is particularly useful for biofilms, or colony forming algae such as *B. braunii* race B. Accordingly, high-resolution chemical imaging has been used to investigate the hydrocarbon storage and distribution of this strong candidate for alternative fuels. Organelles within the hydrocarbon-rich extracellular matrix surrounding the algal cells have been chemically imaged utilizing a 250 nm diameter C60+ primary ion beam on the J105 3D Chemical Imager (Ionoptika, LTD). Organelles of great interest are approximately 1 -2 μm round vesicles within the extracellular matrix, which have been proposed to contain C23 – C37 liquid hydrocarbons, termed "botryococcenes," unique to *B. braunii* race B. They are revealed after depth profiling through the algal cell colonies, thus, enabling the acquisition of chemical information in three dimensions. Unique chemical components of *B. braunii* over m/z 600 may be chemically mapped. Chemical signatures for the polysaccharide network, algal cell wall, and the liquid hydrocarbon extracellular matrix have been identified. The botryococcenes and hydrocarbon-derivative chemical species are located in specific locations dependent on their function within the cell and in the colony. Peak identification has been aided by the use of tandem MS capabilities. With the use of this unique chemical imaging technique, questions surrounding the purpose and function of the *B. braunii* cell's behavior and association with hydrocarbons may be investigated.

ESTABLISHING AXENIC CULTURES: An outcome of the SIMS work was a better understanding of the polysaccharide matrix which appears to function in part by holding the colonies together. Since it is desirable to have axenic cultures of *B. braunii* to carry out chemical feeding and genetic transformation studies, the presence of cellulose in the surrounding matrix suggested we might dissipate the algae colonies with cellulase treatment. This has been substantiated by chemically mapping the presence of cellulose within *B. braunii* colonies. Accordingly, the use of hydrolytic enzymes with endoxylanase and cellulase activity to degrade the polysaccharide wall surrounding these algal colonies resulted in the isolation of single cells. These results indicate that an axenic culture may be obtained through dilution plating or laser capture microdissection of these single cells. The outcome of this work will enable the mixotrophic growth of *B. braunii* resulting in an efficient and expedient means to scaling up for large scale biofuels production.

NILE RED SCREENING: Complementary to the data-intensive chemical profiling provided by SIMS, Nile red fluorescence staining provides a means of rapidly screening for the hydrocarbon content in microalgae. This method has been applied for the determination of the neutral hydrocarbon content in microalgae, where fluorescence intensity is correlated to the quantity of oil present. Conventional solvent extraction methods are time consuming and require large culture samples as compared to the staining of algae in a 96-well plate. With very little sample preparation and low sample volume, a fluorometer using 530 nm excitation and 575 nm emission enables the quick and accurate analysis of oil content in a given algal culture. The ability to sample small amounts of culture volume under a variety of culturing conditions enables high-throughput screening for maximum oil productivity.

(582do) Development of Binding Domains for Improved Cellulose Decrystallization and Hydrolysis

Kang, Y., Georgia Institute of Technology
Bansal, P., Georgia Institute of Technology
Bommarius, B., Georgia Institute of Technology
Realff, M., Georgia Institute of Technology
Bommarius, A. S., Georgia Institute of Technology

The cellulose-binding domain (CBD) has long been recognized as a key moiety in cellulolytic enzymes to bring the full-length enzyme to a close proximity of the cellulose substrate. As one of the most efficient exoglucanases, *T. reesei* Cel7A has attracted enormous interest. The cellulose-binding domain of Cel7A has been reported to be capable of decrystallizing Avicel. In this work, the top nine CBD variants were identified by Principal Component Analysis (PCA), a data-driven protein engineering method, with the concept of consensus. Both wild-type CBD and the identified mutants were expressed as GFPuv fusion proteins in bacterial and yeast expression systems. We identified the mutants with improved functionalities including both cellulose-binding affinity and cellulose decrystallization effect, opening the chance to combine the CBDs with the catalytic domain to yield variant cellulases with improved enzymatic activity on cellulose substrate. Besides, the effects of glycosylation and linker length on substrate binding capacity, reversibility of substrate binding and/or cellulose decrystallization will be discussed.

(582dp) Targeted Chemical Biotinylation of Protein Complexes Using Monomeric Streptavidin

Mann, J., University at Buffalo
DeMonte, D., University at Buffalo
Park, S., University of Pennsylvania

The majority of proteins in the cell interact with other proteins as part of their natural function. Identifying the components of stable and transient protein complexes is therefore important to understand protein function. To this end, mass spectrometry (MS) can significantly accelerate the rate of discovery by analyzing complex protein mixtures through high resolution peptide sequencing. A critical step during MS analysis is sample preparation. The protein complex of interest needs to be separated from other cellular components, many of which are present in high abundance and interact nonspecifically with the target protein. We used monomeric streptavidin (mSA), which was recently designed in our lab, to achieve selective chemical biotinylation of the proteins that interact with a protein of interest. Biotinylation allows the interacting proteins to be subsequently purified on streptavidin beads utilizing well known streptavidin-biotin interaction. Since biotinylation is irreversible, even weakly or transiently associated complexes can be purified using the technique. This technique differs from other existing purification schemes commonly used for MS analysis, e.g. tandem affinity purification (TAP), in that it does not require any prior modification to any part of the protein complex that is being studied. Rather, the chemical crosslinking is catalyzed by mSA, which brings the crosslinking reagent to close proximity of the target molecule. We demonstrate that this technique is specific and works both in vitro and in vivo, and show that it can be implemented in a number of different ways without loss of efficiency.

(582n) Quantitative Tailoring of RNA Device Activity for Conditional Control of Cell Fate
Chang, A. L., Stanford University

RNA is a versatile substrate for encoding sensing, processing, and control functions within cells. Recent advances have demonstrated the use of RNA-based control devices to program sophisticated cellular functions. The ability to quantitatively tailor device regulatory activities will support broader implementation in cellular networks by effectively matching activities of circuit components.

Strategic genes candidates to regulate using programmable controllers possess activities that function as key regulation points in cellular decision making, including threshold activities that trigger phenotypic switching. Modulating gene expression across this threshold using a ligand-responsive RNA switch provides access to both phenotypes, based solely on ligand availability.

We developed a high-throughput, quantitative, cell-based screening strategy to support rapid generation of ribozyme-based RNA switches with user-specified regulatory activities. A dual reporter system was implemented to resolve device regulatory activity from cellular noise, enabling clean isolation of devices with targeted activities. Device libraries were generated through sequence randomization of individual components, from which we isolated switches with enhanced regulatory activities, including greater than tenfold increases in catalytic activity. Furthermore, our screening strategy is capable of enriching a single functional sequence from over a million variants to near purity in two sorting rounds.

We targeted a negative regulator of the yeast mating MAPK pathway to conditionally direct cellular behavior to either a mating or nonmating fate in response to an environmental trigger molecule. A series of RNA devices with varying regulatory activities were tested to locate the negative regulator's threshold activity. Implementation of an RNA switch with improved regulatory stringency that was isolated using our cell-based screen resulted in distinct switching in gene expression, mating pathway activation, and cell cycle arrest phenotype. The ability to match regulatory activities of genetic control elements to thresholds activating phenotypic switching enables rewiring of biological networks and control of cellular decision making in response to user-defined molecular signals.

(582ae) Brownain Dynamics Simulation of Discrete Red Blood Cell Model Based On Kinetic Theory With Constraints

Kim, K. H., University of Illinois at Chicago

The progression of disease is oftentimes accelerated by the changes in the mechanical behavior of cells. The rheological material properties of biological cells such as red blood cells are important factors in the occurrence and onset development of diseases. A better fundamental understanding of flow and its rheological properties will help rational approach to prevention of serious conditions and its costs, and development of new treatments as well.

The empirical data of the human blood, for example, has non-Newtonian behavior and its elastic properties have been measured. The viscosity drops as the shear rate increases instead of remaining constant as in the Newtonian fluids. Often times the pure mathematical Casson model has been used to model the behavior of the blood since it gives the characteristic of power-law slope of the viscosity similar to that of experimental data. However, the model neither predicts the time-dependent behavior nor fit all the range of shear rate empirical data where the viscosity tends to have nearly constant viscosity at the low and high shear limits. More importantly, the Casson model gives us no insight into the dynamics of cells in complex flows.

The Brownian Dynamic simulation will allow predicting the movement of biological cell in methodological way with the consideration of all the forces such as hyodrodynamic, the Brownian, intermolecular, and constraint forces. Instead of considering blood as continuum fluid or using an empirical constitutive equation, we can use kinetic theory where the stress tensor and a stochastic differential equation (SDE) of motion depend on the configuration of the microstructure of the fluid. Blood is considered as a suspension of deformable red blood cells (RBC) in a dilute solution of Newtonian fluid. Our approach is to construct a discrete model of bead-spring RBC based on kinetic theory with a holonomic constraints using Lagrange multiplier method. The constraints such as total area, cell volume, and local curvature describe the effect of the biological cell conserving its overall size throughout the motion of flow while allowing the shape to deform. We present our method and demonstrate its applicability to study rheology of biological cells and further study of blood flow. To validate the capability of the method, the minimalist bead-spring-ring model to represent the RBC was simulated. A simple linear Hookean spring is used to give flexibility to deform. The two geometrical constraints were set to have constant area and constant total length of the connected springs. The two-constraint model gives more restriction to deformation than one-constraint method where the constraint is only an area, and yet still has some degree of flexibility. The model is tested under shear and shear free flow. An assumption was made that the friction tensor is isotropic. Averaged rheological material properties of the fluid can be calculated with the components of stress tensor. The rheological material properties obtained are qualitatively in agreement clearly showing shear-thinning effect of viscosity in shear flow. We can also observe Fahraeus-Lindqvist effect in the capillary flow with this study where the viscosity drops as the vessel diameter decreases.

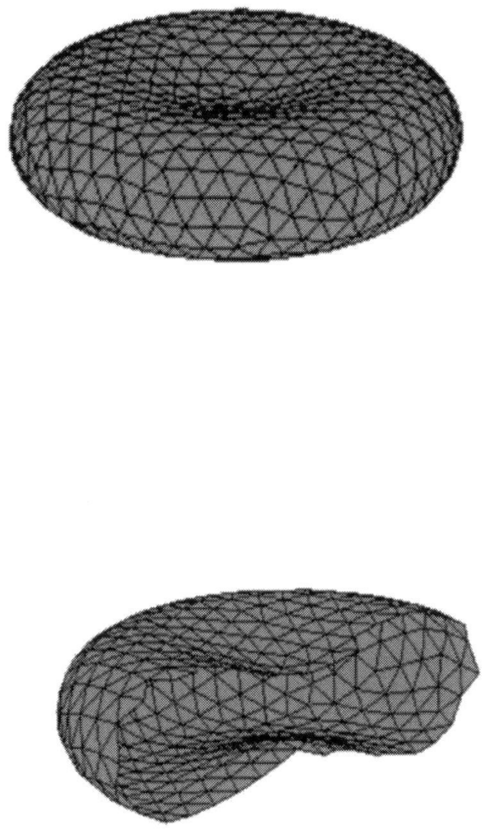

Figure. Biconcave shape of RBC using mathematical expression from Kuchel (1999) in Cartesian coordinate. 3D surface meshing of 558 particles is generated using distance functions and Delaunay triangulation algorithm.

Further study on more complicated geometric modeling of microstructures can be done with hydrodynamic interaction.

(582cb) Engineering Acetogenic Clostridia for Ethanol and n-Butanol Production From CO2

Chen, C. C., The Ohio State University
Yang, S. T., Ohio State University

Some *Clostridia*, including *C. ljungdahlii* and *C. carboxidivorans*, are capable of producing ethanol and butanol from CO_2 and H_2. However, the alcohol productivity, yield, and titer from these native strains are low and uneconomical for industrial applications. In contrast, several acetogenic *Clostridia* can convert CO_2 and H_2 to acetate with high yield, titer, and productivity comparable to those from glucose as the substrate. A high metabolic flux from CO_2 to acetyl-CoA, the immediate precursor for both acetate and ethanol, is the prerequisite for developing a highly productive ethanol and butanol producers. In this study, *C. aceticum* was used as a host to produce ethanol by expressing *adhE2* gene under the control of the thiolase promoter using four different conjugative plasmids (pMTL82151, 83151, 84151, and 85151) each with a different replicon (pBP1 from *C. botulinum* NCTC2916, pCB102 from *C. butyricum*, pCD6 from *C. difficile*, and pIM13 from *Bacillus subtilis*). The effects of different replicons on transformation efficiency, plasmid stability, *adhE2* expression and aldehyde/alcohol dehydrogenase activities, and ethanol production by different mutants of *C. aceticum* were investigated. Ethanol production from diverse electron donors, including CO_2 and H_2, CO_2 and glycerol, cellobiose, xylose, fructose, and glucose were then investigated with the best mutant strain harboring *adhE2*. We also engineered *C. aceticum* for n-butanol production by introducing butanol biosynthesis pathway genes (*thl*, *hbd*, *crt*, *bcd-etfB-etfA*, and *adhE2*) using modular clostridia plasmids. We then knocked out genes (*pta* and *ack*) in acetate biosynthesis pathway to further enhance ethanol and butanol production. The fermentation kinetics of these engineered mutants and their ability to produce ethanol and butanol from diverse electron donors were studied and will be presented in this paper.

(582bh) Hybrid Riboswitch-sRNA for Dual Transcript Control By a Ligand

Lease, R. A., Ohio State University
Lahiry, A.,
Stimple, S. D.,

Metabolic engineering can be used to alter or augment existing enzymatic pathways in microorganisms, in order to turn these cells into chemical plants that produce high-value specialty chemicals such as fuels and pharmaceuticals. Host cell enzymes that lead to undesired metabolites can be deleted or altered at the DNA level. However, outright deletion of native enzyme pathways that produce (for example) organic acid side reactions during *n*-butanol synthesis may compromise the health of fermentation cultures by blocking sources of cellular energy and reduction potential. Here we describe a novel synthetic biology–circuit element that scales the production of enzyme catalysts from their protein-coding mRNA sequences. Native bacterial small regulatory RNA (sRNA) can control gene expression at the level of protein (*e.g.*, enzyme) translation from mRNAs, and can target multiple mRNAs via specific binding (antisense base–pairing) interactions. The sRNA–mRNA pairing may either stabilize or destabilize the target mRNA transcripts, increasing the rate of protein translation or mRNA degradation, respectively. A plasmid DNA is used to deliver the sRNA via an inducible promoter without direct chromosomal modification. By altering certain sRNA sequences, desired target mRNAs can be specified (orthogonal pairing). Our initial experiments describe retargeting the well-characterized *E. coli* DsrA sRNA to reporter gene fusions with the native mRNA targets of DsrA, then re-targeting of the altered–DsrA sRNA to the reporter genes themselves. Ultimately, desired clostridial mRNAs will be targeted, *e.g.*, acetate kinase (*ack*) and phosphotransbutyrylase (*ptb*) that catalyze production of acetate and butyrate 'contaminant' side reactions of n-butanol fermentation, respectively. To enable real-time ligand feedback control of the sRNA and thereby tune the optimal level of side reactions, a ligand-responsive RNA sensor sequence called a riboswitch (RS) will be incorporated, to create a hybrid RS-sRNA synthetic biology regulatory element. Because the RS can be altered to bind different ligands, the RS-sRNA may be used to transduce a desired ligand-binding event into a "portable" and orthogonal RNA regulator. We will present the results of three distinct approaches to RS-sRNA integration, at the level of sRNA synthesis (using an antiterminator-leader RS), sRNA stability (using an RS integrated within the sRNA), and sRNA activity (using a RS–controlled, directed ribozyme cleavage). The ultimate goal of this work is to enable real-time metabolite-to-genetic feedback control of enzyme production for optimization of biofuel fermentation. The RS-sRNA will be portable between different bacterial species for fermentations and other applications as it does not require chromosomal modification of its host cell.

(582a) Activation of Plant Anticancer Compounds By the Gut Microbiota

Diaz, C. A. C., Stanford University
Klein, A. P., Stanford University
Sattely, E., Stanford University

Plant metabolites consumed through the diet have a broad range of chemopreventive activities, yet the effect depends on the composition of the gut microflora. A prominent example is the glucosinolate class of compounds produced by vegetables in the genus *Brassica* (examples include broccoli, kale, and cabbage).[i] These molecules are inactive 'pro-nutrients' that require deglycosylation to release the active nutrient. A well-studied example is sulforaphane, which is derived from the glucosinolate glucoraphanin; this metabolite is a key player in both plant defense and human health.[ii] The average per-person intake of glucosinolates is estimated to be 18 mg / day in the US, 50 mg / day in the UK, and several times higher in some Asian countries.[iii] The role of gut bacteria in liberating sulforaphane is well documented, but no bacterial gene involved in glucosinolate metabolism has been identified to date.

Here we present bioinformatic analysis and biochemical characterization of a prominent gut microbe *Bacteroides thetaiotaomicron* (*B. theta*) capable of catalyzing the activation of glucosinolates. High-resolution mass spectrometry established that sulforaphane is the primary product of glucoraphanin metabolism by *B. theta*. Four independently isolated, sequenced *B. theta* strains exhibited differing abilities to metabolize glucosinolates, spanning 3 orders of magnitude. Comparative genomic and transcriptomic strategies were employed toward identification of the gene responsible for this deglycosylation. Additionally, a high-throughput screening method was developed that uses the inhibition of *E. coli* growth in *B. theta*–spent media to detect glucosinolate metabolism. So far, a small-scale mutant screen has identified two candidate genes involved in glucosinolate activation, which has elucidated biochemical pathways related to sulforaphane release.

We conclude that *B. theta* acquired the ability to deglycosylate glucosinolates and release potent isothiocyanates (e.g. sulforaphane). Further studies aim to test whether the antimicrobial and chemopreventive effects of our glucosinolate–*B. theta* system restrict the growth of enteric pathogens and cancerous cells in a mouse model. The identification of specific microbial genes that process ingested plant metabolites is essential to predict how the gut microbiota influences nutrient uptake. This type of information offers insight on how to best engineer plant-microbe interactions in the gut to lead to optimal health benefit.

[i] Halkier, B.A. and Gershenzon, J. *Annual Review of Plant Biology* **57**, 303–33 (2006).

[ii] Fan, J. et al. *Science* **331**, 1185–8 (2011). Fahey, J.W. et al. *PNAS* **99**, 7610 (2002).

[iii] Krul, C. et al. *Carcinogenesis* **23**, 1009–16 (2002).

(582df) Evaluating a Novel Aquaporin From R. Sphaeroides for Biomimetic Membrane Applications

Erbakan, M., Pennsylvania State University
Shen, Y. X., Pennsylvania State Univerity
Kumar, M., The Pennsylvania State University
Curtis, W. R., The Pennsylvania State University

We demonstrate that a novel aquaporin from *Rhodobacter sphaeroides 2.4.1.*, RsAqpZ, displays superior water transport characteristics in phospholipid vesicles as compared to well-studied *E.coli* AqpZ and most of the other orthodox aquaporins.

BACKGROUND:Most biological cells are surrounded by phospholipid bilayer protecting them from their surroundings and compartmentalizing subcellular organelles to carry out particular processes. Selective permeable nature of the lipid bilayer allows certain organic molecules and ions across the cell with high efficiency and selectivity, which is essential for maintenance of the cellular homeostasis. Membrane proteins are the key players in recognition and transportation of the solutes with a low energy input and excellent specificity. These exceptional features of membrane proteins inspired many researchers from a variety of disciplines to solve difficult separation problems in sensor technology, energy generation and seawater desalination. Aquaporins, one particular family of highly selective water channel proteins integrated in plasma membrane of a wide spectrum of biological cells ranging from bacteria to mammalian cells, emerged as a great potential for water desalination industries. Aquaporins facilitate the high flux of H_2O merely depending on osmotic pressure gradient across the cell membrane without consuming cellular biochemical energy, while they reject ions and organic molecules. High water transport rate and specificity of aquaporins make them potential candidates for utilization in next generation water desalination membranes to reduce the high operational costs required by reverse osmosis membranes in seawater desalination and purification. Since *Rhodobacter* species live in diverse environments, such as consuming organic acids under anaerobic photosynthetic conditions, cloning, expression and characterization of a novel aquaporin from *Rhodobacter sphaeroides 2.4.1.*, RsAqpZ, was undertaken in anticipation of obtaining a more robust aquaporin that is amenable to biomimetic membrane applications with high stability and enhanced water permeability.

RESULTS: RsAqpZ was expressed in *Rhodobacter* under anaerobic photoheterotrophic growth conditions with a yield reaching at 30 mg/L in its homologous expression system. We compared water transport characteristics of RsAqpZ with widely used *E.coli* AqpZ (EcAqpZ) running stopped flow-light scattering experiments using phospholipid vesicles reconstituted with these proteins. A higher water permeability (0.1700 ± 0.0500 cm/s) and lower energy of activation for water transport (2.93 ± 0.60 kcal/mol) was measured for RsAqpZ. Efficiency of incorporation was also significantly higher for RsAqpZ, as we observed a linear relationship between water permeability and protein mass fraction in proteoliposomes until a protein/lipid ratio of 0.04. The highest protein/lipid ratio, we could obtain with EcAqpZ remained to be 0.01.

Initial investigation of incorporation in triblock copolymers resulted a similar trend, as well, which indicates that can be RsAqpZ a good alternative for EcAqpZ for utilization in biomimetic water desalination membranes. For a better understanding of the enhanced incorporation ability of RsAqpZ, we performed 2D crystallization tests with DOPC lipids using microdialysis method. Initial results predicted a tetragonal symmetry for RsAqpZ in phospholipid membranes showing promises for structure determination by electron crystallography. Finally, we constructed fluorescent protein fusions of RsAqpZ to estimate the single channel permeability using a new method based on fluorescence correlation spectroscopy. The presence of the fluorescent protein tag did not change the expression level and water permeability of RsAqpZ as evidenced by Western-blot analysis and stopped-flow light scattering, respectively.

CONCLUSION: A novel water channel protein from purple bacterium Rhodobacter sphaeroides, RsAqpZ was cloned, overexpressed and purified in functional form. Evaluation of functionality demonstrated high water permeability, low activation energy and high incorporation efficiency in liposome and polymersome systems. Our findings suggest that RsAqpZ might be a better alternative to EcAqpZ for utilization in biomimetic water desalination membranes.

(582ch) Metabolic Flux Balances of Mammalian Cho Cells for Investigation of Trace Element Effects On Productivity and Product Quality

Lee, H., University of Massashusetts
Yoon, S., University of Massashusetts

Biopharmaceuticals (defined as recombinant proteins, monoclonal antibody and nucleic acid-based products) is the fastest growing sector within the pharmaceutical industry. About 900 to 1,200 clinical candidate proteins and/or peptides are currently being investigated. Approximately one in every four drugs introduced to the market is a biopharmaceutical. The global biopharmaceutical market was estimated at $80 billion in 2007 with antibodies representing around $28 billion sales and protein products generating almost $52 billion. It is estimated that sales of biopharmaceutical products will exceed $150 billion by 2012. Despite all of the progress and commercial success, the development of production cell lines and fermentation processes with mammalian cells is, to a large extent, still based on empirical knowledge and historical experience. This is mainly because of missing qualitative and especially quantitative data and a lack of understanding of the intracellular mechanisms under bioprocess conditions. The challenges of selectivity, reproducibility and accurately quantitating these products, have seriously handicapped biopharmaceuticals from being commercialized.

Metabolic flux analysis of the mammalian CHO cells for explaining trace element effects for monoclonal antibody product in suspension cultured culture product and quality was assessed. Based on the metabolic flux analysis of the CHO cell culture samples, group of metabolites associated with the trace elements were identified. The Amino Acid metabolism were correspondingly analyzed. A group of amino acids were clustered. For the purpose of the resutlt confirmation, metabolomics was used.

(582dx) Refolding of Laccase With Copper-Based Ionic Liquid

Bae, S. W., INHA university
Koo, Y. M., Inha University

Ionic liquids (ILs) are molten salts which are gaining attention as green solvents to replace organic solvent due to their non-volatile and thermal stable properties and low toxicity. Their physicochemical properties such as hydrophobicity, polarity, density can be easily tuned by the combination of cations and anions. Therefore, these advantages of ILs facilitate their applications in many biological processes.

Laccase is one of the representative metallo-enzyme which needs copper as a cofactor. Therefore, the activity and stability of laccase can be enhanced in copper-induced systems. In general, increased refolding yield could also be achieved by enzyme stabilization, during protein refolding process.

In this study, copper salts and copper-based IL was employed to enhance laccase refolding yield. In order to confirm the effect of copper on the refolding of metallo-enzyme, the same refolding process was also carried with lysozyme which is not metallo-enzyme. When $CuCl_2$ or [EMIM][$CuCl_3$] were added to the refolding buffer instead of urea, the laccase refolding yield was improved more than 1.5 and 2.7 times, respectively, compared with the conventional refolding buffer containing urea without ILs. However, no additive effect was observed in the refolding of lysozyme which is a non metallo-enzmye. From these results, we can conclude that this copper-based IL was only effective to the refolding process of metallo-enzyme.

(582bp) Metabolic Engineering of Escherichia Coli for the Production of Medically Valuable Polysaccharides

Cress, B., Rensselaer Polytechnic Institute
Linhardt, R. J., Rensselaer Polytechnic Institute
Koffas, M. A. G., Rensselaer Polytechnic Institute

Although microbial capsules have traditionally been studied due to the protective effects they confer to their bearers during infection in the face of host immune systems, a specific subset of capsular polysaccharides can serve as precursors to the production of medically important glycosaminoglycans such as heparin and chondroitin sulfate. Our lab has recently sequenced the genomes of two heparosan-producing *Escherichia coli* strains, Nissle 1917 and K5, and a chondroitin producing *E. coli* strain, K4. Genome-scale models have been constructed and used for prediction of genetic interventions—gene deletions, knockdowns, and overexpressions—leading to increased capsular polysaccharide production. Comparative genomics analyses between these capsule producing strains and model acapsular strains are ongoing to understand differences in nucleotide sugar metabolism and carbon utilization that might have evolved between these classes of microbes. Additionally, DNA aptamers have been selected against purified capsular polysaccharides for use in high-throughput screens. Our work covers a broad range of complementary engineering and informatics approaches and demonstrates the feasibility of producing commercially relevant levels of heparin and chondroitin sulfate from bacterial sources.

(582ct) Effect of Coexpression of Cyanobacterial Phosphoenolpyruvate Carboxylase and Carbonic Anhydrase On Succinic Acid Production By Escherichia Coli

Wang, D.,
Zhou, X.,

In order to study the effect of cyanobacterial phosphoenolpyruvate carboxylase (PEPC) and carbonic anhydrase (CA) coexpression on succinic acid production in *Escherichia coli*, artificial operon LY204 with *ppc* and *ecaA* genes tandem cloned onto the plasmid pTrchisB was constructed and firstly introduced into wild type *E. coli* strain W1485. The highest level of succinic acid production was obtained in strains with PEPC and CA expression concurrently, when compared with strains individually overexpressing of either PEPC or CA. Secondly this operon was transformed into strains with lactate and formate pathways inactivation to eliminate the competing pathways for succinic acid formation. Thirdly the glucose phosphotransferase system was inactivated to increase the PEP pool, which functioned as the substrate of the PEPC enzyme, besides the HCO_3^-. Results showed that PEPC and CA coexpression was a novel approach to improve the succinic acid production.

(582ef) Time-Resolved Methodology Development in Biological SAXS

Franklin, J. M., Michigan State University
Spilotros, A.,
Svergun, D.,

With the development in data modeling methods and next-generation synchrotrons, small-angle X-ray scattering (SAXS) has become a powerful technique for structural biology, allowing low-resolution shape determination of solution-state biomaterials. The PETRA III synchrotron (DESY, Hamburg, Germany) provides the opportunity to perform time-resolved biological SAXS measurements. Using microfluidic devices that can perform precise mixing of small volumes, the kinetics of biological reactions can be resolved. An optimized scheme for millisecond synchronization on the P12 beamline is presented. The results from time-resolved studies of protein folding and drug-delivery kinetics are presented.

(582ag) Estimation of Viable Cell Density During Late Growth Phase of CHO Cell Culture Using Dielectric Spectroscopy

Lee, H. W.,
Yoon, S., University of Massashusetts

In mammalian cell culture processes, viable cell density (VCD) is one of the most crucial parameters in monitoring the current status of the cell growth and protein production. Among the various real-time monitoring tools for the VCD in CHO cell culture, dielectric spectroscopy is one of the most reliable tools due to its simplicity, robustness and accuracy, thus enabling efficient monitoring and control of the corresponding processes. However, one of the critical drawbacks of the dielectric spectroscopy is the frequent deteriorations of its accuracy during the late-growth phases of CHO cell culture due to the abrupt changes in the cell physiology as well as their environmental conditions. Therefore, in this study, various kinds of chemometrics tools were employed to understand the spectral dynamics of the dielectric spectroscopy over the entire period of CHO cell culture consisting of different cell growth phases. Then, using the dielectric spectra measured from several batches of the bioreactor operations, the performance of different multivariate regression models was compared to the conventional Cole-Cole equation-based approaches in estimating the VCD in CHO cell culture. The results obtained from this study illustrated the usefulness of the multivariate regression tools in compensating the spectral dynamics observed during the different cell growth phases, inherently into the given estimation models, and the robust prediction performance from the dielectric spectroscopy can be realized by using the regression model based on locally-weighted partial least squares (LWPLS), allowing the optimal and stable estimation of the VCD during CHO cell culture, regardless of the differences in the cell growth phases.

(582y) Characterization of AmiBA2446, a Novel Bacteriolytic Enzyme Active Against Bacillus Species

Mehta, K., RPI

Characterization of AmiBA2446, a novel bacteriolytic enzyme active against *Bacillus* species

As more bacterial pathogens become resistant to available antibiotics, there is a need to develop novel antibacterial agents [1]. A particularly relevant target of high economic and biodefense importance is the spore-forming pathogen *Bacillus anthracis*, the causative agent of anthrax. Given the high cost of decontamination and the recalcitrance of dormant spores [2] there is acute interest in developing an efficient and environmentally-friendly approach to decontamination. Bacteriophage endolysins and lytic enzymes encoded by bacterial genomes are emerging as promising antibacterial agents [3], with highly evolved specificity toward bacterium-specific peptidoglycan cell walls. Moreover, gained resistance against lytic enzymes appears to be extremely low compared to antibiotics, which represent a significant motivation for further investigation. In this work, we analyzed *in silico* the genome of *Bacillus anthracis* Str. *Ames*, and identified a novel lytic enzyme termed AmiBA2446, and demonstrated its bactericidal efficacy against *B. anthracis* and *B. cereus*. We observe that AmiBA2446 retains its activity in solution for at least 24 h at 50 °C, and at least 4 months at room temperature. We also observed significant sporicidal activity in presence of environmental germinants. The approach used to identify and characterize AmiBA2446 showcases our ability to dramatically expand the repertoire of highly selective bactericidal agents. The environmentally-friendly nature and high stability of AmiBA2446 further expands its potential for commercial use. In fact, we are extending the approach used in this work for effective decontamination of other resistant bacterial pathogens by sequencing its genome and using bioinformatics to identify putative lytic enzymes. This approach, therefore, may be important for addressing the growing concern posed by antimicrobial resistance and could be very useful in surface decontamination strategies amongst others [4].

References

1. Menzies, R.E., B.M. Cornere, and D. MacCulloch, *Adaptation of methicillin-resistant Staphylococcus aureus during antibiotic therapy.* J Antimicrob Chemother, 1989. **23**(6): p. 923-7.

2. Driks, A., *The dynamic spore.* Proc Natl Acad Sci U S A, 2003. **100**(6): p. 3007-9.

3. Loessner, M.J., *Bacteriophage endolysins--current state of research and applications.* Curr Opin Microbiol, 2005. **8**(4): p. 480-7.

4. Pangule, R.C., et al., *Antistaphylococcal nanocomposite films based on enzyme-nanotube conjugates.* ACS Nano. **4**(7): p. 3993-4000.

(582aj) Clarification of Mammalian Cell Culture Feed Streams Using Polydiallyldimethylammonium Chloride Flocculation in Conjunction With Clarisolve Depth Filters

Peck, M.,
Singh, N., University of Delaware

The push for disposable technologies, coupled with increasingly higher cell culture densities, has placed a challenge on traditional solid liquid separation technologies. To address this problem, the use of flocculation technologies has been utilized to enhance the clarification throughput and downstream filtration operations. However these very flocculants present unique challenges on downstream purification steps as well as the issue of quantifying residual levels of polymer across major unit operations of downstream purification. In this work EMD Millipore has demonstrated the application of cationic flocculent, Polydiallyldimethylammonium chloride (pDADMAC), as an effective pretreatment method for mammalian cell culture harvest. The effects on depth filtration, downstream purification, polymer toxicity and residual quantification will be discussed.

(582aa) Predication of Sialic Acid Reduction in CHO Culture

Alden, N., Alexion

Sialic acid content is an important quality attribute in the development of many therapeutic glycoproteins. Increased sialic acid content is often linked to an increased half-life in the human body. In a GS-CHO cell line producing a recombinant enzyme, we observed that the total sialic acid content (TSAC) of the glycoprotein decreased over time. As the enzyme is being produced in a second generation process using a new cell line, it is important to determine the primary cause of this decline to help match product quality of the original process. A series of studies were performed to measure TSAC in standard culture conditions as well as in a cell-free harvest hold. In addition, sialidase activity was measured in daily samples, and compared to lactate dehydrogenase (LDH) levels. In a preliminary study, some samples were held at production temperature for 5 days. A sample removed on the fourth day of the culture did not change whereas a sample from the eighth day lost 34% of its TSAC. With this information, a mathematical model was established to estimate TSAC. The model could potentially be used to predict TSAC based on culture parameters including pH, temperature, and viability.

(582z) Investigation of the Impact of Bioprocess Variables on Performance using Multivariate Analysis

Anderson, G.,
Kantardjieff, A., Alexion Pharmaceuticals
Osterman, D.,
Jaluria, P., Alexion Pharmaceuticals

Multivariate analysis of process data can be used to discover new relationships between process variables or to optimize performance of a production cell line. These studies can be used for optimization of late stage manufacturing process or to refine existing platform processes. In this study, we relied on historical data from a variety of small to medium scale cell culture projects and products in order to identify correlations between inputs like lactate production or off-line pH measurements and outputs like specific productivity or titer. The scope of the study includes several Chinese hamster ovary (CHO) cells, producing monoclonal antibodies (mAbs) and fusion proteins. Data from multiple scales was included in the analysis, from the 2 liter scale through the 200 liter scale, as well as data from different stages of the process development timeline. The process data was primarily obtained from off-line samples, including off-line pH, metabolites, and product titer, with additional parameters like mean-adjusted lactate production derived from these values to reveal additional correlations between process variables. The investigation uncovered interesting connections between metabolites like specific lactate productivity and specific productivity. Future multivariate analysis will expand the scope of analysis and investigate the opportunities for improvement at production scale. This study demonstrates the power of multivariate analysis of process parameters through the conclusions that can be drawn from previously existing data, and produces additional value at little cost.

(584az) Microsphere-Collagen Hydrogel With Adipose Derived Stem Cells for Soft Tissue Engineering

Tong, Y. W., National University of Singapore
Kodali, A., National University of Singapore

A composite hydrogel system containing microspheres was developed for tissue engineering applications. Collagen hydrogels (Col) are one of the widely used cell culture matrices as they can closely mimic the in vivo environment. Similarly gelatin microspheres (GMs) are well known for their ability to simultaneously deliver both cells and growth factors for various biomedical applications. Here, we present a system of GMs-incorporated-Col to vary the mechanical strength of the hydrogels for their effects on soft tissue engineering.

GMs were fabricated and further cross-linked before being incorporated into gelatin microsphere-collagen hydrogel composites (Col-GMs). Visco-elastic properties of the scaffolds showed that mechanical strength can be tuned by the composition of GMs. The GMs were then loaded with basic fibroblast growth factor (bFGF) and seeded with human adipose derived stem cells (hADSCs).

hADSCs seeded on the GMs migrated into the relatively softer collagen matrix and seems to populate the whole scaffold within one week of culture period. Osteogenic and adipogenic differentiation of ADSCs were studied. Different amounts of osteogenic genes were expressed, showing higher mechanical strength seeming to enhance the osteogenic differentiation abilities of the ADSCs. Encapsulated bFGF showed that gene expression of osteogenic marker genes was enhanced by the GMs in Col-GMs compared to merely encapsulating bFGF in collagen hydrogels.

Our study showed that co-encapsulation of bFGF loaded Gelatin microspheres along with ADSCs with in a collagen hydrogel matrix enhances the osteogenic differentiation of ADSCs and can be a potential stem cell based therapeutic strategy for different types of tissue regeneration.

(584g) Triple Combination Nanochemoprevention of Breast Cancer

Rai, P., University of Massachusetts, Lowell

According to the American Cancer Society a total of 1,638,910 new cancer cases and 577,190 deaths from cancer are projected to occur in the United States by the end of 2012. There is a dire need for better strategies to prevent, diagnose and treat this deadly disease. Widespread use of chemopreventive agents has been hampered by both, real and perceived toxicities associated with these agents, which are small molecules with poor bioavailability. Nanochemoprevention, which involves use of nanoparticles for delivery of chemopreventive agents, is an emerging field that has shown some promise for cancer prevention by improving the bioavailability and reducing toxicity associated with these drugs. We have developed theranostic nanometer-sized constructs that encapsulate multiple promising chemopreventive agents together with a fluorescent dye for tracking drug delivery. Nanoconstruct synthesis and drug encapsulation were characterized, and the effects of chemopreventive nanoconstructs in both monolayer and 3D cell cultures were studied and compared in normal and cancer cell lines. Our results indicate larger effects on cancer cell lines than on normal and that chemopreventive nanoconstructs inhibit the formation of cancer cell acini in 3D cell culture. We have also investigated the chemopreventive efficacy of the nanoconstructs in mouse models, following pharmacokinetic, toxicity and biodistribution studies that showed extended blood circulation half-life and high tolerability of the nanoconstructs compared to free drugs. Preclinical evaluation of a triple combination nanochemoprevention approach against breast cancer will be presented.

(584h) Novel Biogenic Nanomaterials As Nanocatalysts and Anti-Virulence Agents

Ng, C. K., Interdisciplinary Graduate School, Nanyang Technological University
Mohanty, A., Nanyang Technological University
Yang, L., Nanyang Technological University
Cao, B., Nanyang Technological University

Microorganisms have the potential to change the oxidation state of metal(loid)s and these microbial processes have opened up a new window for novel applications, for example, biosynthesis of metal(loid) nanomaterials. Microbial processes can synthesize nanomaterials in water under gentle and environmentally benign conditions. Hence, biological processes for the production of nanomaterials have become one of the attractive research focuses in current green nanotechnology towards sustainable development. Many microorganisms can be used to synthesize metal nanomaterials. Among them, *Shewanella oneidensis* is able to reduce a diverse range of metal(loid)s extracellularly. One research focus in our group is to explore extracellular synthesis of metal(loid)s nanomaterials by *S. oneidensis* and novel applications of the bacteriogenic nanomaterials. Three examples of our recent research work will be presented.

(i) Using biosynthesis of silver and silver sulfide nanoparticles as a model, we show that surface proteins MtrC and OmcA significantly affect the size and activity of the extracellular bacteriogenic nanoparticles. Our results suggest that it may be possible to control particle size and activity of the extracellular biogenic nanoparticles via controlled expression of the genes encoding surface proteins. We also reveal that the usually neglected non-cell-associated nanoparticles could have high catalytic activity, underlining the need for novel methods that ensure efficient retention of extracellular nanoparticles in biosynthesis processes.

(ii) Biofilms may serve as a promising system for synthesizing and retaining nanomaterials because of the presence of a sticky matrix composed of extracellular polymeric substances. We show that biofilms can produce and immobilize nanocrystals in the biofilm matrix. The *in situ* generated nanocrystals in the biofilms exhibit nanocatalytic activities. Our results demonstrate the development of a novel hybrid biofilm-nanocrystal catalytic system, called nanocatalytic biofilms, which may be used to develop novel multifunctional catalysts with both biocatalytic and nanocatalytic activities.

iii) Bacteria typically require iron with concentrations in the μM, *i.e.*, 10^{-6} M range to support growth. Human serum concentrations of free iron are maintained as low as 10^{-24} M, which pose a serious challenge to pathogenic bacteria that attempt to establish an infection. However, human pathogens can establish infections through self-produced siderophores that can complex Fe(III) to facilitate iron acquisition. We show that an extracellular bacteriogenic nanomaterial significantly inhibits the production of one important siderophore. Using molecular biology tools, we further elucidate the inhibition occurs at the transcriptional level.

(584i) Chitosan/Poly L Lysine (PLL)-Alginate Microspheres for Growth Factor Delivery in Electrospun Scaffolds

Vaidya, P.,
Goldstein, A., Virginia Tech

Chitosan/poly L lysine (PLL)-alginate microspheres for growth factor delivery in electrospun scaffolds

Introduction:

Electrospinning has been widely used to create scaffolds for musculoskeletal, vascular and skin tissue engineering. While electrospun scaffolds are known to provide topographical cues, incorporation of growth factors (GF) in electrospun scaffolds will provide additional biological cues to modulate tissue regeneration. However, electrospinning GF is limited due to instability of GF during electrospinning and influence of fiber degradation on GF release kinetics. We hypothesize that these limitations can be addressed by adopting a two step process. The first step involves fabrication of microspheres (MS) to load and protect the GF; while the second step involves co-axially electrospinning GF loaded MS in outer sacrificial fiber, followed by solvent extraction of sacrificial phase to yield GF loaded MS entrapped on surface of core phase fibers.

Materials and Methods:

Towards the first step, GF loaded chitosan/PLL-alginate MS were manufactured via ionic gelation. Briefly, fluorescein labeled bovine serum albumin (FITC-BSA, model GF) was mixed with alginate solution followed by addition of $CaCl_2$ to form pre-gel. Chitosan/PLL was added to the pre-gel to create MS. The viscosity (mol wt) of alginate, concentration of $CaCl_2$ and ratio of chitosan/PLL to alginate was varied to study the effect of changing processing parameters on MS sizes. Concentration of $CaCl_2$was varied to study its effect on release kinetics of FITC-BSA. Cytotoxicity studies of MS were performed using C3H10T1/2 cells.

With regards to the second step, DiI was added to PLGA and DiO was added to PEO and the fibers were co-axially electrospun with PEO as sheath and PLGA as core phase.

Results and Discussions:

The size of MS increased with increasing viscosity of alginate. For both the ratio of chitosan/PLL to alginate and concentration of $CaCl_2$ the MS size decreased with increase in the processing parameter, reached a minimum and then increased. The release study revealed that different concentrations of $CaCl_2$ released different amounts of FITC-BSA with highest release recorded at highest $CaCl_2$concentration and vice versa. Fluorescent images confirmed the presence of both dyes alluding to fibers being co-axially electrospun.

Conclusions:

Chitosan/PLL-alginate MS were successfully created via ionic gelation and loaded with FITC-BSA. MS sizes can be tuned by tuning the processing parameters such as viscosity of alginates, ratio of chitosan/PLL to alginate and concentration of $CaCl_2$. Release kinetics of FITC-BSA can be tuned by varying the ratio of $CaCl_2$. Polymer fibers were co-axially electrospun with PLGA as core and PEO as sheath phase.

(584w) Human Serum Albumin (HSA) Nanoparticles Stabilized With Intermolecular Disulfide Bonds

Wang, W., Tsinghua University
Zhao, S., Tsinghua University
Shao, T., Tsinghua University
Huang, Y., Tsinghua University
Jin, Y., Tsinghua University
Cheng, Y., Tsinghua University

Proteins as matrices for drug delivery have many advantages including biodegradability, biocompatibility, non-toxic and amenability to surface modification. Albumin nanoparticles have been regarded as an ideal drug carrier material: 1) HSA is an endogenous protein, non-toxic and non-antigenic. 2) many metabolic compounds and therapeutic drugs, such as bilirubin, penicillins, sulphonamides and benzodiazepines, are extensively bound to and transported by this protein in blood circulation, which makes HSA a natural drug carrier. 3) HSA has active targeting ability via gp60 receptor-mediated transcytosis, which may enable tumor targeting.

The intermolecular disulfide bonds formed between the albumin sulfhydryl groups are the ideal stabilization strategy since it can not only stabilize the particles against dilution in *vivo* and the possible organic solvent used in drug loading, but also make the particles have the redox responsiveness in the reducing environment. However, so far there is no satisfying chemistry to crosslink albumin in order to stabilize the nanoparticles against dilution in circulation. Even the one used in clinic, Abraxane, has been shown to dissociate completely after dilution and hence lost its potential advantage in tumor targeting via the enhanced permeation and retention (EPR) effect.

In this work, we proposed and demonstrated the feasibility to use albumin's own thiol groups to form inter-protein disulfide crosslinks and stabilize the albumin nanoparticles against dilution and organic solvents. The formed nanoparticle would dissolve in the physiologically-relevant reducing environment and release albumin monomers. Throughout the fabrication process, only glutathione and ethanol/tertiary butanol were used, and no toxic chemical was involved, making it especially suitable for in *vivo* applications.

In addition, we used our technique to load curcumin and paclitaxel, forming the CCM-HSA-NPs PTX-HSA-NPs to demonstrate the applications of our HSA-NPs. The stability of the drug-HSA-NPs in different solvents and the drug loading efficiency were evaluated.

We believe that our method represents a very valuable progress in nanomedicine materials chemistry and should be interesting to the fields of nanomedicine, drug delivery and biomaterials.

(584x) The Effect of Oxygen and Glucose Stress On the Evolution of a Leukaemia Model System in An in Vitro Bone Marrow Biomimicry

Velliou, E., Imperial College London
Fuentes Gari, M.,
Britos dos Santos, S., imperial College London
Misener, R., Imperial College
Panoskaltsis, N., Northwick Park Hospital
Mantalaris, A., Imperial College London
Pistikopoulos, E. N., Imperial College London

The effect of oxygen and glucose stress on the evolution of a leukaemia model system in an *in vitro* bone marrow biomimicry

E. G. Velliou[1,2], M. Fuentes Garí[1,2], S. Brito Dos Santos[2], R. Misener[1,2],

N. Panoskaltsis[3], A. Mantalaris[2], E.N. Pistikopoulos[1]

[1]*Centre for Process Systems Engineering (CPSE), Department of Chemical Engineering, Imperial Imperial College London, South Kensington Campus, London SW7 2AZ, UK*

[2]*Biological Systems Engineering Laboratory (BSEL), Department of Chemical Engineering, Imperial Imperial College London, South Kensington Campus, London SW7 2AZ, UK*

[3]*Department of Hematology, Imperial College London, Northwick Park & St. Mark's Campus, London HA1 3UJ, UK*

Acute Myeloid Leukaemia (AML) is one of the most common types of leukaemia in adults. AML is a malignant disease of the bone marrow (BM) and blood. Immature white blood cells which are not able to develop into normal functioning blood cells are overproduced and build up in the bone marrow and blood. This inhibits the development of healthy blood and immune cells due to space restrictions as well as inhibitory and clonal factors specific to the disease. Another type of Leukaemia is the Chronic Lymphocytic Leukaemia (CLL) which is also characterized by an accumulation of abnormal white blood cells in the bone marrow, but in a slower rate compared to AML. The most common treatment for AML and CLL is chemotherapy. It is well known that this therapy can result in several life-threatening complications as only few patient-specific factors are taken into consideration in current protocols and the choice of treatment often depends on the physician's experience. Furthermore, inter-patient and intra-leukaemia significantly increase the complexity in determining these treatment protocols. As a consequence, no widely accepted 'optimal' chemotherapy treatment protocol exists, as the evolution of the cancer as well as the chemotherapy (drug) metabolism is highly dependent on patient characteristics such as age, weight, height, medical and family history as well as on a variety of unknown factors/parameters.

Moreover, the absorption/ metabolism of a chemotherapy drug can highly vary within the same patient, depending on the condition of the patient during chemotherapy. For example when the body temperature is higher, i.e., 40°C, than the optimal/normal one, i.e., 37°C, the liquid tumor will experience a temperature stress and both the leukaemia metabolism and the chemotherapy absorption may alter. Another important issue is the variability of parameters such as oxygen or glucose concentration among different body compartments. For example the oxygen concentration in the bone marrow is much lower than the one in peripheral blood. Glucose blood concentration is another crucial factor that may fluctuate within a patient's body. It differs among organs as well as among patients or even within the same patient at different time points (hypoglycemia-hyperglycemia).

The alteration of these environmental factors, i.e., glucose concentration, oxygen concentration, temperature, can be experienced as a severe environmental stress by the cancer population. Several studies have shown that the latter impacts both the cancer growth and/or inactivation kinetics as well as the efficiency of the chemotherapy drug (see as examples Fecteau et al., 2013; Giuntoli et al., 2011; Lodi et al., 2011).

Therefore, understanding the influence of environmental stress such as oxygen and glucose stress is crucial for the optimization of AML chemotherapy treatment. ***A first step towards that direction is the efficient in vitro biomimicry of the human BM***. Haematopoiesis, both normal and abnormal, is regulated by spatially organised cellular microenvironments (niches) within the BM which is a three dimensional (3D) tissue. The deregulation of the niches is implicated in the pathogenesis of cancers of the BM, i.e., leukaemia. Without the provision of growth factors from the niche, the cultivation of leukaemic cells is extremely difficult and requires co-culture with allogeneic/xenogeneic stroma or fibroblasts, or the addition of abnormally high concentrations of exogenous cytokines to the culture. This results in the introduction of artifacts to cell culture kinetics and biasing in outgrowth of cells which are dependent on these signals for survival (Panoskaltsis et al., 2005). An ***"in vitro" 3D BM biomimicry*** for the successful cultivation of leukaemic cells without the use of exogenous factors or allogeneic cell lines would be a powerful tool for the study of different types of treatments in leukaemia, as incorporating molecular, cellular and microenvironmental factors is required for a more efficient and accurate reproduction of the *in vivo* environment *ex vivo*.

The overall aim of the current research is the experimental kinetic study of a leukaemia model system, i.e., leukaemic cell lines, in a 3D polyurethane based collagen coated scaffolding system under oxygen, i.e., ex vivo mimicry of hypoxia, and glucose stress, i.e., ex vivo mimicry of hypo and hyperglycemia.

K-562 and MEC-1 cell lines are selected as a model system for AML and CLL, respectively. The leukaemic cells are seeded in the 3D system that has been developed in our team (Mortera et al., 2010;2011), which is made up of a highly porous polyurethane based scaffold coated with collagen. Hypoxia is imposed by decreasing the oxygen level of the culturing incubator to 5%, which represents the oxygen level of the human BM, as opposed to normoxia at which oxygen is set at 20%. Glucose stress is imposed by alteration of the glucose concentration of the nutrient medium from the normal one, i.e., 4.3 g/L used in most laboratory culture media to 0.6 g/L, i.e., lower level in human peripheral blood or 1.3 g/L, i.e., higher level in human peripheral blood.

In order to evaluate the **cellular growth** within the 3D system, several tests take place. More specifically, the MTS (tetrazolium compound[3-(4,5-dimethylthiazol-2-yl)-5-(carboxymethoxyphenyl)-2-(4sulfophenyl)-

2H-tetrazolium) metabolic assay is applied for the estimation of the cellular proliferation within the scaffolds. The MTS cell proliferation is a colorimetric assay that measures the reduction of the tetrazolium compound by the cells into a coloured formazan product that is soluble in tissue culture media. The amount of formazan produced, as measured by the absorbance (490 nm), is directly proportional to the number of living cells in culture. **Metabolites** present in the growth medium such as lactate, glutamine, glutamate, NH_4^+ are measured with a bioprofiler on a daily basis. Cells are extracted manually with a syringe, from the scaffolds at regular time points in order to determine the percentage of viable cells which is estimated by performing a cell count with a hemocytometer. Moreover, extracted cells are analyzed by Flow Cytometry in order to determine the **cell cycle distribution**. The latter is of great importance, on the one hand as it is an indication of the level of the environmental stress response and adaptation of the cells (higher stress resulting in more cells shifting to the dormant state G_0) and, on the other hand, because knowing the cell cycle distribution is crucial for chemotherapy treatment, as most chemotherapy drugs target cells which are at a specific phase of the cell cycle.

Results indicate that the cellular growth kinetics significantly alter under oxygen and glucose stress. For example extreme glucose stress leads to a higher cellular proliferation the first days of culturing –showing a possible induction of the autophagic mechanism- or in normoxia there is a higher level of glutamine consumption and glutamate synthesis compared to hypoxia. Therefore, our results are pointing out the importance of taking into consideration fluctuations of the micro-environmental conditions (independently and synergistically) of the leukaemic population when designing a chemotherapy protocol and/or when constructing pharmacokinetic and pharmacodynamic mathematical predictive models.

References

Fecteau, J.-F., Messmer, D., Zhang, S., Cui, B., Chen,, L., and Kipps, T.J. (2013) Impact of oxygen concentration on growth of mesenchymal stromal cells from the marrow of patients with chronic lymphocytic leukemia, *Blood*, **121**, 971-974.

Giuntoli, S., Tanturli, M., Gesualdo, F.D., Barbetti, V., Rovida, E., Sbarba, P.D. (2011) Glucose availability in hypoxia regulates the selection of chronic myeloid leukemia progenitor subsets with different resistance to imatinib-mesylate, *Haematologica*, **96(2)**, 204-212.

Lodi, A., Tiziani, S., Khanim, F.L., Drayson, M.T., Gunther, U.L., Bunce, C.M., Viant, M.R. (2011). Hypoxia Triggers Major Metabolic Changes in AML Cells without Altering Indomethacin-Induced TCA Cycle Deregulation, *ACS Chemical Biology*, **6**, 169-175.

Mortera-Blanco T, Mantalaris A, Bismarck A, et al., Long-term cytokine-free expansion of cord blood mononuclear cells in three-dimensional scaffolds (2011). *Biomaterials*, **32**: 9263-9270.

Mortera-Blanco T, A. Mantalaris, A. Bismarck, N. Panoskaltsis (2010). Development of a three-dimensional biomimicry of human acute myeloid leukemia ex vivo. *Biomaterials*, **31**: 2243-2251.

Panoskaltsis, N., Mantalaris, S., Wu, D. (2005). Engineering a mimicry of bone marrow tissue *Ex vivo*. *Journal of Bioscience and Bioengineering*, **100**: 28-35.

Acknowledgements

This work is supported by ERC-Mobile Project (n° 226462), by the European Union Seventh Framework Programme [MULTIMOD Project FP7/2007-2013, n° 238013] and by Richard Thomas Leukaemia Research Fund. Ruth Misener is further thankful for a Royal Academy of Engineering Research Fellowship (Award number: RAEng 10216/118).

(584j) Bacteria Encapsulated Dispersants: Novel Technique for Oil Spill Bioremediation

Gupta, A., Texas A&M International University
Martinez, M.,

Bacteria encapsulated dispersants: novel technique for oil spill bioremediation

Anju Gupta* and Monica Martinez

Texas A&M International University, Laredo, Texas

The most efficient way to clean up an oil spill in oceans is via microbial degradation. This project proposes a novel microbe-encapsulated dispersant system to enhance the natural process of bioremediation of an oil slick in the ocean. The use of dispersants to counter the effects of oil spill has gained a lot of attention due to their lower toxicity to aquatic life and ability to disperse heavy oil. Oil dispersants can reduce the surface tension between oil-water interfaces resulting in oil droplet formation that can easily be accessed by oil-eating-bacteria. Addition of hydrocarbon-degrading bacteria, also known as seeding has often been considered, but has not been reported to be effective. This project proposes a novel idea of combining the seeding and dispersion approaches of oil spill bioremediation. The dispersant system proposed in this project to disperse the oil and increase the population of oil-eating-microbes is both biocompatible and economical. Phospholipids/shear-responsive polymer mixtures will be used to form vesicles with large aqueous core to encapsulate hydrocarbon-degrading microbial culture. Shear produced by the waves, will result in destabilized dispersants releasing the encapsulated bacteria to the oil/water system. Lipids will disperse the layer of oil into tiny droplets that can be easily accessed by bacteria. Lipids from the destabilized dispersants will also serve as a nutrient source of nitrogen and phosphorus to bacteria.

(584y) A Novel PEC Drug Carrier of Nacs-CS-PPS Microcapsules for Colon-Specific Drug Delivery

Wu, Q. X.,
Zhang, Q. L.,
Lin, D. Q.,
Yao, S. J., Zhejiang University

Polyelectrolyte complexes (PEC) prepared with sodium cellulose sulfate (NaCS), chitosan (CS)/water soluble chitosan (WSC) and sodium polyphosphate (PPS) were developed with a simple preparation process, which can be used as a novel colon-specific drug delivery system. NaCS is a polyanion with $–SO_3^-$ groups while CS/WSC is a polycation with $–NH_3^+$groups, and PPS was used as a cross-linking agent. Based on the PEC formed via ionization reaction, spherical microcapsules loaded with drugs were prepared by an orifice-polymerization method. This system was used to successfully load two model drugs: a small molecular drug and a protein drug.

The NaCS-CS-PPS microcapsules loaded with 5-aminosalicylic acid (5-ASA) as a model small molecular drug were prepared. The microcapsules prepared had an average diameter of 1.90 mm with loading efficiency (LE%) of 60.77% and encapsulation efficiency (EE%) of 90.03%. Scanning electron microscope (SEM) results showed that the microcapsules had a double-walled capsule structure with an outer wall thickness of approximately 4.40 μm and inner wall (shell) thickness of approximately 187.14 μm. SEM transection images of the microcapsules showed that 5-ASA entrapped in the microcapsule was in a crystal form. The results of *in vitro*swelling/erosion and release analysis showed that the drug was preferentially and completely released in simulated colonic fluid (SCF, pH 6.4) under the mechanism of anomalous transport.

The NaCS-CS/WSC-PPS microcapsules loaded with lactoferrin (LF) as a model protein drug were also prepared. The average diameter of NaCS-CS-PPS microcapsules and NaCS-WSC-PPS microcapsules were 1.85 and 1.97 mm, and the coefficient of variation (*CV*) value were 5.11% and 4.99%, respectively. SEM studies showed that both of the microcapsules had a typical wall-capsule structure with smooth surfaces. 1H-NMR spectrum confirmed that the WSC used was in the form of chitosan hydrochloride. Fourier transform infrared spectroscopy (FTIR) spectra analysis indicated that $–NH_3^+$ of CS/WSC, $–SO_3^-$ of NaCS and –$[P_2O_5^{4-}]$– of PPS may react and form PEC. Thermo-gravimetric analysis (TGA) studies showed that the weight loss of the NaCS-CS-PPS microcapsules was higher than that of the NaCS-WSC-PPS microcapsules in the same temperature, which indicated that the later one had better thermal stability. Drug loading and encapsulation efficiency studies showed that NaCS-CS-PPS microcapsules had LE% of 49.06% and EE% of 86.3%, respectively, which was higher than that of NaCS-CS-PPS microcapsules of 45.60% and70.72%. *In vitro*release studies showed that both of the microcapsules had a regular drug release behaviors with a trend of slow increasing, and the drug was released sustainably and completely in SCF (pH 5.5-7.0).

All of these results indicated that the microcapsules based on the PEC prepared with NaCS, CS/WSC and PPS can be a good candidate as an enzyme-triggered controlled release drug carrier for colon-specific drug delivery systems.

Acknowledgments: This work was supported by the National Natural Science Foundation of China and the Doctoral Programs Foundation of Ministry of Education of China.

(584s) Dynamics of Mechanosensing in the Bacterial Flagellar Motor

Lele, P., Harvard University
Hosu, B., Harvard University
Berg, H. C., Harvard University

Mechanosensing by the flagellum is thought to trigger bacterial swarmer-cell differentiation, an important step in pathogenesis. How the flagellar motor senses mechanical stimuli is not known. To study this problem, we suddenly increased the viscous drag on single motors, by using optical tweezers to attach latex beads to the flagellum. From the initial speed (after the load change), we inferred that motors running at very low loads are driven by 1 or at most 2 force-generating units. Following the load change, motors gradually adapted by increasing their speeds in a stepwise manner (over a period of a few minutes). Single-motor TIRF imaging of YFP-MotB (part of a stator force-generating unit) confirmed that the response to sudden increments in load occurred by the addition of new force-generating units. We estimate that 6-11 force-generating units drive motors at high loads. Wild-type motors and motors locked in the CW or CCW state behaved in a similar manner, as did motors in cells deleted for *fliL* or for genes in the chemotaxis signaling pathway. Thus it appears that stators themselves act as dynamic mechanosensors. This is a rare example of a biological motor employing a variable number of force generating units to tailor its response to varying external viscous loads.

(584z) Robust In Silico Disease Classification Via Disease- and Procedure-Independent Optimization Models Using Quantitative MS1 Data From High-Throughput Proteomics

Guzman, Y. A., Princeton University
Riley, C. P., Pathology Associates Medical Laboratories
Floudas, C. A., Princeton University

A biomarker is a measurable characteristic that can indicate biological state relative to disease stage or contraction risk. Biomarkers have great potential to transform diagnostic medicine by serving as an early indicator or predictor of developing or oncoming disease. The monitoring of changes in the protein domain has the greatest potential to lead to the discovery of biomarkers due to the close relationship between protein expression and abundance and cellular state [1]. The need for rapid diagnostic turnover in time-sensitive disease systems has yielded an explosion of biomarker discovery research utilizing high-throughput mass spectrometry proteomics protocols within the past decade.

Efforts to discover and define robust protein biomarkers have yielded disappointing results. From 1997 to 2006, about 224,000 biomarker papers were published, while only 15 biomarkers were approved for use by the FDA [2]. The initial stage of biomarker discovery studies typically consists of a small amount of samples, and what seems to be a distinguishing protein biomarker often yields a high amount of false positives and false negatives at later validation stages with larger sample sizes and greater subject heterogeneity [1,3]. Focus has shifted to utilizing panels of biomarkers to create a set of diagnostic rules.

The complexity of biological samples and fluids makes the application of typical high-throughput proteomic analysis exceedingly difficult. The most enticing biofluidic source of protein biomarkers remains the blood and blood plasma for its clinical accessibility, but they display a dynamic concentration range spanning up to 11 orders of magnitude, with 99% of protein mass coming from 22 blood proteins [4-6]. Typical data-dependent acquisition for MS/MS fragmentation will exclude low-abundance proteins whose up- and down-regulation may capture disease response and treatment progression, and its stochastic nature reduces run-to-run reproducibility. These difficulties limit the penetrating depth of untargeted MS/MS protein identification protocols. In response, many biomarker discovery studies have focused on classification using MS^1 features, resulting in very high sensitivity and specificity; these studies have also elicited criticism [7-9], as statistical methods, machine learning techiques, and black-box models are prone to over-training and can magnify features that are actually data artifacts [9,10].

Building on a previous study in which mixed-integer linear optimization models were proposed to classify healthy and diseased samples [11,12], we propose a novel class of robust optimization models that can fingerprint and classify healthy and diseased samples given quantitative MS^1 data. These models can simultaneously select the optimum subset of distinguishing MS^1 peaks while performing parameter estimation. The resulting functions are of diagnostic utility, quantitatively classifying new blind samples given only MS^1 data. The new classification models are general and independent of sample biofluid, experimental protocol, and disease system. The optimal peak subset yields a multiple reaction monitoring protein identification protocol for further sample characterization and biomarker investigation. Results from the proposed models are presented as applied to MS^1 data of proteomics samples collected from different biofluids, subjected to different experimental protocols, and relating to vastly different disease systems, including plasma samples collected from breast cancer patients and gingival crevicular fluid samples collected from patients with chronic periodontitis [13,14].

[1] Rifai N., Gillette M.A., Carr S.A. *Nature Biotechnology*, 24(8):971-983, 2006.
[2] Jin G., Zhou X., Wang H., Wong S.T.C. The Challenges in Blood Proteomic Biomarker Discovery. In Pham T., Computational Biology: Issues and Applications in Oncology. New York: Springer, 2009.
[3] Srinivas P.R., Verma M., Zhao Y., Srivastava S. *Clinical Chemistry*, 48(8):1160-1169, 2002.
[4] Anderson N.L., Anderson N.G. *Molecular & Cellular Proteomics*, 1(11):845-867, 2002.
[5] Schiess R., Wollscheid B., Aebersold R. *Molecular Oncology*, 3(1):33-44, 2009.
[6] Veenstra T.D., Conrads T.P., Hood B.L., Avellino A.M., Ellenbogen R.G., Morrison R.S. *Molecular & Cellular Proteomics*, 4(4):409-418, 2005.
[7] Sorace J.M., Zhan M. *BMC Bioinformatics*, 4:24, 2003.
[8] Poste G. *Nature*, 469(7329):156-157, 2011.
[9] Rogers M.A., Clarke P., Noble J., Munro N.P., Paul A., Selby P.J., Banks R.E. *Cancer Research*, 63(20):6971-6983, 2003.
[10] He Z., Yu W. *Computational Biology and Chemistry*, 34(4):215-225, 2010.
[11] Baliban R.C., Sakellari D., Li Z., Guzman Y.A., Garcia B.A., Floudas C.A. *Journal of Clinical Periodontology*, 40(2):131-139, 2013.
[12] Baliban R.C., Dimaggio P.A., Plazas-Mayorca M.D., Garcia B.A., Floudas C.A. *Journal of Proteome Research*, 11(9):4615-4629, 2012.
[13] Riley C.P., Zhang X., Nakshatri H., Schneider B., Regnier F.E., Adamec J., Buck C. *Journal of Translational Medicine*, 9:80, 2011.
[14] Baliban R.C., Sakellari D., Li Z., DiMaggio P.A., Garcia B.A., Floudas C.A. *Journal of Clinical Periodontology*, 39(3):203-212, 2012.

(584k) "Shake-n-Incubate": Examining the Effects of Orbital Shaking On Iron Oxide Nanoparticle Uptake for D1 Cell Labeling

Clay, N., University of Illinois, Urbana-Champaign
Baek, K., University of Illinois at Urbana-Champaign
Lai, M. H., University of Illinois at Urbana-Champaign
Kong, H., University of Illinois at Urbana-Champaign

Labeling stem cells with superparamagnetic iron oxide nanoparticles (SPIOs) enables clinicians to assess the bioavailability and localization of transplanted cells through MR imaging. To facilitate the cellular uptake of SPIOs for cell therapies, previous protocols optimized the concentration, size, surface charge, and targeting capability of SPIOs, often coupled with a transfection agent. However, these approaches have limited degrees of success, in part due to concerns regarding transfection agent cytotoxicity. In this study, we sought to control cell labeling by manipulating the cellular microenvironment while keeping the intrinsic properties and concentration of the nanoparticles constant . To this end, we prepared SPIO clusters formed by the hydrophobic association between SPIOs and poly(2-hydroxyethyl aspartamide) substituted with hydrophobic octadecyl chains. These water-stable, monodisperse SPIO clusters have a T_2 relaxivity comparable to Feridex, an FDA-approved MR contrast agent. We incubated bone marrow-derived mesenchymal stem cells in cell culture media suspended with SPIO clusters while shaking the cell culture well at varying rates. The cells exposed to the orbital shaking at 50 rpm took up more SPIOs than cells cultured in static culture wells, as quantified through ICP-OES. In addition, cells cultured under shaking at 50 rpm exhibited lower T_2 signal than those cultured without any external mechanical flow. Interestingly, the role of orbital shaking on SPIO uptake became minimal when serum concentration in the cell culture media was decreased. This intriguing protocol will be extended to label a wide array of therapeutic cells, in turn improving transplanted cell tracking and, ultimately, the success of cell therapies.

(584aa) Characterization of Hypromellose 2208 Properties Impacting Drug Delivery Kinetics of a Hydrophilic Matrix Tablet

Rogers, T., The Dow Chemical Company
Zhao, J., The Dow Chemical co
Adden, R., Dow Wolf Cellulosics GMBH & CO. OHG
Theuerkauf, J., The Dow Chemical Co.
Li, Y., The Dow Chemical Co.
Petermann, O., The Dow Chemical Company
Knarr, M., Dow Wolf Cellulosics GMBH & CO. OHG
Meunier, D., The Dow Chemical Co
Broomall, C.,

The purpose of this work is to characterize the key properties of hypromellose 2208 impacting drug release kinetics from a hydrophilic matrix tablet.

Hypromellose (HPMC) from more than thirty HPMC K4M batches was used to formulate matrix tablets containing paracetamol or indapamide. Formulations containing HPMC, API and all fillers and additives were either directly compressed or wet-granulated. Tablets were compressed using a Manesty Beta or Piccola rotary press. Tablet properties and dissolution testing were conducted according to USP methods. Each HPMC batch was assayed using harmonized pharmacopeial methodology. Furthermore, molecular weight distributions were characterized using size-exclusion chromatography (SEC).

The Korsmeyer-Peppas power law equation (equation 1) and Peppas-Sahlin heuristic model (equation 2) below were used to fit the dissolution profiles.

$$M_t / M_\infty = kt^n \qquad (1)$$

$$M_t / M_\infty = k_d t^{0.45} + k_r t^{0.89} \qquad (2)$$

Here, M_t / M_∞ denotes fractional drug release at time t. The constant, k, is a kinetic constant that measures the drug release rate while n is a diffusional exponent that depends on the release mechanism and the geometry of the tested matrix system. The first term of equation (2) represents the Fickian diffusional contribution and the second term is the Case II or the relaxational contribution. Polymer molecular weight and chain overlap concentration, C^*, were obtained via SEC. The viscosity of 2 wt % HPMC in water at 20 °C was predominately impacted by molecular weight and hydroxypropyl substitution level. Correlations were found between M_t, n, k, and C^*. According to Peppas et al., the power law can be seen as superposition of two mechanisms of drug transport, Fickian diffusion and case-II transport. The exponent, n, varied between 0.53 and 0.68, indicating anomalous transport mechanism. In this study, M_t varied with C^* up to t=2 hr. M_t appeared independent of C^* at $t > 4$ h. n was found initially increased with increasing C^*, but subsequently decreased with further increase in C^*.

Chain overlap concentration was used to correlate with modified release performance for the presented HPMC matrix. This modeling strategy facilitates the optimization of sustained API release.

(584b) Cancer Cell Migration in a Three-Dimensional Microfluidic Platform

Bruce, A., West Virginia University
Hindman, B., West Virginia University
Wysomerski, R., West Virginia University
Yang, Y., West Virginia University

Cancer Cell Migration in a Three-dimensional Microfluidic Platform

Metastasis, the migration and spread of tumor cells from a primary tumor to other areas of the body, is the most common cause of cancer patients' death. Knowledge of cancer cell migration is vital to cancer diagnostics and treatment. The conventional *in vitro* cell culture, which uses flat plastic substrates under static condition, does not recapitulate the characteristics of *in vivo* cellular microenvironments where cells reside in a three-dimensional (3-D) extracellular matrix (ECM) and are regulated by the characteristics including ECM mechanical properties and interstitial flows. To this regard, it is essential to develop a 3-D cell culture platform with a fluidic flow to study cancer cell migration. Myosin II, involved in cells ability to migrate and remodel 3-D matrix, is of particular interest. Therefore, we engineered a 3-D microfluidic platform to study cancer cell migration and examined the role of Myosin II in cancer cell migration.

Â A microfluidic chip was fabricated by using photolithography and assembled by applying a microtransfer assembly technique developed. The cells, including metastatic breast cancer cell line (MDA-MB-231 cells) and Myosin IIA and IIB knockdown (KD) MDA-MB-231 cell lines were embedded into Type I rat-tail collagen and then injected into the microfluidic chip.

In the 3-D matrix the cells displayed distinct phenotype compared to the 2-D configurations. On 2-D gels, Myosin IIA knockdown cells had altered actin organization and Myosin IIB knockdown cells exhibited a more irregular shape with prominent stress fibers, compared to the parental control cells. Differently, in 3-D matrix, the parental MDA-MB-231 cells had pyramidal cell bodies with multiple projections. Interestingly, Myosin IIA knockdown cells had rounded cell bodies with highly branched and elongated projections in all directions while IIB knockdown cells where elongated with fewer projections. Moreover, Myosin II isoforms played a significant role in cancer cell migration.Â Loss of Myosin IIA and IIB reduced migratory velocity and altered displacement, and overall track length compared to the parental MDA-MB-231 cells. Blebbistatin (a Myosin II inhibitor) treated MDA-MB-231 also modified migratory velocity, displacement, and track total length. Furthermore, introduction of the fluidic flow had influence on migratory direction and displacement of these cells.

Our study showed that the cancer cells had different migratory patterns in both 2-D and 3-D configurations and Myosin II isoforms played a distinct role in MDA-MB-231 cell migration. Moreover, a 3-D, dynamic cell culture platform mimics the *in vivo* microenvironment and will provide a useful tool for cancer metastasis research.

(584ab) Nanog Transient Overexpression With Optimized Magnitofection to Reverse the Effects of Organismal Aging On Mesenchymal Stem Cells

Son, S., SUNY Buffalo
Liang, M. S., State University of New York at Buffalo
Lei, P., University at Buffalo, State University of New York
Andreadis, S. T., University at Buffalo, The State University of New York

Introduction: Previously, our group demonstrated that overexpression of a single transcription factor, Nanog, can reverse the effects of organismal aging on mesenchymal stem cells (MSCs) proliferation and myogenic differentiation [1]. We also found that transient Nanog overexpression is sufficient to reverse the effects of aging, thereby suggesting the permanent genetic modification may not be necessary. This result prompted us to hypothesize that transient overexpression of Nanog in MSCs by DNA transfection may overcome the effects of cellular senescence. However, it has been proven difficult to deliver genes into MSCs. We hypothesized that we could have efficient gene delivery to achieve our goal by using DNA magnetofection.

Materials and Methods: Human hair follicle derived MSCs (HFMSCs) were used in this study. EGFP- or Nanog-encoding plasmid first formed complexes with polyMAG nanoparticles and pulled down by magnetic field to the cells seeded in 24-well plates. Chloroquine was used to enhance endosomal escape of internalized DNA complexes.

Results and Discussion: Magnetofection on HFMSCs was first optimized with EGFP-expressing plasmid. The optimized dose for transfection without detrimental effects on cell viability was 0.5μgDNA/0.5μLparticles complex. The green positive (G+) cells reached 17.46%±1.21 and the total green fluoresce intensity (GFI) increased by 1.98±0.06 fold comparing to control cells. Repeated addition of optimal magnetofection complex for tree time enhanced the transfection efficiency to G+=47.06%±2.25 (GFI increased by 3.09±0.08 fold) and treatment with choloroquine further promoted the transfection efficiency to 73.64%±2.57; (GFI= 5.19±0.22 fold). This approach is significantly more efficient than the widely used transfection reagent, Lipofectamine 2000, which showed only G+:15.0%±0.51and GFI=2.58±0.07 fold at its optimal DNA/reagent ratio and with lower cell viability. This optimized protocol is currently applied to other MSC such as bone marrow (BM)-MSC as well as for delivery Nanog-expressing plasmid to evaluate the effects of gene transfer on cellular senescence.

Conclusion: This study provides a possible alternative, DNA magnetofection, to overcome the difficulty of DNA delivery to MSCs. Our results imply that this strategy may be employed to reverse the effects of organismal aging or culture senescence on MSC without viral transduction and permanent genome modification.

[1] Han J, Mistriotis P, Lei P, Wang D, Liu S, Andreadis ST. Nanog reverses the effects of organismal aging on mesenchymal stem cell proliferation and myogenic differentiation potential. Stem Cells. 2012;30:2746-59.

(584ap) Aliphatic Lipid Substitution On Aminoglycoside-Based Polymers: Plasmid Delivery and Transgene Expression

Miryala, B., Arizona State University
Potta, T.,
Tian, Y., Arizona State University
Rege*, K., Arizona State University
Pushpavanam, K. S., Arizona State University

Cationic polymers are attractive candidates for transgene delivery and expression since they are considered to be safer alternatives to viral delivery vehicles. Amphiphilic polymers that combine cationic character with hydrophobicity are hypothesized to possess high efficacies for nucleic acid condensation and uptake across cellular membranes. Here, we report the design, synthesis, characterization and transgene expression efficacies of lipid-substituted aminoglycoside polymers. A library of 54 lipid-polymer conjugates was synthesized by N-acylation of the free amines present in the polymers. Aminoglycoside-polymers were derivatized with three different acid chlorides based on C_6, C_{14}, and C_{18} chains. The extent of the lipid substitution was dependent on the lipid-polymer stoichiometry, and the conjugates were characterized using ^{1}H NMR and FTIR spectroscopy. Parallel screening of the conjugates led to identification of several leads that demonstrated higher transgene expression efficacies in several cancer cells, when compared to the parental polymers and poly(ethylene imine), a current polymeric standard for transgene expression. Transgene expression efficacy was dependent on the extent of lipid substitution, as well as the chain length of the lipid substituted. The influence of the various physicochemical properties including polyplex size, zeta potential, and DNA binding on transgene expression efficacies were investigated. This study indicates that lipid substitution on aminoglycoside-based polymers results in high levels of transgene expression compared to unsubstituted polymers. Taken together, these materials show significant promise in non-viral transgene delivery, and may be attractive for gene therapy applications.

(584l) A 3-D Microfluidic Liver Sinusoid Analogue for Drug Screening

Usta, O. B., Massachusetts General Hospital/Harvard Medical School/Shriners Hospital for Children
Bhushan, A.,
Bale, S. S.,
McCarty, W. J.,
Hegde, M.,
Jindal, R., Massachusetts General Hospital, Harvard Medical School
Golberg, I.,
Yarmush, M. L., Center for Engineering in Medicine (CEM) at Massachusetts General Hospital â€“ Harvard Medical School - Shriners Hospital for Children
Senutovich, N., Center for Engineering in Medicine (CEM) at Massachusetts General Hospital â€“ Harvard Medical School - Shriners Hospital for Children
Bakan, A.,
DeBiasio, R.,
Gough, A., University of Pittsburgh
Shun, T.,
Vernetti, L., University of Pittsburgh
Taylor, D. L., University of Pittsburgh Drug Discovery Institute

The liver plays a central role in human drug interactions and is also the most common target for drug-induced toxicity, resulting in costly, late stage drug failures. In this project, we derive inspiration from the building blocks of the liver, the liver sinusoid and the acinus structure, to build a realistic microfluidic liver platform to accelerate drug testing and toxicology studies. The design consists of a microfabricated platform with parallel cords (microfabricated grooves) separated by thin walls. These cords house a three dimensional analog of the liver sinusoid populated with human liver cells. Using a step-by-step approach, we introduce primary hepatocytes into these cords, followed by the non-parenchymal cells (NPCs), including endothelial cells, stellate cells and Kuppfer cells. These NPCs not only support optimal hepatocyte function but also directly contribute to drug metabolism and induced toxicological challenges. Using strategies to mimic in vivo flow characteristics, including nutrient and oxygen zonation, we aim to deliver a platform that closely mimics the in vivo liver and therefore produces data with improved predictive capabilities.

(584ar) Generation of Inhibitory Monoclonal Antibodies Targeting Membrane-Type Matrix Metalloproteinase By Motif Grafting

Nam, D. H., University of California Riverside
Ge, X., University of California Riverside

MMP-14, a membrane-bound zinc endopeptidase, plays an important role in tumor growth, invasion and neovascularization, therefore, it has been recognized as an attractive therapeutic target. However, previous studies of broad-spectrum small molecule inhibitors toward tumorigenic metalloproteinases all failed in clinical trials. It implies that selectivity is the key for a successful proteases inhibition. With exquisite specificity, antibody-based inhibitors are emerging as promising therapeutics. Recently, a cyclic peptide (CFSIAHEC) specifically inhibiting MMP-14 but not other MMPs has been reported. Unfortunately, this peptide inhibitor has a very low affinity of 150 μM, preventing its further applications as a potent therapeutic. We hypothesize that grafting this inhibitory peptide into the CDR-H3 region of antibody scaffolds will confer binding specificity and inhibition function to the antibody. The Fab library carrying the inhibitory motif in CDR-H3 and diversities at other five CDRs was generated by gene assembly. After four rounds of phage panning, several inhibitory Fab clones were identified. Biochemical and enzymatic studies with synthetic substrates were performed to characterize those antibodies. This grafting approach can convert a low affinity cyclic peptide to a more stable and potent therapeutic agent specifically inhibiting MMP-14.

(584c) Quantitative Analysis of Contact Inhibition of Locomotion During Fibrillar-Like Migration

Milano, D., Northeastern University
Asthagiri, A. R., Northeastern University

Metastasis accounts for ninety percent of all cancer-related deaths. This process, driven by cell migration, involves cancer cells escaping from the primary tumor, invading the local tissue and forming secondary tumors in remote areas of the body. Healthy cells do not exhibit this ability because their locomotion is inhibited by contact with other cells. Contact inhibition of locomotion (CIL) is suppressed in cancer cells, one distinctive feature that allows them to become metastatic. In contrast to the 2D culture systems used to study CIL, cancer cells invade through a three-dimensional microenvironment composed of fibers[1]. These fibers are often composed of extracellular matrix proteins including fibronectin and collagen. To date, there has been no work to explore CIL in the context of fibrillar migration. To begin to examine the nature of CIL in confined, fibrillar-like microenvironments, we utilized 1D micropatterned surfaces as an analog to the 3D fibrillar environment in which tumor cells invade. Time-lapse microscopy was used to quantitatively compare homotypic CIL between normal mammary epithelial cells (MCF 10A) and metastatic breast cancer cells (MDA-MB-231). Cells were confined on patterned lines that varied in width from 6 to 33 microns. Pairwise interactions between cells were imaged in real-time and scored as bouncing or sliding collisions using frame-by-frame image analysis. The fraction of sliding collisions was measured as a representative metric of CIL and reported as a function of patterned line width. Our data reveal that cancer cells do overcome CIL even in confined 1D microenvironments. In fact, the ability of cancer cells to evade CIL is significantly enhanced in more confined 1D microenvironments, revealing that the migratory advantage of metastatic cells may be particularly well tuned to the spatial constraints of the local tumor microenvironment. These findings were statistically significant and confirmed this platform as a novel tool to quantitatively measure CIL *ex vivo*. The observed trend in the ability of metastatic cancer cells to undergo sliding collisions on thinner patterns would not have been found using traditional two-dimensional tissue culture dishes.

References:

[1] P. P. P. Provenzano, *BMC Med*, 2006, vol. 4, pp. 38–38.

(584t) Modeling the Dynamics of Acute Phase Proteins During the Heat Shock Response

Giraldo-Wingler, N.,
Huang, Z., Villanova University

The acute phase response is one of the most important and common responses the human body undergoes in response to trauma, injury, or infection. The production of acute phase proteins as a function of time has been investigated comprehensively at regular body temperature [1-4]. However, since the acute phase response is commonly accompanied by fever [5], modeling the production of acute phase proteins under an elevated body temperature (i.e., heat shock condition) is of utmost importance. In particular, the experimental data presented in Karlsson et al., 1998, [2] indicate that the production of acute phase proteins such as haptoglobin and albumin in HepG2 cells is affected by the temperature even if no IL-6 stimulation is imposed on the hepatocyte cells. Furthermore, the research presented in Kanelakis et al., 2002, has shown a correlation between elevated levels of haptoglobin, and the glucocorticoid receptor (GR) encountered at elevated temperatures [6]. These findings suggest that the expression of acute phase proteins is regulated by elevated body temperature during the heat shock response. Although models for heat shock response exist (see [1, 7] for two existing models), no model has been developed to quantify the dynamics of acute phase proteins during the heat shock response. In order to address this, we extend the heat shock response model presented by Petre et al, 2011 [1], to incorporate the reactions linking the heat shock proteins hsp70 to the expression of two acute phase proteins, i.e., haptoglobin and albumin, which represent the positive and negative acute phase proteins, respectively.

During the heat shock response, heat shock proteins (hsps) act as "molecular chaperones" to ensure the proper folding of proteins under stress conditions [3]. This response allows the body to better fight off pathogens by raising the body's own temperature, thus creating unfavorable conditions for the pathogens [3]. Meanwhile, the body's own proteins are kept in proper working conditions by the hsps. It is reported that hsps are activated upon the elevation of body temperature [1]. The activated hsps then activate glucocorticoid receptors, which then translocate into the nucleus of the hepatocyte [7], interact with C/EBPβ to down-regulate the expression of albumin [8], and interact with both STAT3 and C/EBPβ to up-regulate the expression of hatoglobin [6]. In this work, we added these reactions into the heat shock response model presented by Petre et al, 2011 [1]. The ordinary differential equations for these newly added proteins were developed based upon the mass action and Michaelis–Menten kinetics. The newly added parameters were estimated from the experimental data for albumin and haptoglobin at an elevated body temperature that were presented in Karlsson et al., 1998 [2]. On the basis of the developed model, sensitivity analysis was conducted to identify the reactions from the heat shock response that play an important role in regulating the synthesis of albumin and haptoglobin. The sensitivity analysis result showed that the most important parameters in the heat shock part of the model were those involving the interaction between hsp and heat shock factors (hsf). Specifically, the most important parameters were those governing the creation and dissolution of the hsp:hsf complex. This may due to the fact that the sequestration of hsf by hsp eventually leads to the suppression of transcription for hsp, which then leads to lowered levels of hsp but increased levels of misfolded proteins (mfp). This effectively functions as a negative feedback loop regulating the long term dynamics of heat shock proteins.

Reference

1. Petre, I., et al., *A simple mass-action model for the eukaryotic heat shock response and its mathematical validation.* Natural Computing, 2011. **10**(1): p. 595-612.
2. Karlsson, J.O.M., M.L. Yarmush, and M. Toner, *Interaction between heat shock and interleukin 6 stimulation in the acute-phase response of human hepatoma (HepG2) cells.* Hepatology, 1998. **28**(4): p. 994-1004.
3. Pratt, W.B., *The Role of Heat-Shock Proteins in Regulating the Function, Folding, and Trafficking of the Glucocorticoid Receptor.* Journal of Biological Chemistry, 1993. **268**(29): p. 21455-21458.
4. Sanchez, E.R., *Heat-Shock Induces Translocation to the Nucleus of the Unliganded Glucocorticoid Receptor.* Journal of Biological Chemistry, 1992. **267**(1): p. 17-20.
5. Arambasic, J., et al., *Association of the Glucocorticoid Receptor with STAT3, C/EBP beta, and the Hormone-responsive Element Within the Rat Haptoglobin Gene Promoter During the Acute Phase Response.* Iubmb Life, 2010. **62**(3): p. 227-236.
6. Kanelakis, K.C., D.S. Shewach, and W.B. Pratt, *Nucleotide binding states of hsp70 and hsp90 during sequential steps in the process of glucocorticoid receptor hsp90 heterocomplex assembly.* Journal of Biological Chemistry, 2002. **277**(37): p. 33698-33703.
7. Kurata, H., et al., *Module-based analysis of robustness tradeoffs in the heat shock response system.* Plos Computational Biology, 2006. **2**(7): p. 663-675.
8. Wedel, A. and H.W.L. Zieglerheitbrock, *The C/Ebp Family of Transcription Factors.* Immunobiology, 1995. **193**(2-4): p. 171-185.

(584ac) Dual Acid-Responsive Anticancer Polymeric Micelles As New Anticancer Therapeutics

Lee, D.,
Park, S.,
Lee, E.,
Kim, Y.,
Khang, G.,
Kwon, B.,

Cinnamaldehyde is a major component in cinnamon, which is an important dietary factor and food additive and is known induce apoptotic cell death through ROS (reactive oxygen species) generation. Despite its potent anticancer activities, the use of cinnamaldehyde in clinical applications is limited by its poor stability and lack of specificity toward diseased tissues. In order to exploit cinnamaldehyde as anticancer drugs, we developed dual acid-sensitive polymeric prodrug of cinnamaldehyde, poly(cinnamaldehyde-co-β-amino ester)-co-poly(ethylene glycol)(PCAE), which incorporates therapeutic cinnamaldehyde covalently in its backbone through acid-cleavable acetal linkages and release it to induce ROS-mediated apoptotic cell death. PCAE copolymers have average molecular weight ~10,000*Da* and self-assemble to form monodispersed micelles with a mean hydrodynamic diameter of ~90 nm. PCAE micelles have dual acid-sensitive behaviors, micelle dissociation and cinnamaldehyde release due to the presence of amine groups and acid-labile acetal linkages. PCAE micelles induced the ROS-mediated apoptotic cell death in SW620 cells, evidenced by confocal laser scanning micrographs and flow cytomety. We also investigated the ability of PCAE micelles to deliver anticancer drugs, camptothecin (CPT) as a model drug, which triggers ROS generation to induce apoptotic cell death. CPT was loaded at a concentration of 10 wt% of micelles with ~90% encapsulation efficiency. CPT-loaded PCAE micelles induced significantly more ROS generation and apoptotic cell death than free CPT and PCAE micelles, suggesting that PCAE micelles have potential as drug carriers and are able to exert synergistic effects with CPT on ROS-mediated apoptotic cell death generation. In a tumor-bearing mice, PCAE micelles also reduced the tumor growth and showed synergistic anticancer activities with a payload of CPT. We anticipate that dual pH-responsive PCAE micelles are able to serve as anticancer drugs as well as drug carriers and have enormous potential as novel anticancer therapeutics.

(584as) Screening of Functional Aptamers Against Alpha-Methylacyl-CoA Racemase

Yang, D. K., National Taiwan University
Chen, C. S.,
Chen*, L. C., National Taiwan University

Alpha-methylacyl-CoA racemase(AMACR) is an emerging prostate cancer marker, which is known as P504S in cancer research literature. Campared to prostate-specific antigen (PSA)that is a traditional prostate cancer marker, AMACR is more specific and senstive in diagnosis of prostate cancer[1]. Moreover, it is already known that over production of AMACR has been highly associated with not only prostate cancer but also breast, colon, renal and other cancers. Recently, it is proven that inhibition of the AMACR expression can diminish the growth of prostate cancer cell lines[2]. This means that AMACR is a viable chemotherapeutic target in prostate cancer therapy. An aptamer is a manmade protein-binding nucleic acid ligand, and its affinity and specificity can rival an antibody. Moreover, compared to an antibody, an aptamer has many advantages such as high affinity and specificity, easy synthesis and modification, high chemical stability (for DNA and modified RNA), small size, low immunogenicity, and so on. Therefore, this work aims at selection of a panel of functional DNA aptamers that target the catalytic site of AMACR and can thus potentially regulate the catalytic activity of AMACR. To this end, we covalently immobilize AMACR on the surface of micro-epoxy beads and use a DNA library with a centered 35-mer randomized region to interact with the bead-bound AMACR in an *in vitro* selection process called SELEX. After a few rounds of selection, we further use the 2(R)-pristanoyl CoA, a AMACR substrate[3], to compete the binding ligands from the substrate-binding epitope on AMACR, which is considered the catalytic site and collect and amplify the eluted DNA molecules as a new ligand pool for the following selection round. Finally, the functionl aptamer candidates targeting AMACR's catalytic site are evolved. To the best of our knowledge, it is the first time that AMACR's aptamers are selected. Moreover, the AMACR aptamers obtained by an epitope-targeting selection approach can be applied in both affinity-based detection of AMACR and regulation of AMACR's activity.

(1) Sreekumar, A.; Laxman, B.; Rhodes, D. R.; Bhagavathula, S.; Harwood, J.; Giacherio, D.; Ghosh, D.; Sanda, M. G.; Rubin, M. A.; Chinnaiyan, A. M. *J Natl Cancer Inst* **2004**, *96*, 834-43.

(2) Wilson, B. A. P.; Wang, H. F.; Nacev, B. A.; Mease, R. C.; Liu, J. O.; Pomper, M. G.; Isaacs, W. B. *Mol Cancer Ther* **2011**, *10*, 825-838.

(3) Lloyd, M. D.; Darley, D. J.; Wierzbicki, A. S.; Threadgill, M. D. *FEBS J* **2008**, *275*, 1089-102.

(584ad) Identification of Protective Immunogens From Extracellular Secretome of Edwardsiella Tarda

Li, H.,
Song, M.,
Peng, X., Sun Yat-sen University

Edwardsiella tarda is a gram negative bacterium of the Enterobacteriaceae. The bacterium has a strong virulence to many aquatic animal species, which brings a great economic loss in aquaculture. It is also an opportunistic pathogen to humans. Therefore, it is quite important to find out an efficient way to control and prevent the infectious diseases caused by this bacterium.

Vaccination has been known to be the most effective method to prevent infectious diseases caused by pathogenic bacteria for a long time. In view of this, to find out the vaccine candidates with high protective ability against *E. tarda* is increasingly valued by researchers. Until now, only some outer membrane and cytoplasm have been demonstrated to be effective protective antigens in *E. tarda*. Recently, increasing interests have been paid on extracellular secretory proteins. The secretory proteins are bacterial extracellular components and may activate host's immune response. In *E. tarda*, studies related to extracellular secretory proteins includes the mechanisms of type III and type IV secretion system and the secretory proteome of *E. tarda*, but whether the proteins secreted out of the bacterium showed immune protection is largely unknown. Thus, investigation on protective immunogens from extracellular secretory proteins may provide efficient candidates for development of vaccines.

Determination of *E. tarda* extracellular secretory proteins Out of 3575 proteins of *E. tarda* EIB202, 29 proteins were predicted as extracellular secretory proteins and 9 proteins existed in multi-locations with over 1.00 of extracellular scores using PSORTb software. Thus, total 38 proteins were annotated to be extracellular secretory proteins in the present study.

Gene cloning and expression and purification of their recombinant proteins

Then, 20 out of the 38 secretory protein genes were randomly selected and successfully amplified in the present study. They were ETAE_2467, FlgD, ETAE_2130, ETAE_2088, ETAE_2845, ETAE_2132, ETAE_2916, ETAE_1033, PapH, ETAE_2496, ETAE_3382, ETAE_1902, FlgK, FlgL, GcP, FlgG, ETAE_1030, EseC, FlgE, EvpC. Their cDNAs were located in the place with their corresponding molecular mass and then recycled by 1% agorose electrophoresis. The amplification reliability was checked by digestion with restriction endonucleases and sequencing. All of the gene fragments were validated. Then the gene fragments were cloned into expression vector pET-32a (+) plasmid (Novagen). The recombinant plasmids were transformed into *E. coli* BL21 (DH3) and screened by grown on LB plate containing ampicillin. Recombinants were picked and cultured in LB medium, and then IPTG was added to induce the expression of recombinant proteins. The recombinant proteins were purified and checked by SDS-PAGE. Only single bands were detected in all of these samples, which indicate the purity of these proteins. These data show that these recombinant proteins may be used for functional investigation.

Investigation of protective immunogens using active immunization

To find out the candidates of vaccines against *E. tarda*, the twenty recombinant proteins were used to confer immune protection against infection caused by *E. tarda* EIB202 in zebrafish model using the active immunization approach. The zebrafish were grouped and then separately immunized with the 20 recombinant proteins and control was only injected with adjuvent. Bacterial challenge was carried out post 28 days of immunization. Among the 20 secretory proteins, ten, EseC, ETAE_2088, FlgD, ETAE_2130, EvpC, ETAE_2845, ETAE_1902, PapH, ETAE_1030, FlgE, showed significant immune protection, compared with the control group. Very importantly, the first four showed approximately 60% RPS and thus were potential candidates of vaccine against *E.tarda* infections.

Identification of extracellular secretory proteins with highly immune protection

Moreover, we wanted to know whether the four proteins with highly immune protection were detected in extracellular production of the bacterium. Antisera to EseC, ETAE_1902, ETAE_2088, FlgD were commercially from Chengxue Biotech. Corp. Guangzhou and were used as primary antibodies in Western blotting for detection of the four proteins in extracellular fraction and whole-cell lysate fraction. The whole cell lysate fraction was used in this detection because it might be a positive control for the detection of these proteins in the extracellular fraction as well as be indictor of antibody specificity. The antisera separately recognized EseC, ETAE_1902, ETAE_2088, FlgD in the places with corresponding molecular mass using whole-cell lysate fraction, whereas they only recognized EseC and ETAE_2088 in the respective places using the extracellular fraction. These results indicate that EseC and ETAE_2088 were extracellular secretary proteins. ETAE_1902 and FlgD were not detected in the extracellular production in this study, which might be related to the following reasons that they may be low abundance proteins or induced proteins.

In summary, identification of highly protective immunogens is the most key step in the development of efficient subunit vaccines. Here we present a systematic investigation on protective immunogens of extracellular secretory proteins in *E. tarda.* Half out of 20 secretory proteins is verified to be protective immunogens, in which four show approximately 60% of RPS in zebrafish. These results may have significant implications in not only providing efficiently protective immunogens for development of vaccine, but also understanding bacterial stimulation from extracellular secretory proteins to host immune systems.

(584m) The Fractal Dimension of Protein Chain Conformation

Peng, X., Tianjin University
Qi, W., Tianjin University
Su, R., Tianjin University
He, Z., Tianjin University

Proteins are made up of 20 different amino acids linked together by peptide bonds [Petsko, 2004]. From a microscopic scale, a protein is determined by its amino acid sequence which is called primary structure via the process of protein folding. Since a protein's curved shape does not change with the change of observation scale, its backbone conformation obeys statistical laws. From a macroscopic scale, a protein is a three dimensional object which is called tertiary or quaternary structure. On the surface of proteins, there exist a great number of irregular "caves" and "gaps", and with further observation the micro-rugged, very irregular structures are also found. Namely proteins have a strong similarity between local structure and overall structure, which is an obvious characteristic of fractal geometry. The fractal geometry is a fascinating conceptual framework [Mandelbrot, 1982] because of its possibilities to characterize nature irregularities with a single number, a really tempting idea *per se*. Moreover, proteins in nature are irregular, and the most fascinating thing is that irregular objects are the norm in the fractal geometry. So the fractal method can be used to describe the complicated spatial and dynamical structures of proteins [Mandelbrot, 1982].

In this paper we examined the fractal properties of 750 folded proteins from four different structural classes [Mount, 2001], namely (1) the *alpha*-class (dominated by *alpha*-helices), (2) the *beta*-class (dominated by *beta*-pleated sheets), (3) the (*alpha*/*beta*)-class (*alpha*-helices and *beta*-sheets alternately mixed) and (4) the (*alpha*+*beta*)-class (*alpha*-helices and *beta*-sheets largely segregated) by using two fractal dimension methods, i.e. "the local fractal dimension" and "the backbone fractal dimension" (a new and useful quantitative parameter). The results showed that the protein molecules exhibited a fractal character in the range of $1 \leq N \leq 15$ (N is the number of the interval between two adjacent amino acid residues), and the value of backbone fractal dimension was distinctly greater than that of local fractal dimension for the same protein. The average value of two fractal dimensions decreased in order of *alpha*>*alpha*/*beta*>*alpha*+*beta*>*beta*.

All in all, the present results suggest that a protein can be regarded as a fractal object with self-similarity and self-affinity, and the concept of fractal may serve as a useful tool for description of the intrinsic characteristics of protein molecules.

This research was supported by the NSF of China (20976125, 31071509, 51173128) and Tianjin (10JCYBJC05100), the Program for New Century Excellent Talents in Chinese University (NCET-08-

0386), and Beiyang Young Scholar Program (2012).

References

(1) Petsko G., Ringe D.. (2004) Protein structure and function. London: Wiley Blackwell Press.

(2) Mandelbrot B. B.. (1982) The fractal geometry of nature. New York: Freeman Press.

(3) Mount D. W.. (2001) Bioinformatics: Sequence and Genome Analysis. New York: Cold Spring Harbor Laboratory Press.

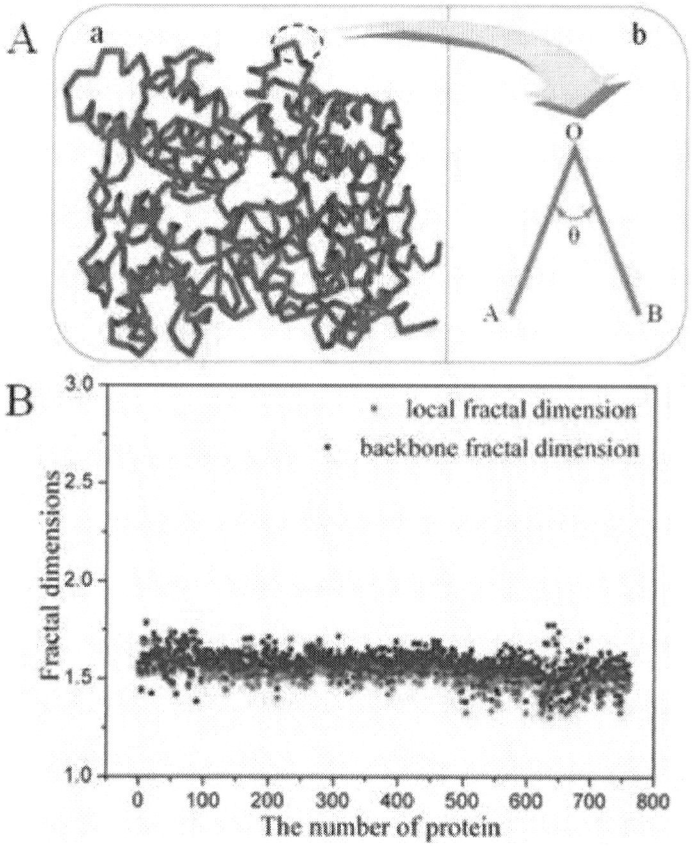

Figure (A) A simple schematic representation of amino acid chain (a) and the bond angle in a dipeptide chain (b). In this protein model the amino acid is denoted by the C*alpha*-atom of amino acid residue. The red thick line represent the length between two adjacent amino acid residues. **(B)** The scatter diagram of fractal dimensions of proteins.

(584at) Temporal Gene Expression Profiling of Transfected Cells to Identify Molecular Factors Guiding Gene Delivery

Martin, T. M., University of Nebraska-Lincoln
Plautz, S. A., University of Nebraska-Lincoln
Kelly, A. M., University of Nebraska - Lincoln
Shea, D., University of Nebraska-Lincoln
Pannier, A. K., University of Nebraska-Lincoln

The design of efficient nonviral gene delivery systems offers potential uses in therapeutic applications such as tissue engineering, biomaterials, and gene therapy. Nonviral gene delivery involves the delivery of exogenous gene(s) to cells, facilitated by the electrostatic complexation of the gene (as plasmid DNA) with a nonviral vector such as a cationic polymer or lipid. Unlike viral based delivery systems, nonviral gene delivery systems can carry a larger genetic payload, overcome several safety concerns, and are easily modified to allow for cellular targeting. However, current nonviral gene delivery systems are inefficient; overcoming the function of cellular barriers to transfection by design of the nonviral gene delivery system is an important strategy for enhancing transfection. While many proposed approaches have resulted in incremental successes to transfection, our rudimentary understanding of the specific molecules that facilitate the DNA transfer process limits the design of better nonviral gene delivery systems.

We have taken a transcriptomics approach to identify up- and down-regulated endogenous genes in transfected HEK293T cells at 8 h, 16 h, and 24 h after delivery of a plasmid encoding the green fluorescent protein (GFP) transgene, complexed by polyethylenimine (PEI) or Lipofectamine® 2000. Bioinformatics analysis using a nonlinear model was used to identify the up (+) and down (-) regulated genes characteristic of transfected cells (GFP+) in comparison to untransfected cells (GFP-). For polyplex transfection, genes differentially expressed between GFP+ and GFP- populations include: 8h (*+RAP1A, +NM_022155, +WDR78, +ACRC*); 16h (*+RAP1A, +CHORDC1, +NM_022155*); and 24h (*+RAP1A, +CHORDC1, +WDR78, -SNHG6, -EIF4A2, -TRA2B, -CKB, -ACTB, -ND2, -AF253979*). The ENRICHR bioinformatics tool showed these genes to be implicated in cell cycle/mitogenesis effectors (*E2F4, EGR1, NFKB, MEF2A*), oncogenic/immunogenic action (*E2F4, FLI1, RFX5, NFKB, MEF2A*), apoptosis (*MEF2A*), gene regulation involved in development/differentiation (*HNF4A, GATA1, NKX2-5, ZFHX3, EGR1, NFKB, FOXC1, SOX2*), xenobiotic metabolism (*ARNT*), the integrin pathway (*RAP1A*), and nucleosome modifications (*PHF8*, H3K23ACH1, H3K23ME2H9, H2BK120ACH9). For lipoplex transfection, genes differentially expressed between GFP+ and GFP- populations include: 8h (*+RAP1A, +PACSIN3, +IFT27, +NDUFA10, -DNAJB11*); 16h (*+RAP1A, +HSPA6, +ZNF548, -AI819238, -BC017896*); and 24h (*+RAP1A, +HSPA6*). The ENRICHR bioinformatics tool showed these genes to be implicated in cell cycle/mitogenesis effectors (*SMARCA4, BRAF, NTRK1, MORF, IL13, NDUFA10),* integrin/cytoskeletal processes (*RAP1A, PAK2*), protein stabilization (HSPA1A, PACSIN3), oncogenic/immunogenic action (*SMARCA4, BRAF, NTRK1, IL13*), apoptosis (*NTRK1, IL13*), gene regulation involved in development/differentiation (*SMARCA4, BRAF, NTRK1, IL13*), DNA repair (*MSH3, MORF*), chromatin/nucleosome modifications (*TRIM28, SMARCA4*), and micro RNA/chromosome overlap (TTTTGAG MIR-373, chr1p13).

These data further support the hypothesis that transfected cells utilize different genes and processes to facilitate DNA transfer, compared to untransfected cells. The results also demonstrate that while some genes and cellular processes are common to both vectors and overlapping time points, there are also time- and DNA-carrier dependent genes identified. Our study suggests a role of these genes and processes as molecular mechanisms that facilitate transfection and offer new targets for improved gene delivery systems.

(584ae) Improved Antitumor Efficacy of Doxorubicin By Lipid Vesicles Integrating Tunable Targeting With Interstitial and Intracellular Release

Bandekar, A., Rutgers University
Sofou, S., Polytechnic Institute of NYU

We hypothesize that the efficacy of chemotherapeutics to vascularized tumors may be improved when delivered by carriers designed to simultaneously address a number of interstitial and intracellular transport barriers. We evaluate a pH-responsive functionalized lipid membrane in the form of vesicles to deliver doxorubicin; these vesicles integrate triggered unmasking of targeting anti-HER2 peptides upon extravasation in the tumor to selectively enhance cell uptake, interstitial release of doxorubicin to improve intratumoral penetration, and endosomal release of doxorubicin to maximize bioexposure. *In vitro*, at the lower limit of interstitial pH, unmasking of anti-HER2 peptides on integrated vesicles results in 175% increase in specific uptake of vesicles by HER2-expressing BT474 cells relative to neutral pH. Exposure of BT474 spheroids to targeted integrated vesicles loaded with doxorubicin results in complete cell killing. *In vivo*, subcutaneous tumor xenografts treated with targeted integrated vesicles are 55% smaller than tumors treated with the non-targeted forms, and 67% smaller than tumors treated with non-targeted vesicles lacking triggered release. Functionalization of the latter improves tumor control, but tumors are 22% larger than these treated with targeted integrated vesicles. This proof-of-concept study supports the significance of designing drug delivery carriers that simultaneously address diverse transport barriers to tumors.

(584e) Role of Water Permeation in Confined Cell Migration and Polarization

Stroka, K. M., Johns Hopkins University
Jiang, H., Johns Hopkins University
Sun, S. X., Johns Hopkins University
Konstantopoulos, K., Johns Hopkins University

Introduction: Cell homeostasis and diverse processes, including migration, are tightly regulated by cell volume. During migration through tissues, metastatic cancer cells experience varying degrees of physical confinement as they migrate through longitudinal tracks within the matrix (1). Intriguingly, cell migration persists in narrow channels even when typical hallmarks of 2D planar migration, such as actin polymerization and myosin II-mediated contractility, are inhibited (2). We therefore hypothesized that an alternate mechanism based on cell volume regulation via ion channels and aquaporins drives cell migration in these confined microenvironments, where cells must deform in order to squeeze through physically restrictive spaces. We used an integrated experimental and theoretical approach, which we have termed the "osmotic engine model", to show that directed water permeation regulates migration in confined microenvironments. Importantly, the theoretical model predicts all key experimental results. Here, our objective was to extend the model by examining the temporal and spatial aspects of cell repolarization in response to an osmotic shock.

Materials and Methods: We engineered a novel microfluidic-based chemotactic device with narrow (3 μm-wide, 10 μm-high, 200 μm-long) extracellular matrix-coated channels. Using timelapse microscopy, we imaged live sarcoma cell migration before and after an osmotic shock at the leading or trailing edge. Pharmacological drugs or siRNA treatments were used to block or knock down proteins relevant to our model, including Na^+/H^+ion channels, aquaporins, myosin II, actin, or microtubules. Cell velocity, volume, and ion channel or aquaporin distribution were quantified and correlated in order to understand cell repolarization and volume changes in response to osmotic shock in confinement.

Results and Discussion: Inhibition of Na^+/H^+ ion channels or knockdown of Aquaporin 5 (AQP5) reduced confined cell migration velocity. These results indicate the importance of ion and water flux across the cell membrane during confined migration and motivated our model. To test the model, osmotic shocks were applied to cells' leading and/or trailing edges within the microfluidic device. In accord with our model, cells reversed migration direction within 5 minutes of applying a hypotonic shock at the leading edge or hypertonic shock at the trailing edge, followed by a two-phase (fast, then slower) migration pattern (Fig. 1). The fast phase correlated with the time during which cell volume was reduced by 40%. Blockage of Na^+/H^+ ion channels, inhibition of actin polymerization, or knockdown of AQP5 interfered with this response. To explore the repolarization mechanism, we quantified the distribution of NHE-1 (a specific Na^+/H^+ion channel) and AQP5 within the cells. As predicted by our model, both NHE-1 and AQP5 polarized to the leading edge of cells migrating in narrow channels. However, they did not begin redistributing to the new leading edge until approximately 30 minutes after the osmotic shock. Thus, cell migration initiated in the opposite direction even before NHE-1 repolarized, indicating that confined cell migration driven by an osmotic shock does not require NHE-1 to be localized at the leading edge, which is a key aspect of our model. Inhibition of actin polymerization completely prevented repolarization of NHE-1.

Conclusions: Together, the results support our "osmotic engine" model of confined cell migration that is driven water permeation via activity of ion channels and aquaporins. This study presents an alternate mechanism of migration in confined spaces that cells exploit in the absence of actin polymerization in confining microchannels. Our osmotic engine model is especially applicable during cell repolarization, where migration in the opposing direction precedes ion channel redistribution. Future work will include live imaging of microtubule, actin, aquaporin, and ion channel dynamics during migration and repolarization in confined channels.

References: (1) Alexander S. *et al*, Histochemistry and Cell Biology 130: 1147-1154 (2008). (2) Balzer E.M. *et al*, FASEB J 26: 4045-4056 (2012).

(584n) Synthesis of Selectively Functionalized Large Pore Silica Materials for Protein and Small Molecule Capture

Schlipf, D., University of Kentucky
Knutson, B. L., University of Kentucky
Rankin, S. E., University of Kentucky

The interaction of proteins with mesoporous silica is of interest for the development of novel platforms for drug delivery, chemical synthesis, and sensing. Recent advances in the synthesis of mesoporous silica via surfactant templating have provided access to a range of pore sizes and particle morphologies that allow for the design of tailored mesoporous structures for protein encapsulation. A major challenge in the design of these mesoporous structures is identifying the location of the protein (surface or pore associated) in order to interpret the effect of the local environment on protein activity or stability. Another challenge is functionalizing the silica surface to provide favorable surface interactions or reduce nonspecific interactions with solutes or proteins at the silica surface. Accessibility of pore loaded proteins with small molecules also needs to be addressed with a goal of optimizing protein availability to interactions with small molecule ligands.

Amino-propyl silanes are a primary functionalization agent for silica, leading to organic-inorganic composites that can be easily chemically modified. Although previous research groups have demonstrated the ability to surface functionalize mesoporous materials, the pore sizes of the functionalized materials were not sufficiently large for protein capture. The silica platforms used in this experiment are synthesized via a mixed surfactant template approach, where the desired pore size (> 7 nm) is achieved through temperature tunable hydrothermal aging during synthesis. A new, facile method for surface functionalization is demonstrated, along with the ability to capture fluorescently tagged proteins within the mesoporous structure of the particles. The confirmation of external functionalization is performed via fluorescein isothiocyanate (FITC) tagging of the selectively functionalized amino propyl silanes, and subsequent imaging in confocal scanning laser microscopy. Pore sizes capable of protein loading are demonstrated via localization of diffused Rhodamine B (RB) tagged proteins within individual particles also using confocal scanning laser microscopy. Finally, Fluorescence Resonance Energy Transfer (FRET) between RB tagged proteins and a small molecule probe (fluorescein) is used to confirm the accessibility of pore loaded proteins within the pore.

(584af) Cofactor Balance By Nicotinamide Nucleotide Transhydrogenase (NNT) Coordinates Reductive Carboxylation and Glucose Catabolism in the Tricarboxylic Acid (TCA) Cycle

Gameiro, P. A., Massachusetts Institute of Technology
Kelleher, J., Massachusetts Institute of Technology
Iliopoulos, O., Massachusetts General Hospital
Stephanopoulos, G. N., Massachusetts Institute of Technology

Cancer and proliferating cells exhibit an increased demand for glutamine-derived carbons to support anabolic processes. In addition, reductive carboxylation of-ketoglutarate by isocitrate dehydrogenase 1 (IDH1) and 2 (IDH2) was recently shown to be a major source of citrate synthesis from glutamine. The role of NAD(P)H/NAD(P) cofactors in coordinating glucose and glutamine utilization in the tricarboxylic acid (TCA) cycle is not well understood, with the source(s) of NADPH for the reductive carboxylation reaction remaining unexplored. Nicotinamide nucleotide transhydrogenase (NNT) is a mitochondrial enzyme that transfers reducing equivalents from NADH to NADPH. Using 13C isotopic tracing methods, we show that knockdown of NNT inhibits the contribution of glutamine to the TCA cycle and activates glucose catabolism in SkMel5 melanoma cells. The increase in glucose oxidation partially occurred through pyruvate carboxylase and rendered NNT knockdown cells more sensitive to glucose deprivation. Importantly, knocking down NNT inhibits reductive carboxylation in SkMel5 and 786-O renal carcinoma cells. Overexpression of NNT is sufficient to stimulate glutamine oxidation and reductive carboxylation, whereas it inhibits glucose catabolism in the TCA cycle. These observations are supported by an impairment of the NAD(P)H/NAD(P) ratios. Our findings underscore the role of NNT in regulating central carbon metabolism via redox balance, calling for other mechanisms that coordinate substrate preference to maintain a functional TCA cycle.

(584ag) Early Prediction of Chemotherapy-Induced Toxicity: A Pharmacometabonomics-Based Approach for Personalized Treatment

Devaraj, J., Purdue University
Skiles, J.,
Renbarger, J.,
Ramkrishna, D., Purdue University

Early Prediction of Chemotherapy-induced Toxicity: A Pharmacometabonomics-based Approach for Personalized Treatment

D Jayachandran[1], Jodi Skiles[2], Jamie Renbarger[2], Doraiswami Ramkrishna[1,*]

[1] School of Chemical Engineering, Purdue University, 480 Stadium Mall Way, West Lafayette, IN 47907, USA

[2] Riley Hospital for Children, Indiana University School of Medicine 705 Riley Hospital Drive, Indianapolis, IN 46202, USA

Vincristine (VCR) is one of the frontline chemotherapeutic agents, which is cytotoxic to cancer cells and effective in a variety of pediatric and adult cancers. While side-effects include neurotoxicity, hyponatremia, mild myelosuppression, and constipation, vincristine-induced peripheral neuropathy (VIPN) is the dose-limiting toxicity. VIPN is highly variable, cumulative and dose-dependent. Although some cases of VIPN are reversible within weeks to months following discontinuation of the treatment, many others suffer chronically from VIPN, even long after treatment is finished. The inability to predict who will develop significant toxicity severely hampers the optimal administration of VCR since the aforementioned side-effects are limited to a subgroup of patients but dosing is a one-size-fits-all approach. This leads to a suboptimal dosing in a subgroup of patients who are not predisposed to VIPN.

Recent advancements in 'omics' and molecular pharmacology technologies have revealed critical information on VCR disposition. VCR is extensively metabolized by cytochrome P450 (CYPs) family of enzymes, specifically CYP 3A4 and 3A5, which are found primarily in human liver microsomes. Studies have shown that tissue distribution of VCR is influenced by the expression of P-Glycoprotein (P-gp), MRP1 and MRP2. Much of the variation in the VCR pharmacokinetics and potentially treatment response can be attributed to genetic polymorphisms observed in these genes. At the molecular level, however, activation or repression of these candidate genes and enzyme activities does not result in accumulation or depletion of corresponding metabolites, which is likely due to the multiplicity and robustness of the metabolic pathway network. Furthermore, human proteome and metabolome are influenced by multiple variables, including epigenetic factors, non-heritable functionally induced (extra-genetic) factors, stochasticity in biochemical reactions, environment, food and lifestyle, micro flora dynamics, and pre- or co-administered drugs, among others. Hence, a focused panel of genetic and phenotypic biomarkers combined with a systems engineering approach is able to effectively predict VIPN.

In this work, we utilized a pharmacometabonomics approach to identify new endogenous metabolites responsible for VIPN in patient samples collected before a dose of VCR was administered. Metabonomics studies the systematic variation in the metabolic profiles due to external stimuli such as genetic modification, biological stimulus and xenobiotic intervention. Pharmacometabonomics, at the intersection of pharmacology and metabonomics, aims to study the global metabolic fingerprints in the pre-dose bio-fluids and characteristic change in the metabolic profiles due to drug dosing in the post-dose bio-fluids. The approach relies on the global metabolic profiling of the pre- and post-dose bio-fluids of the patients using LC/GC-MS and chemometric modeling of the metabolites and VIPN. This provides an unbiased and hypothesis-free analysis of metabolic profile and helps to identify biomarker combinations that enable early prediction of VIPN. This will ultimately make it possible to effectively predict treatment response and optimal administration of VCR according to a patient's metabolic make-up.

(584ah) The Origin of and Cure for Cancer - A Theory

Brinsmade, A. F.,

THE ORIGIN OF AND CURE FOR CANCER - A THEORY

The 96 year old author was prompted to consider this issue when his wife of 69 years was diagnosed with lymphoma and began the battle to regain her health. As an MIT graduate, professional chemical engineer, and member of AIChE for over 60 years, the author has designed and managed plants within the chemical, mining, and commercial industries. He has also worked in the space program for numerous years, where he holds patents and is a published author. Nevertheless, the inter-relationships of the living human systems pose a challenge he is anxious to tackle, especially considering that a full one third of his colleagues may likewise be afflicted with cancer.

The search for the origin of cancer has been intense for the past dozen or more decades but to no avail. The uncertainty is as great as ever. There is a fundamental reason for this lack of success, and that is the assumption that in a biological process there must be either an external biological entity coming into and bearing upon the outcome, or if otherwise in the wholly internal realm, there must be a biological synthesis of sorts necessary to produce the end result.

But there is another more practical approach to the problem of the origin of cancer. And that is to identify the process not as a conventional biological phenomenon but rather, as a plain and simple physical or chemical mishap: a failure to follow in the intended stepwise procedure of a process and therefore to arrive at an unintended and unexpected undesirable or disastrous result.

Though not an astronaut but through careful observation of the facts, Copernicus reached conclusions about the Earth not being at the center of the Universe, which was claimed by Ptolemy 1,500 years earlier. Likewise, though not as immunologists or oncologists, through careful observation of the facts we can reach conclusions about the cause and cure of cancer. Until now, we have known most of the pieces of the puzzle but have not yet put them together.

Cancer is a Self Induced Disease resulting from the dysfunction of the Thymus Gland. The Thymus perhaps has been the most overlooked organ in the human body, the average person even being unaware of its existence. Cancer does not develop as long as the Thymus Gland is functional. Cancer is curable through medications simulating the Thymus Gland's TCell activation.

AKBAR F. BRINSMADE abrinsmade@yahoo.com 486 Channel Mark Dr., Biloxi, MS 39531

(584au) The Search for Future Antispore Molecules

Mi, L., University of Washington
Jiang, S., University of Washington

The bacteria spores have been known to survive under extreme environmental adversities, which include heat, UV, strong acidic/basic/oxidizing conditions as well as exposure to quaternary-ammonium-based antimicrobial compounds. The industrial approach towards spore disinfection using concentrated hydrogen peroxide solution is both dangerous to humans and harmful to the environment. The pathogenic nature of many bacterial spore species, *e.g. B.anthrax*, together with the lack of proper tools to combat this threat has motivated our research to search for novel antispore agents that are both safe and efficient.

To this end, a strategy combining mechanistic study with molecular design was devised to tackle this challenge. Our fundamental study starts with dodecylamine (DDA) lethal germination mechanism. DDA, arguably the strongest antispore molecule known, was reported to trigger spore lethal germination at low concentration and near ambient temperature. Unfortunately, the high toxicity to human has limited its general application outside the realm of academic study. Through our study on DDA mechanism of action, we were able to unveil the molecular principles behind its bioactivity and determine the key structural features essential to this process for the first time in over half a century. Furthermore, by applying this molecular level understanding, we were able to design new molecules from scratch that also show strong antispore activities. Through systematic evaluating these potential antispore agents, a balance can be found between antispore efficacy and environmental toxicity. The results of this study not only shed light on the nature of spore resistance, but also provide practical solutions to the biological threats faced by many clinical and industrial applications.

(584av) Mesoscopic Simulation of DNA - PAMAM Dendrimer Interactions

Yu, S.,
Larson, R. G., University of Michigan

Poly(amido amine) dendrimers (PAMAM dendrimer), which are able to condense semiflexible dsDNA, allowing it to penetrate cell membranes, has many applications in gene therapy and drug delivery. Due to the similarity of the highly charged surface of PAMAM dendrimers and that of histones, the DNA-dendrimer complex mimics chromatin. Since atomistic molecular dynamics simulations of chromatin formation, which is of great interest, is extremely expensive, we developed a coarse-grained model of PAMAM dendrimer-DNA interactions that tracks the dsDNA wrapping around a dendrimer. To speed up our Monte-Carlo simulations of dendrimer-DNA complex formation, we used an implicit solvent model and resolved the DNA double helix structure using the force field developed by de Pablo and coworkers. [PUT IN REF] By using this coarse-grained model, we are able to investigate large systems, without sacrificing important properties of DNA, such as sequence-dependent melting and bending. To eliminate the inherent cytotoxicity of dendrimer, the surface positive charges of PAMAM dendrimer can be partially neutralized by acetylation or PEGylation, which may also reduce its capability to condense DNA. Therefore, we used our simulations to explore the effects of PAMAM dendrimer surface charge density on DNA condensation, in order to help find the optimal charge density of the PAMAM dendrimer.

(584ai) Computational Model to Predict the Effect of Lipids On Pharmacokinetic Profiles in Vivo

Speciner, L., Northeastern University
Buyukozturk, F., Northeastern University
Di Maio, S., Northeastern University
Carrier, R. L., Northeastern University

Dosing lipids, either as a formulation or concurrently with food, has been shown to improve drug bioavailability and absorption for many orally dosed poorly water soluble drugs. While numerous studies have shown this to be the case, there are relatively few commercial lipid-based drug delivery systems, and the effect of food is often not exploited to improve bioavailability, in part because these effects are not currently amenable to quantitative prediction. A computational model based on multiple dynamic processes, such as lipid digestion and drug dissolution kinetics, has been proposed to predict the pharmacokinetic profiles of drugs when dosed either in formulation with lipids or with food. The model was built from *in vitro* experiments that simulate digestion and absorption in the small intestinal environment. In an *in vivo* validation study, blood plasma concentrations of the drug were compared to predicted values from the model. Both drug formulated with lipids as a self-emulsifying drug delivery system (SEDDS) and drug co-dosed with soy bean oil were studied. Results show the feasibility of this model to predict the effect of lipids dosed with drug to increase bioavailability, and thus provide rational guidance for design of lipid-based drug delivery systems.

(584aj) Drug Delivery of Ginsenocides and Polysacharides From North American Ginseng: Immunomodulatory and Angiogenesis Effects

Charpentier, P. A., Western University
Azike, C., Western University
Samimi, R., Western University
Salarian, M., Western University
Akhter, K. F., Western University
Lui, E., Western University

North American ginseng is a medicinal plant used in traditional herbal medicine grown in North America, whose root extracts are being used for commercial over the counter anti-cold formulations such as Cold-FX™. The main components of ginseng are ginsenosides and polysaccharides which provide biological activities such as anti- and pro-inflammatory mediation effects. In this work, development of appropriate delivery systems for oral administration to target colon diseases and bone cements was investigated, with release and bioassay results described. Micro/nanospheres of both ginseng polysaccharide extracts and biodegradable poly (propylene fumarate) (PPF), and poly (DL-lactic-co-glycolic acid) (PLGA) containing ginsenocide extracts were prepared using a microfluidic device and compared to other conventional anti-solvent techniques. The morphology and size of the carriers was characterized by scanning electron microscopy (SEM), ZEISS light microscopy, and dynamic light scattering (DLS). Depending on experimental conditions, microfluidics was found to provide unimodal spheres in the 10-50 um region or down to 20 nm with very narrow particle size distributions. Nanosized extracts were then encapsulated with gelatin, to give 180 nm spheres suitable for integration into capsules for subsequent oral drug delivery. Results showed that the release of ginseng components was controllable with lower initial bursts than those observed with conventional microparticles. Bioassays using murine macrophages were investigated showing significant immunoactivty of extracts while *in vivo* studies in plasma of rat orally fed with NA ginseng root showed measurable concentrations of polysacharides, and significant immunomodulatory effects. These results show the utility of these materials for next-generation biomedical agents as controlled release devices for capsules and bone cements.

(584ak) Evaluating Nanodisks and Nanorods As Carriers for Targeting Diseased Endothelium in Physiological Blood Flows

Thompson, A., University of Michigan
Eniola-Adefeso, O., University of Michigan

Introduction: The development of vascular-targeted carriers (VTC) for the delivery of therapeutics could greatly improve the treatment of many human diseases. Spherical nanoparticles are commonly used as potential drug carriers due to their ability to easily navigate microvasculature and ease of fabrication/drug loading. However, recent literature has shown that spherical nanoparticles do not efficiently marginate and adhere to the vascular wall in model blood vessels compared to larger micron sized particles.[1,2] Therefore spherical nanoparticles may not be effective as VTCs for drug delivery and diagnosis, particularly in targeting diseases which affect medium/large blood vessels (M/LBV), such as atherosclerosis. Recently, particle shape has received attention as a parameter that can be used to improve the performance of VTCs. The purpose of this study is to examine the hemodynamics of rod and disk-shaped particles of different aspect ratios relative to spheres of equal volume both *in vitro* in human blood flow under physiological conditions.

Materials and Methods: Polystyrene rod and disk-shaped particles were fabricated via a polymer film stretch method previously described in literature.[3] Rods, disks, and spherical particles were conjugated with targeting molecules (sialyl Lewis a, sLe[a]) which bind to selectins. The differently shaped particles were tested for their margination efficiency from blood flow both *in vitro* and *in vivo*. The particles explored ranged in diameter from 0.5 – 2 µm (equivalent spherical diameter; ESD, for rods and disks).

Vascular-targeted spheres, rods or disks at a fixed concentration were mixed with RBC and saline flow buffer, and allowed to flow over a layer of IL-1β activated human umbilical vein endothelial cells (HUVEC) using a parallel plate flow chamber. Adhesion of VTCs from steady blood flow from relatively low to relatively high shear rates (200-1000 s^{-1}) was investigated. Since disturbed blood flow profiles are typical in areas of vasculature where atherosclerosis tends to occur, particle adhesion pulsatile and recirculatory blood flow profiles were also examined. The adhesion of rods, disks and spheres to the endothelium from blood flow was imaged/quantified using brightfield and fluorescence microscopy. Further, the localization of these particles to the red blood cell free layer (RBC-FL) while in flow was investigated using confocal microscopy.

Results and Discussion: With *in vitro* assays we find that rod-shaped microparticles at a high aspect ratio have a higher binding efficiency to the HUVEC monolayer from human blood flow than spheres of equivalent volume and targeting ligand site density. Overall, adhesion was greatest for microparticles with ESD=2 μm compared to those with smaller volumes. The improved adhesion pattern for micron-sized rods was not due to better localization of rods to the RBC-FL, as the confocal data showed equivalent levels of microrods/spheres in the RBC-FL. Adhesion of 500 nm ESD spheres and rods was minimal; likely due to poor localization to the RBC-FL. While rods with 500 nm ESD showed no improved adhesion compared to equivalent spheres*, preliminary data suggests that equivalent disk shaped particles display both better localization to the RBC-FL and improved adhesion to the endothelium.*

Conclusions: Our study shows that shape can be a very useful and tunable parameter in the design of vascular-targeted carriers for imaging and treatment of atherosclerosis and other diseases of M/LBVs.

(584aw) A Conformation-Switching, Intrinsically Disordered Protein Probe for Detection of Amyloid Oligomers

Hernandez, M., Polytechnic Institute of New York University
Hu, Y., Polytechnic Institute of New York University
Kim, J. R., Polytechnic Institute of New York University

Amyloid aggregation of intrinsically disordered proteins (IDPs), such as α synuclein (αS) in Parkinson's disease (PD), is a molecular process producing structurally similar yet different oligomeric and fibrillar protein aggregates. In this study, motivated by Nature's use of IDPs as conformation-switching biosensors, we engineered an αS variant together with conformation-sensitive fluorescence to create a novel molecular probe platform for rapid, specific and quantitative detection of αS oligomers, which are the major toxic aggregate forms in PD. Our results suggest that our molecular probe was bound to β sheet-structured αS oligomers, and this binding event was functionally linked to fluorescence signaling. A linkage between binding and signaling could be tuned by binding-induced conformational changes of our molecular probe, enabling selective detection of αS oligomers over unfolded αS monomers, structurally similar αS fibrils and other amyloid oligomers. Our strategy to engineer the structural flexibility of IDP represents a new paradigm for designing rapid-responsive, conformation-switching molecular probes for detection of specific protein forms mediated by not only binding to analytes but also its linkage to signaling.

(584r) Analysis of Complexities in Signal Transduction Though a Study of Signalling Modules

Krishnan, J., Imperial College London

Cellular processes are organized and controlled by signalling networks of considerable complexity. The complexity arises not only from the number of interacting components, but also from the fact that relative small combinations of factors can give rise to subtle and unexpected behaviour. Analyzing these different aspects provides different insights into the behaviour of signalling networks.

In this presentation, I will present multiple examples of analysis of different aspects of signalling through a detailed study of signalling modules.Each of these examples provides insights into how complexity in signal transduction can arise from the combination of relatively simple factors, and is relevant in a broad range of signalling contexts.

In the first case, I discuss multiple scenarios of interacting modules via detailed dynamic analysis, and interaction of pathways and processes via shared components (a ubiquitous feature in signalling, observed in many contexts). In the second case I present and discuss spatial aspects of signal transduction (often ignored in signalling) via the analysis of covalent modification cycles, and representative motifs. These studies show how new complexities in signalling can arise from a relatively small number of factors. This work is done with Daniel Seaton and Aiman Alam-Nazki.

(584al) Tear-Dynamic-Based Pharmacokinetics Model for Ophthalmic Drug Delivery

Peng, C. C., University of California, Berkeley
Radke, C. J., University of California Berkeley

Despite considerable loss inefficiency (> 95 %), topical ophthalmic drug delivery remains the primary application modality to the human eye. Further, topical administration of ocular drugs is limited by uncertainty of where the drug distributes including tear drainage and spill-over loss, dilution in tear fluid, and penetration through cornea and/or conjunctiva (both bulbar and palpebral). Most studies are limited to measuring concentrations in ocular tissues over time following single or multiple administration, which prevents quantitative bioavailability predictions for changes in formulation or dosage regimen. Moreover, the majority of such studies have been conducted by animal subjects such as rabbits and rats. While the physiology properties of these ocular tissues in animal models might be representative, their tear dynamics are significantly different from that of human subjects.

In this study we present a compartmental description of tear dynamics based on human subjects that quantifies drug pharmacokinetics (PK) by topical administration including bioavailability in each compartment of the anterior eye. The tear supply/drainage system is divided into 5 compartments: upper and lower menisci, upper and lower conjunctival sacs, and tear film.[1] A transient description is employed for each compartment for tear volume, salinity, and drug concentration. Compartments are coupled to each other and to drug delivery through kinetic routes.

This tear-dynamic-based PK model was first examined by predicting drug distribution for topical dosing of timolol by eye drop. The results indicate a residence time of drug in precorneal tear film of about 6 min; ocular bioavailability to cornea is about 5% for normal subjects. Furthermore, our model allows us to predict the effect of drop size and dosing sequences on the drug loss to conjunctiva and tear drainage, which is directly related to the systemic side effect of drug delivery. Current delivery models make no distinction on where the drug resides. Another case study predicts precorneal residence time of a fluorescent tracer mixed with commercial artificial tears. The viscosity of solution depends on the concentration of active agent (biocompatible polymers). Although dilution reduces the initial high viscosity, thickness (volume) of precorneal tear film increases and tear drainage rate decrease, thus extending drug residence time. These predictions are consistent with clinical observations.[2]

The proposed tear-dynamics model is the first attempt to develop a topical ocular drug-delivery model based on compartmental tear dynamics. The model is especially useful for optimizing drug formulations and dose regimens for both normal and dry-eye subjects and, thus, for streamlining FDA approval. Further, quantitative prediction of conjunctival uptake can be integrated with available PK models to improve drug uptake prediction accuracy in both the anterior and posterior segments of eye. The new model is suitable for quantifying ocular drug delivery by other vehicles including contact lens, and emulsion and nanoparticle suspensions.

1. Cerretani C, Radke CJ. ***Current Eye Research*** 2013; *submitted*

2. Paugh JR, Nauyen AL, Ketelson HA, Christensen MT, Meadows DL. ***Optom Vis Sci***. 2008;85:725-731.

(584am) Uptake, Co-Localization and Cytotoxicity of Acid-Labile Pegylated Poly(amidoamine)-Doxorubicin Conjugates in An in Vitro Model of the Lung Adenocarcinoma

Zhong, Q.,
Dimovski, R., Wayne State University
da Rocha, S. R. P., Wayne State University

Lung cancer is the leading cause of cancer death among both men and women in the United States. Adenocarcinoma accounts for more than 40% of all lung cancer cases. Doxorubicin (DOX) is an FDA-approved chemotherapeutic that has been widely used as primary anticancer drug in the treatment of a variety of cancers, including lung adenocarcinoma. However, rapid elimination, uncontrolled release, systemic distribution, and life-threatening cardiotoxicity, reduce tumor cytotoxicity and thereafter curb the application of DOX and other anti-cancer therapeutics in the treatment of several patient populations suffering from lung cancer. Therefore, the development of DOX delivery carriers (and carriers for other anti-cancer therapeutics for that matter) that can achieve controllable release, reduce toxicity to healthy tissue, and prolonged blood circulation times would be of great relevance in the treatment of lung adenocarcinoma.

In this work, a series of PEGylated poly(amidoamine) dendrimer nanocarriers with acid-labile and acid-nonlabile DOX conjugates were synthesized and characterized. We employed a two-step PEGylation strategy (pre-PEGylation followed by supplementary PEGylation) to increase the DOX payload. We investigated the impact of pH (neutral and acidic pH), PEGylation density (low, medium and high) and number of DOX conjugates (low and medium) on the release of DOX from the dendrimer nanocarrier, the kinetics of carrier uptake, intracellular release kinetics of DOX from the nanocarrier, and toxicity in an alveolar adenocarcinoma cell line (A549). PEGylation retards the release of DOX in acidic medium, and also intracellularly, as determined by co-localization studies with confocal microscopy. We also observed that the kinetics of cellular entry of the nanocarrier with DOX increased significantly compared to free DOX, as determined by flow cytometry. At the highest PEGylation density, the rate of internalization of the nanocarrier containing DOX was even higher than that of free DOX. PEGylation density also affects cytotoxicity as seen by an increase in IC_{50} for DOX-conjugated dendrimer compared to free DOX, due to the controlled release of the therapeutic.

(584u) In Vivo Imaging of Anticancer Drug Activity At the Cellular Level

Dubach, J. M., Massachusetts General Hospital/Harvard Medical School
Vinegoni, C.,
Weissleder, R., Massachusetts General Hospital/ Harvard Medical School

Intravital fluorescence microscopy provides cellular information under physiological conditions in vivo. We have developed a novel optical imaging modality based on fluorescence detection that enables the study of cellular response to anticancer drug treatment in tumor mouse models. Fluorescently labeled drug can be tracked in the tumor over time and the cellular uptake in vivo can be determined at subcellular resolution. This imaging modality provides both pharmacokinetic and pharmacodynamic information of drug effects on tumors in vivo. Here we demonstrate the ability to image drug activity at the cellular level in vivo. These results can be used in the future to study the cellular mechanism of both clinically successful and unsuccessful drugs as well as tumor resistance heterogeneity during drug treatment.

(584o) Fabrication of Protein Dots Pattern Via Template Assisted Particle Lithography

Li, R.,
Taylor, Z.,
Schmidtke, D. W., University of Oklahoma

Fabrication of protein patterns in micro-scale and nano-scale has gained tremendous strides over the years due to their wide range of applications in the field of tissue engineering, biosensors, and live cellular microfluidics experiments. Previously, our lab had developed an inexpensive and effective protein patterning technique based on particle lithography to fabricate hexagonal patterns of protein dots (450-1200 nm diameter). These substrates have been utilized to investigate the role of protein spatial organization on neutrophil spreading. One potential limitration of this particle lithography technique along with other self-assembled methods is the limited control on particle separation and overall particle patterns. The center-to-center separations along with the size of protein dots formed are intrinsically related to each other based on the diameters of the beads used to form the monolayer, severely limiting in choice feature of size-to-separation. In this work, we demonstrated that photolithography can be used to direct the deposition location of PS beads via template assisted self-assembly (TASA) mechanism. As a result, by controlling the beads' deposition patterns, the protein dot patterns in subsequent steps can then be easily manipulated without changing the dot size.

(584ba) Multimodal Contrast Agent for Prostate Cancer Diagnosis

Rogers, H. B., Auburn University
Anani, T., Auburn University
David, A. E., The University of Michigan

One of the most prominent issues that have arisen due to increased life expectancy in developed countries is the increased rate of prostate cancer. Prostate cancer is currently the most common type of cancer in men and is the second leading cause of cancer related deaths. Despite decades of effort in the fight against cancer, current diagnostic techniques are inaccurate and therapies for prostate cancer are often ineffective. Nanomedicine, however, has the potential to solve the current shortcomings of prostate cancer treatments.

Much of our failure in treating cancer stems from the fact that, despite all the knowledge gained, still very little is known about the factors affecting cancer progression. Recently, there has been a great interest in combining data from multiple diagnostic modes – for example, the physiological information from magnetic resonance imaging (MRI) and metabolic data from positron emission tomography (PET) – to improve our understanding of cancer growth. In response to these developments, this project looks to develop multi-modal (MRI/PET) contrast imaging agents that would provide qualitative anatomical information and quantitative data on the biological activities within tumor lesions. We use magnetic iron oxide nanoparticles as our multifunctional system due to their intrinsic diagnostic capabilities as MRI contrast agents and the ability to add functionalities, through appropriate surface modifications, to include targeting moieties, fluorescence dyes and radioactive agents. In this work, we demonstrate the successful synthesis and fractionation of iron oxide nanoparticles that can be used as multimodal imaging agents. Each fractionated size distribution was then used in several *in vitro* experiments to determine what effect nanoparticle size has on particle size stability, protein binding, and macrophage uptake, which all affect *in vivo* pharmacokinetics and biodistribution of the nanoparticles. The results of this work allow us to identify an optimal size to be used *in* vivo. Moving forward, peptides identified thorough phage display technology will be investigated as viable moieties to target molecules specific to prostate cancer and to monitor the proteolytic activity in the tumor's microenvironment. These targeted contrast agents can then be utilized to optimize nanomedicines for cancer diagnosis and therapy.

(584ax) Encapsulation of Adenovirus With Pegylated Ternary Cationic Liposome Shows Improved Transduction of CAR-Negative Cells

Vupputuri, S., Oklahoma State University
Tayebi, L., Oklahoma State University
Ramsey, J. D., Oklahoma State University

Encapsulation of Adenovirus with PEGylated Ternary Cationic Liposome Shows Improved Transduction of CAR-Negative Cells

Sravanthi Vupputuri, Lobat Tayebi and Joshua D. Ramsey

School of Chemical Engineering, Oklahoma State University, Stillwater, OK 74078

Adenovirus (Ad) has been widely used in gene therapy clinical trials because of its ability to transduce different human cells and to accommodate large transgenes. Ad-based gene delivery vectors, however, have drawbacks, such as the inability to efficiently infect cells lacking the coxsackie adenovirus receptor and a tendency to elicit an inflammatory response, which initiates clearance of the virus. To address these drawbacks we explored encapsulating the virus in a PEGylated ternary cationic liposome (DDCP) composed of 1,2-dioleoyl-3-trimethylammonium-propane (DOTAP), 1,2-dioleoyl-*sn*-glycero-3-phosphoethanolamine (DOPE), cholesterol and polyethylene glycol (PEG). We also investigated the effect of non-ionic surfactants and amino acids on the physical and chemical stability of the PEGylated liposomes.

Our results showed that encapsulation of Ad with PEGylated cationic liposomes enabled the virus to infect CAR-negative cells, providing an almost 300-fold improvement in gene expression compared to the native virus alone. We also found that gene expression depended on the ratio of the different lipids, with the optimum molar ratio of DDCP being 1.0:0.7:0.3:0.02. The level of proinflammatory cytokine IL-6 produced by mouse macrophage cells exposed to Ad encapsulated within the DDCP liposome was significantly lower than for cells exposed to the native virus. Further, measurements of cytotoxicity indicated that PEGylation provided a significant reduction in toxicity. Nanoparticle size, zeta-potential and polydispersity were characterized using dynamic light scattering. Overall, our results show that encapsulating Ad within PEGylated liposomes is an attractive option for producing an improved adenoviral vector with improved transduction of CAR-negative cells and reduced cytokine production.

(584ay) Role of Urinary Proteins in Moderating Calcium Oxalate Kidney Stone Formation

Farmanesh, S., University of Houston
Rimer, J. D., University of Houston

Proteomic studies of organic constituents occluded within the matrix of human kidney stones reveal the presence of more than 60 different proteins, glycosaminoglycans, and other molecular species. Past studies have identified a small fraction of these constituents as putative inhibitors of calcium oxalate monohydrate (COM) crystallization. COM is the most prevalent crystalline component of human stones, accounting for more than 80% of nephrolithiasis incidence. Here we will discuss *in vitro* studies of COM crystallization in the presence of urinary proteins and biomolecules ranging from anion-rich species (*e.g.* serum albumin and transferrin) to cation-rich species (*e.g.* lactoferrin and lysozyme). These studies reveal that urinary proteins function as either inhibitors or promoters of COM crystallization with variable efficacy and specificity for binding to different COM crystal surfaces. We used a combination of experimental techniques to quantify the effects of these urinary constituents on COM crystal size, habit, surface architecture, and growth kinetics. Results of these studies reveal that certain binary combinations of urinary components exhibit a synergetic enhancement of their efficacy, while other combinations yielded antagonistic effects. This work explores the physicochemical properties of effective protein inhibitors of COM crystallization towards an improved understanding of disease pathogenesis; and a generalized framework for the rational design of biomimetic analogues (*i.e.*peptide and organic growth inhibitors) as potential drug candidates for preventative therapies of kidney stone disease [1,2].

[1] Farmanesh, S., Chung, J., Chandra, D., Sosa, R.D., Karande, P., Rimer, J.D., *J. Cryst. Growth* (2013) In Press.

[2] Rimer, J.D., An, Z., Zhu, Z., Lee, M.H., Goldfarb, D.S., Wesson, J.A., Ward, M.D., *Science*

330 (2010) 337-341.

(584an) Characterization of the Size, Shape, and Drug Encapsulation Efficiency of PLGA Microcapsules Produced Via Electrojetting for Anticancer Agent Delivery

Fattahi, P., The Pennsylvania State University
Borhan, A., The Pennsylvania State University
Abidian, M. R., The Pennsylvania State University

Despite significant progress in the development of new chemotherapeutic agents and drug delivery methods for brain tumors, malignant gliomas (high grade brain tumor) remains deadly with a median survival period of only about a year. The high dosage of chemotherapeutic agents required for penetration through the blood brain barrier during chemotherapy not only kills cancer cells but also damages healthy tissues and causes adverse side effects. Hence, a major unmet challenge in the treatment of malignant gliomas is the development of effective and targeted local delivery of chemotherapeutic agents at the cellular level. Drug-loaded PLGA microcapsules with high drug encapsulation efficiency and controlled shape and size are attractive candidates for a more precise control of anticancer agent delivery at the tumor sites. Here, we report the results of a systematic study of the size, shape, and drug release profiles of Poly(lactic-co glycolic) (PLGA) microcapsules produced and loaded with the anticancer agent 1,3-bis(2-chloroethyl)-1-nitrosourea (BCNU) using an electrojetting technique. We quantify the shape and size distribution of BCNU-loaded PLGA microcapsules as a function of the polymer concentration and flow rate used during electrojetting, and measure drug release profiles for microcapsules of three different morphologies: flattened microspheres, microspheres, and microfibers. The BCNU release profiles for all three microcapsule morphologies are found to be in good agreement with model predictions for drug release as a result of drug diffusion and degradation of PLGA microcapsules.

(584v) Modeling Motor-Assisted Intracellular Cargo Transport

Loewenberg, M., Yale University
Pascal, J. A., Tennessee Tech University
Cristini, V., University of New Mexico
Chauviere, A., University of New Mexico

Modeling Motor-Assisted Intracellular Cargo Transport

Michael Loewenberg, Jennifer Pascal, Arnaud Chauviere, Vittorio Cristini

Yale University, New Haven, CT, USA

University of New Mexico, Albuquerque, NM, USA

Intracellular transport of cargo, including macromolecules, vesicles and organelles, through the attachment to microtubules via molecular motors, such as kinesin and dynein, is a complex process that plays a significant role in cellular function. Here, we present an analysis and experimental validation of the unidirectional traffic-type partial differential equations that describe motor-assisted intracellular cargo transport (Smith and Simmons 2001). A new analytical solution is obtained for the case of a single type of molecular motor. The solution exhibits wave-like hyperbolic behavior at short times and a convective/diffusive behavior at longer times. The long-time behavior is characterized by two parameters: the mean velocity of the cargo and a dispersion coefficient that results from the saltatory mechanism by which cargo binds and unbinds from the motors. The solution and thus the traffic-type model is tested using stochastic sampling of nanoparticle trajectories in the squid giant axon (doi 10.1088/1478-3975/9/5/055005). We present a generalized analysis of motor-assisted cargo transport for an arbitrary number of motors, providing numerical results and analytical formulas that describe the long-time evolution of the system.

(584ao) Safety and Biodistribution of Carbohydrate-Functionalized Polyanhydride Nanoparticles

Vela Ramirez, J., Iowa State University
Goodman, J., Iowa State University
Roychoudhury, R., Indiana University
Boggiato, P.,
Pohl, N., Indiana University
Wannemuehler, M. J., Iowa State University
Narasimhan, B., Iowa State University

An emerging area of research in vaccine development is the design of adjuvant formulations that can be targeted to specific receptors and activate specific cell populations. Biodegradable polymers have shown promising characteristics as adjuvants and/or delivery vehicles by enhancing antigen presentation and processing compared with soluble protein. The properties of these materials can be tailored to enhance their interactions with specific receptors on a variety of cells. In particular, polyanhydride nanoparticles exhibit these desirable characteristics both as adjuvants, including immune response enhancement, modulate inflammation, activation of antigen presenting cells, and induction of high titer and highly avid antibodies; and as delivery vehicles in which they sustain antigen release, and stabilize protein antigens. Pattern recognition receptors such as C-type lectin receptors (CLRs) are key components of innate immunity, and also important for the modulation of adaptive immune responses. In these studies, the effect of carbohydrate functionalization on the safety and biodistribution of polyanhydride nanoparticles was analyzed.

Upon *in vivo* administration, nanoparticle distribution depends on size, chemistry, interactions with a variety of cells, and route of administration. In this context, toxicity studies are a necessary first step to assess the safety of the carbohydrate-functionalized particles prior to future efficacy studies. Analysis of the distribution of targeted polyanhydride nanoparticles is necessary to rationally select and design particles targeted to specific cells and/or locations in order to induce-effective immune responses.

In this work, the effect of carbohydrate functionalization on the toxicity and biodistribution of polyanhydride nanoparticles following intranasal administration of mice was investigated. Dye-loaded polyanhydride nanoparticles based on 1,6-bis-(*p*-carboxyphenoxy)hexane (CPH), and 1,8-bis-(*p*-carboxyphenoxy)-3,6-dioxaoctane (CPTEG) were synthesized using anti-solvent nanoencapsulation. Surface functionalization with di-mannose, galactose, or glycolic acid (as a "linker" control) was performed using an amine-carboxylic acid coupling reaction. These studies have important implications for the *in vivo* performance of carbohydrate-functionalized nanoparticle-based systems for vaccine delivery.

(584p) Model-Based Decision Support Tool for Personalized Treatment of Cancer

Devaraj, J., Purdue University
Ghanty, T.,
Ramkrishna, D., Purdue University

Model-based Decision Support Tool for Personalized Treatment of Cancer

Jayachandran Devaraj[a], *Tuhina Ghanty*[b], *Doraiswami Ramkrishna*[a,*]

[a] School of Chemical Engineering, Purdue University, 480 Stadium Mall Way, West Lafayette, IN 47907, USA

[b]Department of Chemical Engineering, Indian Institute of Technology, Mumbai, India

[*]Corresponding author at: School of Chemical Engineering, Purdue University, 480 Stadium Mall Way, West Lafayette, IN 47907, USA. Tel: +1 765 494 4066; fax: +1 765 494 0805; email: ramkrish@purdue.edu

Abstract

Personalized medicine has been in practice in some form or the other ever since medicine has been in existence. With the discovery of new complex diseases, elevated incidence and mortality rate of existing diseases, and advancements in high throughput technologies, the focus and importance on personalized medicine has been gaining momentum in the recent times. Every individual is unique both genetically and phenotypically. Hence, the gist of personalized medicine lies in evaluating genetic and biomolecular traits specific to a patient, followed by clinical decision making based upon the patient's genetic and phenotypic make-up. Since several important processes interfere during the manifestation of gene sequence to molecular phenotypes and further into cellular responses, a holistic approach, integrating many levels of biomolecular entities and events, provides superior information for clinical decision making. Given the complexity of the biological processes and amount of information, it is impossible to reach a conclusion through simple deductive reasoning. Quantitative tools, suitably empowered by systems theoretic approach, could serve as a decision-support mechanism to physicians to quantitatively predict the response and adjust the dose for a given patient.

In this work, we present a model-based standalone GUI to quantitatively predict the clinical outcome for a specific patient and determine the optimal chemotherapeutic dose to achieve a desired clinical response. We demonstrate an application of such a tool using an important chemotherapeutic drug known as 6-Mercaptopurine in tailoring treatment for Leukemia patients.

(584q) Development of a Dual Fluorescence Whole-Cell Biosensor to Detect N-Acyl Homoserine Lactones

Zhuang, G.,
Ma, A.,

Detection of *N*-acyl homoserine lactones (AHLs) is useful for understanding quorum sensing (QS) behaviors, including biofilm formation, virulence and metabolism. In this study, to not only detect AHLs but also to indicate the host cells *in situ*, the plasmid pUCGMA2$T_{1\text{-}4}$ was constructed to make a dual fluorescent whole-cell biosensor based on the AhlI/R AHL system of *Pseudomonas syringae* pv. *syringae* B728a. The plasmid contains three components: constitutively expressed P_{nptII}::*gfp* for indicating host cells, P_{ahlI}::*mcherry* that produces red fluorescence in response to AHL, and the *ahlR* gene that encodes an AHL regulatory protein. Meanwhile, two copies of $T_{1\text{-}4}$ (four tandem copies of a transcriptional terminator) were added into the plasmid to reduce background. The results showed that when the plasmid was placed into *Escherichia coli*, the dual fluorescence whole-cell biosensor was able to respond with red fluorescence within 6 hr to 5×10^{-8}–10^{-5} mol/L of 3OC6-HSL. Bright green fluorescence indicated the host cells. Furthermore, when the plasmid was transferred to wild-type *Pseudomonas* PhTA125 (an AHL-producing bacterium), it also showed both green and red fluorescence. This result demonstrates that this plasmid can be used to construct whole-cell indicators that can indicate the AHL response and spatial behaviors of microbes in a microenvironmental niche.

(584aq) Multifunctional Bacterial Lyases: Roles in Disease and Therapeutic Applications

MacDonald, L., Lehigh University
Wong, E. L., Lehigh Valley Hospital
Berger, B. W., Lehigh University

Stenotrophomonas maltophilia is a gram-negative, obligate aerobic bacteria widely distributed throughout the environment, particularly in water sources or damp environments. *S. maltophilia* is among the most common, emerging multi-drug resistant (MDR) organisms found in the lungs of people with chronic infections as well as with cystic fibrosis (CF). The prevalence of *S. maltophilia* in lung infections is increasing, with multiple studies pointing to *S. maltophilia* as an independent predictor of pulmonary exacerbation requiring hospitalization. A common mechanism of persistence during chronic infection is through biofilm formation, in which exopolysaccharide (EPS), DNA and other components are secreted into the extracellular space to form a barrier preventing access for antimicrobials and other therapeutics. We recently identified a family of unique polysaccharide lyases that exhibit broad substrate specificity against a range of EPS components. Strikingly, in contrast to other lyases, we find substrate specificity is highly dependent on pH, with activity against both mammalian polysaccharides as well as bacterial substrates. The possible role of dual substrate specificity in virluence and approaches for therapeutic intervention will be discussed.

(585c) Isolation, Identification and Quantification of Two Major Antioxidants From Sarang Semut (Myrmecodia pendan)

Engida, A.,

Sarang semut was extracted with methanol at 45 °C for 3 h. The extract was re-dissolved in distilled and successively partitioned with hexane (HEX), dichloromethane, (DCM), ethyl acetate (EtOAc) and n-butanol (BuOH). EtOAc fraction showed the highest DPPH scavenging activity. Two major phenolic compounds, rosmarinic acid and procyanidin B1, were isolated from EtOAc fraction using silica gel column chromatography and gradient solvent system. Rosmarinic acid and procyanidin B1 were obtained with purities of 99.01%, and 98.44%, respectively as determined by HPLC. Their chemical structures were characterized using HPLC–ESI/MS/MS and NMR spectroscopic techniques. The DPPH inhibition power of the two compounds was evaluated. The IC_{50} value of proanthocyanidin B1 and rosmarinic acid was found to be 27.59 ±13.50 µg/mL and 35.80 ± 3.79 µg/mL, respectively. The results of this study suggested that the antioxidant activities of these phenolic compounds may be influenced by the number of hydroxyls in their structure. The contents of proanthocyanidin B1 and rosmarinic acid were quantified using HPLC and found to be 3.236 ± 0.280 and 20.688 ± 1.573 mg/g dry sample, respectively.

(585ad) Spray Drying of Uniform Microparticles for Encapsulation and Controlled Release

Selomulya, C., Monash University

The Biotechnology and Food Engineering group at Monash University has developed a low temperature assembly approach to synthesise monodisperse microparticles in a scalable, waste-free process, stemming from their research on thermal sensitive dairy powders. Using a microfluidic jet spray drying method, functional particles with uniform properties, including fluorescent and magnetic composites, polymeric microparticles for encapsulation and controlled release, and mesoporous microspheres, can be synthesised. The fabrication route is applicable to other self-assembling systems and could be a cost effective, energy and material-efficient route to produce bioactive powders with potential technology transfer to functional foods and therapeutics industry. The talk will give an overview of this assembly method with examples of several applications, and the view to invite future collaboration opportunities.

Related publications:

1. Liu, W., Wu, W., **Selomulya, C.***, Chen, X. D., **2013**, On designing particulate carriers for encapsulation and controlled release applications, *Powder Technology*, 236, 188-196.

2. Liu, W., Wu, W., **Selomulya, C.***, Chen, X. D., **2012**, Spray drying of monodispersed microencapsulates: implications of process parameters on microstructural properties and controlled release functionality, *Journal of Microencapsulation*, 29: 7, 677-684.

3. Rogers, S., Fang, Y., Lin, S., **Selomulya, C.**, Chen, X. D., **2012**, A Monodisperse Spray Dryer for milk powder: Modelling the formation of insoluble material, *Chemical Engineering Science*, 71, 75-84.

4. Liu, W., Wu, W., **Selomulya, C.***, Chen, X. D. **2011**, Facile spray-drying assembly of uniform microencapsulates with tunable core-shell structures and controlled release properties, *Langmuir*, 2011, 27 (21), pp 12910–12915.

5. Wu, W., Liu, W., **Selomulya, C.**, Chen X. C., **2011**, On Spray Drying of Uniform Silica-Based Microencapsulates for Controlled Release, *Soft Matter*, 7, 11416-11424.

6. Fang, Y., Rogers, S., **Selomulya, C.***, Chen, X. D., **2011**, Functionality of milk protein concentrate: Effect of spray drying temperature, *Biochemical Engineering Journal*, Volume 62, 101-105.

7. Liu, W., Wu, W., **Selomulya, C.***, Chen, X. D., **2011**, A single step assembly of uniform microparticles for controlled release applications, *Soft Matter*, 7, 3323 - 3330.

8. Wu, W., Amelia, R., Hao, N, **Selomulya, C.**, Zhao, D., Chiu, Y., Chen, X. D., **2011**, Assembly of Uniform Photoluminescent Microcomposites Using a Novel Micro-Fluidic-Jet-Spray-Dryer, *AIChE Journal*, 57:10, 2726–2737.

*: *C. Selomulya as the corresponding author*

(585b) Antioxidant Activity of Limnophila Aromatica Extracts

Do, Q. D., National Taiwan University of Science and Technology

In recent years, the ingestion of fresh fruits and vegetables rich in natural antioxidants has been associated with the prevention of cancer and cardiovascular diseases. Many medicinal plants have been extensively studied for their antioxidant activity. *Limnophila aromatica* (Lamk.) Merr. (syn. *L. gratissima* Blume) is used as a spice and medicinal herb in Southeast Asia. Some studies on this species have found that *L. aromatica* (cited as *L. gratissima*) yielded 0.1% of an antibacterial, colourless essential oil, containing limonene and perillaldehyde as the principal constituents. Other minor constituents such as 7-desmethylartemetin (5,7-dihydroxy-3,6,30,40-tetramethoxyflavone, but later was revised as a mixture of nevadensin (5,7-dihydroxy-6,8,40-trimethoxy-flavone) and salvigenin (5-hydroxy-6,7,40-trimethoxyflavone)) and 8-oxygenated flavonoids were isolated from aerial parts of *L. aromatica*. Chlorogenic and ca eic acids were also reported as constituents of this species.

L. aromatica used in this study was bought at a Vietnamese store in Taiwan. In this study, water and various concentrations (50%, 75 % and 100%) of methanol, ethanol and acetone in water, were used as solvent in extraction of L. aromatica. The antioxidant activity, total phenolic content and total flavonoid content of the freeze dried L. aromatica extracts were investigated by using various in vitro assays. The extract obtained by 100% ethanol showed the highest total antioxidant activity, reducing power and DPPH radial scavenging activity. The same extract also exhibited the highest phenolic content (40.5 mg GAE/g DFLA) and the highest flavonoid content (31.11 mg QCE/g DFLA). The highest extraction yield was obtained by using 50% aqueous acetone. These results indicate that L. aromatica can be used in dietary applications with potential to reduce oxidative stress

(585aa) Survival of Oral and Faecal Bacteria in Beverages

Mbachu, C., Middlesex University London

Escherichia coli, Enterococci, Coli forms and *Pseudomonas aeruginosa* are common known bacteria that can be found in beverages such as fruit juices and bottled water and these organisms can thrive well depending on the conditions (e.g. temperature and pH). Bacterial contamination of beverages has been a major concern and can cause many health issues. These beverages sold at all public places and roadside shops and they can become a health concern depending on handling and storage or if the manufacturing process is unsafe. In this research, five different beverages (spring water, fountain water, fruit juice, lucozade and coke) were inoculated with *Pseudomonas aeruginosa* and *Escherichia coli* and survival rate monitored over time under two different temperature (4°C and 22°C) conditions. In this, the effect of storage time and temperature of different beverages on the survival growth of bacteria were studied to assess for the potential risk of food borne illnesses on different beverages. Analysis of Variance (ANOVA) of the data was conducted to determine and assess the differences in growth rate and for comparison purposes and to know which condition was significant. It was found that temperature was significant in the growth rate of *Escherichia coli* and both temperature and type of beverage used for the study was significant in the growth rate and survival of *Pseudomonas aeruginosa*.

(585ae) Modeling and Validation of the Continuous Htst Thermal Processing of Liquid Foods With Plate Heat Exchangers

Benze, R. V., University of Sao Paulo, Escola Politecnica
Gut, J. A. W., University of Sao Paulo, Escola Politecnica

Thermal processing of liquid foods aims the inactivation of micro-organisms and enzymes that compromise the safety or the shelf life of the product. The design of continuous processes using heat exchangers usually rely on simplifying assumptions that provide a wide safety margin and facilitate calculations, such as isothermal processing only at the holding tube at the minimum residence time. These assumptions can impair sensory or nutritional attributes of the product because of over-processing. The aim of this work was to develop and validate a phenomenological model to simulate the continuous HTST pasteurization with plate heat exchangers in order to determine the distribution of the temperature of the fluids and of the concentration of the target micro-organism or activity of the target enzyme throughout the process.

The considered pasteurizer comprises the inlet tank, a plate heat exchanger (PHE) with three sections (heating, cooling and heat regeneration), the holding tube, the tubular connections between the exchanger and the holding tube, the hot water circuit and the cold water circuit (Figure 1). The model includes equations for flow, heat exchange and lethality. Main assumptions are: steady-state operation, plug-flow in the channels of the PHE and tubes, uniform flow distribution among PHE channels, heat exchange with ambient air only in the tubular sections, no axial diffusion and uniform thermo-physical properties in each section.

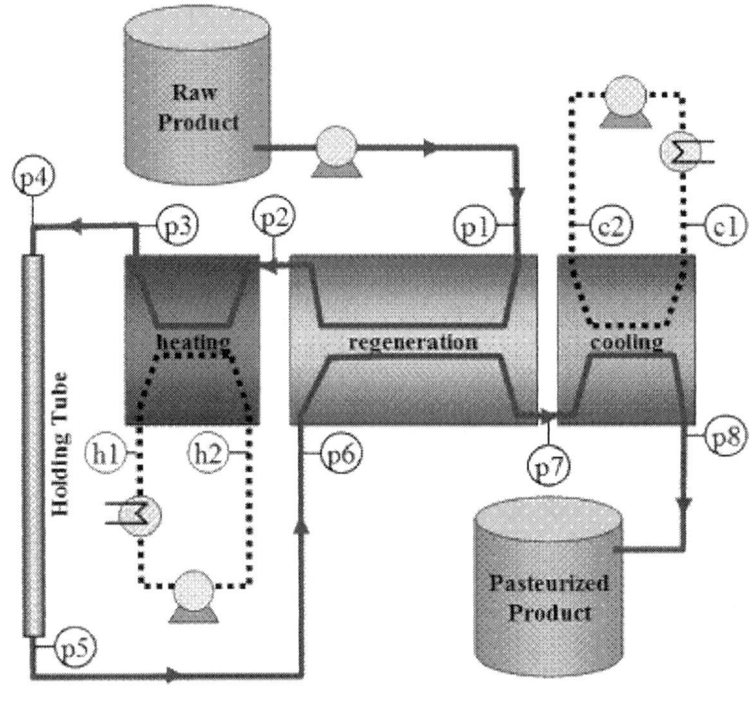

Figure 1: Representation of the HTST thermal process unit

Each PHE channel or tube provides one differential equation for heat transfer and one for thermal inactivation with the corresponding boundary conditions. The sterilization value was evaluated as the number of decimal reductions on the target micro-organism. Space time (length velocity ratio) can be replaced in the thermal inactivation equations by the experimental mean residence time or by the minimum residence time, depending on the desired study (evaluation, validation, design or optimization). Model simulation provides axial distribution of product temperature and the associated lethality, allowing the assessment of the contribution of each section of the process.

In order to test and validate the model, a FT-43A laboratory scale plate pasteurizer (Armfield) was used. The sections of the PHE were configured for series flow (one channel per pass) with 11 thermal plates in the heating section, 7 thermal plates in the cooling section and 19 thermal plates in the heat regeneration section. Product flow rate was 20 L/h and the flow rate of the service fluids was 1,0 L/min (hot water and cold water). The tested product was phosphate buffer (pH = 6.6) with alkaline phosphatase (EC 3.1.3.1) acting as an enzymatic time temperature integrator (TTI), as proposed by Aguiar et al. (2012, DOI: 10.1016/j.lwt.2011.12.027). Enzymatic activity was measured using a reflectometric method and the inactivation kinetic model used was first order with two isoenzymes.

Processing temperatures were 70, 75, 80 and 85 Â°C and the controller manipulated the inlet temperature of the hot water in order to achieve the desired temperature at the end of the holding tube. Temperatures at 12 points along the process were registered using thermocouples and a data acquisition system (National Instruments). Samples of the TTI were collected at the exit of the heating section of the PHE (p3 in Figure 1) and after the cooling section (p8 in Figure 1). Samples were immersed in ice-water bath to stop the inactivation, prior to activity assessment.

The process was simulated using the proposed mathematical model in order to compare experimental and predicted results of temperature and lethality. Parameters provided for model simulation were: geometrical parameters of plates and tubes, mean thermo-physical properties, inlet temperatures and flow rates, inlet concentration or activity of target, kinetic parameters of thermal inactivation, heat exchange parameters and mean residence times. The model was solved using a finite difference method in software gPROMS 3.2 (Process System Enterprise). Discretized variables were the temperature of the fluids and target concentration.

The mean residence time in the channels of the PHE were obtained from Gutierrez et al. (2011, DOI: 10.1016/j.applthermaleng.2011.02.015) and the mean residence time in the holding tube were obtained from Gutierrez et al. (2010, DOI: 10.1016/j.jfoodeng.2010.01.004). Space time was calculated for the tubular connections. The Nusselt x Reynolds correlation for the heat transfer in the PHE was obtained according to Gut et al. (2004, DOI: 10.1016/j.ces.2004.07.025) for series pass arrangements. Mean overall heat transfer coefficients for the tubular section were acquired from experimental data.

The model simulation provided the temperature history of the product along the process and the distributed residual activity of the enzymatic TTI. Figure 2 presents the obtained results for temperature and residual activity of the TTI. There was an excellent agreement between predicted and experimental temperatures at the registered points with a mean absolute error of 1.1+?1.0 Â°C for all four processing conditions.

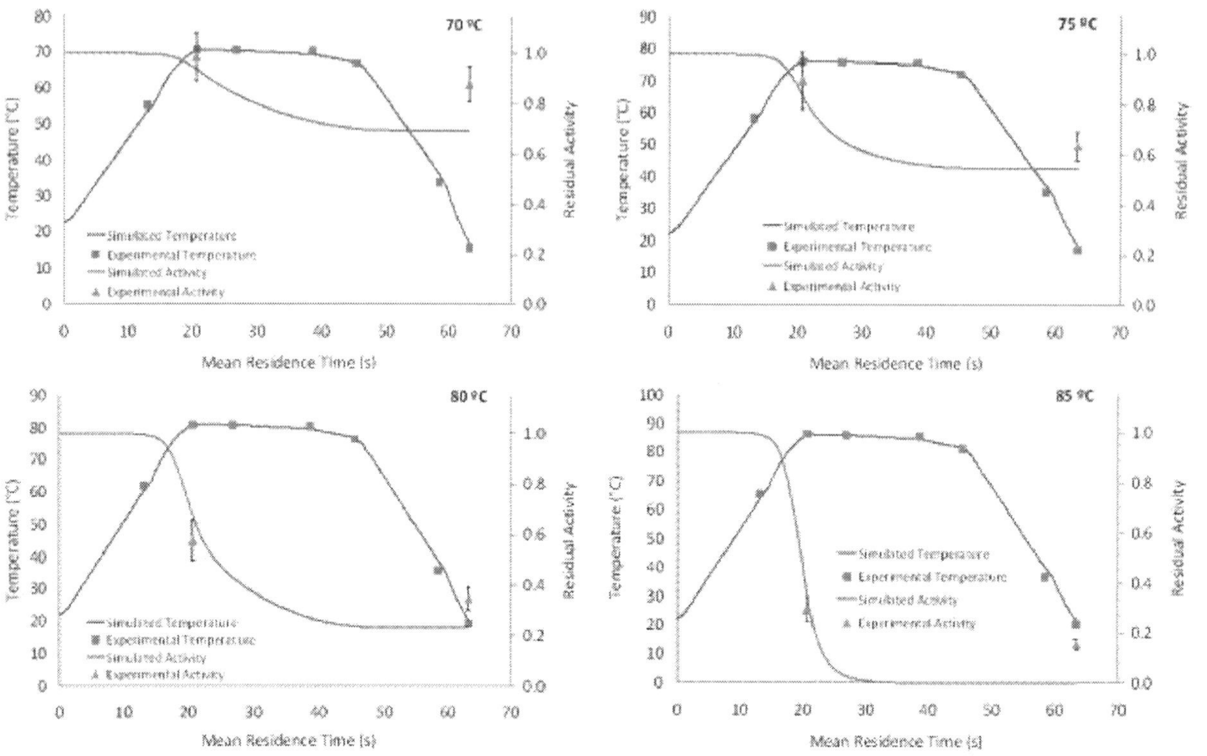

Figure 2: Obtained experimental and simulated results for temperature distribution and residual activity of the TTI for the four processing temperatures

The results of temperature history and residual activity of the enzyme showed that the inlet and outlet connections of the holding tube play an important role on the process lethality. The lethality associated with the heat exchangers was also significant. There was a very good agreement between predicted and measured residual activities at the end of the heating section because the simulated values were in the confidence interval of the experimental values. However, the simulated values of residual activity at the end of the process were consistently inferior to the measured values. This deviation can be explained by the use of the space time instead of the mean residence for the tubular connections. Since the experimental mean residence time is shorter than the calculated space time, the predicted residual activity was smaller.

The proposed model allows the simulation of a continuous HTST thermal processing with plate heat exchangers in order to determine the temperature history and process lethality. With the experimental data used in this work it was possible to validate this model, which can be used for analysis, design or optimization problems, depending on the residence time used for each process step (space time, experimental mean residence time or minimum residence time).

(585y) Study Of Spray Dried Yoghurt Production In A Pilot Scale Equipment

Triboli, E. P. D. R.,
Gut, J. A. W., University of Sao Paulo, Escola Politecnica

The search for new solutions in both products and processes guide the development of the food industry. When compared to their hydrated forms dehydrated products have advantages for ease of handling, transportation and storage costs, and usually an extended shelf life. Yogurt is very well accepted by Brazilian society and is consumed in its natural state in various formulations. If dehydrated it may be used as an ingredient in many processed foods, stimulating the development of new products. This paper attempts to establish suitable conditions for the production of yogurt powder using the technique of spray drying in a pilot scale equipment, studying the influence of the rotation speed of the atomizer disk (25,000, 27,500 and 30,000 rpm), the temperature of the air entering the drying chamber (160, 170 and 180 Â°C) and the feeding flow of the product (30, 40 and 50% of the rated capacity of the feed pump, respectively, 5.2, 6.9 and 8.7 kg/h) by a randomized complete 23 factorial design with central point. The results showed that there are significant interactions between the three studied parameters. The effect of the feed flow rate of yogurt in the dryer is very important to control the particle size distribution of the dried product, affecting powder homogeneity.

The design of experiments technique proved to be a suitable choice for the study of the process, since there are interactions between the parameters in system responses, a fact that is hardly observed by using a strategy of changing one condition at a time.

Plain yoghurt from Fazenda Bela Vista (Brazil) was used in the experiments (88.2% water, 5.8% carbohydrates, 3.7% proteins, 2.0% fats). Equipment was a spray dryer from Tecnape (Brazil) with nominal evaporation capacity of 5 L of water per hour (Figure 1). Hot fumes from a LGP burner with temperature control is feed co-currently in the drying chamber (0.63 m3). Powdered product is collected from the bottom of the chamber and also from a cyclone separator. The process conditions are suitable to be scaled-up to equipment of larger size.

Figure 1. Spray dryer SD-05 from Tecnape (Brazil)

There was considerable deposit (20 to 30%) of product in the wall of the drying chamber, indicating that the optimum conditions for drying have not been determined yet. This is due to the low value of the glass transition temperature of yoghurt with respect to drying conditions of the process. It was possible to know the relative importance of each parameter in the drying process, however, the sticking of material on the wall of the drying chamber is probably the major challenge to be addressed for deemed satisfactory drying process as a whole. Results suggest that a reduction of inlet air temperature in the drying chamber should contribute to the reduction of the deposit. Nevertheless, equipment operating limits do not allow temperatures lower than 160 Â°C for stable steady state operation.

The feed flow rate of yoghurt proved to be the most important parameter since any excess compromise the performance of the whole process. The best conditions to produce a powder having greater homogeneity and less amount of material deposited on the wall occurred at 30% of the capacity of the pump (5.2 kg/h), the lowest feed flow value studied, and at the upper value of the atomizer disk speed, rpm 30,000. With the inlet air temperature of 160 or 180 Â°C results were satisfactory. However, because of the material that adheres to the drying chamber wall one cannot consider that the drying process has been successfully carried out.

The particle diameter analysis was done by laser diffraction using a Mastersizer 2000 (Malvern, UK) and the difference [d(0.9) ? d(0.1)] can be used as an index of homogeneity of the distribution of particle diameters. It is understood that the more homogeneous the sample with respect to particle size, the better the performance of the process.

In an ideal dryer, the drying is an isenthalpic process, so that the energy required for the evaporation of water from the wet material is achieved by reducing the air temperature entering the device. If satisfactory moisture level in the final product is achieved, the lower the temperature of the outlet air, the smaller the heat effects such as color change and microorganisms destruction.

The sticking on the wall of the drying chamber is a very undesirable effect and suggests that the chosen material to dry or process conditions chosen are not the right ones. The sticked material on the wall was evaluated by weight percentage of brushed powder from the drying chamber with respect to the total mass of product recovered in the process. It is understood that the smaller the amount of material sticked on the wall, the better the process of spray drying.

The color difference analysis was established against Molico skimmed milk powder (NestlÃ©) by means of CIELAB color values under D65 illuminant and 10Â° angle using a ColorQuest XE (Hunter Lab, EUA). The smaller the difference ΔE^*, the better drying process performance. All variables studied were related to process parameters to identify key trends and effects and the main results of the experiments are presented in Table 1.

Table 1. Process performance indexes for plain shaken yoghurt spray drying for tests in coded variables

Condition	Rotation speed	*T* inlet air	Feed rate	[d(0,9)-d(0,1)] (µm)	*T* outlet air (Â°C)	Sticked material (%)	ΔE^*
1	?	?	?	49.8	102.3	41.1	14.11
2	+	?	?	35.2	108.0	**22.1**	6.14
3	?	+	?	50.9	125.6	34.2	**3.79**
4	+	+	?	**33.8**	121.1	29.3	9.36
5	?	?	+	151.8	**73.8**	54.4	17.16

6	+	?	+	177.1	80.6	39.6	33.11
7	?	+	+	98.0	94.8	40.5	27.34
8	+	+	+	119.7	99.3	43.9	3.92
9	0	0	0	77.4	94.5	27.7	7.37
10	0	0	0	62.0	101.2	27.4	6.15
11	0	0	0	59.9	102.0	26.8	5.10

By inspection of Table 1 it is possible to verify that the desirable minimum values of process performance indexes appear in four different study conditions (2, 3, 4 and 5). Thus, for the study conditions there was a not a condition that would accommodate the various desired goals. Thus, to define which conditions proved the best, with the goal of guidance for further study will be necessary to find a compromise solution, which should be verified in practice later. With this aim other values besides the minimum ones were also considered to settle the best possible process conditions. So, the conditions of test number 2 showed a minimum percentage of material sticked on the wall and were reasonably satisfactory in two other parameters: sample homogeneity and color difference. Therefore, one can conclude that, in this project, the best working conditions were: speed of rotation of the atomizer: 30,000 rpm; inlet temperature of the air in the drying chamber: 160 Â°C and feed flow rate of pump: 30% of the rated capacity, or 5.2 kg/h. These conditions will be used as basis for further study.

The sticking on the wall phenomenon is occurring, probably, but maybe not exclusively, on account of having exceeded the glass transition temperature of the yogurt. This problem can be solved partly or wholly with the addition of maltodextrins in the shaken plain yogurt. Maltodextrins with DE = 5 or 10 have high glass transition temperatures and thus can help on decreasing the gumminess condition of yoghurt, hindering its grip on the wall and on other particles also. Based on literature review it is evident the importance of obtaining a product with viable microorganisms and, accordingly, both the addition of skimmed milk powder and gum acacia suggest the need for research.

(585z) Drinking Water Disinfection By Low Voltage Micro-Mini Pulsed Electric Field Device

Lam, K. W., The Hong Kong University of Science and Technology
Hung, P. S., The Hong Kong University of Science and Technology
Lee, O. W.,
Kwan, S. M., Hong Kong University of Science and Technology
Sim, V., The Hong Kong University of Science and Technology
Kwan, J. K. C.,
Yeung, K. L., The Hong Kong University of Science and Technology

The microbial contamination of building water supply systems is very common all over the world. Among the 29 common waterborne pathogens, Legionella pneumophila, Campylobacter jejuni, Shigella spp, and E. Coli have a relatively high infectivity, while non-tuberculous mycobacteria is also reported to be a high chlorine-resisting bacteria. This puts public heath in serious jeopardy. Water disinfection technologies such as chlorination, ozone disinfection, UV exposure and reverse osmosis are now widely used in drinking water disinfection. However, these technologies are not effective. In 2012, the discovery of Legionella in the water from the taps of Hong Kong's brand new government headquarters illustrates the gravity of the situation and its potential threat to public health. Chemicals such as chlorine and ozone are powerful oxidizing agents that can kill most of the waterborne organisms, at the cost of producing unwanted by-products which are toxic. Reverse osmosis will remove all the minerals and nutrients inside the drinking water while and UV exposure may foster high infectivity microorganisms due to cell mutation. The current disinfection technologies bring a lot of questions other than disinfection effectiveness .

Pulse electric field disinfection is an effective, environmentally benign and safe technology for point-of-use disinfection of tap water. Microbial cell wall will be elongated and suppressed under a pulse electric field. This causes irreversible damage to the microbial cell wall. The inner structures such as cytoplasm of microbial will diffuse out of the cell through the enlarged pores. Thus, the microbial is no longer able survive. This disinfection technology avoids the use of chemical disinfectants that can alter the quality of the drinking water and induce resistance in microorganisms. It requires no heating so the energy consumption is far less than that of current disinfection technology such as ozone generation and reverse osmosis. The technology was first used for non-thermal sterilization pharmaceuticals and food products, eventually extending to sewage and clinical waste disinfection. However, the current technology uses high-input voltage pulse electric field as the key disinfection parameter. This creates a potential electrical hazard to operators. The design in this project is able to use low-input voltage with micro-engineered electrode to attain 99.9% bacteria reduction in the running tap water. The pulse electric field device is portable and small enough for home and public tap water. It believed that this device should be used as the last step for disinfection of drinking water from the water distribution system.

(585a) The Use of Dynamic Vapor Sorption to Characterize Surface Properties of Amorphous Valnemulin Hydrochloride

Ouyang, J., National Engineering Research Center and Industrial Crystallization Technology, Tianjin, 300072, China;
Hao, H., National Engineering Research Center for Industrial Crystallization Technology, School of Chemical Engineering and Technology, Tianjin University

Amorphous systems have attracted considerable attention due to their favorable properties; however, their stability issues still pose a major challenge resulting from their strong hygroscopicity. The interaction of vapor with these pharmaceuticals is an important topic, from the aspect of stability as well as processing treatment. Dynamic vapor sorption method is used to characterize hygroscopicity properties of solid powder, especially in the field of pharmaceuticals. In the present investigation we used dynamic vapor sorption to study polar vapor and nonpolar vapor uptake by amorphous valnemulin hydrochloride powder as a function of time and relative humidity. We model the data obtained to suggest sorption mechanisms, which are based on surface adsorption or bulk adsorption. In addition, the surface properties, specific surface area and surface energy are investigated using related equations through dynamic vapor sorption analysis. In conclusion this study allows insights into both vapor adsorption/desorption behavior and surface properties of amorphous valnemulin hydrochloride.

(585x) Dielectric Properties Of Simulated Green Coconut WATER From 500 To 3,000 Mhz At Temperatures Between 10 and 80 °c

Franco, A. P., University of SÃ£o Paulo
Tadini, C. C., University of SÃ£o Paulo
Gut, J. A. W., University of Sao Paulo, Escola Politecnica

Introduction

Green coconut water is a known refreshing beverage and is distinguished for its content of minerals, vitamins and sugars. It can be consumed naturally from the fruit or processed in order to extend its shelf life. Nowadays, conventional thermal treatments such as pasteurization and sterilization are used to preserve this kind of product (Zhu, et al., 2012). However, high temperature is required to inactivate the target enzymes peroxidase and polyphenol oxidase, resulting in degradation of color, taste and nutritional value (Matsui et al., 2008; Zhu, et al., 2012).

Liquid foods can also be heated by electromagnetic waves, such as microwaves (Decareau, 1985; RyynÃ¤nen, 1995). Microwave heating is supported on the ability of a dielectric material to absorb electromagnetic energy and convert part of it to heat (Datta et al., 2005; Nelson and Datta, 2001). This method offers advantages in pasteurization and sterilization processes when compared to conventional heating methods. For instance, this treatment allow more rapid heating and shorter processing time as a consequence of the molecular level interactions of the food and the alternating electric field (Decareau, 1985; Venkatesh and Raghavan, 2004). Microwave heating has also disadvantages; the main limitation of this technology is the non-uniform temperature distribution during the process as a result of the limited power penetration of electromagnetic waves in liquid water (Giese, 1992; Thostenson and Chou, 1999).

Knowledge about dielectric properties of liquid foods is required in order to design thermal treatments for preservation using microwave energy (Giese, 1992). Dielectric properties characterize the interaction of electromagnetic waves with the media (Datta et al., 2005; RyynÃ¤nen, 1995) and provide information regarding heating rate and penetration depth.

Dielectric properties of materials are defined in terms of relative permittivity e' and dielectric loss factor e?. The relative permittivity represents the material energy storage capability in response to an applied electric field and the dielectric loss factor refers to energy dissipation as heat (Datta et al., 2005; Sosa-Morales et al., 2010). The power penetration depth of electromagnetic waves in materials is defined as the depth in which the power is reduced to 1/e (e = 2.7183) (RyynÃ¤nen, 1995; Venkatesh and Raghavan, 2004). These properties are affected for many factors; the main are composition, temperature and the electric field oscillation frequency (IÃ§ier and Baysal, 2004).

In this work, dielectric properties of simulated green coconut water were studied at frequencies from 500 to 3,000 MHz and temperatures between 10 and 80 Â°C in view of continuous thermal processing of green coconut water.

Materials and methods

Simulated green coconut water was obtained dissolving fructose, glucose, sucrose, calcium chloride, magnesium chloride, monobasic potassium phosphate, sodium sulfate and potassium sulfate in distillated water to mimic the average composition of this product. The samples were cold stored and temperature was adjusted from 10 to 80 Â°C using an oil bath TC-550 (Brookfield, USA).

The dielectric properties measurements were carried out using the open-ended coaxial line probe technique with a Dielectric Probe Kit 85070E connected to an E5061B Network Analyzer (Agilent Technologies, Malaysia). An Electronic Calibration Module 85093C (Agilent Technologies, Malaysia) was used for improving the measurement quality. Before the measurements, the network analyzer was allowed to warm up for at least 90 min. The system was calibrated using the equipment standard configuration: air, short-block and deionized water. The measurements were performed for field frequencies between 500 and 3,000 MHz at temperature intervals of 10 Â°C from 10 to 80 Â°C. In this work, the dielectric properties were analyzed at 915 and 2,450 MHz which are used industrially and household, respectively. The measurements were carried out in triplicates at each temperature and the standard deviations were calculated. Results were correlated with temperature and the penetration depths were calculated using proper equations for electromagnetic wave penetration (Datta et al., 2005; RyynÃ¤nen, 1995; Zhu et al., 2012).

Results and discussion

Figure 1 shows the relative permittivity and the loss factor of the simulated green coconut water at temperatures from 10 to 80 Â°C as a function of the frequency. It can be seen that these dielectric properties significantly depend on both temperature and frequency. For all testing temperatures, the relative permittivity e' decreased with increasing frequency between 500 and 3,000 MHz, which can be justified by the decreased ability of the water molecule dipole in accompanying the oscillating electric field at increasing frequency. In addition, this property also decreased with temperature increase at any given frequency, because the thermal agitation disturbs the dipole alignment with the electric field.

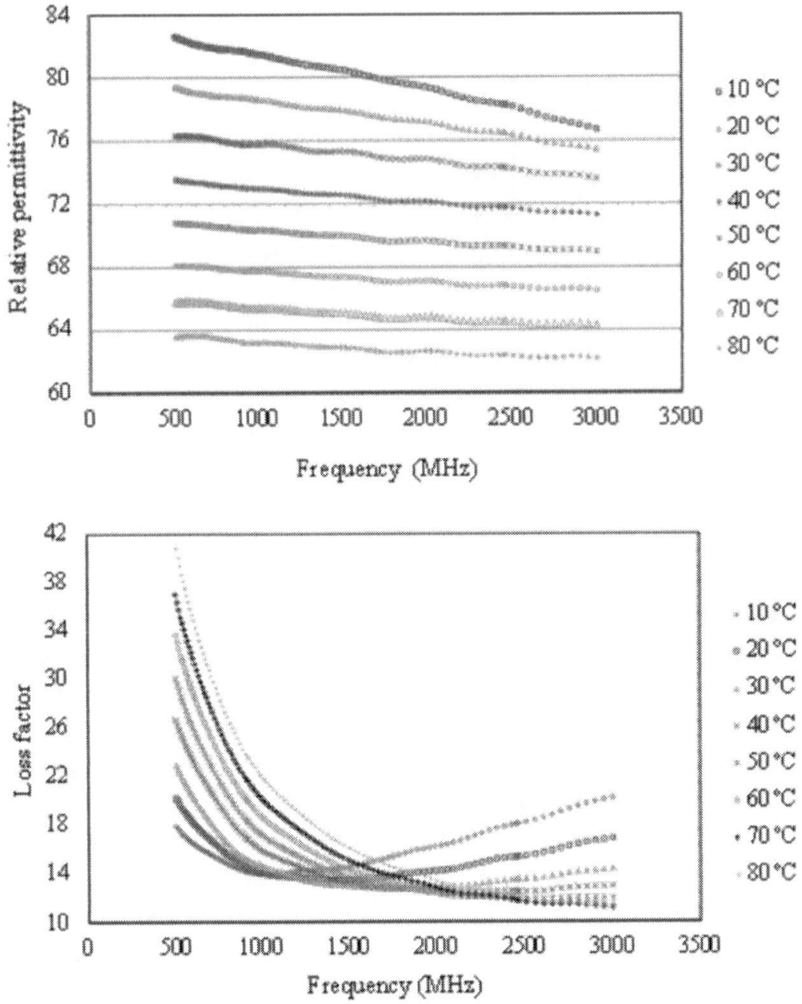

Figure 1: Dielectric propeties of simulated green coconut water as affected by frequency and temperature

The dielectric loss factor e? showed in interesting dependence with temperature and frequency, as can be seen in Figure 1. The drop on the loss factor for frequencies between 500 and 1,000 MHz is associated with the ionic conduction in the solution. This mechanism does not contribute with the polarization of the media and only generate energy loss due to heating. For increasing field frequency, there is less ionic movement and, therefore, a decrease in the loss factor.

For higher frequencies, it can be seen In Figure 1 that there is an increase in the loss factor, which as associated with the dielectric relaxation of the water molecule. The dipole of the water molecule is unable to follow very rapid field reversals, generating heat and increasing the loss factor. The relaxation frequency of water at 20 Â°C is 17,004 MHz (peak of the relaxation curve) and this value increases with the temperature, in accordance to the trend of the curves in Figure 1 (lower temperatures show a stronger increase in the loss factor). It is known that for frequencies beyond the relaxation frequency, the water dipole gradually losses the ability to follow the oscillating field, reducing both the relative permittivity and the dielectric loss factor (Datta et al., 2005).

The relative permittivity and the dielectric loss factor of the simulated green coconut water were successfully correlated with temperature for the commercial frequencies of 915 and 2,450 MHz. The coefficients of the linear correlation for the relative permittivity are a0 = 8.392E+01, a1 = ?2.648E-01 for 915 MHz (R2 = 0.9975) and a0 = 8.099E+01, a1 = ?2.335E-01 for 2,450 MHz (R2 = 0.9986) and of the polynomials adjusted for dielectric loss factor are a0 = 1.340E+01, a1 = 4.058E?02, a2 = 1.118E?03 for 915 MHz (R2 =

0.9956) and a0 = +2.037E+01, a1 = ?2.873E?01, a2 = 2.326E?03 for 2,450 MHz (R2 = 0.9873). The power penetration depth decreased with the temperature increment at lower frequencies. At frequencies above 2,450 MHz, penetration depth increased with temperature increment from 10 to 50 Â°C and decreased at temperatures between 60 and 80 Â°C.

Conclusions

Dielectric properties of simulated green coconut showed a temperature and frequency dependent behavior, which was characterized. The relative permittivity decreased with temperature increasing from 10 to 80 Â°C at any given frequencies in the studied range. The loss factor and penetration depth are generally lower at high frequencies from 500 to 3,000 MHz. At temperatures from 50 to 80 Â°C and frequencies between 2,000 and 3,000 MHz the loss factor has the lowest values.

Acknowledgements

The authors would like to acknowledge financial support from FAPESP 2012/04073-0 (The State of SÃ£o Paulo Research Foundation) and to CNPq (National Council for Scientific and Technological Development).

References

Datta, A. K.; Sumnu, G. and Raghavan, G. (2005). Dielectric Properties of Food. In: M. A. Rao, S. Rizvi and A. K. Datta (Eds.). *Engineering properties of foods* (pp. 101?147). USA: CRC Press.

Decareau, R. V. (1985). *Microwaves in the Food Processing Industry*. Orlando: Academic Press.

Giese, J. (1992). *Food Technology*. 46(9), 118-123.

IÃ§ier, F. and Baysal, T. (2004). *Critical Reviews in Food Science and Nutrition*. v. 44, 465-471,

Matsui, K. N.; Gut, J. A. W.; De Oliveira, P. V. and Tadini, C. C. (2008). *Journal of Food Engineering*. 88, 169-176.

Nelson, S.O; Datta A. K. (2001). Dielectric Properties of Food Materials and Electric Fields Interactions. In R. C. Anantheswaran and A. K. Datta. *Handbook of Microwave Technology for Food Applications*. New York: Marcel Dekker, Inc.

RyynÃ¤nen, S. (1995). *Journal of Food Engineering*. 26, 409-429.

Sosa-Morales, M. E.; Valerio-Junco, L.; Lopez-Malo, A. and Garcia, H. S. (2010). *Food Science and Technology*. 43, 1169-1179.

Thostenson, E. and Chou, T. (1999). *Composites Part A: Applied Science and Manufacturing*. 30, 1055-1071.

Venkatesh, M. and Raghavan, G. (2004). *Biosystem Engineering*. 88(1) 1-18.

Zhu, X.; Guo, W. and Wu, X. (2012). *Journal of Food Engineering*. 109, 258-266.

(585f) Statistical Medium Optimization for High Biomass Production of Trichoderma Harzianum: A Potential Biocontrol Agent for Ganodermal Oil Palm Disease

Othman, N. Z., Institute Bioproduct Development, Universiti Teknologi Malaysia
Rahaeifard, M., Universiti Teknologi Malaysia
Then, C., All Cosmos Industries Sdn. Bhd.
H'ng, W. C.,
Abd Malek, R., Institute Bioproduct Development, Universiti Teknologi Malaysia
Wan Mustapha, W. A., All Cosmos Industries Sdn Bhd
Jalal, Y., Institute Bioproduct Development, Universiti Teknologi Malaysia
Sarmidi, M. R., Universiti Teknologi Malaysia
Aziz, R., Institute Bioproduct Development, Universiti Teknologi Malaysia
El Enshasy, H. A., Institute Bioproduct Development, Universiti Teknologi Malaysia

Trichoderma harzianum plays significant role in the future of sustainable agriculture since it can act as biological control agent against broad range of plant pathogens. Basal Stem Rot (BSR) caused by *Ganoderma boninense* which is a white rot fungus has counted as significant defeat in various regions around the world including South East Asia. In this study, statistical medium optimization in shake flask cultivation applied as an approach to optimize medium composition for high biomass production. Medium optimization was carried out using statistical method of response surface methodology (RSM) based on Box-Behnken design in shake flask culture. The interaction was studied between four different levels of variables that were glucose, peptone, K_2HPO_4, and KH_2PO4. Second order polynomial models were used to correlate the factors to the cell biomass production. The models were found to be significant with the cell biomass production in optimized media reached to approximately 20.0 g L^{-1} increment of 80 % while compared in un-optimized media only 4.0 g L^{-1}. Consequently, the yield coefficient was increased from 0.08 (g g^{-1}) in un-optimized media to 0.215 (g g^{-1}) in optimized media. Based on medium statistical design study, results indicated that the optimal medium supporting maximal biomass production had the following composition (g L^{-1}): glucose, 121.0; peptone, 12.3; KH2PO4, 4.36; and K2HPO4, 8.73. Furthermore, the cell biomass cultivation in the 16-L stirred-tank bioreactor under un-controlled and controlled pH showing no differences between both cultivation as the maximal biomass produced was 25.0 gL^{-1} and 24 gL^{-1}, respectively.

(585g) Bioprocess Development for High Cell Density Cultivation of Rhizobium Trifolii in Semi-Industrial Scale Stirred Tank Bioreactor

Othman, N. Z., Institute Bioproduct Development, Universiti Teknologi Malaysia
Wei Chang, H., All Cosmos Industries Sdn. Bhd.
Then, C., All Cosmos Industries Sdn. Bhd.
Abd Malek, R., Institute Bioproduct Development, Universiti Teknologi Malaysia
Sarmidi, M. R., Universiti Teknologi Malaysia
Aziz, R., Institute Bioproduct Development, Universiti Teknologi Malaysia
Wan Mustapha, W. A., All Cosmos Industries Sdn Bhd
El Enshasy, H. A., Institute Bioproduct Development, Universiti Teknologi Malaysia

Rhizobium trifolii is a symbiotic N-fixing bacterium that is able to facilitate the conversion of atmospheric nitrogen into ammonia in legumes through the formation of root nodules. Although the use of Rhizobia as microbial inoculants is well-established, there is a lack of information regarding mass production of this type of bacteria in large scale. Therefore, the objective of this research is to develop an optimum cultivation condition for high cell mass production of *R. trifolii* in liquid culture, for 16-L and 150-L stirred-tank bioreactors. Batch cultivation under controlled and uncontrolled pH in 16-L stirred tank bioreactors yielded cell mass concentrations of 8.05 g L^{-1} and 16.87 g L^{-1}, respectively. To study the limiting factors in Rhizobia cultivation, subsequent bioreactor experiments were carried out in different fed-batch mode such as constant feeding and intermittent addition method. Intermittent addition method yielded 37.95 g L^{-1} of cell mass which was lower compared to 69.10 g L^{-1} of cell mass using constant feeding strategy. Fed-batch cultivation of *R. trifolli* in a 150-L stirred tank bioreactor using the optimum cultivation strategy yielded a maximum cell mass of 27.32 g L^{-1}. This was due to the decrease in oxygen transfer in a larger bioreactor. Under oxygen limited condition, exopolysaccharides (EPS) was accumulated in culture medium and reached about 9.5 g L^{-1}.

(585q) The Role of Simulation and Scheduling Tools in Bioprocess Development and Manufacturing

Petrides, D., INTELLIGEN, INC.
Siletti, C., INTELLIGEN, INC.
Carmichael, D., Intelligen, Inc

The successful scale up and commercialization of biopharmaceuticals is a challenging task that requires collaboration of professionals from many disciplines. Process simulators can facilitate this task by assisting scientists and engineers to answer the following and other related questions: What is the impact of product titer increase on the capacity load of the downstream section, the overall throughput of a plant, and the cost of goods? What changes are required in an existing multi-product facility to accommodate the process of a new product? What is the range of variability that a process can accommodate if it operates under a tight cycle time? What is the impact of single-use systems on the demand for utilities, the environment and the cost of goods? Our experience in addressing the above questions will be presented using industrial examples in which we evaluated alternative technologies for producing therapeutic monoclonal antibodies and vaccines. In bioprocessing, the balance of upstream and downstream capacity has become increasingly important. Upstream improvements have lead to high yield batches which may lead to bottlenecks in purification.

(585e) Improving Thermostable Phytase Production in Recombinant Escherichia Coli During Fed-Batch Cultivation in Semi-Industrial Scale Stirred Tank Bioreactor

Othman, N. Z., Institute Bioproduct Development, Universiti Teknologi Malaysia
El Enshasy, H. A., Institute Bioproduct Development, Universiti Teknologi Malaysia
Abd Malek, R., Institute Bioproduct Development, Universiti Teknologi Malaysia
Ramli, S., Universiti Teknologi Malaysia
Sarmidi, M. R., Universiti Teknologi Malaysia
Aziz, R., Institute Bioproduct Development, Universiti Teknologi Malaysia
Hatti-Kaul, R., Lund University

The expression of thermophile phytase gene from *Bacillus* sp. MD2 under the control of *lac*-0 promoter by *E. coli* BL21 (DE3) was studied in 16L bioreactor. This is to investigate the expression of thermostable phytase in relation to the effect of continuous glycerol feeding in the post-induction phase for high cell density cultivation when one pulse induction with lactose was applied. The expression of recombinant protein was induced at the mid of exponential phase by one pulse induction strategies of 30 mM lactose. The challenge of a conventional single induction with lactose is decreased of cell growth after induction that effect lactose uptake as inducer for foreign protein expression. To overcome this problem the strategies should focusing on maintaining the exponential growth at the post-induction phase. Therefore the mid-optimum continuous glycerol feeding consisted of 46 g L^{-1} hr^{-1} glycerol can maintained the exponential cell growth with the specific growth rate of 0.20 h^{-1}. The expression of total phytase activity reached 47.97 U mL^{-1} with the specific phytase activity of 3,646.61 U g^{-1}. Moreover, the excretion as extracellular phytase activity was predominantly excreted outside the cell membrane with the maximum excretion after 7 hour of induction with the volumetric phytase production rate of 6,850.00 U L^{-1} h^{-1}. When compared with maximum and low glycerol feeding rate during the post-induction phase, both showing 50 % decreased of cell growth immediately after induced with lactose and the expression level was very low with the volumetric total phytase production rate only 2,300.00 and 1,310.28 U L^{-1} h^{-1} respectively.

Keywords: fed-batch cultivation, *lac* promoter, phytase production, recombinant *E. coli*.

(585d) Optimization of Xylanase Enzyme Production Using Recombinant Escherichia Coli By Response Surface Methodology (RSM)

Othman, N. Z., Institute Bioproduct Development, Universiti Teknologi Malaysia
Kunhi Kandiyil, S., Universiti Teknologi Malaysia
Abd Malek, R., Institute Bioproduct Development, Universiti Teknologi Malaysia
Jalal, Y., Institute Bioproduct Development, Universiti Teknologi Malaysia
Aziz, R., Institute Bioproduct Development, Universiti Teknologi Malaysia
Hatti-Kaul, R., Lund University
El Enshasy, H. A., Institute Bioproduct Development, Universiti Teknologi Malaysia

Escherichia coli BL21 (DE3) with a plasmid vector pET-22b(+) carrying xylanase coding gene isolated from an extreme thermophilic bacteria, *Thermotoga sp.* is used in this study for the enhanced production of recombinant xylanase. The study was focused on optimization of various physiochemical factors effecting cell growth and xylanase expression. The statistical approaches based on Response Surface Methodology (RSM) were used to optimize the fermentation media and induction strategy in the shake flask experiments. A General Linear full factorial model was used to screen various nutrient sources in the media. A maximum xylanase activity of 48 IU mL^{-1} was observed in maltose and $(NH_4)_2$ HPO_4 based defined media at pH 7 and temperature 37°C. The concentration of each screened nutrients in the media were then optimized by a Box-Behnken Design and an enhanced cell growth was observed at 16.2 gL^{-1} of maltose and 12.7 gL^{-1} of $(NH_4)_2$ HPO_4. Xylanase gene expression was further optimized using RSM based experimental design. Factors such as inducer type, inducer concentration, induction initiation time, post induction incubation temperature were optimized at this phase. Induction at mid log phase with 3.4 gL^{-1} lactose with post induction incubation temperature at 32°C was found the optimum for enhanced xylanase gene expression. The Xylanase activity was found improved up to 65 IU mL^{-1} with statistically optimized lactose based induction strategy and was found relatively higher yield and economic way of production than the IPTG based induction (38 IU mL^{-1}) strategies used in previous studies with unoptimized media.

(585r) Impact of Enzymatic Hydrolysis On Solubility of a Modified Soy-Maize Protein

Soria-Hern ndez, C. G., Tecnol gico de Monterrey
Serna-Sald var, S. R. O., Tecnol gico de Monterrey
Chuck-Hern ndez, C. E., Tecnol gico de Monterrey

Plant proteins, particularly those from cereals and legumes are nutritive and less expensive than counterparts from animal sources. Its use as ingredients in food products rely on various types of functional properties. Solubility maybe the most relevant because of its significant influence on other functional characteristics such as hydration and foaming. These features affect the stability and behavior of the food system. Some plants proteins display a limited solubility, making necessary structural modification, through enzymatic, physical and/or chemical treatments. Enzymatic hydrolysis is a more likeable treatment compared to the chemical counterpart because of the employment of milder process conditions. In addition, enzyme catalysis is more specific and reduces the yield of byproducts. The peptides obtained by proteolysis have smaller molecular size and less secondary structure than native protein and therefore they have an enhanced solubility. In the present study, a soy–maize modified protein was hydrolyzed in order to increase its solubility. Two point four percent protein dispersions were hydrolyzed with Neutrase® using a concentration and hydrolysis time of $8.48x10^{-3}$ to 0.424 g of enzyme/g of protein and either 30, 60 and 90 min., respectively. The optimum solubility of the modified soy-maize protein was obtained at the enzyme level of 0.191 g of enzyme/g of protein and 30 min. Under these conditions, free alpha-amino nitrogen increased 191% and protein solubility increased 21.6% compared to the native protein. Higher solubility would allow the development of a modified soy-maize protein more useful as ingredient for food applications.

(585h) Proteomics: Mechanism of Biosimilar Quality Regulation in Three CHO Host Cells

Xu, N., The University of Alabama
Ma, C., University of Alabama
Liu, X., The University of Alabama

The interest in developing "biosimilars" (generic or follow-on versions of biologics) has been increased with the patents expiration of many biologics. However, this development effort requires a comprehensive understanding of mammalian cell based protein expression and production. The construction of a high-yield protein producing cell line is time-consuming work, so the choice of an appropriate protein expression system based on the requirements of the characteristics and the timeframe of the expressed products is critical in biopharmaceutical industry. In this study, three CHO cells (i.e. CHO K1, CHO DG44, CHO S) were investigated to evaluate the effect of host cell regulation on therapeutic protein expression and quality control. First, appropriate chemical defined media (CD CHO, CD FortiCHO and CD M4CHO) were applied to cultivate the suspension culture of these three CHO cells. Second, the global proteomics profiling of host cells, especially the enzymes in glycosylation and sialylation pathways, were analyzed using Proteomics analysis. Third, both GS and dhfr- expression vectors were used to construct biosimilar IgG 1 expression cell lines. The comparison of protein glycosylation by different host cell in different production media could provide us a direction in the development of biosimilar expression and production. With the developed fundamental understanding of host cell regulation of protein glycosylation, it is feasible to rationally design a robust engineered CHO host cell in future.

(585j) Comprehensive Evaluation of Two Genome-Wide Metabolic Network Models On Scheffersomyces Stipitis

Damiani, A., Auburn University
He, Q. P., Tuskegee University
Wang, J., Auburn University

In the past few decades, significant progress has been made in the development of computational and high-throughput experimental approaches. These advancements enable the reconstruction of genome-scale metabolic network models, which provides a holistic view of the organism's metabolism and the genotype with phenotype of the organism. These models, and constraint-based metabolic flux analysis methods, such as flux balance analysis (FBA), have been used extensively to study genome-wide cellular metabolic networks as they provide quantitative descriptions of the cellular metabolism without the knowledge of kinetic information which is often difficult to obtain[1].

Various applications of genome-scale models have been reported, such as hypothesis-driven discovery, genetic identification for overproduction of bio-value chemicals, and drug targeting. For example, the genome scale model for the bacterium *Mannheimia succiniciproducens* was used to guide the improved production of succinic acid by genetic engineering[2]. The network structure of genome-scale models can also be used to provide mechanistic knowledge under different conditions, as variation in fluxes is often coupled to changes in transcription level of genes. A recent study showed that genome-scale models can be used to determine reactions that are transcriptional controlled[3].

Despite the genome-scale metabolic networks models enormous potentials, there is a large gap between the *in silico* results and comprehensible biological information that can be easily understood by biologists for genetic engineering, which have limited their more extensive use in many application. Because these models typically predict over thousands of fluxes, there is a need for an effective method to decipher the complex network and to determine whether a model provides an accurate description of the organism. In this work, instead of examining the massive array of intracellular fluxes, we have developed a subspace identification based approach to analyze and study cellular metabolism where genome scale models are used as high-fidelity simulators to investigate how a perturbation propagates through the whole network. The proposed framework is developed to study how a specific factor, such as increases oxygen pick-up rate, regulates a particular enzyme and subsequently affects the metabolic network, by carrying out the following three steps: 1. Design and conduct a series of in silico experiments to reflect the desired change that would be introduce to the cell; 2. Apply multivariable analysis tools to analyze the high dimensional in silico experimental results, and identify the key component of the metabolic network that are affected by the factor; 3. Visualize the analysis results to show how the network is affected.

This approach is carried out on *Scheffersomyces stipitis*, which is a yeast strain that has the highest native capacity for xylose fermentation, and the potential to significantly impact the production of biomass-derived biofuels and other value-added chemicals[4]. Two recently published genome-scale metabolic network models[5,6] have been evaluated comprehensively, which for simplicity will be named Model A[5] and B[6] respectively. The manual examination shows that there are explicit differences between the models in terms of the model structure and reactions included, but both models predicted very similar phenotypes on growths under various oxygen and xylose conditions. Therefore, it is difficult to pin-point how those differences contribute to the model prediction, and it is difficult to determine which model provides a better description of the cellular metabolism of the strain.

By applying the proposed approach, we found that the two models have significant discrepancies mechanistically when responding to oxygen pickup rate under both oxygen-limited and aerobic conditions. These results provide a clear perspective on the metabolic rearrangement when varying oxygenation levels, and what reactions are being up-regulated or de-regulated under certain signaling. Our analysis results clearly indicate that Model B is better than Model A, because the model predictions comply better with general qualitative knowledge of the strain. This work demonstrates that the proposed system identification based approach has the capability to dissect large scale genome-wide metabolic network models, and can significantly expedite the understanding of complex large scale metabolic networks.

References

1. Lewis, N.E. et al. (2012) Constraining the metabolic genotype–phenotype relationship using a phylogeny of in silico methods. *Nature Reviews Microbiology* , **10**, 291.
2. Oberhardt, M.A. et al. (2009) Applications of genome-scale metabolic reconstructions. Molecular System Biology, **5**, 320
3. Osterlund, T. et al. (2013) Mapping condition-dependent regulation of metabolism in yeast through genome-scale modeling, BMC Systems Biology, **7**, 36
4. Rumbold, K et al. (2009) Microbial production host selection for converting second-generation feedstocks into bioproducts. Microb Cell Fact, **8**, 64
5. Caspeta, L et al. (2012) Genome-scale metabolic reconstructions of Pichia stipitis and Pichia pastoris and in silico evaluation of their potentials. BMC System Biology, **6**, 24
6. Balagurunathan, B et al. (2012) Reconstruction and analysis of a genome-scale metabolic model for Scheffersomyces stipites. Microb Cell Fact **11**, 27

(585p) Simultaneous Production of Multi-Functional Peptides in An Enzymatic Membrane Reactor Via Combinational Chromatography

Wu, S.,
Qi, W., Tianjin University
Su, R., Tianjin University
He, Z., Tianjin University

Bioactive peptides derived from bovine casein exert various physiological functions. These peptides ususally included angiotensin converting enzyme inhibitory peptides (ACEIPs), phosphopeptides (CPPs) and antimicrobial peptides (AMPs)[1]. ACEIPs possess the capacity to lower blood pressure in hypertensive subjects. CPPs, containing the cluster sequence Ser(P)-Ser(P)-Ser(P)-Glu-Glu, can stabilize calcium and phosphate ions, thus enhancing calcium absorption. AMPs are cationic and typically consisting of 12 to 50 residues. They can slow the growth of bacteria and play an important part in the host defense system of many animal species.

Although numerous techniques have been investigated for the preparation of these peptides, the commercial production has been limited due to the lack of suitable large-scale technologies as well as the high cost of enzyme and raw materials. Enzymatic hydrolysis seems to be the most appropriate method for preparation of peptides in nutritional applications, which could be performed in a conventional batch reactor (CBR) or in an enzymatic membrane reactor (EMR) for continuous production of peptides. Several studies have concluded that EMR could result in higher productivities and more uniform products than CBR. However, EMR has been only applied to the preparation of one kind of specific peptide until now.

On the other hand, as reviewed[2], various potential bioactive peptides have been derived from casein hydrolysates, indicating that screening diverse bioactive peptides simultaneously from one kind of raw material would be achievable. However, most reports only focused on the preparation of one specific peptide, while simultaneous preparation of various peptides was rarely investigated.

Based on the above analysis and considering of high productivity & low cost, three kinds of bioactive peptides (ACEIPs, AMPs and CPPs) were simultaneously prepared from casein by pancreatic hydrolysis via an enzymatic membrane reactor (EMR) (see Fig. 1a, Integrated System) and combinational chromatography (see Fig. 1a, Simultaneously preparation) in this work. The reaction was performed at 37 °C and pH 8.0 for 3h followed by ultrafiltration. Strong cation exchange high-performance liquid chromatography (SCE-HPLC) was used to isolate AMPs and CPPs from retentate fractions ranging from 1kDa to 5kDa. Concurrently, ACEIPs were purified by size exclusion chromatography (SEC) from permeate fractions with molecular weight below 1kDa. Fraction 5 corresponding AMPs from SCE exhibited the highest activity against *S.aureus* (97.49% in 1.0mg/ml) (Fig. 1b). The ACEIPs in fraction B from SEC had the highest ACE-inhibiting-activity (73.5% in 1.0mg/ml) (Fig. 1c). Following reverse-phase high performance liquid chromatography-electrospray ionization tandem mass spectrometry (RP-HPLC-ESI-MS/MS) analysis, potential ACEIPs and a total of 34 CPPs were identified (18 corresponded to alpha$_{s1}$-casein, 2 to alpha$_{s2}$-casein, 10 to beta-casein and 4 to kappa-casein) (Fig. 1d). This work provided a novel method to Simultaneously prepare multi-functional peptides from casein hydrolysates via enzymatic membrane reactor and combinational chromatography.

This research was supported by the NSF of China (20976125, 31071509, 51173128, 21206113) and Tianjin (10JCYBJC05100), the Ministry of Science and Technology of China (Nos. 2012YQ090194, 2013AA102204, 2012BAD29B05), the Program for New Century Excellent Talents in Chinese University (NCET-08-0386), and Beiyang Young Scholar Program (2012).

References

1. Silva, S.V. and F.X. Malcata. (2005). Caseins as source of bioactive peptides. *International Dairy Journal*, *15*(1), 1-15.

2. Phelan, M., Aherne, A., FitzGerald, R. J., & O'Brien, N. M. (2009). Casein-derived bioactive peptides: Biological effects, industrial uses, safety aspects and regulatory status. *International DairyJournal, 19*(11), 643-654

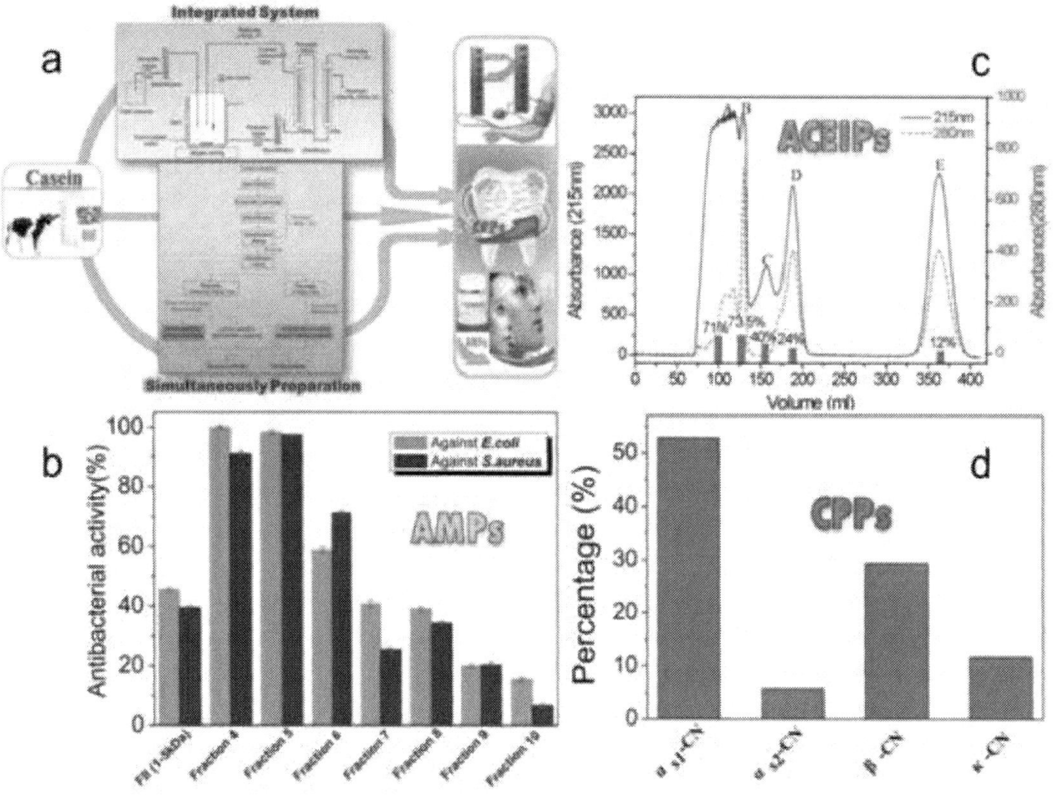

Figure 1 Simultaneous preparation of AMPs, ACEIPs and CPPs. a) Schematic diagram of the production of AMPs, ACEIPs and CPPs in an enzymatic membrane reactor via combinational chromatography, b) The antibacterial activity against *E.coli* and *S.aureus* strains of fractions 4-10 in cation exchange chromatogram; c) Size exclusion chromatogram of FIII on a SephadexG-15 gel column and ACE-inhibitory activity (%) of each fractions with green bars, d) The percentage of the number of phosphorylated peptides.

(585o) Synthesis and Characterization of Thermo-Responsive Bioconjugates: Immobilized Enzyme for Galacto-Oligosaccharide Formation

Palai, T., Indian Institute of Technology Kanpur

Kumar, A., Indian Institute of Technology Kanpur

Bhattacharya, P. K., Indian Institute of Technology, Kanpur

Synthesis and characterization of thermo-responsive bioconjugates: immobilized enzyme for galacto-oligosaccharide formation

Tapas Palai[1], Ashok Kumar[2] and Prashant K. Bhattacharya[1*]

1 Department of Chemical Engineering, Indian Institute of Technology Kanpur, Kanpur-208016, India.

2 Department of Biological Sciences and Bioengineering, Indian Institute of Technology Kanpur, Kanpur-208016, India.

* Corresponding author: E-mail: pkbhatta@iitk.ac.in Tel.: +91-512-2597093; Fax: +91-512-2590104

Abstract

Thermo-responsive polymers, also known as smart/intelligent/environmentally sensitive/and reversible polymer, have received much attention in biotechnology, medicine and engineering applications. With a change in temperature around its lower critical solution temperature (LCST), the polymer undergoes a reversible phase change form soluble to insoluble form or vice versa [1]. This property was considered in immobilizing enzymes on such reversible polymers which are known as polymer-protein bioconjugate and the attachment of enzymes may be via physical or chemical means. Further, such bioconjugates may effectively carry out enzymatic reactions under immobilized states [2]. In order to test one such enzymatic reaction, the formation of galacto-oligosaccharides from lactose [3] was considered with β-galactosidase (EC 3.2.1.23; commercial name: Biolacta FN5) on reversible polymer, poly-N-isopropylacrylamide (PNIPAAM). In the present study, apart from PNIPAAM, a non-thermo-responsive polymer poly-acrylamide (PAAM), was chosen as control polymer [4]. Experimental evidences too confirm the formations of PNIPAAM-β-galactosidase bioconjugates, exhibiting thermo-responsive behavior, whereas PAAM-β-galactosidase bioconjugates do not exhibit such thermo-response. Oligosaccharides and their derivatives are important nutraceuticals and prebiotic food ingredient, having a range of important functions in biological systems. The utility of thermo-responsive bioconjugates help in carrying out the reaction under homogeneous condition (below LCST) and effectively help in separating enzymes above LCST.

The synthesis of bioconjugates was carried out in three steps [4] using commercial grade β-galactosidase from *Bacillus circulans*. Firstly, acryl group was introduced to enzyme by treating with itaconic anhydride in phosphate buffer (pH 6.0). Furthermore, unreacted itaconic anhydride was removed by dialysis. The extent of enzyme modification was estimated by 2,4,6-trinitrobenzenesulphonic acid (TNBS) method [5].

Then, PNIPAAM and PAAM-β-galactosidase bioconjugates were copolymerized from their respective monomers (N-isopropylacrylamide (NIPAAM) and acrylamide (AAM)). The reaction was catalyzed and initiated with N,N,N',N'-tetramethylethylenediamine (TEMED) and ammonium persulphate (APS), respectively. Finally, the bioconjugates thus formed were separated from their respective monomers by salt-induced thermal precipitation (for PNIPAAM) and dialysis (for PAAM). The activities of native enzyme, PNIPAAM and PAAM bioconjugates were determined using *o*-nitro-phenyl-β-D-galactopyranoside (ONPG) as substrate at 40°C and pH 6.0. [3]. Bioconjugation yield was estimated by measuring the amount of conjugated protein using bicinchoninic acid method. The yield of PNIPAAM- β-galactosidase bioconjugate was measured as 75% at an enzyme modification level of 50%. A decreased LCST of 29.5°C, in comparison to 32.5°C for the pure PNIPAAM polymer, was observed for PNIPAAM-enzyme bioconjugate. This decrease may be due to the increase in hydrophobicity of the polymer during bioconjugation with enzyme [4].

The catalytic activities of PAAM, PNIPAAM-enzyme bioconjugates and native enzyme were measured at varying temperature (20-80°C) and pH (4.0-6.0) using ONPG as substrate. The activities of all the three preparations were observed to increase with increasing temperature till 60°C and thereafter it a decrease was observed only for PAAM-enzyme bioconjugate and native enzyme. On the other hand, the activity of PNIPAAM-enzyme bioconjugate increased till 70°C and thereafter it starts decreasing. Furthermore, it was observed that the activities of both the PAAM, PNIPAAM bioconjugates were less sensitive to pH variation than its native form. Thermal stability was also performed; in two different ways. In the first case, enzyme solutions were incubated at different temperature (20-80°C) for 15 min, followed by residual activity measurement. PNIPAAM-enzyme bioconjugates retained ~100% activity at 50°C; however, it is around 50% of initial activity at 80°C. On the other hand, PAAM-enzyme retained ~60% and ~8% activity at 50°C and 80°C, respectively; whereas, the native enzyme retained ~60% and 0% at 50°C and 80°C, respectively. In the second case, the enzyme solutions were heated continuously at 55°C for 90 min and residual activity was measured at regular time intervals. The PNIPAAM-bioconjugate offered a maximum resistance to thermal deactivation by keeping 67.5% of initial activity after 90 min of incubation as compared to PAAM-enzyme (24%) and native enzyme (16.5%). Furthermore, study on operational stability showed the residual activities of PNIPAAM-enzyme, PAAM-enzyme and native enzyme after 24 cycles were 91%, 88.5% and 86%, respectively. The effect of organic solvents (ethylene glycol and ethanol) on activity of PNIPAAM-bioconjugate was studied by keeping it at varying concentrations of organic solvents (2-25% v/v) for 24 h at 30°C. Residual activity was measured at a regular interval of time. It was observed that the activity of PNIPAAM-bioconjugate got increased by 45% with 20% (v/v) ethylene glycol; whereas, with ethanol, it showed negative trend.

The storage stability was monitored by keeping the enzyme solution at 20°C for 30 days. PNIPAAM-bioconjugate retained ~90% of its initial activity; whereas, PAAM-bioconjugate retained ~80% till 30 days. The native enzyme however lost its full activity within 20 days.

Further, Michaelis-Menten constant (K_m) and maximum reaction rate (V_m) values were also estimated (from Lineweaver-Burk method) using ONPG as substrate at 30°C. The values were found to be: K_m = 1.1 mM, V_m = 2.61×10^{-5} mM s^{-1}.

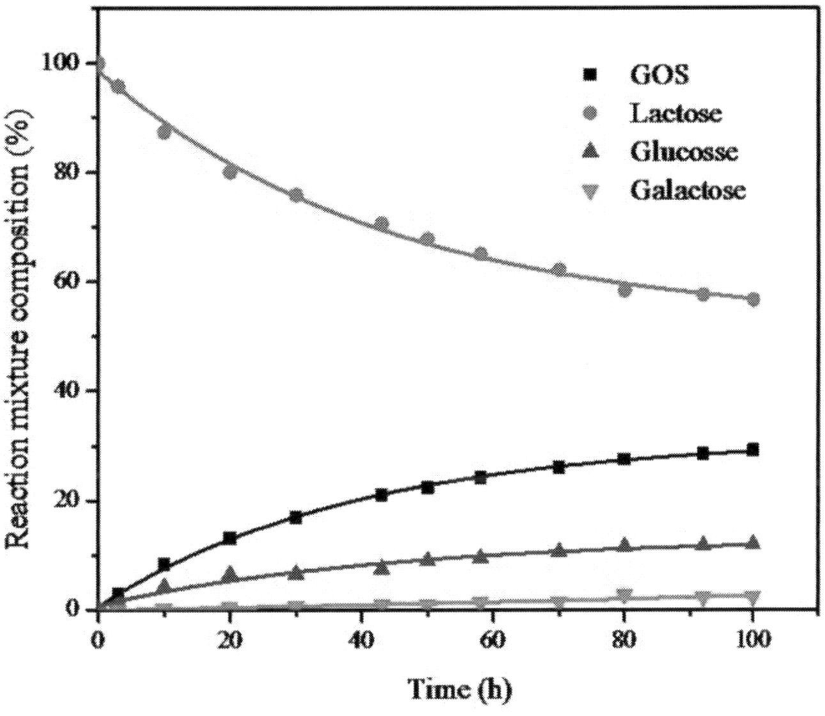

Fig. 1: Time-course reaction mixture composition

Finally, the synthesis of GOS from lactose was carried out with the application of PNIPAAM-β-galactosidase bioconjugates. Utilizing such bioconjugates for the enzymatic formation of GOS, thus considered the system would provide not only a homogeneous system also an immobilized system too. Accordingly, the enzyme β-galactosidase converts lactose via two simultaneous reactions; trans-galactosylation reaction, which produces GOS and hydrolysis reaction, which forms glucose and galactose [2]. The reaction was carried out under simple batch mode at 30°C and pH 6.0 for 100 h. Samples were collected at a regular interval and analyzed through high performance liquid chromatography (HPLC). A maximum of 29.5% GOS (dry basis) with 44.0% of lactose conversion was obtained after 100 h reaction time at 50 g/L initial lactose (Fig. 1). The equilibrium reaction mixture contained 7.0% tetra-saccharide, 22.5% tri-saccharide, 12.0% glucose, 2.5% galactose along with 56.0% unreacted lactose.

Keywords: Bioconjugate, Responsive polymer, enzyme immobilization, Galacto-oligosaccharide, Lactose

References:

[1] A. Kumar, A. Srivastava, I.Y. Galaev, B. Mattiasson, Smart polymers: Physical forms and bioengineering applications, Progress in Polymer Science 32 (2007) 1205-1237.

[2] T. Palai, P.K. Bhattacharya, Kinetics of lactose conversion to galacto-oligosaccharides by β-galactosidase immobilized on PVDF membrane, Journal of Bioscience and Bioengineering 115 (2013) 668-673.

[3] T. Palai, S. Mitra, P.K. Bhattacharya, Kinetics and design relation for enzymatic conversion of lactose into galacto-oligosaccharides using commercial grade β-galactosidase, Journal of Bioscience and Bioengineering 114 (2012) 418-423.

[4] A.K. Shakya, P. Sharma, A. Kumar, Synthesis and characterization of thermo-responsive poly(N-isopropylacrylamide)-bovine liver catalase bioconjugate, Enzyme and Microbial Technology 47 (2010) 277-282.

(585n) An Integrated Mechanistic Model for Nucleotide Sugar Metabolism and Monoclonal Antibody Glycosylation

Jimenez del Val, I., Imperial College London
Constantinou, A., Imperial College London
Dell, A., Imperial College London
Haslam, S., Imperial College London
Jedrzejewski, P., Imperial College London
Polizzi, K. M., Imperial College London
Kontoravdi, C., Imperial College London

Pharmaceutical quality by design (QbD) is a framework for the development and approval of manufacturing processes that aims to build quality into drugs at every stage of process development. Implementation of QbD requires identification of the critical attributes that define product quality (CQAs) which, in the case of pharmaceuticals, are the properties that define a drug's safety and therapeutic efficacy. Implementation of QbD relies on all available knowledge regarding the mechanisms that quantitatively relate manufacturing process inputs with product CQAs. These mechanistic relationships are then used to control the manufacturing process so that product CQAs are maintained and end-product quality is ensured. Within this framework, mathematical models that mechanistically describe drug product quality as a function of process inputs are of paramount importance.

Monoclonal antibodies (mAbs) are one of the leading products of the pharmaceutical industry having had worldwide sales of over $43 billion in 2010. All approved mAbs contain a consensus N-linked glycosylation site on their constant fragments. It has been widely reported that the complex carbohydrates (glycans) bound to these sites directly impact the safety and efficacy of mAbs when administered to patients. This has led to defining N-linked glycosylation as a CQA of mAbs. It has also been reported that many bioprocess conditions (e.g. nutrient availability, dissolved oxygen, pH, temperature, stirring speed and culture medium supplementation) influence the composition and distribution of glycans bound to mAbs. Within the context of QbD, we have defined a mathematical model that mechanistically and quantitatively describes the glycosylation profiles of mAbs as a function of nutrient availability during cell culture.

The mathematical model consists of two elements, the first of which describes cell culture dynamics and has been adapted to include intracellular accumulation of nucleotide sugars (NSDs) as a function of extracellular nutrient availability. The metabolic pathway for nucleotide sugars has been reduced from 32 to 8 reactions, yet the regulatory motifs of the full pathway have been included in "lumped" rate expressions. The second element is a model for N-linked glycosylation, which approximates the Golgi apparatus to a plug-flow reactor and includes NSDs as co-substrates for the glycosylation reactions that occur therein.[1] Both elements are linked through mAb specific productivity and intracellular availability of NSDs.

The unknown parameters of the model were estimated with data obtained from murine hybridoma (CRL-1606, ATCC) culture. Typical data (viable cell density, extracellular glucose, glutamine, lactate, ammonia and mAb titre) was collected along with intracellular measurements of nucleotides (ATP, CTP, GTP and UTP) and nucleotide sugars (CMP-Neu5Ac, GDP-Fuc, GDP-Man, UDP-GalNAc, UDP-Glc and UDP-GlcNAc). In addition, the mAb glycan profiles were obtained using MALDI mass spectrometry. With the estimated parameters, the model reproduces the experimental data accurately. Finally, the full model was used to perform simulations of glucose and glutamine depletion. These studies predict accumulation of high-mannose glycans under glutamine starvation, a phenomenon that is consistent with previous observations.[2]

The final product of this work is a mathematical model that mechanistically and quantitatively describes how extracellular glucose and glutamine availability impact the glycosylation profiles of mAbs by using intracellular NSD availability as the fundamental link between both elements. Mathematical models that mechanistically relate process inputs with product CQAs may prove successful in aiding process development, control and optimisation under the QbD scope.

References:

1. del Val, I.J., Nagy, J.M. & Kontoravdi, C. A dynamic mathematical model for monoclonal antibody N-linked glycosylation and nucleotide sugar donor transport within a maturing Golgi apparatus. *Biotechnol Progr* **27**, 1730-1743 (2011).
2. Wong, D.C.F., Wong, K.T.K., Goh, L.T., Heng, C.K. & Yap, M.G.S. Impact of dynamic online fed-batch strategies on metabolism, productivity and N-glycosylation quality in CHO cell cultures. *Biotechnol Bioeng* **89**, 164-177 (2005).

(585ac) Two Different Processes to Obtain Antimicrobial Packaging Containing Natural Oils

Valderrama Solano, A., TecnolÃ³gico de Monterrey
Rojas-de Gante, C., TecnolÃ³gico de Monterrey
trujillo De Santiago, G., TecnolÃ²gico de Monterrey

Abstract

In this study, antimicrobial packaging materials were developed by incorporating known concentrations (w/w) of essentials oils of oregano (Origanum vulgare) and thyme (Thymus vulgaris) into low-density polyethylene (LDPE), suitable for use as food packaging, via two different methods: ionizing treatment and directly by extrusion. The mechanical, barrier, and antimicrobial properties of the packaging were evaluated against the following foodborne pathogens: Salmonella typhimurium, Listeria monocytogenes, and Escherichia coli O157:H7. The results demonstrate that films developed by extrusion incorporating 4% (w/w) of essential oils had a higher inhibitory effect than those obtained using the ionizing treatment. The packaging developed by extrusion containing 1% (w/w) showed a positive inhibitory effect, while those obtained by the ionizing treatment had no inhibitory effect against any of the test microorganisms. The incorporation of essential oils on the LDPE films generated a plasticizer effect, whereas the ones obtained by means of ionizing treatment did significantly affect the barrier properties of the films. The results of this study showed that plantderived essential oils could be incorporated in active films for food packaging.

(585t) Enzymatic Hydrolysis Changes Induced for Thermoplastic Extrusion and Protease Supplementation in Red Sorghum, Decorticated Sorghum and Maize: Reducing Sugars Yield, Free Alpha Amino-Nitrogen Compounds and Proteins Fractions Distribution

Peralta-Contreras, M., ITESM
Perez-Carrillo, E., ITESM
Garc a-Escamilla, E., Center for Research and Development in Health Sciences (CIDICS)
Serna-Sald var, S. R. O., TecnolÃ³gico de Monterrey
Soria-Hern ndez, C. G., TecnolÃ³gico de Monterrey

The growing demands for efficient processes and a variety of cereal commodities have been the driving force for the implementation of thermoplastic extrusion in food and bio-fuel industries. This technology allows a quick starch gelatinization and promotes important structural changes in the food matrix. The objective of this research was to compare the extruded cereal with dried ground meals in terms of reducing sugars yield, free alpha amino-nitrogen (FAN) compounds profile and protein fractions distribution during the starch hydrolysis to produce sorghum worts. A split plot design was proposed with three replications to evaluate the enzymatic hydrolysis. The main plot treatment comprised the cereal type while the sub plot treatments were milling, extrusion conditions, and the protease supplementation. The proteins fractions were studied with a modified Landry-Moreaux protocol changing the polarity of solvent applied to the hydrolyzed samples The protein fractions considered in this study were globulins, prolamins, crosslinked prolamins and glutelins; these fractions are important to differentiate the protein matrix that entraps the starch granules and cause certain differences in hydrolysis efficiency.

The extruded treatments increased sugar production (185 g/L) compared with the ground treatments (165g/L), maize presented a better yield (185g/L) respect to red sorghum (168g/L).However, the FAN presents higher production in the ground treatments (130mg/L)compared to the ground ones (90mg/L) during the first 50 minutes of hydrolysis (140 g/L) in comparison with the grounded treatments (45 g/L). Preliminary results have shown that extrusion yields a decrement in the protein content in red sorghum. Protease increases the FAN compounds in extruded treatments which can be correlated with starch breakdown in the first 25 minutes of the process. The research has demonstrated that the use of extrusion modifies the protein matrix, increases the hydrolysis yield and affects the FAN profile.

(585u) Supercritical CO2 and Ethanol Extraction of Phospholipids From Natural Products

Dyer, P., Callaghan Innovation
Tallon, S. J., Industrial Research Ltd

Phospholipids are of increasing interest for their dietary benefits. They are soluble in supercritical fluid carbon dioxide with an ethanol co-solvent and this is an attractive method for preparing extracts for use in food ingredients and dietary supplements. In this paper, solubility data are presented along with extraction profiles from selected natural sources of marine, animal, and plant phospholipids.

(585v) In Vitro Cell-Based Assays for Assessing the Impact of Thermal Processing On the Allergenic Potential of Milk Allergens

Zhao, Y.,
Fu, T. J., U.S. Food and Drug Administration

A better understanding of how processing may affect the allergenicity of proteins will allow a more accurate allergen risk assessment and achieve a greater success in food allergy management. To determine whether food processing can affect the allergenic potential of proteins, immunochemical methods such as ELISA are commonly used, however, these methods are not able to determine changes in the proteins' ability to elicit allergic reactions. Although food challenge studies are the gold standard for determining the allergenicity of proteins, such studies are seldom performed due to ethical reasons and the difficulty in recruiting a sufficient number of study participants. In recent years, a number of mediator release assays have been developed to measure the allergenic potential of food proteins in vitro. How these cell-based assays can be used to assess the impact of thermal or other processing conditions on the allergenic potential of proteins remains to be determined. This study evaluated the use of sensitized rat basophilic leukemia (RBL) cells for assessing the impact of boiling on the allergenic potential of milk allergens by measuring changes in the proteins' ability to induce cell degranulation and the release of β-hexosaminidase. RBL-2H3 cells sensitized with rat IgEs that were specifically raised against NIST non-fat milk standard reference material, β-lactoglobulin (BLG), α-lactalbumin (ALA) and α-casein (αCAS) were challenged with various concentrations of each allergen in its native and processed (boiled in PBS for 10 min) forms. Cell degranulation caused by the allergen was monitored by measuring the release of β-hexosaminidase. All allergens (NIST milk, BLG, ALA, and αCAS) were able to induce the release of β-hexosaminidase release, ranging from 7.90% to 39.02%. In general, a decrease in % release of β-hexosaminidase was observed when the cells were challenged with a decreasing concentration of the allergens. Boiling of ALA caused an increase in the % release of β-hexosaminidase from sensitized cells across antigen concentrations from 10^{-9} mg/mL to 1 mg/mL. In conclusion, this study demonstrated that the in vitro mediator release assay method may serve as a useful model to evaluate the impact of thermal processing on the allergenic potential of milk allergens.

(585i) L-Lactic Acid Production With Simultaneous Saccharification and Fermentation of Xylo-Oligosaccharides Waste Residue By Rhizopus Oryzae

Zhang, L., The Ohio State University
Yong, Q.,
Yang, S. T., The Ohio State University
Li, X.,

Abstract: High substrate cost is one of the biggest challenges in fermentative L-lactic acid production. The waste residue of xylo-oligosaccharides (XOS) manufacturing process has been considered as an attractive alternative substrate for cost-effective L-lactic acid production because of its high enzymatic digestibility, various environmental benefits, and extremely low cost. Simultaneous saccharification and fermentation (SSF) technology is an advanced strategy for bioconversion of carbohydrate materials to lactic acid by coupling the enzymatic hydrolysis and microbial fermentation into a single step. In this study, L-lactic acid production from XOS waste residue by *Rhizopus oryzae* in a SSF process was investigated with a focus on improving substrate utilization efficiency and reducing production cost. Compared to separate hydrolysis and fermentation (SHF), a higher L-lactic acid titer and yield (22.39 g/L vs. 15.40 g/L and 44.77 g/100 g dry waste residue vs. 30.81 g/100 g dry waste residue) was achieved in the SSF process at a high temperature (40 Â°C) and 5 % substrate loading. When the substrate loading was increased to 10 %, 32.67 g/L L-lactic acid with a yield of 32.67 g/100 g dry waste residue was obtained in the SSF process. This is the first report on cost-effective L-lactic acid production from XOS waste residue by *R. oryzae* at a high temperature in a SSF process.

(585k) Metabolomic Study Of Propionibacterium Acidipropionici ATCC 4875 Revealed The Metabolic Regulations In Response To FBB Adaptation and Different pH Values

Lu, N., The Ohio State University
Wang, Z., The Ohio State University
Yang, S. T., Ohio State University

Drastic alterations in FBB adaptation are known to cause large changes in metabolism in *Propionibacterium acidipropionici*. The pH value could affect the propionic acid production in *P. acidipropionici* fermentation. Metabolomics was used to study the possible metabolic regulations leading to the enhanced propionic acid production in the FBB adapted cells and how pH could have affected the metabolic pathway in propionic acid biosynthesis. As a result of the rapid sampling, extraction, and analysis by gas chromatography coupled to time-of-flight mass spectrometry (GC-TOF/MS), more than 572 analytes from a single sample could be measured, of which 140 metabolites of their chemical structures and names have been identified. Wild-type, FBB adapted and cells at pH 6.5 and pH 5.5 after high cell density fermentation were clustered into four groups after orthogonal projection on latent structure discriminant analysis (OPLS-DA) pattern recognition due to their highly distinctive metabolite profiles. Compared to the wild type, FBB adapted mutant biosynthesized less polysaccharide which contributed to that more C sources were directed toward propionic acid biosynthesis. The N-compounds mainly nucleotides decreased in FBB adapted cells, indicating reduced cell growth activity. Amino acids, whose biosynthesis requires large amounts of ATP and NADPH, also decreased in FBB-adapted cells. The cell membrane composition also changed in the adapted cells with more long-chain saturated fatty acids and less unsaturated fatty acids compared to the wild type. Cells after high cell density fermentation at pH 5.5 showed lower concentrations of pyruvate, succinate and lactate, confirming that more propionic acid was produced from the substrate carbon with less diverted to other byproducts as observed in the fermentation. Overall, metabolic alterations could be assigned to central metabolic pathways such as glycolysis, Wood-Werkman cycle, TCA cycle, nucleotide and amino acids metabolism, which will be discussed in this paper.

(585w) Evaluation of Mixing Using Extension Rate Distribution in a Co-Rotating Twin Screw Mixer

Rathod, M., Rutgers, The State University of New Jersey
Kokini, J., University of Illinois

Many food processes involve mixing to incorporate ingredients, introduce air and form desirable structural attributes. When examining mixing efficiencies, one is concerned with both distributive and dispersive mixing. One common measure of dispersive mixing is the Manas-Zloczower mixing index (λ). Although this incorporates elongational flow, there is additional information that is not captured in this parameter. Another way of examining mixing is by mapping extension rate distribution.

This work examined a nine-paddle mixer geometry with a fluid inflow rate of 55.31 cc/s and a mixer speed of 100 rpm. The material was modeled as a power law fluid using parameters previously obtained for a 2% aqueous CMC (carboxymethylcellulose) solution. Numerical simulation was performed using Polyflow (ANSYS, Inc.). Five xy planes were chosen to compare extension rate as axial distance from the mixer entrance increased.

The initial plane had significant extension rate peaks slightly left of the mixer center due to the vertical position of the left-hand paddle. Additional features seen in later planes were not evident in this location due to its position relative to the mixer entrance. Beyond the initial plane, remaining planes had similar maximum extension rate values. Greatest extension rate values were seen in the mixer center and in the region of the vertical left-hand paddle, with peaks at the upper and lower paddle tips. No gradual increase was seen along the paddle length. Maximum extension rate values were seen in areas with high shear, between the two paddles and between paddle tips and the barrel wall.

The highest extension rate values were seen between λ values of 0.4 and 0.75, a much wider range than the predicted value for efficient dispersive mixing at λ=0.7. More elongational flow did not indicate more material stretching above a certain λ value. A more equal division of rotational and elongational flow produced more stretching.

(585m) Construction and Characterization Of Ack and Pta Gene-Deleted Mutants Of Propionibacterium Freudenreichii For Enhanced Propionic Acid Fermentation

Lin, M., The Ohio State University
Wang, Z., The Ohio State University
Yang, S. T., Ohio State University

Propionic acid is a widely used industrial product currently produced mainly via petrochemical processes. As the crude oil price has surpassed US$100 per barrel, there have been increasing interests in propionic acid production from renewable bioresources by fermentation using propionibacteria. The main metabolic product of Propionibacterium freudenreichii subsp. shermanii is propionic acid with acetic and succinic acids as two by-products. Propionic and succinic acids are generated in the dicarboxylic acid pathway, in which the formation of both acids is directly correlated with each other. On the other hand, acetic acid is formed in a separate pathway, first from pyruvate decarboxylation to acetyl-CoA and then to acetyl phosphate and finally to acetic acid. The last two steps of acetic acid formation are catalyzed by phosphotransacetylase (PTA) and acetate kinase (ACK), respectively. In order to enhance propionic acid production, *pta* and/or *ack* were mutated through homologous recombination. Gene mutations were confirmed by PCR. Fermentations in serum bottles and fermentors using different carbon sources were conducted for mutants and wild type. Fermentation results showed differences regarding acetic and propionic acids formation between the wild type and single knock-out mutants. The effects of these gene knockouts on the flux distributions between acetic acid and propionic acid biosynthesis pathways were investigated and will be presented in this paper.

(585l) High-Cell Density Fermentation With An Acid-Tolerant Strain Of Propionibacterium Acidipropionici For Propionic Acid Production

Jin, Y., Ohio State University
Wang, Z., The Ohio State University
Yang, S. T., Ohio State University

Propionic acid is an important short-chain fatty acid with wide industrial applications. With the depletion of petroleum, rising oil price and market demanding, bio-based propionic acid has gained more and more attention. *Propionibacterium acidpropionici*are Gram-positive, nonspore-forming, rod-shaped, anaerobic bacteria. It can be used to produce propionic acid with acetate, succinate and carbon dioxide as byproducts. To compete with petrochemical processes, a *Propionibacterium acidipropionici* acid-tolerant mutant strain was first developed by adaptation in a fibrous bed bioreactor (FBB) and serious sub-culturings on agar plates. Compared to the wild-type strain, the mutant strain gave higher productivity and product purity. Then a high-cell-density (HCD) fermentation process, with high propionic acid yield and productivity was developed. A productivity of 2.89 g/L·h (vs. 0.5 g/L·h) was attained in repeated-batch HCD fermentation at pH 6.5. At lower pH values of 5.5 and 5.0, higher propionic acid yield and slightly lower productivity were obtained. The HCD fermentation process with the acid-tolerant strain can be used to produce propionic acid from glucose and other biomass feedstocks, which will be presented in this paper.

(597a) Experimental and Computational Analysis of Chinese Hamster Ovary Stable Transfectants Grown in Fed-Batch Culture

Sou, S. N., Imperial College London
Sellick, C.,
Lee, K.,
Mason, A.,
Polizzi, K. M., Imperial College London
Kontoravdi, C., Imperial College London
Originally presented on: 11/7/2013 8:30:00 - 8:48:00

With positive effects in medical treatments, therapeutic drugs that are derived from cells (biologics) have become one of the fast growing drug groups in the pharmaceutical market [1]. To date, Chinese Hamster Ovary (CHO) cells are the most common industrial hosts for therapeutic recombinant protein (rProtein) production. Despite the high specificity and medical efficacy of biologics, there is limited information on the metabolic activities of the cultured organisms during rProtein synthesis, while their production yields are often low. An increasing demand for the production of rProteins such as monoclonal antibodies (mAbs) places a necessity to understand the intracellular events, as well as to search for ways to control cell viability and improve productivity.

To increase the cell-specific production rate of mAbs (q_{mAb}), variations in culture environments are being integrated into fed-batch cultures. Mild hypothermic conditions (30 – 34 °C) in particular have been shown to enhance protein titres in CHO cell cultures [2]. Despite the increase in protein productivity in mild hypothermic cultures, this effect is dependent on host cell type, temperature variation and time of temperature shift. Therefore thorough understanding of the way host cells regulate their dynamic cellular processes at lower temperatures is necessary to enhance bioprocess design and optimisation. In this study, CHO cells expressing a mAb were cultured at physiological temperature (36.5°C) and with a temperature shift to 32 °C at 3 days or 6 days post inoculation. We then compared culture performance with respect to cell growth, extracellular metabolite concentrations, stability of mAb mRNA and assembly intermediates, as well as secretion rates.

We found that at hypothermic conditions, cells grow at a reduced rate compared to those at physiological temperature. Cell metabolic activities also vary at hypothermic culture conditions. Regardless of the induction time of temperature shift, there is an overall reduction in glucose consumption. Decrease in lactate production was only observed in CHO cells that were cultured at 32°C from day 6 onwards. Our results correlate with those observed in *Nam et al.* [3] where consumption and production of metabolites are reduced.

In the case of mAb synthesis, our experimental results show that the specific mAb productivity of cells that are cultured under hypothermic conditions is highly dependent on the induction time of the temperature shift. Reducing culture temperature early in exponential phase (day 3) has no beneficial effect on the overall mAb titre. However a small increase in q_{mAb} is observed when the temperature is reduced to 32 °C at late exponential phase (Day 6) when compared to those at 36.5 °C. Similar results are observed at mRNA level. CHO cells that experience temperature shift on day 6 achieve an overall higher mRNA gene copy number of heavy chain and an increase in intracellular H_2L_2 intermediates when compared to the other two cases. This is in agreement with *Marchant et al.* [4], who have shown that when cells are cultured at a cold-stress condition, the residence time of the protein in the ER increases and proteins are processed more efficiently. However, concentrations of intracellular H_2 and H_2L assembly intermediates remain lower at 32 °C when compared to those at physiological temperature.

In addition to experimental analysis of fed-batch CHO cultures, model-based techniques were introduced with an aim to identify bottlenecks and guide future experiments. In this respect, a hybrid unstructured/structured cell culture model was being developed for CHO cell culture grown in commercially-available media and supplemented with commercial feeds. Model parameters were estimated using datasets for culture under physiological temperature and with the switch to mild hypothermia. Simulation results provide good fits for cell growth, metabolite concentrations, concentrations of intracellular mRNA and mAb assembly intermediates, as well as secreted monoclonal antibody titres, when compared to experimental data for the three scenarios examined. We envisage that this model-based tool will provide a means of design of experiments, which aids in bioprocess optimisation.

Reference

[1]. Harris M. Market-Leading Biotechnology Drugs 2009: Blockbuster Dynamics in an Ailing Economy. *BioWorld,* Atlanta, GA. 2009.

[2]. Wulhfard S *et al.* Mild Hypothermia Improves Transient Gene Expression Yields Several Fold in Chinese Hamster Ovary Cells. *Biotechnol. Prog.* 2008; 24: 458-465.

[3]. Nam JH et al. The effects microcarrier culture on recombinant CHO cells under biphasic hypothermic culture conditions. *Cytotech.* 2009; 59: 81-81.

[4]. Marchant, RJ *et al.* Metabolic rates, growth phase, and mRNA levels influence cell-specific antibody production levels from in vitro-cultured mammalian cells at sub-physiological temperatures. *Mol. Biotech.* 2008; 39 (1): 69-77.

(597b) Match-Maker's Dilemma: Developing a Novel Cell Culture Process for a Late-Stage Program for Which Every Parameter Changes, But Product Quality Remains Unchanged

Jaluria, P., Alexion Pharmaceuticals
Sui, S., Alexion
Facenda, J., Alexion
Alden, N., Alexion
Leslie, B., Alexion
Lucka, A., Alexion
Originally presented on: 11/7/2013 8:48:00 - 9:06:00

Following the acquisition of a late-stage program with clearly defined product quality attributes, we were tasked with developing an entirely new cell culture process from cell line to harvest capable of meeting a pre-defined increase in productivity
(> 5-fold) while still matching existing product quality attributes (purity, glycosylation patterns and enzyme activity). Initially, we mapped out a series of experiments across multiple scales evaluating many different clones. As we proceeded to distinguish -clones from each other on the basis of productivity and product quality, we began to identify limitations in our existing cell culture platform process. For instance, necessary co-factors for the desired product were at low levels in the basal media with respect to the product concentration, and supplementation of these co-factors was found to be beneficial to productivity. The timing and exact amount of supplementation was determined experimentally along with modifications to the feeding strategy. Once the production clone was selected, additional experiments were performed to fine tune operating conditions with both product quality and scale-up in mind. Of particular interest were glycosylation patterns and maintaining similar sialylation levels. Ultimately, modifications to feed delivery lines, timing of supplementation, feed composition, gassing strategy and harvest criteria led to a successful program capable of meeting all pre-defined criteria.

(597c) Managing Simultaneous Manufacturing, Process Development, and Technology Transfer Challenges for a Sensitive CHO Process With Rapid Timelines

Doyle, B.,
Derfus, G., Gilead Sciences
Krishnan, R., Gilead Sciences
Originally presented on: 11/7/2013 9:06:00 - 9:24:00

For monoclonal antibody products, early stage process development typically focuses on rapid cell line development, clone selection, and production of material for clinical trials, without the final large scale manufacturing site in mind. For robust Chinese hamster ovary (CHO) cell lines, the risk associated with this approach is usually minimal, as the cells typically perform similarly under a moderate range of bioreactor operating parameters and facility configurations. However, for a more sensitive cell line, relatively small changes in mixing and sparging strategy may significantly impact overall performance. This case study highlights the challenges encountered for a sensitive CHO cell line which led to decreased viability and 35% lower titer upon scale-up. Subsequent lab scale experiments and statistical analysis demonstrated that the reduced titer was correlated to high power input per unit volume, high gas sparge rates, and high dissolved CO_2 in the culture medium. To address these limitations, the relationships among power input, sparge rate, and kLa were modeled to identify the feasible process design space for a target manufacturing site. Based on the model, a comprehensive DOE was performed to investigate and decouple these parameters at lab scale and identify those with the highest overall impact to dissolved CO_2, viability, and titer. As a result, competing factors, including multiple process sensitivities and facility fit restrictions, were mitigated in a short timeframe to maintain supply for ongoing clinical trials and improve productivity.

(597d) Method for Improved Control of Anti-Tumor Effector Function of MAb By Modification of NS0 Cell Culture Medium

Cuenca, J.,

Originally presented on: 11/7/2013 9:24:00 - 9:42:00

Fc effector function is a key anti-tumor mechanism of action for recombinant therapeutic monoclonal antibodies (mAbs). In particular, antibody-dependent cellular cytotoxicity (ADCC) mediated by natural killer (NK) cells correlates directly with Fc binding of the mAb to the human CD16a receptor, which is in turn affected by the type of Fc *N*-linked glycan moiety. More specifically, an increased level of a-fucosylated glycans (including mannosylated species) present on the mAb Fc, leads to increased Fc binding to CD16a receptor and enhanced ADCC activity. This study investigated the effect of increased concentrations of the amino acid glycine in the chemically-defined culture media of NS0 murine cells engineered to express five different mAbs. When cells were cultured in media with a 7-fold increase in glycine concentration, there was an increased prevalence of a-fucosylation and increased ADCC activity across all five monoclonal antibodies tested. This study shows that elevated levels of glycine in cell culture media can be used to potentially improve the efficacy of possible therapeutic mAb candidates by enhancing ADCC and the Fc-directed killing of target tumor cells.

(597e) Evaluation and Characterization of Medium Factors Impacting the Fc Glycan Structure of Recombinant Antibodies Using Chemostat Cultures

Huang, J. C. J., Amgen
Luo, Y., Amgen
Richardson, J., Amgen
Lin, H., Amgen
Originally presented on: 11/7/2013 9:42:00 - 10:00:00

Glycan structures attached to the CH2 domain of recombinant antibody produced by Chinese Hamster Ovary (CHO) cells have been shown to affect antibody effector function, clearance, and therapeutic efficacy. Hence, glycosylation heterogeneity on the Fc moiety is an important critical product attribute for commercially-manufactured antibodies. During process development, multiple medium and process factors have been demonstrated to affect Fc glycan structures. However, these factors tend to be cell line specific and vary in their magnitude of effect. To develop a consistent Fc glycoform for a therapeutic antibody, a robust system to study and evaluate the impact of various factors on Fc glycosylation is necessary. A steady-state chemostat culture features constant specific cell growth rate, high viability, prolonged cell culture duration, and stable metabolite profiles, therefore, making it an ideal experimental tool for studying product quality of recombinant antibodies. In this study, a CHO chemostat producing recombinant antibody was operated for several months with high viability, and used to study the effects of various media and supplements on Fc glycosylation, including high mannose content and galactosylation percentage. NMR and mass spectroscopy were also used to characterize cell culture processes to correlate and elucidate the mechanisms of Fc glycan modulation. The results indicate that culture medium profoundly affects Fc glycan profile during antibody production. Individual supplements that impact protein glycosylation were also evaluated in the chemostat and their impact on Fc glycosyaltion were assessed. From this study, we demonstrate that chemostat culture can be a useful tool to complement cell culture process development by determining Fc product quality of a therapeutic antibody under varying cell culture medium and supplement conditions.

(597f) Improved Process Control Through Advanced Analytics During Media Development

McElearney, K., Biogen Idec
Gilbert, A., Biogen Idec
Zang, L., Biogen Idec
Kshirsagar, R., Biogen Idec
Ryll, T., Biogen Idec
Originally presented on: 11/7/2013 10:00:00 - 10:18:00

Robust media are critical for the development of biologic molecules to maintain process consistency and control product quality attributes. Recent advances in analytics have provided a deeper understanding of the medium which has led to a more robust composition. Light exposure of the medium has been shown to lead to chemical degradant products which have a negative impact on the cell culture performance, specifically with regards to the specific growth rate and viable cell density. As previously described[1] , a liquid chromatography-mass spectrometry (LC-MS) was used to characterize multiple lots of media to determine the root cause for poor cell culture performance. It was shown that the medium that had performance below expectations had an increased level of tryptophan oxidation products and a riboflavin degradant, lumichrome, present. In this study we evaluated chemically- defined, proprietary media and commercially available medium using multiple cell types to gain a better understanding of the cell culture behavior and the effect of medium. We evaluated cell culture performance based on viable cell density and viability to determine a robust strategy for media development to ensure high product yields. Based on the results, we identified a medium formulation that mitigates the risk associated with light exposure of the medium

[1] Zang, L., Frenkel, R., Simeone, J., Lanan, M., Byers, M., & Lyubarskaya, Y. (2011). Metabolomics Profiling of Cell Culture Media Leading to the Identification of Riboflavin Photosensitized Degradation of Tryptophan Causing Slow Growth in Cell Culture. *Anal. Chem.*, 5422–5430.

(597g) Essential Factors Promoting Long-Term Cell Survival and Growth of Drosophila Cells in Chemically Defined Media

Zartman, J. J., University of Zurich
Burnette, M., University of Notre Dame
Chen, J.,
Originally presented on: 11/7/2013 10:18:00 - 10:36:00

Drosophila cell culture finds multiple applications as a screening platform to identify novel targets and therapeutics to treat human diseases and as an expression system for generating recombinant proteins. In an effort to define and characterize the minimal essential requirements for maintaining long-term growth of Drosophila cell lines, we have performed a medium-throughput screen for small molecule compounds that promote cell proliferation in a chemically defined medium. Multifactorial secondary screening provided an efficient method for validation of the candidate growth supplements and identification of important synergies between medium components. For the screening efforts, we focused on an epithelial cell line, Cl.8, which is well characterized and has been previously used to identify novel insect-specific growth factors. We have characterized candidate molecules relevant to stimulating growth and survival of both imaginal disc-derived Cl.8 cells and the industrially relevant cell line Drosophila S2. We found that one factor promoting autophagy in particular is critical for enabling multiple passages in a chemically defined medium. Expansion of our pilot screen will enable identification an optimal combination of small molecules that can support the maintenance and generation of new Drosophila cell lines, increasing the versatility of Drosophila as both a genetic and biochemical model system.

(597h) A Novel Patented Open Cultivation System for Microalgae Using Recycled Water

Ogden, K. L., University of Arizona

Originally presented on: 11/7/2013 10:36:00 - 10:54:00

In order to make production of algal biofuels a reality, it is necessary to recycle water and nutrients. Furthermore, large scale cultivation systems must have temperature, nutrient addition, pH, and agitation control to enhance productivity. A novel reactor system has been developed, patented, and tested at the pilot scale to achieve better systems control, minimize operational cost, and allow for year round cultivation even when evening temperatures are below freezing. The overall systems design will be presented as well as cultivation results that are compared to traditional raceways. Finally, results using recycled water will be discussed. There are two recycle strategies: one is to recycle the water or spent media only after harvesting the algae via centrifugation or filtration or other more cost effective technologies, and the other is to harvest a part of the culture, recycle the water, add more nutrients, and operate the system as a fed-batch reactor. This presentation will show the benefits and consequences of batch versus fedbatch operation with recycled water. Data suggests that after 3 to 5 fed batch operations, organics and nitrogen build up in the spent media and inhibit growth and productivity. Recycling the water or spent media when fresh inoculum is supplied is feasible for extended recycle periods, however fed batch operation is more limited. Growth rates, productivities, and the effects of recycled media will be presented. Nitrogen and phosphorous as well as total organic content is monitored as a function of recycle period. Although the results are dependent on algal species, overall it is feasible to recycle the water routinely, but the algal inoculum must be "refreshed" after a few fed batch cycles.

(623a) Control of Mitochondrial Substrate Utilization By the Mitochondrial Pyruvate Carrier

Vacanti, N. M., University of California, San Diego
Divakaruni, A. S., University of California, San Diego
Murphy, A. N., University of California, San Diego
Metallo, C. M., University of California San Diego
Originally presented on: 11/7/2013 8:30:00 - 8:48:00

Pyruvate oxidation in the mitochondria marks the node connecting the two most well studied and arguably most important pathways in catabolic and anabolic metabolism, glycolysis and the tricarboxylic acid (TCA) cycle. Deregulation of this process contributes to the pathogenesis of numerous diseases, including diabetes, obesity, and cancer. Existence of a protein carrier to facilitate pyruvate transport into mitochondria has been recognized for decades; however the identity of the genes encoding this complex eluded researchers until last year. In this study we evaluated the function of the mitochondrial pyruvate carrier (MPC) in skeletal muscle cells using ^{13}C metabolic flux analysis. *Mpc1* or *Mpc2* were knocked down with lentviral infection of hairpins coding for shRNA or pharmacologically inhibited in myoblasts and differentiated C2C12 mouse skeletal muscle cells. Proliferating myoblasts maintained growth and oxygen consumption with *Mpc1/2* knocked down or the protein products inhibited; however their reliance on substrates for energy and biosynthetic metabolism completely shifted from glucose to amino acid oxidation. TCA cycle flux and fatty acid synthesis were maintained through glutamine anaplerosis, oxidation, and reductive carboxylation. Similar shifts were seen in differentiated myotubule energy metabolism in response to *Mpc1/2* knockdown and pharmacological inhibition. The identification of the major shifts in metabolic substrate utilization in response to *Mpc1/2* regulation provides new insights into the treatment of metabolic disease.

(623b) DNA CpG Methylation Patterns Distinctively Modulate Chromatin Structure

Jimenez-Useche, I., Purdue University
Ke, J.,
Yu, J.,
Tian, Y.,
Shim, D.,
Yuan, C., Purdue University

Originally presented on: 11/7/2013 8:48:00 - 9:06:00

DNA CpG methylation plays a key role in gene regulation. It induces compaction of the chromatin fiber and leads to gene silencing. DNA methylation is a potential biomarker for early detection of cancer since abnormal DNA methylation patterns have been observed in the promoter regions of tumor suppressor genes in early stages of tumorigenesis. Therefore, it is critical to understand the underlying mechanism as to how DNA methylation directly modulates chromatin structure and gene expression. In this study, we constructed chromatin-like molecules, i.e. nucleosomes and nucleosome arrays, containing defined unmethylated and methylated CpG patterns. Using various biophysical assays, we elucidated how different CpG patterns and methylation levels can modulate chromosome dynamics and conformation. Unmethylated CpG dinucleotides at 10bp intervals induced a more open nucleosome conformation. In addition, when in contact with the histone proteins, this CpG pattern destabilized the nucleosome structure. As we increased the DNA methylation level to 100%, we found that methylated CpG dinucleotides can distinctively alter nucleosome and nucleosome array conformation, depending on their relative positions within the nucleosomal DNA. A more open conformation was observed in nucleosomes with a stretch of methylated CpG dinucleotides in the central dyad. The opposite effect was found when the methylated CpG dinucleotides were located at 10bp intervals. In the case of nucleosome arrays, DNA methylation had a different effect on array conformation depending on the position of the CpG sites. These observations give us molecular insights into the regulatory mechanisms of gene expression driven by DNA methylation. In addition, identifying CpG patterns that are important for modulation of chromosome structure associated with DNA methylation is instrumental in the identification of relevant DNA methylation biomarkers for cancer.

(623c) RNA/Protein Phase Transitions and Cell Growth

Brangwynne, C., Princeton University

Originally presented on: 11/7/2013 9:06:00 - 9:24:00

Ribonucleoprotein (RNP) bodies are non membrane bound macromolecular assemblies. They are crucial for many fundamental biological processes, including RNA splicing, translational control, embryonic development, and cell growth. Among dozens of examples include germ granules and P bodies in the cytoplasm, and Cajal bodies and nucleoli in the nucleus. We have shown that these structures behave as liquid droplet phases of RNA and protein. We are particularly interested in the nucleolus, an RNP droplet which functions as a micro-reactor to catalyze the RNA processing steps in ribosome biogenesis, critical for cell growth. Using quantitative live-cell imaging and molecular biology approaches, we show that the nucleolus assembles by an intracellular phase transition resulting from protein-protein and protein-RNA interactions; perturbing these interactions using RNA interference (RNAi) leads to altered nucleolar structure and function. Moreover, we demonstrate that this phase transition is sensitive to geometric cues within the cell. Finally, we use microfabrication approaches to investigate the consequences of these perturbations on cell growth within a development context, using the model organism *C.elegans*.

(623d) Effects of Ozone and Complex Urban Air Pollution On Post-Transcriptional Modification of Total RNA

Baldridge, K., University of Texas at Austin
Zavala, J.,
Li, W.,
Sexton, K.,
Surratt, J.,
Bryant-Friedrich, A.,
Contreras, L. M., University of Texas at Austin
Originally presented on: 11/7/2013 9:24:00 - 9:42:00

Ribonucleic Acid (RNA) is involved in a wide variety of cellular processes, including the translation of genetic information into proteins (messenger RNA), the synthesis of proteins in the ribosome (transfer RNA, tRNA; ribosomal RNA, rRNA), and the regulation of cellular metabolism (non-coding RNA, ncRNA)[1–3]. The integrity of RNA molecules is hypothesized to be of central importance to cellular health given that the vast majority of RNA functions depend on accurate base pairing with other nucleic acids and proteins[1,4–6]. Therefore, changes in the sequence or chemical identity of base pairs through mutations or post-transcriptional modifications can significantly alter regulation of all cellular processes. In this study, chemical modifications to RNA were induced by reactive oxygen and nitrogen species (ROS) in the context of various air pollution sources. We will discuss our treatment of purified RNA samples and of human A549 lung epithelium cells with various air exposures at an air-liquid interface that include ozone, simple and complex gas-only pollutant mixtures, and particulate diesel exhaust. We will also discuss our molecular characterization of both of these exposed systems using a variety of techniques that include ELISA, MALDI-TOF and circular dichroism. Our observations reveal that levels of chemical modification and tertiary structure breakdown correlate with complexity of pollution mixtures. We will also discuss implications of these observations on cellular regulation and human health.

References

1. Huang, Y. *et al.* Molecular functions of small regulatory noncoding RNA. *Biochemistry (Moscow)* **78**, 221–230 (2013).

2. Mercer, T., Dinger, M. & Mattick, J. Long non-coding RNAs: insights into functions. *Nature Reviews Genetics* (2009).at <http://www.nature.com/nrg/journal/v10/n3/abs/nrg2521.html>

3. Storz, G., Vogel, J. & Wassarman, K. M. Regulation by small RNAs in bacteria: expanding frontiers. *Molecular cell* **43**, 880–91 (2011).

4. Vogel, J. & Luisi, B. F. Hfq and its constellation of RNA. *Nature reviews. Microbiology* **9**, 578–89 (2011).

5. Waters, L. S. & Storz, G. Regulatory RNAs in bacteria. *Cell* **136**, 615–28 (2009).

6. Esteller, M. Non-coding RNAs in human disease. *Nature reviews. Genetics* **12**, 861–74 (2011).

(623e) Engineering Exosomes to Probe and Program Loading of RNA Cargo Into Biological Nanoparticles

Marcus, M. E.,
Leonard, J. N., Northwestern University
Originally presented on: 11/7/2013 9:42:00 - 10:00:00

Exosomes are emerging as important new players in the landscape of intercellular communication. These extracellular vesicles serve as biological nanoparticles (~30-200 nm in diameter) that transfer proteins, mRNA, and microRNA from a "sender" cell to a "receiver" cell, wherein these cargo molecules functionally modulate cell state. This mechanism appears to play a particularly important role in cancer, and tumor-derived exosomes are a newly appreciated modality by which cancer cells modulate host function to promote tumor survival and metastasis. Given this capacity to functionally deliver biomolecules, exosomes are also potentially attractive vehicles for the therapeutic delivery of biomolecules [1], and recent reports have demonstrated the promise of this approach. To date, however, the mechanisms by which a subset of cellular proteins and RNAs are sorted into exosomes are poorly understood. Therefore, in order to better understand the roles of exosomes in normal and disease processes and to enable effective exosome-based therapeutics, we are developing novel biochemical tools for interrogating and directing the incorporation of specific RNA cargo molecules into exosomes. We have developed a system for engineering exosomal membrane proteins that bind to defined "packaging" sequences within engineered RNA cargo molecules. Using this system, we showed that by engineering defined protein-RNA interactions, we can direct the incorporation of specific proteins and RNA into exosomes. We also identified RNA motifs and features that may impact loading in a manner that cannot be explained by simple bulk flow or mass action driving forces. These results suggest that novel active sorting mechanisms and/or biophysical constraints dramatically modulate RNA loading into exosomes. We are using this platform to investigate the constraints that regulate packaging of RNAs into exosomes and the features of exosomal RNA cargo that affect function in receiver cells, in order to develop a quantitative framework for understanding and modulating exosome-mediated delivery. Together, these investigations serve to both improve our understanding of natural exosome biology and enable translation of this understanding into novel therapeutic capabilities.

[1] Marcus, M.E.; Leonard, J.N. FedExosomes: Engineering Therapeutic Biological Nanoparticles that Truly Deliver. *Pharmaceuticals* **2013**, *6*, 659-680.

(623f) Intracellular Cell Cycle Model of the Embryonic Stem Cell G1-S Transit During Self-Renewal and Differentiation

Task, K., University of Pittsburgh
Banerjee, I., University of Pittsburgh
Originally presented on: 11/7/2013 10:00:00 - 10:18:00

Embryonic stem cells (ESC) exhibit numerous unique traits which allow them to maintain self-renewal and differentiation potential. The cell cycle is one important trait showing vastly different behavior between ESC and adult cells. In particular, the G1-S checkpoint, a main regulator maintaining proper growth of somatic cells, is structured differently in ESC. Because ESC are characterized by a shortened G1 phase compared to mature cells, this checkpoint must give limited hindrance to transit to S phase while also being adaptable enough to allow lengthening of the G1 phase during differentiation . Although these unique cell cycle traits are central to ESC, limited work has been done in mechanistically describing this checkpoint during the transition from undifferentiated to mature cells. While mechanistic information for this complex, non-linear behavior would be difficult to obtain with a purely experimental approach, mathematical modeling can aid in obtaining a more comprehensive understanding of the system. In the current work, we develop a mathematical model to describe the unique stem cell behavior during G1-S transit and during differentiation.

Molecular dynamics are a major factor in governing cycle phase transitions in individual cells. Hence, an intracellular protein network model was developed to describe hESC at the single cell level. To develop our single-cell model, we employ a system of ordinary differential equations (ODE) to describe molecular dynamics during the G1 phase and G1-S transition. The system of ODE consist of 17 state variables representing key cell cycle molecules governing cell cycle dynamics and hESC cycle behavior, including cyclins, CDK inhibitors (CKI), and Rb/E2F. A model of adult G1-S transit was used as a base model, with parameter estimation and structure modifications included to capture characteristics/signatures unique to the hESC system. A major inclusion to the model is miRNA, with multiple miRNA complexes being combined and associated with inhibition of cyclin D and CKI.

Our ODE model is able to capture the unique hESC G1-S transition, including shortened G1 time, which gradually lengthens, but does not arrest, upon differentiation. hESC molecular dynamics are appropriately captured, including low levels of CKI and cyclin D. Inclusion of miRNA regulation is necessary to capture the system's unique behavior. miRNA levels remain elevated during pluripotency and decrease upon differentiation (as shown by miRNA qPCR data), leading to increased cyclin D and CKI. These latter two molecules perform antagonistic roles (promoting and inhibiting G1-S transition, respectively), thereby leading to a "balancing act" during differentiation which helps G1 lengthen but not arrest. At a critical threshold, the CKI remain elevated, leading to growth arrest and terminal differentiation.

The developed ODE model can accurately describe hESC cycle behavior at the single-cell level. This model of the G1-S restriction point can mechanistically explain cell cycle observations made at the population level. These non-linear molecular dynamics of the ESC cell cycle cannot be fully comprehended without computational support, and it has been shown herein that mathematical modeling of the system extracts more information than would be possible with an experimental approach alone.

(623g) Exploring the Role of Sirtuin Deacetylases in Regulating Megakaryocyte and Erythrocyte Differentiation

Duncan, M. T., Northwestern University
Yi, M., Northwestern University
Miller, W. M., Northwestern University
Originally presented on: 11/7/2013 10:18:00 - 10:36:00

Exploring the role of sirtuin deacetylases in regulating megakaryocyte and erythrocyte differentiation

Mark T. Duncan, Minchang Yi, William M. Miller

Department of Chemical and Biological Engineering, Northwestern University, Evanston, IL, United States of America

Previously, our group has reported that nicotinamide potently enhances the polyploidization of megakaryocytes (MK) in *ex vivo* culture, and that this phenomenon is at least in part a consequence of its inhibition of sirtuin deacetylase activity.[1,2]Â We further investigated the functional role of three of the sirtuins that are found in the nucleus (SIRT1, SIRT2, and SIRT7) by silencing them in two leukemic cell lines, K562 and CHRF-288-11 (CHRF), that undergo MK differentiation in response to PMA treatment.Â Despite their similar response to PMA, CHRF cells are already committed to the MK lineage, while K562 cells are bipotent for the erythroid (E) and MK lineages.Â In K562 cells, only SIRT1 silencing significantly repressed MK differentiation, as indicated by moderately reduced CD41 expression and greatly reduced polyploidization.Â However, since this effect was not recapitulated in CHRF cells, we considered three possible explanations for the differential dependence on SIRT1:Â 1) that the effect was dependent on the absence of p53 (K562 cells are p53 null, while CHRF cells express functional p53), 2) that SIRT1 is important for commitment of uncommitted progenitor cells, or 3) that K562 cells' dependence on SIRT1 was a consequence of its bcr-abl+ status.Â To test the importance of the presence of p53, we performed SIRT1 silencing experiments in K562 cells with a stably integrated p53 expression vector.Â Silencing of SIRT1 in these cells had a similar effect as in wildtype cells, suggesting the observed effects were not related to p53-status.Â To test the possibility that SIRT1 was involved in the commitment of immature progenitor cells, we transduced CD34+ cells from mobilized peripheral blood with a SIRT1 silencing vector.Â We did not observe any significant difference in the commitment or rate of differentiation of the transduced cells, as indicated by acquisition of the identifying MK antigen CD41 or CD42, which is expressed by mature MKs.Â However, we did find a slight (typically ~25%) but consistent enhancement in the polyploidization of SIRT1-silenced MKs, suggesting a fundamental difference between the endomitotic programs of primary MKs and K562 cells.Â Thus, our current results suggest the SIRT1-dependence of K562 cell differentiation by PMA is related to its expression of bcr-abl.

Since megakaryocytes and erythrocytes are closely related lineages, derived from a common progenitor cell, we are further investigating whether SIRT1 has a role in regulating aspects of erythropoiesis.Â Preliminary experiments performed in K562 cells demonstrated that SIRT1 overexpression strongly suppresses cell proliferation, while stimulating spontaneous erythroid differentiation (indicated by increased glycophorin A expression and hemoglobinization).Â This observation could have potential clinical implications for diseases including chronic myeloid leukemia, and hereditary hemoglobin disorders such as sickle cell anemia and beta-thalassemia.Â We are currently investigating the mechanism of this effect, and also exploring whether it is recapitulated in primary erythrocyte differentiation.

References

1 Giammona, L. M., Fuhrken, P. G., Papoutsakis, E. T. & Miller, W. M. Nicotinamide (vitamin B3) increases the polyploidisation and proplatelet formation of cultured primary human megakaryocytes. *British Journal of Haematology* **135**, 554-566, doi:10.1111/j.1365-2141.2006.06341.x (2006).

2 Giammona, L. M. *et al.* Mechanistic studies on the effects of nicotinamide on megakaryocytic polyploidization and the roles of NAD+ levels and SIRT inhibition. *Experimental Hematology* **37**, 1340-1352.e1343, doi:10.1016/j.exphem.2009.08.004 (2009).

(623h) Nucleocytoplasmic Translocation of Proteins Detected By a Novel Flow Cytometry Assay Without Imaging

Cao, Z., Virginia Tech
Geng, S.,
Li, L.,
Lu, C., Virginia Tech
Originally presented on: 11/7/2013 10:36:00 - 10:54:00

Nucleocytoplasmic transport (the protein translocation between the cytosol and the nucleus) is often regulated via upstream signal transduction pathways that lead to protein modifications. As expected, alterations in any of the regulatory steps can result in the mislocalization of a protein that may lead to cancers. [1] Most traditional research on nucleocytoplasmic translocation focuses on population-average assays (e.g. immunoblotting), which mask the distinct responses by cellular subpopulations. Recent studies with live cell imaging can achieve single cell resolution and track signal dynamics for a period of time with fused fluorescent protein markers. However, intracellular signaling can be changed by the fusion of fluorescent protein markers [2]. Furthermore, the low number of cells interrogated by imaging-based analysis does efficiently reflect population statistics [3, 4]. In contrast, flow cytometry provides both single cell resolution and high throughput. However, traditional protocol with flow cytometry lacks subcellular resolution (i.e. only the fluorescence from the entire cell is quantified) and does not detect nuclear translocation [3].

In this work, we demonstrate a novel and simple protocol for studying nucleocytoplasmic transport based on preferential release using classic flow cytometry combined with immunofluorescence staining. We demonstrate the proof-of-principle using the translocation of a transcription factor NF-κB. It is known that NF-κB translocates from the cytosol to the nuclear under stimulation by reagents such as TNF a. In our experiment, we permeabilized the plasma membrane using saponin which dissolved its cholesterol content. Such permeabilization preferentially allowed NF-κB in the cytosol to release out of the cells while keeping the nuclear fraction of the same protein intact. After fluorescent labeling using an antibody that was specific to NF-κB, the fluorescence intensity of cells was screened by a classic flow cytometer and the fluorescence intensity detected can be correlated with the original subcellular localization of the protein.

Our method provides a simple solution to study protein translocation at the single cell level without imaging at high speed. This approach is ideal for studying a large population of cells. With predominant localization of NF-κB being found in breast, ovary, colon, pancreas and thyroid tumor cells5], we envision that our method will provide an effective approach for a range of clinical investigations.

(626a) 13C Metabolic Flux Analysis of Xylose Metabolism in Saccharomyces Cerevisiae

Feng, X., University of Illinois at Urbana-Champaign
Zhao, H., University of Illinois at Urbana-Champaign
Originally presented on: 11/7/2013 8:30:00 - 8:48:00

Cellulosic biofuel plays an increasingly important role in sustainable energy supply and greenhouse gas emissions reduction. Among all the endeavors for producing cellulosic biofuels, one of the most promising strategies is to engineer yeast strains to utilize xylose for bioethanol production. To this end, a heterologous xylose pathway identified from fungal species is often introduced into *Saccharomyces cerevisiae* by functionally expressing xylose reductase (XR), xylitol dehydrogenase (XDH), and xylulose kinase (XKS). Despite extensive *in vitro* studies on the fungal xylose pathway, the intracellular metabolism rewiring in response to the heterologous xylose pathway remains largely unknown. In this study, we applied ^{13}C metabolic flux analysis to systemically investigate the flux distributions in a series of recombinant *S. cerevisiae* strains. It was found that the oxidative pentose phosphate pathway was active and played a dominant role in producing NADPH for the heterologous xylose pathway. The TCA cycle activity was strong and tightly correlated with the requirements of maintenance energy and biomass yield. Compared to *Scheffersomyces stipitis*, more carbon fluxes were diverted into the oxidative pentose phosphate pathway and the TCA cycle in the recombinant *S. cerevisiae* strains. Based on *in silico* simulations, reducing the maintenance energy can be an important strategy to improve ethanol production. Meanwhile, the cofactor-balanced xylose pathway could be more beneficial to the xylose-based ethanol production than its counterparts, by providing a wider range of fermentation conditions for optimal fermentation conditions. As indicated by the *in silico* simulations, the advantages of using cofactor-balanced pathway are more significant when oxygen uptake level is lower.

(626b) Optimizing the Biomass Equation of a Genome-Scale Model Using Simple Measurements Obtained By Raman Spectroscopy

Senger, R. S., Virginia Tech
Nazem-Bokaee, H., Virginia Tech
Originally presented on: 11/7/2013 8:48:00 - 9:06:00

The utility of genome-scale metabolic flux modeling has been demonstrated over the past decade. However, it has also been shown fairly recently that the chemical composition of the cell (i.e., the biomass equation) exerts major influence on the flux results returned by flux balance analysis (FBA). Studies have also shown that flux solutions returned by 13C-labeling experiments and metabolic flux analysis (13C-MFA) do not necessarily agree with FBA results. Here, we present some of the first research designed to improve FBA predictions to match measured flux values obtained by 13C-MFA. We hypothesized that more accurate representations of the cell chemical composition in the biomass equation of a genome-scale metabolic flux model would produce flux predictions that agree with measured values. We have tested this hypothesis with two approaches. The first involves the use of Raman spectroscopy to obtain critical cell composition data that can be used in the biomass equation of a genome-scale model. Raman spectroscopy is advantageous for this application because (i) data can be obtained in near real-time, (ii) minimal sample preparation is required, and (iii) a single Raman spectra contains a snapshot of the entire chemical composition of a single cell or entire culture. The second approach involves non-linear optimization of the biomass equation by minimizing uptake/secretion processes not observed in the laboratory but required by the model to satisfy the mass balance. Using these techniques, we successfully modeled the growth of the cellulose utilizer *Clostridium cellulolyticum* ATCC 35319 through several distinct growth phases using a newly developed genome-scale model for this organism. We have also reconciled FBA predictions with 13C-MFA measurements in *Escherichia coli* K12. This research demonstrates the discrepancies that arise when using approximated biomass equations in genome-scale modeling. The new tools presented here are applicable to all genome-scale models and provide researchers with a means of minimizing the bias in flux predictions arising from sub-optimal biomass equations.

(626c) Thermodynamics-Based Significance Ranking of Candidates for Metabolomics

Kiparissides, A., Imperial College
Hatzimanikatis, V., Swiss Federal Institute of Technology of Lausanne (EPFL)

Originally presented on: 11/7/2013 9:06:00 - 9:24:00

Cellular metabolism is the driving force of all living cells and organisms. Numerous research projects undertaken over the past century have attempted to decipher the function and operating principles of metabolism with varying degrees of success. A wide spectrum of scientific disciplines have studied metabolic function and regulation from a variety of perspectives including, Biology, (Bio-) Chemistry, (Bio-) Engineering, Mathematics, Analytical Chemistry, Physics and Control Theory. These efforts have led to the development, amongst others, of a number of high throughput, –omics, techniques able to conduct measurements at the subcellular level and a number of accompanying mathematical and statistical tools for the refinement and evaluation of the collected information.

However the sheer amount of incoming information often obscures our vision by hiding significant systemic properties behind a multitude of seemingly less meaningful correlations. The ongoing increase in the popularity of Systems Biology over the past 15 years is a testament of the need for a systems level approach that can handle large datasets. Moreover the complete genome sequencing of an increasing number of cell types has increased the availability of genome scale metabolic reconstructions further promoting the expansion of the "systems" approach. Genome scale models, comprising the entirety of the metabolic reactions occurring within the cell, provide a system's level overview of central carbon metabolism and offer valuable insight regarding optimal resource allocation towards a desired macroscopic behavior, usually maximizing the production of a valuable metabolite. Proper integration of experimental data to genome scale models still poses a significant challenge both from a data collection and a computational point of view.

Integration of experimental measurements of intracellular metabolite concentrations in genome scale models restricts the thermodynamically feasible flux space and reduces uncertainty regarding the net outcome of by-directional reactions. By combining Thermodynamics based Flux Balance Analysis (TFBA), Marcov Chain sampling, Experimental Design and Global Sensitivity Analysis we present an efficient algorithm to quantify the effect intracellular metabolites have on the thermodynamic flexibility of cellular metabolism. Metabolites are ranked based on the extent by which they reduce the thermodynamically feasible flux space when fixed at any value within their respective feasible bounds. The proposed methodology effectively defines the minimal amount of experimental information required to precisely describe the state of cellular metabolism (i.e the flux distribution) by providing a ranked list of targets for metabolomics.

(626d) Improving Glycosylation Yield in Escherichia Coli Using An Adapted Genome-Scale Metabolic Model and Metaheuristic Optimization Techniques

Wayman, J. A., Cornell University
Mansell, T. J., Johns Hopkins University
DeLisa, M. P., Cornell University
Varner, J. D., Cornell University
Originally presented on: 11/7/2013 9:24:00 - 9:42:00

Protein glycosylation, the post-translational covalent attachment of oligosaccharide groups to specific amino acid residues, is required of many therapeutic proteins. This modification affects various protein properties including pharmacokinetic activity and immunogenicity. Currently, eukaryotes possessing native glycosylation machinery serve as the preferred production host of therapeutic glycoproteins. The discovery of bacterial glycosylation in the pathogen *Campylobacter jejuni* and the subsequent transfer of these pathways into *E. coli* has spurred interest in producing non-native glycans in a more genetically tractable host. Recently, human-like glycan production was achieved in *E. coli* with the recombinant expression of a synthetic glycosylation pathway. Producing glycoprotein from prokaryotic hosts continues to suffer from several limitations including insufficient yield. Glycosylation in *E. coli* partly depends upon the available pool of nucleotide sugars making up the desired oligosaccharide. Removal of competing metabolic pathways and bottlenecks that diminish this pool may improve glycosylation efficiency. The design of strains displaying improved glycoprotein production may be aided greatly by the incorporation of a metabolic network modeling strategy. Genome-scale metabolic reconstructions of industrial microbes are commonly used to identify genetic perturbations that will produce a desired biochemical production phenotype. In this study, we modify the existing genome-scale *E. coli* model iAF1260 (1260 open reading frames) to include a variety of glycosylation pathways, including those for *C. jejuni* and human-like glycans. Added reactions include the biochemical transformations associated with glycan biosynthesis and flipping into the periplasm, as well as the expression and glycan conjugation of a target protein. The adapted network consists of 3601 reactions, 1290 open reading frames, and 2233 species segregated into cytosolic, periplasmic, and extracellular compartments. Optimal gene knockout designs were identified by solving a bilevel optimization problem which sought to couple glycan production to a cellular growth objective. For a given set of gene knockouts, flux balance analysis was used to calculate the flux distribution that optimized the cellular objective of maximizing growth rate. A genetic algorithm was applied to generate strains and select those that maximized the engineering objective, a function of glycan production. This approach has the advantage of generating strains with variable number of knockouts and has the ability to accommodate non-linear engineering objectives. We discuss characteristics of strains that couple growth to glycan production and explore metabolic engineering strategies that may improve glycosylation efficiency in a recombinant prokaryotic host.

(626e) Improving UDP-Sugar Precursor Availability for Heparosan Biosynthesis in Escherichia Coli

Cress, B., Rensselaer Polytechnic Institute
Linhardt, R. J., Rensselaer Polytechnic Institute
Koffas, M. A. G., Rensselaer Polytechnic Institute
Originally presented on: 11/7/2013 9:42:00 - 10:00:00

E. coli Nissle 1917, an archetypal probiotic agent, naturally produces and secretes heparosan as the primary component of its cell capsule. Heparosan is an unsulfated polysaccharide precursor to the glycosaminoglycan heparin, the oldest and most successful polysaccharide pharmaceutical of the 20th century that is still in widespread use today. Due to an international crisis stemming from adulteration of the feed stream of this highly sulfated anticoagulant pharmaceutical, increasing costs and concern over quality control led our group to develop a process for producing commercially relevant quantities of bioengineered heparin starting from *E. coli* derived heparosan. Supplementing *E. coli* cultures with the nucleotide activated sugar precursors UDP-*N*-acetylglucosamine and UDP-glucuronic acid improved heparosan titers, suggesting that intracellular availabilities of these valuable metabolites were limiting. We have recently sequenced the genomes of Nissle 1917 and related strain K5 and constructed genome-scale models enabling computational predictions for simultaneous gene overexpressions, downregulations, and deletions to improve availability of UDP-sugar precursors. In conjunction with metabolic remodeling, transcription of the heparosan biosynthetic pathway was modulated using ePathBrick, a combinatorial pathway assembly method developed in our lab to evaluate the effect of regulatory control element configuration. We demonstrate that coupling computational modeling with transcriptional modulation is an effective strategy to explore metabolic space for production of microbial polysaccharides. Engineered strains capable of generating high intracellular concentrations of UDP-activated sugar donors will be critical for *in vivo* glycosylation of natural products and for production of other medically important glycosaminoglycans like chondroitin sulfate, dermatan sulfate, keratan sulfate, and hyaluronan.

(626f) A Systems Biology Approach to Quantify the Metabolism of Escherichia Coli for Generating 3-Hydroxypropionic Acid From Glycerol

Lee, K., University of Michigan
Huang, Z., Villanova University
Originally presented on: 11/7/2013 10:00:00 - 10:18:00

A Systems Biology Approach to quantify the metabolism of *Escherichia coli* for Generating 3-Hydroxypropionic Acid from Glycerol

Kilho Lee, Zuyi (Jacky) Huang

Abstract

3-hydroxypropionic acid (3-HP; $C_3H_6O_3$? MW 90.08) has the potential for a broad range of industrial applications like deriving acrylic acid, 1,3-propanediol, methyl acrylate, propiolactone, malonic acid, acrylamide, and hydroxyamides (Paster et al., 2003; USITC, 2008). However, 3-HP production via chemical synthesis is not commercially feasible due to numerous technological and environmental concerns (Wolff, 1991). Instead, genetically modified *Escherichia coli* strains have been engineered that produce 3-HP from glycerol (Rathnasingh et al., 2009). Due to a significant increase in the amount of crude glycerol generated as a by-product during the biodiesel production process, the market value of crude glycerol is low and its value is expected to further diminish in correspondence to its increased supply. Therefore, a comprehensive understanding of the recombinant *E. coli* strains to further enhance the production level and provide a cost effective strategy for producing 3-HP. Rathnasingh et al., 2009 has reported that two recombinant *E. coli* strains, SH-BGA and SH-BGK, has capability of successfully converting glycerol into 3-HP. Both strains of *E. coli* were genetically modified to express glycerol dehydratase (DhaB) and glycerol dehydratase reactivase (GDR). The protein expression of DhaB and GDR enables the conversion from glycerol to 3-hydroxypropionaldehyde, or 3-HPA. SH-BGA contains aldehyde dehydrogenase (ALDH), while SH-BGK includes a-ketoglutaric semialdehyde dehydrogenase (KGSADH). Both ALDH and KGSADH converts 3-HPA into 3-HP. The experimental approach demonstrated biological 3-HP production from glycerol feedstock. However, the production level of 3-HP is not high enough to be commercially or industrially feasible. Since the metabolism of *E. coli* depends on hundreds to thousands of highly-interacted metabolic reactions, a mathematical modeling approach is required to identify the reactions that can be used to further engineer *E. coli* for a higher yield of 3-HP. Â

In this work, we developed the first mathematical model to quantify 3-HP produced by engineered *E. coli* strain from glycerol. The model can be further used to suggest experimental strategies, such as gene-knockout, for increasing 3-HP production level. The proposed approach is based on a previously published

E. coli kinetic model that includes glycerol metabolic network (Cintolesi et al., 2011). This kinetic model consists of the glycerol fermentation pathway with ethanol as the primary product. The model has been extended to incorporate the metabolic reactions that produce 3-HP from glycerol. Specifically, three reactions have been added into the model presented by Cintolesi et al., 2011: 1) glycerol dehydratase dependent reaction that converts glycerol to 3-HPA, 2) aldehyde dehydrogenase dependent reaction that converts 3-HPA to 3-HP, and 3) a reaction converting 3-HPA to 1,3-propanediol. 1,3-propanediol, or 1,3-PDO, is a by-product that is produced in a significant amount due to the accumulation of NADH from 3-HPA (Rathnasingh et al., 2009). Differential equations for the newly added metabolites were developed based on Michaelis-Menten kinetics and added to the model.

Appropriate experimental data were provided by Dr. Park's group at Pusan University for their initially developed *E. coli* strains. The data were used to estimate the newly added parameters via a nonlinear least square approach. The outputs of the estimated model match the experimental data for the glycerol consumption and the production of several metabolites, such as 3-HP and 1,3-PDO. Based upon the developed model, sensitivity analysis was conducted to identify the metabolic reactions that have influence the 3-HP production. Furthermore, we analyzed the flux change of each enzymatic reaction in the glycerol fermentation pathway for *E. coli* to produce 3-HP. It was found that the fluxes through the pathway from glycerol to pyruvate were decreased in the engineered *E. coli* that can produce 3-HP.

This work presents a computational approach to quantify the metabolism of a genetically modified *E. coli* strain that produces non-native metabolite, 3-HP. In-depth experimental analysis of a recombinant *E. coli* is often difficult to carry out due to technical limitations as well as the viability of the microorganism. The proposed modeling approach can overcome these limitations and provide valuable information on how to enhance the 3-HP production level that can extend to the experimental strategies.

References

Cintolesi, A., Clomburg, J. M., Rigou, V., Zygourakis, K. and Gonzalez, R. 2012, Quantitative analysis of the fermentative metabolism of glycerol in *Escherichia coli*. Biotechnol. Bioeng., 109: 187?198.

Paster M, Pellegrino JL, Carole TM. 2003. Industrial bioproducts: Today and tomorrow, US DOE report. http://www.brdisolutions.com/pdfs/ BioProductsOpportunitiesReportFinal.pdf.

Rathnasingh, C., Raj, S. M., Jo, J.-E. and Park, S. 2009, Development and evaluation of efficient recombinant *Escherichia coli* strains for the production of 3-hydroxypropionic acid from glycerol. Biotechnol. Bioeng., 104: 729?739.

USITC. 2008. Industrial biotechnology: Development and adoption by the U.S. Chemical and biofuel industries, Investigation No.332?481.Publ.4020. July 2008. http://hotdocs.usitc.gov/docs/pubs/332/pub4020.pdf.

Wolff GT. 1991. Kirk-Othmer encyclopedia of chemical technology. In: Kroschwitz J I, Howe-Grant M, editors. Air pollution. Vol. 1, 4th edn. New York: Wiley. p 711?749.

(626g) Optimizing Metabolic Dynamics for Increased Metabolite Production

Sowa, S., University of Texas at Austin
Baldea, M., University of Texas
Contreras, L. M., University of Texas at Austin
Originally presented on: 11/7/2013 10:18:00 - 10:36:00

Optimizing Metabolic Dynamics for Increased Metabolite Production

Steven Sowa1, M. Baldea2, L. Contreras2

1Microbiology Graduate Program, University of Texas at Austin, 100 E. 24th Street, STOP A6500 Austin, Texas 78712
2McKetta Department of Chemical Engineering, University of Texas at Austin, 200 E. Dean Keeton St., Stop C0400, Austin TX 78712
Abstract

The development of in silico metabolic models has received extensive attention in the past decades. Static models, which are at present well understood and have been extensively validated for several organisms, have proven effective at screening and identifying genetic knockout strains that maximize production of a desired metabolite. More recent research efforts have focused on constructing and validating dynamic models, capable of predicting the transient behavior of specific pathways or entire organisms.
In the present work, we describe a new avenue for exploiting the newly acquired knowledge on dynamics of cellular metabolism. Specifically, we propose to tune the rates of metabolic reactions by varying relevant enzyme concentrations over time, with the objective of maximizing production of targeted metabolites. Equivalently, we seek time-varying enzyme concentration levels that can direct a wild-type cell towards optimal metabolic state on par with those that are engineered via traditional metabolic engineering approaches (e.g., gene knockouts). We formulate this concept as a dynamic optimization problem over a finite time horizon, using the total production of the metabolite of interest as the objective function, and the time-varying concentrations of the cellular enzymes as decision variables. We propose solving this problem using a feasible path approach, discretizing the decision variables using either piece-wise continuous functions or sinusoids with user-defined periods and amplitudes.
We illustrate this theoretical framework using an established kinetic model of Escherichia coli central carbon metabolism (Chassagnole 2002). We demonstrate that the production of phosenolpyruvate (PEP) can be significantly increased by optimizing specific sets of time-dependent enzyme levels. The optimal enzyme levels vary periodically in time, and we link this finding to classical results in chemical reaction engineering (Kevrekidis 1986, Ozgulsen 1993), which indicate that the time-integral of the quantity of a reaction product can be increased by operating the reactor in an unsteady regime. We conclude by discussing an experimental

approach for tuning existing cellular regulatory networks and by validating these theoretical results.

Chassagnole, C., Noisommit-Rizzi, N., Schmid, J. W., Mauch, K. & Reuss, M. Dynamic modeling of the central carbon metabolism of Escherichia coli. Biotechnology and Bioengineering 79, 53-73, doi:10.1002/bit.10288 (2002).
I. G. Kevrekidis, L. D. Schmidt, and R. Aris, Some common features of periodically forced reacting systems. Chemical engineering science 41, 1263 (1986).
S. J. K. Ozgulsen, A. Cinar, Nonlinear predictive control of periodically forced chemical reactors. AIChE J. 39, 589 (1993).

(626h) Kinetic Optforce: Integrating Kinetics With FBA for Strain Design

Chowdhury, A., Pennsylvania State University
Zomorrodi, A. R., The Pennsylvania State Univeristy
Maranas, C. D., The Pennsylvania State University
Originally presented on: 11/7/2013 10:36:00 - 10:54:00

In this talk we will introduce Kinetic OptForce, building upon our previous stoichiometric-based strain design protocol, OptForce, by integrating available kinetic models of metabolism with constraint-based models to sharpen intervention strategies in the enzyme and reaction levels for improving the production of the chemical of interest. Incorporation of the kinetic information contracts the feasible space for reaction fluxes, thereby leading to more streamlined identification of reactions that must be altered from their reference state in the face of overproduction target (i.e., MUST sets). The MUST sets are subsequently used to identify a minimal set of interventions, comprising of enzymatic changes (for reactions with available kinetic information) and flux alternations (for reactions lacking such information), forcing the network to meet a pre-specified overproduction target (i.e., FORCE set). This procedure was used first to study the overproduction of L-serine in *E. coli*. Kinetic OptForce revealed interventions necessary to remove the substrate-level inhibition of key enzymes in the central carbon metabolism to enhance the flux flow towards L-serine, which regular OptForce relying solely on stoichiometry cannot identify. In addition, Kinetic OptForce points at the cause of infeasibility (usually concentration bound violation) of intervention strategies found by regular OptForce. For the second case study, we explored the overproduction of triacetic acid lactone (TAL) in *S. cerevisiae*. Here a higher yield of TAL was predicted with fewer interventions using Kinetic OptForce. In this example kinetics are consistent with the flux redirections needed for overproduction. Preliminary experience with the use of kinetics in strain design reveals that both the number and nature of interventions are altered. Interventions tend to cause less dramatic rearrangements of metabolism so as not to violate concentration bounds. Importantly, the impact of interventions on concentration pools is quantified and additional model improvements can be gleaned by contrasting with experimental results.

(636a) Development and Comparison of Computer-Aided Methods for the Optimal Design of Lyophilized Protein Formulations

Roughton, B. C., University of Kansas
Camarda, K. V., University of Kansas
Topp, E. M., Purdue University
Originally presented on: 11/7/2013 8:30:00 - 8:50:00

Protein drugs are increasingly important, both therapeutically and commercially. Of the top 100 drugs by U.S. sales in the fourth quarter of 2012, 28 were protein drugs or other biologics. Protein drug products are often lyophilized to ensure stability and prolong shelf-life. Lyophilization, or freeze-drying, aims to produce an amorphous solid with minimal water content. Despite the benefits to chemical stability, the lyophilization process itself can cause protein aggregation in the final drug product. Protein aggregation is undesirable due to product loss during manufacturing, reduced efficacy and increased potential for adverse immune response in patients. Formulation aims to minimize aggregation through the inclusion of excipients or additives. Excipient selection decisions are often based on prior history and/or generate-and-test methods. Neither approach guarantees that a best or even good solution will be found. Computer-aided molecular design (CAMD) offers a methodology for the rational selection of excipients based on the optimization of target formulation properties of a lyophilized drug product. The work here focuses on the properties and design of carbohydrate excipients, which represent the most common class of excipients in lyophilized protein products.

CAMD requires predictive property models for all target excipient properties. Properties of interest include glass-transition temperatures, water content in the lyophilized solid and contribution to protein stability. A statistical framework is presented for descriptor selection and cross-validation of property models to insure that the final developed model correlates well with known data and provides reasonable prediction accuracy for new data. Both linear and non-linear property models are considered. For linear models, descriptor selection is performed using Mallow's Cp statistic and the model quality is assessed using R^2 and Q^2. For non-linear models, descriptor selection is guided by parameter sensitivity analysis and model quality is assessed using %AAD and reduced χ^2. Prediction accuracy for a designed excipient is quantified through the use of prediction intervals. For further validation of property models, prediction accuracy is compared to experimental error.

Given a set of predictive property models, CAMD next requires a solution method. Stochastic methods are considered here due to the presence of non-linearities in descriptors and property models. Stochastic methods are also desirable due to the ability to quickly generate locally optimal solutions. Due to error in experimental measurement and property prediction, locally optimal solutions may have statistically similar values and may be statistically similar to the global optimal solution. Thus, multiple solutions may be equally valid and worthy of further consideration. Prediction intervals are used to compare solutions. The two stochastic methods used are Tabu search and genetic algorithms. Tabu search is a metaheuristic method that uses memory of previous solutions to enhance the search for a new solution. Genetic algorithms use the principles of evolution to guide an initial population of solutions to optimality. Both methods are tuned for the CAMD application and compared to determine if one is better suited as a stochastic CAMD solution method. Comparison results are presented specifically for the excipient design application and also in general for CAMD methods. After tuning, the stochastic methods are used to design carbohydrate excipients for several example proteins. Results between methods and proteins are compared to assess the performance of the CAMD method. The final CAMD method provides a novel lyophilized formulation design tool for the selection of optimal carbohydrate candidates given a protein drug of interest. The final design candidates can be further tested post-design through molecular modeling and/or experimental analysis.

(636b) System Identification and the "Omics" - Challenges in the Time of Big Data

Shoemaker, J. E., Japan Science and Technology Agency
Kawaoka, Y., Japan Science and Technology Agency
Kitano, H., Systems Biology Institute
Originally presented on: 11/7/2013 8:50:00 - 9:10:00

System Identification and the "Omics" – Challenges in the Time of Big Data

Network and systems biology combined with "omics" technologies, e.g. genomics and proteomics, have often been cited as the key to understanding complex diseases and changing the fundamentals of drug discovery from large scale screening processes to streamlined, predicted protein targeting. But the current approaches to model identification are not well suited to handle big data as they attempt to overlay small subsets of the data on to incomplete models that have been mined from literature. Here, we present a novel approach to modeling biological systems in which we attempted to reconstruct the innate immune response using only gene expression data. From the gene expression in the lungs of mice infected with influenza virus, we constructed modules of co-expressed genes which on further analysis were found to represent distinct immune-related pathways and functions. Each module was then considered to represent the output of its associated pathway/function and the gene expression was summarized with the scaled average of the expression of all genes in the module. We then applied a collection of network inference algorithms, e.g, time-delay ARACNE, to the modules to predict relationships between them and compared the resultant networks to the expectations of immunology experts. Interestingly, we were able to accurately reconstruct several aspects of the immune response as well as identify some unexpected trends. Key events such as the timing and magnitude of interferon production and the influx of macrophages and natural killer cells were all accurately predicted from gene expression data. The most counterintuitive prediction that was later validated by FACS analysis was that the number of B cells decline early in the infection. Of the network inference algorithms applied, the time-delay ARACNE performed the best by identifying events known to occur early during an infection and correctly identifying interferon's regulatory role in leukocyte migration. In conclusion, the work presented here supports the use of omics data to create medically relevant biological models as an alternative to time consuming approaches such as literature mining. This work is now being extended so as to construct a complete, dynamic mapping between a large spectrum of pathogens and our body's immune response.

(636c) Advanced Modeling of Oral Drug Delivery -- Innovation in Tablet Design

Pavurala, N., Virginia Tech
Achenie, L. E. K., Virginia Polytechnic Institute and State University
Originally presented on: 11/7/2013 9:10:00 - 9:30:00

Advanced modeling of oral drug delivery -- innovation in tablet design

Naresh Pavurala, Luke E.K Achenie

Department of Chemical Engineering, Virginia Polytechnic Institute and State University, Blacksburg, Virginia 24060

The oral administration route is by far the most common way of administering pharmacological substances. In oral drug delivery the challenge is for the drug active to be released in a controlled way from the solid tablet into the blood circulatory system in order to maintain the therapeutic blood plasma concentration levels following a time schedule. Good computational models are needed to predict the drug release and transport into the blood plasma. This would help in reducing the expense, time and effort involved in drug design. Several models are available in literature describing either the release of drug from the tablet into the gastrointestinal (GI) tract or the transport of drug through GI tract into the blood plasma. Our first objective was to develop a composite computational model for an oral drug delivery system (involving a drug active within a solid polymer matrix), which describes the release as well as the transport of drug from the tablet through GI tract into the blood plasma.

We employ a modified form of the drug release model proposed by Balaji and Peppas [1] based on the following assumptions. As the polymer matrix comes into contact with bodily fluid (modeled as water) inside the digestive system, it penetrates the polymer matrix; the latter then swells forming a gel layer. Drug molecules are released through the gel layer into the digestive system, crossing the enterocyte membrane into the blood stream. In our strategy, we envision two stages in the transport of drug into the blood. Therefore we couple the drug release model with a compartmental absorption and transit model (CAT), see for example Yu and Amidon [2]. The drug release model (Stage 1) is a moving boundary problem consisting of partial differential equations (PDEs) and ordinary differential equations (ODEs) which describe the diffusion of (i) bodily fluid into the drug tablet; (ii) drug out of the tablet; and (iii) polymer chains through a boundary layer. The model accounts for two moving interfaces (namely, tablet/gel interface and gel/bodily fluids interface in the digestive system).

In the CAT model (Stage 2), the GI tract is divided into a number of compartments representing the stomach, small intestine, colon and the enterocyte membrane. These are represented as a system of ODE's with linear transfer kinetics and nonlinear metabolism/transport kinetics, which describe absorption and transit of drug through three kinds of compartments, namely unreleased, released and dissolved. The model takes into account physicochemical properties such as pKa, solubility, particle density, and permeability. It also accounts for physiological factors such as gastric emptying, intestinal transit rate, first pass metabolism, and luminal transport. Finally the model accounts for factors such as the shape and size of the tablet and the dosage.

The coupling of Stage 1 and Stage 2 would be useful for predicting the release, distribution, absorption and elimination of an oral drug through the GI tract to obtain plasma concentration profile and pharmacokinetic characteristics. The developed model can also be used to inform drug tablet design for target pharmacokinetic characteristics.

The geometric characteristics and the make-up of the drug tablet have a large impact on the drug release profile in the GI tract. Our second objective was to develop a strategy for tablet customization, namely designing the dosage form of the tablet based on a desired release profile inspired by a desired plasma concentration profile (therapeutic effect). Our efforts include the systematic identification of tablet geometry and designs that could give specific desired drug release profiles such as a constant release profile or a pulsatile release profile. We expect the proposed approach to aid in the drug development process and contribute to the development of personalized medicine.

1. Narasimhan, B. and N.A. Peppas, *Molecular analysis of drug delivery systems controlled by dissolution of the polymer carrier.* Journal of Pharmaceutical Sciences, 1997. **86**(3): p. 297-304.

2. Yu, L.X. and G.L. Amidon, *A compartmental absorption and transit model for estimating oral drug absorption.* International Journal of Pharmaceutics, 1999. **186**(2): p. 119-125.

(636d) Regulatory T-Cell Differentiation Plasticity Model

Chaar, D., Texas A&M University
Hahn, J., Rensselaer Polytechnic Institute
Jayaraman, A., Texas A&M University
Alaniz, R., Texas A&M Health Science Center
Steinmeyer, S.,
Originally presented on: 11/7/2013 9:30:00 - 9:50:00

Cells of the immune system undergo various responses that include differentiation into effector cells to combat a detected antigen. Particularly, T-lymphocytes are of interest because of their diverse range of possible immune responses. The differentiation of T-cells depends on the cytokine environment [1]. In vivo studies have established the plasticity of differentiation of T-cells to better allocate resources, reduce collateral tissue damage, and more effectively combat antigens [1-2]. T-cells can differentiate into many effector cells, but the focus of this work is upon Th1, Th2, Th17. Th1 cells produce IFN-γ and contribute to cellular immunity against microbes. Th2 cells produce IL-4, IL-5, and IL-13 where their immunity role is primarily humoral. Th17 produce IL-17 that is vital for combating extracellular bacteria and fungi. More importantly, Th17 differentiation is dependent on the induction of the orphan receptor (ROR)- γ which is stimulated by TGF-β, and IL-6 among other pro-inflammatory cytokines [1,3]. In summary, over-active Th1 and Th17 cells are responsible for organ-specific autoimmunity while Th2 cells are responsible for asthma attacks and allergies.

The immune system also regulates the strength of the response by causing a portion of the differentiated effector cells to revert to Regulatory T-cells (Treg). Studies have shown that cytokine as well as bacterial metabolite stimulation of a T-cell population has the potential of shifting the population towards Treg cells. FoxP3 is determined to be the key regulatory protein for Treg differentiation, as well as its functionality in immune suppression. This protein is utilized to identify the differentiated Treg cells among the other effector cells. Being able to control the population composition of the immune system cells has important therapeutic applications that include control over infections and combating of autoimmune diseases through preventing the effector T-cells from being overly active.

This research proposes a differentiation model of T-cells upon simulation with different combinations of IL-2, IL-4, IL-6, TGF-β, and indole. The model essentially fits the change in the T-cell population composition in response to external cytokine stimulation. As the biological system is not well understood, the prediction of how the composition of the T-cells will shift in response to cytokine stimulation is a non-trivial problem. More importantly, it is even more challenging to predict the population after exposure to a combination of cytokines which are very likely to interact and influence the outcome of the population.

Therefore, a data driven approach to this complex biological system is necessary to develop a quantitative understanding of the mechanisms at work that govern the differentiation plasticity of T-cells. By constructing an experimental dataset that transverses a variety of cytokine concentrations and combinations, the resulting model parameters are capable of predicting the response of the cell population to an unfamiliar cytokine stimulation more robustly.

The levels of FoxP3 and IL-17 within a cell population were measured. Since Treg cells predominantly express FoxP3, while Th17 cells express IL-17, the presence and detection of those proteins as well as their relative compositions is translated to reflect the T-cell population composition that results in the detected cytokine profile. A neural network model was used to capture the effect that exposure to the various stimulants has on composition of the population. Neural networks were chosen based on their flexibility for fitting data and their ability to accommodate a large degree of nonlinearity within the model [4-5]. However, it is important to note that an inherent challenge in neural networks is ensuring that the model does not over-fit the data through over-parametrization. The procedure used is that the developed model receives the concentration of the cytokines as input, and predicts the T-cell population compositions of Th17 and Treg. The network is trained using a Bayesian Regulation algorithm to generalize the model to all the training data. This method inherently embodies the task of avoiding over-parameterization of the model [6]. Different network architectures are constructed that include different transfer functions within the neurons of the network. The training data is applied to the different network architectures, and the output of the networks is analyzed. In addition, the network models are also challenged to predict the T-cell population upon input of cytokine concentration combinations that the model has not been trained with. The models are then compared and based on their overall performance with respect to the training dataset, the testing dataset, and their complexity in terms of number of parameters.

The identified model contains a total of 22 parameters, and 4 processing neurons–2 neurons within each hidden layer–that are able to capture the nonlinearity and correlate the inputs and outputs from the dataset. The model suggests an optimal, indole concentration that can be utilized to obtain the largest shift in the T-cell population composition to Treg cells. In addition, the model suggests a novel combination of cytokine compositions that display important dynamics in the metabolic response of the T-cells to the cytokine stimulations. This work is unique in its investigation of the effect of cytokines as well as bacterial metabolites on T-cell population and transcription factors as well as the modeling of the biological system.

Reference

[1] Zhou L, Chong M.M.W, Littman D.R. Plasticity of CD4+ T cell lineage differentiation. *Immunity* 2009;30:646-655.

[2] Josefowicz S.Z, Rudensky A. Control of regulatory T cell Lineage commitment and maintenance. *Immunity*2009;30:616-625.

[3] Opal S.M, DePalo V.A. Impact of basic research on tomorrow's medicine: Anti-inflammatory cytokines. *CHEST* 2000;117:1162-72.

[4] Qian N. Sejnowski T.J. Predicting the secondary structure of globular proteins using neural network models. *Am J Mol Biol.* 1988;202:865.

[5] de Freitas J.F.G. Bayesian methods for neural networks. *Trinity College University of Cambridge*2003.

[6] MacKay D.J.C. Bayesian interpolation. *Neural comput* 1994;4:415-419.

(636e) Effects of Scaffold Degradation and Geometrical Structure On Vascularization: A Multi-Agent Systems Approach

Mehdizadeh, H., Illinois Institute of Technology
Bayrak, E. S., Illinois Institute of Technology
Somo, S., Illinois Institute of Technology
Brey, E. M., Illinois Institute of Technology
Cinar, A., Illinois Institute of Technology
Originally presented on: 11/7/2013 9:50:00 - 10:10:00

Abstract

Modeling and simulation are used to gain a better understanding of the underlying mechanisms of biomedical systems, resulting in improved theoretical knowledge and facilitating the design of experiments. In this work, degradation of biomaterial scaffolds and its effects on scaffold vascularization have been investigated using a multi-layer agent-based model (ABM). Tissue engineering scaffolds are used as a physical support structure and regulator of cellular activities in a wide range of biomedical applications, including production of functional implants for regenerative medicine. Scaffold vascularization is of essential importance for providing the growing tissue cells within the scaffold structure with sufficient oxygen and nutrients.

The computational models investigating the effect of biomaterial scaffold architecture on vascularization assume that the scaffold structure remains constant over time and do not account for the effects of scaffold degradation on vascularization. The degradation behavior of the scaffold is important for predicting the lifetime of biomaterials. When designing a scaffold for tissue engineering, it is important that the scaffold degrades at a balanced rate to provide mechanical support so that scaffold integrity is preserved during tissue development until enough tissue has formed. As a result, the microstructure of biomedical scaffolds is characterized by parameters such as porosity, average pore size, interconnectivity, and pore shape as well as mechanical properties such as Young's modulus, which change dynamically with the degradation of scaffold. Predicting the degradation behavior of biomaterials allows optimization of the tissue engineering scaffolds.

We have included a structural-kinetic model describing the bulk degradation of PEG hydrogels in our models for vascularization of scaffolds to predict the degradation of the porous hydrogel scaffolds and study the effects of scaffold degradation on angiogenesis and tissue growth. Since the degradation model is based on fundamental parameters that are taken directly from the physical system, it would be possible to use the model for different types of hydrogels that are degraded via similar bulk erosion mechanism.

The developed framework is implemented in Java, using Repast (Recursive Porous Agent Simulation

Toolkit), which is an open-source agent-based modeling and simulation platform. Software agents are developed as independent computational entities to represent endothelial cells (ECs). EC agents interact with their neighbors and with their microenvironment, leading to formation of new capillaries and invasion into deeper parts of scaffolds. As the scaffold degrades, more space becomes available for EC invasion, facilitating capillary ingrowth.

An embedded rule base governs the behavior of individual EC agents. EC agents are capable of sensing their microenvironment to perceive the location of neighboring agents as well as the geometry of surrounding scaffold and concentrations and gradients of the soluble and insoluble factors. They perform migration, elongation, proliferation, and sprouting, leading to capillary extension and branching. In each capillary branch, only the leading EC, referred to as tip cell, is active and capable of performing EC actions. Other ECs in each branch are referred to as stalk cells, and are only activated randomly during simulation to perform sprouting, if other conditions in their environment allow them to do so.

Three-dimensional scaffold models with heterogeneous spherical pores, designed previously in our group, have been utilized to investigate the combined effects of scaffold architecture and degradation on rate of capillary growth and invasion. Simulation results illustrate that various scaffold degradation dynamics affect scaffold vascularization in a complex manner. These results are used to identify the optimal combination of scaffold porous structure and degradation dynamics, yielding improved scaffold vascularization. It is observed that the effect of scaffold degradation becomes more significant at lower porosities and pore sizes. Experimental studies are being undertaken to investigate these findings. Controlling scaffold vascularization through modulating EC behavior with changes in the characteristics of the extracellular environment enables development of improved functional replacement tissues. The model results help us better understand the complex interactions between the growing blood vessel network and a degrading scaffold structure and identify optimal combinations of geometric and degradation characteristics of tissue engineering scaffolds.

(636f) Thrombogenesis & Intra-Clot Molecular Transport: Modeling Vs Reality

Voronov, R. S., University of Pennsylvania
Welsh, J., University of Pennsylvania
Stalker, T. J., University of Pennsylvania
Brass, L. F., University of Pennsylvania
Diamond, S. L., University of Pennsylvania
Originally presented on: 11/7/2013 10:10:00 - 10:30:00

The assembly of platelet deposits and fibrin polymerization results in over one million heart attacks and strokes each year in the US. Conversely, deficiencies in these processes result in bleeding risks that confront surgeons on a regular basis. Therefore, a unified model explaining these phenomena would provide a greater understanding of thromboembolism and coagulopathy, as well as allow for better risk assessment and enhanced therapeutics. Unfortunately, no such model exists to this day. This talk will give an overview of the disconnect that exists between the current thrombogenesis modeling methods and experimental observations. Additionally, it will summarize the results of a three year study of thrombus formation hemodynamics and intra-thrombic molecular transport. The investigation utilized fluorescently labeled intravital confocal microscopy coupled with laser injury experiments in mice, as well as Lattice Boltzmann Method fluid dynamics / Lagrangian Scalar Transport image-based simulations. The reported findings are expected to contribute to the development of future thrombogenesis and drug delivery simulation methods.

(636g) Whole-Brain Computational Fluid Dynamic Analysis

Marrinan, T.,
Linninger, A., University of Illinois at Chicago
Originally presented on: 11/7/2013 10:30:00 - 10:50:00

Whole-Brain Computational Fluid Dynamic Analysis

The human brain is a large, metabolically sensitive organ requiring an uninterrupted supply of oxygen delivered through the blood stream. In order to properly perfuse cerebral tissue with oxygen, an extensive vascular network encompasses the brain. Cortical blood supply starts at the base of the brain with three major arteries, the left and right internal carotid artery (ICA) and the basilar artery (BA). These three arteries combine to form the Circle of Willis, which is a cyclic structure capable of redistributing blood throughout the brain in the case of occlusions or stenoses [1,2,3,4]. Six main cerebral arteries branch out from the Circle of Willis – left and right anterior, posterior, and middle cerebral arteries. Each of these arteries branch as they leave the Circle of Willis and travel along the gyrations of the cortical surface to create pial networks in independent territories of the brain. The pial vessels then penetrate the surface of the brain to form column-like microvascular structures that perfuse local regions with oxygen [5,6,7]. In a reverse process, blood is drained through the veins as they converge to the confluence of sinuses.

In order to detect morphologically consistent vascular architecture and blood flow properties of a specific patient, physicians commonly use non-invasive imaging modalities. These modalities produce three-dimensional images and structural reconstructions that can be visualized using volume and surface rendering respectively to facilitate interpretation by physicians. Various imaging modalities are used in clinical practice depending on what physiological properties are deemed relevant. In order to gather information regarding blood vessels, either Computed Tomography Angiography (CTA), Magnetic Resonance Angiography (MRA), or Digital Subtraction Angiography (DSA) is utilized. CTA is fast to acquire, MRA provides flow measurements, and DSA can be performed during surgeries. However, performing multiple scans on a patient is not feasible due to the limited resources in emergency centers and the time it takes to obtain a single medical image. Because of these limitations, developing a computational model of the vasculature that can accurately simulate physiological phenomena would be useful to physicians when investigating blood flow dynamics and devising a surgical plan.

While the expansive field of intracranial vasculature simulation has made great progress in the area of constructing computational fluid dynamics models capable of predicting hemodynamic patterns [8,9], many of these models do not address the flow dynamics in an entire, patient-specific brain. Models that treat the brain as a lump Windkessel or Balloon type model [10] are useful for predicting the aggregate behavior of the brain but are unable to address the growing need for quantitative predictions of cerebral blood flow, blood velocity, pressure distributions and angiographic transit times. In order to make these predictions, a complete cerebral vasculature model is needed that captures large arteries and veins as well as the microvasculature.

Based on images and reconstructions output from medical scanners, centerline extraction and radius detection can be performed on the blood vessel structure to represent a network of tubular objects [11]. We then perform data reduction and smoothing in order to limit the number of segments necessary to store accurate information about a blood vessel and to enforce G1 (positional and tangential) continuity between vessel segments. This technique yields a morphologically consistent patient-specific model of large cerebral vessels. However, hemodynamic models of the entire cerebral vasculature require an accurate reconstruction of not only the major vessels, but also require reconstruction of smaller vessels that exist beyond the scope of angiographic resolution. Detailed morphological studies have uncovered the hidden architecture of the microvasculature that bridges the gap between the major arterial and venous structures [5,6]. Vascular growth algorithms have been proven to be an accurate representation and have been thoroughly explored and validated against measured morphological and hemodynamic studies [12]. Once a complete tubular network of the major vessels and the microvasculature is constructed, it can be used to perform dynamic blood flow simulations.

The final, and most important, step in the entire process is analyzing the resulting data. In order to display the information so that fine details can be viewed along with the context of the entire vascular structure, our patient-specific vascular models are rendered in the 72 megapixel, stereoscopic, cylindrical CAVE2™ System. This immersive environment allows neurosurgeons, bioengineers, and researchers to walk through the unique three-dimensional structure of a specific patient. This affords unprecedented ability to assess morphology, spatial relationships, and simulation results of both macro and micro vessels.

References

[1] S. Hirsch, et al., "Topology and Hemodynamics of the Cortical Cerebrovascular System," J Cereb Blood Flow Metab, vol. 32, pp. 952-67, Jun 2012.

[2] P. Blinder, et al., "Topological Basis for the Robust Distribution of Blood to Rodent Neocortex," Proc Natl Acad Sci USA, vol. 107, pp. 12670-5, Jul 13 2010.

[3] M. Brozici, et al., "Anatomy and Functionality of Leptomeningeal Anastomoses - A Review," Stroke, vol. 34, pp. 2750-2762, Nov 2003.

[4] A. Shuaib, et al., "Collateral Blood Vessels in Acute Ischaemic Stroke: a Potential Therapeutic Target," Lancet Neurology, vol. 10, pp. 909-921, Oct 2011.

[5] H. M. Duvernoy, et al., "Cortical Blood vessels of the Human Brain," Brain Res Bull, vol. 7, pp. 519-79, Nov 1981.

[6] F. Cassot, et al., "Branching Patterns for Arterioles and Venules of the Human Cerebral Cortex," Brain Research, vol. 1313, pp. 62-78, Feb 8 2010.

[7] F. Lauwers, et al., "Morphometry of the Human Cerebral Cortex Microcirculation: General Characteristics and Space-Related Profiles," NeuroImage, vol. 39, pp. 936-48, Feb 1 2008.

[8] A. Bui, et al., "Dynamics of pulsatile flow in fractal models of vascular branching networks," Med Biol Eng Comput, vol. 47, pp. 763-72, 2009.

[9] L. Grinberg, et al., "Modeling Blood Flow Circulation in Intracranial Arterial Networks: A Comparative

3D/1D Simulation Study," Annals of Biomedical Engineering, vol 39, pp. 297-309, 2011.

[10] D.A. Boas, et al., "A Vascular Anatomical Network Model of the Spatio-Temporal Response to Brain Activation," NeuroImage, vol. 40, pp. 1116-29, 2008.

[11] F. Benmansour, et al., "Tubular Structure Segmentation Based on Minimal Path Method and Anisotropic Enhancement," Int J Comput Vis, vol. 92, pp. 192-210, 2011.

[12] R. Karch, et al., "Staged growth of optimized arterial model trees," Annals of Biomedical Engineering, vol. 28, pp. 495-511, 2000.

(656a) Encapsulation of Thermotropic Liquid Crystals in Hollow Polymer Microcapsules: Design of Colloidal Liquid Crystal-Based Chemical Sensors That Function in Cellular Environments

Zayas-Gonzalez, Y. M., University of Wisconsin-Madison
Manna, U., University of Wisconsin-Madison
Carlton, R. J., University of Wisconsin-Madison
Abbott, N. L., University of Wisconsin-Madison
Lynn, D. M., University of Wisconsin-Madison
Originally presented on: 11/7/2013 12:30:00 - 12:48:00

This presentation will focus on the design of ultrathin, covalently crosslinked, and amine-reactive polymer microcapsules, and on fundamental investigations of these microscale objects as containers for the confinement and control of droplets of thermotropic liquid crystals (LCs) in aqueous environments. The goals of this work are motivated broadly by potential applications of microscale LC droplets as environmental sensors in biological contexts, and oriented specifically toward the design of hollow microscale containers that are (i) stable in physiologically relevant environments (e.g., in culture media or inside cells), and (ii) chemically reactive in ways that allow the inner and outer walls of the polymer capsules to be functionalized independently (e.g., to create microenvironments inside the capsules that interface with and direct the orientational order of the LC, and outer surfaces that confer water solubility or the ability to enter into cells or attach to their surfaces).

Our approach builds on the results of past studies demonstrating methods for the encapsulation of thermotropic LCs within "polyelectrolyte multilayer" microcapsules, but exploits new methods for the "reactive/covalent" assembly of azlactone-containing polymer multilayers developed recently by our group. We used this reactive "layer-by-layer" approach to design monodisperse, amine-reactive microcapsules (~5 micrometers in size) using a sacrificial colloidal template. Subsequent treatment of these reactive capsules with the small-molecule dimethylaminopropylamine lead to water-soluble, amine-functionalized capsules that could be filled with LC by direct immersion in mixtures of LC and ethanol. This general approach could also be used to design bi-functional capsules having hydrophobic (amine-functionalized) and hydrophilic (alkyl-functionalized) inner- and outer-walls to explore the impacts of these parameters on both capsule solubility and the defect structure of the LC contained within it.

Partial or complete filling of these permeable capsules with thermotropic LCs yielded encapsulated LC droplets with defect structures (e.g., either radial or bipolar defects) that can respond to the presence or absence of analytes (e.g., surfactants and other amphiphiles) present in surrounding aqueous phases. Our results demonstrate that the walls of the surrounding capsules are semi-permeable, and that they can be used to discriminate between model analytes (and thus tune or predict the responses of the LC droplets) based on parameters such as size and charge. These observations suggest opportunities to use these LC-filled containers as selective reporters of molecular events in complex media (e.g., by protecting the aqueous/LC interface from contact with proteins, but permitting the transport of small-molecule substrates or products to the interface).

Our results demonstrate that these LC-filled capsules can be incubated in the presence of mammalian cells without inducing cytotoxic effects observed when cells are exposed to bare (uncoated) LC droplets. Capsules functionalized with cationic outer surfaces adhere to the outer surfaces of mammalian fibroblast cells, and these cell surface-bound LC-filled capsules retain their ability to respond rapidly to the addition of cytotoxic amphiphiles to culture media. Experiments aimed at understanding (i) the influence of the surrounding capsule on the orientation of the LC (e.g., defect structures), (ii) the sensitivity of orientation of the LC to the presence of surfactant (e.g., induced changes in defect structures), and (iii) the ability of these LC-filled microcapsules to function as sensors of toxic agents in biological/cellular environments will be discussed.

(656b) Decoupling Contributions to Cell Adhesion in the Context of Laterally Mobile Ligands

Kourouklis, A., Umass, Amherst
Lerum, R. V., University of Massachusetts Amherst
Bermudez, H., University of Massachusetts Amherst
Originally presented on: 11/7/2013 12:48:00 - 13:06:00

Cells employ a set of subcellular structures including integrin receptors and focal adhesions (FAs) in order to transmit, sense, and respond the properties of their extracellular environment. Previous works attempted to understand the role of ligand mobility on cell adhesion by using receptor-bearing vesicles as a model system. Their findings show that ligand mobility enhances force-induced cell adhesion, interpreted as being due to the energy of vesicle membrane deformation. However, these model systems obviously neglect the roles of FAs, the cytoskeletal network, and receptor diversity in living cells. To improve upon these early descriptions of receptor-ligand adhesion, we have studied cell (receptor) interactions on polymer films presenting laterally mobile ligands. The polymer films were created by an interfacial self-assembly process, whereby a fraction of the polymer chains bear RGD ligands to allow for integrin-specific adhesion. We find that NIH 3T3 fibroblasts seeded on our polymer films exhibit biphasic responses in spreading and adhesion strength as a function of the ligand mobility, with a minimal response for intermediate mobility values. These results highlight ligand mobility as a key property of biologically active materials, and may also point to pathophysiological roles. Towards decoupling the contributions of cellular components such as FAs and actin cytoskeleton, we independently administer pharmacological agents (e.g., Y-27632, cytochalasin, and trypsin) to cells seeded on films with different ligand mobility. This method has previously been used to estimate the mechanical properties of adherent cells, with the rate of cell de-adhesion corresponding to cellular viscoelasticity (i.e., storage and loss moduli). Our work will give experimental insight into the contributions of both cell and substrate mechanical properties on cell adhesion in the context of mobile ligands. Furthermore, our results will assist the development of more complete theoretical representations of cell adhesion.

(656c) Large Amyloid Fibers As Robust Biomaterials

Ridgley, D., Virginia Tech
Barone, J. R., Virginia Tech
Originally presented on: 11/7/2013 13:06:00 - 13:24:00

Amyloid structures are found throughout nature and have been shown to be robust biomaterials with a specific modulus comparable to dragline silk. Amyloids have been self-assembled *in vitro* into nanometer-sized fibrils from a host of proteins at primarily denaturing conditions. We have developed a mechanism to grow amyloid fibers 10-20 μm in diameter and several millimeters in length at near physiological conditions.[1] A short, hydrophobic template peptide interacts with an α-helical adder peptide to cooperatively self-assemble into large amyloid fibers.[2] By altering the *in vitro* solution conditions and peptides in the mixture it is possible to engineer a fiber with a tailored morphology (rectangular or cylindrical cross-section) and modulus (0.1-2.5 GPa).[3] Here, we show that it is possible to encode micron-sized self-assembled structures at the genetic level. A template peptide is expressed in *Escherichia coli* and combined with an adder protein to grow amyloid fibrils (width 200-500 nm) and fibers (width 7-20 μm) in solution. Beginning at the genetic level, the system could be used to build custom protein fibers, program cells to build their own scaffold for tissue engineering, immobilize cells for bioprocessing, program cell death, or reconstruct or recompartmentalize cells.

References

1 Ridgley, D. M., Ebanks, K. C. & Barone, J. R. Peptide Mixtures Can Self-Assemble into Large Amyloid Fibers of Varying Size and Morphology. *Biomacromolecules* **12**, 3770-3779, doi:10.1021/bm201005k (2011).

2 Ridgley, D. M. & Barone, J. R. Evolution of the Amyloid Fiber over Multiple Length Scales. *ACS nano* **7**, 1006-1015, doi:10.1021/nn303489a (2012).

3 Ridgley, D. M., Claunch, E. C. & Barone, J. R. The effect of processing on large, self-assembled amyloid fibers. *Soft Matter* **8**, 10298-10306, doi:10.1039/c2sm26496j (2012).

(656d) OB-Cadherin Regulates Mesenchymal Stem Cell Differentiation Into Smooth Muscle Cells and Development of Contractile Function in Vivo

Alimperti, S., State University of New York at Buffalo
You, H., University at Buffalo, State University of New York
George, T. A., Baylor College of Medicine
Agarwal, S., Baylor College of Medicine
Andreadis, S., State University of New York at Buffalo
Originally presented on: 11/7/2013 13:24:00 - 13:42:00

Mesenchymal stem cells are multi-potent cells that can be differentiated into fat, bone, cartilage and muscle. Although soluble factors (e.g. TGF-β1) induced mesenchymal stem cell (MSC) differentiation toward smooth muscle (SMC) lineage, the role of adherent junctions in this process is not well understood. To this end, we examined how cell-cell communication controls the fate of MSC towards smooth muscle cells (SMC). Specifically, we identified the molecular mechanism of MSC differentiation into vascular fate during cell-cell contact through a molecule called OB-cadherin. We further applied this study in-vivo and we confirmed our hypothesis in a mice model. Initially, we observed that increased cell density increased adherens junction (AJ) formation and at the same time, high cell density increased expression of myogenic markers such as αSMA as well as N-cadherin and OB-cadherin. Interestingly, we found that OB-cadherin mediated SMC differentiation through activation of serum response factor (SRF) via ROCK pathway. Also, myogenic differentiation proceeded even in the absence of TGF-β receptor, while exogenous TGF-β1 failed to promote SMC differentiation in the absence of OB-cadherin. Engagement of OB-cadherin increased its own expression through SRF, suggesting the presence of an auto regulatory feedback loop that regulated SMC differentiation. Notably, SMC-containing tissues such as bladder and arteries from $OB^{-/-}$ mice showed significantly reduced levels of SMC proteins and exhibited diminished contractility and mechanical properties. To conclude, we revealed a previously unknown role of OB-cadherin based AJs in the process of SMC regeneration and shed light into the mechanism of MSC differentiation with potential implications for the development of strategies to understand and control stem cell fate decisions.

(656e) Fabrication of Chitosan-Poly(ethylene glycol) Hybrid Microparticles Via Replica Molding and Its Application Toward Facile Biomolecular Conjugation Via Copper-Free Click Chemistry

Jung, S., Tufts University
Yi, H., Tufts University
Originally presented on: 11/7/2013 13:42:00 - 14:00:00

Facile conjugation of biological molecules on robust platforms is important in many application areas such as therapeutics and diagnostics. One such example is nano and microparticles based on polysaccharide chitosan due to the abundant primary amine groups that offer covalent binding sites via nucleophilic reactions. However, poor physical properties of chitosan such as high viscosity, poor solubility in solvents and low mechanical strength due to its rigid crystalline structure have limited its utility in the fabrication of biofunctionalized platforms. While the chitosan-based particles can be formed with water-in-oil emulsion method, the shape of the particles is usually limited to spheres. Meanwhile, standard amine-reactive conjugation chemistries for the biomolecular conjugation (e.g. glutaraldehyde or homobifunctional *N*-hydroxysuccinimidyl (NHS) ester) have drawbacks such as limited stability, nonselective conjugation, and self-blocking of active amine sites by the chemical agents. In addition, reactive Cu(I) catalysts in the Cu-catalyzed click chemistry can cause structural damage to the biomolecules during the conjugation reaction.

In this work, we exploited replica molding (RM) to fabricate non-spherical, shape-controlled hybrid microparticles containing chitosan moieties for facile biomolecular conjugation in a simple, robust, and inexpensive manner. We also exploited Cu-free click chemistry between strain-promoted alkyne of dybenzylcyclooctyne (DBCO) and azide for selective and efficient conjugation of biomolecules with the chitosan-containing microparticle platforms under mild reaction conditions.

Specifically, we utilized preparticle solution composed of chitosan oligomer and poly(ethylene glycol) diacrylate (PEGDA) to fabricate chitosan-PEG hybrid microparticles with uniform and well-defined dimensions and shapes via RM. The incorporation and distribution of chitosan molecules within the PEG networks as well as the chemical reactivity of chitosan's amine groups were confirmed via fluorescent labeling. The results show that the chitosan molecules are mostly located near the particle surfaces, while particles with high PEG concentration yielded retention of chitosan molecules within the core area.

Next, the primary amines in the chitosan-PEG microparticles were exploited for conjugation of biomolecules such as single-stranded (ss) DNAs and fluorescent proteins (R-phycoerythrin, R-PE) via Cu-free click chemistry. The results show selective conjugation of ssDNAs and R-PE near the particle surface where mass transfer limitation is minimal for rapid biosensing. In this presentation, our recent results on protein conjugation kinetics of the Cu-free click chemistry will be highlighted.

We then examined antibody-conjugated chitosan-PEG microparticles as a protein sensing platform through Cu-free click chemistry. The results show that the target proteins are selectively captured with the antibody-conjugated particles, and subpicomolar quantity of the proteins can be readily detected with standard imaging conditions. In this presentation, our recent results on substantially improved antibody conjugation capacity via genetically modified tobacco mosaic virus (TMV1cys) as a nanotubular template will also be introduced.

Combined these results demonstrate a facile fabrication-conjugation strategy of hybrid bio-/synthetic polymer microparticles via RM and Cu-free click chemistry, enabling efficient use of precious biomolecules. We envision that our methodologies may find wide application areas where efficient fabrication and conjugation of precious materials are highly desired.

(660a) Development of High Throughput Methods for Mammalian Cell System

Wang, W., Massachusetts Institute of Technology
Wang, D. I. C., Massachusetts Institute of Technology
Originally presented on: 11/7/2013 12:30:00 - 12:48:00

For the past decade, biopharmaceutical industry had a fast growth, creating a hundred billion dollar market worldwide with the approval of recombinant proteins, monoclonal antibodies and nucleic acid–based drugs. Among all these production systems, the Chinese Hamster Ovary (CHO) cells are the most widely used mammalian cell systems for transfection, expression, and large-scale recombinant protein production, due to its advantage in accurate glycosylation for production of natural proteins *in vitro*. In recent CHO cell culture technology, significant improvement in recombinant protein production leading to a titer up to 10 g/L was achieved, to meet the high market demand by using stable and high producer together with the optimization of culture process. However, rapid selection for high production of clones is still needed.

In current study, we applied a new screening concept for on line and *in situ* high throughput screening for mammalian cell system. CHO cell lines producing Herceptin were used as the model mammalian cell system. By measuring the key parameters of the cell culture system on line, the performance of cell line can be predicted, and the high producers can be identified without performing invasive assays. The impact of the research will be very significant in upstream bioprocesses. In cell line development, product titer is a crucial parameter for clone selection. Our method does not require any external reporter, and it is a non-invasive liquid-based assay. We only need to monitor basic information of cell culture in order to calculate the product quantity. High and low producers can be easily identified in a high-throughput manner on-line and in situ using automated handling and aforehand programming. This will be a revolutionary change to current mammalian cell screening system.

(660b) High Throughput Screening of Cyanobacteria Cultures for Maximized PHA Production

Lane, C. E.,
Benton, M. G., Louisiana State University
Originally presented on: 11/7/2013 12:48:00 - 13:06:00

Dr. Michael Benton benton@lsu.edu Courtney Lane clane6@tigers.lsu.edu

High Throughput Screening of Cyanobacteria Cultures for Maximized Bioplastic Production

Polyhydroxyalkanoates (PHAs), biodegradable and biocompatible biopolymers used for carbon storage by many microorganisms, are receiving much attention as possible replacements for petroleum-based polymers. Typical PHA fermentation processes utilize heterotrophic organisms such as *Escherichia coli* or *Cupriavidus necator* and show high yields, approaching 80-90% dry cell weight in some cases. Still, such production remains economically uncompetitive with petroleum-based plastics due to high production costs, 50% of which can be attributed to the carbon feedstock. Cyanobacteria are phototrophic prokaryotes and have the potential to make PHA bioproduction economically viable as they can fix atmospheric carbon through photosynthesis, lessening the cost of carbon feedstock. Unfortunately, typical PHA yields in cyanobacteria are typically around 10% dry cell weight, far below the level needed for such viability. However, some atypical PHA producing cyanobacteria have been found which can accumulate up to 60% dry cell weight of PHA. Therefore, strain choice is paramount in the design of a cyanobacteria-based PHA production process and more high-yield strains of cyanobacteria must be identified if this bioprocess is to become viable. Cyanobacteria are an extremely diverse phylum ranging from simple unicellular organisms to fully differentiating vegetative colonies and span over 10,000 taxonomic entries in the National Center for Biotechnology Information database. With so many strains to categorize, a rapid, robust, and efficient screening process is needed to maximize the opportunity of identifying species capable of economically feasible PHA production. In this work, a high-throughput method to determine the genotypic and phenotypic PHA accumulation characteristics in cyanobacteria via PCR amplification and fluorescent staining was developed and tested on a diverse range of cyanobacteria: *Plectonema sp.* UTEX 1541, *Spirulina platensis* UTEX LB 2340, *Synechococcus sp.* UTEX 2434, *Synechocystis sp.* PCC 6803, and *Synechocystis sp.* UTEX 2470. Positive PHA accumulation was observed via staining and PCR amplification in PCC 6803, UTEX LB 2340, and UTEX 2470, while no signs of PHA accumulation were observed in UTEX 1541 or UTEX 2434. This novel assay will rapidly identify those cyanobacteria which could (based on culture conditions) accumulate relatively high amounts of PHA. The characterization of these outlier strains would lead to new insights and a better understanding of PHA accumulation in cyanobacteria. The atypical producers detected utilizing this test could ultimately be implemented in a more economically and environmentally sustainable cyanobacteria-based PHA bioprocess, creating a sustainable production system more competitive with petroleum-based processes.

(660c) Continuous Monoclonal Antibody Production On a Chip

Garza-Garc a, L. D., TecnolÃ³gico de Monterrey
Carrillo-Cocom, L. M., TecnolÃ³gico de Monterrey
Garc a-L pez, E., TecnolÃ³gico de Monterrey
L pez-Pacheco, F., TecnolÃ³gico de Monterrey at Monterrey
Ara z-Hern ndez, D.,
Camacho-Le n, S., TecnolÃ³gico de Monterrey
Tapia-MejÃa, E. J., TecnolÃ³gico de Monterrey
Rodr guez-Gonz lez, C. A., TecnolÃ³gico de Monterrey
Alvarez, M. M., TecnolÃ³gico de Monterrey
Originally presented on: 11/7/2013 13:06:00 - 13:24:00

An important number of experimental reports are now available on the use of micro-devices in cell culture applications, including a few examples of continuous flow micro-devices for cell culture. However, reports on the use of micro-devices for the production of recombinant proteins are still scarce. To our knowledge, no previous study has documented the production of biopharmaceuticals in continuous micro-devices.

We report a proof-of-principle for the use of micro-devices as continuous bioreactors for the production of a glycosylated pharmaceutical product (a monoclonal antibody). We culture CHO cells on the surface of PMMA/PDMS micro-channels textured with semi-spherical cavities and coated with fibronectin. Three different micro-channel geometries were tested in a continuous flow regime in the range of 0.1 to 6.0 microL/min. Both, the geometry of the micro-device and the flow rate had a significant effect on cell adhesion, cell proliferation, and monoclonal antibody production. In the most proficient configuration, we observe a steady-state productivity above 7,000 mg L^{-1} day^{-1}, at least 100 times higher than that observed in commercial scale fed-batch systems.

Immediate applications of these micro-devices are the optimization of process conditions for CHO cell production, the comparison of CHO cell clones, or the precise evaluation of productivity of a cell line.

(660d) Traversing Six Logs in Scale-Down Modeling: Two Case Studies in Developing a Scale-Down Model for the Advanced Microscale Bioreactor System From a 15,000L Production Bioreactor

Kwiatkowski, C., Biogen Idec
Huang, Y. M., Biogen Idec
Kshirsagar, R., Biogen Idec
Ryll, T., Biogen Idec
Originally presented on: 11/7/2013 13:24:00 - 13:42:00

Traversing six logs in scale-down modeling: Â two case studies in developing a scale-down model for the advanced microscale bioreactor system from a 15,000L production bioreactor

The use of TAP Biosystem's advanced microscale bioreactor system (AMBR) has gained popularity in recent years as a high throughput cell culture process development tool. The following presents two case studies in which a scale-down model was developed from 15,000L production bioreactor data and scaled down to the 15mL AMBR using specific scale-down methodologies and strategies. Multivariate data analysis was then performed on representative data across AMBR, bench top, pilot, and manufacturing scales to demonstrate comparability across the scales. We will highlight the major variations across scales and discuss their contributions to cell culture performance and product quality.Â We will also discuss the methods and strategies used to develop the scale-down model in addition to operational problems and successes using the AMBR system as a cell culture process development tool.

(660e) Systematic Development of a Scale-Down Model for a 25,000 L Fed-Batch CHO Culture Process

Xu, S., Bristol-Myers Squibb, Co.
Abu-Absi, S., Bristol-Myers Squibb, Co.
Maranga, L., Bristol-Myers Squibb, Co.
Originally presented on: 11/7/2013 13:42:00 - 14:00:00

We will present a typical process of developing and fine-tuning a 5 L bioreactor model for a 25,000 L scale CHO fed-batch process. The process was first scaled-down according to engineering principles (same power input per volume and vvm). Â We were able to reproduce most of the culture parameters in 5 L bioreactors such as specific growth rate, peak viable cell density, titer and specific productivity. However differences in certain culture indicators (such as viability and lactate metabolic shift) and critical quality attributes (such as glycoform profile) were observed between 5 L and 25,000 L bioreactors.

We carried out a systematic analysis and step-wise experimentations to identify key cell culture parameters (such as CO_2 stripping and base consumption) that contributed to the observed differences. Intrinsically, small scale bioreactors are superior at mixing and O_2/CO_2 transfer and base required for pH control is often a few fold less than that in large scale bioreactors. In experiments, cell inoculum and media were obtained from manufacturing operations to avoid culture source and materials variation. Controlled experiments showed that the glycoform profile differences at 25,000 L scale was mainly due to decrease in cell viability. We were able to fine tune the scale-down model to match the large scale culture process closely by artificially re-creating the 25,000 L osmolality trend in 5 L bioreactors using 1 N NaOH. Protein quality attributes of specific interest, glycoform profiles, were produced at the same levels as seen in 25,000 L bioreactors after the culture processes were aligned. We will discuss the impact of osmolality and pCO_2 on cell viability and protein quality attributes. Relationships between $pCO_{2,}$ lactate, osmolality and base consumption will be presented. Applications of the improved scale-down model in cell culture performance evaluations (such as the impact of cell population doubling length) will also be discussed.

(660f) Gas Transfer Characterisation Methodology to Improve Bioreactor Scale-Up/Scale-Down

Mollet, M.,
Rose, S. T., Eli Lilly and Company
Kenty, B.,
Chitta, R.,
Originally presented on: 11/7/2013 14:00:00 - 14:18:00

During cell culture process development, bench scale bioreactors are routinely used to optimize and characterize the operating conditions. These vessels can be 1:100 to 1:10000 the scale of the manufacturing vessels. Regularly we observe that the dissolved carbon dioxide concentration is lower in the scale down reactors compared to that in the manufacturing scale. This difference in carbon dioxide levels has resulted in shifts in productivity or metabolism and may impact the representativeness of, and hence the conclusions drawn from the scale-down runs. To address this gap in the scale down model, we developed a systematic and mathematical method to conserve the dissolved carbon dioxide concentration across the scales. The method uses a mass balance-based approach to characterizing the gas transfer coefficients (kLa) in bench- and intermediate- scale bioreactors that incorporates gas transfer from both the head space and sparged gas bubbles. We found that the headspace contribution to gas transfer which is significant at small scale can be predicted by using parameters such as temperature, working volume, dissolved oxygen, agitation rate, and gas flow rates. After the bioreactors are fully characterized with respect to gas/liquid mass transfer, we can make rational decisions on how to operate the bioreactors to conserve the dissolved carbon dioxide levels. Example data from a CHO-based process comparing bench-scale to production scale bioreactors will be shown.

(660g) Signal Transduction Engineering As a Powerful Tool for Enhancing Valuable Secondary Metabolite Production By Plant Cells and Higher Fungi

Zhong, J. J., Shanghai Jiao Tong University
Xu, Y. N., Shanghai Jiao Tong University
Huang, C., Shanghai Jiao Tong University
Originally presented on: 11/7/2013 14:18:00 - 14:36:00

Plant cells and medicinal mushrooms are a treasury of therapeutically bioactive agents. Ginsenosides are an important group of tetracyclic triterpene glycosides produced by *Panax ginseng*, a well-known traditional Chinese medicinal plant, and the versatile pharmacological activities of ginsenosides include hypocholesterolemic, antitumor, immunoadjuvant and anti-inflammatory functions. Ganoderic acids (GAs) produced by *Ganoderma lucidum*, a traditional Chinese medicinal mushroom, have significant pharmacological activities such as anti-tumor, anti-metastasis and anti-HIV. Both ginsenosides and GAs are synthesized through the mevalonate/isopenoid pathway, but the details of later steps in their biosynthetic pathways are yet unclear. The fact implies that construction of heterologous biosynthetic pathway of them in other hosts is not feasible at this stage, although a few di/tri-terpenoids-derived natural products have been produced in microorganisms.

Signal transduction engineering is important to improve secondary metabolite production. In this work, the effects of heavy metals and other elicitors on the ginsenoside and GAs biosynthesis in liquid cultures of *P. ginseng* and *G. lucidum* were investigated. The optimal concentration and time of vanadate addition for ginsenoside biosynthesis and Mn^{2+} addition for GAs biosynthesis were identified. The increased activities of key enzymes and up-regulation of transcriptional levels of genes encoding enzymes in the triterpenes' biosynthesis were observed. Further investigation by intracellular Ca^{2+} imaging and calcineurin inhibitor study revealed that addition of Mn^{2+} led to Ca^{2+} influx from medium to the cells to trigger calcineurin signals. The regulation mechanism of Mn^{2+} on the GA biosynthesis was considered through calcineurin signaling transduction. For the ginseng cell culture, our results indicated that vanadate stimulated the ginsenoside biosynthesis by inducing the signal molecule jasmonic acid. This work demonstrated that elicitors manipulation, as efficient induction strategy, significantly improved the production of ginsenosides and GAs in the cell cultures. The approach of signal transduction engineering proposed here is very useful for large-scale production of ginsenosides, GAs, and other important secondary metabolites in plant-cell/mushroom cultures.

(660h) Genetic and Environmental Manipulation of Somatic Embryo Formation for Plant Improvement

Curtis, M. S., The Pennsylvania State University
Florez, S.,
Maximova, S., Pennsylvania State University
Guiltinan, M., Pennsylvania State University
Curtis, W. R., The Pennsylvania State University
Originally presented on: 11/7/2013 14:36:00 - 14:54:00

We present our progress in developing technology for plant improvement based on transient genetic manipulation of the developmental process of embryo formation as well as the environmental manipulation of the bioreactor in which the tissue culture takes place.

Somatic embryogenesis (SE) is a way to make superior plants with a desired trait. SE allows the reprogramming of a somatic cell to form a new plant. This extremely complex system is tightly controlled by a network of regulating factors in the plant's signaling mechanism. This network can be tweaked by environmental changes or more recently by the introduction of transcription factors to activate or repress specific pathways. This manipulation can be done through an agrobacterium mediated gene transfer however, very precise timing and control is necessary to prevent abnormal development or the production of transgenic plants. As a result, transient expression, the introduction of a gene into the plant for a short period of time, is a critical tool that can be used to increase productivity of SE. In addition to the modified genetic environment, an optimal culture environment is essential for the successful implementation of this process at a scaled-up or commercial scale. Plant tissue bioreactors provide a way to control this environment but are often prohibitively expensive. A bioreactor system has been fabricated for plant tissue culture propagation which utilizes low-cost plastic bags combined with gravity driven flows as an approach to achieving scale-up and reducing costs. These temporary immersion bioreactors (TIB) provide for complete media-tissue contacting while minimizing boundary layer mass-transfer resistances and water-logging the plant tissue.

Using cacao (chocolate tree) as a plant model, we are mapping the gene profile expression during the process of SE. Using bioinformatics, transcription factors involved in regulation of SE have been identified and their interaction between these genes are being studied with the goal of developing a dynamic model to better understand the genes responsible for the somatic to embryo transition. Preliminary results of these genes overexpressed in Arabidopsis have shown regenerative capacity resulting in the development of new embryo like structures. These new structures developed abnormalities due to the continued expression of these transcription factors. Using a transient expression strategy however, cacao embryo production was doubled when the transcription factor Babyboom was expressed without stable integration into the genome.

Despite the success observed using transcription factors, the process remains labor intensive at a small scale.

In order to scale up the production the process must be automated while providing appropriate growing conditions. Bioreactors provide means to dynamically control the environmental conditions as well as incorporate transient gene expression. To minimize the cost of production we are developing a low cost TIB. The rationale behind the design focuses on the unique needs that come from plant propagation. The slow growth rates of plant embryos require much less stringent operating conditions than typical cell culture allowing the use of low cost disposable plastic bags as vessels allowing this control to be economically feasible. The different gas compositions required by the different stages of plant propagation can easily be adjusted by decoupling of gas and liquid flow, something that gravity driven flow makes possible. This low cost design has introduced minor setbacks including an increased risk of contamination and mechanical failure. These problems have been addressed through iterative design improvement. As a result the TIB has already been shown to improve the growth of N. benthamiana roots under elevated oxygen conditions as well as successfully grow watermelon photoautotrophically when supplemented with carbon dioxide.

Due to the growing prevalence of generic bio-products in the market, the future of bioengineering relies in increasing productivity and decreasing the cost of production. Using a transient gene expression approach to increase yield alongside a low cost TIB system provides for a plant tissue culture propagation system designed to meet these requirements.

(666a) Mechanistic Modeling of PLGA Microsphere Drug Delivery: Analytical Autocatalytic Degradation of Polymer and Hindered Diffusion of Drug

Ford Versypt, A. N., Massachusetts Institute of Technology
Braatz, R. D., Massachusetts Institute of Technology
Pack, D. W., University of Illinois Urbana-Champaign
Originally presented on: 11/7/2013 12:30:00 - 12:50:00

Controlled release drug delivery is a category of pharmaceutical administration techniques where the concentration of medicine in a patient's body is sustained for an extended period of time with a single dose. Controlled-release drug delivery devices such as polymer microspheres composed of poly(lactic-co-glycolic acid) (PLGA) are alternatives to conventional pharmaceutical dosage forms such as pills or injections that require repeated doses or complicated dosing regimens. Controlled-release drug delivery alleviates the pressure on patients to adhere to strict dosing schedules, thus improving the compliance with the prescribed treatment. Controlled-release drug delivery also can enhance the control of drug concentrations in a patient and reduce side effects associated with frequent doses, missing doses, or peaks and valleys in drug concentration.

Biodegradable PLGA microspheres have been extensively studied as drug delivery devices [1-4]. Modeling of the drug release from the devices must consider the interdependent phenomena that contribute to drug release [1,5]; the most significant of the phenomena are autocatalytic polymer degradation and drug diffusion. The microspheres encapsulate drug molecules dispersed throughout the polymer. The polymer undergoes autocatalytic hydrolysis, breaking the polymer bonds and generating smaller polymer chains with acidic end groups that catalyze further hydrolysis of the degradation products. Autocatalytic hydrolysis is more severe in the interior of large microspheres where the diffusion of degradation products is more limited [5-9]. Sufficiently small oligomers produced by the degradation are water-soluble and can diffuse out of the polymer microspheres through water-filled pores. The resulting polymer mass loss increases the pore volume in the microspheres. Encapsulated drug molecules diffuse through the aqueous pores in the polymer by hindered diffusion [10-11]. As the pore network in the microspheres evolves, the effective diffusivity of the drug increases, thus enhancing the diffusion of drug through the microspheres and accelerating drug release.

A mathematical reaction-diffusion model [12] is defined and solved analytically for the concentration of carboxylic acid end groups of the polymer chains in PLGA microspheres as a function of radial position and time. The model treats simultaneous diffusion and first-order degradation to capture the microsphere-size-dependent effects of autocatalysis on PLGA erosion. The analytical expression is useful for predicting the conditions under which drug release from PLGA microspheres transitions from diffusion-controlled to erosion-controlled release, for understanding the dynamic coupling between the PLGA degradation and erosion mechanisms, and for designing drug release particles. The model is the first to provide an analytical prediction for the dynamics and spatial heterogeneities of PLGA degradation and erosion within a spherical particle. The analytical solution is general and can be applied to other spherical systems with simultaneous diffusive transport and first-order generation by reaction.

The analytical expression for the autocatalytic polymer degradation and erosion is coupled to a model of pore evolution, which is related to the effective diffusivity through hindered diffusion theory, and subsequent diffusive drug release [12-13]. The unique contribution of the modeling work is that it combines autocatalytic PLGA degradation mechanisms [10-11] with hindered diffusion in aqueous pores [14-15] having variable pore sizes. The model performance for the case of drug release from microspheres of different sizes is presented to highlight the capability of the model for predicting size-dependent, autocatalytic effects on the polymer and the release of drug.

References

[1] A. N. Ford Versypt, D. W. Pack, R. D. Braatz, Mathematical modeling of drug delivery from autocatalytically degradable PLGA microspheres—A review, Journal of Controlled Release 165(1):29–37, 2013.

[2] J. M. Anderson and M. S. Shive. Biodegradation and Biocompatibility of PLA and PLGA Microspheres. Advanced Drug Delivery Reviews, 28(1):5-24, 1997.

[3] N. K. Varde and D. W. Pack. Microspheres for Controlled Release Drug Delivery. Expert Opinions on Biological Therapy, 4(1):35-51, 2004.

[4] S. Freiberg and X. X. Zhu. Polymer Microspheres for Controlled Drug Release. International Journal of Pharmaceutics, 282(1-2):1-18, 2004.

[5] A. N. Ford Versypt. Modeling of Controlled-Release Drug Delivery from Autocatalytically Degrading Polymer Microspheres. Ph.D. Dissertation, University of Illinois at Urbana-Champaign, Urbana, IL, 2012.

[6] J. Siepmann, K. Elkharraz, F. Siepmann, and D. Klose. How Autocatalysis Accelerates Drug Release from PLGA-Based Microparticles: A Quantitative Treatment. Biomacromolecules, 6(4):2312-2319, 2005.

[7] C. Berkland, E. Pollauf, C. Raman, R. Silverman, K. Kim, and D. W. Pack. Macromolecule Release from Monodisperse PLG Microspheres: Control of Release Rates and Investigation of Release Mechanism. Journal of Pharmaceutical Sciences, 96(5):1176-1191, 2007.

[8] D. Klose, F. Siepmann, K. Elkharraz, and J. Siepmann. PLGA-Based Drug Delivery Systems: Importance of the Type of Drug and Device Geometry. International Journal of Pharmaceutics, 354(1-2): 95-103, 2008.

[9] A. N. Ford Versypt, D. W. Pack, and R. D. Braatz, Analytical Expression for Predicting Concentration of Acidic Polymer End Groups in PLGA Microspheres during Simultaneous Diffusion and Autocatalytic Degradation, In Preparation.

[10] P. Dechadilok and W. M. Deen. Hindrance Factors for Diffusion and Convection in Pores. Industrial and Engineering Chemistry Research, 45(21):6953-6959, 2006.

[11] A. Zhao, S. K. Hunter, and V. G. J. Rodgers. Theoretical Prediction of Induction Period from Transient Pore Evolvement in Polyester-Based Microparticles. Journal of Pharmaceutical Sciences, 99(11):4477-4487, 2010.

[12] A. N. Ford Versypt, D. W. Pack, and R. D. Braatz, Theoretical Drug Release from PLGA Microspheres during Simultaneous Hindered Diffusion and Autocatalytic Degradation Coupled by Analytical Acidic Polymer End Group Concentration Profiles, In Preparation.

[13] A. N. Ford, D. W. Pack, and R. D. Braatz. Multi-Scale Modeling of PLGA Microparticle Drug Delivery Systems. In E. N. Pistikopoulos, M. C. Georgiadis, and A. C. Kokossis, editors, 21st European Symposium on Computer Aided Process Engineering: Part B, pages 1475–1479. Interscience, New York, 2011.

[14] H. Antheunis, J. C. van der Meer, M. de Geus, W. Kingma, and C. E. Koning. Improved Mathematical Model for the Hydrolytic Degradation of Aliphatic Polyesters. Macromolecules, 42(7):2462-2471, 2009.

[15] S. Lyu, R. Sparer, and D. Untereker. Analytical Solutions to Mathematical Models of the Surface and Bulk Erosion of Solid Polymers. Journal of Polymer Science Part B: Polymer Physics, 43(4):383-397, 2005.

(666b) Hybrid 1D/3D Blood Flow Simulations of the Arterial Human Network

Apostolidis, A., University of Delaware
Beris, A. N., University of Delaware
Johnson, D. A., Merck & Co., Inc.
Originally presented on: 11/7/2013 12:50:00 - 13:10:00

Continuing our previous work of blood flow simulations within the human arterial network [1,2] we present here an investigation of the effects of the non-Newtonian character of blood rheology. This involves the implementation of a recently developed viscoplastic description of the steady-state blood flow rheology into the previously developed 1D model of the flow within the arterial network. This new rheological model is still based on the Casson equation but involves a new parameterization of its coefficients in terms of important physiological parameters, such as the blood hematocrit and fibrinogen concentration, so that it much better fits available experimental data. A similar re-parameterization is also performed for the apparent Newtonian viscosity used in the description of the transient component of the flow.

Those newly described aspects of non-Newtonian rheology of blood are then examined with respect to their influence to the predictions of the 1D approximate arterial flow model as well as with respect to their effect to a hybrid 1D/3D simulation, following [2], of the time-periodic blood flow through a non-axisymmetric (coronary) arterial bifurcation. These simulations involve, as explained in [2], an efficient implementation of the proper (*in vivo*) outlet boundary conditions, as provided in the form of Fourier frequency impedance coefficients by the 1D model, to a commercial 3D (FLUENT) code. In contrast to our previous simulations of the left coronary artery bifurcation [2] a much more faithful to the actual geometry mesh representation is used at this time, in addition to the implementation of non-Newtonian rheology characteristics both in FLUENT and in the 1D model (as described above). The non-Newtonian effects are examined in two cases: First a healthy system, patterned after the left main coronary arterial system, and second a diseased case where an occlusion has developed in one of the daughter vessels, resulting in strengthening the asymmetry of the bifurcation.

References

[1] Johnson, D.A., Spaeth, J.R., Rose, W.C., Naik, U.P., Beris, A.N., An Impedance model for blood flow in the human arterial system, Part I: Model Development and Matlab Implementation. Computers in Chemical Engineering, 35(7): 1304-1316, (2011).

[2] Johnson, D.A., Naik, U.P., Beris, A.N., Efficient implementation of the proper outlet flow conditions in blood flow simulations through asymmetric arterial bifucations. Int. J. Numer. Meth. Fluids, 66(11): 1383-1408, (2011).

Acknowledgement

This work is supported by NSF, Grant Award # NSF CBET 1033296

(666c) Predicting the Mitochondrial Toxicity of Nucleoside-Analogue Antiretroviral Drugs Via Stochastic Modeling

Alathur Srinivasan, P. A., The Pennsylvania State University
Armaou, A., The Pennsylvania State University
Originally presented on: 11/7/2013 13:10:00 - 13:30:00

Nucleoside-analogue Reverse Transcriptase Inhibitors (NRTIs) are an important class of drugs used for the treatment of HIV infection. Although NRTIs have been effective in controlling viral populations below the detection limit, their prolonged use leads to several medical complications. Clinical and molecular biological evidences suggest that NRTIs cause toxic side-effects by interfering with normal mitochondrial function. Since NRTIs are structurally similar to the natural deoxynucleotides (dNTPs), they competitively inhibit the mitochondrial DNA (mtDNA) replication and transcription processes. The objective of this work is to investigate the individual and synergistic effects of NRTIs on these two processes using a stochastic computational model.

Human DNA polymerase-γ is the enzyme responsible for the replication of mtDNA. Early studies on NRTI-induced toxicity led to the 'DNA pol-γ hypothesis', which postulates that the inhibition of pol-γ by NRTIs causes mtDNA depletion and thereby mitochondrial dysfunction. Since the NRTIs lack the 3'-OH group required for the elongation of mtDNA, their incorporation into the strand by pol-γ leads to termination of mtDNA replication. However, recent findings indicate that the inhibition of the human mitochondrial RNA polymerase (POLRMT) is as important as that of pol-γ, if not more (Mallon et al, 2005). POLRMT is the enzyme responsible for the transcription of mtDNA into three unique preliminary RNA strands. The effects of NRTIs and AVRNs (antiviral ribonucleosides) on these two polymerases in different types of cells are crucial for the understanding of clinical toxicity.

The known mechanism of polymerase chain reaction is modeled and simulated using Gillespie's algorithm, a kinetic Monte Carlo method. At each location on the mtDNA strand, the polymerases can perform one of these actions: (1) Addition of the correct nucleotide (2) Addition of an incorrect nucleotide (non-Watson-Crick pair) (3) Addition of an incorrect sugar (4) Addition of a NRTI (5) Addition of an AVRN (6) Excision of the previously added nucleotide (7) Disassociation of the polymerase from the strand (8) Association of the polymerase to the strand. The simulation stops either when a strand is completely replicated or transcribed, or when the polymerase disassociates and fails to re-associate. The occurrence of the latter is more probable when a NRTI or AVRN was added in the previous step and was failed to be removed.

Apart from obtaining a quantitative measure of clinical toxicity, the model is used to determine the sensitivity of drug response to various parameters. The model serves to help experimentalists in identifying key factors and parameters for measurement. In future, the model will be extended to include the mitochondrial translation machinery to predict the under-expression of OXPHOS proteins.

(666d) Comparison of Different Methods for Predicting Customized Drug Dosage in Superovulation Stage of In-Vitro Fertilization

Yenkie, K. M., University of Illinois, Chicago

Diwekar, U., Vishwamitra Research Institute, Center for Uncertain Systems: Tools for Optimization and Management

Originally presented on: 11/7/2013 13:30:00 - 13:50:00

Comparison of Different Methods for Predicting Customized Drug Dosage in Superovulation stage of In-vitro fertilization

Kirti M. Yenkie, [a,b] and Urmila M. Diwekar [a,b]

[a]Department of Bio Engineering, University of Illinois, Chicago, IL 60607 - USA

[b]Center for uncertain Systems: Tools for Optimization & Management (CUSTOM),

Vishwamitra Research Institute, Clarendon Hills, IL 60514 ? USA

Abstract

In vitro fertilization (IVF) is one of the highly pursued assisted reproductive technologies worldwide. The IVF procedure is divided into four stages: Superovulation, Egg-retrieval, Insemination/Fertilization and Embryo transfer. Superovulation is the most crucial stage in IVF, since it involves external injection of hormones to stimulate development and maturation of multiple oocytes. The maximum amount of effort and money for IVF procedure goes into superovulation. Although numerous advancements have been made in IVF procedures, little attention has been given to modifying the standard protocols based on a predictive model. Currently, the same protocol is followed for every patient undergoing the IVF superovulation procedure. In reality every patient reponds differently and hence the proposition to modify the amounts of drug administered based on the patient's initial treatment response is a reasonable approach. The modification of drug dose if based on a well developed mathematical model which takes into account the variability in the follicle growth dynamics as well as the desired outcome thus increasing the predictive value of the method.

A model for the follicle growth dynamics and number as a function of the injected hormones and patient characteristics has been developed and validated. The modeling basics have been adapted from batch crystallization moment model, since moments are representatives of specific properties like number, shape and size of the particles under consideration. Based on this model, the dosage of the hormones to stimulate multiple ovulation or follicle growth is predicted by using the theory of optimal control. The objective of successful superovulation is to obtain maximum number of mature oocytes/follicles within a particular size range. Using the mathematical model involving follicle growth dynamics and the optimal control theory, optimal dose and frequency of medication customized for each patient is predicted for obtaining the desired result.

The optimal control problem is solved by different methods like the maximum principle and dicretized non-linear programming. The problem is solved with and without constraints to check the variation in the dosage amounts and size of follicles at the retrieval time. The results from different approaches are compared. Thus, a systematic comparison of the different methods will help in deciding the best solution strategy for customized drug dosage. It will also provide information about the sensitivity of the model parameters and hence the source of uncertainty in the system.

Keywords: multiple ovulation, follicle growth, maximum principle, non-linear programming

1. Introduction:

Fig. 1 shows a schematic diagram of the overall IVF procedure. Superovulation is a method to retrieve multiple eggs using drug induced stimulation of the ovaries. In normal female body only one egg is ovulated per menstrual cycle, but with the use of fertility drugs and hormones, more number of eggs can be ovulated per cycle.

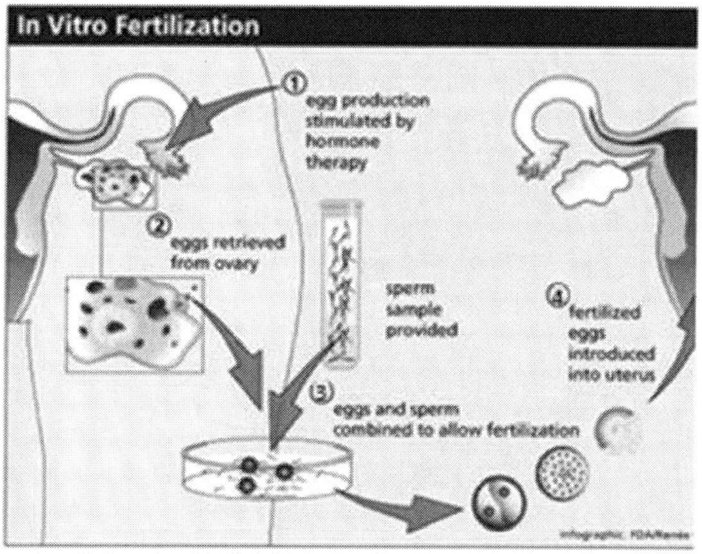

Figure 1. Schematic diagram of the IVF procedure

The success of superovulation is critical in proceeding with the further stages of the IVF. The fertility drugs used for inducing multiple ovulations are external hormones which are very expensive. The superovulation protocols follow a standard for hormonal dosage and after the initial dose of follicle stimulating hormone the patient requires daily testing and monitoring to keep a check on the patient's response and thus modify the dose. Improper dose may result into life threatening complications like ovarian hyperstimulation syndrome (OHSS) or at times may not cause any response in the patient. Thus, existing protocols lack planned treatment initiation and are highly dependent upon the equipments for monitoring.

The current work aims at developing a model predictive method for the drug dosage in superovulation. The drug dosage prediction method will take into account the treatment which is highly cost intensive and lacks individualized treatment variations will get a strong base. The comparisons between different method will help in deciding the best predictive method. The information regarding the single or multiple solutions possible by the different methods used will also provide insight into the system uncertainties.

2. Model Equations

Superovulation model is developed on the lines of batch crystallization modeling using the method of moments. The follicle size is converted into mathematical moments by assuming the follicles to be spherical in shape. The eq. 1 is used for converting follicle size to moments. Each moment has its own significance and corresponds to a feature of the follicles. The zeroth moment is a representative of the number, first moment represents the size, second the area, etc.

$\mu i=njr,trjiDrj$ $\mu_i = \sum n_j(r,t) r_j^i \Delta r_j$ (1)

Here, $\mu_i = i^{th}$ moment

$n_j(r,t)$ = number of follicles in bin 'j' of mean radius 'r' at time 't'.

r_j = mean radius of jth bin

Δr = range of radii variation in each bin

It is assumed that follicle growth is dependent on follicle stimulating hormone (FSH) administered. The growth term is written as shown in equation (2) ;

$$G = kC_{fsh}^{\alpha} \quad (2)$$

Here, G - follicle growth rate

k - rate constant

C_{fsh} - amount of FSH injected

α - rate exponent

From the literature by Baird, 1997 it can be assumed that the number of follicles activated for growth are always constant (for a particular protocol initiation) for a particular patient, hence the zeroth moment has a constant value. We use the 0th to 6th order moments since they help in better prediction of moment values as well as help in efficient recovery of the size distributions as against the lower order moments. The moment equations for the follicle dynamics can be written as in equations (3) and (4).

$$\mu_0 = constant \quad (3)$$

$$\frac{d\mu_i}{dt} = iG(t)\mu_{i-1}(t);\ (i = 1,2,\ldots 6) \quad (4)$$

Here, G - follicle growth rate

μ_i - i^{th} moment

3. Optimal Control Problem

The moment model for follicle size distribution prediction and the method for deriving normal distribution parameters have been used as the basis for deriving expressions for the mean (eq. 5) and coefficient of variation (eq. 6) for the follicle size distribution. The paper by John *et. al.* 2007 highlights the techniques for reconstructing distributions from moments and they obtain good approximation of realistic distributions using some finite order of moments. Since, the data clearly reflects a normal distribution it is quite reasonable to assume it as an apriori distribution for follicles.

$$\bar{x} = \frac{\mu_1}{\mu_0} \quad (5)$$

$$CV = \sqrt{\frac{\mu_2\mu_0}{\mu_1^2} - 1} \quad (6)$$

Here, $\bar{x}$ - mean follicle size

CV - coefficient of variation

μ_0, μ_1 and μ_2 - zeroth, first and second order moment respectively.

Thus, the objective of superovulation can be stated as; to minimize the coefficient of variation at the final time ($CV(t_f)$)and the control variable for this shall be the dosage of FSH with time ($C_{fsh}(t)$).

4. Solution Methods

The optimal control problem is solved by unconstrained and constrained maximum principle and dicretized non-linear programming. Initial guess for the optimization variables are varied to check for multiple solutions. The results from all the methods are compared for better understanding of the parametric uncertainty as well as the best solution strategy. The fig. 2 shows the flowchart for the maximum priciple method.

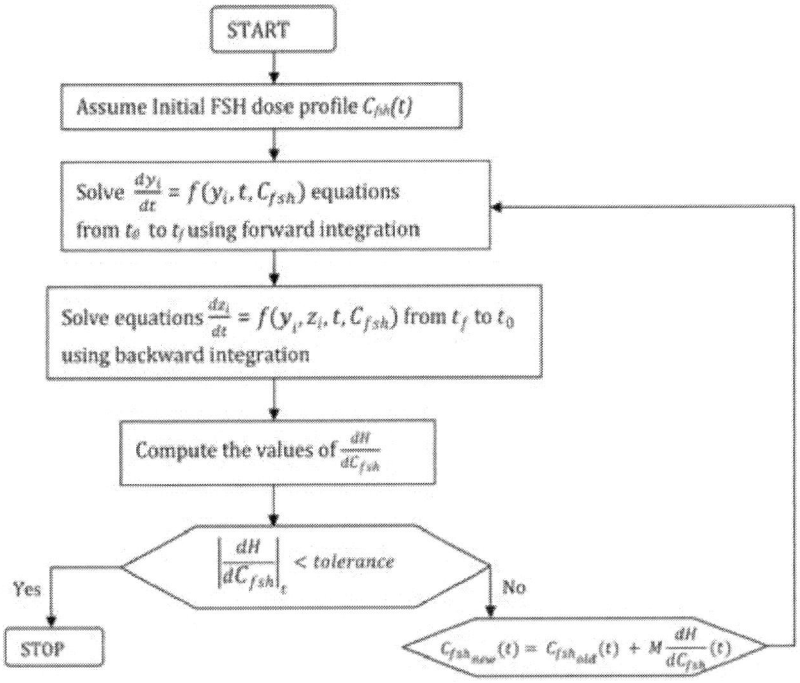

Figure 2. Flowchart for optimal FSH dosage evaluation using maximum principle

5. Some Results

The optimal control results show the comparison of optimized prediction with the actual observations on the final day of the follicle size and number measured. The figures for two patients B and C are shown and observations for 4 patients are summarized in the Table 1.

Table 1. Summary of Results for Patient A, B, C & D

Patient (No. of Growing Follicles)	FSH (used)	FSH (opt)	No. of Follicles ($9 \leq$ Mean ≤ 12) (used)	(opt)	% reduction in FSH
A (8)	1650	1388	4	6	18.88
B (23)	525	463.5	12	15	13.02
C (18)	750	617	10	13	17.73

D (5)	2250	1855	0	1	17.56

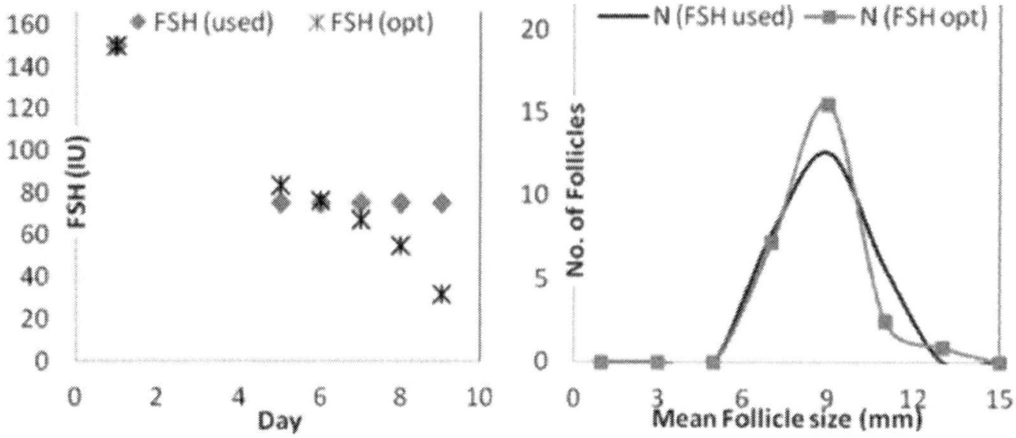

Figure 3a. Optimal FSH dose against the administered dose

3b. Final day follicle number prediction (Patient B)

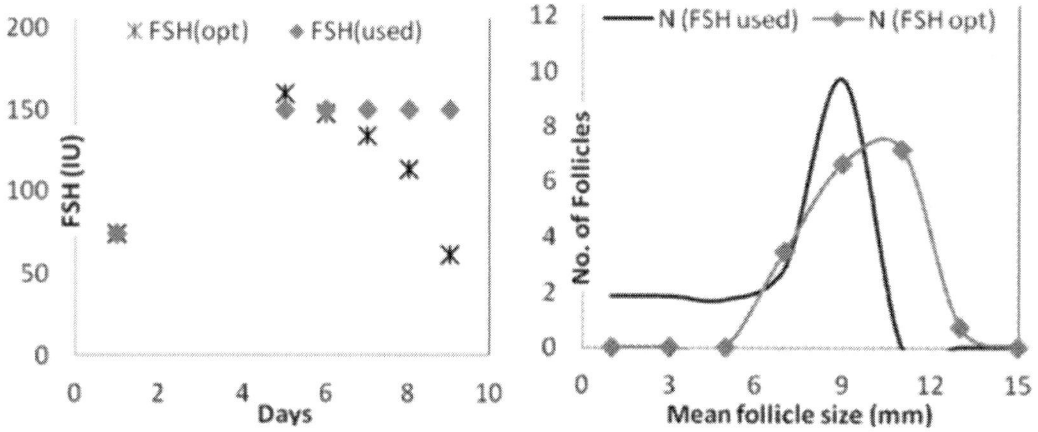

Figure 4a. Optimal FSH dose against the administered dose

4b. Final day follicle number prediction (Patient C)

6. Conclusions

The moment model developed for IVF superovulation predicts the follicle size distribution which is in good agreement with the actual size distribution. The optimal control theory application to superovulation stage provides a new approach for model predictive drug delivery in IVF. The mathematical formulation of the objective function in terms of the coefficient of variation by utilizing the concepts of normal distribution provides a reasonably good measure of the final outcome. The predetermined dosage saves the cost of excess medicines and also the requirements of daily monitoring and testing. It can be said that the applications of control principles to a medical treatment procedure like IVF which was previously based on trial and error gets a good basis for planned treatment initiation. Also the current work has been done in collaboration with clinicians and real patient data has been used; which makes the study more emphatic when compared to theoretical work.

(666e) Parametric Cerebral Blood Flow Predictions

Gould, I., University of Illinois at Chicago
Hsu, C. Y., University of Illinois at Chicago
Linninger, A., University of Illinois at Chicago
Originally presented on: 11/7/2013 13:50:00 - 14:10:00

Abstract

The control of intracellular oxygen tension in neural tissue is the result of a complex interaction of transport across the vasculature endothelial wall, diffusion through intracellular spaces, neuronal cellular metabolism, and the release of vasomodulating signals. In order to test the various theories regarding this tightly controlled system, it is necessary to construct a model that can accurately model these dynamically integrated elements. Such a model must be able to quickly solve large sets of differential equations, capture the true nature of the intracellular oxygen problem utilizing a 3D approach, and provide physiologic results for blood pressure, blood flow, red blood cell velocity and oxygen tension.

1. Introduction

The need for parametrically driven clinical models for predicting the perfusion of oxygen in brain tissue is apparent in the patients suffering from cerebral ischemia in the emergency room, the angiography suite and the intensive care unit. Few models give context to oxygen supply sensitive to blood pressure, pH and hemoglobin oxygenation, and fewer models still incorporate patient-specific morphology that can widely skew the result. These topological parameters are critical for assessing quantitative results, allowing for the differentiation between healthy and ischemic tissue. Conversely, the abundance of morphologically accurate models of a single vessel structure within the Circle of Willis or a handful of capillaries in the microvasculature suffer from boundary condition estimations and give a highly accurate but narrow view.

2. Methods

A holistic model of the human cerebral vasculature was generated which couples the large vessel morphology with the microvasculature structures in the pial vessels and capillary bed, a hemodynamic equivalent emerges which can be selectively focused and reassessed to determine a patients' cerebral perfusion based on the solution of first-principle scientific computing effort.

2.1 Topological Construction

The large vessel anatomy of the cerebral vasculature model including both arteries and veins was captured from MRA and MRV images using a centerline extraction approach and diameter extrusion to create a network representation of the cerebral vasculature. As the resolution magnetic resonance images does not extend to any vessel smaller than 100μm in the large viewing scope necessary to capture the entire brain domain, a geometric vessel sprouting algorithm was employed to complete the pial and capillary vessels that complete the cerebral vasculature structure. This 3D topologically constrained vessel growth algorithm was modified from the Constrained Construction Optimization (CCO) algorithm introduced Schreiner and Karch (1999). By expanding the image-reconstructed network with these geometric space-filling methods, a morphologically accurate representation of the arterial and venous trees of the cerebral vasculature was constructed, containing the Circle of Willis, the left and right middle, anterior and posterior cerebral arteries, the Superior and Inferior Sagittal Sinus. Secondary branches arising from these major vessels were sprouted along the surface of the cerebral cortex geometry, which was obtained from T_2-weighted MRI. A modified Voronoi mesh constructed from this geometry, and was connected to these acyclic bifurcating trees as a microvasculature surrogate, completing the connectivity of the model. The final model consisted of over 100,000 vessel bifurcations and over 80,000 vessels with three inlet boundary vessels (left and right internal carotid, basilar artery) and two outlet boundary vessels (left and right jugular veins).

An extensive literature review was conducted to determine the physiologic and topological parameters for constructing a human cerebral microvasculature model. Though many of these models incorporated topologically accurate dense microvasculature structures, venous, arterial or capillary vessels, consisting structure of extravascular space, computational blood flow and pressure predictions, or oxygen transport - no model existed which portrayed all of these components. What has been established is that random and tortuous blood vessel networks enhance oxygen distribution and mixing compared to evenly spaced, ordered vessel configurations. Additionally, several studies of cerebral microvasculature using confocal microscopy and synchrotron-CT methods have been undertaken in human, rat and macaque models to rigorously establish topological features. By matching our model to these studies it can be shown that a length scale, vessel lumen fractions, diameter distributions, and angle bifurcations can be rigorously matched.

2.2 Blood Flow Equations

Blood flow was described using established hemodynamic network models established by Pries (1990). Flow resistance of each cylindrical segment was approximated by the Poiseuille law, where the pressure drop across a single vessel is equivalent to the product of the bulk blood flow through that segment and the segment's vascular resistance. The continuity equation was applied at each vessel connection. The pulsatile nature of the microcirculation was ignored, and a quasi-steady flow assumed as low Womersley numbers ($Wo<0.1$) dominate in the microvasculature. Inertia is neglected in our simplified fluid flow model due to low Reynolds numbers ($Re<1$). The effect of biphasic blood rheology was approximated by the Fahraeus-Lindqvist equation, which predicts the apparent viscosity as a function of vessel segment diameter and the volumetric ratio of erythrocytes to plasma volume, called hematocrit.

To parametrically predict the transport of oxygen as a function of systemic blood thinning, or a reduction of hematocrit, it is necessary to depict the biphasic nature of blood flow. A global method was implemented in this work to capture the relationship between the hematocrit, the vascular resistance and the oxygen delivery to tissue.

2.3 Oxygen Delivery to Tissue

After blood flow was solved, convective transport equations were constructed across each vasculature bifurcation. Using the hematocrit phase alone for the blood flow driven transportation of oxygen throughout the cerebral vasculature, an inlet Neumann boundary condition was assigned to the arterial inlets and a draining flux was assigned at the venous outlet. Oxygen molar flux was computed across each vessel segment, oxygen tension was assigned to the node of each.

A dual mesh technique was employed using the above described vasculature network of cylindrical segments and a volumetric tetrahedral brain mesh constructed from the T_2-weighted MRI of the same patient. A tetrahedral tissue volume may contain multiple vascular bifurcation nodes; however each vascular bifurcation node is uniquely assigned to one tetrahedral tissue volume. The species conservation for each network node was balanced, and the overall mass transfer coefficient is a function of wall thickness and the oxygen permeability of epithelial tissue. Oxygen mass exchange is driven by the concentration difference between oxygen tension in the hematocrit and oxygen tension in the brain tissue.

The extravascular space modeled by the volumetric mesh is a modeled is a single continuum with extracellular and intracellular space lumped together, which was consumed at a constant rate. Finally, diffusion equations were assigned in this extravascular tissue space and solved to predict the transport of oxygen throughout the tissue domain. Though many variations and improvements of the analytical Krogh cylinder model have been proposed, these analytical models are frequently hampered by their inability to capture axial diffusion of oxygen in the tissue compartment and are limited to simple structures. For these reasons, the discretized approach described above was employed.

2.4 Parallel Solvers

With the flow field established, linearization of the nonlinear biphasic fluid flow eqiautions was proposed to determine the distribution of hematocrit. Decoupling the flow and hematocrit transport equations allows for solution by linear algebra approaches, for which iterative parallel solvers implemented in PETSc excel in decreasing the solution time of this nonlinear set of ordinary differential equations. With the flow field established, conservation of mass to each node was applied to determine the distribution of hematocrit throughout the network.

3. Results

The pial arterial inlet pressure was set to 90mmHg, the pial venous drain had 10mmHg, as obtained from literature values, giving a volumetric flow rate of 750mL/min, which is comparable to measured values. The oxygen tension in this whole brain model gradually dropped from 90mmHg in the internal carotid arteries down to 30mmHg in jugular veins. The redirected oxygen extraction compares well to cortical measurements, and radial oxygen perfusion profiles in the tissue were also in good agreement to measured values.

Radial oxygen tension in the molecular layer ranged from 80mmHg at a distance of 10μm, to 45mmHg at 50μm distance. These oxygen gradients seem to be a result of the capillary-depleted regions adjacent to larger arterioles, which were captured in the detailed microvasculature surrogate. Our simulations for regions without larger vessels show shallow oxygen profiles, completely flattening about 30μm away from the nearest capillary, equivalent to healthy results measured in rat models.

4. Conclusion

This morphologically accurate approach is necessary for accurately capturing the fate of oxygen delivered in the perfusion of cerebral tissue.These models of topologically complex biological system such as the cerebral microvasculature and the predictions of first-principle computations can be used to (a) improve the interpretation of *in vivo* open cranial measurements acquired in both human and rodent models, (b) better understand the role of topological features of the cerebral microvasculature that promote inferior collateral to shed light on perfusion clinical imaging modalities, (c) exploring intercellular transport phenomena at a length scale and scope that is not available to imaging methods, and (d) gain insight into the transport of oxygen in the intracortical regions of human patients for whom invasive measurements are not an option.

(666f) Design of Patient-Specific Inter-Colonoscopy Interval Based On a Natural History Model of Colorectal Neoplasia

Sherer, E., Roudebush VA Medical Center

Originally presented on: 11/7/2013 14:10:00 - 14:30:00

A model framework is suggested, in which an individual patient's risk for colonic neoplasia varies based on findings from his previous colonoscopies, to predict longitudinal colonoscopy results. The neoplasia natural history model describes progression through four neoplasia development states with patient age. Multiple natural history model parameter sets are assumed to act concurrently on the colon and parameter set prevalence combinations, whose *a priori* likelihoods are a function of patient sex, provide a basis set for patient-level predictions. The novelty in this approach is that after a colonoscopy, both the parameter set combination likelihoods and their model predictions can adjust in a Bayesian manner based on the results and conditions of the colonoscopy. The adjustment of model predictions operationalizes the clinical knowledge that multiple or advanced neoplasia at baseline colonoscopy is an independent predictor of multiple or advanced neoplasia at follow-up colonoscopy - and vice versa for negative colonoscopies – and the adjustment of parameter set combination likelihoods accounts for the possibility that patients may have different neoplasia development rates. A model that accurately captures serial colonoscopy results could potentially be used to design and evaluate post-up colonoscopy treatment strategies based on the risk of individual patient.

To support model identification, observational longitudinal colonoscopy results, procedure details, and patient characteristics were collected for 4,084 patients. At least two parameter sets specific to each sex with model adjustments was required to capture the longitudinal colonoscopy data and inclusion of multiple possible parameter set combinations, which account for random variations within the population, was necessary to accurately predict the second-time colonoscopy findings for patients with a history of advanced adenomas. Application of this model to predict CRC risks for patients adhering to guideline recommended follow-up colonoscopy intervals found that there are significant differences in risk with patient age, gender, and preparation quality and demonstrates the need for a more rigorous investigation into these recommendations. In this talk, the colorectal neoplasia natural history model is presented and used in a Markov decision process to identify the inter-colonoscopy interval that minimizes the cost per quality adjusted life year depending on the age, gender, and previous colonoscopy results of a patient.

(666g) The Interplay Among De Novo Formation, Fission, and Lipolysis in Lipid Droplet Organelle Distribution in Dividing Eukaryotes

Dalhaimer, P., University of Tennessee

Originally presented on: 11/7/2013 14:30:00 - 14:50:00

One of the key questions in cell biology is how daughter cells inherit organelles from the parent cell. To help address this question, I simulated lipid droplets as model organelles in dividing fission yeast cells. Lipid droplets are dynamic bodies that store neutral lipids and can form both *de novo* (biogenesis) and by fission, they can also be broken down (lipolysis) for energy or phospholipid needs. The quantitative interplay between formation and breakdown mechanisms is unknown but would seem crucial for controlling neutral lipid distribution, especially in dividing cells. Because long-term microscopy experiments over times that would cover multiple division cycles would be challenging, I instead used Brownian Dynamics simulations to quantify droplet distribution in populations of dividing cells. Simulation parameters were obtained from time-lapse confocal microscopy experiments of lipid droplets in live fission yeast cells. Of the three main events studied: biogenesis, fission, and lipolysis, the third plays the key role. Droplets must persist (not be completely broken down) for the number and volume of droplets to be stable in dividing cells even with very high *de novo* formation and fission rates. Thus, the sustained presence of droplets – most likely through connections with the endoplasmic reticulum – seems necessary for this organelle's experimentally observed distribution in dividing eukaryotes.

(683a) Metabolic Control of Persister Formation

Amato, S. M., Princeton University
Orman, M. A., Rutgers, The State University of New Jersey
Brynildsen, M. P., Princeton University
Originally presented on: 11/7/2013 12:30:00 - 12:48:00

Bacterial persisters comprise of a subpopulation of cells from a genetically identical population that phenotypically exhibit multidrug tolerance. Persisters survive supralethal antibiotic stress and upon reinoculation, their progeny possess the same antibiotic sensitivity as the original population [1, 2]. Persisters are particularly significant for biofilm infections, where they are thought to be responsible for the propensity of such infections to relapse after the conclusion of antibiotic therapy [1]. Therefore, understanding the mechanisms that give rise to persisters can provide therapeutic targets to combat relapse infections. Current research suggests that the terminal effectors of persistence are toxin-antitoxin (TA) modules, due to their ability to inhibit growth, increase antibiotic tolerance, and exhibit bistability. However, the molecular mechanisms that mediate this tolerance upstream of TA modules remain elusive with only a few signaling cascades defined [3-5]. Previous work has demonstrated that the majority of persister formation during normal growth in rich media occurs during mid- to late-exponential phase [6], which is characterized by numerous metabolic perturbations, the onset of oxygen limitation, and the appearance of higher density phenotypes, such as quorum sensing. Inspired by this work, we sought to identify molecular mechanisms by which persisters form during normal growth. Since the extent of stresses experienced during growth suggests that diverse mechanisms may be present, we focused on a single native stress in isolation. In particular, we sought to answer the fundamental question of whether carbon source transitions stimulate persister formation, and if so, determine the intracellular pathway associated with formation. To accomplish this, we analyzed batch growth of *Escherichia coli* in defined minimal media containing one or more carbon sources. We measured persister levels using the fluoroquinolone antibiotic, ofloxacin, quantified metabolite concentrations, and used genetic perturbations, flow cytometry, and a mathematical model to define a mechanistic persister formation pathway from initial stress (glucose exhaustion) to a metabolic TA module (guanosine tetraphosphate and its biochemical network) and its inhibition of the antibiotic's primary target (DNA gyrase). The results presented here identify a comprehensive signaling pathway responsible for persister formation from a native stress and suggest that TA-like metabolic networks may play a significant role in persister formation.

1. Lewis, K., *Persister cells.* Annu Rev Microbiol, 2010. **64**: p. 357-72.

2. Balaban, N.Q., et al., *Bacterial persistence as a phenotypic switch.* Science, 2004. **305**(5690): p. 1622-5.

3. Dorr, T., M. Vulic, and K. Lewis, *Ciprofloxacin causes persister formation by inducing the TisB toxin in Escherichia coli.* PLoS Biol, 2010. **8**(2): p. e1000317.

4. Nguyen, D., et al., *Active starvation responses mediate antibiotic tolerance in biofilms and nutrient-limited bacteria.* Science, 2011. **334**(6058): p. 982-6.

5. Vega, N.M., et al., *Signaling-mediated bacterial persister formation.* Nat Chem Biol, 2012. **8**(5): p. 431-3.

6. Keren, I., et al., *Persister cells and tolerance to antimicrobials.* FEMS Microbiol Lett, 2004. **230**(1): p. 13-8.

(683b) Phosphoplipid Alterations and ER Calcium Efflux Promote Metabolic Dysfunction in the Context of Hepatic Lipotoxicity

Egnatchik, R. A., Vanderbilt University
Leamy, A. K., Vanderbilt University
Young, J. D., Vanderbilt University
Jacobson, D.,
Jacobson, D. A, Molecular Physiology and Biophysics, Vanderbilt University
Masakazu, S., Molecular Physiology and Biophysics, Vanderbilt University
Shiota, M.,
Originally presented on: 11/7/2013 12:48:00 - 13:06:00

Lipotoxicity from free fatty acid overload has been implicated in a variety of disease states including type II diabetes and non-alcoholic fatty liver disease (NAFLD). Incubation of hepatic cells with saturated fatty acids (SFAs) induces apoptosis characterized by the accumulation of reactive oxygen species (ROS), metabolic dysfunction, and apoptosis. However, a consensus mechanism linking SFA-induced metabolic alterations to apoptosis has been difficult to establish. Lack of such knowledge represents an important problem, because it limits the ability of researchers to develop novel nutritional and/or pharmacologic interventions to combat the effects of NAFLD. Our central hypothesis is that SFA overexposure leads to increased saturation of ER membrane phospholipids, which alters membrane stability, induces endoplasmic reticulum (ER) calcium efflux leading to metabolic dysfunction, and ultimately causes lipoapoptosis.

Our objective was to first determine how perturbations of lipid metabolic pathways influence SFA-induced lipotoxicity. We measured tritiated palmitate incorporation into total cellular phospholipids in H4IIEC3 rat hepatoma cells. We found increased saturation of phospholipids which correlated to decreased ER calcium stores. Interestingly, co-treatment with the mono-unsaturated fatty acid oleate prevented phospholipid oversaturation by redirecting palmitate carbon to triglyceride stores, which blocked apoptosis. We hypothesized that the observed increases in ER saturation could disrupt membrane homeostasis and enhance calcium efflux. In fact, after six hours of palmitate treatment we observed decreases in ER calcium indicating a net efflux of calcium from the ER.

To determine if intracellular calcium flux affects ROS accumulation and apoptosis, we co-treated H4IIEC3 cells with both palmitate and the intracellular calcium chelator BAPTA. BAPTA co-treatment suppressed ROS accumulation and caspase activation, indicating that these events are calcium dependent. To identify metabolic pathways causing hepatic lipotoxicity, we applied metabolic flux analysis (MFA) using [U-$^{13}C_5$]glutamine as an isotopic tracer to quantify phenotypic changes in H4IIEC3 cells treated with either palmitate alone or both palmitate and BAPTA in combination. Our results indicate that palmitate treatment leads to increased channeling of pyruvate into TCA cycle and increased mitochondrial oxidation of glutamine as an energy substrate. We hypothesizd that these metabolic alterations are the primary cause of toxic ROS accumulation. Co-feeding BAPTA restored most mitochondrial fluxes to their control levels while suppressing ROS and apoptosis. Based on the timing of these events, we hypothesized that the observed uncoupling between glycolytic and TCA cycle fluxes is responsible for triggering mitochondrial ROS generation and apoptotic cell death in palmitate-treated cells. Additionally, our results indicate that the observed metabolic uncoupling is a primary effect of palmitate overexposure due to lipid-induced ER calcium efflux. These findings shed light on the mechanisms of hepatic lipotoxicity and may suggest novel strategies for the treatment and prevention of NAFLD.

(683c) Development of a Single-Cell Clonogenic Assay

Maass, K., Massachusetts Institute of Technology
Wittrup, K. D., Massachusetts Institute of Technology
Originally presented on: 11/7/2013 13:06:00 - 13:24:00

Standard cell proliferation assays use the bulk media drug concentration to determine a drug's potency. A more clinically relevant quantity is the amount of drug that is actually taken up by the cell. In this work, we developed an *in vitro* single-cell clonogenic assay based on flow cytometry. The assay uses a fluorescent drug and the CellTrace^{TM} Cell Proliferation Kit (Molecular Probes) to evaluate the cytostatic effects of drug exposure. The amount of drug taken up by individual cells can be correlated to the proliferation of that cell. Doxorubicin, a naturally fluorescent chemotherapeutic drug, was used in the development and validation of this assay. Uptake of drug by treated cells was quantified via fluorescence. Various drug concentrations in the media and lengths of exposure to the drug were tested in the following cancer cell lines: BT474, HCT-15, HT-29, IGROV-1, MDA-MB-231, and T47D. Results were also compared to a standard WST-1 assay, which measures bulk levels of cell proliferation based on metabolic activity. In order to demonstrate the utility of measuring actual cellular drug uptake, the effect of inhibiting the multi-drug resistance pump in HCT-15 cells that express the multi-drug resistance pump was shown. This assay can also be generalized to study any fluorescent drug including antibody-drug conjugates which are labeled fluorescently or which carry fluorescent drugs such as Doxorubicin.

(683d) Ability of Polyphenols to Attenuate Alzheimer's Disease By Reducing Nuclear Factor- ÎšB Activation Through Inhibition of Both Amyloid- Î' Oligomer Formation and Reactive Oxygen Species

Pate, K., University of South Carolina
Rogers, M., University of South Carolina
Moss, M. A., University of South Carolina
Originally presented on: 11/7/2013 13:24:00 - 13:42:00

Alzheimer's disease (AD) is the most common form of dementia and is characterized by extracellular plaques in the brain. Â These plaques are created when monomeric amyloid-β (Aβ) protein aggregates into fibrillar structures. Â Soluble Aβ aggregates, including oligomers that form along the aggregation pathway, have the ability to induce production of reactive oxygen species (ROS) resulting in various cellular responses, including increased cerebrovascular inflammatory responses associated with AD pathology.Â Specifically, ROS serve as messengers for increased NF-κB activation, which has been shown previously by our laboratory to regulate endothelial monolayer permeability as well as endothelial adhesion of monocytes and subsequent transendothelial monocyte migration.Â Epidemiological studies have shown that diets rich in polyphenols, compounds found in fruits, vegetables, and herbs, result in a reduced incidence of AD.Â Many polyphenols of the flavonoid subclass can inhibit Aβ aggregation thereby neutralizing the protein's damaging effects.Â Alternatively, polyphenols have been shown to inhibit Aβ induced cellular responses by neutralizing ROS through their antioxidant properties.Â This study sought to identify polyphenols that can either inhibit Aβ aggregation or reduce ROS in order to determine the role of these two activities in reducing NF-κB activation. Â Additionally, polyphenols that have both of these activities were identified to investigate possible synergy between the two mechanisms and the potential of polyphenols to serve as dual action drugs for AD.

Polyphenols investigated were from the subclass flavonoids and included various flavones, isoflavones, flavanols, and catechins in order to explore correlations between structure and inhibitory effects.Â To identify polyphenols capable of inhibiting the formation of oligomers, polyphenols were added in 10-fold excess to monomeric $A\beta_{1-42}$ prior to induction of oligomerization.Â The resulting size of oligomers formed was analyzed via SDS-PAGE and Western blot.Â To identify polyphenols capable of neutralizing ROS stimulated by aggregated Aβ, antioxidant capacity was evaluated using the OxiSelect? Oxygen Radical Antioxidant Capacity Activity Assay.Â A Vitamin E analog, Trolox, was used as a standard.Â To verify that high antioxidant capability led to a reduction in intracellular ROS, the compounds were also tested for their ability to decrease ROS production in a neuronal cell culture model.Â Â To determine which mechanism, oligomer formation inhibition or ROS reduction, has the greatest effect on NF-κB activation, SH-SY5Y human neuroblastoma cells were treated with oligomers formed in the presence of polyphenols or treated with oligomers formed in the absence of polyphenols followed by addition of polyphenols capable of reducing ROS.Â Cells were fixed and stained for NF-kB (rabbit polyclonal anti-NF-κB?p65 antibody) and nuclear location (DAPI).Â Localization of NF-κB-p65 subunits within the nucleus, indicating NF-kB activation, was quantified.

Flavones tested showed a correlation between the number of hydroxyl groups present and the ability to inhibit Aβ oligomer formation.Â Results from the flavanols further suggest that not only the number of hydroxyl groups but also their placement is crucial for oligomer inhibition.Â Isoflavones and catechins were shown to lack the ability to inhibit oligomer formation even with the presence of several hydroxyl groups.Â The correlation between hydroxyl groups and antioxidant capacity was weaker, and at least one compound from each group exhibited antioxidant capability similar to that of Trolox.Â Polyphenols with either anti-aggregation or antioxidant activity were capable of reducing the activation of NF-kB, and polyphenols possessing both activities were the most effective inhibitors of NF-kB activation.

Characterization of polyphenols for anti-aggregation and antioxidant activities led to the identification of polyphenols with each of these activities as well as polyphenols with both activities.Â Distinct correlations were observed between the structure of the polyphenols and the ability to inhibit oligomer formation. This panel of compounds facilitated elucidation that both of these characteristics of polyphenols can protect against Ab-induced NF-kB activation associated with AD.Â Furthermore, polyphenols that posses both activities can exert synergism to function as dual-action drugs.

(683e) Mechanical Enhancement of VEGF Intracellular Signaling in Endothelial Cells

Spagnol, S. T., Carnegie Mellon University
Dahl, K. N., Carnegie Mellon University
Originally presented on: 11/7/2013 13:42:00 - 14:00:00

A majority of intracellular signaling that triggers changes in gene expression is propagated by chemical cascades that occur as distinct temporal and spatial events within the cell. However, there are other aspects of DNA structure which regulate genome expression including epigenetic modifications, higher order chromatin compaction as well as gene positioning and movement. Our work is aimed at investigating global genome scale reorganization during physiological processes and its role in genome regulation. Specifically, we focus on intracellular mechanical signaling processes that influence gene expression by altering nuclear organization. We examine changes in cytoskeletal force generation, propagation of force through the nuclear envelope and changes to the chromatin in the nuclear interior. To do this, we visualize stimulated actin reorganization in live cells and the associated active nuclear deformations through image analysis of nuclear envelope fluctuations. Further, the propagation of these forces to the nuclear interior is examined using particle tracking of fluorescently-tagged sub-nuclear markers, where we utilize the mechanical-functional coupling of chromatin to monitor mechanical changes as a means of inferring functional changes. Additionally, we investigate the associated changes in chromatin condensation state using fluorescence lifetime imaging microscopy and DNA texture analysis.

Our results indicate that we can decouple the effects of chromatin condensation and intranuclear active stresses arising from intracellular force generation using the particle tracking experiments through calculation of the mean squared displacement (MSD) fit to a power-law rheological model $MSD(\tau) = D_{eff}\tau^{\alpha}$, as is consistent with cell rheological modeling. These results demonstrate that intracellular force generation through the cytoskeleton and its propagation to the nuclear interior is a dominant means of chromatin agitation for nuclear reorganization. Additionally, our work indicates these effects are largely dominated by motor activity, with the actin network serving as a primary driver of intranuclear agitation via myosin-II. These findings are extended to VEGF-stimulated angiogenesis, which is known to induce a dramatic transition in gene expression. The results provide mechanistic details underlying the nuclear reorganization and subsequent turnover in gene expression, where early intracellular force generation induces chromatin agitation leading to chromatin decondensation and reorganization at the long time scales coinciding with changes in expression. Thus, our work highlights the effects of intracellular force generation on chromatin rearrangement and agitation as a means of altering higher order genome organization during physiological changes. These mechanical events occur as part of the underlying intracellular chemical signaling processes and act to enhance the efficacy of the response that brings about changes in gene expression.

(683f) Image Mining and the Temporal Evolution of Protein Spatial Patterns: A Case Study in Drosophila Embryogenesis

Dsilva, C., Princeton University
Lim, B., Princeton University
Shvartsman, S. Y., Princeton University
Kevrekidis, I. G., Princeton University
Originally presented on: 11/7/2013 14:00:00 - 14:18:00

The Extracellular Signal Regulated Kinase (ERK) pathway controls a wide range of processes in cells and tissues. Multiple lines of evidence suggest that biological effects of ERK depend on dynamics of its activation, but experimental studies of ERK dynamics have been largely limited to cells in culture. We study the dynamics of the activation of the epidermal growth factor receptor (EGFR) in the ERK pathway in the Drosophila embryo. Because ERK is double-phosphorylated (dpERK) directly downstream of the EGFR pathway, we follow the dynamics of dpERK in order to study the activation of the EGFR pathway.

Microfluidic arrays are used to collect dpERK concentration profiles from multiple embryos; because each embryo must be fixed prior to staining for dpERK, it is impossible to obtain dynamic trajectories of the dpERK concentration profiles from a single organism followed over time. Instead, we obtain concentration profiles for many different embryos, with each embryo fixed at a slightly different developmental time. We are interested in reconstructing dynamics from such "cross-sectional" data. We use linear (PCA) as well as nonlinear (diffusion maps) dimensionality reduction techniques to extract the relevant dynamics. Spatial symmetries within the embryo have important implications in this process: an "angular synchronization" problem must be solved to correctly align the data before extracting the dynamics. We discuss the algorithms involved in the process, and we illustrate their use in analyzing the evolution of concentration profiles along the circumference of the embryo, as well as in direct embryo images.

(683g) Systems Analysis of Components of PI3K/AKT Pathway Involved in Maintenance of Self-Renewal State of Human Embryonic Stem Cells

Mathew, S., University of Pittsburgh
Sundararaj, S., University of Pittsburgh
Mamiya, H., University of PIttsburgh
Banerjee, I., University of Pittsburgh
Originally presented on: 11/7/2013 14:18:00 - 14:36:00

Introduction: Differentiation of human embryonic stem cells (hESCs) to the definitive endoderm lineage is a critical first step towards deriving important functional cell types of the pancreas and liver. The process of lineage specific differentiation is under the control of complex and non-linearly interacting parallel signaling pathways. The complexity of these interactions restricts a purely experimental analysis and necessitates a systems level mathematical approach. Recent investigations into the signaling pathways have identified the key role of insulin mediated PI3K/AKT pathway in maintaining self-renewal and influencing differentiation [1]. It is established that the levels of the molecule p-AKT correlate well to the state of self-renewal (pAKT ON) and differentiation (pAKT OFF). However most of these studies analyze key molecules of the pathway in isolation. What is still lacking is a systems level understanding of the regulatory interactions of molecules as a component of a pathway. In particular, the PI3K pathway involves several regulatory modules; namely (i) insulin receptor activation and trafficking in the endosomes, (ii) receptor mediated activation of intracellular IRS1/PI3K/AKT, (iii) negative feedback by serine phosphorylation of IRS1 by molecules like PKC-ζ (iv) negative regulators like PTEN, SHIP and PTP and (iv) positive feedback by AKT. In the current work, we have analyzed the dynamics of the pathway and the sensitivity of the pathway to the various regulatory interactions to identify the mechanism of maintenance of p-AKT levels in self-renewal and its implications during differentiation.

Materials and Methods: The dynamics of the key components of the PI3K/AKT pathway in hESCs was first analyzed by experimentally stimulating H1 cells with 100 nM insulin after growth factor starvation. The dynamics was then compared to *in silico* simulations using a detailed mechanistic differential equation model of insulin mediated PI3K/AKT pathway (20 state variables, 25 parameters in the current form) by Sedaghat *et al* [2]. Using clustering techniques, the dynamics predicted by the model was compared with the experimental profiles to identify the feasible parameter ranges. The steady-states of the model were further analyzed using global sensitivity analysis (GSA) to identify the key interactions in the pathway that affect the levels of molecules like p-AKT. We utilized a computationally efficient algorithm called Random Sampling High Dimensional Model Representation (RS-HDMR) to capture sensitivity under non-linear interactions between the large numbers of model parameters. The results from GSA were validated in hESCs using small molecule inhibitors of the sensitive nodes in the self-renewal state. Finally, we combined the validated GSA results with global parameter optimization using evolutionary strategies to track the pathway dynamics in hESCs. These methods allowed us to capture detailed dynamics and steady-states of the pathway nodes as well as the sensitive biochemical processes that are involved in self-renewal and differentiation.

Results and Discussion: We analyzed the dynamics and steady states of four major nodes of the PI3K/AKT pathway. These include the active insulin receptors (p-IR), tyrosine IRS1 (p-IRS1 (Y)), serine IRS1 (p-IRS1 (S)) and p-AKT. Dynamics from insulin stimulation experiments showed that intracellular components of PI3K/AKT in hESCs show the typical overshoot with steady-state behavior influenced by negative feedback by p-IRS1 (S). From a detailed rate parameter based GSA of the steady-states, we identified that many of the negative regulators of the pathway sensitized the pathway more than the positive regulators. These processes kept p-AKT levels considerably low in the explored high dimensional parameter space. We validated the influence of negative regulators in hESCs where it was found that inhibition of negative feedback by p-PKC-ζ increased p-AKT levels considerably and PTEN and PTP inhibition had a secondary effect. During parameter optimization, we found parameter ensembles that favor a low rate of negative regulation by PTEN and PTP but simultaneously favor a strong negative feedback by p-IRS1 (S) to best explain the hESC dynamics. Thus our analysis allowed us to identify the mechanism that regulates p-AKT levels in hESCs during self-renewal and which should be down-regulated during differentiation.

Conclusions: Our detailed mechanistic model and experimental analysis identified the mechanism of control of p-AKT levels in hESCs under self-renewal. During self-renewal, negative feedback by pIRS1 (S) controls p-AKT levels at a sufficient threshold. Therefore, our work has identified an important role for negative feedback to maintain the PI3K/AKT pathway in hESCs under homeostatic conditions which was previously unappreciated. We are currently investigating the role of this negative feedback in the context of cross-talks with TGF-β induced endoderm differentiation of hESCs.

References:

1. Singh *et al.* Cell Stem Cell (2012) 10: 312-326

2. Sedaghat *et al.* American J of Physiol-Endocrinol and Metabol (2002) 283: E1084-E1101

(683h) Directing Stem Cell Differentiation By Engineering Cell-Cell Adhesion Pathways

Alimperti, S., State University of New York at Buffalo
Row, S., State University of New York at Buffalo
George, T. A., Baylor College of Medicine
Agarwal, S., Baylor College of Medicine
Andreadis, S., State University of New York at Buffalo
Originally presented on: 11/7/2013 14:36:00 - 14:54:00

The differentiation fate of a stem cell is controlled by its microenvironment. Although the role of soluble factors (e.g. TGF-β1) in mesenchymal stem cell (MSC) differentiation towards smooth muscle (SMC) lineage has been studied, the role of intracellular adhesion in this process remains elusive. Our previous studies have shown that cell-cell adhesion is necessary to induce MSC differentiation into vascular fate, particularly through the cell surface molecule, OB-cadherin. To further investigate the role of OB-cadherin in MSC proliferation and differentiation potential, we employed a fusion protein of OB-cadherin with the Fc domain (OB-Fc), which allowed us to study cadherin-cadherin interactions in an isolated manner. In particular, OB-Fc is a chimeric homo-dimer consisting of CH2 and CH3 (Constant Heavy-chain) region of FC and is linked through the region of the antibody to all five extracellular domains of OB cadherin. Initially, IgG-1 which preferentially bound to FC fragment was coated on a hydrophobic plate and afterwards, Bone Marrow Mesenchymal Stem Cells (BM-MSCs) were plated on OB-FC surface. To this end, these cells exhibited increased proliferation (~5-fold) and suppressed expression of senescence-associated proteins such as p21 over several passages. At the same time, culture on OB-Fc surface enhanced the BM-MSCs differentiation potential towards the SMC lineage as evidenced by the levels of key myogenic markers such as αSMA and CNN1. In particular, engagement of OB-cadherin increased αSMA expression through ROCK pathway and its downstream effectors such as SRF. On the other hand, knocking down OB-cadherin using shRNA reversed these effects. Finally, tissue constructs prepared from cells that were propagated on OB-Fc exhibited significantly increased vascular contractility and improved mechanical properties. Notably, these results were confirmed *in vivo* using OB-cadherin knockout mice ($OB^{-/-}$). In particular, smooth muscle containing organs from $OB^{-/-}$ animals e.g. bladder exhibited diminished vascular reactivity, decreased collagen content and significantly decreased mechanical properties. To this end, our results demonstrate that engineering OB-FC containing substrates that mimic the in vivo microenvironment promotes MSC proliferation and differentiation potential into SMC, thereby suggesting that this may be a novel and potentially clinically applicable strategy for directing stem cell fate.

(686a) Rapid Evolution of Enzymes and Regulatory Factors Through An in Vivo Continuous Evolution (ICE) Approach in Yeast

Crook, N., The University of Texas at Austin
Abatemarco, J.,
Sun, J., University of Illinois at Urbana-Champaign
Schmitz, A., The University of Texas at Austin
Alper, H., The University of Texas at Austin
Originally presented on: 11/7/2013 12:30:00 - 12:48:00

Engineered cellular systems have the potential to solve pressing issues facing society, including the production of chemicals from renewable sources, degradation of environmental toxins, and the treatment of disease. However, the engineering efforts leading to these outcomes almost always rely upon the optimization of enzymes and regulatory factors. Typically, a directed evolution approach is undertaken to improve performance at both the individual component and the final strain level. However, traditional directed evolution techniques introduce mutations through *in vitro* procedures, which fundamentally limit the overall library size and thus can constrain success. Here, we describe a new technique for continuous evolution of desired genes and synthetic parts that reduces the practice of library generation to a simple cloning step followed by cell outgrowth. In this technique, library sizes are not limited by transformation efficiency and instead scale with culture volume, which is the fundamental upper limit to the size of any evolving population. The utility of this approach is demonstrated with both model proteins as well as key metabolic and regulatory factors to demonstrate the benefits of continuous evolution over traditional approaches. These results show the capability of ICE to significantly accelerate strain engineering efforts and enable advanced studies of gene evolution in eukaryotes.

(686b) Syncretic Adaptive Evolution: Combining Multiple Tolerance Phenotypes With Conjugation

Winkler, J. D., Texas A&M University
Kao, K., Texas A&M University
Originally presented on: 11/7/2013 12:48:00 - 13:06:00

Evolutionary engineering is a widely applicable method to improve strain tolerance and production phenotypes. Typically, after the conclusion of an adaptive laboratory evolution experiment, resistant clones will be isolated and characterized on a genetic, regulatory, and phenotypic level to determine the underlying mechanisms of adaptation. Engineering the discovered tolerance phenotype into another strain then requires extensive manual reconstruction, depending on the number and type of mutations involved. This process must be repeated for every required strain, further slowing the process of strain development. Here, in order to circumvent this laborious process, we present a conjugation-based method to transfer furfural resistance from an evolved donor population to arbitrary recipient strains through simple selection and screening. The feasibility of this technique is demonstrated by endowing previously evolved acetate or n-butanol resistant strains with furfural tolerance, followed by characterization of the multi-tolerant recombinants. Overall, this technique should be a useful new tool for strain engineers to rapidly transfer and examine complex tolerance phenotypes.

(686c) Automation and Miniaturization Towards High-Throughput Genome Engineering

Zeitoun, R. I., University of Michigan
Gill, R. T., University of Colorado Boulder
Originally presented on: 11/7/2013 13:06:00 - 13:24:00

Engineering organisms typical requires a three-step design cycle of creating a genetically diverse library, testing that library's phenotype and finally learning which trait of combinations of modifications contribute towards a desired phenotype. However, the complexity and non-linearity associated with genome engineering coupled with the low throughput of the design cycles remains a barrier towards thorough and efficient genome engineering. We describe approaches to improve the rate of library generation and testing through automation and miniaturaztion, respectively. The rate of library generation is improved with a bench-top cell-culture, liquid handling and electroporation system. This allows for the rapid and targeted generation of a diverse library population. Screening and testing of this diverse population is adapted in a microfluidic setup to improve screening throughput while reducing associated screening costs. These approaches are applied towards olefin and alcohol production in *E.coli*.

(686d) Connecting Anthranilate Catabolism and Anabolism Leads to a Novel Muconic Acid Biosynthetic Approach

Lin, Y., the University of Georgia
Sun, X., Beijing University of Chemical Technology
Huang, Q.,
Yuan, Q., Beijing University of Chemical Technology
Yan, Y., the University of Georgia
Originally presented on: 11/7/2013 13:24:00 - 13:42:00

Adipic acid is an important platform chemical that can be used for plastic and nylon production. The global demand for adipic acid is over 2 billion kilograms annually. Currently, the major approach to adipic acid production relies on chemical synthesis using benzene as the starting material. However, this approach is generally considered nonrenewable and environmentally incompatible due to the toxicity of the starting material and intermediates and the release of greenhouse gas N_2O. Microbial synthesis from renewable carbon sources provides a feasible and promising alternative in the circumstance of environment deterioration and petroleum depletion. Indeed, adipic acid can be easily generated by hydrogenation of muconic acid, a naturally existing intermediate in aromatic compounds degradation by some soil bacteria. Draths and Frost first reported the constitution of an artificial pathway in *Escherichia coli* by shunting 3-dehydroshikimate from shikimate pathway, affording the production of muconic acid from glucose. Here we designed a novel muconic acid biosynthetic pathway by connecting anthranilate catabolism with anabolism. Specifically, anthranilate, an *E. coli* endogenous intermediate in the tryptophan biosynthetic branch was sequentially converted into catechol and muconic acid by anthranilate 1,2-dioxygenase (ADO) and catechol 1,2-dioxygenase (CDO). First of all, screening for efficient ADO and CDO from different microorganisms led to the convertion of anthranilate to gram per liter-level of muconic acid. Further, to achieve the *de novo* muconic acid biosynthesis, anthranilate overproducing strains were constructed by blocking tryptophan biosynthesis and over-expressing the key enzymes in shikimate pathway. Interestingly, introduction of a strengthened glutamine regeneration system by over-expressing glutamine synthase dramatically enhanced anthranilate production. Finally, the engineered strain carrying the full pathway produced 390 mg/L of muconic acid from simple carbon sources in shake flask. This approach will demonstrate scale-up potential for microbial production of muconic acid with further condition and pathway optimizations.

(686e) Installing the Wood-Ljungdahl Pathway in Clostridium Acetobutylicum

Fast, A. G., University of Delaware
Al-Hinai, M. A., Sultan Qaboos University
Schmidt, E. D., University of Delaware
Sandoval, N. R., University of Delaware
Papoutsakis, E. T., University of Delaware
Originally presented on: 11/7/2013 13:42:00 - 14:00:00

The ability to express the genes necessary for instating a functional Wood-Ljungdahl (WL) pathway into *C. acetobutylicum* constitutes a major advance in that the WL pathway has never before been instated into a heterologous organism. In the specific case of *C. acetobutylicum,* it would allow one to grow this organism to high cell densities and then switch the available carbon substrate from a sugar to CO_2/H_2, CO/CO_2, or a mixotrophic culture with both sugars and gases, thereby enabling the production of carboxylic acids and solvents from syngas in a flexible and scalable fermentation system. From the perspective of biological fundamentals, the ability to install this complex, primordial pathway into a heterologous host would constitute a major advance in cell/metabolic engineering that would open new horizons for pathway engineering and synthetic approaches in the genus *Clostridium.* Comparative genetic analysis of three sequenced acetogens, *C. ljungdahlii*, *C. carboxidivorans*, and *C. difficile*, showed that the WL pathway genes of these clostridial acetogens are concentrated in a highly conserved, 18 kilobase region on their respective genomes. By examining the *C. acetobutylicum* genomes, we focused on expressing 11 core genes coding for enzymes and accessory proteins. To achieve this goal, we developed a system of two co-existing plasmids combined with a method to integrate some of these genes, starting with the formate dehydrogenase gene, into the chromosome. We will present the new chromosomal integration method and the step-by-step verification of the expression and function of the cloned genes/proteins as well as culture experiments to test the functional installation of the WL pathway in *C. acetobutylicum*.

(686f) Driving Force and Oxygen Tolerance As Design Principles to Enable Efficient Photosynthetic n-Butanol Biosynthesis in Cyanobacteria

Lan, E. I., University of California Los Angeles
Liao, J. C., University of California Los Angeles
Originally presented on: 11/7/2013 14:00:00 - 14:18:00

Direct chemical and fuel production from CO_2 has emerged as a promising approach for sustainability and CO_2 mitigation. However, metabolic engineering of photosynthetic microorganism such as cyanobacteria for the production of fuels or chemicals is challenging, particularly when the pathway involves oxygen-sensitive enzymes and/or unfavorable thermodynamics. We aim to take *Clostridium* n-butanol fermentation pathway as an example to demonstrate some of the design principles for engineering cyanobacteria. Two of the major obstacles for using *Clostridium* pathway to produce n-butanol in cyanobacteria are the thermodynamically unfavorable condensation of acetyl-CoA and the oxygen sensitivity of its enzymes. To address the issue of unfavorable acetyl-CoA condensation to acetoacetyl-CoA, the first and committed step of the *Clostridium* pathway, we incorporated energy of ATP hydrolysis into this reaction by utilizing malonyl-CoA dependent formation of acetoacetyl-CoA. Using this strategy, we effectively push carbon flux against the natural thermodynamic barrier of the *Clostridium* pathway and achieved n-butanol synthesis in cyanobacterium *Synechococcus elongatus* PCC 9742. To address the difficulty of oxygen sensitivity, we characterized six oxygen tolerant aldehyde dehydrogenases for their aerobic activity toward butyryl-CoA. Replacing *Clostridium* aldehyde dehydrogenase with its oxygen tolerant counterpart in the n-butanol synthesis pathway resulted in 20 fold increase of n-butanol production. Together, these results demonstrate the importance of driving force and oxygen tolerance for a successful pathway design in cyanobacteria.

(686g) Engineering Styrene Biosynthesis in Yeast Via Rational and Combinatorial Approaches

McKenna, R., Arizona State University
Pugh, S., Arizona State University
Thompson, B., Arizona State University
Nielsen, D. R., Arizona State University
Originally presented on: 11/7/2013 14:18:00 - 14:36:00

Renewable production of styrene, a versatile commodity chemical and monomer, has to date been limited by the low toxicity threshold of *E. coli* (~300 mg/L). To address this, multiple alternative methods have been investigated in parallel, including the development of more robust yeast platforms, as well as via multiple product removal strategies. *Saccharomyces cerevisiae* BY4741 as well as more robust industrial yeast strains have been investigated as initial strains for the development of novel styrene producers. As the immediate endogenous precursor to the styrene pathway is the proteinogenic amino acid phenylalanine, the first aim of this study was to engineer a feedback-resistant mutant capable of its overproduction. Multiple approaches of both rational and combinatorial nature were explored to this end. This included engineering known feedback-resistant point mutations in the DAHP synthase (ARO4 ^{K229L}) and the chorismate mutase (ARO7 ^{G141S}) as well as the use of classical mutagenesis and anti-metabolites to increase flux through the shikimic acid pathway. Multiple rounds of mutagenesis coupled with increasing growth challenges ultimately elucidated a strain with increased flux through the phenylalanine pathway. This approach had led to the development of yeast strains capable of producing over 350 mg/L phenylalanine from glucose. To achieve styrene biosynthesis, it was found that only over-expression of phenylalanine ammonia lyase (PAL2 from *Arabidopsis thaliana*) to deaminate phenylalanine to *trans*-cinnamate was required as *trans*-cinnamate was then naturally decarboxylated to styrene by the native FDC1. Factors influencing FDC1 transcription and the impact of its expression level were investigated to ensure ample flux through the pathway was possible. At last, multiple approaches for *in situ* styrene removal from cultures including via liquid-liquid extraction and vapor stripping/collection were also investigated. These approaches led to improvements in overall productivity.

(686h) Improving D-Glucaric Acid Production in E. Coli Through Directed Evolution and Delayed Induction

Shiue, E., Massachusetts Institute of Technology
Prather, K. J., Massachusetts Institute of Technology
Originally presented on: 11/7/2013 14:36:00 - 14:54:00

D-glucaric acid is a naturally occurring compound that has been investigated for a number of potential uses, including cancer treatment, cholesterol reduction, and biopolymer synthesis. A pathway for the biosynthesis of D-glucaric acid from glucose has been constructed in *E. coli*, consisting of three heterologously expressed natural enzymes: myo-inositol-1-phosphate synthase (INO1) from *Saccharomyces cerevisiae*, myo-inositol oxygenase (MIOX) from *Mus musculus*, and uronate dehydrogenase (Udh) from *Pseudomonas syringae*. Characterization of the pathway's constituent enzymes revealed MIOX to be the rate-limiting pathway enzyme. Achievement of industrially relevant titers, therefore, hinges on improving the specific activity and/or stability of MIOX. Several synthetic biology and metabolic engineering strategies have been explored to improve MIOX activity and stability, including time-delayed induction of pathway enzymes, directed evolution of MIOX for improved activity, and protein fusions for enhancement of soluble expression. We will present experimental results from the application of these strategies towards enhancing MIOX activity and glucaric acid productivity.

(715a) Biomaterial Properties of Aggregating Tripeptides Designed Using a De Novo Multimeric Protein Design Framework

Smadbeck, J., Princeton University
Khoury, G. A., Pennsylvania State University-University Park
Chan, K. H., Nanyang Technological University
Hauser, C. A. E., Nanyang Technological University
Floudas, C. A., Princeton University
Originally presented on: 11/7/2013 15:15:00 - 15:33:00

Computational protein design has had great success in the development of novel therapeutics, novel protein folds, and protein-based biomaterials. Much of this design has focused on the design of protein tertiary structure, but as computing power improves it becomes increasingly possible to shift this focus to the quaternary structure of protein complexes. In nature any number of identical or non-identical proteins can self-assemble into ordered (amyloid) or disordered (hydrogel) complexes. Biologically these complexes can be either essential or disrupting to cell function. Additionally, they have application in the design of self-assembling biomaterial. There has been much effort in the design of protein-protein interfaces, which is generally restricted to the design of a single monomer in a multimeric complex [1-9]. Recently, we have developed a novel de novo design method capable of designing for multimeric peptides/proteins simultaneously. This method has been applied to the design of aggregating peptides with success and we present the designed peptides and the properties of the resulting designed hydrogel and crystal structures.

The novel protein design method involves three stages and is capable of improving the aggregation affinity of a multimeric system. The first stage is a Mixed Integer Linear Optimization model [10-12] Sequence Selection stage capable of designing multiple identical or non-identical protein chains simultaneously. A flexible template structure for the protein complex is initially determined through Molecular Dynamics (MD) simulations. The potential energy of the system is minimized to generate a rank-ordered list of designed peptide sequences. The second stage calculates the Fold Specificity of each designed sequence for the flexible aggregation template. The third stage employs MD simulations in CHARMM [13] to predict the assembly of multiple peptides into a macromolecular structure. From these MD generated structure ensembles, we calculate approximate molecular partition functions of the structures, which are used to calculate an approximate aggregation affinity of the designed proteins [14]. This approximate aggregation affinity is used as a matric in choosing the peptides that will be tested experimentally.

The work detailed in the presentation shows the application of the novel Multimeric Protein Design method to the design of aggregating peptide systems. Experimental validation of the designed aggregating tripeptides has been carried out successfully and demonstrates aggregation into hydrogel and crystal structures. Several of the hydrogel aggregating peptides demonstrate higher stability, aggregation rate, and storage moduli than the template tripeptide. Additionally, one of the designed peptides demonstrates rapid self-assembly into large crystals that lend themselves well to crystallographic analysis. The biomaterial properties of the designed peptides provide important information for the understanding of which properties control peptide hydrogel aggregation and crystallization.

[1] Kortemme, T. and D. Baker. Computational design of protein–protein interactions. *Current Opinion in Chemical Biology*, 8:91–97, 2004.

[2] AndrÃ©, I., C. E. M. Strauss, D. B. Kaplan, P. Bradley, and D. Baker. Emergence of symmetry in homooligomeric biological assemblies. *PNAS* 105 (42) 16148-16152, 2008.

[3] Mandell, D. J. and T. Kortemme. Computer-aided design of functional protein interactions. *Nat. Chem. Biol.* 5, 797–807, 2009

[4] Huang, P-S., J. J. Love and S. L. Mayo. A de novo designed protein protein interface. *Protein Sci.* 16:2770-2774, 2007.

[5] Fu, X.R., H. Kono, J. G. Saven. Probabilistic approach to the design of symmetric protein quaternary structures. *Protein Eng.* 16:971-977, 2003.

[6] Colombo, G., P. Soto, and E. Gazit. Peptide self-assembly at the nanoscale: a challenging target for computational and experimental biotechnology. *Trends in Biotechnology*, 25(5):211-218, 2007.

[7] Bellows, M. L., H. K. Fung, C. A. Floudas, A. Lopez de Victoria, and D. Morikis. New Compstatin Variants Through Two De Novo Protein Design Frameworks. *Biophysical Journal,* 98:2337–2346, 2010.

[8] Bellows, M. L., M. S. Taylor, P. A. Cole, L. Shen, R. F. Siliciano, H. K. Fung, and C. A. Floudas. Discovery of entry inhibitors for HIV-1 via a new de novo protein design framework. *Biophysical Journal,* 99:3445–3453, 2010.

[9] Bellows-Peterson, M. L., H. K. Fung, C. A. Floudas, C. A. Kieslich, L. Zhang, D. Morikis, K. J. Wareham, P. N. Monk, O. A. Hawksworth, and T. M. Woodruff. De Novo Peptide Design with C3a Receptor Agonist and Antagonist Activities: Theoretical Predictions and Experimental Validation. *Journal of Medicinal Chemistry 55* (9), 4159-4168, 2012.

[10] Fung, H. K., M. S. Taylor, and C. A. Floudas. Novel formulations for the sequence selection problem in de novo protein design with flexible templates. *Optimization Methods and Software*, 22:51-71, 2007.

[11] Fung, H. K., C. A. Floudas, M. S. Taylor, L. Zhang, and D. Morikis. Toward Full-Sequence De Novo Protein Design with Flexible Templates for Human Beta-Defensin-2. *Biophysical Journal* 94:584-599, 2008.

[12] Fung, H. K., W. J. Welsh, and C. A. Floudas. Computational De Novo Peptide and Protein Design: Rigid Templates versus Flexible Templates. *Industrial & Engineering Chemistry Research*, 47:993-1001, 2008.

[13] Brooks, B. R., et al. CHARMM: the biomolecular simulation program. *Journal of Computational Chemistry*, 30(10):1545-1614, 2009.

[14] Lilien, R. H., B. W. Stevens, A. C. Anderson, and B. R. Donald. A novel ensemble-based scoring and search algorithm for protein redesign and its application to modify the substrate specificity of the Gramicidin Synthetase A Phenylalanine Adenylation enzyme. *Journal of Computational Biology*, 12:740-761, 2005.

(715b) Production of Polymalic Acid (PMA) From Biomass Hydrolysates By Aureobasidium Pullulans

Zhou, Y., The Ohio State University
Cheng, C.,
Yang, S. T., The Ohio State University
Originally presented on: 11/7/2013 15:33:00 - 15:51:00

PMA is a biocompatible, biodegradable and bioresorbable water-soluble polymer composed of malic acid. PMA can be readily functionalized by the large number of carboxyl groups present in the molecule. Therefore, PMA has great potential to be used as prodrugs or drug carriers. Currently, PMA can be produced by either chemical synthesis or microbial fermentation. Compared to chemical ways, bioprocess represents a better PMA source due to the advantages of higher enantiomerical purity in products, flexibility to feedstock and being more environmentally friendly. Previously, we have developed an efficient fermentative process for high-level production of PMA from *Aureobasidium pullulans* strain ZX-10 with high titer of 123 g/L. However, the increasing cost of raw materials such as glucose and water may significantly hamper the economical competitiveness of PMA fermentation.

In this work, we achieved efficient PMA production with decent titer (>20g/L) and yield (0.5~0.8g/g sugar) from agricultural biomass substrates such as cotton stalk, cassava bagasse and soybean hull hydrolysates and soybean molasses. Among these substrates, oligosaccharide contained in soybean molasses including stachyose and raffinose can be directly converted to PMA without pretreatment while soybean hull hydrolysate gave the highest titer (38 g/L) and yield (0.8g/g sugar). Therefore, these results showed that using inexpensive substrates such as agricultural biomass hydrolysate provides an efficient and economical way for PMA production and is promising for industrial application.

(715c) Development of Functional Proteinaceous Hydrogels for Biotechnology Applications

Banta, S., Columbia University
Dooley, K., Columbia University
Bulutoglu, B.,
Tu, R., City College of New York (of CUNY)
Originally presented on: 11/7/2013 15:51:00 - 16:09:00

Hydrogels consisting of three-dimensional cross-linked polymer networks have emerged as interesting scaffolds for a host of applications including regulated drug delivery, tissue engineering, and biosensors. Self-assembly of these supramolecular networks is driven by the incorporation of a cross-linking domain into a monomeric building block. We have used protein engineering to create hydrogels by genetically appending cross-linking domains to globular proteins, which results in unique proteinacous biomaterials that incorporate biological activity [1]. In our previous work, we have investigated α-helical leucine zipper domains to facilitate cross-linking between proteins. By appending these helices to the N and C termini of globular proteins, we have created hydrogels with benchmark proteins such as GFP [2], and more interestingly, with catalytic enzymes such as organophosphate hydrolase (OPH) [3] and alcohol dehydrogenase (AdhD) [4]. More recently, we have created a hydrogel out of three self-assembling dehygrogenase enzymes which results in a hydrogel that contains a synthetic metabolic pathway that is capable of oxidizing methanol to carbon dioxide [5]. We have now been expanding this approach as we have been extensively characterized a calcium responsive repeats-in-toxin (RTX) domain and evaluated its potential as an alternative cross-linker. In the absence of calcium, the RTX domain is largely unstructured. In calcium rich environments, it will reversibly fold into a beta helix secondary structure [6]. By rationally designing one face of the folded beta helix to contain leucine residues, we have created an environmentally-responsive cross-linking domain capable of self-assembly only in the presence of calcium [7]. The folded RTX domain contains a second face amenable to mutation, which we have also designed to contain leucine residues. We are currently characterizing this "double-face" leucine-rich RTX domain and evaluating is mechanical and biophysical properties using circular dichroism, michrorheology, and multi-angle light scattering. The RTX domains represent a truly "smart" cross-linking strategy in that the addition of a specific effector molecule (calcium) enables the reversible formation of the hydrogel network.

1. Banta, S., I.R. Wheeldon, and M.A. Blenner, *Protein engineering in the development of functional hydrogels.* Annu Rev Biomed Eng, 2010. **12**: p. 176-186.

2. Wheeldon, I.R., S.C. Barton, and S. Banta, *Bioactive proteinaceous hydrogels from designed bifunctional building blocks.* Biomacromolecules, 2007. **8**(10): p. 2990-4.

3. Lu, H.D., I.R. Wheeldon, and S. Banta, *Catalytic biomaterials: engineering organophosphate*

hydrolase to form self-assembling enzymatic hydrogels. Protein Eng Des Sel, 2010. **23**(7): p. 559-566.

4. Wheeldon, I.R., E. Campbell, and S. Banta, *A chimeric fusion protein engineered with disparate functionalities-enzymatic activity and self-assembly.* J Mol Biol, 2009. **392**(1): p. 129-42.

5. Kim, Y.H., et al., *Complete oxidation of methanol in biobattery devices using a hydrogel created from three modified dehydrogenases.* Angewandte Chemie, 2013. **52**(5): p. 1437-40.

6. Blenner, M.A., et al., *Calcium-Induced Folding of a Beta Roll Motif Requires C-Terminal Entropic Stabilization.* J Mol Biol, 2010. **400**(2): p. 244-256.

7. Dooley, K., et al., *Engineering of an environmentally responsive beta roll peptide for use as a calcium-dependent cross-linking domain for peptide hydrogel formation.* Biomacromolecules, 2012. **13**(6): p. 1758-64.

(715d) Recombinant Spider Silk Protein Production in Escherichia Coli Cell-Free Protein Synthesis

Hong, S. H., Northwestern University
Ye, X., Virginia Tech
Jewett, M. C., Northwestern University
Originally presented on: 11/7/2013 16:09:00 - 16:27:00

Spider dragline silk is a tough natural fiber that promises numerous industrial and biomedical applications. Producing native-sized recombinant spider silk (>250 kDa) has been limited because of difficulties in translation due to unbalanced tRNA pools and the extremely long template. Recently, these challenges were addressed for the first time *in vivo* by engineering an *Escherichia coli* strain with an elevated glycyl-tRNA pool. Here, we sought to express native-sized spider silk using cell-free protein synthesis (CFPS). In CFPS, components can be tuned to unnatural levels, new components (natural and unnatural) can be added or synthesized and can be maintained at precise ratios, and the chemical environment can be controlled, actively monitored, and rapidly sampled. We targeted the synthesis of two spider silk-like proteins, the major ampullate spidroins 1 and 2 (MaSp1 and MaSp2). We constructed plasmid templates of different repeat sizes of spider silk gene ranging 4 to 128 repeats via multiple rounds of recursive cloning and tested in CFPS using *E. coli* cell extract. Synthesis of spider silk protein was detected up to 64 repeats (170 kDa). To our knowledge, this is largest polypeptide expressed using bacterial CFPS. The addition of excess glycyl-tRNA and glycine improved spider silk protein production. Beyond improvements afforded by tuning the components of translation, I will also discuss our efforts to incorporate unnatural amino acids into native-sized spider silk proteins by amber suppression in extracts from genomically recoded organisms (GROs) for introducing new functional properties. Looking forward, we anticipate that protein-like sequence defined polymers hold promise as a new class of high performance, lightweight, and innovative materials. Moreover, we expect that CFPS will provide a highly tunable approach for the production of genetically encoded materials that are difficult to produce *in vivo*.

(715e) Sticky Adeno-Associated Viral Vectors for Spatially Controlled Gene Delivery

Kim, E., Yonsei University
Song, I. T., KAIST
Lee, S., Yonsei University
Kim, J. S., Yonsei University
Jang, J. H., Yonsei University
Lee, H., Korea Advanced Institute of Science and Technology
Originally presented on: 11/7/2013 16:27:00 - 16:45:00

Viral gene delivery is a powerful approach for tissue engineering applications. It can modulate cellular behaviors to grow specific tissues to replace those have damaged organs. In growing tissues, spatially controlled secretion of inductive factors is critical to elucidate a variety of biological mechanisms. It can control spatial distribution of extra cellular molecules on substrates and overcome limitations found in protein delivery such as systemic toxicity and short protein half-life. AAV is one of the safe and highly efficient viral vehicles for the delivery of therapeutic genes for tissue engineering applications. Hence, we developed spatially controlled AAV-mediated gene delivery system in order to intensively control the neurotropic factors, such as nerve growth factor (NGF). In this study, we developed patterned gene delivery system with adhesive materials, PEI-catechol which were used to modify surfaces of AAV viruses for sticky viruses. The negative surface charge of AAVr3.45 was successfully complexed with PEI-catechol which is often used to non-viral gene delivery as polycation polymers. Because of the adhesive characteristic of sticky virus (AAVr3.45/PEI-C) we were able to generate spatially controlled viral patterns on substrates by the simple "gene-vector drawing" technique without any additional steps. Therefore, this study developed a simple, cost effective and versatile platform for the spatially patterned gene expressions that can be utilized in the functional tissue organization. *(ref: Angew. Chem. Int. Ed. 2012, 51, 5598 –5601)*

(715f) Engineering Patterned Signalling Molecule Release Within Hydrogels Via Programmed Reaction and Diffusion Processes

Schulman, R., Johns Hopkins University
Scalise, D.,
Originally presented on: 11/7/2013 16:45:00 - 17:03:00

A fundamental challenge in bioreactor design and tissue engineering is that when cells are grown outside of their normal environment, portions of the complex communications cells expect to receive during growth may be missing, leading to changes in function. While some aspects of cellular microenvironment can be synthetically emulated using tools such as microfluidics, the reconstruction of complex signalling gradients and gradients in 3 dimensions remain largely unaddressed.

We present a mechanism for the rational design of complex 2- or 3-dimensional spatial gradients for sustained release of nucleic acids within a hydrogel: the output molecule is released only within regions of the gel that are part of a programmed pattern. This pattern, which can have feature sizes on the scale of microns to centimeters, emerges from a designed network of interactions between interacting synthetic nucleic acid species embedded in the gel. It requires as input to generate the pattern only a small number of "spots" of specific molecules that control pattern size. As a result, no specialized equipment is required to produce these patterns. We demonstrate the design of simple patterns such as points and lines and then how these structures can be modularly combined to produce more complex patterns.

(715h) Genetic Screening of a Saccharomyces Cerevisiae Strain Provides Insight Into the Cellular Molecular Response Towards Artificial Synthetic Peptides

Mora, C.,
Chaparro, F.,
Nieves, J. C., University of Puerto Rico
Gellman, S. H., University of Wisconsin-Madison
Palecek, S. P., University of Wisconsin-Madison
Ortiz-Berm dez, P., University of Puerto Rico
Originally presented on: 11/7/2013 17:21:00 - 17:39:00

The development of new drugs to treat fungal infections has become an urgent necessity due to the increasing resistant strains of fungi and bacteria, the currently limited selection of treatment and its adverse side effects along with the increasing numbers of immunocompromised patients. A possible solution to this problem is the development of β-peptides. These foldamers, are short nanorods made from artificial amino acids, that have been designed to mimic natural antimicrobial peptides known as host-defense peptides (HDPs). HDPs are made by most living organisms as a way to counteract a broad spectrum of microbial attacks, which include bacteria and fungi. Some advantages of using β-peptides compared to natural HDPs are: their conformational stability, predictable folding behavior, resistance to proteases and activity at physiological salt concentrations. From previous studies, the selected β-peptide sequence, $(ACHC\text{-}\beta^3V\text{-}\beta^3K)_3$ nicknamed as $(ACHC\text{-}V\text{-}K)_3$, because of its decent antifungal potency compared to amphotericin B properties and its low toxicity against erythrocytes. The mechanism of action of β-peptides still remains unknown and in order to find possible routes of action we hypothesized that, using a genetic molecular approach, we can identify genes that are actively involved in the phenotypic response the genetic model *S. cerevisiae*. Using this approach, 15 genes have been identified, three of which are being discussed in this work (HSP33, FEN2, SRN2).

(716a) Bioprospecting the Cellulases and Xylanases Thermozymes for the Production of Biofuels

Kumar, S., Sardar Swaran Singh National Institute of Renewable Energy
Behera, S.,
Arora, R.,

Originally presented on: 11/7/2013 15:15:00 - 15:40:00

Bioprospecting the cellulases and xylanases thermozymes for the production of biofuels

Shuvashish Behera, Richa Arora, Sachin Kumar*

Sardar Swaran Singh National Institute of Renewable Energy, Jalandhar-Kapurthala Road, Wadala Kalan, Kapurthala, Punjab-144601, India

Abstract

A wide variety of microorganisms are present in the environment, which can be harnessed for the purpose of mankind. The isolation and screening of the efficient enzymes like cellulases and xylanases producing microbes can be helpful to overcome the current challenges in biofuel production. This study focuses on isolation of efficient cellulase and xylanase producing thermophilic bacteria found in different soil samples, which could be used in enzymatic saccharification of biomasses. The bacterial isolates with appreciable hydrolytic zones were isolated from soil sample based on zone of clearance on solid plates. All isolates were studied by cultural morphological characteristics and microscopic examinations. About 17 bacterial strains showed the cellulose and/or xylanase enzyme activity. Some isolates showed individual enzyme activity i.e. either cellulase or xylanase and rest showed the activity for both the enzymes. Four isolates designated NIRE C3, NIRE C16, NIRE U1 and NIRE U2 showed superior cellulase activity with more than 15 mm zone of clearance and two isolates NIRE B7 and NIRE B8 shows highest xylanase activity producing more than 16 mm zone of clearance after 24 h of incubation at 50°C. These cultures represent excellent candidates for further enzyme production and characterization.

*Corresponding author. Tel.: +91 1822 255543; fax: +91 1822 255544.

E-mail address: sachin.biotech@gmail.com (S. Kumar)

(716b) Novel Thermostable Xylanase Production From Termite Gut Microbes and Its Application in Biofuels

Dheeran, P., 1SRM Research Institute, SRM University
Kumar, S., Sardar Swaran Singh National Institute of Renewable Energy
Adhikari, D., Indian Institute of Petroleum
Originally presented on: 11/7/2013 15:40:00 - 16:05:00

Novel thermostable xylanase production from termite gut microbes and its application in biofuels

Pratibha Dheeran[1,*] , Sachin Kumar2 and Dilip K. Adhikari[3]

[1]SRM Research Institute, SRM University, Kattankulathur-603 203, Tamilnadu, India

[2]Sardar Swaran Singh National Institute of Renewable Energy, Kapurthala-144601, India

[3]Biotechnology Area, Indian Institute of Petroleum, Dehradun-248 005, India

Abstract

The termites harbor more than 200 species of microbes that produce the novel cellulosic and hemicellulosic degrading enzymes, which in turn can be efficiently applied for the production of ethanol from lignocellulosic biomass in biofuel industries. In the current work, wood feeding termites were collected from the forest of Rajaji National Park Range, Dehradun (India) and they were used to isolate the thermostable xylanse producing microbes. As far as literature concern, very few information is available on using the termite gut microbes for xylanase production.

The wood feeding termite gut's extract were used for screening of xylanase producing microbes at 50°C. A novel thermophilic strain *Panebacillus macerans* IIPSP3 was isolated which was found to be potent thermostable xylanase producer. Maximum xylanase production was observed in late exponential phase at 50°C and pH 6.0. The isolated thermostable xylanase worked at broad temperature and pH range 40-90°C and 3.5-9.5, respectively, with optimum activity at temperature of 60°C and pH 4.5. This novel strain *P. macerans* IIPSP3 from wood feeding termites could utilize agro wastes such as corn cob as a substrate for growth and produce high level of xylanase, which has benefits for its economical production and industrial applications. IIPSP3 xylanase is a novel enzyme, which can be used in biofuel and paper & pulp industries as it could efficiently be hydrolyzed the lignocellulosic biomass at pH 4.5 and 9.0. The isolated xylan is cellulase free, which can be used in biobleaching of pulp.

Key Words: Higher Termite, Thermostable xylanase, Biofuels.

*Author for correspondence: E-mail: pratibhadheeran@gmail.com

(716c) Inducing Cellulase Activity of Microbulbifer Hydrolyticus Using Hot-Water Wood Extract As Substrate

Arens, K., SUNY College of Environmental Science and Forestry
Liu, S., SUNY College of Environmental Science and Forestry
Originally presented on: 11/7/2013 16:05:00 - 16:30:00

This study used wood extract from *Paulownia elongata* (WE) and hydrolisate of sugar maple (hydrolysate) produced during hot-water extraction as substrates to stimulate the production of cellulase from *Microbulbifer hydrolyticus* IRE-31(ATCC 700072). Selective strain medium adaptation of the original wild *M. hydrolyticus* IRE31 was performed in order to improve bacterial growth in mediums supplemented with hydrolysate and WE. Comparative studies were performed to establish better bacterial growth conditions and enzymatic activity at a given temperature, agitation and medium pH. The effect of different substrates, growth supplements and bases utilized to control the medium pH were among the studies evaluated. Flask tests where pure marine broth medium with an enzymatic activity of 0.009 U/ml broth medium was used as control and compared it to mediums supplemented with xylose, glucose, cellobiose, hydrolisate, WE or yeast extract, showed higher cellulase activity in the sample broth containing either WE or hydrolisate. No discernible activity was revealed in broths containing either glucose or xylose. Scale up from 100ml flask incubations to 1L stirred tank bioreactor set at pH 7.5, 37Â¢ÂºC and 300 rpm, and supplemented with 15% (v/v) *P. elongata* wood extract or 10% (v/v) sugar maple hydrolisate, yield a maximum enzymatic activity of 1.417 U/ml broth medium and 0.678U/ml broth medium respectively. These results reveal a significant increase in cellulase activity with growth mediums supplemented with complex substrates containing xylooligosaccharides such as *P. elongata* wood extract and sugar maple hydrolysate.

(716d) Synthesis and Performance of Cellulase-Mimetic Polymeric Solid Catalysts for Hydrolysis of Cellulose

Yang, Q., University of Wisconsin, Madison
Pan, X., University of Wisconsin, Madison
Originally presented on: 11/7/2013 16:30:00 - 16:55:00

Hydrolysis of cellulose to glucose is the most critical step in bioconversion of lignocellulosic biomass to fuels and chemicals. Cellulases and homogenous acids are widely used for cellulose hydrolysis. However, cellulases can only be used at moderate conditions and need longer time to achieve satisfactory cellulose hydrolysis. Homogenous acids can tolerate higher temperature, but they have limitations and issues such as equipment corrosion, recycling and wastewater treatment. To address these issues, heterogeneous solid acids have recently drawn a lot of attention for hydrolyzing cellulose. Traditional solid acids such as sulfonated carbon and resins, however, are not as effective as homogenous acids and cellulases in hydrolyzing cellulose because they have poor access/affinity to cellulose. In this study, a series of porous polymeric solid acids were synthesized for cellulose hydrolysis. These cellulase-mimetic solid acids have hydroxyl, halide, or boronic acid as cellulose-binding group in addition to sulfonic acid as cellulose-hydrolytic group. Their performance in hydrolyzing cellulose and real lignocellulose was evaluated. The properties of the synthesized catalysts such as particle size, surface area, and functional groups and their effect on cellulose hydrolysis were investigated.

(716e) Engineering Strategies to Improve Cellulase Production: Trichoderma reesei as a Pellet

Callow, N., The University of Akron
Ju, L. K., The University of Akron
Kelbly, M., The University of Akron
Originally presented on: 11/7/2013 16:55:00 - 17:20:00

In this study, we demonstrate the use of surfactants and specific nutrient control to regulate the growth and behavior of *Trichoderma reesei* Rut C-30. Although *T. reesei* is the ubiquitous cellulase producing fungus, many practical engineering issues remain. Typical processing conditions utilize a high growth density complicated with high nutrient demand in a viscous filamentous biomass filled fluid phase. This work shows that surfactants can be used to alleviate these processing difficulties by leveraging the inherent cell biology to our advantage. Surfactants have long been used to reduce cellular agglomeration, lyse cells, and improve cellular membrane permeability. In fungal cultures the application the surfactants tends to produce variable, species dependent, responses. In the case of the normally filamentous *T. reesei* Rut C-30, certain surfactants were found to promote growth into a compact pellet formation (diameter 1 ? 4 mm). These physical changes to the morphological structure have an overall positive effect on fermentation, allowing for improved mixing and mass transport through the continuous fluid phase. To improve yields and further decouple cellulase production from the growth phase, the restriction of key nutrients were tested. The specific control of phosphorus was found to successfully halt cell growth without affecting cellulase production. These results indicate that long-term cellulase can be continually produced from a pelleted, stationary phase, *T. reesei* culture.

(717a) Towards a Personalised Treatment of Acute Myeloid Leukaemia: The Impact of Considering the Cell Cycle

Fuentes-Gari, M., Imperial College London
Velliou, E., Imperial College London
Misener, R., Imperial College
Brito Dos Santos, S.,
Panoskaltsis, N., Northwick Park Hospital
Mantalaris, A., Imperial College London
Pistikopoulos, E. N., Imperial College London
Originally presented on: 11/7/2013 15:15:00 - 15:33:00

Towards a personalised treatment of Acute Myeloid Leukaemia:

The impact of considering the cell cycle

M. Fuentes Garí[1,2], E. G. Velliou[1,2] , R. Misener[1,2], S. Brito Dos Santos[2], N. Panoskaltsis[3], A. Mantalaris[2], E.N. Pistikopoulos[1]

[1]*Centre for Process Systems Engineering (CPSE), Department of Chemical Engineering, Imperial Imperial College London, South Kensington Campus, London SW7 2AZ, UK*

[2]*Biological Systems Engineering Laboratory (BSEL), Department of Chemical Engineering, Imperial Imperial College London, South Kensington Campus, London SW7 2AZ, UK*

[3]*Department of Hematology, Imperial College London, Northwick Park & St. Mark's Campus, London HA1 3UJ, UK*

Acute Myeloid Leukaemia (AML) is a type of blood cancer characterized by the release into the blood of abnormal amounts of immature cells of the myeloid lineage that preserve the ability to produce new cells. Patients diagnosed with AML are typically treated with chemotherapeutic drugs. These drugs (cytarabine, daunorubicin) prevent highly proliferative cells from progressing in their duplication and irreversibly program those cells to undergo apoptosis. A major challenge with chemotherapy is to eradicate a big proportion of the malignant cells while keeping enough healthy cells for the patient to survive. Indeed, these drugs are only selective for proliferative cells, but make no distinction between normal and abnormal cell types. Therefore, it is crucial to assess the response of both cell types to chemotherapy **to be able to target** (at least partially) **the malignant cells by their kinetic behaviour in favour of the healthy cells**. Our challenge is two-fold: *in vitro*, we need to reproduce an environment where the cells will grow as they would in the bone marrow (BM); and *in silico*, we need to model the system keeping in mind the real processes occurring in the cell and the existence of techniques to quantify them.

The BM is a highly porous and vascularized environment that supports the growth and differentiation of hematopoietic stem cells (HSCs). These cells are sensitive to both chemical and mechanical cues. They are stimulated chemically by the presentation of integrins (attachment proteins) by other cells or the BM matrix, and by the presence/absence of nutrients and metabolites. HSCs reside in "niche" arrangements, which provide the mechanical signals for their settlement. These niches consist of a high wall surface for cell attachment as well as an extensive network of interconnected channels, for the transport of signals and the accommodation of daughter cells (Baker and Chen 2012). The BM microenvironment is maintained by the production and regulation of growth factors and extracellular matrix components from the stromal cells (Panoskaltsis et al., 2005; (Kim 2005)). To assess cell growth kinetics, it is important to mimic as closely as possible this environment, by keeping a 3D structure with similar mechanical characteristics and reproducing the extracellular matrix properties, such as the presence of collagen type I for cell attachment. Such a system has previously been developed by our group in the form of a polyurethane (PU) porous scaffold coated with collagen type I, and has proved to be successful in recapitulating the BM properties (Mortera et al., 2010). Mononuclear stem cells from cord blood have successfully been cultured in this 3D platform for periods of over a month, in the absence of exogenous growth factors (Mortera et al., 2011).

Proliferation kinetics has its roots in the cell cycle. The cell cycle is an organized process by which cells duplicate their DNA and divide into two new cells. It consists of four different phases, highly regulated by the timed expression of cell cycle proteins (cyclins) and their partner cdks (cyclin-dependent kinases) (Darzynkiewicz et al., 1996). By following experimentally the cyclin expression over time, the cell cycle kinetics can be predicted mathematically, and optimal treatment strategies can be derived. **Our goal is to capture individual cell cycle kinetics in the 3D environment and introduce them as parameters into the chemotherapy model.**

In this work, we compare 2D and 3D kinetics and the changes in cell cycle and metabolic response, as a first step to justify the use of the 3D platform. K-562 (a widely used leukemic cell line) cells are cultured in parallel under 2D (liquid suspension contained in wells/flasks) and 3D (collagen-coated PU scaffolds dipped in growth medium) conditions for up to 4 weeks. Growth kinetics are captured indirectly *in situ* with the MTS assay ((3-(4,5-dimethylthiazol-2-yl)-5-(3-carboxymethoxyphenyl)-2-(4-sulfophenyl)-2H-tetrazolium, which is a substance that can be degraded enzymatically by cells and changes colour in the process), and by DNA quantification of the lysate. All of the *in situ*analyses are done in triplicates by three different operators (9 total replicates per condition). In addition, cell counts are performed weekly (2 replicates x 3 operators per condition) on 2D cells without further processing, and on 3D cells extracted from the scaffolds, giving the cell numbers and viability. Those cells are further analysed via flow cytometry for the characterization of: intracellular cyclin proteins, DNA content and proliferation markers. Bi- and tri-variate analysis provides information on: (A) the cell cycle regulatory patterns in 2D or 3D conditions, and (B) the cell cycle distribution. Finally, the metabolic activity of the cells is obtained by analysing the levels of glucose, lactate, glutamine and glutamate.

Faster proliferation kinetics were observed in the 2D samples, possibly due to the 2D-to-3D adaptation period (after all, cell lines are 2D-adapted populations) but also to intrinsic slower kinetics in the three-dimensional environment, as previously reported for other cancer types (Chitcholtan K. et al., 2013). This was not due to overcrowding of the cells, as the space availability in the 3D environment was high even after several weeks of culture (as observed in scanning electron microscopy images). We have to consider whether slower kinetics might indicate a closer behaviour to the one the cells have inside the patient. The cyclin expression profiles in both 2D and 3D looked similar, suggesting that the levels of protein expression stay the same for a particular cell line.

References

Baker, B. M. and C. S. Chen (2012). Deconstructing the third dimension: how 3D culture microenvironments alter cellular cues. *Journal of Cell Science* **125**(Pt 13): 3015-3024.

Chitcholtan K, E. Asselin, et al. (2013). Differences in growth properties of endometrial cancer in three dimensional (3D) culture and 2D cell monolayer. *Exp Cell Res* **319**(1): 75-87.

Darzynkiewicz, Z., J. P. Gong, et al. (1996). Cytometry of cyclin proteins. *Cytometry* **25**(1): 1-13.

Kim, J. B. (2005). Three-dimensional tissue culture models in cancer biology. *Seminars in Cancer Biology* **15**(5): 365-377.

Mortera-Blanco T, A. Mantalaris, A. Bismarck, N. Panoskaltsis (2010). Development of a three-dimensional biomimicry of human acute myeloid leukemia ex vivo. *Biomaterials*, **31**: 2243-2251.

Mortera-Blanco T, Mantalaris A, Bismarck A, et al., Long-term cytokine-free expansion of cord blood mononuclear cells in three-dimensional scaffolds (2011). *Biomaterials*, **32**: 9263-9270.

Panoskaltsis, N., Mantalaris, S., Wu, D. (2005). Engineering a mimicry of bone marrow tissue *Ex vivo*. *Journal of Bioscience and Bioengineering*, **100**: 28-35.

Acknowledgements

The research leading to these results has received funding from the European Union Seventh Framework Programme [FP7/2007-2013] under grant agreement n° 238013 (MULTIMOD ITN). It has additionally been supported by ERC-Mobile Project (n° 226462) and the Richard Thomas Leukemia Research Fund. R.M. is further thankful for a Royal Academy of Engineering Research Fellowship (RAEng 10216/118).

(717b) An Agent Based Model of Angiogenesis and Tissue Growth Within Porous Scaffolds

Bayrak, E. S., Illinois Institute of Technology
Mehdizadeh, H., Illinois Institute of Technology
Sumo, S., Illinois Institute of Technology
Brey, E. M., Illinois Institute of Technology
Cinar, A., Illinois Institute of Technology
Akar, B., Research Service, Edward Hines, Jr. V.A. Hospital, Hines
Originally presented on: 11/7/2013 15:33:00 - 15:51:00

Repair and replacement of large tissue defects involve the combination of many complex interacting factors and provide challenges to tissue engineers. Experiments alone are not sufficient to study the effects and interactions of various components involved in tissue formation such as biodegradable scaffolds, signaling molecules and different types of cells. A computational model that can realistically simulate the behavior of such systems over time would be beneficial for many different theoretical and practical purposes. These purposes include rapid testing of alternative hypotheses, better understanding the mechanisms involved, estimation of the important variables of the system and prediction of their effects on system level behaviors. A simulation based on the computational model will provide a tool for rapidly screening alternatives, determining the most promising experimental search space and investigating ways of intervening in such systems for functional tissue formation.

Researchers in computational biology and biomedical engineering have started using multi-agent modeling techniques to study complex systems with many interacting elements. The central idea in agent-based models (ABM) is to define agents that represent the building blocks of a system and to develop rules that regulate their interactions. The rules originate from the vast qualitative and quantitative knowledge gained through many years of studying the individual components of these biological systems. Agent-based systems are naturally suitable for modeling biological systems as they are comprised of individual constituents like cells that interact with each other to form macro-scale bodies like tissue.

In this study, a multi-layer ABM is presented to simulate tissue growth and vascular network formation within porous biodegradable scaffold. For cells to survive and form functional tissue, they should receive necessary nutrition and oxygen from blood vessels and this requires the scaffold to be well vascularized. Consequently, development of blood vessels should be considered simultaneously with tissue formation. In our model, one layer describes the event occurring within the scaffold, the second layer describes the angiogenesis, focusing the individual behavior of endothelial cells, and finally the third layer simulates the behavior of tissue cells.

Endothelial cells, the cells lining the blood vessel wall, and tissue cells are represented by different agents. These agents are capable of migration, growth (or elongation), proliferation and differentiation into desired tissue. Cells go through the cell cycle during their life time, and depending on their stage in cell cycle, they can perform different actions. Tissue cells are homogenously seeded in the pores scaffold and they undergo severe hypoxia until anatomized blood vessels reach their vicinity. On the other hand, the cells under hypoxia secrete the growth factor which stimulates the endothelial cell proliferation.

This computational framework couples two different cell types interacting each other and with their environment. Using this simulation framework, we can quantify the dynamic impact of cellular level events and actions on a complex multi-cellular system. This framework can be used to predict *in vivo* and *in vitro* tissue formation from individual seeded cells and host blood vessels and find the optimal environmental conditions for functional tissue.

(717c) Impact of Mechanical Pre-Stretch On Axonal Growth and Myelination

Liu, C., Michigan State University
Baek, S., Michigan State University
Sakamoto, J.,
Tuszynski, M.,
Chan, C., Michigan State University
Originally presented on: 11/7/2013 15:51:00 - 16:09:00

Spinal cord injury (SCI) is a disease of the central nervous system. Current therapeutic approaches focus on axonal regeneration, remyelination, and inhibition of inflammatory responses. Much of the research has focused on axonal regeneration, which has met with some success. However, remyelination has not been as extensively studied and remains a challenge in SCI. Remyelination is critical for successful functioniong of regenerated nerves. In this study, we discuss how an anisotropic surface arising from mechanical pre-stretch affects axonal regeneration and myelination..

Initiation of myelination has been shown to depend highly on the thickness of the axon. Using engineered nanofibers, it was found that above a certain fiber thickness, myelination initiates on the fibers independent of axonal signaling [1]. Concomitantly dorsal root ganglion (DRG) cells seeded on a stretch-growth system aligned in the stretched direction [2]. These DRG cells formed longer and thicker axons as the stretch on the surface increased. Their results showed some degree of myelination. We previously demonstrated that mesenchymal stem cells aligned on pre-stretch induced anisotropic surface, because the cells can sense a larger effective stiffness in the stretched direction than in the perpendicular direction. We explore how axons on a pre-stretched surface are impacted by the degree of pre-stretch (10%, 20%, or 30% pre-stretch) on the orientation, thickness and length of the axons as well as myelination.

[1] S. Lee, M.K. Leach, S.A. Redmond, S.Y.C. Chong, S.H. Mellon, S.J. Tuck, Z.Q. Feng, J.M. Corey, J.R. Chan, A culture system to study oligodendrocyte myelination processes using engineered nanofibers, Nature Methods 9 (2012) 917-+.

[2] B.J. Pfister, A. Iwata, A.G. Taylor, J.A. Wolf, D.F. Meaney, D.H. Smith, Development of transplantable nervous tissue constructs comprised of stretch-grown axons, Journal of Neuroscience Methods 153 (2006) 95-103.

(717d) Generation of a Heart Patch Using a Blend of Native Heart Matrix, Chitosan and Polycaprolactone

Pok, S., Rice university
Jacot, J. G., Rice university
Originally presented on: 11/7/2013 16:09:00 - 16:27:00

Introduction: One common cardiac defect, tetralogy of Fallot, generally requires surgical placement of a patch or baffle across the right ventricular outflow tract (RVOT) in an area that consists of myocardial tissue. Various types of biocompatible polymers or decellularized matrices have been attractive candidates for cardiac patch applications, including Dacron, Gore-Tex®, and autologous or bovine pericardium[1-2]. However, these constructs have significant drawbacks in cardiac patching, including the absence of growth potential, loss of mechanical strength over time, and inability to grow with the patient as well as an increased risk of infection and aneurysm. To improve these drawbacks, we have investigated a multi-layered scaffold manufactured by forming a gelatin (typeA)-chitosan hydrogel around a self-assembled polycaprolactone core[3]. In this study, decellularized porcine heart myocardium (HM) powder was blended with chitosan and formed a hydrogel. Cardiac specific decellularized matrices can provide the improvement of mimicking the native biochemical and composition of native extracellular matrix[4]. Chitosan allows structural stability, mechanical properties conducive to cell invasion and function, and controllable biodegradation rate[3].

Materials and Methods: Polycaprolactone matrix (10% in glacial acetic acid) was formed by self-assembly in aqueous media. Chitosan (C) solutions (2% in ddH_2O with 0.5 M acetic acid) were mixed with decellularized porcine heart myocardium solution (HM) (2% in ddH_2O) or type-A gelatin (G) solutions (2% in ddH_2O) respectively and emulsified using sonicator at 37°C for 30 min. 250 mL blended solutions were poured into custom-made Teflon moulds in between PCL matrices and lyophilized at -50°C for 24 hours followed by freezing with the dry ice. Formed hydrogels were neutralized using 100% ethanol and rehydrated using phosphate buffered saline (PBS). Hydrogels were assessed to test surface characteristics, mechanical properties, and cellular attachment, gene expressions and electrophysiological characteristics of neonatal rat ventricular myocytes (NRVM) isolated from 1 to 3 day old Sprague-Dawley rats.

Results and Discussion: Effects of different blending ratios of HM to chitosan (1:1, 1:2, 1:3, 1:4) on pore size, compressive modulus, NRVM adhesion and viability were characterized to identify the optimal blending ratio. 25% HM and 75% chitosan blended hydrogel (HM/C13) showed similar pore sizes and stiffness as G/C11 hydrogels which had the optimal environment for NRVM from our previous research[3]. Further, NRVM cultured on HM/C13 formed a 3-D formation after 7 days of culturing with ~82% of viability. Gene expression analysis results indicated that NRVM cultured on HM/C13 hydrogel had significantly higher a-myosin heavy chain (MYH6) expression (Fig. B), and faster conduction velocity

(4.05±2.59 cm/s of G/C11and 9.30±3.99 cm/s of HM/C13; $p<0.05$, n=5) compared to cells on G/C11 samples.

Conclusions: The results of this study demonstrate that mixtures of heart myocardium and chitosan hydrogel has sufficient mechanical strength, can maintain cardiomyocyte viability, and improved cell maturation for use in a cardiac patch application. The best results in terms of cell spreading and viability and scaffold integrity resulted from a blend of 25% HM-75% chitosan. In summary, this novel hydrogel shows significant potential for use of cardiac patch to repair full-thickness RVOT defects.

Acknowledgements: We would like to thank Dr. Jessica DeQuach and Dr. Karen Christman at University of California, San Diego for donating decellularized porcine heart myocardium powder.

References: [1]Jenkins, K. J. et al., *Circulation* 2007, *115* (23), 2995-3014. [2]Kochupura, P. et al., *Circulation* 2005, *112* (9), I144-I149. [3]Pok S, etc. Acta Biomater. 2013;9:5630-42. [4]Singelyn JM, etc. J Cardiovasc Transl Res 2010;3:478-86.

(717e) Engineering Clustered VEGF to Promote Angiogenesis in Vitro

Gojgini, S., UCLA
Segura, T., University of California, Los Angeles
Originally presented on: 11/7/2013 16:27:00 - 16:45:00

The process of angiogenesis, defined by the development of new blood vessels from pre-existing vessels, is essential in tissue remodeling and regeneration. This complex process involves extensive interplay between cells, soluble factors and extracellular matrix (ECM) components. In our laboratory, we are interested in designing biomaterials that can deliver pro-angiogenic agents such as Vascular Endothelial Growth Factor (VEGF) to activate the vascular process and stimulate the formation of a vascular network at ischemic tissue sites. The presence of VEGF at the site of ischemic damage enables specific endothelial cells (ECs) to undergo a phenotypic switch to become tip cells, which lead the way in a branching vessel. Previously our laboratory found that the physical presentation of VEGF (e.g. covalently bound, clustered or soluble) affected the way in which ECs became activated both at the molecular level (receptor phosphorylation) and at the multi-cellular level (tube formation). In particular, we observed that clustered VEGF enhanced the number of branching points and the total network length compared to less clustered VEGF and soluble VEGF, using a bead HUVEC (Human Umbilical Vascular Endothelail Cells) branching assay. Interestingly, the VEGF presentation didn't affect the total number of sprouts and the tube thickness. To understand the mechanism by which clustered VEGF resulted in enhanced tube branching points, we optimized the formation of VEGF clusters, studied VEGF receptor-2 (VEGFR-2) activation by its phosphorylation at two different tyrosine residues (1175 and 1214), and studied the activation of p38, Erk1/2 and Akt downstream signaling pathways. Briefly, Huvec cells were plated on 2D and were exposed to basic media containing 20 ng/mL of soluble or clustered VEGF and were studied at different time points. We evidenced that VEGF clusters could be generated with a uniform Z-average size was 95.97 ± 0.5 nm with a polydispersity index of 0.246 ± 0.005 using dynamic light scattering (DLS) and transmission electron microscopy (TEM). Interestingly, although the level of receptor activation was at highest for soluble, VEGF clusters achieved sustained phosphorylation. These results indicate that clustered VEGF causes extended activation when compared to soluble VEGF. Moreover, we studied the tip cell phenotype by evaluating their gene expression profile, and evidenced an increased expression of tip cells specific genes Delta-like 4 (Dll4), Jagged-1, CXCL12 and VEGFR3. Dll4 and Jagged-1 are the two ligands in the Notch signaling pathway that regulates the shaping of a branched vascular network. In addition, CXCL12 and VEGFR3 have been associated with the tip cell phenotype. Moreover, the analysis of mRNA levels by qPCR at different time exposure showed an increased expression of these genes at 4 hours in the group exposed to clustered VEGF compared to soluble VEGF. At the cellular level, quantification of DNA content demonstrated that clustered VEGF resulted in enhanced proliferation compared to soluble VEGF at different concentration of VEGF. To continue to study the mechanism by which clustered VEGF induced enhanced branching, we aim to investigate differences in protein expression and distribution as a result of exposure to clustered VEGF and soluble VEGF.

(717f) Acrylated Hyaluronic Acid Hydrogels to Study Cancer Angiogenesis

Shen, T., Johns Hopkins University
Abaci, E.,
Burdick, J. A., University of Pennsylvania
Gerecht, S., Johns Hopkins University
Originally presented on: 11/7/2013 16:45:00 - 17:03:00

Many have adapted three dimensional (3D) tissue culture models that recapitulate aspects of the tumorogenic *in vivo* microenvironment to study cancer progression *in vitro*. For example, both hypoxia and matrix stiffness were shown to promote tumor growth. We have previously engineered a modular culture system using hyaluronic acid (HA) hydrogel to engineer a functional microvascular network from endothelial progenitor cells (EPCs). The acrylated HA hydrogel is modified with binding adhesive domain (RGD) and matrix metallopeptidase (MMP) cleavable sites. Here we hypothesize that our tunable acrylated HA hydrogel system can be used to study cancer cell fate and its adaptive angiogenic responses.

Using mathematical modeling we first found that within 30 minutes of culture of HA-encapsulated HT1080 fibrosarcoma cells ($5x10^6$ cells/ml) in 5% oxygen, oxygen tension gradient is generated from the HA hydrogel edge (~4%) to the center (~1%). Analyzing cell cycle, we discovered that the there is a reduced percentage of Sub G1 population during three days period compared to atmospheric conditions, suggesting that HT1080 cells rapidly recover from encapsulation-induced apoptosis at 5% oxygen. Interestingly these were not significantly affected by altering the viscoelasticity of the HA hydrogel from 200 Pa to 30 Pa. Moreover, gene expression for MMP1, VEGF, ANG1 were altered across different viscoelasticity of gel only when cultured in 5% oxygen. This suggests that HT1080 cells in hypoxic conditions are more sensitive in responding to matrix stiffness compared to HT1080 cells cultured in atmospheric condition. Finally, based on these results, we also developed an angiogenesis assay to showcase cancer cell recruitment of EPC under different stiffness and oxygen tension. Here we are reporting a hydrogel system that can be incorporated with multidimensional parameters in investigating the cancer cell fate and its adaptive angiogenic responses in a 3D *in vitro* culture condition.

(717g) Imparting Hepatoprotection to Engineered Liver Tissues Using Anti-Microbial Peptides

Vu, L., Virginia Tech
Barron, A. E., Stanford University
Rajagopalan, P., Virginia Tech
Originally presented on: 11/7/2013 17:03:00 - 17:21:00

Every day the liver is confronted with a wide range of toxins resulting in the apoptosis of hepatic cells which if left uncontrolled, could potentially cause significant liver failure. While antimicrobial peptides (AMP) are typically known for their bactericidal activity, recent studies have shown they also possess a hepatoprotective effect [1]. However, there is limited information on how this effect is conferred by the administration of AMPs. Human cathelicidins, such as LL-37 (LLGDFFRKSKEKIGKEFKRIVQRIKDFLRNLVPRTES), are peptide sequences that also exhibit anti-microbial properties. In this study, we have investigated the hepatoprotective effect of LL-37 on organotypic liver models by administration with transforming growth factor β (TGF-β), a pro-inflammatory signaling molecule typically associated with the activation of liver fibrosis and hepatocyte apoptosis. To the best of our knowledge, this is the first study that demonstrates that the human cathelicidin, LL-37, can prevent hepatic cell death. Our goal is to obtain information on its protective effect on each of three major hepatic cell types (hepatocytes, liver sinusoidal endothelial cells, and Kupffer cells) as well as changes in their inter-cellular communications.

Organotypic liver models were assembled using primary hepatic cells obtained through the surgical excision of rat livers. Cellular models that closely mimic the three dimensional architecture and in vivo composition of the liver were assembled using detachable polyelectrolyte multilayers that serve as a mimic for the space of disse [2]. Hepatocyte monolayers (HMs) served as controls. All hepatic cultures were maintained for up to 12 days. TGF-β was added to cultures at different time points and its effects on cell death and hepatic phenotype were monitored. LL-37 was administered at a physiological concentration of 1 μg/mL at these same time points. Apoptosis was monitored through fluorescence imaging for FITC-Annexin V and assaying for caspase 3 activity.

Studies conducted so far have demonstrated that LL-37 alone does not result in significant cell death. Furthermore, apoptosis that results due to the addition of exogenous TGF-β can be reduced by administering LL-37. When LL-37 was administered to HMs 30 min prior to TGF-β, the protective effect was maximized when investigated by FITC-Annexin staining. This was further confirmed by caspase 3 activity where TGF-β treatment exhibited a decrease only when hepatocytes had already been exposed to LL-37. However, this trend was not observed at later time points in HMs, possibly due to dedifferentiation of hepatocytes. .

Since the organotypic liver model is a multi-cellular system, our current efforts are focused on unearthing the functional changes of each hepatic cell type and perturbations in inter-cellular signaling.. We have demonstrated for the first time that human cathelicidins such as LL-37 confer a protective effect against TGF-β induced apoptosis in hepatocytes.

References:

1. Lee, W.R., and coworkers, Toxicology and Applied Pharmacology, 2011. **256**(2): p. 209-215.

2. Larkin A. L. and coworkers, Tissue Engineering Part C: Methods, In Press, 2013.

(717h) Nanofiber Based Cellular Protrusions of Cancerous Cells
Koons, B., Virginia Tech
Sharma, P., Virginia Tech
Nain, A. S., Virginia Tech
Originally presented on: 11/7/2013 17:21:00 - 17:39:00

Nanofiber based Cellular Protrusions of Cancerous Cells

Brian Koons[1], Puja Sharma[2], Amrinder S. Nain[1,2]

[1]*Mechanical Engineering Department, Virginia Tech, Blacksburg, USA*

[2]*School of Biomedical Engineering and Sciences, Virginia Tech, Blacksburg, USA*

Introduction

As cells migrate and interact with their mechanical environment of the extracellular matrix (ECM), they sense their surroundings using cytoplasmic protrusions. Generally, these protrusions can be classified as either lamellipodia or filopodia, where actin within the protrusion is in the form of a broad branched sheet or thin long bundle, respectively [1]. Although lamellipodia have been known to play a pivotal role on 2D flat substrates, filopodia proves to be the dominant sensing mechanism in 3D microenvironments [2]. In cancerous cells, filopodia-like projections known as invadopodia are used to break down the surrounding ECM and infiltrate the proximate blood vessels. Current studies consist of using chemical gradients to force cancer cell protrusions across a porous membrane. Here, alternatively, we demonstrate a biophysical method using nanofibers to selectively position areas of contrasting structural stiffness (N/m) in order to elicit cellular protrusions. Using the previously reported STEP technique, non-electrospun suspended polymer nanofibers are deposited on substrates with varying fiber diameters in aligned configurations to create stiffness and geometric gradients sensed by the cell [3]. Cells interacting with this system are seen to migrate on the stiff fibers while sending protrusions towards and on the less stiff environment of nanofibers. Therefore, this method when coupled with immunostaining, can provide an understanding of the surrounding biophysical influence on actin polymerization and depolymerization. Use of this method provides a protrusion-based comparison for the invasiveness and metastatic potential of various cancer cell types including brain, breast, and prostate.

Materials and Methods

Nanofibers were fabricated from polymer solutions of polystyrene dissolved in xylene using various concentrations and molecular weights. The fibers were suspended in a highly controlled and aligned fashion across polystyrene frames via STEP technique to create a fibrous grid where crossing fibers are perpendicular to one another. Vertical fibers averaged 2 μm in diameter while horizontal fibers were 250 nm or 500 nm in diameter for each sample. Diameter measurements were confirmed using scanning electron microscopy. Fusing fibers at their intersections created unified fiber meshes with both fibers coexisting on the same geometric plane. The fiber meshes were coated with ECM protein fibronectin (2 μg/ml) to promote cellular adhesion prior to cell seeding with Denver Based Tumor Research Group (DBTRG-05MG) glioma cell line. Time-lapse imaging and measurements were performed using Carl Zeiss microscope and Axiovision software respectively. As cells initiated protrusions throughout the nanofiber mesh, protrusion length (μm), extension rate (μm/min), and retraction rate (μm/min) were measured using aforementioned software.

Results

Protrusion Behavior

DBTRG-05MG favors the 2 fibers, and mostly stayed wrapped around the 2 μm fibers due to high structural stiffness. Cells rarely migrated on less stiff nanoscale fibers. Instead, the cells extended protrusions from the stiff microfibers onto the less stiff, smaller diameter nanofibers, as shown in Figure 1. These protrusions were observed to be thin rod-like cytoplasmic projections, and their occurrence was observed to depend on the fiber diameter upon which they extended. Cells were observed to extend multiple protrusions in opposing directions, sensing potential migratory pathways within distance of contact.

Protrusion Analysis

The average protrusion length of 23 μm was constant between both 250 nm and 500 nm fibers. In addition, the protrusion extension rate of 3 μm/min did not differ between diameters. However, the rate of protrusion retraction for the larger diameter fibers (3.3 μm/min) was slower than when compared to retraction rates on smaller diameter fibers (4.2 μm/min) as shown in Figure 1.

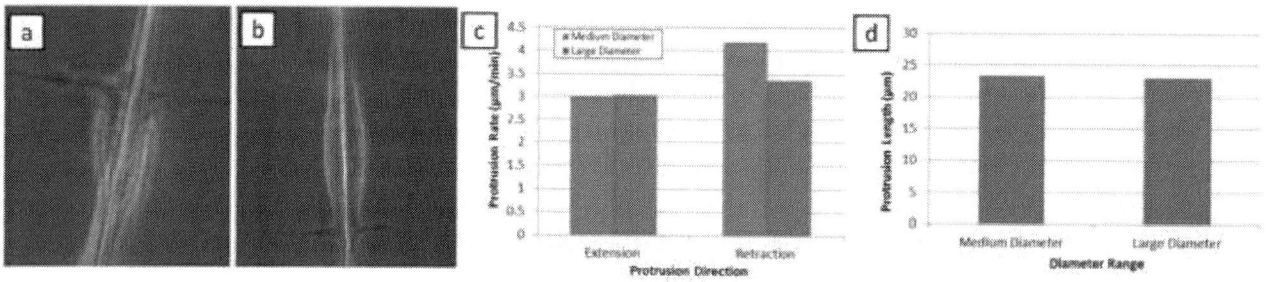

Figure 1. (a) Cell protrusion on large, 500 nm fiber, (b) bi-directional cell protrusion on medium, 250 nm fiber,Â (c) protrusion extension and retraction rates between large and medium fiber diameters, and (d) total length of protrusion compared between large and medium fibers.

Discussion

The study shows that biophysical characteristics of the extracellular microenvironment can control cell attachment and migration preference. Preliminary data illustrates that retraction can be regulated through the variance of fiber diameter. However, protrusion length and extension rates were seen to be constant on the diameters tested. Although the protrusions are sensing the mechanical characteristics of the extracellular environment, they can also be seen as migratory attempts to overcome the prohibitory stiffness gradient which is keeping the cell migrating along the large, stiffer fiber. Their protrusion extending potential can quantify the invasiveness and metastatic potential of cancerous cells. Our future experiments are aimed at exploring protrusion dynamics for a variety of cancer and normal cell lines on fibers of different diameters and structural stiffness.

References

[1] Faix, Jan Breitsprecher, Dennis Stradal, Theresia E B Rottner, Klemens. Filopodia: Complex models for simple rods. The international journal of biochemistry & cell biology. 2009

[2] Albuschies, J rg Vogel, Viola. The role of filopodia in the recognition of nanotopographies. Scientific Reports. 2013

[3] Nain AS, Sitti M, Jacobson A, Kowalewski T, Amon C. Dry Spinning Based Spinneret Based Tunable Engineered Parameters (STEP) Technique for Controlled and Aligned Deposition of Polymeric Nanofibers. Macromolecular Rapid Communications. 2009;30:1406-12.

(719a) Invited Speaker: Next Generation Controlled Release Systems for Immunoregulation

Little, S. R., University of Pittsburgh

Originally presented on: 11/7/2013 15:15:00 - 15:33:00

Currently, aberrant inflammation (responsible for autoimmune disease and transplant rejection) affects over 10 million people in the US. It is well documented that a potential treatment option for these diseases is to increase the presence of regulatory T cells (Treg) at local sites, which leads to reduced severity of immune responses. At present, an increase in local numbers of Treg is achieved through the systemic infusion of ex vivo cultured cells, a method plagued with problems. In this talk, we discuss the development and testing of formulations that are inspired by our own sophisticated immunoregulatory mechanisms that can be used to increase local populations of Treg, with the potential to ameliorate aberrant inflammation, autoimmunity and transplant rejection. For example, one interesting new strategy is a mimetic formulation that focuses on increasing local numbers of regulatory T cells through recruitment of these cells from distal sites (CCL22 microparticles). Another is a formulation that focuses on establishing an immunosuppressive environment and inducing regulatory T cells (IL-2, TGF-β and rapamycin microparticles; termed as Drug/CytokineMP). These formulations have been explored for various in vitro Treg indications such as the treatment of periodontitis and composite tissue allotransplantation (CTA).

(719b) Therapeutic Protein Nanoparticles That Subvert Intracellular Pathways for Immunomodulation As a Treatment of IBD

Herrera Estrada, L., Georgia Institute of Technology
Wu, H.,
Zhang, G.,
Ling, K., Lafayette College
Sumagin, R.,
Parkos, C. A.,
Neish, A. S.,
Champion, J., Georgia Institute of Technology
Originally presented on: 11/7/2013 15:33:00 - 15:51:00

Certain bacterial pathogens use needle-like structures, known as the type III secretion system (T3SS) to inject effector proteins directly into the cytoplasm of host cells. These effectors have evolved to subvert a variety of eukaryotic signaling pathways, including MAPK and NF-κB, and hence, modulate immune signaling and fate of the host cell. We have identified one of these effectors, acetyltransferase-A (ATA), as a potential therapeutic agent for acute and chronic inflammation such as that observed in inflammatory bowel disease (IBD). However, significant challenges remain to deliver ATA *in vivo* without bacterial T3SS in order to enable its use as an immunotherapeutic.

In prior work, we established that ATA decreases immune and inflammatory signals *in vitro*. However, delivering ATA to cells in the gut is difficult due to the low solubility of ATA and the harsh conditions of the gastrointestinal system and the mucosal barrier. In the present study, we replace the T3SS delivery mechanism by producing protein nanoparticles made directly from ATA and enhanced-green fluorescent protein. We have demonstrated nanoparticle size control through modification of multiple desolvation parameters including imidazole concentration, protein concentration, cross-linker type and cross-linking time. Tuning the size of these particles is critical for facilitating penetration and crossing of the mucosal layer, as well as endocytosis by intestinal epithelial or immune cells.

We observe that ATA nanoparticles are internalized *in vitro* at significantly higher rates than soluble protein and that ATA in the nanoparticles retains its bioactivity when delivered. Furthermore, we demonstrate anti-inflammatory activity in two different *in vivo* models. In a murine peritonitis model of acute inflammation, ATA nanoparticle pretreatment decreases neutrophil influx. In a mouse colitis model, we observe decreased inflammation measured by clinical parameters and histological indices. Therefore, this bacterial effector protein nanoparticle is a novel alternative that can replace the T3SS of bacterial pathogens for immunomodulation as a treatment for IBD, and potentially inflammatory and autoimmune diseases in general.

(719c) Selectively Targeting the TLR9-IRF7 Signaling Pathway By Polymer Blend Particles

Chen, H. C., University of Washington
Zhan, X., University of Washington
Tran, K. K., University of Washington
Shen, H., University of Washington
Originally presented on: 11/7/2013 15:51:00 - 16:09:00

Signaling through toll-like receptor 9 (TLR9) has been exploited for cancer therapy. The stimulation of TLR9 leads to two bifurcating signaling pathways – NF-κB-dependent pro-inflammatory cytokines pathway and IRF-7-dependent type I interferons (IFNs) pathway. In this study, we employ polymer blend particles to present the synthetic ligand, CpG oligonucleotides (CpG ODNs) to TLR9. The polymer blend particles are made from the blend of a pH-insensitive and pH-sensitive copolymer. By tailoring the composition of the pH-sensitive polymer, CpG ODNs are presented to TLR9 in a way that only activates the IRF-7 signaling pathway. CpG ODNs have been used for cancer therapy in both preclinical and clinical studies. The selective activation of IRF-7 could potentially enhance the apoptosis of tumor cells and immunological control of tumor progression without inadvertently activating NF-κB-dependent oncogenesis.

(719d) Development of Antiphagocytic CD47-Tagged Polymeric Prodrug Micelles to Target Alpha(v)Beta(3) Integrin-Bearing Cells

Sawdon, A. J., Michigan Technological University
Peng, C. A., Michigan Technological University
Originally presented on: 11/7/2013 16:09:00 - 16:27:00

Prodrug 5'-DFUR was used as the initiator in ring-opening polymerization of ε-caprolactone to form hydrophobic 5'-DFUR-polycaprolactone, which was then grafted with hydrophilic cationic polymer (i.e., chitosan or polyethyleneimine) to form amphiphilic copolymers for the preparation of stable micellar nanoparticles. The formed polymeric prodrug micelles were biotinylated via the interaction of biotin-NHS with the amines on cationic polymers. CD47-streptavidin (extracellular domain of self-marker CD47) fusion protein was expressed in lysY/I bacteria and then purified by biotinylated agarose column. Through biotin-streptavidin affinity, CD47 was bound to biotinylated polymeric prodrug micelles and showed antiphagocytic efficacy when CD47-tagged polymeric prodrug micelles exposed to J774A.1 macrophages. Since CD47 is not only an antiphagocytic ligand but also an integrin-associated protein, it can be employed to target integrin $\alpha_v\beta_3$, which is overexpressed on tumor-activated neovascular endothelial cells. In this study, CD47-tagged polymeric prodrug micelles were employed to treat two cells lines - one expressing integrin $\alpha_v\beta_3$ and one without integrin $\alpha_v\beta_3$ expression. Results showed that functionalized polymeric drug nanocarriers were successfully targeted to integrin $\alpha_v\beta_3$ by using CD47 as a targeting moiety.

(719e) Single-Cell Analysis of Specific B Cell Binding and Uptake of Peptide-Targeted Liposomes for Vaccine Formulations

Tokatlian, T., University of California, Los Angeles
Irvine, D. J., Massachusetts Institute of Technology
Originally presented on: 11/7/2013 16:27:00 - 16:45:00

Lipid-based nanoparticles have been widely characterized for the delivery of small molecules, peptides and proteins. Liposomes containing surface-bound antigens are similarly being developed as vaccines and have been shown to induce strong immune responses when compared to soluble antigens by facilitating cross-presentation and enhancement of humoral responses. Yet so far no studies have focused on how specific liposome characteristics modulate immune responses at the single cell level to enhance CD4+ T cell help and CD8+ CTLs through interaction and activation of specific B cells. Using a 3-83Ig transgenic mouse model, we evaluated the role of liposome composition on the direct binding and internalization of liposomes with surface-displayed peptide antigens to antigen-specific B cells. Initial studies were focused on the role of surface PEGylation. We hypothesized that while surface PEGylation has been shown to improve liposome stability and prolong circulation in vivo, that it may hinder binding and internalization through specific BCRs. Unilamellar PEGylated liposomes containing maleimide-functionalized PEG-lipids were synthesized by membrane extrusion, followed by post-synthesis conjugation to cysteine-containing antigenic peptides (p31 and p0 peptides with high and no binding affinities, respectively). Through flow cytometry analysis of liposomes with 10, 5, and 1% PEG-DSPE and 5% MPB (0% PEG) but identical antigen density, we found that 5% PEG liposomes bound with the highest intensity to B cells. There was, however, only a slight difference in the total number of B cells with bound liposomes between 10, 5, and 1% PEG liposomes. Only in the case of 0% PEG liposomes, where the peptide antigen was anchored directly to the bilayer surface, was binding significantly reduced, likely due to steric hinderance. Surprisingly, binding and internalization were inversely related, with 5% PEG liposomes having only 13% internalization after 22 hours at 37°C compared to almost 75 and 90% internalization of 1 and 0% PEG liposomes, respectively. Future studies will focus on the relationship between internalization and B cell activation by assessing B cell proliferation, Ig class switch, presentation of activation markers, and the ability to present helper peptides to CD4+ T cells. We believe the results obtained in these studies will further the development and understanding of effective particle based vaccines.

(719f) Improved Vaccination Through the Use of Self-Assembled Peptide Amphiphile Micelle Delivery Devices

Ulery, B. D., University of Chicago
Trent, A., University of California, Santa Barbara
Black, M. J., University of California, Santa Barbara
Tirrell, M., University of Chicago
Originally presented on: 11/7/2013 16:45:00 - 17:03:00

Whole-killed and live-attenuated vaccines have been tremendously effective in preventing pathogenic infections, but can be associated with undesirable side effects. Subunit vaccines that deliver just the peptide antigen of interest have been shown capable of stimulating an immune response, but are generally weak immunogens on their own requiring strong adjuvants (non-specific immunostimulants) to be effective. In order to enhance the immunogenicity of peptide vaccines, new delivery systems must be designed.

Peptide amphiphiles are unique biomaterials consisting of a hydrophilic peptide conjugated to a hydrophobic lipid or fatty acid tail that undergo self-assembly into micelles in water. Micellization is driven by the need to shield the hydrophobic tail from the aqueous surroundings leading to a micelle comprised of a hydrophobic tail core and a hydrophilic peptide corona. Peptide amphiphile micelles allow for the presentation of hundreds to tens of thousands of peptide on the surface of each micelle. Peptide amphiphile micelles also crowd peptides within their coronas providing an artificial tertiary structure which induces desirable peptide secondary structure conformations. Since peptide vaccines are often delivered in low local concentrations, peptide amphiphile micelles provide a novel platform to improve the host immune response to peptide vaccines.

A conformational B cell epitope (J8) of Group A Streptococcus, the bacterial causative agent of diseases ranging from mild pharyngitis to severe toxic shock syndrome, was chosen to evaluate the immunostimulatory effect of peptide amphiphile micelles. J8 was covalently tethered to a di-palmitic acid tail (J8-diC_{16}) and fabricated into peptide amphiphile micelles in water. When delivered to mice subcutaneously, J8-diC_{16} was found to induce J8-specific high antibody titers greater than soluble J8 delivered with a commercially available adjuvant. To further enhance the antibody response, mixed micelles comprised of J8-diC_{16} and amphiphilic adjuvants were synthesized. Mixed micelles induced a strong immune response after a single vaccination and higher titers than all other formulations regardless of the vaccine regimen. These results provide significant evidence that peptide amphiphile micelles are innovative constructs capable of enhancing the immunogenicity of peptide vaccine candidates.

(719g) Design of HIV-1 Nanovaccine Using Active Targeting Mechanisms

Vela Ramirez, J., Iowa State University
Tygrett, L., University of Iowa
Roychoudhury, R., Indiana University
Hao, J., Case Western Reserve University
Habte, H., Iowa State University
Cho, M.,
Greenspan, N.,
Pohl, N., Indiana University
Waldschmidt, T.,
Narasimhan, B., Iowa State University
Originally presented on: 11/7/2013 17:03:00 - 17:21:00

Infection caused by Human Immunodeficiency Virus (HIV) is one of the greatest challenges for researchers around the world. Since its outbreak in 1981, more than 25 million people have died of Acquired Immunodeficiency Deficiency Syndrome (AIDS). An ideal strategy to stop the spread of the virus is to develop a prophylactic or therapeutic intervention against HIV-1that can be easily administered and distributed and elicits a robust and balanced immune response in order to avoid the infection of the host. Due to the high mutation rate of the virus and multiple mechanisms used to evade the immune response, HIV remains a daunting challenge.

A current focus in vaccine development is the design of novel adjuvant formulations and delivery systems that can elicit strong and balanced immune responses towards a variety of pathogens. Active targeting of these adjuvants can help in the design of efficacious vaccines since the use of subunit proteins as antigens requires developing delivery platforms that will provide the appropriate environment to preserve the structure of these novel antigens, and at the same time, offers the prospect of enhancing the typically modest ability of these immunogens to elicit specific immune responses. Our work is focused on the gp41-54Q-GHC immunogen, which is based on the membrane proximal external region of an HIV-1 glycoprotein (gp41) that elicits broadly neutralizing antibodies for epitopes conserved in different virus isolates. The main goal of this study is to design a biodegradable nanoparticle-based delivery platform for gp41-54Q-GHC that stimulates a robust immune response. In particular, polyanhydride nanoparticles exhibit desirable characteristics as adjuvants and delivery vehicles, including immune modulation, activation of antigen presenting cells, induction of high titer and highly avid antibodies, sustained antigen release, and stabilization of protein antigens.

In this work, polyanhydride copolymers based on 1,6-bis(*p*-carboxyphenoxy) hexane (CPH) and 1,8-bis(*p*-carboxyphenoxy)-3,6-dioxaoctane (CPTEG) and CPH were synthesized via melt polycondensation. CPTEG:CPH nanoparticles encapsulating gp41-54Q-GHC protein were fabricated using an anti-solvent nanoencapsulation method. Surface functionalization of the antigen-loaded nanoparticles was carried out using an amine-carboxylic acid coupling reaction. Di-mannose (i.e. alpha-D-mannopyranosyl-(1,2)-D-mannopyranoside) and glycolic acid (linker) modified nanoparticles were characterized using zeta potential measurements and a phenol-sulfuric acid assay. Subcutaneous administration of these particles using footpad injections in BALB/c mice was performed. Germinal center formation and isotype switching analysis were analyzed following injection of these particles. These studies lay the groundwork for using the targeted nanovaccine platform as a dual adjuvant/delivery system for gp41-54Q-GHC in future studies.

(719h) On Designing Magnetic Nanoparticles for Targeted Delivery of Malaria DNA Vaccine

Selomulya, C., Monash University

Originally presented on: 11/7/2013 17:21:00 - 17:39:00

Malaria is a deadly infectious disease, while to date an effective vaccine still remains elusive. This paper provides an overview of works that we did to develop a gene delivery system for malaria vaccine using magnetic nanoparticles tested under a range of conditions *in vitro* and *in vivo*. The examples include the design of magnetic polyplexes stable in physiological conditions with dendritic-cell targeting functionality, and a comparison of efficacy and types of immune responses from plasmid malaria DNA vaccine encoding VR1020-PyMSP1$_{19}$ delivered via different administration routes in *in vivo* mouse studies. Higher serum antibody titers against PyMSP1$_{19}$ were observed with intraperitoneal and intramuscular injections, with robust IgG2a and IgG1 responses observed for intraperitoneal administration, possibly due to the physiology of peritoneum as a major reservoir of macrophages and dendritic cells. Heterologous DNA prime followed by single protein boost vaccination regime also enhanced IgG2a, IgG1, and IgG2b responses, indicating the induction of appropriate memory immunity that can be elicited by protein on recall. These outcomes support the possibility to design superparamagnetic nanoparticle-based DNA vaccines to optimally evoke desired antibody responses, useful for a variety of diseases including malaria.

Related publications:

Al-Deen, F.N., Ma, C., Xiang, S. **Selomulya, C.***, Plebanski, M., Coppel, R., **2013**, On the efficacy of malaria DNA vaccination with magnetic gene vectors, *Journal of Controlled Release*, 168:1, 10-17.
Al-Deen, F. N., **Selomulya, C.***, Williams, T., **2013**, On designing stable magnetic vectors as carriers for malaria DNA vaccine, *Colloids and Surfaces B: Biointerfaces*, 102, 492– 503.
Al-Deen, F., Ho, J., **Selomulya, C.***, Ma, C., Coppel, R., **2011**, Superparamagnetic Nanoparticles for Effective Delivery of Malaria DNA Vaccine, *Langmuir*, 27 (7), pp 3703–3712.

Ho, J., Al-Deen, F., Aboodi, A., **Selomulya, C.**, Xiang, S., Plebanski, M., Forde, G., **2011**, N,N'-Carbonyldiimidazole-mediated functionalization of superparamagnetic nanoparticles as vaccine carrier, *Colloids and Surfaces B: Biointerfaces*, 83, 83-90.
Xiang, S. D., **Selomulya, C.**, Ho, J., Apostolopoulos, V., Plebanski, M., **2010**, Delivery of DNA vaccines: an overview on the use of biodegradable polymeric and magnetic nanoparticles, *Wiley Interdisciplinary Reviews: Nanomedicine and Nanobiotechnology*, vol 2, John Wiley & Sons, United States, pp. 205-218.

**: C. Selomulya as the corresponding author*

(723a) Evaluating the Use of HL60-Derived Neutrophilic Cells to Examine the Molecular Dynamics of the Regulation of CD18 Surface Expression By Shear Stress Exposure

Akenhead, M. L., University of Kentucky
Zhang, X.,
Shin, H. Y., University of Kentucky
Originally presented on: 11/7/2013 15:15:00 - 15:33:00

Introduction

Chronic inflammation is a common denominator for a wide range of cardiovascular diseases and is associated with an impaired neutrophil response to fluid shear stress exposure [1]. In fact, fluid flow mechanotransduction by neutrophils has been shown to be part of a control mechanism to minimize inflammation in the blood under physiologic (i.e. non-pathogenic) conditions [1]. Specifically, shear stress exposure derived from blood flow deactivates neutrophils, even in the presence of low quantities of cell agonists. A key aspect of this mechanism is the mechanosensitive cleavage of CD18 cell-cell adhesion molecules off the surface of neutrophils, which prevents their heterotypic binding to other cells (e.g., endothelium, platelets) [2,3]. Recently, cathepsin B (catB) has been identified as a key protease involved in CD18 cleavage [2,4], but it is unclear how shear regulates the release of this protease into the extracellular milieu. To elucidate the dynamics of shear-induced catB release, we seek to overexpress a catB-GFP (i.e. green fluorescent protein) construct in neutrophils and track its release into the extracellular environment. Since transfection of human neutrophils with a catB-GFP construct is not feasible due to their short lifespan (<24 hours) outside the body, we tested the possibility of using HL60-derived neutrophilic cells as a transfectable cell culture model. For this study, we hypothesized that shear stress-induced CD18 cleavage is a characteristic of the neutrophilic phenotype.

Methods

HL60 cells were purchased from the American Type Culture collection (ATCC; CCL-240) and subcultured in RPMI 1640 media supplemented with 10% FSB plus 1% penicillin/streptomycin/L-glutamine solution. To induce a neutrophilic phenotype, HL60 cells were treated with 1.25% DMSO for 5 days (referred to as dHL60 cells). Immediately after neutrophil differentiation, dHL60 cells were resuspended in HBSS media. Some dHL60 cell populations were left untreated, while others were preincubated with 10 nM to 1 μM fMLP for 10 minutes. Following preincubation, the cells were exposed to 5 dynes/cm^2 steady laminar shear stress for 10 minutes using a custom cone-and-plate viscometer as described [3]. For some experiments, both unstimulated and fMLP-stimulated dHL60 cells were also pretreated with 30 μM E64, a known blocker of shear-induced CD18 cleavage [2], prior to shear exposure in the cone-and-plate viscometer. Controls were dHL60 cells maintained under static (no-flow, no-shear), but otherwise, similar experimental conditions. After shear and control experiments, dHL60s were fixed with 1% paraformaldehyde for 20 minutes.

Immunofluorescence in conjunction with flow cytometric analyses was then conducted to assess dHL60 binding to fluorescent antibodies for CD18, CD11a, CD11b, and catB.

Results

Based on our analyses, sheared cells exhibited reduced expression of CD18 and CD11a, but not CD11b. E64, a cysteine protease inhibitor capable of blocking catB, attenuated the shear-induced CD18 and CD11a cleavage responses of cells. Furthermore, by increasing fMLP concentrations up to 1 μM, the shear-induced cleavage of CD18 and CD11a was dose-dependently attenuated. Notably, unlike human primary neutrophils, dHL60 cells did not appear to express appreciable quantities of catB.

Discussion

Based on our results, dHL60 neutrophilic cells exhibit preferential shear-induced cleavage of CD11a over CD11b isotypes, which differs from that of human neutrophils. Primary neutrophils have been shown to exhibit preferential cleavage of CD11b over CD11a isotypes [3]. This is consistent with the near absence of catB expression by dHL60 cells and, in conjunction with previous studies [2,4], further points to catB as responsible for the CD11b cleavage observed for sheared human neutrophils exhibiting abundant catB expression levels. Despite these differences, our results indicate that the CD18 cleavage response to shear is a distinct trait of the neutrophil phenotype. Additionally, this shear response appears to serve as an anti-inflammatory measure below 1 μM fMLP. It should be noted that this threshold concentration of fMLP is similar to that below which shear-induced pseudopod retraction, another neutrophil-deactivating effect of fluid flow, remains intact.

Conclusion

The results of the present study confirm the potential utility of the dHL60 neutrophilic cells in characterizing catB release dynamics using molecular cloning approaches. Moreover, our results further point to shear-induced CD18 cleavage as an important control mechanism that serves to restrict the neutrophil activation under physiological (i.e. non-inflamed, non-pathogenic) conditions.

References

[1] Makino A, Shin HY, Komai Y, Fukuda S, Coughlin M, Sugihara-Seki M, et al. Mechanotransduction in leukocyte activation: a review. Biorheology. 2007;44(4):221-49.

[2] Fukuda S, Schmid-Schonbein GW. Regulation of CD18 expression on neutrophils in response to fluid shear stress. Proc Natl Acad Sci U S A. 2003;100(23):13152-7.

[3] Zhang X, Zhan D, Shin HY. Integrin subtype-dependent CD18 cleavage under shear and its influence on leukocyte-platelet binding. J Leukoc Biol. 2013;93(2):251-8.

[4] Shin HY, Simon SI, Schmid-Schonbein GW. Fluid shear-induced activation and cleavage of CD18 during pseudopod retraction by human neutrophils. J Cell Physiol. 2008;214(2):528-36.

(723b) Characterization and Modeling of Cancer Cell Motility in 3-D Extra-Cellular Matrix

Wu, P. H., Johns Hopkins University
Wirtz, D., Johns Hopkins University
Giri, A., Johns Hopkins Physical Sciences - Oncology Center
Sun, S. X., Johns Hopkins University
Originally presented on: 11/7/2013 15:33:00 - 15:51:00

Cancer cell migration through the extra-cellular matrix (ECM) is one of the most important processes during metastasis. Cell migration is a multi-scale process, which integrates molecular kinetics and signaling; cell mechanics and microenvironment. A quantitative understanding of cell motion in three-dimensions will facilitate the development of a system-level understanding of cellular processes in cancer metastasis and further promote identification of new therapeutic targets. Though various mathematical models were proposed to measure cell migration on 2-D surfaces, adequate tools to characterize and model migration in 3-D are lacking. In this work, we observed the migration patterns of individual HT1080 fibrosarcoma cells on 2-D substrate and in 3-D collagen matrices for more than 8 hours. We found that non-Gaussian behavior of cell velocity distribution at various time scales in population level is a universal characteristic for both cell 2-D and 3-D migration and it is as result of the highly varied cell-to-cell motility behaviors. Further, cell velocity in 3-D matrices is anisotropic and velocity profiles display different speed and self-correlation processes at different directions. Together, by including cell-level variation, anisotropic velocity and observation noise in persistent random walk model, we were able to predict the population level of cell motility system in 3-D matrices which corresponding well to experimental observation. Our model further allows the estimation of speed for HT1080 cells in 3-D matrices at more pathophysiological relevant time scale (months) and we found there are small populations of cells that display ~10x higher speed than the general population and could imply that metastatic cells come from small populations of cells within a tumor.

(723c) Mechanisms of Three-Dimensional Glioma Cell Motility in Non-Fibrillar Matrices

Ananthanarayanan, B., University of California, Berkeley
Singh, G., University of California, Berkeley
MacKay, J. L., University of California, Berkeley
Chang, C. W., University of California, Berkeley
Kim, Y., University of California, Berkeley
Kumar, S., University of California, Berkeley
Originally presented on: 11/7/2013 15:51:00 - 16:09:00

Diffuse infiltration of single cells into brain parenchyma is a hallmark of malignant glioma. This extreme invasiveness makes complete tumor resection difficult and contributes to the high mortality rate associated with this disease. Brain parenchyma has a distinct physical structure characterized by densely packed neural cell processes and sub-micron extracellular space, and is largely devoid of the fibrillar collagen scaffolding typically found in stromal tissue. Consequently, glioma cells migrating in brain slices exhibit a distinct type of motility, with branched protrusions and hourglass-shaped cell-body deformations that help squeeze cells through tight spaces (Beadle *et al., Mol. Biol. Cell* 2008; 19(8):3357-68). However, the mechanistic details of this unique mode of motility remain incompletely understood. To address this question, we synthesized brain-mimetic nanoporous, non-fibrillar extracellular matrices (ECMs) based on cross-linked hyaluronic acid (HA), the major component of brain ECM, and functionalized them with Arg-Gly-Asp (RGD)-containing peptides to facilitate integrin-mediated cell adhesion. In a three-dimensional (3D) spheroid invasion paradigm, glioma cell motility was highly reminiscent of that seen in brain slices, validating the use of these ECMs a model system (Ananthanarayanan *et al. Biomaterials* 2011; 32(31):7913-23). Eliminating RGD peptides from the ECM abolished invasion, suggesting a requirement for integrin-mediated adhesions for this mode of motility. Increasing matrix density, or inhibiting myosin-based cellular contractility by blebbistatin or shRNA-induced knockdown of Myosin IIA also severely impaired motility. We also report preliminary results from two sets of studies: First, we investigate the balance of protrusive and contractile forces in glioma cell motility by pharmacologically and genetically manipulating the Rho GTPases RhoA, Rac1, and Cdc42. Second, we explore the involvement of actin-nucleating proteins such as Arp2/3 and formins in the protrusive dynamics observed in this mode of motility. Our studies collectively help define the mechanotransductive signaling mechanisms that underlie the distinct mode of glioma cell motility observed in non-fibrillar matrices.

(723d) Molecular Tension Sensors Measure Forces Generated By Single Integrin Molecules in Living Cells

Mekhdjian, A. H., Stanford University
Morimatsu, M., Stanford University
Adhikari, A. S., Stanford University
Dunn, A. R., Stanford University
Originally presented on: 11/7/2013 16:09:00 - 16:27:00

Disruption in the ability of cells to sense the mechanical properties of their surroundings represents a hallmark of many diseases, including muscular dystrophy, arteriosclerosis, cardiomyopathies, and cancer. Although cells have numerous mechanisms for detecting mechanical inputs, one of the most prominent is through integrins, a class of heterodimeric transmembrane proteins that cluster into micron-sized assemblies termed focal adhesions (FAs).Â FAs link the cell cytoskeleton to the surrounding extracellular matrix (ECM), transmitting and responding to mechanical force. Force transmission through integrins is essential for cell migration and adhesion, while force sensing at FAs regulates numerous cellular processes including cancer cell proliferation and stem cell differentiation. Despite intense interest, how force is generated within individual FAs remains poorly understood due to a lack of methods that directly visualize molecular-scale forces in living cells.

We engineered FÃ¶rster resonance energy transfer (FRET)-based molecular tension sensors to measure cell-generated forces at single integrins. This approach allows us to directly visualize the distribution of mechanical forces within FAs, a measurement that is not feasible using conventional approaches such as traction force microscopy. We find that regions of low FRET localize primarily to the cell periphery and coincide with paxillin recruitment, a canonical FA marker, showing that force generation is largely localized to FAs. However, we observe strikingly complex distributions of tensions within individual FAs. In addition, FRET values at single probe molecules indicate that the majority of integrins generate tensions in the range of 1-4 pN, considerably less than the forces required to break integrin-ECM bonds. These observations are consistent with a model in which the number of engaged integrins within a FA controls the total force exerted while the force per integrin remains bracketed in a relatively narrow range. In ongoing work, we have generalized the molecular tension sensor design for use in wide variety of cell adhesion and cell-ligand interactions, providing a potentially powerful tool for elucidating the molecular mechanisms underlying cellular mechanotransduction.

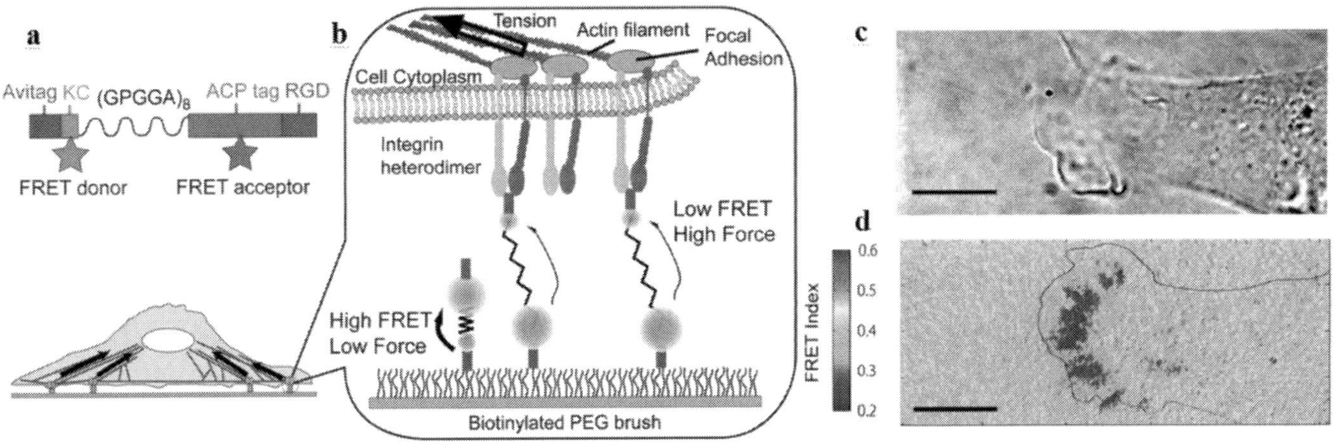

Figure 1

site-specifically labeled with biotin (Avitag), a FRET donor (KC tag), and FRET acceptor (ACP tag), and present the RGD sequence from fibronectin to promote adhesion. The $(GPGGA)_8$ sequence acts as an entropic spring that is stretched upon the application of force. **(b)** Sensors are attached to a coverslip via biotin and Neutravidin; the biotinylated PEG brush prevents nonspecific cell and sensor attachment. Integrin heterodimers attach to the RGD domain and apply load generated by the cell cytoskeleton. **(c)** Brightfield image of a spreading fibroblast on a sensor-functionalized surface. **(d)** Corresponding FRET map showing areas of high force localized at the cell periphery. Color bar represents FRET index (blue: low FRET/high force; red: high FRET/low force). Scale bar: 10 μm.

(723e) Prostate Cancer Cell Migration Is Influenced By Suspended Fiber Structural Stiffness

Sharma, P., Virginia Tech
Bhatia, S.,
Hughes, C.,
Nain, A. S., Virginia Tech
Originally presented on: 11/7/2013 16:27:00 - 16:45:00

Introduction: Â

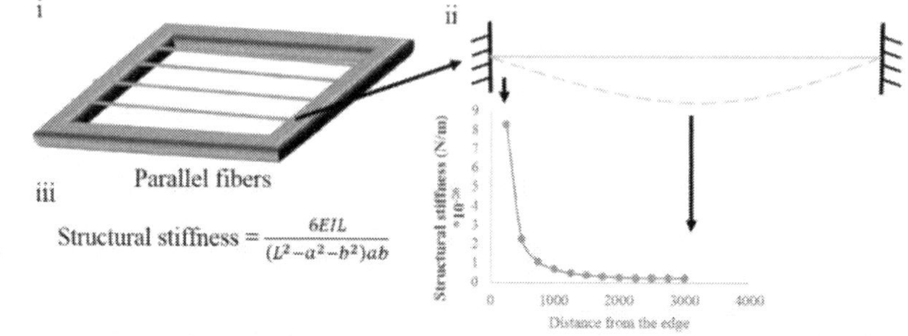

Figure 1. (i) Schematic of STEP aligned nanofibers, (ii) Structural stiffness of a single nanofiber, (iii) Structural stiffness formula using Euler beam mechanics

Cancer is the second leading cause of death in the U.S.[1], and accounts for over $200 billion in health and morbidity costs [2]. The major cause of death in cancer patients is metastasis [3]. Tumors are classified in terms of grades by the World Health Organization. While grades I-II represent tumors that are mostly contained in the periphery of the primary tumor, grades III-IV represent tumors that have migrated from the primary site of occurrence to distant tissues and organs. When higher grade tumors invade other tissues, it starts compromising the functionality of the organ of origin and the organs affected via metastasis, ultimately leading to the death of the person [4, 5].

One of the most effective ways to control tumors is by administering anti-metastatic drugs [6]. Some of the drugs that compromise the ability of cancer cells to migrate include paclitaxel, nocodazole, and marimastat [7, 8]. These drugs control metastasis by primarily depolymerizing cytoskeletal components and severely compromising the ability of cells to migrate. While this is shown to be effective, another potential strategy to minimize metastasis can include the conditioning of the tumor associated extracellular matrix (ECM) such that it does not encourage cell migration. In order to design such strategies, it is important to understand how the biochemical and biophysical properties of the ECM influences single cancer cell migration.

It has been well established that bulk mechanical properties like substrate elasticity alters integrin expression in cancer cells facilitating a migratory phenotype [9, 10]. However, it has also been observed that the local and cell specific mechanical properties might have a stronger influence on cell migration behavior than bulk properties [11]. It is still not completely understood how cancer cell migration behavior is influenced by biophysical properties of their extracellular matrix. One such property is structural stiffness measured in

N/m. Unlike substrate elasticity, structural stiffness accounts for local changes in the material property, and from our preliminary investigation, have been shown to influence cancer cell migration.

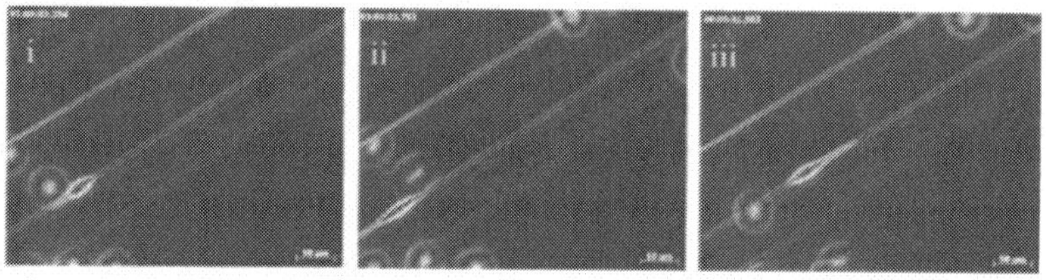

Figure 2. Time lapse images of PC-3 cells migrating along a single nanofiber

In this study, we utilized a previously explained Spinneret based Tunable Engineering Parameters (STEP) pseudo-dry spinning technique to manufacture aligned, parallel, and suspended nanofibers [12, 13]. This platform offers design of: a) nanofibers of diameters close to the native fibrous ECM are manufactured, b) a non-2D environment of suspended fibers on which the cells wrap around, which closely represent the *in vivo* conditions, and c) change in the structural stiffness (N/m) of the fibers through different Â material, diameter and length of fibers. A schematic of the suspended nanofiber substrate with change in structural stiffness along its distance from the edge is shown in Fig. 1.

Materials and Methods:

Polystyrene (*E= 3GPa [14]*), nanofibers were manufactured using a non-electrospinning, STEP technique. Highly aligned nanofibers of diameter 500nm and length 6 mm were deposited on plastic frames to obtain suspended nanofibers (Fig 1). The fibers were sterilized with 70% ethanol, coated with fibronectin to facilitate cell attachment, and seeded with prostate cancer cells (PC-3, purchased from ATCC) in F-12K media (ATCC). After attachment, time lapse images were obtained every 10 minutes for 6 hours using a Zeiss microscope with incubating capacity. The position of cell in terms of its distance from the edge of the substrate was calculated using Axio Vision software. Also the migration (maximum displacement within the 6 hour period) of single prostate cancer cells was measured and the distance of the cell from the substrate edge was recorded.

Results and Discussion:

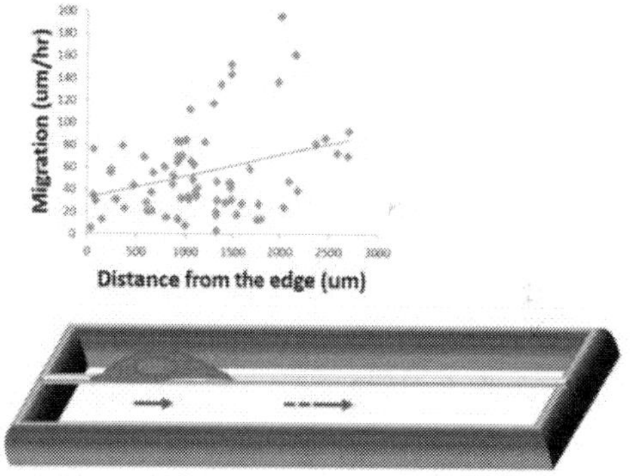

Figure 3. PC-3 cell migration as a function of distance from the edge of the substrate

Using Euler beam mechanics, the structural stiffness of the nanofiber can be calculated. Using the equation for beam deflection, it can be said that the structural stiffness decreases towards the middle span length of the fiber (Fig. 1 (ii) and (iii)). The cells attached to the fibronectin coated fibers within 2-6 hours after seeding, and formed spindle morphologies as they migrated along a single nanofiber (Fig. 2). While some cells demonstrated persistent migration, cells that migrated back and forth along the fiber were also observed. Prostate cancer cells exhibited different migration speeds along the fiber span length (Fig.3). While the migration speed of PC-3 ranged from 3-195µm/hr, linear regression analysis showed that cells migrated faster when they were migrating at the center of the fiber span length. This observation suggests that as structural stiffness decreases, prostate cancer cell migration speed increases.

Structural stiffness is a measure of local mechanical property of the nanofiber. This observation shows that even when the elastic modulus of the substrate (N/m^2) was constant for polystyrene, prostate cancer cells exhibited different migratory behavior due to the change in structural stiffness (N/m), a local mechanical property. This suggests that prostate cancer cells are constantly probing their local microenvironment, and are able to manipulate their migration speeds in response to mechanical changes in the microenvironment.

Conclusion:

Cell migration is a complex phenomenon that is influenced by a myriad of factors within and outside the cell. Here, we demonstrate that by changing the structural stiffness of the immediate microenvironment of the cell, its migration speed changes. Moreover, this study shows how stiffness differences within millimeter distances can impact the migration dynamics of prostate cancer cells. Hence, drugs that alter the local mechanical property of the immediate ECM could potentially be coupled with the already existing anti-metastatic drugs to achieve lower cancer metastasis rates.

References:

1. Stewart, S.L., et al., *1990-2000.* MMWR Surveill Summ, 2004. **53**(3): p. 1-108.

2. Lyman, G.H., J Oncol Pract, 2007. **3**(3): p. 113-4.

3. Mehlen, P. and A. Puisieux, Nat Rev Cancer, 2006. **6**(6): p. 449-58.

4. McGovern, S.L., et al., J Neurosurg, 2010. **112**(5): p. 925-33.

5. Lilleby, W., et al., Cancer, 2001. **92**(2): p. 311-9.

6. Chay, C.H., et al., Clin Prostate Cancer, 2002. **1**(1): p. 14-9.

7. Nukatsuka, M., et al., Â Int J Oncol, 2004. **25**(6): p. 1531-6.

8. Haier, J., M. Nasralla, and G.L. Nicolson, Â Br J Cancer, 1999. **80**(12): p. 1867-74.

9. Caccavari, F., et al., Cell Adh Migr, 2010. **4**(1): p. 124-9.

10. Levental, K.R., et al., Cell, 2009. **139**(5): p. 891-906.

11. Carey, S.P., et al., Biomaterials, 2012. **33**(16): p. 4157-65.

12. Sheets, K., et al., Acta Biomater, 2013.

13. Nain, A.S., et al., Small, 2008. **4**(8): p. 1153-9.

14. Dokukin, M.E. and I. Sokolov, Langmuir, 2012. **28**(46): p. 16060-71.

(723f) Investigating The Critical Dimensions Of Bacterial Transport

Tandogan, N., Northeastern University
Goluch, E. D., Northeastern University
Originally presented on: 11/7/2013 16:45:00 - 17:03:00

We investigate the transport of bacteria through sub-micron channels by using polydimethylsiloxane (PDMS) microfluidic devices. It is important to understand how bacteria move through confined spaces, toward certain chemicals and away from others, and whether they will continue to advance toward the source when the confinement becomes smaller than their own dimensions. The results obtained here will reveal more about bacterial behavior in soil environments where the surroundings resemble micro- and nano-pores.

To determine the critical dimension for bacterial migration, PDMS devices containing micro- and nanochannels were made using replica molding. Initial master molds were created using a combination of electron-beam and photolithography. The devices are very simple in their design. They include a series of sub-micron channels in various sizes and geometric shapes that connect large (approximately 10 x 100 micrometer) microfluidic channels to a central main channel. By varying constriction geometries it is possible to control where bacteria go in the devices. The bacteria are loaded into the main channel while fresh food is loaded into the side channels, creating a gradient along the sub-micron constrictions.

In this study, we used fluorescently labeled *Bacillus subtilis* and *Escherichia coli,* which are prevalent in soil and studied extensively in research laboratories. A chemotactic response was imposed on the cells by filling the main channel containing bacteria with phosphate buffered saline while the side channels were filled with Lysogeny Broth (LB). Images taken with fluorescent microscope show that bacteria quickly move toward the food source and form biofilms when the dimensions of the constrictions approach the diameter of the bacteria cells. When the cross-sectional area of the constrictions is significantly smaller than the cell dimensions, cells attach themselves at the constriction entrance. An interesting observation is that the cells were able to deform and squeeze through constrictions where only one dimension is smaller than their diameter, for example 0.7 x 1.5 micrometer.

These simple devices can be used to elucidate and understand the complex interactions between microbial species in the environment. We will next investigate the chemotactic response of microbes to other chemical compounds.

(723g) Biofilm Formation of Candida Albicans On Nanofiber Textured Surfaces

Nain, A. S., Virginia Tech

Behkam, B., Virginia Tech

Originally presented on: 11/7/2013 17:03:00 - 17:21:00

Introduction: Medical implants are susceptible to infection caused by microorganisms that adhere to the surface and form biofilms. Biofilms consist of extracellular polymeric substance which protects the embedded microbial cells from the antimicrobials. Due to the high cost of surgical replacements and ineffectiveness of long-term antibiotic as treatments, a new method is needed. The new research direction for anti-biofilm surfaces have been prompted by the anti-fouling topographies of marine organisms. Our research involves interaction between microorganism and bio-inspired submicron topographic features on surface. In the work presented here, we investigate the effect of surface topography feature size and shear stress on biofilm formation and differentiation of *Candida albicans*, a model organism for fungal infection.

Materials and Methods:

Sample Fabrication- Nanofiber-textured model surfaces were fabricated by depositing a single layer of highly aligned and isodiametric polystyrene nanofibers on 45 mm^2 flat polystyrene substrates. The Spinneret based Tunable Engineered Parameters (STEP) nanofiber manufacturing platform developed by Nain *et al.* was utilized to make the polystyrene nanofibers (diameter = 750 nm; separation distance = 2000 nm – 3000 nm). Three different types of surfaces were manufactured for the experiment; smooth (control), single layer, and crisscross double layer. The scanning electron microscopy (SEM) images of single and double layer fiber surfaces are shown in Figure 1.

Biofilm Growth Assay- All samples were soaked in fetal bovine serum (FBS) overnight prior to the biofilm growth assay in order to condition the surface to mimic the natural environment of the microorganism. *Candida albicans* (*C. albicans*) SC5314, a model human fungal pathogen, was used in this study. Dynamic retention assay was conducted by placing the smooth (control) and the nanofiber-textured surfaces on the sample holder bars of a center for disease control (CDC) biofilm reactor and exposing them to a suspension of $1 \bullet 10^5$ CFU/ml of *C. albicans* in yeast nitrogen base (YNB) media with 50 mM dextrose. After 24 hours of growth at 37°Æ C and 80 RPM (1.6 $dyne/cm^2$), the cells are detached from the surfaces and counted using serial dilution and standard plate count method.

Results and Discussion: Our preliminary experimental results shown in Figure 2 suggest that the minimum experimental adhesion density occurs for samples with single layer nanofiber texture. When compared to the smooth surface, the adhesion density for these samples is reduced by approximately 24%. Unlike the single layer surface, the double layer nanofiber surface increased the cell adhesion by 170% when compared to the smooth surface.

Conclusions: Our results indicate that a highly aligned single layer of nanofiber texture significantly delays biofilm formation of *C. albicans*, and is therefore a potentially effective way to minimize the pathogenic effect of this organism. The shear stress of common implant devices that are affected by *C. albicans* are shown in Table 1. Our current and future experiments are focused on conducting pathogenicity assays and optimizing the nanofiber texture parameters for maximum delay in biofilm formation under different physiologically relevant shear stresses.

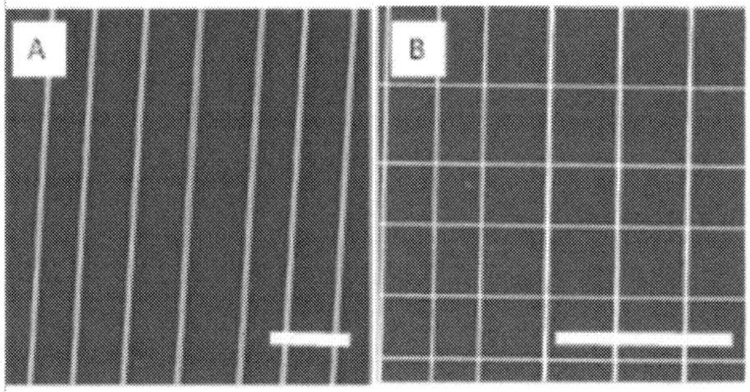

Figure 1. Scanning electron microscopy images of A- single layer fibrous surface and B- double layer fibrous surface is shown. The double layer surface is manufactured by repeating the STEP method on a single layer fibrous surface in a perpendicular direction. The scale bar represents A- 10mm, B- 50mm.

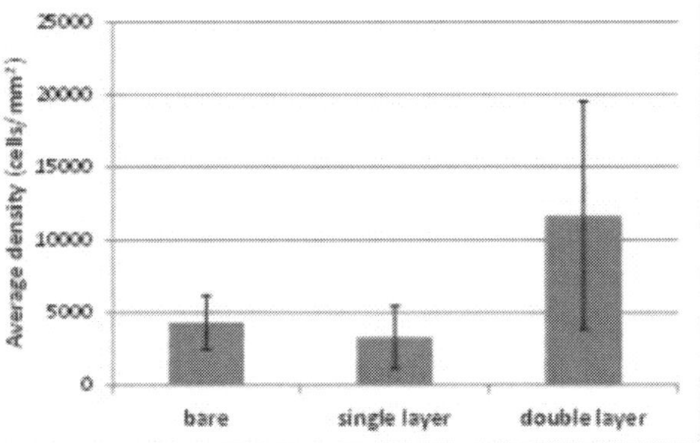

Figure 2. The average cell density (cells per squared mm) is shown for each type of surface. The single layer fiber coated surface shows approximately 24% reduction in cell density when compared to the bare (control) surface. The crisscross double layer fiber coated surface, unlike single layer, showed increase in adhesion by approximately 170%.

Table 1. Implant devices affected by yeast infections

Device	Physiological shear stress (dyne/cm^2)	Corresponding rotational velocity (RPM)
Oral environment	0.6	8
Pacemaker	25	322
Catheter	200	2580
Artificial heart valve	1000	12900

(723h) Elucidating the Roles of Chemical and Mechanical Stimuli in Cell Migration

Jain, G.,
Rajagopalan, P., Virginia Tech
Originally presented on: 11/7/2013 17:21:00 - 17:39:00

Cell migration plays a critical role from morphogenesis to cell death. For example, it plays a prominent role during gastrulation, tissue repair and in mediating the response. *In vivo*, cells exhibit directed migration when presented with single or multiple stimuli that are chemical, mechanical, electrical, or optical in nature[1]. Prior reports have focused on investigating cellular migration only when a single cue is presented in the form of a gradient. In the present study, our goal was to investigate migratory behavior when cells are presented with a choice between increasing substrate rigidity (mechanical) or higher protein concentration (chemical). We have chosen to focus on this unique environment since it recapitulates the wound microenvironment *in vivo*.

Polyacrylamide (PAAM) hydrogels exhibiting chemical and mechanical profiles in opposing directions were created by photomasks and UV polymerization. The Young's modulus and collagen concentrations were modulated in a manner such that they increased in opposing directions. AFM measurements were conducted on hydrated PAAM gels to calculate their elastic modulus. The surface concentration of collagen was measured using an indirect ELISA assay[2]. PAAM gels were seeded with Balb/c 3T3 fibroblasts and time lapse microscopy was used to obtain phase contrast and fluorescence images. Cellular parameters were measured using Nikon NIS Elements software.

Hydrogels were designed exhibiting mechanical gradient values ranging from 34-120kPa. Initial studies were conducted to validate durotaxis on two different mechanical gradient profiles (119kPa/57kPa) and (95kPa/34kPa) while keeping the surface concentration of immobilized collagen constant. We observed that greater than 85% of the cells on both mechanical gradient profiles exhibited a preference towards the rigid side of the substrate. Hence, we concentrated on 119kPa-57kPa gels for future studies since statistically significant, higher displacement values were observed on these substrates. Next, we designed hydrogels with opposing dual chemical and mechanical gradient profiles to investigate which cue would play a dominant role in directing cell locomotion. The collagen concentration was increased either ~ 4-fold or 7-fold. In dual gradient gels, 86% and 100% of cells exhibited directed migration towards the soft (57kPa) side of the gel when the collagen concentration increased 4-fold and 7-fold respectively. A shallow (~ 4-fold) increase in immobilized collagen elicited statistically higher displacements and speeds in comparison to a steep (~ 7-fold) gradient.

We have designed a novel dual gradient hydrogel substrate that can provide new insights into cellular locomotion. Cellular locomotion on these interfaces demonstrated that durotaxis can be reversed when a chemical cue was presented in the opposing direction. These results can be used to design novel biomaterials for interfacial tissue engineering.

References:

(1) Hale, N. A.; Yang, Y.; Rajagopalan, P. *ACS Applied Materials & Interfaces* **2010**, *2*, 2317.

(2) Gaudet, C.; Marganski, W. A.; Kim, S.; Brown, C. T.; Gunderia, V.; Dembo, M.; Wong, J. Y. *Biophysical journal* **2003**, *85*, 3329.

(723i) Loss of Giant Obscurins Promotes a Metastatic Phenotype in Breast Epithelium

Stroka, K. M., Johns Hopkins University
Shriver, M.,
Konstantopoulos, K., Johns Hopkins University
Kontrogianni-Konstantopoulos, A.,
Originally presented on: 11/7/2013 17:39:00 - 17:57:00

Introduction: Obscurins, encoded by the single *OBSCN* gene, are a family of giant proteins that are expressed in striated muscle cells where they play key roles in their structural organization and contractile activity. In a recent systematic study of 13,023 genes in breast and colorectal cancers, *OBSCN* was identified as one of 189 "candidate cancer genes" due to its high mutational frequency (1). Indeed, findings from our laboratory indicate that loss of giant obscurins provides breast epithelial cells with a survival and growth advantage after exposure to DNA damaging agents (2). Here, we explore the role of obscurins in other processes related to cancer development, including epithelial-to-mesenchymal transition, cell invasion and migration, and cytoskeletal dynamics.

Materials and Methods: Stable clones of MCF10A breast epithelial cells depleted of giant obscurins were created using small hairpin RNA (shRNA) technology (2). A combination of immunoblotting, semi-quantitative RT-PCR and immunostaining was used to quantify the expression levels and localization of adherens junctions proteins, as well as proteins and transcriptional regulators associated with an epithelial-to-mesenchymal transition (EMT) in cells depleted of obscurins. Furthermore, a novel microfluidic chemotaxis-based device was engineered and used to quantify cell migration velocity and persistence within narrow and wide (3-50mm) channels. Finally, fluorescence recovery after photobleaching (FRAP) was used to measure actin cytoskeletal dynamics upon depletion of obscurins.

Results and Discussion: We first demonstrated that giant obscurins are down-regulated in human breast cancer biopsies, in comparison with their matched normal samples. Furthermore, depletion of obscurins promoted epithelial-to-mesenchymal transition, as evidenced by (1) disruption of the localization and reduction of the amounts of the adherens junctions proteins E-cadherin, β-catenin, and α-catenin, (2) increased expression of the mesenchymal proteins N-cadherin and vimentin and of the transcriptional regulators Slug and Twist, (3) decreased expression of the junctional epithelial proteins claudin-1, ZO-1, connexin-43, and plakoglobin. Loss of obscurins also promoted a more migratory phenotype, as demonstrated using our microfluidic chemotaxis-based device. Cell migration velocity and chemotactic index (a measure of persistence) were both elevated in obscurin-depleted cells compared to controls (Fig. 1A-B). This trend held for cell migrating in channels of width 3, 6, 10, 20 and 50 μm. Furthermore, FRAP experiments measuring Lifeact-GFP dynamics at cell-cell junctions revealed an increase in the mobile actin fraction and recovery time in cells depleted of obscurin, in comparison with control cells (Fig. 1C-D). This increase in actin cytoskeletal dynamics likely contributes to the enhanced ability of obscurin-depleted cells to transition from an epithelial to mesenchymal phenotype.

Conclusions: Loss of obscurins leads to epithelial-to-mesenchymal transition, enhanced actin dynamics at cell-cell borders, and increased cell migration and invasion into confined spaces. Collectively, our results suggest that obscurins play a role in preventing breast epithelial cells from progressing to a metastatic phenotype. In future work, we plan to explore the possible role of obscurins in cell mechanosensing of matrix mechanical properties.

References: (1) Sjoblom T. *et al*, Science 314: 268-274 (2006). (2) Perry N.A. *et al*, FASEB J 26(7): 2764-2775 (2012).

(724a) Redirection of Glycolytic Metabolism By Overexpression of Bcl-2Î" in CHO Cells

Templeton, N., Vanderbilt University
Lewis, A., Johns Hopkins University
Smith, K., Janssen Pharmaceuticals (J&J)
Dorai, H., Janssen Pharmaceuticals Inc.
Lang, S., Janssen Pharmaceuticals (J&J)
Betenbaugh, M. J., Johns Hopkins University
Young, J. D., Vanderbilt University
Originally presented on: 11/7/2013 15:15:00 - 15:33:00

Previous work has shown that overexpressing anti-apoptotic genes in CHO cells leads to lower lactate accumulation and more rapid consumption. Multiple studies have found that there is a correlation between reduced lactate accumulation and increased antibody titer. However, little is known as to how apoptotic pathway modulations affect central metabolism. Accordingly, we conducted a study where Bcl-2Δ was overexpressed by transfecting a commercially available CHO-S line. Following transfection, we measured Caspase 3/7 expression as a proxy for the overall effectiveness of apoptotic resistance (ApoR) in selected clones. Two clones were selected, each overexpressing Bcl-2Δ as determined by Western blot. To better understand how lactate metabolism was perturbed, we conducted metabolic flux analysis (MFA), in an attempt to elucidate the connection between apoptotic and metabolic pathways. Through MFA, we determined that there was a considerable redirection of glycolytic metabolism at the pyruvate node. In both ApoR clones, a more substantial portion of pyruvate (nearly 40% in one ApoR clone, whereas only 25% in the control) was directed to the TCA cycle and oxidative metabolism. Consequently, less pyruvate was diverted to lactate. Interestingly, in the ApoR cell lines, enhanced NADH oxidase enzymatic activity (a 2 to 3-fold increase) was measured, an enzyme known to oxidize NADH in the electron transport chain. This result, in conjunction with the MFA study, provides a potential explanation for how Bcl-2Δ suppresses lactate production. We will present the results from this study, as well as the impact of Bcl-2Δ on increasing the integrated viable cell density (IVCD) and improving the overall biomass yield on glucose.

(724b) Applying Topological Metabolic Analysis to Industrially Relevant CHO Cell Cultures

Sharfstein, S. T., University at Albany
Shen, X.,
Mitchell, J., Rensselaer Polytechnic Institute
Schaefer, E., Janssen Pharmaceutical Companies of Johnson and Johnson
Migliore, N., Janssen Pharmaceutical Companies of Johnson and Johnson
Tomlin-Verni, K., Janssen Pharmaceutical Companies of Johnson and Johnson
Originally presented on: 11/7/2013 15:33:00 - 15:51:00

Despite extensive efforts to model mammalian cell metabolism in academic research labs, dating back over 30 years and similar levels of bioprocess analysis in industrial research laboratories, much about mammalian cell metabolism in industrial biprocesses remains poorly understood. Recent advances in metabolomics in conjunction with the sequencing of the CHO genome have generated the ability to create and analyze genome-scale models of CHO cell metabolism. However, this level of analysis is impractical for analysis of day-to-day bioprocesses or even for selection of clones that are compatible with platform technologies.

To address this modeling and analysis "disconnect", in this academic-industrial collaboration, we applied an existing, large-scale metabolic model[1,2] to analyze bioprocess data with the following objectives:

- To identify the minimum amount of cellular metabolite data that needs to be collected to sufficiently model cellular metabolism (*e.g.* amino acids, TCA cycle intermediates, etc.).
- To determine whether metabolite data from supernatant samples is sufficient for this purpose or if cell pellet samples would need to be analyzed as well.
- To develop and implement a model that gives a reasonable overview of the cellular state, in terms of specific metabolic signatures, which can be used to identify clones that do or do not fit in with the platform process and/or to determine when a cellular state will adversely affect product quality attributes.
- To determine whether these cellular states can be used to identify changes in cell culture performance due to changes in inputs, *e.g.*, raw materials, which may affect protein quality attributes.

In this presentation, we discuss our success in this effort and provide guidance for the use of large-scale models for analysis of industrial bioprocesses. In addition, we compare our approach with other multivariate analysis approaches to identify optimal approaches for applying metabolic analysis to industrial bioprocesses.

(1) A.C. Baughman, S.T. Sharfstein, and L.L. Martin, *A Flexible State-space Approach for the Modeling of Metabolic Networks I: Development of Mathematical Methods,* Metabolic Engineering, 13: 125-137 (2011). doi:10.1016/j.ymben.2010.12.002

(2) A.C. Baughman, S.T. Sharfstein, and L.L. Martin, *A Flexible State-space Approach for the Modeling of Metabolic Networks II: Advanced Interrogation of Hybridoma Metabolism,* Metabolic Engineering, 13: 138-149 (2011). doi:10.1016/j.ymben.2010.12.003

(724c) Metabolomics and Network Analysis for Sensitive Physiological Monitoring in Industrial Cell Culture Engineering

Vernardis, S., University of Patras
Goudar, C. T., Bayer HealthCare Pharmaceuticals
Klapa, M., Foundation for Research and Technology-Hellas (FORTH)
Originally presented on: 11/7/2013 15:51:00 - 16:09:00

Mammalian cell cultures have been widely used for the production of therapeutic proteins. The primary objective of most current programs for the development of therapeutic protein production processes is the rapid development of bioreactor cultures that are characterized by high product yield and consistent product quality. In addition, due to the high manufacturing cost of these processes, the identification and use of accurate and sensitive controls for cell cultivation robustness is desirable. These controls could provide early warnings of problems in protein productivity and/or final quality before the end of the cultivation. Today, both bioreactor monitoring and process improvements are based primarily on cell growth, metabolic activity and protein productivity data. While useful, the limitations of this cell specific rate-based approach have been recognized. There is a clear need for the development and application of methods that enable the comprehensive characterization of the physiological state of mammalian cell cultures. Moreover, utilized in the context of programs for industrial process improvement, which allow for experimentation with various cell lines and physiological conditions, these methods could contribute to enhancing our overall understanding of the protein production and manufacturing processes. Metabolomics, referring to the quantification of the (relative) concentration profile of the free small metabolites, is the most recently introduced high-throughput method for the measurement of the metabolic fingerprint of a biological system. Between the molecular levels of the cell function, metabolism is directly and very dynamically affected by the environmental changes. By quantifying a complete and accurate metabolic profile is foreseen to have a major positive impact in cell culture engineering research.

To investigate this hypothesis, baby hamster kidney (BHK) cells were cultivated in high cell density perfusion bioreactors. In the first experimental series we examined the effect of the cell age on the metabolic physiology of the cell culture [2], while in a second large experimental process the pH, temperature (T), dissolved oxygen (DO) and/or perfusion rate (CSPR) were varied over the course of the experiments using a Design of Experiments (DOE) strategy. Samples were acquired at the end of each experimental condition and subsequently dried, as described in [2]. The metabolite extracts from the dry biomass samples were obtained by methanol/water extraction and, after derivatization, the metabolic profiles were acquired using the Varian Saturn 2200 Gas Chromatograph (GC) - (ion trap) Mass Spectrometer (MS), as described in [2]. After appropriate normalization and filtering from experimental biases and artifacts [3,4], the profiles were analyzed using multivariate statistical analysis algorithms, as implemented in the TM4 MeV software (www.tm4.org/mev). The results were visualized in the context of a BHK cell line metabolic network that was reconstructed based on the metabolic databases KEGG (*www.kegg.com*), EXPASY (*enzyme.expasy.org*), MetaCyc (metacyc.org) and available literature. Multivariate statistical analysis of the metabolomics data showed the differences between the cell culture samples, which were based on cell age and subtle changes in cell culture growth parameters, while these differences are not directly observable based on the conventional monitoring toolbox.

REFERENCES

[1] Vernardis SI., Goudar CT, Klapa MI (2013) Metabolic profiling reveals that time related physiological changes in mammalian cell perfusion cultures are bioreactor scale Independent *Metabolic Engineering* (in press)

[2] Chrysanthopoulos PK., Goudar CT. and Klapa MI. (2009) Metabolomics for high-resolution monitoring of the cellular physiological state in cell culture engineering. *Metabolic Engineering* **12**: 212-222

[3] Kanani H., Klapa MI. (2007) Data correction strategy for metabolomics analysis using gas chromatography-mass spectrometry. *Metabolic Engineering* **9**:39 – 51

[4] Kanani H., Chrysanthopoulos PK., Klapa MI. (2008) Standardizing GC-MS metabolomics. *Journal of Chromatography B: Biomedical Applications* **871**:191 - 201

(724d) Investigation of pH-Induced Metabolic Switch By Flux Balance Analysis and Gene Expression Analysis

Sarenac, M., ETH Zurich
Noh, H., ETH Zurich
Morbidelli, M., ETH Zurich
Soos, M., ETH Zurich
Originally presented on: 11/7/2013 16:09:00 - 16:27:00

Mammalian cells, used for the production of large, complex and post-translationally modified proteins, generally rely on glucose and glutamine as main carbon and nitrogen sources and consequently produce lactate and ammonia as waste products. The accumulation of lactate and ammonia results in reduced cell growth in batch and fed-batch culture. Additionally the high production of lactate also demonstrates the inefficient metabolism of mammalian cells marked by the strong dependence on the glycolytic pathway as energy source, thus resulting in lower net energy production [1]. Different bioprocess strategies to reduce lactate production by keeping low glucose concentrations [2], [3] or by feeding alternative carbon sources such as galactose [4], have been described.

Our goal is to systematically study the effect of process parameters (pH, osmolarity, dissolved gases and hydrodynamic forces) on cell metabolism, in particular the metabolic switch from lactate production to lactate consumption. Under pH-shifted culture conditions we observed a switch to lactate consumption at pH 6.8 and an increased lactate production above pH 7.5. For comparison, we also performed glucose-limited cultivations, which also resulted in a switch to lactate consumption. By applying flux balance analysis, we investigated changes in the metabolic pathways due to shifted process conditions and compared them to glucose-limited conditions. Furthermore we also aim to find intracellular triggers of metabolic switches and therefore apply transcriptome analysis to evaluate the over- or underexpression of certain metabolic genes as well as cell and mitochondrial transporter genes. With this we aim to validate our flux balance analysis as well as to identify possible flux constraints.

Finally, the knowledge gained will be used to construct a kinetic metabolic model that is able to describe the effect of operating parameters, in particular pH, as well as glucose limitation on cell metabolism.

[1] O. Warburg, "On the origin of cancer cells.," *Science (New York, N.Y.)*, vol. 123, no. 3191, pp. 309–14, Feb. 1956.

[2] H. J. Cruz, a S. Ferreira, C. M. Freitas, J. L. Moreira, and M. J. Carrondo, "Metabolic responses to different glucose and glutamine levels in baby hamster kidney cell culture.," *Applied microbiology and biotechnology*, vol. 51, no. 5, pp. 579–85, May 1999.

[3] W. Zhou, J. Rehm, and W. S. Hu, "High viable cell concentration fed-batch cultures of hybridoma cells through on-line nutrient feeding.," *Biotechnology and bioengineering*, vol. 46, no. 6, pp. 579–87, Jun. 1995.

[4] C. Altamirano, C. Paredes, J. J. Cairó, and F. Gòdia, "Improvement of CHO cell culture medium formulation: simultaneous substitution of glucose and glutamine.," *Biotechnology progress*, vol. 16, no. 1, pp. 69–75.

(724e) Lactate Metabolism in Cell Culture Processes: Model Prediction of the Trajectory to Lactate Consumption

Yongky, A., University of Minnesota
Shu, C. C., University of Minnesota
Mulukutla, B. C., University of Minnesota
Grimm, S.,
Hu, W. S., University of Minnesota
Originally presented on: 11/7/2013 16:27:00 - 16:45:00

Mammalian cells in culture consume glucose and produce a large quantity of lactate. However, under certain culture conditions, glucose flux decreases and the metabolism shifts to lactate consumption. The transition to lactate consumption has been associated with an enhanced productivity as evidence by mining a trove of process data. However, the shift of metabolism is elusive in nature; under similar conditions some cultures metabolically shift to lactate consumption state while others continue to produce lactate. A mechanistic kinetic model based on reactions in glycolysis, pentose phosphate pathway and TCA cycle was developed. The allosteric regulations, acting on the various glycolysis isozymes, together form regulatory loops that give rise to a multiplicity of states in the glycolytic behavior. Such a behavior is profoundly affected by culture lactate levels as well as cell growth rate resulting is a complex topology of the steady state fluxes, in a multi-dimensional space. The switch from the lactate production state to the consumption state can occur only at a certain region in the steady state topology. To further understand such complex behavior, the model was used to simulate the transient behavior in the critical period of metabolic shift and compared to the trajectories of the runs in the archived data. Insights on how to achieve a control of cell metabolism will be presented.

(724f) Using Metabolic Modeling to Examine Differences in the Metabolism of Chlorella Protothecoides Grown On Glucose Or Glycerol for the Production of Lipids for Biodiesel

Morgan, J. A., Purdue University
O'Grady, J., Purdue University
Originally presented on: 11/7/2013 16:45:00 - 17:03:00

Algae are a promising source of lipids for the production of biodiesel due to their high rates of lipid production, the ability for continuous culturing, and their use of non-arable land. The microalga *Chlorella protothecoides* is known to grow to high cell densities with high lipid content when cultured on glucose. We have previously shown that it maintains this high lipid content when grown on glycerol and exhibits an increased yield, albeit at the cost of slower productivity, showing glycerol to be a less expensive and higher yielding substrate. In this work we have employed 13C-metabolic flux analysis (13C-MFA) to further examine the differences between its metabolism on glucose and its metabolism on glycerol during exponential growth. The flux maps generated from 13C-MFA show a significantly higher flux rate through the oxidative pentose phosphate pathway during growth on glucose as well as a lower flux through the TCA cycle. These results were compared to predictions generated by flux balance analysis, which had effectively demonstrated the increased yield of glycerol, but predicted it to have a lower flux through the TCA cycle than glucose, contrary to our 13C-MFA results. We also conducted corollary maintenance studies using a CSTR system to determine the relative efficiencies of each substrate and the underlying factors influencing these differences.

(724g) Isotope-Assisted Metabolic Flux Analysis Reveals Metabolic Landscape Alterations in Poplar Induced By Carbon-Nitrogen Interactions

Zhang, X., University of Maryland, College Park
Nargund, S., University of Maryland
Misra, A., University of Maryland
Coleman, G. D., University of Maryland, College Park
Sriram, G., University of Maryland
Originally presented on: 11/7/2013 17:03:00 - 17:21:00

Nitrogen is an indispensable element for life because it is present in numerous biochemical compounds including amino acids, nucleotides and cofactors such as NAD(P)H. However, nitrogen is usually supplied in the form of fertilizers that are produced mainly by high-energy industrial processes. Fortunately, most trees are capable of conserving and recycling nitrogen. Poplar, specifically, is capable of efficiently managing its nitrogen reserves by recycling nitrogen from leaves during senescence in fall, storing this recycled nitrogen in the form of a protein (bark storage protein [BSP]) during winter and mobilizing this stored nitrogen for *de novo* protein synthesis in spring. Previous research shows that this nitrogen recycling mechanism is strongly correlated with carbon availability that influences BSP accumulation. Therefore, it is desirable to study how carbon and nitrogen nutrient levels affect metabolic landscapes in poplar.

Isotope-assisted metabolic flux analysis (MFA) is a powerful tool in understanding metabolism with its ability to accurately quantify intracellular chemical reaction rates (fluxes). We performed isotope-assisted MFA to calculate fluxes of central carbon metabolism in cells from poplar trees under different carbon and nitrogen nutrient levels. Besides, in order to improve the statistical quality of the flux results, we employed multiple types of labeled carbon sources in parallel experiments and consolidated the ensuing mass isotopomer data into a single flux evaluation. Moreover, we conducted transcriptome analysis to measure the expression levels of all metabolic genes and compared them with the flux map obtained from MFA. The consistency of fluxome and transcriptome clearly shows the metabolic significance of the availability of carbon and nitrogen sources. We anticipate that this study can shed light on future research on nitrogen recycling mechanism and experiment design in poplar plants.

(724h) Analysis of Central Metabolic Reaction Rates Via Non-Stationary 13C Isotope Study Combined With Kinetic Model in Yeast

Jung, J. Y., Korea University
Kim, T. Y., Korea University
Im, D. K.,
Oh, M. K., Korea University
Originally presented on: 11/7/2013 17:21:00 - 17:39:00

Calculation of metabolic reaction rates in the central carbon metabolic pathways (glycolysis, pentose phosphate pathway and TCA cycle) can improve our understanding about microbial physiology of metabolically engineered strains. This has become possible by recently introduced non-staionary ^{13}C metabolite profiling, which directly measures isotope tracer conversion rates of metabolites. For this purpose, a few critical experimental steps should be established and examined, such as proper sampling protocol including quenching and extraction, the verification of metabolite leakage, and the assessment of appropriate mass peaks corresponding to the derivatized metabolites. Also, enzyme kinetic models should be hired for the calculation of metabolic conversion rates from the dynamic ^{13}C isotope data.

In this study, we performed comprehensive analysis of intermediate metabolites with quadruple gas chromatography mass spectrometry (GC-MS) and calculation of metabolic reaction rates in glycolysis, pentose phosphate pathway and TCA cycle in yeast. Firstly, non-stationary metabolite profiling methods were set up for central carbon metabolites analysis from *Saccharomyces cerevisiae* and *Kluyveromyces marxianus*. For example, a bioreactor with fast sampling device was designed, metabolite leakage was tested by mass balance analysis among different types of sample, and the proper mass peaks corresponding target metabolites were assessed via statistical approach. Secondly, isotope-based dynamic studies were performed with uniformly labeled ^{13}C glucose. From time series isotopic studies, conversion of glucose into the central metabolites were well demonstrated. Lastly, kinetic model of each pathway enzymes were used to calculate metabolic reaction rates. The results well demonstrated the difference of glucose utilizing metabolism between *S. cerevisiae* and *K. marxianus*.

(746a) Development of New Reagents and Procedures to Facilitate Cyanobacterial Genetic Engineering

Mehta, K., Stanford University
Evitt, N., Stanford University
Davison, M.,
Bhaya, D., Carnegie Institution, Stanford University
Swartz, J., Stanford University
Originally presented on: 11/7/2013 15:15:00 - 15:33:00

Cyanobacteria and other photosynthetic organisms have tremendous potential as host strains for industrial chemical production driven by energy from the sun. Two important engineering tasks - transformation with exogenous DNA, and measurement of intracellular samples - are difficult in cyanobacteria because they have particularly strong outer peptidoglycan and exopolysaccharide layers that impede DNA access and cell lysis. Currently, large amounts of exogenous DNA are required for reliable transformation of cyanobacteria, which makes it cumbersome to, for example, screen a large library of constructs. Cells must be lysed by disintegration with beads or treatment with harsh solvents, which could damage important cellular components and severely limits the throughput of intracellular measurements.

We are developing a series of novel lysozymes as tools to make both of these tasks easier. Lysozymes, which hydrolyze peptidoglycan chains, could be used to soften the cell wall to improve transformation efficiency and, in a stronger treatment, completely degrade the cell wall to gently lyse cells. Originally discovered in a cyanobacterial community, our lysozymes appears to be particularly suited to treating cyanobacteria. We have used *in vivo* and cell-free platforms to optimize expression conditions for these enzymes, and have developed procedures for cell lysis and pretreatment during transformation. These engineered enzymes together with the new procedures should serve as valuable tools for the cyanobacterial research and engineering community to improve transformation efficiency and enable biologically gentle, high-throughput cell lysis for intracellular measurement.

(746b) Design Principles of Bacterial Organelles

Savage, D.,

Originally presented on: 11/7/2013 15:33:00 - 15:51:00

Carbon fixation is the defining biosynthetic reaction of an autotrophic organism. It also presents numerous catalytic challenges to the host, which must possess specific mechanisms for concentrating carbon and catalyzing fixation in the proper place and time. One such example is the carboxysome, a microcompartment structure that encapsulates the key enzyme of the Calvin Cycle, RuBisCO, and facilitates carbon fixation in cyanobacteria and chemoautotrophs. Although there is considerable genetic and structural data regarding the carboxysome, there are still numerous unanswered questions about how this complex assembles *in vivo* and actually facilitates carbon fixation. We present here a comparative cell biological and biochemical analysis of carboxysome function in two organisms, *S. elongatus* and *H. neapolitanus*. Using these systems, we are investigating the biophysical properties of the carboxysome, including the nature and specificity of self-assembly and what endows its organelle-like function. In addition, we have recently demonstrated that carboxysomes may be transgenically expressed in heterologous hosts. To this end, we will discuss future bioengineering opportunities for using this structure to facilitate carbon fixation in novel organisms, and ultimately, how microcompartments can be used in the construction of synthetic organelles.

(746c) Isobutanol Pathway As a Metabolic Sink for the Glgc Mutant of Synechococcus Elongatus PCC 7942

Li, X., UCLA
Shen, C. R., University of California Los Angeles
Liao, J. C., University of California Los Angeles
Originally presented on: 11/7/2013 15:51:00 - 16:09:00

Glycogen synthesis catalyzed by the glucose-1-phosphate adenylyltransferase (*glgC*) represents a major carbon storage route in cyanobacteria which could divert significant portion of the carbon flux. Here we present the construction of *glgC* knock-out in the *S. elongatus* PCC7942 isobutanol producer and the analysis of its effects on carbon distribution. Deletion of *glgC* increased the normalized isobutanol productivity under constant light during the log-phase. Interestingly, whereas wild type Δ*glgC* suffered severe growth inhibition and photo-bleaching, overexpression of isobutanol synthesis pathway in the Δ*glgC* strain prolonged cell growth and led to higher chlorophyll content. In addition, the growth rescue effect caused by isobutanol pathway activity is positively correlated to the capacity of isobutanol production. Bashed on the carbon 14 incorporation analysis, while wild type Δ*glgC* mutant displayed significant amount of carbon fixation loss due to inability of glycogen synthesis, overexpression of the isobutanol pathway in the Δ*glgC* strain restored the total carbon fixation rate back to the level of wild type. In addition, deletion of *glgC* successfully channeled more carbon flux into the isobutanol pathway. This high percentage suggests that cyanobacteria based biofuel productivity may be limited by its intrinsic carbon fixation capacity.

(746d) Cyanobacterial Conversion of Carbon Dioxide to 2,3-Butanediol

Oliver, J. W. K., UC Davis
Machado, I. M. P., UC Davis
Yoneda, H., Asahi Kasei Corporation
Atsumi, S., UC Davis
Originally presented on: 11/7/2013 16:09:00 - 16:27:00

Conversion of CO_2 for the synthesis of chemicals by photosynthetic organisms is an attractive target for establishing independence from fossil reserves. However, synthetic pathway construction in cyanobacteria is still in its infancy compared with model fermentative organisms. Here we systematically developed the 2,3-butanediol (23BD) biosynthetic pathway in *Synechococcus elongatus* PCC7942 (*S. elongatus*) as a model system to establish design methods for efficient exogenous chemical production in cyanobacteria. We identified 23BD as a target chemical with low host toxicity, and designed an oxygen-insensitive, cofactor-matched biosynthetic pathway coupled with irreversible enzymatic steps to create a driving force toward the target. Production of 23BD from CO2 reached 2.38 g/L, which is a significant increase for chemical production from exogenous pathways in cyanobacteria. This work demonstrates that developing strong design methods can continue to increase chemical production in cyanobacteria. In this talk I will discuss the effect of these design techniques on 2,3-butandiol pathways, screening 10 genes between in *E. coli* and *S. elongatus* hosts.

(746e) Carbon and Nitrogen Instationary Metabolic Flux Analysis On the Source Recovery Urea Cycle of a Unicellular Diatom

Zheng, Y., University of Maryland
Sriram, G., University of Maryland
Originally presented on: 11/7/2013 16:27:00 - 16:45:00

The urea cycle is known to produce urea from ammonia in vertebrates, thus enabling organs to remove excess nitrogen. It was previously assumed that this cycle operates only unidirectionally, always eliminating metabolized nitrogen. However, recent studies have found that in the unicellular, marine diatom *Phaeodactylum tricornutum*, this cycle may perform a biosynthetic role by assimilating nitrogen under nitrogen-depleted conditions. Previous studies have suggested (i) the operation of an incomplete ornithine-urea cycle in this diatom, and (ii) a role for amino acids such as aspartic acid and glutamic acid in nitrogen assimilation, perhaps in conjunction with the tricarboxylic acid cycle.

Diatoms are already known to be photosynthetically efficient. This, combined with an unusual role for the urea cycle, may explain the ecological success of diatoms in adapting to challenging environment. Yet, very little is known about fluxes through diatom metabolic pathways that facilitate carbon and nitrogen assimilation. In this presentation, we will report isotopes labeling experiments and metabolic flux analysis to elucidate fluxes through these pathways. Our metabolic flux analysis involves design of labeling experiments, isotopomer measurements using mass spectrometry and mathematical modeling of metabolic networks. We expect this analysis to reveal pathways of inorganic carbon and nitrogen assimilation including fluxes through the urea cycle and the interrelationship between the urea and tricarboxylic acid cycles. The coordination of nitrogen assimilation and photosynthesis will also be discussed, as well as the roles of these processes on biomass production.

(746f) Isotopically Nonstationary 13C Flux Analysis of Arabidopsis Thaliana Rosettes At Varying Light Intensities

Jazmin, L. J., Vanderbilt University
Ma, F.,
Allen, D. K., Donald Danforth Plant Science Center
Young, J. D., Vanderbilt University
Originally presented on: 11/7/2013 16:45:00 - 17:03:00

Photoautotrophic metabolism represents the primary source of all food on earth as well as raw materials for bio-based production of fuels and chemicals. Efforts to improve photosynthetic efficiency would be aided by more quantitative descriptions of primary metabolism in plants. We have conducted isotopically nonstationary ^{13}C metabolic flux analysis (INST-MFA) studies in *Arabidopsis thaliana* rosettes to quantify metabolism under photoautotrophic conditions using two different light intensities of 200 (normal light) and 500 $\mu mol/m^2/s$ (high light). Using LC-MS/MS and GC-MS profiling techniques on isotopically labeled intracellular metabolites, along with measurements of net photosynthetic CO_2 uptake and starch production rate, we have created comprehensive flux maps of the central carbon metabolism in *Arabidopsis* rosettes under both normal and high light conditions. The resulting flux maps provide a subcellular compartmentalized description of the carbon fixation pathways, which is the first application of INST-MFA to a terrestrial plant system *in planta*. We show that modeling subcellular compartmentation is necessary to account for differences in glucose-6-phosphate labeling patterns in the chloroplast and cytosol, which give rise to starch and sucrose, respectively. The INST-MFA results reveal that Calvin cycle and sucrose synthesis fluxes were increased two-fold in high light relative to normal light conditions, while flux towards starch synthesis remained the same. However, photorespiration increased from 13% of photosynthesis in normal light to 28% in high light. We hypothesize that this increase in photorespiration under high light serves to dissipate surplus ATP and NADPH. This study shows that INST-MFA is a feasible approach for quantifying photosynthetic metabolism in *Arabidopsis* rosettes and provides a quantitative framework to examine genetic and environmental perturbations that impact leaf metabolism.

(746g) Towards the Elucidation and Metabolic Engineering of the Podophyllotoxin Biosynthetic Pathway

Lau, W.,
Sattely, E., Stanford University
Originally presented on: 11/7/2013 17:03:00 - 17:21:00

Towards the elucidation and metabolic engineering of the podophyllotoxin biosynthetic pathway

Warren Lau[1], Elizabeth Sattely[1]
(1) Department of Chemical Engineering, Stanford University, 381 North-South Mall, Stanford, CA, 94305-5025

The lignan compound, podophyllotoxin, is an isolated secondary metabolite used for the semi-synthesis of successfully utilized anti-tumor drugs etoposide, etopophos and teniposide. The growing demand for podophyllotoxin as a precursor for anti-cancer drugs has endangered the traditional source, Podophyllum emodi, which provides 0.1-1% by mass of the biologically active compound. Though other Podophyllum sources have been identified, there is an urgency to relieve the severe pressure on natural sources. Chemical synthesis is one possible solution, but current synthetic routes are not economically feasible for large scale production. Another possible alternative is the overproduction of the molecule in a tractable host organism by pathway engineering. However, there are currently an estimated number of five unknown genes missing for the complete biosynthesis of podophyllotoxin. Also, traditional hosts, such as E. coli, do not carry the complex machinery necessary to produce many plant proteins in their soluble and functional form. Herein, we present the development of an in planta screening platform that combines rapid, transient expression in the model plant, Arabidopsis thaliana (Arabidopsis) with liquid chromatography-mass spectrometry (LC-MS) metabolomics that will enable the functional characterization of candidate genes that are potentially involved in podophyllotoxin biosynthesis. Additionally, we will discuss the usage of this platform as an effective means of evaluating expression constructs for metabolic engineering efforts with a specific focus on building the initial pathway towards podophyllotoxin in Arabidopsis.

(746h) Examining the Transcriptional Response of Overexpressing Anthranilate Synthase in Catharanthus Roseus Hairy Roots

Sun, J.,
Peebles, C. A. M., Colorodo State University
Originally presented on: 11/7/2013 17:21:00 - 17:39:00

Catharanthus roseus produces a variety of terpenoid indole alkaloids (TIAs) including valuable anti-cancer drugs vinblastine and vincristine. Previous study has shown over expression of a feedback insensitive anthranilate synthase αβ unit (ASαβ) behind an inducible promoter in *C. roseus* hairy roots increase the levels of some TIA metabolites, but decrease the levels of other TIA metabolites.

Although the genetic modification is targeted to ASαβ, we hypothesized that expression of other genes in the TIA pathway would be altered. In order to determine the transcriptional response caused by the overexpression of ASαβ, we utilized RT-qPCR to analyze the transcriptional response of all known TIA genes and regulators after 12, 24, and 48 hours of induction (i.e. increased expression of ASαβ). The results indicate an increase in mRNA abundance for some TIA genes and regulators while the mRNA abundance decreases for others. The results and their implications for engineering increased TIA production in *C. roseus* hairy roots will be discussed.

(752a) Controlled Synthesis of Nanoparticles Using a Coaxial Turbulent Jet Mixer

Lim, J. M., Massachusetts Institute of Technology
Gilson, L. M.,
Chopra, S.,
Langer, R. S., Massachusetts Institute of Technology
Farokhzad, O. C., Brigham and Women's Hospital - Harvard Medical School
Karnik, R., MIT
Originally presented on: 11/7/2013 15:15:00 - 15:35:00

Nanoparticles have shown great promise for the diagnosis and therapy of major diseases such as cancer, cardiovascular, and infectious diseases. Use of microfluidics for synthesis of these nanoparticles has been of great interest due to controllability and reproducibility in their physicochemical properties. However, there are intrinsic limitations in conventional polydimethylsiloxane (PDMS) microfluidic systems for the synthesis of nanoparticles. First, only a handful of organic solvents are compatible with conventional PDMS microfluidic systems, which hinders PDMS microfluidic systems from serving as versatile platforms for synthesis of various types of nanoparticles. Second, productivity of microfluidic systems is lower than that of batch reactors due to low flow rates, making it challenging to produce large quantities of nanoparticles required for *in vivo* or clinical studies. Here, we report a novel coaxial turbulent jet mixer capable of synthesizing various types of functional nanoparticles with high production rates suitable for *in vivo* studies and clinical trials, while maintaining the controllability and reproducibility of PDMS microfluidic systems.

In the coaxial turbulent jet mixer, the inner flow stream containing raw materials is mixed with outer stream of non-solvent by turbulent jet flow for generation of uniform nanoparticles by rapid solvent exchange method called nanoprecipitation. The mixing of the inner and outer streams was visualized using phenolphthalein, a pH indicator. Two dimensionless parameters, flow velocity ratio and Reynolds number (Re), were used to classify the mixing behavior into flow regimes corresponding to laminar, transition, and turbulence regimes. Operating in the turbulence regime, the mixing time was tunable in the range of 1 - 40 ms by changing the Re. Nanoparticles obtained using the coaxial turbulent jet mixer at high Re (rapid mixing) were more homogeneous and smaller than those synthesized by bulk mixing, because the mixing time scale is more controllable and shorter than the characteristic aggregation time scale. Since the coaxial turbulent jet mixer is compatible with various organic solvents, it is versatile system where various types of nanoparticles including poly(lactide-co-glycolide)-b-polyethyleneglycol (PLGA-PEG) nanoparticles, iron oxide nanoparticles, lipid vesicles, and polystyrene nanoparticles can be made. In each case, the nanoparticle size decreased as mixing time decreased (*i.e.*, as Re increased). Various functional agents including anticancer drug, insulin, and fluorescent dye could be loaded in the nanoparticles for diagnostic and therapeutic applications during the nanoprecipitation. The coaxial turbulent jet mixer developed in these studies can be used to make functional nanoparticles with high production rates and reproducible manner suitable for clinical studies and mass production.

(752b) Hierarchically Porous Microparticles VIA Spray Drying

Waldron, K., Monash University
Wu, W. D., Monash University
Wu, Z., Monash University
Liu, W., Monash University
Selomulya, C., Monash University
Zhao, D., Fudan University
Chen, X. D., Monash University
Originally presented on: 11/7/2013 15:35:00 - 15:55:00

The versatility of mesoporous silica with tailored pore size and functionality renders it ideal for a wide range of applications such as bio-adsorption and as catalyst substrates. A rapid synthesis method combining evaporation induced self-assembly and surfactant templating via spray-drying can overcome drawbacks associated with traditional synthesis routes, namely long processing time, non-scalable batch operations, and that most powders are sub-micron polydisperse particles. A micro-fluidic jet spray-dryer, developed at Monash University, was used to produce monodisperse microparticles with mesoporous characteristics. We demonstrated that hierarchical structures could be generated through the addition of co-templates in the precursors. Microparticles produced with pluronic block copolymer (F127) with the addition of Eudragit RS 30D ®, a polymer-colloidal dispersion of ~ 120 nm nanoparticles as a co-template, produced microparticles with very open, porous structure with macropores of the ~ 120 nm, along with ~9 nm mesopores. These particles exhibited highly ordered hexagonal mesostructures in addition to the macropores created by the removal of the polymer, with surface areas up to 460 m^2/g and pore volumes up to 0.68 cm^3/g. Through the variation in weight ratio of the co-templates and the silica different hierarchical structures can be obtained. These macro-meso-structures can be used for a range of applications such as bioseparation and bioadsorption in dynamic systems and for immobilisation of bio-catalysts.

(752c) Nano Encapsulation of Active Ingredients

Lakshminarayanan, V.,
Meesters, G., DSM
Mendes, E.,
Originally presented on: 11/7/2013 15:55:00 - 16:15:00

In this work the production of polymeric micelles is studied. Their size and size distributions are measured. Also the stability of the micelles will be measured and discussed. Secondly the polymeric micelles are loaded with an active ingredient. The loading of these polymeric micelles with the active is measured by HPLC methods. Next to this several other techniques will be discussed to elucidate in what state, e.g. crystaline or amorphous, the active is present in the polymeric micel. Also the loaction of the acive in the polymeric micel will be discussed. Techniques like TEM and Cryo TEM are used for this.As a final step, the quantification of the active is carried out to realize the complete characterization of the system. With complete information of the encapsulation system, it is possible to control the and optimize the variables of the system to better tune the nano-system towards the application for which it is developed

(752d) Effect of Nanosmearing On the Rate of Nanostructure Formation in Individual Grains and Tablet Manufacturing

Kick, C., Western Michigan University
Ari-Gur, P., Western Michigan University
Pingali, K., Western Michigan University
Originally presented on: 11/7/2013 16:15:00 - 16:35:00

The influence of mixing order, flowing agent formulation, and compaction on tablet hardness, hydrophobicity, coating uniformity, extent of coating, micoporosity, mesoporosity, and nanoporosity was studied to understand how nanostructures affect bio-availability of the active pharmaceutical ingredient (API). Multiple pharmaceutical formulations of both lubricated and un-lubricated blends were prepared and mixed in a v-blender under fixed shear and strain conditions to characterize the formation of microstructures and nanostructures on the API. The changes in powder properties and surface topography throughout the various blends can be attributed to these formations of nanostructures, which was the focus of our study. These nanostructures were characterized using gas absorption and Scanning Electron Microscopy (SEM) with Electron Dispersive Spectroscopy (EDS) mapping. While the uniformity and extent of coverage on the API particle interface controls the powder flowability, only a fraction of the MgSt lubricant used in the flowing agent formulation coats the API surface and forms the nanostructure. The powder flowability of the blends was quantified in a gravitational displacement rheometer (GDR), which showed that the flow index did improve when the extent of the coating area was increased. However, the extent of coating for the Cab-O-Sil was non-uniformly distributed on the API and much weaker than MgSt, which can be attributed to the interaction between the coating and the API particle surface. When tablets were produced on a rotary tablet press, mercury intrusion, gas absorption, and SEM imaging analysis software were used to characterize the size and distribution of the nanostructures formed. The best flow properties in a blend were exhibited with a formulation consisting of 1% MgSt and 0.5% Cab-O-Sil. The distribution or presence of Cab-O-Sil was seen to enhance the distribution and smearing of MgSt on the particle surface only when the Cab-O-Sil concentration did not exceed 0.5% of the formulation. The variations in uniformity, extent of coating, and characteristic nanostructure formation in the powders are key parameters that influence tablet performance and bioavailability, which has wide applications for bio-medicine manufacturing.

(752e) Cell Free Production of Flagellin As a Vaccine Adjuvant and Its Ordered Display On Virus-Like Particles

Lu, Y., Stanford University
Swartz, J. R., Stanford University
Originally presented on: 11/7/2013 16:35:00 - 16:55:00

Bacterial flagellin has been explored as a potential vaccine adjuvant for enhancing immune responses. In this report, we describe *Escherichia coli*-based cell-free protein synthesis (CFPS) as a method to rapidly produce soluble phase 1 flagellin (FliC) protein from *Salmonella typhimurium*. Especially when flagellin is urgently needed for new and highly potent vaccines at short notice, for example, to combat pandemic influenza threats, CFPS can provide such a means for rapid, cost-effective and high-level production of protein products in a few hours. The synthesized flagellin product had high affinity for the Toll-Like Receptor 5 (hTLR5) (EC50 = 2.4 ± 1.4 pM); approximately 10 to 50 fold higher than previously reported. Observations from protein purification, SDS-PAGE, alanine scanning and multiple mutations indicated that the C-terminal D0 domain of flagellin has disordered tertiary structure, making it susceptible to proteolytic degradation. This proteolysis could be reduced by protease inhibitors, the use of protease-deficient cell extracts, or deletion of the flagellin D0 domain.

We hypothesized that the TLR5 receptor activation could be enhanced if flagellin were presented in an ordered array on the surface of a virus-like particle (VLP). We incorporated a non-natural amino acid displaying an alkyne moiety into flagellin using the CFPS system and attached flagellin to hepatitis B core virus-like particles (VLPs) using bioorthogonal azide-alkyne cycloaddition reactions. The ordered and oriented VLP display of flagellin increased its specific TLR5 stimulation activity by approximately 10-fold.

In this study, we set out to produce more potent vaccine designs. Unexpectedly, we also gained an important insight into our innate immune systems. Cell-free protein synthesis overcame serious proteolytic challenges, and the intact flagellin then exhibited unexpectedly high affinity to the toll-like receptor, TLR5. It stimulated cellular responses at 2 pM (10^{-12}M) concentrations, at least a thousand fold lower than for most biological interactions. When attached to VLPs, the flagellin was even more effective. When bacteria invade our bodies, they find an extremely hospitable environment. In order to overcome this threat, our immune systems must respond very rapidly and very strongly. Our measurements indicate that immune cells can detect incredibly low concentrations of flagellin so they can quickly mobilize to win the battle.

(752f) Polycarboxybetaine Esters for Gene Delivery: Molecular Tuning, Controlled Degradation and Vaccine Applications

Sinclair, A.,
Bai, T., University of Washington
Zhang, L., University of Washington
Jiang, S., University of Washington
Nowinski, A.,
Originally presented on: 11/7/2013 16:55:00 - 17:15:00

The safe and controlled delivery of genes to targeted cells promises exceptional advancements in clinical disease treatment, next-generation vaccines, and tissue engineering. [1-3] An ideal polymeric vector would condense nucleic acids into stable polyplexes that can successfully navigate the extracellular environment and pass through the cell membrane, as well as protect genes from degradation inside and outside the target cells and assist with escape from the endosome or trafficking vesicle. We have developed a gene delivery platform based on polycarboxybetaines (PCBs) modified with degradable esters, with the goal of optimizing each function without detriment to the others.

When the anionic carboxylate groups in zwitterionic PCB side chains are esterified, the polymer is rendered cationic and thus able to bind and condense nucleic acids. [4,5] A tertiary amine can be introduced in place of the quaternary ammonium to make the polymer pH-responsive, buffering pH changes during polyplex trafficking and providing a mechanism via the "proton sponge" for endosomal or phagosomal escape. Another key function of this platform is the hydrolytic conversion of cationic PCB esters to zwitterionic PCB. The rate of this charge shifting can be easily modified by the incorporation of different ester leaving groups and different distances between the ester and the tertiary amine groups. This appends a "smart" characteristic to this polymer; DNA is first strongly complexed by cationic PCB-esters, but is efficiently released when ester degradation reveals the anionic carboxylate groups, rendering PCB zwitterionic and nontoxic.

In our recent study [6], the side chain length was varied to include either one or two alkyl spacers denoted between the amine and the carboxylate/ester groups. Additionally, each chain length was synthesized with both an ethyl ester and a tert-butyl ester hydrolytic group. These molecular modifications result in monomers and resulting polymers with varying proton buffering capacities and hydrolytic profiles. An alkyl spacer length of one brings the pKa of each amine into the optimal range for endosomal buffering, and the hydrolytic ester can be easily modified to speed or slow charge switching. An optimized polymer is able to deliver plasmid DNA an order of magnitude more effectively than polyethyleneimine, without demonstrating any cytotoxicity.

We have also developed a novel photolabile o-nitrobenzyl ester of polycarboxybetaine (PCB-NBE) to give this platform a UV- sensitive "switch" for active degradation control, to study how the charge neutralization caused by ester degradation directly catalyzes DNA release from a polyplex. Photoinitiated ester degradation precipitates the rapid release of over 70% of complexed DNA from PCB-NBE polyplexes. This molecule is promising for triggered release of nucleic acid therapies in vitro.

Extending the functionality of this polymer class, we have applied a similar platform to the passively targeted delivery of DNA vaccines with polycarboxybetaine micro-"gels". [7] These hydrogel microparticles target the phagocytosis pathway of immunostimulating cell types to delivery nucleic acid vaccines adsorbed to their surface. Tertiary amines and hydrolytic esters confer similar benefits of attraction, buffering, and release as they do in nanosized polyplexes. A model vaccine delivered to macrophage cells has resulted in over 10x higher expression than that delivered from a cationic CTAB microparticle control. This platform is also nontoxic and easily tunable.

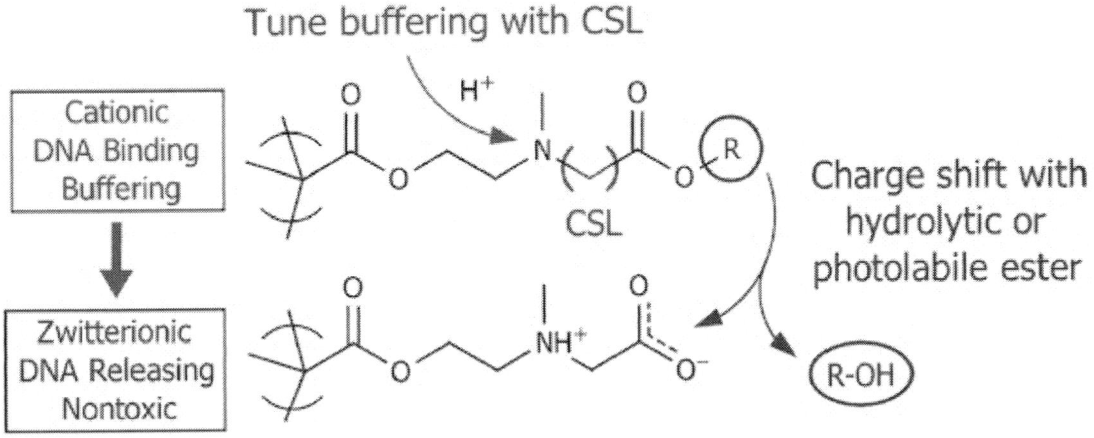

References:

[1] Sheridan, C. *Nat. Biotechnol.* **2011**, 29, 121-128.

[2] Ferraro, B.; Morrow, M. P.; Hutnick, N. A.; Shin, T. H.; Lucke, C. E.; Weiner, D. B. *Clin. Infect. Dis.* ***2011**, 53, 296-302.*

[3] Vallet-Regi, M.; Ruiz-Hernandez, E.; Gonzalez, B.; Baeza, A. J. *Biomater. Tissue Eng.* ***2011**, 1, 6-29.*

[4] Zhang, Z.; Cheng, G.; Carr, L. R.; Vaisocherova, H.; Chen, S.; Jiang, S. *Biomaterials* **2008**, 29, 4719-4725.

[5] Carr, L. R.; Jiang, S. *Biomaterials* **2010**, 31, 4186-4193.

[6] Sinclair, A.; Bai, T.; Carr, L. R.; Ella-Menye, J-R.; Zhang, L.; Jiang, S. *Biomacromolecules* **2013** (DOI: 10.1021/bm400221r)

[7] Zhang, L.; Sinclair, A.; Cao, Z.; Ella-Menye, J-R.; Xu, X.; Carr, L. R.; Pun, S. H.; Jiang, S. *Small* **2013** (DOI: 10.1002/smll.201202727)

(752g) Hollow Stimuli-Responsive Magnetic Silica Microparticles for Controlled Release and MRI Imaging

Sarvasova, N., Institute of Chemical Technology
Stepanek, F., Institute of Chemical Technology, Prague
Originally presented on: 11/7/2013 17:15:00 - 17:35:00

The possibility to control the release of active substances represents one of the most important attributes of targeted drug delivery (TDD). Here we are presenting one of the possible means for such delivery, which could be provided by microscopic, synthetically created particles from composite material based on silica, thermoresponsive polymer poly-N-isopropylacrylamide (PNIPAM, LCST 32°C) and magnetite nanoparticles. All of the materials mentioned above provide desired microparticles with unique characteristics such as endurance and hardness of silica, the ability of thermoresponsive polymers to change its volume in dependence on temperature and the possibility of induction heating as a result of the presence of magnetite nanoparticles. By combining these attributes, the final microparticles could act as a vessel for TDD able to transport the delivered substance into its destination and then release it there as a reaction to an external stimulus being the application of alternating magnetic field.

The synthesis of such microparticles consists of several steps. Firstly, it is essential to prepare macroporous hollow silica microparticles with incorporated magnetite nanoparticles using a solid template method. This is achieved by following a chain of reactions starting with the production of solid polystyrene (PS) template in the form of monodisperse spheres with relatively uniform size in the range of 1 mm. Then the silica/PS/Fe_3O_4 composite microparticles are obtained using modified Stober method based on the reactions of condensation and hydrolysis. During this process, Fe_3O_4 nanoparticles are gradually absorbed into the growing silica seeds, thus creating continuous composite silica/ Fe_3O_4 network on the surface of the PS templates. The last step, constituting of the PS core removal, is easily done by dissolving in toluene and leads to the creation of hollow silica/Fe_3O_4 microparticles with pores of the approximate size near 100 nm. In order to proceed to the final product being silica/PNIPAM/Fe_3O_4 hybrid microparticles, the already existing silica/Fe_3O_4 microparticles have to be functionalized by introducing the double bond onto its surface utilizing 3-(trimethoxysilyl)propyl methacrylate (3-MOP), allowing PNIPAM to be subsequently attached onto the microparticles as a result of radical copolymerization of 3-MOP silica with NIPAM. The as prepared hybrid microparticles were characterized in terms of its composition and morphology. Additionally, also the possibility of radiofrequency heating and release process of a model substance (Nile Red) were investigated. Since it would be highly beneficial to be able to observe the behavior of microparticles containing magnetite in 3D surroundings, the method of MRI was used.

(752h) Iodinated Nanoparticles for Diagnostic Medical Imaging Applications

Tang, C., North Carolina State University
Chacko, A. M.,
Mikitsh, J.,
Alavi, A.,
Prud'homme, R. K., Princeton University
Originally presented on: 11/7/2013 17:35:00 - 17:55:00

Non-invasive medical imaging techniques such as X-ray-computed tomography (CT) and positron emission tomography (PET) are powerful tools in diagnostic medicine. In these applications, contrast agents are needed to enhance the diagnostic quality of the images. Ideally, contrast agents would be biocompatible and contain high payloads of contrast agent thus nanoparticle formulations of radiopaque or radioactive materials offer opportunities for improvements in diagnostic imaging. In this work, we explore the use of Flash NanoPrecipitation (FNP) as a platform technology to produce such nanoparticle formulations of iodine (radiopaque) and its radioactive isotopes. In FNP, amphiphilic block copolymers direct self-assembly of biocompatible nanoparticles with high concentrations of encapsulated components. The precipitation process is controlled by carefully tuning the time scales of micromixing, self-assembly, and nucleation and growth. Using this method, various routes to iodinated nanoparticles have been explored. In one approach, an iodinated polymer can be encapsulated within the nanoparticles during the FNP process. Alternatively, the FNP can be used to encapsulate polyphenolic compounds and the phenolic moieties within the nanoparticles can be iodinated. Comparison using TEM and elemental analysis of these two approaches using sodium iodide with sodium hypochlorite as a model chemistry for iodination will be presented. Radioiodination of various size nanoparticles (80 to 300 nm) for PET imaging applications will also be discussed.

(766a) Unlocking Yeast Metabolism: Identification of Key Metabolites Affecting Reactions Near Thermodynamic Equilibrium

Kiparissides, A., Imperial College
Hatzimanikatis, V., Swiss Federal Institute of Technology of Lausanne (EPFL)
Originally presented on: 11/8/2013 8:30:00 - 8:48:00

The industrial demand for biological processes has been significantly increasing over the last decades (Biotechnology Industry Organisation, 2010; http://bio.org). This sustained growth can be partially attributed to advancements in biology and medicine and partially to external factors such as the need for cleaner, environmentally friendlier processes. Despite the economic turmoil of recent years (2008-10) Thomson&Reuters concur that the bio-industry is a viable platform for low risk investments with a good profit margin. However, industrial production is struggling to keep up with increasing global demand for products such as biologics, biofuels and biodegradable polymers leaving room for further growth. Common problems associated with bioprocessing are batch-to-batch variability, low product yields, poor product quality and long time to market. Process optimization and control remains fundamentally heuristic in the biological industry due to both the complex nature of cell culture and a limited understanding of cellular metabolism.

The ongoing increase in the popularity of Systems Biology over the past 15 years is a testament of the need for a systems level approach that can identify meaningful correlations amongst the multitude of interactions that constitute cellular metabolism. Moreover the complete genome sequencing of an increasing number of organisms has increased the availability of genome scale metabolic reconstructions further promoting the expansion of the "systems" approach. Genome scale models, comprising the entirety of the metabolic reactions known to occur within the cell, provide a "bird's eye view" of central carbon metabolism and offer valuable insight regarding optimal resource allocation towards a desired macroscopic behavior, usually maximizing the production of a valuable metabolite.

Integration of thermodynamic constraints in genome scale models (Henry et al. 2007) restricts the flux space to include only thermodynamically feasible profiles and is a significant first step in reducing uncertainty regarding the net distribution of flux. Thermodynamics based Flux Balance Analysis (TFBA) allows for the identification of bottlenecks in the metabolic network by finding reactions that are at or near thermodynamic equilibrium. However the problem of allocating the (remaining) uncertainty and identifying input factors with a strong effect on the output of the model is far from trivial even for TFBA. A number of studies have attempted to study the sensitivity of FBA outputs. The applied approaches are commonly one-factor-a-time numerical perturbation simulations and/or derivative based local measures, which almost exclusively study the effect of the model constituents on the value of the objective function. Whilst valuable qualitative information has often been the product of the proposed approaches they are not able to identify which metabolite(s) affect the qualitative and/or quantitative behavior of a particular reaction or a subset of reactions which is vital information for the metabolic optimization of any cellular process.

By combining TFBA, Marcov Chain sampling, Experimental Design and Global Sensitivity Analysis we present an efficient algorithm to quantify the effect intracellular metabolites have on the thermodynamic flexibility of cellular metabolism. Metabolites are ranked based on their ability to affect reactions at or near thermodynamic equilibrium when fixed at any value within their respective feasible bounds. The proposed methodology effectively identifies actuators for the optimization of cellular metabolism by providing a ranked list of metabolites that can un(b)lock thermodynamic bottlenecks.

(766b) Combinatorial and Evolutionary Engineering of Pentose Sugar Catabolism in Saccharomyces Cerevisiae for Efficient Biofuels Production

Lee, S. M., Korea Institute of Science and Technology
Jellison, T., The University of Texas at Austin
Alper, H., The University of Texas at Austin
Originally presented on: 11/8/2013 8:48:00 - 9:06:00

Biofuels production from lignocellulosic biomass can be both sustainable and economical when all available carbon sources are completely utilized. Pentose sugars, such as xylose and arabinose, constitute significant portion of lignocellulosic biomass hydrolysates. However, these sugars are poorly utilized by the yeast *Saccharomyces cerevisiae* despite decades of research. Here, we discuss the advantages of both combinatorial and evolutionary engineering approaches for the improvement of pentose sugar utilization. Specifically, we discuss the directed evolution of a xylose isomerase-based pathway to achieve a 61 fold improvement in growth rate and 8 fold improvement in ethanol production from xylose. We next discuss combinatorial and evolutionary engineering efforts to establish functional pathways. In particular, key catabolic enzymes were either randomly mutated from a previously reported enzyme or newly obtained from a pentose sugar utilizing fungi. We show that alternative pentose sugar catabolic pathways in *S. cerevisiae* can perform better compared with previously reported pentose sugar pathways. Moreover, key design principles for catabolic enzyme pathways can be extracted from these studies. These strains can be further improved through adaptive or evolutionary engineering efforts. Finally, these studies describe an effective combinatorial engineering strategy to develop efficient heterologous pathway in *S. cerevisiae*.

(766c) Investigating Host Dependence of Xylose Utilization in Recombinant Saccharomyces Cerevisiae Strains Using RNA-Seq Analysis

Feng, X., University of Illinois at Urbana-Champaign
Zhao, H., University of Illinois at Urbana-Champaign
Originally presented on: 11/8/2013 9:06:00 - 9:24:00

Xylose-based ethanol production by recombinant *Saccharomyces cerevisiae* is of great interest to basic and applied bioenergy research. By expressing three different fungal pathways in two *S. cerevisiae* hosts respectively, we found that the xylose utilization efficiency by recombinant *S. cerevisiae* depends not only on the choice of xylose pathway but also on the choice of host, exhibiting an obvious host or context dependence. To investigate molecular mechanisms of this context dependence, we applied RNA-seq analysis in this study for a systematic characterization of the xylose utilization via different pathways in different *S. cerevisiae* hosts. The transcripts that were regulated during xylose utilization have been identified in this study. Three transcription factors involved in regulation of amino acid metabolism, responses to oxidative stresses, and degradation of aggregated proteins, respectively, were found to participate in xylose metabolism regulation regardless of which pathway was expressed and which host the xylose pathway was expressed in. Nine transcription factors, involved in homeostasis, regulation of amino acid metabolism, and stress responses, were identified as the key modules responsible for the host-specific responses to the same xylose pathway. In addition, the transcriptional regulations of xylose utilization in different yeast hosts were compared to two reference regulation patterns, which suggested that diverse regulation strategies were adopted by different hosts for improved xylose utilization. As indicated by the functions of the conserved transcription factors involved in xylose metabolism regulation, the xylose utilization in recombinant *S. cerevisiae* may be affected by both carbohydrate metabolism regulation and stress responses. Based on the comparison of transcriptional regulation patterns, the metabolic optimizations of xylose utilization in different hosts went toward different directions, which may explain the host dependence observed in this study. The knowledge revealed by this study could provide valuable insights towards the improvement of metabolic engineering strategies for cellulosic ethanol production.

(766d) Expanding the Biofuel Portfolio By Integrating Chemistry and Biology

Baer, Z., UC Berkeley
Sreekumar, S., UC Berkeley
Anbarsan, P., IIT-Madras
TostÃ©, D., University of California
Blanch, H. W., University of California at Berkeley
Clark, D. S., University of California
Originally presented on: 11/8/2013 9:24:00 - 9:42:00

Acetone, a product of solventogenic *Clostridia*'s acetone-butanol-ethanol (ABE) fermentation, harbors nucleophilic α-carbons, which are amenable to C-C bond formation with the electrophilic α-carbon of the alcohols (ethanol and n-butanol). These inherent functionalities enable coupling chemistry to form higher molecular weight hydrocarbons, similar to those found in current jet and diesel fuels, by a transition-metal-catalyzed alkylation reaction. Here we describe the integration of biological and chemocatalytic routes to efficiently convert ABE fermentation products into ketones, ranging from 2-pentanone to 6-undecanone, by a palladium or copper-catalyzed alkylation reaction. Tuning of the reaction conditions permits production of either predominately gasoline or predominately jet and diesel blend stocks. The integration of the two technologies, ABE fermentation and the alkylation reaction, is made possible by *in situ* liquid-liquid extraction (LLE) of the ABE with a non-toxic extractant. Glyceryl tributyrate was used for *in situ* selective extraction of both acetone and n-butanol to enable simple integration of ABE fermentation and chemical catalysis, while reducing the energy demand of the overall process. A 2L 60-hr extractive fermentation of wild-type *C. acetobutylicum* in a 1:1 volume ratio of medium:extractant produced 40.8 g of ABE from 105 g of glucose (90% theoretical yield). Similar titers were observed when sucrose was the major carbon source. Solvents distilled from the extractant phase were as reactive as the pure chemicals, achieving an overall molar yield of 93% and 97% conversion of acetone. This process provides a means to selectively produce gasoline, jet, and diesel blendstocks from lignocellulosic and cane sugars at yields near their theoretical maxima. Replacing acetone with isopropanol in the ABE mixture predictably increases the concentration of diesel-range components in the blendstock. *C. acetobutylicum* was metabolically engineered to produce isopropanol-butanol-ethanol (IBE), achieving an alcohol titer of 21.5 g/L from 54 g/L glucose in 35 hours. Computational modeling of solute/extractant LLE using Cosmo-RS™ identified oleyl alcohol as a superior extractant for isopropanol-butanol-ethanol extractive fermentation on the basis of isopropanol removal. Extractive fermentation with the metabolically engineered *C. acetobutylicum* SACE strain produced 39.6 g of IBE from 117 g glucose. Alcohols distilled from the extractant phase were as reactive in the alkylation reaction as the corresponding synthetic mixtures. In conclusion, we have leveraged advantages from both theromchemical and biological processes for the efficient conversion of sugars to biodiesel blendstocks.

(766e) Raman Spectroscopy to Monitor the Phenotypic Responses of Bacteria to Environmental Stresses

Zu, T. K.,
Athamneh, A. I., Virginia Tech
Senger, R. S., Virginia Tech
Originally presented on: 11/8/2013 9:42:00 - 10:00:00

The chemical composition of a bacterial cell is determined by a complex function involving the genotype of the organism and its environmental conditions. One reason for studying cellular chemical composition is to determine the physiological response to a toxic compound. Indeed, this knowledge will help guide future metabolic engineering approaches to making highly resistant cells. Producing bacteria that can withstand high concentrations of alcohols, alkane biofuels, or commodity chemicals would be advantageous. However, physically determining the chemical composition of a cell routinely involves mass spectrometry and is labor and time-intensive. Furthermore, chemical composition of a bacteria cell can change continuously, demonstrating the need for real-time measurements. Here, we demonstrate a new technology combining Raman spectroscopy and multivariate statistics to effectively monitor cell chemical composition in real-time. To demonstrate our novel approach, we exposed *Escherichia coli* K12 to various growth inhibitory levels of 1-butanol, isobutanol, 2-butanol, and 1,4-butanediol. We performed mass spectrometry-based analyses of membrane fatty acids, chromatographic analysis of amino acids, fluorescence anisotropy determination of membrane fluidity, and routine analyses of protein, nucleotide, and carbohydrate content. These measurements were all in agreement with specific Raman signatures obtained in near real-time and illuminated by comparative peak analysis and discriminant analysis. Furthermore, results of the Raman study were used to identify differences in the mechanisms of toxicity among these 4-carbon alcohols to *E. coli*. This approach has very broad applications for studying phenotype response and ultimately deriving metabolic engineerng strategies to address specific mechanisms of toxicity. In addition, we have made our Raman Diagnostic Analysis (RDA) tool, containing all relevant multivariate statistical models, publicly available as a stand-alone application for the analysis of Raman data.

(766f) A Reverse Glyoxylate Shunt to Build a Non-Native Route From C4 to C2 in E. Coli

Wong, S. S., University of California Los Angeles
Mainguet, S. E., University of California, Los Angeles
Gronenberg, L., University of California
Liao, J. C., University of California Los Angeles
Originally presented on: 11/8/2013 10:00:00 - 10:18:00

Most central metabolic pathways such as glycolysis, fatty acid synthesis, and the TCA cycle have complementary pathways that run in the reverse direction to allow flexible storage and utilization of resources. However, the glyoxylate shunt, which allows for the synthesis of four-carbon TCA cycle intermediates from acetyl-CoA, has not been found to be reversible to date. As a result, acetyl-CoA can only be produced by decarboxylation of the three-carbon molecule pyruvate in heterotrophs. A reverse glyoxylate shunt (rGS) could be extended into a pathway that converts C_4 carboxylates into two molecules of acetyl-CoA without loss of CO_2. Here, as a proof of concept, we engineered in *E. coli* such a pathway to convert malate and succinate to oxaloacetate and two molecules of acetyl-CoA. We introduced ATP-coupled heterologous enzymes at the thermodynamically unfavorable steps to drive the pathway in the desired direction. This synthetic pathway in essence reverses the glyoxylate shunt at the expense of ATP. When integrated with central metabolism, this pathway has the potential to increase the carbon yield of acetate and biofuels from many carbon sources in heterotrophic microorganisms, and could be the basis of novel carbon fixation cycles.

(766g) A Genetic Circuit Approach to Alleviating Membrane Transporter Stress

Boyarskiy, S., University of California, Berkeley
Davis, S. A., University of Californa, Berkeley
Tullman-Ercek, D., University of California, Berkeley
Originally presented on: 11/8/2013 10:18:00 - 10:36:00

A key component in microbial production of biofuels and other chemicals is the control of the selective permeability of the cell's membrane. There is a large research effort in engineering membrane transporters for influx of feed sugars and for efflux of toxic molecules and fermentation products. One limiting factor is the increased stress associated with over-expression of membrane-associated proteins and the difficulty in controlling their expression without the addition of a costly inducer molecule. Therefore expression of transporters must be tightly regulated for maximum productivity and yield. To address this issue, we chose the *Escherichia coli* protein AcrB, a member of a multi-drug resistance pump, which can be engineered to secrete small biofuel molecules such as alcohols. To control the expression of the somewhat toxic AcrB, we have designed and implemented a feedback circuit. Our circuit utilizes a native *E. coli* stress promoter that downregulates pump expression in response to stress induced by the pump but not the metabolic product secreted by the pumps. With this presentation, we will discuss how we identified this promoter in a screen of regions upstream of differentially expressed genes in response to various stresses, how the overall pump concentration can be controlled via the ribosome binding site, and how the circuit behaves under various conditions. This approach can also be extended towards stable production of other growth inhibiting or toxic proteins, and can be used to achieve stable, tightly controlled expression, without the drawbacks of small molecule inducer systems.

(766h) Rational Design of Metabolic Engineering Strategies for Xylose Metabolism With Scheffersomyces Stipitis and Saccharomyces Cerevisiae

Liang, M., Auburn University
He, Q., Tuskegee University
Wang, J., Auburn University
Originally presented on: 11/8/2013 10:36:00 - 10:54:00

Rational design of metabolic engineering strategies for xylose metabolism with *Scheffersomyces stipitis* and *Saccharomyces cerevisiae*

Meng Liang[1], Q. Peter He[2], Jin Wang[1]

1. Department of Chemical Engineering, Auburn University
2. Department of Chemical Engineering, Tuskegee University

Lignocellulosic ethanol production represents an attractive alternative source for long-term renewable energy supply due to the rising concerns over energy sustainability, global warming and feed stock availability[1]. However, many barriers exist for industrializing lignocellulosic ethanol processes and one of them is the effective conversion of xylose, the second abundant mono-saccharide and representative pentose in the hydrolysate of lignocellulosic biomass.

Saccharomyces cerevisiae is the most promising candidate for lignocellulosic ethanol production due to its excellent glucose fermentation capability, high ethanol tolerance, and resistance to inhibitors presented in lignocellulosic hydrolysate[2]. However, native *S. cerevisiae* strains cannot utilize xylose for either growth or ethanol production. Despite of the inability to utilize xylose, *S. cerevisiae* has the metabolic pathway to convert xylulose, an isomerized product of xylose, into ethanol. Thus, numerous studies have been carried out to link extracellular xylose with intracellular xylulose and to enhance the capacity of xylulose fermentation so as to further increase the efficiency of xylose fermentation[3].

As the most promising native strain for xylose fermentation[4], *Scheffersomyces stipitis* (previously named *Pichia stipitis*) has been chosen as one of the most common gene providers for recombinant *S. cerevisiae*. Besides working as a gene provider, *S. stipitis* itself also shows good overall performance on lignocellulosic hydrolysate[5]. Understanding its metabolism, especially the central carbon metabolism, is very important to improve the strain, or to provide hints for metabolic adjustment of other strains. However, limited results have been published on the metabolic engineering of *S. stipitis*.

Redox couples NAD+/NADH and NADP+/NADPH have been reported to play an important role in energy metabolism and product formation of yeasts[6]. For *S. stipitis*, it has been reported that its product distribution is very sensitive to oxygen transfer rate due to redox imbalance. Meanwhile, it has been recognized that the heterologous expression of xylose reductase (XR) and xylitol dehydrogenase (XDH) in recombinant *S. cerevisiae* would cause perturbation to the intrinsic redox balance, which has been considered as one of the reasons of some recombinant *S. cerevisiae*'s subordinate performance on xylose fermentation[7]. Therefore, understanding the mechanism of redox balance and tuning up the redox conditions by cofactor engineering or adjusting other NADH- or NADPH-related steps have significant impacts on ethanol production for xylose fermentation.

Besides the redox balance, there are other bottlenecks reported for recombinant *S. cerevisiae* [7, 8], such as xylose transporter with low efficiency, insufficient capacity from xylulose to xylulose-5-P, limited flux of non-oxidative pentose phosphate pathway. The metabolic changes caused by the heterologous expression of XR and XDH will propagate through the metabolic network and interact with other bottlenecks. Therefore, the design of metabolic engineering strategy should be considered systematically.

In this work, we reconstructed the central carbon metabolic network model of *S. stipitis* and modified published genome-scale metabolic network model of *S. cerevisiae* to mimic real genetic modifications. The two models were used to gain better understanding of xylose metabolism of the two strains, as well as to evaluate the various metabolic engineering strategies for xylose fermentation to ethanol and their influences to glucose fermentation. By analyzing the modeling results we propose the candidate strategies for optimizing ethanol production from xylose with the two strains.

Specifically, we first reconstructed the central carbon metabolic network model of *S. stipitis* by integrating genomic, biochemical and physiological information available for this microorganism and other related yeast. Meanwhile a published genome-scale metabolic network model of *S. cerevisiae* was modified by adding additional reactions and constraints to simulate the recombinant *S. cerevisiae* with introduced *XYL1* and *XYL2* from *S. stipitis*. Then we propose a system-identification based framework to characterize the xylose metabolism under various conditions, such as different oxygen conditions or varied cofactor preferences of the enzymes. The reactions most influenced by introducing various perturbations, whose capacities would also affect the introduced perturbations most, as well as the desired regulations directions (e.g., depress or up-regulating reaction activity), have been identified. The reactions chosen by other researchers as the targets of the metabolic engineering for xylose metabolism in recombinant *S. cerevisiae* have all been identified by the proposed approach.

With the reactions identified, the published metabolic engineering strategies were evaluated using the models through constraint-based FBA. The modification, deletion, depression or overexpression of the genes have been simulated by applying hard constraints or flux ratio constraints. The simulation predictions agreed well with the reported experimental results. In addition, the influences of the strategies on glucose fermentation were evaluated as well. Cofactor engineering by varying cofactor preferences of the enzymes through protein engineering has been characterized. Based on these results, the optimal metabolic strategies for both strains have been proposed which predicted significantly increased ethanol yield compared to the previously published strategies.

References

1. Cherubini, F., The biorefinery concept: Using biomass instead of oil for producing energy and chemicals. Energy Conversion and Management 2010, 51 (7), 1412-1421.
2. Olsson, L.; Nielsen, J., The role of metabolic engineering in the improvement of Saccharomyces cerevisiae: utilization of industrial media. Enzyme and microbial technology 2000, 26 (9-10), 785-792.
3. Jeffries, T. W., Engineering yeasts for xylose metabolism. Current opinion in biotechnology 2006, 17 (3), 320-6; (b) Matsushika, A.; Inoue, H.; Kodaki, T.; Sawayama, S., Ethanol production from xylose

in engineered Saccharomyces cerevisiae strains: current state and perspectives. Applied microbiology and biotechnology 2009, 84 (1), 37-53.

4. Jeffries, T. W.; Van Vleet, J. R. H., Pichia stipitis genomics, transcriptomics, and gene clusters. FEMS yeast research 2009, 9 (6), 793-807.
5. Rumbold, K.; van Buijsen, H. J. J.; Gray, V. M.; van Groenestijn, J. W.; Overkamp, K. M.; Slomp, R. S.; van der Werf, M. J.; Punt, P. J., Microbial renewable feedstock utilization: a substrate-oriented approach. Bioengineered bugs 2010, 1 (5), 359-66.
6. Bakker, B. M.; Overkamp, K. M.; van Maris, A. J.; Kötter, P.; Luttik, M. a.; van Dijken, J. P.; Pronk, J. T., Stoichiometry and compartmentation of NADH metabolism in Saccharomyces cerevisiae. FEMS microbiology reviews 2001, 25 (1), 15-37.
7. Cai, Z.; Zhang, B.; Li, Y., Engineering Saccharomyces cerevisiae for efficient anaerobic xylose fermentation: reflections and perspectives. Biotechnology journal 2012, 7 (1), 34-46.
8. Kim, S. R.; Park, Y.-C.; Jin, Y.-S.; Seo, J.-H., Strain engineering of Saccharomyces cerevisiae for enhanced xylose metabolism. Biotechnology advances 2013.

(770a) A Simplified Route to Prepare Biomass-Derived Terephthalic Acid (TA)

Tong, Z., University of Florida
Wang, F.,
Originally presented on: 11/8/2013 8:30:00 - 8:55:00

Abstract : In response to diminishing oil reserves and increasing costs for petroleum-based products, there is a great need to develop new routes to chemicals intermediates from renewable resources. Here we reported a novel synthesis route to directly transform biomass-derived chemicals (isoprene and acrylic acid) to terephthalic acid (TA) through p-toluic acid (PTA). TA is an important monomer for synthesizing polyesters including polyethylene terephthalate (PET), polypropanediol terephthalate (PPT) or polybutylene terephthalate (PBT), which have broad applications from drink bottles to polyester fibers. The starting chemicals can be produced from biomass through established chemical processes. Bio-based isoprene can be produced by the fermentation of biomass sugars. Acrylic acid can be produced either from dehydration of glucose-based lactic acid or dehydration and oxidation of glycerol (available in large quantity from biodiesel production). In this study, TA was converted from isoprene and acrylic acid through three simple reactions including Diels-Alder reaction, dehydro-aromatization, and oxidation. The simplified process is based on the efficient dehydro-aromatization reaction of cyclohexene carboxylic acid to PTA using a cost-effective catalyst (sulfuric acid). All three reactions were conducted under mild conditions. Finally, pure TA was successfully synthesized with the overall yield of approximately 60% and its structure was confirmed by ^{1}H-NMR (Nuclear Magnetic Resonance). Theoretically, at the first time, we elucidated the dehydro-aromatization reaction mechanism of cyclohexene carboxylic acids using sulfuric acid as the catalyst.

(770b) Co-Fermentation of Glucose and Xylose From Steam Pretreated Wheat Straw With Recombinant S. Cerevisiae Strains

Nielsen, F.,
Zacchi, G., Lund University
Originally presented on: 11/8/2013 8:55:00 - 9:20:00

Advanced bioethanol produced by fermentative conversion of lignocellulosic agricultural residues relies on efficient hydrolysis and fermentation, as well as extensive utilization of available substrates in order to obtain high yields and render the process economically viable. Another important aspect is high ethanol concentration in the fermentation broth, as this reduces the energy demand. Obtaining a high ethanol concentration requires high water insoluble solids (WIS) content in simultaneous saccharification and fermentation (SSF) or in the enzymatic hydrolysis in separate hydrolysis and fermentation (SHF). This usually has an adverse effect on ethanol yield, due to the increased inhibition and poor mass transfer. Opting for either configuration is essentially a trade-off between optimal temperatures and inhibitory glucose concentrations on the one hand (SHF) and sub-optimal temperatures and ethanol-inhibited cellulolysis on the other (SSF). However, the ethanol yield and concentration can also be improved by efficient co-fermentation of glucose and xylose; especially from agricultural residues, such as wheat straw, due to the high amount of hemicelluloses in these materials.

In the present study, SSF and SHF of steam pretreated wheat straw was performed in a variety of process configurations in 2 to 30 L scale. Two different recombinant S. cerevisiae strains, which has the ability to co-ferment glucose and xylose, and two steam-pretreated wheat straw slurries, one pretreated at mild conditions and one pretreated at more sever conditions, were utilized. Substrate feeding and enzyme dosage were varied in order to maximize the ethanol yield. The highest overall ethanol yield achieved, based on the amount of glucose and xylose present in the material fed to the system, was 92% of theoretical for the milder pretreated straw and 84% of theoretical for the more severe pretreated straw. The ethanol concentration was 45.5 and 41.9 g/L, respectively. In both cases the xylose consumption was higher than 90% and the xylitol produced was 1.8 and 3.3 g/L respectively.

(770c) Nutrient Supplementation Key to Inhibitor-Tolerant Yeast Development and Fermentation Performance On Switchgrass Hydrolyzates

Slininger, P. J., National Center for Agricultural Utilization Research, USDA-ARS
Liu, Z. L., National Center for Agricultural Utilization Research, USDA-ARS
Dien, B. S., USDA-ARS NCAUR
Originally presented on: 11/8/2013 9:20:00 - 9:45:00

Inhibitory compounds generated during acid hydrolysis pretreatment of lignocellulosic biomass interfere with subsequent fermentation to ethanol. Tolerant yeast strains have recently been developed by targeted evolution in the presence of biomass hydrolyzate inhibitors. In order to optimize performance, comparative data were collected on both the tolerant and parent strains in experiments testing the impact of nitrogen sources, vitamins, and minerals on ethanol production potential. With respect to optimal ethanol production, a striking difference in the nutritional requirements of parent and adapted strains was observed in both synthetic media and dilute acid hydrolyzates of switchgrass that were supplemented with a variety of commercially available nitrogen sources. Since nutrient availability significantly impacted both ethanol yield and production kinetics, and was strain-dependent, nutrient supplementation requirements were considered in the evaluation and ranking of tolerant isolates as well as in the subsequent optimization and design of hydrolyzate fermentation processes.

(770d) Alkaline Hydrogen Peroxide Pretreatment of Softwood

Alvarez-Vasco, C., Bioproducts Science Engineering Laboratory, Washington State University
Zhang, X., Bioproducts Science Engineering Laboratory, Washington State University
Originally presented on: 11/8/2013 9:45:00 - 10:10:00

Pretreatment of softwood is most challenging largely due to the high lignin content and difficulty in break down lignin compared to hardwood and grass. Acid or SO_2 pre-impregnated steam explosion and organosolv pretreatments have been employed. Both are facing high capital cost and low recovery of hemicellulose. Hydrogen peroxide (HP) is an effective delignification agent and has been widely employed in lignin remove/degradation in paper industry. We also recently found that hemicellulose can be recovered as valuable chemicals from HP pretreatment. This study demonstrated an integrated HP treatment of Douglas fir softwood with the opportunity to simultaneously produce valuable chemicals along with biomass pretreatment in order to maximize the value of both hemicellulose and cellulose.

(770e) Bioconversion of Water Hyacinth (Eichhornia crassipes) At Its Optimum Age Into Bioethanol Via Separate Hydrolysis and Fermentation With Continuous Overliming

Caparanga, A. R., Mapua Institute of Technology
Bathan, R. S.,
Originally presented on: 11/8/2013 10:10:00 - 10:35:00

This study explored the conversion of water hyacinth (*Eichhornia crassipes*) into bioethanol. Prior to bioconversion into ethanol, the plant's optimum age, or age (in days) at which its holocellulose to lignin content was highest, was determined to be 5 days on the average (leaf area ~ 64 cm^2; average stem height ~ 10 cm). At this stage, holocellulose content was highest at 76.3% and lignin content was lowest at 12.3%. Therefore, only five-day old plants were harvested for use in the rest of the experiments. Ethanol was produced from the lignocellulosic materials from the plant by hydrolysis, which breaks the cellulose and hemicellulose polymers to fermentable sugar, followed by fermentation which converts the sugar into ethanol. However, some byproducts such as furan compounds are released during hydrolysis and inhibit the yeast during fermentation. Hydrolysate detoxification using lime ('overliming') was employed. During batch detoxification, sudden addition of lime shows severe degradation of sugar simultaneously with furan compounds. Continuous detoxification was introduced in this study. Its optimum condition was determined to be at pH 12 and dilution rate of 0.667 hr^{-1}. Gradual addition of lime in continuous detoxification could keep the sugar concentration high enough at more than 90% with approximately similar detoxification levels of furan compounds. Using a traditional strain of *S. cerevisiae,* 12.7 g/L ethanol was produced. The ethanol yield per unit biomass was comparable to those reported in other agricultural biomasses: 0.151g/g dry biomass.

(772a) Traction Stress As a Modulator for Protease Dependent Invasion of Cancer Cells

Aung, A., University of California, San Diego
Varghese, S., University Of California, San Diego
Originally presented on: 11/8/2013 8:30:00 - 8:48:00

Introduction: The ability of cancer cells to invade into and migrate within the extracellular matrix (ECM) is central to metastatic dissemination. Conventionally, this process was thought to occur via proteolytic degradation of the pericellular ECM network. However, studies demonstrating that cell invasion and migration can be only partially attenuated by protease inhibitors suggest the existence of protease independent mode of migration as a compensation mechanism. Although recent findings have indicated the role of physicochemical cues of the ECM in dictating the particular mode of migration, the intracellular signal by which a particular mode of migration occurs remains elusive. In this study, we demonstrate that cell generated traction stress modulates the transition from protease independent to dependent mode during invasion of cancer cells into 3D matrices through the activation of a regulatory Golgi transport pathway known as CARTS.

Materials and Methods: Single cell invasion assay into Matrigel was utilized to probe the transition from protease independent to dependent mode of invasion and the role of ECM properties on this. To this end, Matrigel, embedded with fluorescent particles, were spin coated onto glass surfaces to obtain thicknesses of 10, 20, and 30 μm. Cancer cells, MDA-MB-231, were plated onto these matrices to initiate the invasion process. As a function of invasion depth, individual cells were analyzed for the traction stress utilized to deform the matrix and whether these forces result in any permanent deformation of the Matrigel due to proteolysis. Furthermore, we examined the location of a key transmembrane matrix metalloproteinase (MT1-MMP) as well as its colocalization with the Trans-Golgi Network Marker 46 (TGN46) and a secretory protein involved in the CARTS pathway, Pancreatic Adenocarcinoma Up-regulated Factor (PAUF).

Results and Discussion: By decreasing the thicknesses of the Matrigel on the length scale of a single cell, we increased the mechanical resistance encountered by the invading cells while keeping other factors such as ligand density and other interfacial properties unaltered. We observed that the protease induced permanent deformation of the Matrigel occurs at traction stresses of approximately 120 Pa irrespective of Matrigel thickness. We also applied mechanical stresses up to 200 Pa via glass bead modified probes using atomic force microscopy and failed to observe the mechanical yielding of Matrigel. Furthermore, at stresses below 120 Pa, we observed that MT1-MMP remains within the cytoplasm as the cell utilize bleb formation to penetrate into the Matrigel. At larger forces, we observed the translocation of MT1-MMP to the cellular periphery as well as actin and cortactin rich protrusions at the leading edge of the cell indicating the formation of invadopodia. We next examined whether the translocation of the MT1-MMP occurs through CARTS pathway during this process. We observed that the depletion of the TGN46 through inhibition of Protein Kinase D abrogated the invasion processes. In addition, we observed the colocalization of TGN46 and PAUF with MT1-MMP within cells generating traction stresses larger than 120 Pa.

Conclusions: Traction stresses generated by the cells during their invasion triggers the transition from protease independent to dependent mode of migration. During protease indepedent mode, cells utilize bleb formation devoid of MT1-MMP on the cellular periphery to deform the matrix while during protease dependent mode, cells utilize invadopodia-like structures with MT1-MMP localized to the cellular periphery to degarde the matrix. We aslo shows that the transport of MT1-MMP during this process occurs through a regulatory Golgi transport pathway known as CARTS.

(772b) Mechanical Tropism in Metastatic Ovarian Cancer Cells

McGrail, D. J., Georgia Institute of Technology
Kieu, Q., Georgia Institute of Technology
Dawson, M. R., Georgia Institute of Technology
Originally presented on: 11/8/2013 8:48:00 - 9:06:00

Ovarian cancer is the fifth leading cause of cancer deaths among women, possibly because it is often diagnosed at late stages after metastasis (1). In contrast to following the normal metastatic process of intravasation to the vascular system and extravasation at a distal site, ovarian cancer is more likely to disseminate through the intraperitoneal fluids. From there, it preferentially accumulates in soft tissues such as the adipocyte-rich omentum (2). Previous work suggests this is because adipocytes act as a rich energy source and actively promote ovarian cancer cell homing via cytokines such as interleukin-8 (2). However, these studies were based solely on chemical factors, whereas the burgeoning field of physical oncology has recently shown the mechanical environment a cell experiences can be of equal importance. Based on these results, we hypothesized that the preferential accumulation of ovarian cancer cells in soft tissues may be due to intrinsic mechanical properties of the environment as opposed to the presence of soluble signaling molecules.

To test this hypothesis, we utilized a series of biophysical and biochemical techniques to understand the response of ovarian cancer cells to extracellular matrices of varying rigidities. We found that ovarian cancer cells (OCCs) show increased adhesion to compliant matrices at short time scales coupled with increased spreading. After engraftment, OCCs on soft matrices are more proliferative and resistant to standard chemotherapeutic drugs. In addition to these increases in growth, cells also displayed increased migratory capacity. Further immunocytochemistry and gene expression analysis revealed a shift from a more epithelial phenotype on stiff substrates to a more mesenchymal phenotype on soft matrices. Finally, biophysical characterization with small molecule inhibitors suggests the role of a Rho/ROCK mechanosensitive pathway in determining this tissue tropism.

1. Landen, C.N., M.J. Birrer, and A.K. Sood. 2008. Early events in the pathogenesis of epithelial ovarian cancer. Journal of clinical oncology□: official journal of the American Society of Clinical Oncology. 26: 995–1005.

2. Nieman, K.M., H. a Kenny, C. V Penicka, A. Ladanyi, R. Buell-Gutbrod, et al. 2011. Adipocytes promote ovarian cancer metastasis and provide energy for rapid tumor growth. Nature medicine. 17: 1498–503.

(772c) A Microfluidic Device to Measure Traction Forces During Confined Chemotactic Migration

Paul, C. D., University of Arkansas
Raman, P. S., Johns Hopkins University
Stroka, K. M., Johns Hopkins University
Konstantopoulos, K., Johns Hopkins University
Originally presented on: 11/8/2013 9:06:00 - 9:24:00

Introduction: There is a wide heterogeneity of environments in which tumor cells migrate *in vivo*. In many cases, alignment of fibers in the ECM or cell-mediated matrix remodeling creates microchannel-like tunnels in the tumor stroma[1,2]. Previous work has suggested a reduced importance of actomyosin contractility for tumor cell migration in highly confining microchannels[3]. Traction forces are essential for two-dimensional cell migration and can be measured using deflectable PDMS microposts[4] or deformable gels[5]. However, force measurement in three-dimensional environments, particularly those showing the architectural heterogeneity seen in the body, is much more difficult. Therefore, we engineered a device to measure traction force exertion during confined migration.

Materials and Methods: To quantify force exertion during chemotactic migration, we designed and built a microfluidic device containing a ladder-like array of microchannels with deflectable PDMS microposts at their bases. NIH-3T3 fibroblasts and HOS cells were induced to migrate on top of the posts, and the post deflections were quantified and converted to time-dependent forces using the beam bending theory. To explore the effects of modulating actomyosin contractility on force exertion in confining and non-confining environments, we treated cells with 50 μM blebbistatin, a myosin II inhibitor, or 0.1 nM calyculin A, which promotes myosin light chain phosphorylation. Migration speeds in microchannels containing flat bases were measured under these same conditions to explore migration efficiency. Statistical significance was assessed with non-paired Student's t-test.

Results and Discussion: Initial experiments were performed with 3T3 fibroblasts, a common model used for cell migration studies, to demonstrate the utility of our device. In wide channels, 3T3 force exertion was consistent with measurements made on beds of microposts[4]. Remarkably, measured traction forces were significantly reduced in narrow channels vs. wide channels for both 3T3 and HOS cells, as shown in Figure 1. To explore the extent of actomyosin participation in the confined migration of cancer cells, drug treatments were used to increase or decrease cell contractility. Blebbistatin treatment significantly reduced migration speed and exerted force in wide channels but did not affect either in narrow channels. Treatment with calyculin A increased forces and migration speed in wide channels but not narrow channels. Together, these data strongly suggest that actomyosin contraction plays a reduced role in migration through confined spaces.

Conclusions: Understanding the diverse mechanisms of cell migration is vital for abrogating cancer cell migration. Here, we present the fabrication of a novel device to measure force exertion in well-defined, heterogeneous environments and quantitatively demonstrate that confined cells use a migration mechanism that is less dependent on actomyosin contractility than was previously appreciated. Future work will focus on determining how modulation of cell contractility affects migration through confined spaces for a variety of cell types and how this modulation could provide clues for inhibiting tumor cell migration.

References: (1) Wolf, K. *et al.*, Sem Cell Devel Biol, 2009, 20, 931-941; (2) Fisher, K.E. *et al.*, J Cell Sci, 2009, 122, 4558-4569; (3) Balzer, E.M. *et al.*, FASEB J, 2012, 26, 4045-4056; (4) Han, S.J. *et al.*, Biophys J, 2012, 103, 640-648; (5) Dembo, M. *et al.*, Biophys J, 1999, 76, 2307-2316.

(772d) Regulation of Melanoma Induced Endothelial Gap Formation By Contractile Forces

Pohler, S. E., The Pennsylvania State University
Fu, C.,
Dong, C.,
Gomez, E. W., The Pennsylvania State University
Originally presented on: 11/8/2013 9:24:00 - 9:42:00

Metastasis is a complex process whereby cancer cells detach from the primary tumor and migrate to remote locations within the body through the vascular and lymphatic systems. During metastasis, breakdown of endothelial adherens junctions is critical for tumor cell extravasation through blood vessel walls. Extravasation is mediated by a combination of tumor secreted soluble factors and receptor-ligand interactions which induce VE-cadherin disassembly and actin-myosin-mediated endothelial cell contraction. Here, we investigate the roles of key intracellular signaling molecules and cytoskeletal contractility in regulating melanoma-induced VE-cadherin disassembly. We find that pharmacological inhibition of cytoskeletal contractility reduces melanoma-mediated gap formation in endothelial monolayers. Likewise, through the use of traction force microscopy we find that endothelial cell contractility increases when cells are exposed to soluble factors secreted by melanoma cells. These results provide insight into how biochemical and biophysical signals act in concert during extravasation and may aid in future identification of therapeutic targets to block metastasis.

(772e) Assessment of Microgroove-Aligned Endothelial Cells for in Vitro Simulations of Cancer Cell Adhesion

Fischer, J., University of Kentucky
Gray, L., University of Kentucky
Eitel, R. E., University of Kentucky
Trinkle, C., University of Kentucky
Anderson, K. W., University of Kentucky
Originally presented on: 11/8/2013 9:42:00 - 10:00:00

A thin layer of endothelial cells line the entirety of circulatory vessel walls. Adhesion of cancer cells to this endothelial cell lining is an important step in the metastatic cascade and researchers are currently using *in vitro* techniques to investigate these interactions. Under static culture conditions, endothelial cells grow in a characteristic cobblestone pattern rather than growing in straight lines due to the absence of shear stresses that would normally be present. It has recently been suggested that such changes in cell morphology may also affect surface expression profiles, which can alter how frequently or strongly cancer cells bind to endothelial cells. While flow adapting endothelial cells is important prior to studying cancer cell binding *in vitro*, traditional methods can be cumbersome due to the fact that the cells have to be exposed to flow for an extended period of time under controlled environmental conditions. In this study, we investigated the efficacy of a microgroove-aligned method for acquiring a flow adapted phenotype. Human Umbilical Vein Endothelial Cells (HUVECs) were cultured on glass slides that were fabricated using UV lithography which employed SU-8 2000.5, an epoxy-based negative photoresist, such that microscale grooves were present on the slides. The portions of the SU-8 photoresist exposed to UV light polymerized and remained on the slide while the unexposed areas were removed with developer, leaving only the hardened grooves. The resulting microgrooves were approximately 0.5 microns in both width and depth and were separated by 0.5 micron spaces. Actin staining and fluorescent microscopy were used to compare the morphology of human umbilical vein endothelial cells (HUVECs) grown under static conditions on unmodified glass slides versus micropatterned glass slides. Results demonstrated that desirable elongation and unidirectional growth was achieved using the microgroove-aligned system. Surface expression analyses of the HUVECs were also conducted to evaluate expression of four common cellular adhesion molecules involved in cancer metastasis (ICAM-1, VCAM-1, E-Selectin and P-Selectin) for HUVECs cultured on unmodified glass slides versus micropatterned slides. HUVECs grown on unmodified glass slides and stimulated with TNF-alpha (tumor necrosis factor alpha) were also employed as a positive control in these analyses, as TNF-alpha is commonly used in adhesion assays to stimulate adhesion molecule expression. Fluorescent imaging was used to obtain histograms of image pixel intensities corresponding to adhesion molecule expression as well as spectrophotometer readings for further comparisons. Preliminary results indicate upregulation of VCAM-1 on HUVECs cultured on microgrooves as compared to unmodified controls, more reminiscent of *in vivo* expression. Also presented will be a proof of concept cancer adhesion study in a parallel plate flow system employing unmodified and microgroove-aligned statically grown HUVECs versus HUVECs grown under flow conditions.

(772f) Utilizing Multiple Particle Tracking Microrheology and Traction Force Microscopy to Monitor Mechanical Progression of Differentiating Mesenchymal Stem Cells

McAndrews, K., Georgia Institute of Technology
McGrail, D. J., Georgia Institute of Technology
Quach, N.,
Dawson, M. R., Georgia Institute of Technology
Originally presented on: 11/8/2013 10:00:00 - 10:18:00

Bone marrow derived mesenchymal stem cells (MSCs) are highly regenerative multipotent stem cells that differentiate into a variety of non-hematopoietic cells, including adipocytes, osteoblasts, chondrocytes, muscle cells and neurons. Recent studies have shown that mechanical properties have a significant effect on stem cell fate [1]; however, the complex interaction between mechanical and chemical cues has not been completely elucidated. In addition, traditional methods of determining if a cell has differentiated typically are endpoint assays performed at late time points. We hypothesized that changes in intracellular rheology and traction forces would correlate with early lineage commitment.

In this study, human MSCs (hMSCs) were cultured in mixed differentiation media on soft (10 kPa) or intermediate (26kPa) polyacrylamide gels. The effect of mechanical and chemical cues on hMSC fate was characterized by histological staining and quantitative real time PCR (qRT-PCR). In addition, multiple particle tracking microrheology (MPT) and traction force microscopy (TFM) were used to determine the intracellular rheology and traction forces associated with lineage commitment and stem

cell differentiation. Lineage commitment and cell differentiation were characterized with traditional

methods, including qRT-PCR and histological staining, after 21 and 35 days exposure to culture

conditions. Cells on soft substrates had an increased propensity to differentiate into fat, while cells on

hard substrates were more likely to differentiate along the osteogenic lineage. Differences in rheology

and traction forces were detected by day 7 in differentiation media. In addition, a correlation between

local rheology and local traction forces was developed. Changes in lineage commitment were associated

with changes in intracellular rheology and extracellular forces, suggesting these methods may be useful

for determining if a cell has differentiated at early time points without having to lyse or fix the cell,

which would result in cell death.

References:

1. Engler, A.J., et al., Matrix Elasticity Directs Stem Cell Lineage Specification. Cell, 2006. 126(4): p. 677-689.

(772g) The Role of Actomyosin Generated Tension in Coordinating Cell Movements During Zebrafish Development

Chai, J., Stanford University
Hamilton, A.,
Krieg, M., Stanford University
Buckley, C.,
Dunn, A. R., Stanford University
Originally presented on: 11/8/2013 10:18:00 - 10:36:00

Introduction

Cells engage in collective migration in circumstances ranging from wound healing to embryonic development. While much is known about how single cells migrate, it is unclear how groups of cells coordinate movements across large distances. In this study we use zebrafish embryos as a model system to investigate the role of cell-generated line tensions in generating long-range, coordinated motion. Early zebrafish development exhibits intricately orchestrated cell migration and patterning. Within the first 24 hours a mass of cells termed the blastoderm spreads uniformly over a spherical yolk, starting at one end and moving toward the opposite pole in a process termed epiboly. During epiboly, the cells at the blastoderm margin (edge) migrate at a uniform rate and thus reach the opposite pole of the yolk at the same time (**Figure 1A-C**). This long-range coordination is essential in establishing the axes that define the front, back, top and bottom of the resulting organism. A band of actin and myosin, termed the actin band, forms near the blastoderm margin. It has been previously shown that the actin band produces a contractile tension; however its functional role is not known.

Results

We used mechanical shape perturbations to probe the function of the actin band during epiboly. In shape-perturbed embryos, we find that the rate of blastoderm migration is no longer uniform: lagging areas of the margin speed up, while leading areas slow down. This equalizes progress along the blastoderm margin perimeter such that epiboly reorients to coincide with the long axis of the embryo. When the tension within the actin band is disrupted using chemical treatments such as blebbistatin or reduction of calcium this reorientation no longer occurs. This observation suggests that the tension within the actin band is a necessary component in coordinating blastoderm migration along the length of the margin. We additionally observed that cells along the blastoderm margin align parallel to the actin band. Treatment with either blebbistatin or calcium depletion resulted in a markedly rougher margin, suggesting that the actin band additionally coordinates motion between neighboring cells.

Discussion

Based upon these results, we developed an analytical model in which the actin band acts like a pre-tensioned rubber band, introducing a line tension that coordinates blastoderm migration over millimeter length scales (**Figure 1D**). This line tension also provides a resistive force against instabilities at the margin of the migrating blastoderm, ensuring that neighboring cells progress at similar rates (**Figure 1E**). Our model is in quantitative agreement with margin migration rates observed in a wide variety of embryo shapes, suggesting that a simple physical parameter (actin band tension) is sufficient to account for the long-range cell coordination that we observe. The physical properties of the zebrafish blastoderm margin are similar to those of spreading cell sheets observed in a wide variety of circumstances. We suggest that similar quantitative, physical models may account for collective cell movements observed in a wide variety of circumstances including other developmental processes, wound healing, and a variety of tissue engineering applications.

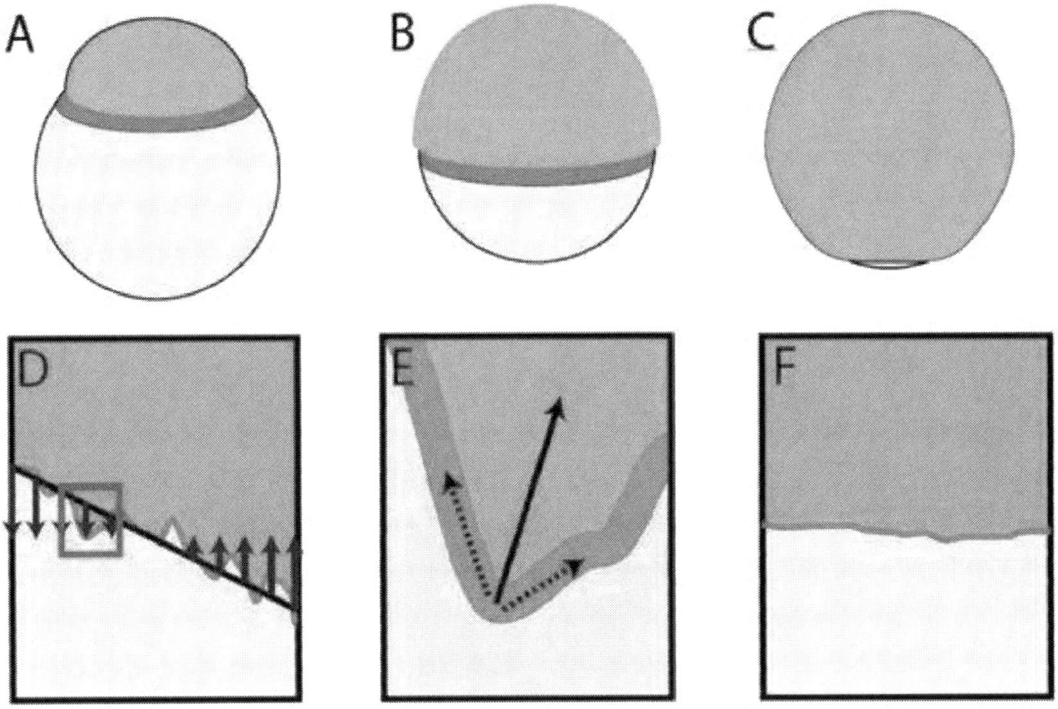

Figure 1. Mechanical tension leads to long-range coordination between blastoderm cells in the zebrafish embryo. **A-C**) Progression of epiboly (green ? blastoderm, yellow ? yolk, orange ? actin band). **A**) Epiboly begins with the blastoderm sitting on one pole of the yolk cell. **B**) The blastoderm migrates over the yolk cell until it finishes epiboly by converging at the pole opposite the initial blastoderm pole (**C**). **D-F**) Description of the roles of actomyosin generated line tension. **D**) Line tension (black line) along the boundary of a migrating cell mass (green) generates local forces (blue) that reorient the cell boundary. **E)** Zoomed in view of an instability at the boundary (red square in D). Tension at the margin (black dotted arrows) generates a net restorative force (black solid arrow). **F**) As a result, the cell mass reorients and the boundary maintains a smooth profile that is parallel to the actin band.

(772h) Physical Role of E-Cadherin Mediated Cell-Cell Adherent Junctions in Multicellular Motility

Wu, P. H., Johns Hopkins University
Wirtz, D., Johns Hopkins University
Marchand, J., Johns Hopkins University
Lee, M. H., SUNY Buffalo
Originally presented on: 11/8/2013 10:36:00 - 10:54:00

Cell migration is one of the hallmarks for the metastatic process in cancer. Single cell migration has been extensively studied and different models have been proposed to understand cell motility processes. It has been suggested that metastasis may involve collective cell migration and invasion. However, the physical mechanisms of multicellular cell migration remain to be elucidated. In this work, we adopt both experimental and computational approaches to elucidate the physical role of E-Cadherin mediated cell-cell adherent junction in multicellular migration. To study the cell migration with contacts with other cells, we maintained human breast epithelial cells, the MCF-10A, and transformed metastatic breast cancer cells, MDA-MB-231, in a semi-confluent condition and observed their motility patterns. Thousands of cell movements were tracked for each condition. Our result shows that cell velocity decreases with increase of cell-cell contacts for MCF-10A cells. The same phenomenon was not found on MDA-MB-231 cells. Further, cells present the elevated correlated velocity when they are in proximity. Both these multicellular associated phenomena are associated with E-Cadherin mediated adherent junctions since overexpression of E-cadherin in MDA-MB-231 or knockdown of a-catenin in MCF-10A will change these profiles accordingly. Based on these observations, we developed a computational model for multicellular cell motility in which we introduces intercellular drag force and velocity association caused by neighboring cells into the persistent random walk model. We validated our model by predicting motility behavior for monolayer of MCF-10A cells. Together, we have identified the principal characteristics of multicellular motility for normal and malignant breast epithelium cells. Our results reveal fundamental physical mechanism of multicellular migration and how they arise from cell-cell interactions.

(792a) Engineering Synthetic Organelles: Encapsulating Heterologous Proteins Into Bacterial Microcompartments

Kim, E. Y., University of California
Tullman-Ercek, D., University of California
Originally presented on: 11/8/2013 8:30:00 - 8:48:00

Biological systems often use compartmentalization as a strategy for optimization. In bacteria, naturally occurring protein microcompartments (MCPs) of approximately 125nm in diameter are used to compartmentalize metabolic pathways, thereby concentrating enzymes and their substrates, sequestering potential toxic intermediates, and insulating the reactions within from competing enzymes or metabolites. For the propanediol utilization (Pdu) MCP of *Salmonella enterica*, it has been shown that an 18 amino acid N-terminal signal peptide is sufficient for localizing heterologous proteins into the Pdu MCP. We aim to use the Pdu MCP as a template to create customizable nanobioreactors in bacteria.

A key step towards encapsulating biochemical pathways is to gain the ability to tune the enzyme stoichiometry within the MCP. Towards this goal, we have constructed a library of single amino acid substitution mutants in the signal peptide and characterized its strength for encapsulation by using a flow cytometry based assay. We also use this flow cytometry based assay to investigate whether two different naturally occurring N-terminal signal peptides compete for encapsulation, or have orthogonal mechanisms for encapsulation. These results help us gain an understanding of the requirements for localizing enzymes to the interior of the Pdu MCP and gain tunability in enzyme stoichiometry within the MCP by manipulating the signal sequence.

(792b) Synthetic Co-Assembly of Metabolic Enzymes On Protein Scaffolds for Improved Flux

Connell, K. B.,
Dueber, J. E., U.C. Berkeley

Originally presented on: 11/8/2013 8:48:00 - 9:06:00

Enzyme complex formation is extremely common in natural metabolism. In fact, recently been found that there are many more dynamic complexes dependent on environmental conditions than previously appreciated. These enzyme-enzyme interactions can be important for allosteric regulation of catalytic activities, increased efficiency of intermediate hand-off, and subcellular localization. Gaining an ability to engineer similar synthetic complexes in a programmable manner would provide several advantages to metabolic engineers for increasing pathway performance with lower enzyme expression levels, minimizing intermediate metabolite accumulation and resultant toxicity, and/or reducing undesired flux through side pathways. We are taking a synthetic biology approach in co-localizing pathway enzymes into scaffolded complexes in an effort to probe the key design principles relating flux improvements achieved from formation of these complexes. In particular, we are interested in testing our hypothesis that formation of multimeric complexes – an assembly of multiple scaffolded complexes – is critical to increased pathway performance. I will also discuss some of the unexpected challenges we have encountered in building these scaffolded systems will be discussed.

(792c) Post-Translational Modification of Artificial Cellulosomes for Efficient Cellulose Hydrolysis

Sun, Q., University of Delaware
Chen, W., University of Delaware
Originally presented on: 11/8/2013 9:06:00 - 9:24:00

Efficient hydrolysis of cellulose to glucose is gaining momentum because of its significant role in biofuel production. Cellulosome is a naturally occurring multi-enzyme system with a substantially enhanced ability for cellulose hydrolysis due to the effect of substrate targeting and enzyme clustering. The major component of cellulosome is a structural scaffoldin that consists of at least one cellulose binding module (CBM) and repeating cohesin domains that are docked individually with an enzyme tagged with a corresponding dockerin domain. Although artificial scaffoldins containing three or four orthogonal cohesins have been constructed, larger designer scaffoldins tend to aggregate and degrade. In this talk, we will highlight our recent strategies in engineering more complex artificial cellulosome structures through post-translational recombination for complex artificial cellulosome construction. The effect of enzyme loading on the overall enhancement in cellulose hydrolysis will be discussed.

(792d) Correlating Biomass Degradation With the Dynamic Composition of Fungal Cellulosomes

O'Malley, M. A., University of California, Santa Barbara
Haitjema, C., University of California, Santa Barbara
Solomon, K., University of California, Santa Barbara
Henske, J.,
Theodorou, M. K., Durham University
Originally presented on: 11/8/2013 9:24:00 - 9:42:00

Anaerobic gut fungi thrive in the digestive tract of large herbivores, where they have evolved to break down plant biomass through filamentous growth and the secretion of powerful enzymes. In particular, gut fungi secrete complexes of tethered cellulases, hemicellulases, and carbohydrate binding modules to facilitate synergistic biomass hydrolysis. These scaffolded complexes, referred to as fungal cellulosomes, differ considerably from their bacterial counterparts as they display a unique array of glycosyl hydrolases. In addition, catalytic enzymes are fused to tandem repeat fungal dockerin domains, which are speculated to mediate assembly of the large multi-protein complex. However, the mechanisms that govern the structure, enzymatic reactivity, and adaptability of fungal cellulosomes remain elusive.

We hypothesize that fungal cellulosomes are regulated at the post-transcriptional and post-translational level, leading to dynamic, tunable cellulosomes that optimize biomass breakdown. In order to correlate expression and cellulosome architecture with substrate availability, we have isolated a panel of novel gut fungi from the *Piromyces, Neocallimastix, and Anaeromyces* genera. Proliferation of the fungal isolates was monitored via fermentation gas production, and cellulosomes from each species were isolated through cellulose-precipitation. All fungi exhibited high enzymatic reactivity against a range of cellulosic and lignocellulosic substrates (filter paper, Avicel, reed canary grass), which was repressed in the presence of simple sugars. Within isolated cellulosomes, striking similarities are observed for certain dockerin-fused glycosyl hydrolases, and these proteins are not secreted from fungi when simple sugars are present. This suggests that the secretion of cellulosomal enzymes is highly regulated in response to simple sugars. We will further link this behavior to the transcriptional regulation patterns observed for important enzyme families under catabolic regulatory conditions via RNAseq. In addition, we will discuss a new model for fungal cellulosome assembly and evolution during lignocellulose degradation.

(792e) High-Yielding Escherichia Coli-Based Cell-Free Protein Synthesis of Commercially Relevant Proteins

Smith, M. T., Brigham Young University
Hawes, A. K., Brigham Young University
Rainsdon, J. M., Brigham Young University
Shrestha, P., Brigham Young University
Bundy, B. C., Brigham Young University
Originally presented on: 11/8/2013 9:42:00 - 10:00:00

Cell-free protein synthesis (CFPS) is an open and accessible *in vitro* protein production system that is highly controllable, easily probed and economical. These traits have proved useful and widely applicable for the high-throughput production, characterization and engineering of novel proteins such as oxygen-tolerant hydrogenases, signaling peptides, kinases and membrane-anchored proteins. *Escherichia coli*-based CFPS (eCFPS) is the most cost-efficient and is therefore commonly employed. Unfortunately, eCFPS lacks the ability to produce many commercially relevant proteins due to the absence of cofactors necessary for proper synthesis. However, commercially relevant proteins can be properly produced in eCFPS by: 1) adding proper cofactors that are exogenously produced and purified and/or 2) directly expressing proper cofactors during cell-extract preparation. Direct expression of cofactors during cell growth avoids the tedious and time-consuming steps of exogenous production and purification, and therefore lends itself to speedier and more economical high-throughput applications. However, previous reports demonstrate that yields in such systems are regrettably low. We hypothesized that yields could be increased by adjusting factors during cell extract growth and protein synthesis. By circumspectly adjusting these factors, we were able to produce two commercially relevant proteins at levels up to 7 times higher than previously reported yields. This finding further enables the economical *in vitro* production of complex proteins and high-throughput screening. In addition, insights gained could help identify the optimal levels of enzyme expression for *in vivo* synthetic biology applications.

(792f) Cell-Free Synthesis of Membrane Proteins and Glycoproteins
Kubick, S.,
Originally presented on: 11/8/2013 10:00:00 - 10:18:00

Membrane proteins have become an important focus of the current efforts in structural and functional genomics and the rapid progress of various genome sequencing projects has greatly accelerated the discovery of novel genes encoding membrane proteins. In contrast, the molecular analysis of membrane proteins lags far behind that of cytosolic soluble proteins. Preparing high quality samples of functionally folded proteins represents a major bottleneck that restricts further structural and functional studies. Cell-free protein expression systems, in particular those of eukaryotic origin, have recently been developed as promising tools for the rapid and efficient production of a wide variety of membrane proteins. A large number of these proteins, however, require posttranslational modifications for optimum function. Several membrane proteins have been expressed *in vivo* to date, most of them being functionally, antigenically, and immunogenically similar to their authentic counterparts. This is mainly due to the properties of cultured eukaryotic cells, which are able to carry out many types of posttranslational modifications such as the addition of N- and O- linked oligosaccharides, but also palmitoylation, myristylation, and phosphorylation. Based on these versatile properties of cultured cell lines, we have developed a technique for the standardized production of translationally active eukaryotic lysates from insect cells. In contrast to other cell-free protein synthesis systems (e.g., rabbit reticulocyte lysates and wheat germ extracts) our homogenization procedure avoids any serious breakdown of membrane vesicles already existing in the cytoplasm of the prepared eukaryotic cells. We have demonstrated the functional integrity of these subcellular components by showing signal peptide cleavage as well as glycosylation of *in vitro* expressed membrane proteins.

The development of this novel eukaryotic *in vitro* translation system now expands the possibilities of cell-free protein synthesis, since posttranslational modifications significantly alter the physical and chemical properties of proteins, including their folding and conformational distribution and these modifications are frequently a fundamental prerequisite for functional activity.

Cell-Free Systems: Functional modules for Synthetic and Chemical Biology

STECH, M., BRÖDEL, A.K., QUAST, R.B., SACHSE, R. and KUBICK, S.

In: Advances in Biochemical Engineering / Biotechnology 2013, April 11

Functional evaluation of candidate ice structuring proteins using cell-free expression systems

BRÖDEL, A. K., RAYMOND, J. A., DUMAN, J: G., BIER, F. F. and KUBICK, S.

In: Journal of Biotechnology 2013, 163, 301-310

Synthesis of membrane proteins in eukaryotic cell-free systems

Sachse, R., Wüstenhagen, D., ŠAMALIKOVA, M., Gerrits, M., Bier, F. F. and Kubick S.

In: Eng. Life Sci. 2013, 13, No. 1, 39–48

Production of functional antibody fragments in a vesicle-based eukaryotic cell-free translation system

STECH, M., MERK, H., SCHENK, J. A., STÖCKLEIN, W., WÜSTENHAGEN, D., MICHEEL, B., DUSCHL, C., BIER, F. F. and KUBICK, S.

In: Journal of Biotechnology 2012, 164, 220–231

The protease-activated receptor 1 possesses a functional and cleavable signal peptide which is necessary

for receptor expression

ZAMPATIS.DE, RUTZ, C., FURKERT, J., SCHMIDT, A., WÜSTENHAGEN, D., KUBICK, S.,TSOPANOGLOU, NE. and SCHÜLEIN, R.

In: FEBS Lett. 2012 Jul 30;586(16):2351-9.

Cell-free synthesis and characterization of a novel cytotoxic pierisin-like protein from

the cabbage butterfly *Pieris rapae*

Orth JHC , Schorch B, Boundry S, Ffrench-Constant R, Kubick S, Aktories K

In: Toxicon 57, Issue 2, 199-207 (2011)

Towards an artificial cell: Protein incorporation in giant unilamellar vesicles under physiological

conditions

Paige M. Shaklee, Stefan Semrau, Maurits Malkus, Stefan Kubick, Marileen Dogterom, Thomas Schmidt

In: ChemBioChem (2010)

***In vitro* Synthesis of posttranslationally modified Membrane Proteins**

Kubick, S., Gerrits, M., Merk, H., Stiege, W., Erdmann, V.A.

In: 'Current Topics In Membranes' Vol. 63, Elsevier, Chapter 2, (2009)

(792g) Characterizing and Alleviating Substrate and By-Product Limitations for Improved Yeast Cell-Free Protein Synthesis

Schoborg, J. A., Iowa State University
Hodgman, C. E., Northwestern University
Anderson, M. J., Northwestern University
Jewett, M. C., Northwestern University
Originally presented on: 11/8/2013 10:18:00 - 10:36:00

Cell-free protein synthesis (CFPS) has become a powerful technology for high-throughput, rapid production of proteins. Towards the development of a cost-effective, high-yielding eukaryotic CFPS platform, we recently reported a system derived from *Saccharomyces cerevisiae*. Here, we study factors causing the early cessation of protein synthesis in a yeast CFPS batch reaction. Specifically, high-performance liquid chromatography and chemical assays were used to characterize the substrate and by-product concentrations during the CFPS reaction over time. We discovered that creatine phosphate, the secondary energy substrate, as well as nucleoside triphosphates were rapidly degraded during the cell-free reaction. This led to a significant drop in the energy charge of the reaction (from 1 to 0.4 in 3 hr), and as a consequence, the cessation of protein synthesis. Of the 20 amino acids, aspartic acid, asparagine, alanine, and proline were consumed. Additionally, inorganic phosphate accumulated, which is a toxic by-product. Semi-continuous exchange reactions, where passive diffusion enables substrates to be replenished and by-products to be removed, led to over a 100% increase in active superfolder green fluorescent protein (sfGFP) expressed as compared to the batch system. This yielded 17.9 ± 3.7 μg sfGFP mL^{-1} in a 10 hr semi-continuous reaction. In addition to increasing sfGFP production, this characterization identified promising targets for future improvements to the batch reaction format.

(792h) Production of Chlamydia Trachomatis Protein Antigens for Vaccine Development

Chan, W., Stanford University
Swartz, J. R., Stanford University
Originally presented on: 11/8/2013 10:36:00 - 10:54:00

Chlamydia trachomatis is an obligate intracellular bacterium that primarily infects mucosal membranes of the eyes and reproductive organs. Though the disease is treatable using antibiotics, infected individuals are often asymptomatic and thus do not seek treatment. The most effective way to reduce the number of infections is a vaccine. Three proteins have been identified as potential vaccine targets: CT681 (Major Outer Membrane Protein), CT443 (Outer Membrane Protein B), and CT043. These *C. trachomatis* proteins are associated with the outer membrane of the organism and can be used to elicit an immune response against the organism. However, due to the intracellular nature of the organism, culturing the cells to produce sufficient quantities of these proteins is laborious and difficult. Using an *E. coli* based *in vitro* protein synthesis platform, we are able to produce these three *C. trachomatis* proteins as soluble proteins with high yields. We are also able to incorporate azide and alkyne functional non-natural amino acids into the *C. trachomatis* protein, thus allowing for site specific conjugation. The ability to conjugate these proteins is valuable because it allows us to link adjuvants that stimulate Th1 response, which has been shown to be the dominant pathway to clear a current infection and to prevent future infections. We are currently working on conjugating these proteins, as well as potential adjuvants, onto the surface of a virus-like particle (VLP). By linking the *C. trachomatis* proteins with the adjuvants, we believe we can elicit a much stronger immune response compared with injecting the proteins and adjuvants separately.

(795a) Attenuated Bistability in Bacterial Sugar Utilization

Afroz, T., North Carolina State University
Biliouris, K., University of Minnesota
Kaznessis, Y. N., University of Minnesota
Beisel, C. L., North Carolina State University
Originally presented on: 11/8/2013 8:30:00 - 8:48:00

Sugar utilization in bacteria has served as a powerful model to understand gene regulation and the emergence of single-cell behaviors. The prototypic example is the 'all-or-none' response in lactose utilization, where cells are either uninduced or fully induced in the presence of a lactose analog. The utilization of lactose and other sugars predominantly has been studied in the absence of sugar catabolism--either through the use of stable sugar analogs or deletion of the catabolic genes. In contrast, natural utilization pathways inherently catabolize the sugar as part of the response. This discrepancy begs the question: how does sugar catabolism shape the response of individual cells, especially across the myriad of sugar utilization pathways present in bacteria?

Using a combination of transcriptional reporters and stochastic modeling, we are investigating the single-cell response of representative sugar utilization pathways in *Escherichia coli*. Our results indicate that examined pathways exhibit ranging responses that can be generally described as attenuated bistability. This response is bimodal yet exhibits reduced hysteresis in comparison to bistable systems. Deletion of the catabolic genes reverts the exhibited response to an 'all-or-none' response with greater hysteresis, in line with the behavior of pathways lacking sugar catabolism. Mathematical modeling predicted these same responses and indicated the importance of the balance between the induction and activity of the transporters and catabolic enzymes. Our overall results suggest that bacterial sugar utilization pathways exhibit complex, single-cell responses, potentially representing common foraging strategies in sparse and unpredictable environments. With continued efforts, insights into these responses may reshape our fundamental understanding of how microorganisms utilize nutrients in their surrounding environment.

(795b) Dormancy Is Not Necessary Or Sufficient for Bacterial Persistence to Antibiotics

Orman, M. A., Rutgers, The State University of New Jersey
Brynildsen, M. P., Princeton University
Originally presented on: 11/8/2013 8:48:00 - 9:06:00

Persisters are phenotypic variants that are tolerant to high concentrations of antibiotics, and this phenomena has been attributed to transient dormancy. Evidence in support of this model has been obtained from time-lapse fluorescence microscopy (1-3) and fluorescence-activated cell sorting (FACS) experiments that traced cell division (4) or used markers of growth-rate (5). Although these studies demonstrated that persisters can originate from growth-inhibited cells, it had not been determined to what extent dormancy is a characteristic of persisters, or if persister antibiotic tolerances require growth inhibition. In this context, we sought to determine whether dormancy in replication or metabolism prior to antibiotic treatment was necessary or sufficient for persistence. To quantify the metabolic and cell division distributions of persisters within an exponentially-growing *E. coli* culture at single cell-level, we used fluorescence-activated cell sorting, a fluorescent indicator of cell division, a fluorescent measure of metabolic activity, and persistence assays. FACS was used to segregate cultures into quantiles based on a fluorescent signal providing information about the cell division or metabolic states, and each quantile was then treated with antibiotics to determine the number of persisters present. First, we verified that fluorescent labeling, flow cytometry, and segregation did not change the level of persisters in the population. Using the cell division reporter, we observed that although the non-growing sub-population was more enriched with persisters (~1%) than the growing sub-population (~0.01%), approximately 20% of all the persisters originated from normally-growing cells. Since the most rapidly-growing sub-population also included persisters, these data demonstrated that growth inhibition prior to antibiotic exposure is not required for persistence. When we employed a fluorescent measure of metabolism, we found that cells with low metabolic activity were ~40 times more likely to be persisters than cells with high metabolic activity. These data, which include the first direct measurements of persister metabolic activity, suggest that persister sub-populations are highly heterogeneous, and in spite of the fact that inhibition of replication or metabolism enhances the likelihood of a cell to be a persister, persistence is far more complex than dormancy and additional characteristics are needed to define the persister phenotype.

References

1. **Balaban NQ, Merrin J, Chait R, Kowalik L, Leibler S.** 2004. Bacterial persistence as a phenotypic switch. Science **305:**1622-1625.

2. **Rotem E, Loinger A, Ronin I, Levin-Reisman I, Gabay C, Shoresh N, Biham O, Balaban NQ.** 2010. Regulation of phenotypic variability by a threshold-based mechanism underlies bacterial persistence.

Proc. Natl. Acad. Sci. U. S. A. **107:**12541-12546.

3. **Gefen O, Gabay C, Mumcuoglu M, Engel G, Balaban NQ.** 2008. Single-cell protein induction dynamics reveals a period of vulnerability to antibiotics in persister bacteria. Proc. Natl. Acad. Sci. U. S. A. **105:**6145-6149.

4. **Roostalu J, Joers A, Luidalepp H, Kaldalu N, Tenson T.** 2008. Cell division in Escherichia coli cultures monitored at single cell resolution. BMC Microbiol. **8**.

5. **Shah D, Zhang ZG, Khodursky A, Kaldalu N, Kurg K, Lewis K.** 2006. Persisters: a distinct physiological state of E-coli. BMC Microbiol. **6**.

(795c) Super-Resolution Imaging of the Bacterial Chemotaxis System in Bacillus Subtilis

Agrawal, U., University of Illinois at Urbana-Champaign
Walukiewicz, H. E.,
Rao, C. V., University of Illinois, Urbana Champaign
Schroeder, C. M., University of Illinois at Urbana-Champaign
Originally presented on: 11/8/2013 9:06:00 - 9:24:00

In this work, we apply fluorescence nanoscopy to elucidate a near molecular scale view of key proteins involved in bacterial motility. The ability to probe cellular structures and events using fluorescence microscopy is of paramount importance for understanding dynamic biological processes. However, optical diffraction limits the spatial resolution of conventional fluorescence microscopy to ≈ 250 nm, which precludes spatially resolved analysis of nanometer-scale cellular protein assemblies that mediate key biological processes, such as microbial motility. Recent advances in super-resolution imaging have enabled imaging at 20-25 nm spatial resolution[1].

Chemotaxis refers to the movement of microorganisms under the influence of chemical gradients used by microbes to search for nutrients for survival. Chemotaxis is exploited industrially for designing microbial biosensors for detection and degradation of a broad spectrum of chemicals including toxins, environmental pollutants, as well as bio-warfare agents. In general, bacteria perform chemotaxis via run-and-tumble strategy using an intricate internal signaling network. Chemotaxis is mediated by protein clusters, which display dynamic changes in spatial organization and architecture in response to chemical cues. These chemotactic receptor clusters play a key role in response to stimulants, eventually eliciting behavioral response post signal transduction via the chemotactic network. As chemotactic clusters are several fold smaller than the diffraction limit of light (200 nm), the application of fluorescence nanoscopy is imperative to develop an accurate model for chemotaxis.

We are employing single molecule localization based super-resolution imaging to directly visualize dynamic changes in nano architectures of key chemotactic proteins in response to chemical stimulation in *B. subtilis* to elucidate near molecular scale view on signal amplification and signal transduction. Recent studies on *B. subtilis* have revealed that receptor lattices are not static and undergo significant structural rearrangement when exposed to ligands[2]. However, the understanding of mechanism of this rearrangement, activity modulation by receptors and eventual signal transduction remains unclear. Our work has revealed marked differences in the subcellular localization and clustering patterns of receptor proteins upon chemical stimulation in individual bacterial cells. In first-of-its-kind super-resolution visualization in *B. subtilis*, we observed that receptor rearrangement is characterized by a largely polar localization in unstimulated cells to a more diffused polar-lateral configuration in cells that have been exposed to ligand. Our results conform to the previous diffraction limited studies[2]. We achieved a spatial resolution of ≈ 25 nm using immunofluorescence based stochastic optical reconstruction microscopy (STORM). Our super-resolution data also provides crucial information on changes in structure and composition of the polar and lateral receptor clusters in *B. subtilis* during chemotaxis towards asparagine. This work provides a novel platform to study how chemotaxis clusters form, reorganize and localize during chemotaxis. Our data also embodies individual receptor molecules or small cluster sizes, which were previously inaccessible using conventional fluorescence microscopy. Overall, this work presents tremendous potential to provide complete molecular-scale understanding of chemotactic response using receptor dynamics by tapping information on cluster densities, cluster sizes, cluster shapes and number of receptors forming a cluster. Our work will provide novel insights to help discern long-standing mystery of mechanistic details of receptor organization and signal transduction during chemotaxis in *B. subtilis*.

References:

1. Agrawal, U., Reilly, D. T. & Schroeder, C. M. *Current opinion in biotechnology* 1–8 (2013).

2. Wu, K., Walukiewicz, H. E., Glekas, G. D., Ordal, G. W. & Rao, C. V. *The Journal of biological chemistry* **286**, 2587–95 (2011).

(795e) Profiling Cellular Populations At the Single Cell Level With Drop-Based Microfluidics

Rotem, A., Harvard University
Ram, O., Broad Institute
Shoresh, N.,
Sperling, R.,
Bernstein, B.,
Weitz, D. A., Harvard University
Originally presented on: 11/8/2013 9:42:00 - 10:00:00

Populations of cells have substantial heterogeneity that is important for their function and understanding. This variability is reflected in cell to cell variations of genetic and epigenetic features that are quickly averaged when characterized without single-cell resolution. We developed a method for single-cell profiling that combines drop-based microfluidics with genomics and DNA barcoding. The method is based on encapsulating individual cells in micron-size droplets, lysing the cells inside the droplets, and labeling the cellular content with molecular barcodes unique to each droplet. When sequenced, each barcode is used to group DNA fragments that originate in the same cell, yielding multiple density maps for individual cells. The platform is compatible with both DNA and RNA, uses ligation or hybridization to attach the barcodes and can be scaled up to a large number of cells ultimately limited only by the capacity and cost of Next-Gen sequencing.

(795f) Quantitative Single-Cell Functional Characterization of CD19-Specific Chimeric Antigen Receptor T Cells

Liadi, I., University of Houston
Singh, H., The University of Texas MD Anderson Cancer Center
Romain, G., The University of Houston
Rey-Villamizar, N., The University of Houston
Merouane, A., The University of Houston
Roysam, B., The University of Houston
Cooper, L., The University of Texas MD Anderson Cancer Center
Varadarajan, N., The University of Houston
Originally presented on: 11/8/2013 10:00:00 - 10:18:00

Adoptive cell therapy (ACT) utilizing chimeric antigen receptor (CAR) T cells rendered specific for CD19 have demonstrated significant anti-tumor effects in patients with CD19+ chronic lymphocytic leukemia (CLL). In spite of the clinical promise of ACT in achieving complete responses, their efficacy remains unpredictable and new approaches are needed to a priori define the therapeutic potential of T-cell based therapies. In our current work, we characterize the in vitro functionality of CD19-specific (CD19RCD28) CAR+ T cells propagated using artificial antigen presenting cells, by employing a novel methodology: single-cell nanowell screening (SNS). We show that CAR+ T cells exert specific cytotoxicity against NALM6 cells (31 ± 8 %) when co-incubated at a 1:1 ratio in nanowell containers. Furthermore, single CAR+ T cells were capable of engaging and killing multiple targets; 17 ± 8% of T cells killed two target cells within the 6 hour window of observation. Hierarchical clustering indicated that interferon-gamma (IFNγ) secretion is not correlated to cytotoxicity of T cells. CAR+ T cells that secreted IFNγ upon target ligation did not undergo activation-induced cell death (AICD) whereas T cells that engaged in repeated killing showed an increased propensity to undergo AICD ($p = 0.04$). In summary, our SNS-based methodology allows the deep functional characterization of clinical grade CAR+ T cells and can be used to determine in vitro functions of CAR+ T cells.

(795g) On-Chip Magnetic CELL Separation and Subsequent Molecuar Analysis

Chalmers, J., The Ohio State University
Sooryakumar, R., The Ohio State University
Chen, A., Ohio State University
Byvank, T., The Ohio State University
Wu, Y., The Ohio State University
Lustberg, M., The Ohio State University
Miller, B. L., The Ohio State University
Originally presented on: 11/8/2013 10:18:00 - 10:36:00

The demand for high-throughput single cell assays is gaining importance with the recognition of the heterogeneity of cell suspensions. These suspensions may display cell-to-cell variability at the gene expression level through protein expression level which can have impact in our study of functional genomics, cancer, stem-cell research and drug screening. The on-chip separation and analysis of individual cells in an isolated environment can reduce cross-contamination, provide high recovery, and enable the study of biological traits at a single cell level.

In this presentation we will report on our next generation of cell separation chip, which we introduce recently in a paper by Chen et al. (2013). This devices integrates mobile magnetic trap array with microfluidic technology (Henighan et al. 2010; Vieira et al. 2009) to separate magnetically labeled cells into streams that allow further analysis. In parallel with this on chip separation, we are further developing and optimizing commercial RNA "ish" technology which allows single cell RNA analysis without the need for PCR combined with antibody labeling to allow the potential, multiparameter analysis on a single chip.

(795h) Droplet Microfluidic Systems For Single Cell Whole Genome Amplificiation

Rhee, M., Sandia National Labs
Meagher, R. J., Sandia National Laboratories
Light, Y., Sandia National Laboratories
Yilmaz, S. A., Sandia National Laboratories
Sustarich, J.,
Singh, A. K., Sandia National Laboratories
Originally presented on: 11/8/2013 10:36:00 - 10:54:00

DROPLET Microfluidic systems FOR SINGLE CELL WHOLE GENOME AMPLIFICIATION

M. Rhee, R.J. Meagher, P. Liu, Y.K. Light, S. Yilmaz, J. Sustarich and A.K. Singh

Sandia National Laboratories, USA

The diversity of microbes in environments ranging from soil and water to the human body has been long recognized while there are big challenges for the study of microbial diversity and their roles, including extremely low cell density and uncultivable microbes. The majority of such microbial organisms are uncultivable, making it challenging to perform genome sequencing and functional characterization. Uncultured microbes in the human body may play important roles in processes related to pathologic development, and uncultured environmental microbes may also hold key metabolic pathways enabling more efficient biofuel processing and bioremediation. Even a slight change of this microbial composition may be a critical indicator of a variety of the progression of diseases; however, its complex pathway to the change is still uncertain yet. Although Metagenomics has contributed to the discovery of a great diversity of bacterial species, the complexity of microbial communities is highly beyond the tractable range of standard metagenomic techniques where relatively large amounts of starting genetic material is required. In addition, population-averaged statistical data from metagemonic study can be hardly used to identify the link between the genotype and phenotype of each individual cell. Thus, a targeted approach to identify, sort and isolate individual bacterial cells of interest from a pool of microbial samples is required to perform genome amplification and sequencing without culture. We present here droplet-based microfluidic tools to isolate individual microbial cells and amplify their whole genomes in a non-cultivation manner as a route to single-cell genome sequencing. The system consists of flow cytometry for identifying bacteria and on-demand droplet generation for cell encapsulation followed by injection of amplification reagents.

We have introduced a microfluidic platform for fluorescence in situ hybridization (FISH) on chips in our previous paper [1]. In order to identify or label specific bacteria with molecular specificity, we have used fluorescently labeled sequence of 16S rRNA for hybridization which has long been used for determining phylogenetic relationships between bacteria [2]. The following fluorescence-activated cell sorting (FACS) offers a convenient way to physically isolate individual cells. The individually isolated bacterial cell is then contained in a safe environment where crosstalk between cells or other contamination is strictly prevented. Aqueous emulsion droplets in oil phase serves as ideal containers for isolated single cells since they provide physical water/oil barriers to prevent contamination between cells. Microfluidic environments provide an ideal platform to generate and manipulate aqueous droplets in oil for various preparation steps prior to amplification. Evaporation of oil phase during long incubation time may become another challenge for successful amplification. Decrease in the amount of carrier oil due to evaporation can cause significant reduction of physical oil barriers between droplets or make the droplet exposed to atmosphere, resulting in non-specific droplet merging or breakage. To minimize evaporation during amplification, a fast and non-thermal amplification method is preferred. Multiple displacement amplification (MDA) using φ29 polymerase yields relatively uniform coverage when performed in a very small volume.

We present here a microfluidic processor with three modules constructed on a single chip, capable of performing (*i*) fluorescence-based flow cytometry and cell sorting, (*ii*) encapsulation of selected cells in sub-picoliter droplets, and (*iii*) injection of amplification reagents into the droplets for single-cell genome amplification. Especially, to our best knowledge, this is the first report in the field that proposes on-demand droplet generation for single cell sorting. Encapsulation of bacteria in picoliter plugs in particular allows us to scale down conventional assays into much smaller reaction volumes better suited to the size of an individual microbe. By dramatically reducing the reaction volume, the effective concentration of template is also increased, reducing amplification artifacts that often arise in single-cell reactions carried out at a conventional scale. Fluorescently labeled bacterial cells in the sample stream are hydrodynamically focused by two aqueous side sheath streams to minimize the detection area for more accurate reading. Multiple laser beams with desired wavelengths spot the detection point to determine the presence of desired cells using computerized PMT based on the fluorescence and scattered light. Upon the successful detection of bacterial cells, the control software triggers the microdispensing pump connected to a dispensing fluid. The dispensing fluid is injected, gets mixed with the sample stream and forcedly generates a droplet through a small orifice into the adjacent oil stream. The dispensing is programmed based on calculated delay time between detection and dispensing and the size of the droplets can be fine-tuned by dispensing pressure and pumping time. These parameters all together not only determine whether the cell or particles of interest can be encapsulated in the generated droplet or escape to the waste collection but also affect the recovery time for the next droplet generation. Typically, as the size of droplets decreases, the pump requires less recovery time. For droplets of ~50 μm in diameter, the full recovery time is approximately 200 ms, which enables droplet generation at the rate of 5 Hz. This generation rate is not comparable to that of high-throughput droplet generation systems that typically produce hundreds to thousands of droplets per second, but the on-demand nature of our system inevitably has a limitation of speed in compensation of versatile controls over droplet generation. For our microbial studies, our candidate samples of interest typically are rare and have a very low cell density, which makes this platform most suitable for the purpose. In case of high concentration samples, prior dilution will facilitate single cell encapsulation.

An encapsulated cell in the aqueous droplets is physically protected by water/oil interface. However, this physical barrier imposes another challenge of adding lysis and amplification reagents into the flowing droplets. Having a specially designed injection module down the stream, we successfully demonstrated that various reagents and buffer solutions could be merged into the cell containing droplets without losing or contaminating encapsulated cells. When the flowing droplets encounter concurrent reagent stream, by applying voltage of 100 V at the merging site, the concurrent flow injects an adjustable amount of reagents into the droplets. Because of internal circulation inside the injected droplets, almost instantaneous mixing occurs and the merged droplets become ready for the following amplification. After careful collection of droplets, the isolated single-cell genomes were amplified by multiple displacement amplification (MDA). using high purity φ29 DNA polymerase and their 16s rRNA genes were amplified and sequenced for identification.

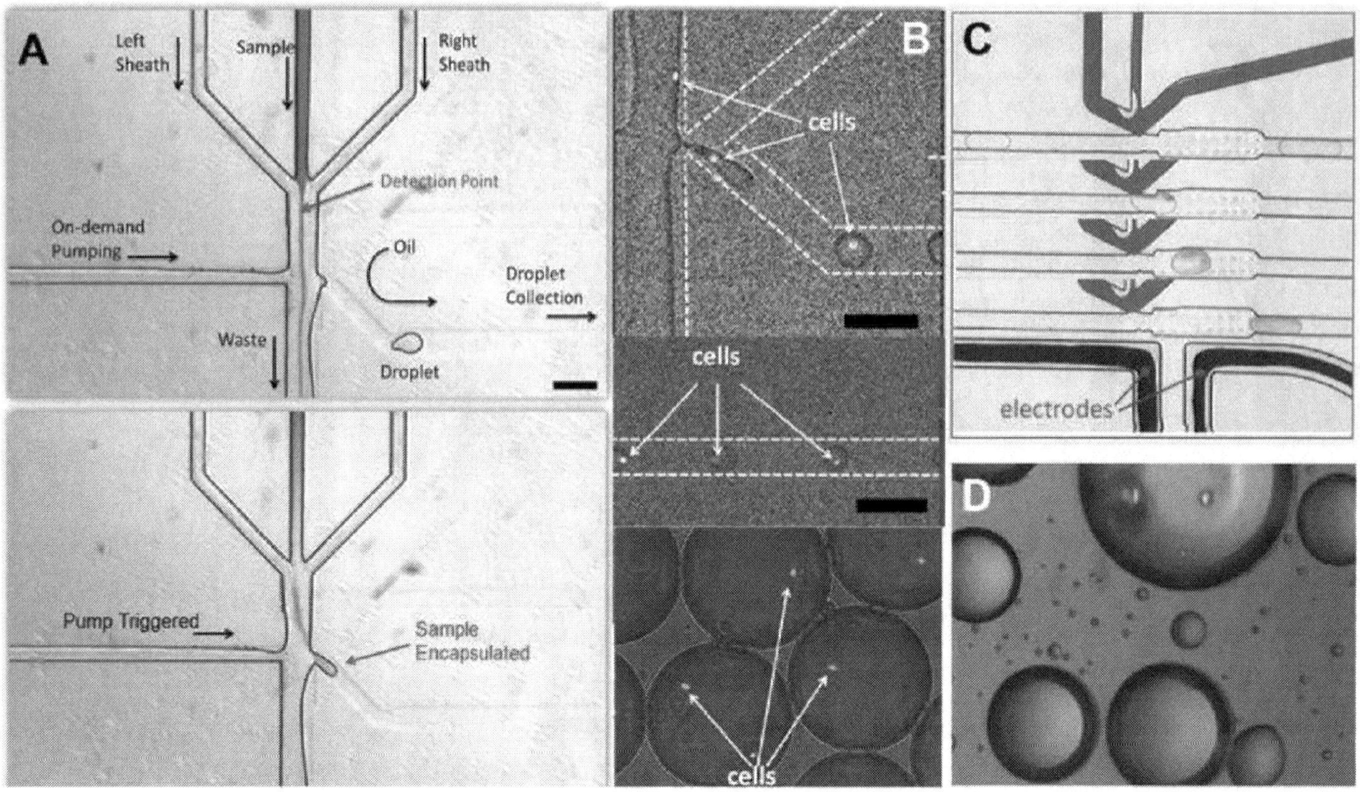

Figure Captions

A. (top) Upon the detection of bacterial cells, the control software actuates the micro-dispensing pump connected to a dispensing fluid. (bottom) The dispensing fluid is injected into the channel and forcibly reroutes the sample stream into adjacent oil stream through a small orifice, forming an emulsion droplet.

B. (top) GFP expressing *E. coli* cells are detected and encapsulated in aqueous droplets in oil one by one. (middle) Single *E. coli* cells contained in emulsion droplets flow down the channel for reagent injection. (bottom) Microdroplets serve as discrete, non-communicating containers or reaction vessels.

C. Injection of Fixed amount of reagents enabled by con-current pico-injector

D. A microscope photo of droplets containing amplified genomic DNA spread inside the droplet.

REFERENCES

[1] P. Liu et al., "Microfluidic fluorescence in situ hybridization and flow cytometry (μFlowFISH)", *Lab on a Chip*, vol. 11, pp. 2673-2679, 2011.

[2] R.I. Amann et al., "Combination of 16S rRNA-targeted oligonucleotide probes with flow cytometry for analyzing mixed microbial populations", *Applied and Environmental Microbiology*, vol. 56, pp. 1919-1925, 1990.

[3] T.D. Perroud et al., "Microfluidic-Based Cell Sorting of Francisella tularensis Infected Macrophages Using Optical Forces", *Analytical Chemistry*, vol. 80, pp. 6365-6372, 2008.

(796a) Probiotics Engineered to Fight Against Human Pathogens

Hwang, I. Y., Nanyang Technological University
Tan, M. H., Nanyang Technological University
Koh, W. C. E., Nanyang Technological University
Wong, C. K., Nanyang Technological University
Lo, T. M., Nanyang Technological University
Poh, C. L., Nanyang Technological University
Chang, M. W., Nanyang Technological University
Originally presented on: 11/8/2013 8:30:00 - 8:48:00

There is a growing sense of urgency in searching for a new effective treatment regimen for infectious diseases as the emergence of multi-drug resistant pathogens continue to pose threats. In our previous study, we engineered *Escherichia coli* to effectively sense and kill *Pseudomonas aeruginosa*, as a proof-of-concept of a novel microbe-based antimicrobial strategy. In this study, we engineered *E. coli* Nissle, a probiotic strain, to be equipped with clinically relevant functionalities to fight against *P. aeruginosa* infection. *Nissle* is a non-pathogenic and therapeutically relevant bacterium that is currently used for gastrointestinal disorder treatment. We developed novel genetic circuits that allow *Nissle* to move towards *P. aeruginosa* and produce bacteriocin and biofilm-degrading enzyme, thereby targeting two prominent states of the pathogen; planktonic and biofilm. With enhanced sensitivity and added functionalities, the killing efficiencies of the engineered *Nissle* were further demonstrated in *Caenorhabditis elegans* and mice infection models. This study extends the proof-of-concept design of engineered microbes to a novel synthetic biology-driven antimicrobial strategy, that can potentially be applied in fighting *P. aeruginosa* and other infectious pathogens.

(796b) Construction of a Cell-Cell Communication-Dependent and Gate

Shong, J., Rensselaer Polytechnic Institute
Collins, C. H., Rensselaer Polytechnic Institute
Originally presented on: 11/8/2013 8:48:00 - 9:06:00

The integration of cell-cell communication signals with other chemical inputs at a single promoter has the potential to provide a new set of tools for applications that require cells to respond to both population density and the presence of a second, exogenous signal. We have contstructed and characterized a set of promoters that behave as AND-gates, where gene expression is turned on only in the presence of a quorum-sensing (QS) signal and an exogenous inducer. In addition to widely used transcription factor operator pairs (LacI/LacO, TetR/TetO), we have employed the EsaR protein and its DNA target, the *esa* box. Unlike LuxR and other common QS regulators, EsaR can act as a repressor, where binding an acyl-homoserine lactone (AHL) molecule causes it to release its DNA target. We have shown that expression from the AND-gate promoters occurs only in the presence of both AHL and either anhydrotetracycine or IPTG. Enabling endogenous AHL production via the expression of an AHL synthase, EsaI, allowed us to control gene expression in a density-dependent manner, but only in the presence of the exogenous inducer. This system provides a unique alternative to typical autoinduction systems and provides a framework for the generation of AND-gates that respond to both cell-cell communication and any metabolite for which there is an available regulator.

(796c) Cooperative Co-Culture of Escherichia Coli and Saccharomyces Cerevisiae for Overproduction of Paclitaxel Precursors

Zhou, K., Singapore MIT Alliance
Qiao, K.,
Edgar, S.,
Stephanopoulos, G., Massachusetts Institute of Technology
Originally presented on: 11/8/2013 9:06:00 - 9:24:00

In nature, there are many examples of microbial consortia that can efficiently accomplish chemically difficult processes through the division of labor among different species. To date, such strategy has not been fully explored in engineering microbes for production of valuable metabolites. In this study, we demonstrated a synthetic cooperative microbial system to produce precursors of anti-cancer drug paclitaxel by using two model laboratory microbes, *E. coli* and *S. cerevisiae*. The two species with distinct advantages cooperated synergistically in two levels, cell growth and taxane production. This design could also make the engineering of paclitaxel biosynthetic pathway to be more modular, because each species carrying part of the pathway can be optimized individually and then simply mixed together to complete the whole pathway. Because of these advantages, the concept demonstrated in this project, cooperation of multiple cells with their own specialties, should be generally applicable to many other similar processes.

(796d) Design and Construction of Synthetic Fungal-Bacterial Consortia for Direct Production of Biofuels and Commodity Chemicals From Cellulosic Feedstocks

Minty, J. J., University of Michigan
Singer, M. E., University of Michigan
Bae, C. H., University of Michigan
Ahn, J., University of Michigan
Scholz, S., University of Michigan
Foster, C., Michigan State University
Liao, J. C., University of California Los Angeles
Lin, X. N., University of Michigan
Originally presented on: 11/8/2013 9:24:00 - 9:42:00

In nature microbes usually live in synergistic communities in which individual species with specialized roles cooperate to perform complex tasks. The powerful features of natural consortia have inspired great interest in engineering synthetic consortia for biotechnology applications. However, there have been few reports of synthetic consortia being deployed in real-world applications, and stability, robustness, and regulation of population composition remain key barriers. In this work we bridge ecology theory with engineering principles to develop robust synthetic fungi-bacteria consortia for efficient biosynthesis of valuable products from lignocellulosic feedstocks. The required biological functions are divided between two specialists: the fungus *Trichoderma reesei*, which secretes cellulase enzymes to hydrolyze lignocellulosic biomass into soluble saccharides, and the bacterium *Escherichia coli*, which metabolizes soluble saccharides into desired products. We developed and experimentally validated a comprehensive modeling framework for *T. reesei* / *E. coli* (TrEc) consortia, providing insights on key determinants of performance. We show that ecological dynamics within the TrEc consortium lead to stable coexistence between *T. reesei* and *E. coli*, and provide mechanisms for tuning population composition. To illustrate bioprocessing applications, we demonstrate direct conversion of microcrystalline cellulose and AFEX pre-treated corn stover to isobutanol with the TrEc consortium, achieving titers up to 1.86 g/L and 62% of theoretical yield, which represent the highest reported to date for conversion of cellulosic substrates to next-generation biofuels. While we offer isobutanol production as a proof-of-concept application, the modularity of our system will allow it to be readily adapted to the large portfolio of existing microbial strains metabolically engineered to produce valuable products, and we are currently validating the consortia platform for production of specialty and commodity chemicals.

(796e) Use of Oxygen Transport Limitations to Mediate a Fuel Producing Symbiosis

Zuroff, T., Montana State University
Xiques, S., Neol, Inc.
Curtis, W. R., The Pennsylvania State University
Originally presented on: 11/8/2013 9:42:00 - 10:00:00

We have demonstrated an oxygen transport mediated symbiosis between the cellulolytic anaerobe, *Clostridium phytofermentans*, and the ethanol producing yeast *Saccharomyces cerevisiae* cdt-1 or *Candida molischiana* for ethanol production from cellulose. The consortium produce over two times more ethanol from α-cellulose when grown under simultaneous saccharificaiton and fermentation conditions. Oxygen transport induces a symbiotic cooperation between the two organisms which is both stable and scalable. These results demonstrate the power of microbial consortia for lignocellulosic biofuel production processes and the use of environmentally mediate symbioses presents a fresh look on a natural paradigm for regulating consortium function.

The process of consolidated bioprocessing (i.e., simultaneous biological hydrolysis and fermentation) of lignocellulosic is an increasingly feasible solution for the production of sustainable, liquid transportation fuels. However, the lack of a suitable organism capable of efficient hydrolysis and fermentation has limited the implementation of consolidated bioprocessing. As an alternative to single organism bioprocessing, mixed cultures represent a potential solution to achieve the appropriate combination of hydrolytic and fermentative capabilities. The stability of mixed consortia and lack of simplistic bioprocessing are often cited as challenges to this approach. This work presents the development and characterization of a semi-synthetic consortium of the cellulolytic mesophile, *Clostridium phytofermentans*, and a cellodextrin fermenting yeast, *Candida molischiana* or *Saccharomyces cerevisiae* cdt-1, for direct ethanol production from cellulose. The symbiotic cooperation is induced by the diffusion of oxygen into the culture medium. Oxygen acts to inhibit the growth of the obligate anaerobe, *C. phytofermentans* but when provided a soluble carbon source from *C. phytofermentans* hydrolysis, the yeast metabolizes oxygen relieving the inhibitory effect. When fermenting cellobiose, a mutually accessible carbon source, under semi-aerobic conditions *C. phytofermentans* is "rescued" from oxygen inhibition by yeast growth. This paradigm was extended to cellulose hydrolysis in static, aerobic filter paper degradation which was only possible using co-cultures. We then applied this symbiotic consortium to cellulose fermentation and found that the rate of oxygen transfer had a profound impact on population control as well as maintenance of the desired fuel product, ethanol. Using oxygen diffusion through submerged neoprene tubing we achieved efficient consortia-mediated cellulose fermentation and found that populations were stable up to 2 months while the system appeared to be hydrolysis rate limited. In order to overcome this rate limitation, we simulated improved hydrolysis capability by adding exogenous endoglucanase from *Trichoderma viride* at a moderate level of 400 mg/L. Using this simultaneous saccharification and fermentation approach the consortium produced greater than two times more ethanol than either mono-culture. The ethanol production level was greater than the sum of the mono-culture titers suggesting a synergism in the system. This work represents a viable demonstration of the utility of environmentally controlled microbial consortia in the production of liquid transportation fuels from cellulosic biomass. Our future work will focus on characterizing the physical interactions in the consortium system as well as developing mathematical models to predict dynamics to improve productivity. Ultimately, we aim to expand the paradigm of symbiotic biofuels production to more advanced biofuel products and true lignocellulosic substrates. In this presentation we will present our previous work in light of new advancements in understanding the consortium bioprocess.

(796f) Dynamic Model-Based Analysis of Furfural and HMF Detoxification By Pure and Mixed Batch Cultures of S. Cerevisiae and S. Stipitis

Hanly, T., University of Massachusetts
Henson, M. A., University of Massachusetts Amherst
Originally presented on: 11/8/2013 10:00:00 - 10:18:00

The conversion of lignocellulosic materials to commodity chemicals requires the efficient fermentation of sugar monomers to products. The chemical and physical processes commonly used in the pretreatment and hydrolysis of sugar polymers typically produce dehydration products such as furfural and 5-hydroxymethyl furfural (HMF) that can have a deleterious effect on fermentative microorganisms. Many microbes have the capacity to reduce these furan compounds to less toxic alcohols. Of interest in the present study is the reduction of furfural to furfuryl alcohol and the conversion of HMF to 2,5-bis-hydroxymethylfuran. Furan detoxification in yeast is achieved using a combination of enzymes from the alcohol dehydrogenase, aldehyde dehydrogenase and pyruvate dehydrogenase families. Depending on the specific enzyme used, furan aldehyde reduction requires the use of either NADH or NADPH. Thus, furfural and HMF induce changes to the redox balance by sequestering these cofactors in the detoxification process. Furan aldehydes can also reduce the rates of sugar uptake and ethanol synthesis by fermentative microbes.

Co-culturing different species of yeast has been shown to be an effective strategy for fermentation of sugar mixtures to ethanol. In contrast to monocultures with engineered organisms, mixed cultures allow the innate conversion strengths of two or more microbes to be combined. A variety of microbial co-cultures have been proposed for fermentation of lignocellulosic hydrolysates. For example, co-culturing a respiratory deficient mutant of the ethanologic yeast *Saccharomyces cerevisiae* with the xylose-fermenting yeast *Scheffersomyces stipitis* has been shown to produce larger amounts of ethanol from glucose and xylose mixtures than either yeast in monoculture. The use of respiratory deficient *S. cerevisiae* prevents the reassimilation of ethanol and allows for more precise regulation of oxygen levels needed to establish microaerobic conditions under which *S. stipitis* has high ethanol synthesis capabilities. Previous studies have shown that *S. cerevisiae* and *S. stipitis* respond differently to the presence of furan inhibitors in batch monoculture. HMF is reduced by *S. cerevisiae* at approximately a quarter of the rate that furfural is reduced. Although *S. stipitis* can detoxify HMF at a faster rate than *S. cerevisiae*, furan aldehydes have a more deleterious effect on *S. stipitis* growth and ethanol production.

In this paper, genome-scale reconstructions of *S. cerevisiae* and *S. stipitis* metabolism are combined with batch monoculture and co-culture experiments to develop improved understanding of furaldehyde detoxification by these yeast species. Uptake kinetics and stoichiometric equations for the intracellular reduction reactions associated with each inhibitor were added to genome-scale metabolic reconstructions of the two yeasts. Further modification of the *S. cerevisiae* metabolic network was necessary to satisfactorily predict the amount of acetate synthesized during HMF reduction. Inhibitory terms that captured the adverse effects of the furan aldehydes and their corresponding alcohols on cell growth and ethanol production were added to attain qualitative agreement with batch experiments. When the two yeasts were co-cultured in the presence of the furan aldehydes, inoculums that reduced the synthesis of highly toxic acetate produced by *S. cerevisiae* yielded the highest ethanol productivities. The model described here can be used to generate optimal cellular engineering and fermentation strategies for the simultaneous detoxification and fermentation of lignocellulosic hydrolysates by *S. cerevisiae* and/or *S. stipitis*.

(796g) Dynamic Multi-Level Analysis and Metabolic Modeling of Microbial Communities Using Optcom

Zomorrodi, A. R., The Pennsylvania State Univeristy
Islam, M. M.,
Mahadevan, R., University of Toronto
Maranas, C. D., The Pennsylvania State University
Originally presented on: 11/8/2013 10:18:00 - 10:36:00

The species within microbial communities communicate through unidirectional or bidirectional exchange of biochemical cues. These inter-species interactions and their temporal changes in response to environmental stimuli are known to significantly affect the structure and function of microbial communities and play a pivotal role in species evolution. Despite the growing availability of high-throughput experimental techniques and data, we still know very little about the metabolic contributions of individual microbial players within an ecological niche and the extent and directionality of metabolic interactions among them. This calls for development of efficient modeling frameworks to elucidate less understood aspects of metabolism in microbial communities. We have recently developed a comprehensive framework, called OptCom, for the flux balance analysis of microbial communities using genome-scale metabolic models, which relies on a multi-level and multi-objective optimization formulation to properly describe trade-offs between individual vs. community level fitness criteria. Given that microbial communities change with time (e.g., day/night cycle) new approaches that would be able to capture temporal interspecies metabolic interactions and tradeoffs are needed. To this end, we extend the OptCom procedure for the dynamic metabolic modeling of microbial communities by combining it with the DMMM (Dynamic Multi-species Metabolic Modeling) framework previously proposed by Zhuang *et al.* This is done by replacing the separate FBA problems for each individual species in DMMM with the OptCom's multi-level optimization structure to account for interspecies interactions and temporal driving forces. In addition, the OptCom procedure was further extended to incorporate the uptake kinetics whenever available. These approaches were used for modeling the competition between *Rhodoferax* and *Geobacter* in subsurface anaerobic environment, and to examine addition of a new member to this community in order to assess the newly formed inter-species metabolic interactions and driving forces, and their dynamics under different environmental conditions.

(796h) Label-Free, Strain-Resolved, Shotgun Proteomics of a Defined Bacterial Co-Culture

Chignell, J.,
Reardon, K. F., Colorado State University
Park, S., Colorado State University
Originally presented on: 11/8/2013 10:36:00 - 10:54:00

Systems biology approaches can provide new, detailed insights into the interactions among members of communities of microorganisms that are unavailable through conventional methods. An "-omics" approach provides metabolic information about the effects of interactions on different community members, thereby suggesting functional mechanisms to explain dynamics of the system as a whole. Recently, analysis of defined co-culture systems has been identified as a horizon for metabolic modeling and bioprocessing. Strain-resolved proteomics can track proteome changes of each species of a co-culture. To evaluate the complexity of molecular-level responses associated with growth in community, a quantitative, shotgun, label-free proteomics study was conducted on a defined bacterial co-culture consisting of two well-characterized soil species: Gram-negative *Pseudomonas putida* KT2440 and Gram-positive *Bacillus atrophaeus* ATCC 9372, the standard surrogate for the pathogen *Bacillus anthracis*. A total of 1151 proteins were identified across pure and co-culture samples. Differentially-expressed proteins were quantified by two methods: precursor ion intensity and spectral counts. Proteins were categorized according to Gene Ontology categories and KEGG pathways. Twenty-seven *B. atrophaeus* proteins and 50 *P. putida* proteins were significantly more abundant in the co-culture than in the respective pure culture ($p<0.05$; fold-change > 2), while 137 *B. atrophaeus* and 42 *P.putida* proteins were significantly less abundant in the co-culture, compared to the pure culture ($p<0.05$; fold-change < 2). For both species, the most common Gene Ontology biological process category for upregulated proteins was regulation of transcription. For *B. atrophaeus*, nearly 15% of the proteins upregulated in coculture were related to stress response. These proteins included GrpE (4. fold higher in co-culture compared to pure culture), penicillin-binding lipoprotein 3 (2 fold higher in co-culture) and glycosyltransferase (3.7 fold higher in co-culture), the latter two of which are a response to antibiotic or toxin. For *P.putida*, several proteins were related to toxin biosynthesis, including PPM/P mutase (2.2 fold higher in co-culture than in pure culture), 2-dehydro-3-deoxyphosphooctonate aldolase 2 (2. fold higher in co-culture), phenazine biosynthesis protein (6.5 fold higher than in co-culture), and GTP cyclohydrolase-2 (2.3 fold higher in co-culture). Futhermore, for each species several differentially-expressed proteins related to iron uptake. For *B.atrophaeus*, 2,3-dihydroxybenzoate-AMP ligase, an enzyme in the siderophore biosynthesis pathway, was 7-fold higher in the co-culture than in pure culture. Bacterioferritin was down-regulated in *P.putida*. Follow-up growth studies involving growth in culture supernatant and at different levels of iron availability were conducted to verify interactions suggested by the proteomics data. This study contributes to a growing literature that seeks to discover and characterize patterns of microbial interactions through powerful "-omics" analysis of co-culture systems of model species.

(798a) Genome Evolution and Engineering for Elucidating the Genetic Architecture of Isobutanol Tolerance In Escherichia Coli

Minty, J. J., University of Michigan
Lai, L., University of Michigan
Kennedy, L., University of Michigan
Boyer, D.,
Zhu, L.,
Zaroff, T. A. III, University of Michigan
Wang, H., Harvard University
Church, G. M., Harvard Medical School
Lin, X. N., University of Michigan
Originally presented on: 11/8/2013 12:30:00 - 12:48:00

Understanding the genetic architecture underlying complex phenotypes is of great fundamental interest and also has important ramifications in biotechnology. Metabolic engineering efforts have enabled microbial production of many fuels and commodity chemicals, but frequently toxicity limits production. Microbial stress tolerance is a complex multigenic trait that is intractable to traditional genetic study and rational engineering efforts. We are developing an innovative evolutionary-genomics methodology for elucidating and improving complex phenotypes. Our approach entails experimental evolution of stress tolerance followed by genome re-sequencing to identify acquired mutations, genomic and functional dissection to reverse engineer mechanisms of tolerance, and targeted genome engineering for further phenotype improvement. As a proof-of-concept, we have been studying *E. coli* tolerance to isobutanol, a promising next-generation biofuel. Through our evolution and genome re-sequencing work (Minty et al. *Microb Cell Fact* 2011 10:18), we identified 247 genetic loci correlated with isobutanol tolerance. We are currently performing targeted mutagenesis of selected genetic loci using Multiplex Automated Genome Engineering (MAGE), a recently developed technology that entails repeated cycles of high efficiency recombination using libraries of mutagenic DNA oligonucleotides (Wang et al. *Nature* 2009 460). This strategy enables rapid exploration of vast genotype space without being constrained to adaptive walks. Improved strains isolated from mutant libraries are further characterized via detailed phenotyping and genotype analysis, allowing for systematic mapping of isobutanol tolerance phenotypes and genotypes. Our work reveals prevalent epistasis between genetic loci and provides insights into biochemical mechanisms of tolerance, as well as generating improved strains of *E. coli* that may be immediately useful in the production of isobutanol.

(798b) Eukaryotic Reprogramming By Directed Genome Evolution

Si, T., University of Illinois at Champaign-Urbana
Zhao, H., University of Illinois at Urbana-Champaign
Originally presented on: 11/8/2013 12:48:00 - 13:06:00

Genome-wide cellular reprogramming will not only uncover fundamental biological mechanisms of cell differentiation and development but also result in a broad range of applications such as gene therapy, regenerative medicine, microbial production of chemicals and fuels, and transgenic crops with herbicide, salt, and drought tolerance. However, the current ability to reprogram a cell on the genome scale is limited to bacterial cells and in most cases is confined to modifications of a limited number of gene targets. Here we report an efficient, genome-scale and generally applicable method for eukaryotic reprogramming, by applying directed evolution strategies to genome engineering. Iterative cycles of genome-wide trackable mutagenesis coupled with high-throughput selection or screening enable the accumulation of beneficial modifications in an evolving eukaryotic genome. We show the application of directed genome evolution to *Saccharomyces cerevisiae* for improved acetic acid (HAc) tolerance, a key trait for microbial production of chemicals and fuels. Three genetic modification targets were identified which acted synergistically to confer the highest acetic acid tolerance ever achieved in an engineered yeast strain. Our strategy may greatly accelerate the design and engineering of organisms with desired traits and provide new insights on genome structure, function, and evolution.

(798c) Novel Variants of Self-Excising Beta-Recombinase/Six Cassettes Facilitating Efficient Sequential Repetitive Gene Deletions in Neurospora Crassa

Szewczyk, E., University of California
Kasuga, T., University of California
Fan, Z., University of California, Davis
Originally presented on: 11/8/2013 13:06:00 - 13:24:00

For over 70 years, *Neurospora crassa*, a model filamentous ascomycete, has been studied to understand biological processes and phenomena relevant to higher eukaryotes. It has received increasing interest because of its ability to convert lignocellulosic biomass to sugars, which can be converted into biofuels or platform chemicals. Development of efficient genetic engineering tools including efficient targeted gene deletions methods are essential to meet the requirements of various tailor-made applications.

Recently we validated functionality of a bacterial recombination system employing β serine recombinase acting on *six* recognition sequences (β-rec/*six*) in the filamentous fungus *Neurospora crassa*. This self-excising resistance marker cassette allows repetitive deletions of genes with only one round of transformation per gene deletion. Moreover, the β serine recombinase acts in a strict *cis*-action manner, which enables only intramolecular recombination events and eliminates the risk of chromosomal rearrangements as brought by the Cre/*loxP* or FLP/*FRT* systems. Whereas the screening process is straightforward in Aspergilli, whose mononucleate conidia allow automatic homokaryon purification, the multinuclear nature of *N. crassa* macroconidia used for transformation makes the process more complicated. The promoter used for the induction of the cassette eviction is also of very low efficiency in *N. crassa*.

Here we report on the construction and validation of new variants of β-rec/*six* system cassettes that facilitate screening for self-excision events in *N. crassa*. Introduction of the *tk* (thymidine kinase) gene allows fast-forward, one-step, positive screening for the excision event and automatic homokaryon purification. Use of the regulatable promoter native to b-*xylanase A* of *N. crassa*, also significantly improves the efficiency of cassette eviction in this fungus.

(798d) Combinatorial Metabolic Engineering of Clostridium Cellulolyticum Using Clonable Genomic Library Fragments Generated From Degenerate Oligonucleotide Primed PCR

Freedman, B. G., Virginia Tech
Senger, R. S., Virginia Tech
Originally presented on: 11/8/2013 13:24:00 - 13:42:00

The anaerobic mesophile *Clostridium cellulolyticum* ATCC 35319 is an efficient cellulose utilizer that is characterized by pyruvate secretion and a slow growth rate. Resolving a critical metabolic bottleneck in *C. cellulolyticum* has been performed using an over-expression genomic library to identify gene fragments that impart a growth advantage on cellulosic substrates. A novel, template-independent approach involving degenerated oligonucleotide primed PCR was used to generate genomic libraries reliably and rapidly from the genomic DNA of multiple bacteria with different methylation characteristics. This approach allowed for isolation and amplification of genomic fragments from environmental sources rich in biological and cellulolytic activity. Also unique to our new method is that error-prone PCR is supported in library generation, dramatically expanding the search space for improved biological activity. We show that the use of single-stranded degenerate oligonucleotide primers consisting of a 10-15bp 3' tail, a 6-15bp degenerate interior, and a 2-6bp 5' anchor allow for random binding and amplification of pure or mixed DNA samples using two separate PCR cycling parameters. These DNA fragments clone at high efficiency to commercial TOPO™ cloning vectors and are easily transposed into other vectors using the Gateway™ cassette system. This new methodology has made generating and cloning a genomic library a routine and easily repeatable procedure in our laboratory. A genomic library was generated from *C. cellulolyticum* DNA as well as DNA amplified from several environmental sampling sites. Expressing the library in continuously sub-cultured *C. cellulolyticum* grown on different cellulosic carbon sources (including recycled paper) allowed for identification of unique gene fragments that resolved metabolic bottlenecking and improved cellulose degradation rates. Ultimately, this new strain will enable the conversion of cellulose into valuable chemicals and fuels.

(798e) Expanding the Searchable Genomic Space: Heterologous Sigma Factor Expression for Screening of Metagenomic Libraries

Gaida, S. M., University of Delaware
Sandoval, N. R., University of Delaware
Nicolaou, S. A., University of Delaware
Papoutsakis, E. T., University of Delaware
Originally presented on: 11/8/2013 13:42:00 - 14:00:00

Most bacteria found in nature cannot be cultured in a laboratory but are able to perform many interesting and valuable biocatalytic functions. Discovery of the underlying genetics requires functional screening of heterologous and metagenomic libraries. A major limitation in the application of this screening is the low-level heterologous gene expression within the screening host, such as *Escherichia coli*. To overcome this limitation, we engineered *E. coli* strains that possess non-native transcription machinery capable of recognizing heterologous promoters. Such strains can be employed for screening heterologous DNA libraries, thus greatly enlarging the genomic space that can be functionally sampled. Specifically, we expressed in *E. coli* sigma factors from phylogenetically-distant bacteria (e.g. *Lactobacillus plantarum*, *Bacillus subtilis*). Strains expressing foreign sigma factors recognized a significantly greater fraction of promoters from multiple heterologous genomic libraries compared to the native *E. coli* transcription machinery alone (as quantified using a flow cytometry-based *gfp*-trap method). Importantly, high levels of cross-recognition were observed between foreign sigma factors and libraries of differing species. Effect of simultaneous expression of multiple heterologous sigma factors on mixed genomic libraries was explored and optimized. Screening fosmid-based genomic libraries of *L. plantarum* in an *E. coli* strain expressing the *L. plantarum rpoD* gene (coding for its major sigma factor) allowed the identification of *L. plantarum* genetic elements imparting ethanol tolerance in *E. coli*.

(798f) Facile Generation of Knockdown Phenotypes in Yeast Through RNA Interference

Crook, N., The University of Texas at Austin
Schmitz, A., The University of Texas at Austin
Alper, H., The University of Texas at Austin
Originally presented on: 11/8/2013 14:00:00 - 14:18:00

Microbial synthesis offers a promising approach for the sustainable production of fuels and chemicals. These metabolic engineering efforts often rely upon a finite set of common genetic approaches such as gene deletions or overexpressions. However, knockouts are time-consuming to construct and difficult to rapidly transfer to other strains, which greatly lowers the throughput of strain engineering programs. Therefore, in order to facilitate the identification of valuable knockout candidates, a method enabling the rapid construction of gene knockdowns is needed. Here, we develop a new technique for the facile generation of knockdown phenotypes in yeast through RNA interference, which is able to confer gene knockdowns using a simple, portable, plasmid-based system. By importing the RNA interference pathway from *S. castelli* and optimizing the design of the siRNA cassette, we have been able to achieve up to 93% downregulation of genes from a wide variety of initial expression levels. Finally, we demonstrated the utility of this technique for metabolic production phenotypes in yeast. This approach will significantly expedite yeast strain development and greatly expand our understanding of how native metabolism interfaces with heterologous pathways.

(798g) Metabolic Pathway Inactivation Via Post Translation Modifications

Sekar, K., Northwestern University
Tyo, K. E. J., Northwestern University
Originally presented on: 11/8/2013 14:18:00 - 14:36:00

Metabolic engineering relies on recombinant DNA technology to optimize a biological system toward the production of commodity chemicals. Generally, this technology can be classified into three methods: overexpression of product-producing pathways, knockouts of competing pathways, and rewiring of regulatory networks. One important feature of these technologies is that they take place only on the transcriptional and translation synthesis level. This is problematic because biomass-producing enzymes cannot be knocked out lest the organism doesn't grow. Even when synthesis is modulated on the transcriptional or translation level, protein stability ensures continued biomass anabolism. Protein half-lives are long, with only 2% to 7% degraded under half an hour in *E. coli*. Other studies indicate generally long half-lives for the majority of proteins in *E. coli* (>20 hours).

We have established a tool to inactivate proteins directly. Our technology has been successfully applied to two unique proteins in *E. coli*: mCherry and beta-galactosidase. We are investigating the techology on enzymes citrate synthase and
isocitrate dehydrogenase. Inactivation is fast and efficient.

Implications of this technology mean that metabolic engineers can reach unprecedented yields, where resources are not wasted toward the cell's biomass, and potentially longer fed batch processes where chemical catalysis continues without cell accumulation problems such as oxygen transfer.

(798h) Isotope Labeling Experiments Enable Dissection of Parallel Carbon-Reducing Pathways in the Marine Diatom Phaeodactylum Tricornutum

Quinn, A., University of Maryland
Kreeger, E.,
Sriram, G., University of Maryland
Originally presented on: 11/8/2013 14:36:00 - 14:54:00

In algae and plants, the 2-C-methyl-D-erythritol phosphate (MEP) and mevalonate (MVA) pathways independently produce isopentenyl pyrophosphate (IPP), an important precursor to a spectrum of highly reduced molecules that are biofuel and biochemical precursors. For example, numerous photosynthetic species convert IPP into the branched C_5 alkene isoprene. Once extracted from a cell, isoprene can be polymerized into true diesel fuel or natural rubber. Therefore, an organism engineered to produce high quantities of IPP could serve as a robust bio-factory for highly reduced fuels far superior to the current generation of bio-alcohols and biodiesel.

Photosynthetic organisms offer a significant advantage over heterotrophic bacteria and yeast due to their ability to grow on atmospheric CO_2 in addition to reduced carbon substrates. This eliminates the necessity of costly sugars derived from terrestrial crops or cellulosic biomass. One promising organism is the unicellular marine diatom *Phaeodactylum tricornutum* (Pt). Pt is highly prized for its ability to efficiently sequester carbon dioxide using a unique C4 pathway, as well as for its propensity to convert a large proportion of its biomass into reduced lipids.

Like many photosynthetic organisms, Pt operates a plastid-localized MEP pathway and a cytosolic MVA pathway. Using isotope labeling techniques, researchers have previously shown that the organism utilizes the MEP pathway to synthesize phytol, a precursor to photosynthetic pigments, whereas the MVA pathway synthesizes steroids such as brassicasterol. The plastidic and cytosolic pools of the IPP precursor are considered separate from one another; however, investigations with *Arabidopsis thaliana* and *Nicotiana tabacum* have hinted at some degree of mixing between the two compartments in those organisms. Any effort to overproduce IPP in Pt will require an understanding of the degree of mixing between these two pathways so that the metabolic pathways can be rewired to both maximize the flow of carbon towards a large pool of available IPP, while also synthesizing sufficient steroids and carotenoids to maintain high cell viability and growth.

In this presentation, we will report the dissection of the metabolic cross-talk between the compartmentalized MEP and MVA pathways in Pt based upon *in vivo* experiments employing enzyme inhibition and isotope-assisted metabolic flux analysis. By inhibiting one pathway and measuring the abundances of key downstream metabolites, we will determine whether the organism is able to reroute carbon through the uninhibited isoprenoid pathway and across the plastid-cytosol membrane. We will also feed various carbon substrates strategically labeled with the ^{13}C isotope of carbon. By measuring the labeling patterns of intracellular metabolites, we will construct metabolic flux maps highlighting how the organism alters its' central carbon metabolism to compensate for the inhibition of the isoprenoid pathways. We anticipate that these results will reveal efficient and effective strategies to increase IPP production in Pt whilst also maintaining adequate levels of essential, secondary isoprenoids.

(803a) Mixing During High Consistency Enzymatic Hydrolysis - Influence of Raw Material, Viscosity and Particle Size Distribution

Palmqvist, B., Lund University
Kadic, A., Lund University
Lid n, G., Lund University
Originally presented on: 11/8/2013 12:30:00 - 12:55:00

In order for commercial scale lignocellulosic bioethanol to be economically feasible, high solid operation throughout the whole process is regarded as a necessity. This will not only reduce the need for high amounts of water and waste water treatment, but also increase ethanol titers after the fermentation step, which will lower energy demands in the following distillation step.

However, since concentrated biomass slurries are associated with very high non-Newtonian viscosities, achieving sufficient mixing during high consistency enzymatic hydrolysis will be a challenging task. Poor mixing can lead to mass- and/or heat transfer problems as well as inhomogeneous enzyme distribution, all with possible yield reduction as an effect. Moreover, we have previously shown that stirrer speed (or more accurately – mixing-power) strongly affects the hydrolysis yield of certain lignocellulosic materials, in particular spruce, where an almost 100 % yield increase was achieved by increasing the stirrer speed from 75 to 500 rpm. On the contrary, other materials such as for example the energy crop Arundo Donax (Giant Reed) seems rather unaffected by mixing power input during enzymatic hydrolysis.

This presentation summarizes what we have learned during the last years about rheology and mixing of high consistency biomass during enzymatic hydrolysis. The fact that different materials respond completely different to mixing power input during hydrolysis will be discussed and seems to be related to the viscosities of the materials. Furthermore, a discussion about liquefaction of biomass versus hydrolysis yield and the effect of particle size distribution on hydrolysis rates will be presented.

(803b) Evolution of Yield Stress During High-Solids Fed-Batch Enzymatic Liquefaction of Cellulose Measured Using a Non-Invasive Real-Time Rheometer

Cardona, M., University of California, Davis
Tozzi, E. J., University of California, Davis
Karuna, N., University of California, Davis
Jeoh, T., University of Califonrnia, Davis
Powell, R. L., University of California-Davis
McCarthy, M. J., University of Calfiornia Davis
Originally presented on: 11/8/2013 12:55:00 - 13:20:00

Enzymatic hydrolysis of cellulosic biomass is a key step of biological routes for production of fuels and chemicals. Economic considerations for large-scale implementation of the process demand operation in the high solids regime. In this regime, the biomass is processed at concentrations of 15% or higher by weight, forming a high viscosity slurry which introduces processing challenges especially in the initial stages of hydrolysis (liquefaction). Rheological measurements with rotational rheometers are challenging in these materials due to rapid changes in rheology that require high time resolution, and the large particle sizes which cause gap effects and settling. We employed a non-invasive real-time rheometer (FlowScan(TM), Aspect Imaging) to characterize the real-time evolution of yield stress of cellulose undergoing enzymatic hydrolysis in a recycle-flow reactor. The hydrolysis was performed in fed-batch mode, with fibers being added at various time points. The suspension exhibited sustained changes in rheology upon initial dispersion of fibers in water (prior to the first addition of enzyme), which was interpreted as hydration or swelling. Each addition of fibers caused a rapid increase followed by a decay in yield stress due to the enzymatic action. The decay of the yield stress was more rapid for the initial loads, becoming slower as the hydrolysis progressed. The influence of fiber loading on the rheology of the slurry undergoing hydrolysis is discussed.

(803c) Maximize Enzymatic Saccharification and Minimize Sugar Degradation in SPORL Pretreatment of Douglas-Fir At a Low Temperature: A Kinetic Approach

Zhang, C., Tianjin University
Houtman, C. J., USDA Forest Service, Forest Products Laboratory
Zhu, J. Y., USDA Forest Service, Forest Products Laboratory
Originally presented on: 11/8/2013 13:20:00 - 13:45:00

Maximize sugar yield and minimize inhibitor formation are often difficult to achieve simultaneously in pretreatment of lignocelluloses. The objective of this work was to evaluate the hemicellulose hydrolysis and degradation models with the goal of predicting performance of both hydrolysis and degradation process and optimizing the pretreatment conditions. Various investigations have developed reaction rate expressions based on Arrhenius and first order rate law. Based on the experiment data, the hemicellulose hydrolysis and degradation kinetic models were developed and the best-fit kinetic parameters were determined. The kinetic models showed that the hemicellulose degradation reaction was more sensitive to pretreatment temperature compared with hydrolysis reaction during pretreatment process. Under the same pretreatment severity expressed by the Combined Hydrolysis Factor (CHF) using SPORL (sulfite pretreatment to overcome recalcitrance of lignocelluloses), a low temperature pretreatment at 165°C for 75min can reduced the inhibitor generation compared with a pretreatment at 180°C for 30min while the substrate enzymatic digestibility was almost identical under these two conditions.

(803d) Comparison of Different Process Configurations Using SSF At High Solid Loadings of Steam-Pretreated Corn Stover

Bondesson, P. M., Lund University
Zacchi, G., Lund University
Galbe, M., Lund University
Originally presented on: 11/8/2013 13:45:00 - 14:10:00

The use of lignocellulosic biomass for ethanol reduces the dependency of fossil fuel and lowers the greenhouse gas emissions. A high ethanol concentration in the broth improves the economics of the ethanol production process. However, when working with high solid loadings, the ethanol yield may decrease due to difficulties to mix, which causes mass transfer limitations. Also, the effect of higher inhibitor concentrations becomes more pronounced. Some of the problems can be overcome by selecting proper configurations for enzymatic hydrolysis and the fermentation process.

In this work washed steam-pretreated fibres from corn stover was used to study the impact of a high solids loading using different process configurations with negligible amounts of inhibiting compounds. A prehydrolysis step prior to SSF or fed-batch SSF was investigated and compared with ordinary SSF using a water-insoluble solids (WIS) loading of 20%. Different variables, such as prehydrolysis time, temperature, feeding time and feeding interval as well as enzyme loading were investigated. The results showed that the inclusion of a prehydrolysis step did not increase the ethanol yield (70-75%) compared with SSF performed batchwise (75%). On the other hand, an improvement was found utilizing fed-batch, which increased the yield (79-82%). Results from this study will be presented.

(804a) A Convenient Homogeneous Enzyme Immunoassay for Both Small Molecules and Proteins

Chiu, M. L., University of California, Los Angeles
Tseng, T. T. C., University of California, Los Angeles
Lai, D., University of California, Los Angeles
Monbouquette, H. G., University of California, Los Angeles
Originally presented on: 11/8/2013 12:30:00 - 12:48:00

The enzyme-multiplied immunoassay technique (EMIT) is a little studied cousin to ELISA that normally is used for small-molecule drug testing. Unlike ELISA, however, EMIT is a simple, homogeneous assay that also can be performed quickly without expensive instrumentation. The EMIT approach entails covalent attachment of an antigen-bearing molecule to a reporter enzyme such that when antibody binds to the enzyme-bound antigen, reporter enzyme activity is repressed substantially. Yet when competing free antigen (the analyte) is present in an analytical sample, antibody binds the analyte and the reporter enzyme is "switched on" to an extent related to the analyte concentration. We have created an EMIT-based protein assay by conjugating a cysteine-modified HA peptide (from influenza hemagglutinin A) to the reporter enzyme, glucose-6-phosphate dehydrogenase. The 13-minute assay gave a free HA limit of detection of 10 nM and proved effective for detection of a high molecular weight model protein tagged with HA. We also have prepared an EMIT assay system for estradiol by covalently binding β-estradiol-6-(O-carboxymethyl)oxime (E2-6-CMO) to glucose-6-phosphate dehydrogenase (G6PDH). The lower detection limit of the estradiol assay is 1 nM in aqueous solution, and the standard curve is linear on logit-log scale up to 6.7 mM estradiol. A detection limit of 11.5 nM in estradiol-spiked human serum samples suggests the feasibility of applying this assay to monitoring estradiol levels for the prediction and prevention of ovarian hyperstimulation syndrome. Toward understanding the antibody-induced inhibition mechanism of EMIT, characterization of morphine-glucose-6-phosphate dehydrogenase (G6PDH) conjugates was pursued. The conjugates were characterized using a combination of methods including tryptic digestion, immunoprecipitation, matrix-assisted laser desorption/ionization mass spectrometry, and electrospray ionization tandem mass spectrometry. Twenty-six conjugate sites were identified including some in the vicinity of the enzyme active site. EMIT assays are convenient, versatile and powerful, and may be developed for a variety of applications including biotoxin and infectious disease detection.

(804b) Supported Biomembrane Microenvironments for the Protease Functional Analysis and Screening of Gamma-Secretase, the Notch Pathway Drug Target

Gilchrist, M. L., City College and the Graduate Center of the City University of New York
Martin, J., City College of New York of CUNY
Hur, J. Y., Molecular Pharmacology and Chemistry Program, Memorial Sloan-Kettering Cancer Center, New York, NY
Ahn, K., Memorial Sloan-Kettering Cancer Center
Li, Y., Memorial Sloan-Kettering Cancer Center
Originally presented on: 11/8/2013 12:48:00 - 13:06:00

The intramembrane protease gamma-secretase is a current target of therapeutic intervention, with pivotal pathological functions within Alzheimer's disease and cancer. Our primary objective is to expand methods for *in vitro* studies of the intramembrane proteases (IMPs) with the development of a microsphere-supported biomembrane (proteolipobead) platform. Despite extensive studies, understanding of IMP regulation and catalysis has been hampered by membrane-associated enzymology. The precise reaction mechanisms of the intramembrane proteases remain to be elucidated and furthermore, rigorous analysis of the kinetics of interfacial catalysis in these systems has not yet been undertaken.

High throughput of screening of gamma-secretase drug candidates has been carried out in bulk assay systems using cell membrane fragments, solubilized enzymes and is underway in proteoliposomes. A critical barrier to further progress in the study and screening of gamma-secretase is that such bulk systems do not allow for the direct in situ quantification of enzyme, substrates, or inhibitors or their relative distributions within the structures under assay in native lipids. Due to these factors and the instability of these systems there are limitations in the scope of assays possible using solubilized systems and proteoliposomes in bulk solution. We have expanded in vitro models of gamma-secretase to include proteolipobeads (microsphere-supported biomembranes), using novel substrates such as Notch1-Sb1 (truncated Notch 1 with FLAG tag) to probe the enzyme. Using confocal and super-resolution imaging in tandem with flow cytometry, we will present findings from i) direct measurements of substrate and gamma-secretase loading and in-membrane orientation within supported biomembranes, ii) the localization and quantification of enzyme:substrate and enzyme:inhibitor complexes within supported biomembranes, and iii) the introduction of new protein-orthogonal biomembrane tethering strategies that include membrane fusion.

(804c) Simultaneous, m-RNA and/Or Micro-RNA, and Protein Expression On Individual, Circulating Tumor Cells

Chalmers, J., The Ohio State University
Wu, Y., The Ohio State University
Deighan, C.,
Lustberg, M., The Ohio State University
Amaya, P.,
Miller, B. L., The Ohio State University
Park, K. J.,
Originally presented on: 11/8/2013 13:06:00 - 13:24:00

Human Epidermal Growth Factor Receptor family includes four members: EGFR/HER1, ErbB2/HER2, HER3, and HER4. This signaling cascade has important prognostic and predictive significance in a variety of solid tumors and has several targeted therapies that are currently approved or in development. EGFR and Her 2 currently have approved therapeutic that can block such pathways. Therefore, characterization of CTCs for presences of these drugable targets is of significant clinical interest since CTCs may give the most dynamic snapshot of the state of tumor progression.

Since it is challenging to obtain serial metastatic site biopsies, a significant promise of CTCs is in utilizing them as liquid biopsies and evaluate them for various targets including the Her family. A significant challenge in using CTCs as predictive markers is that more than a third of patients with metastatic disease do not have detectable CTCs by EpCAM-based technology. One hypothesis is that these patients are more likely to have CTCs not captured by EpCAM based positive enrichment, such as those that have undergone epithelial to mesenchymal transition, EMT, which down regulates epithelial markers, such as EpCAM. EMT-like characteristics of CTCs has been reported recently by several groups.

As opposed to using positive EpCAM selection we are using a negative enrichment system, we have identified CTCs in all breast cancer stages. However, to further characterize these cells, we need more than the traditional four color epi-flourescent and confocal approaches. In this presentation, we will demonstrated how we are able to obtain 6, and possibly more, specifically targeted cell markers, including mRNA, microRNA, as well as cellular proteins on each cell. Markers targeted include: cytokeratin proteins, vimentin proteins, HER and EGRF surface proteins, and messenger RNA, and microRNA for HER, EGFR, and other relevant markers. In addition, we will present the potential to quantify this presence of these targets using spectral deconvolution technology.

(804d) Electronic Platform To Quantify The Anticancer Mechanisms Of Cardiac Glycoside and Synthetic Analog

Eldawud, R., West Virginia University
Dong, C., West Virginia University
Stueckle, T., National Institute for Occupational Safety and Health
Rojanasakul, Y., West Virginia University
Chen, M.,
Dinu, C. Z., West Virginia University
Originally presented on: 11/8/2013 13:24:00 - 13:42:00

The clinical applications of digitoxin as a cardiac glycoside have been halted by many concerns associated with its cardiac toxicity. Recently it was shown that the toxic profiles induced by therapeutic ranges of digitoxin exposure are selective to many types of neoplastic cells from breast to prostate and lung cancer, making this drug an attractive candidate for chemotherapies. The anticancer potency of digitoxin lays in its trisaccharide moiety linked to a steroid core. Further research has shown that synthetic manipulation of digitoxin structure leads to more potent analogues. In this research, a new approach is used to characterize and quantify the cellular behavior of immortalized and tumorigenic human lung cells (BEAS-2B and H460 respectively) upon exposure to different concentrations of digitoxin and synthetic monosaccharide analogue (D6-MA) in real time. The approach relies on an electrical cell impedance sensing system (ECIS) used as a proxy to measure the morphological changes and cell-cell interactions *in vitro*. Such measurements are supplemented with biophysical techniques for multiparametric analysis to derive structure function relations that correlate the cellular dynamics with molecular pathways upon drug or analogue exposure. Our results provided novel means to investigate the underlying anticancer mechanisms associated with drug exposure at a single cells level and promise to extend such cellular-based interface studies to other analysts and toxins.

Disclaimer: The findings and conclusions in this abstract are those of the authors and do not necessarily represent the views of the National Institute for Occupational Safety and Health.

(804e) Phenotypic Signatures for Metastatic Cancer Cells

Wu, P. H., Johns Hopkins University
Wirtz, D., Johns Hopkins University
Philip, J.,
Chen, W. C., Johns Hopkins University
Maitra, A., Johns Hopkins University School of Medicine
Leek, J. T., Johns Hopkins University
Originally presented on: 11/8/2013 13:42:00 - 14:00:00

Cancer is the leading cause of death in U.S with ninety percent of cancer patients dying from the metastasis. Genomic analysis has been vigorous used to study cancer samples, however no consensus on molecular signatures has been found that are specific to metastasis. Accumulating evidences has suggested that cell mechanics could be a robust marker of metastasis. Therefore, the physical principles underlying metastasis could be translated into effective diagnostic and prognosis tools, as well as drug screening platforms. Here, we developed a comprehensive morphological analysis, visual aided morpho-phenotyping recognition (VAMPIRE), to classify irregular cellular and nuclear shapes using a set of common shape modes through eigenshape decomposition and clustering analysis. We collected a cohort of more than 20 different cancer cell lines including prostate, breast and pancreatic cancer cell lines which are either derived from primary tumors or from metastatic lesions. Using an automated high-throughput microscopy assay we collected more than data from more than 100,000 cells to examine the phenotypes of metastasis using VAMPIRE assay. Our results show that the lack of cell/nuclear morphological heterogeneity is a highly predictive feature of metastatic cancer cells and suggest that metastasis is associated with a selection process for biophysical features.

(804f) Detecting Pseudomonas Sp. Strain ADP in An Atrazine Enriched Environment With Raman Spectroscopy

Henry, V. A., The University of Iowa
Peeples, T. L., The University of Iowa
Jessop, J. L., The University of Iowa
Originally presented on: 11/8/2013 14:00:00 - 14:18:00

Raman spectroscopy is a technique that is seeing wider application for the analysis of biological samples because it offers the potential to distinguish cells in a non-destructive manner. Through measurement of vibrational, rotational, and other low-frequency modes in molecules, Raman spectroscopy enables rapid detection of chemical bond changes. Using confocal Raman microscopy, efforts are made to obtain three dimensional distributions of substances with high spatial resolution. For the purposes of this research, cellular and chemical changes occurring during the metabolic degradation of an environmental pollutant are traced with Raman spectroscopy. To allow for immobilization of contaminant, improved cell life, and increased horizontal gene transfer, organisms are grown as biofilms and examined with 532 and 785nm excitation lasers. Proposed hypotheses are: 1) Raman scattering can be used to identify and distinguish bacteria in the free cell state and in a biofilm and 2) Raman scattering can be used to evaluate the concentration of the substrate and metabolites in flow systems.

To create a model system *Pseudomonas* sp. strain ADP is grown in a nutrient limited medium with atrazine as its sole nitrogen source. *Pseudomonas* sp. strain ADP mineralizes atrazine through a series of hydrolytic reactions to carbon dioxide and ammonia. Along the degradation pathway five intermediates are produced: hydroxyatrazine, N-isopropylammelide, cyanuric acid, biuret, and allophanate. The accumulation and disappearance of these intermediates depend on the expression of atrazine pathway genes, which may be differentially expressed from the large atrazine degradation plasmid retained within the microorganism. For Raman analysis, microbial biofilms are grown in shake flasks and flow reactors under varying shear stress to allow for selection of optimal residence time required for cellular adhesion to atrazine.

Raman analysis of cells as well as atrazine and metabolites are being used to characterize the active biofilm. Success of this research will contribute to development of non-invasive techniques to examine both internal composition and spatial distribution of compounds present in biofilms. This will also lead to a firmer grasp of scientific fundamentals associated with pollutant degradation by biofilms, to improved remediation applications, and ultimately, to reduced pollutant-associated illness. Gaining insight to conditions required to create systems conducive for efficient transformation by biofilm may have many useful applications environmentally, industrially, and medically.

(804g) Engineering Virus-Based Nanoparticles As Adjustable Platforms for High-Density Site-Directed Decoration

Smith, M. T., Brigham Young University
Bundy, B. C., Brigham Young University
Originally presented on: 11/8/2013 14:18:00 - 14:36:00

Nanoparticle technologies become increasingly valuable as demand for their applicable technologies escalates in fields such as nanosensing, imaging, drug and therapeutic delivery, biocatalysis, and electronics. A fundamental difficulty of nanotechnology is controlling the traits of synthetic nanoparticles, such as monodispersity, assembly, stability, size, and morphology. One compelling group of candidates that may address these fundamental challenges is that of viruses. Viruses derive portions of their structures from peptide-based complexes. These complexes form nanoparticles that are often monodispere, self-assembling, symmetric and stable under general physiological conditions. Virus-based nanoparticles (VNPs) may be engineered to be noninfectious and innocuous while maintaining desirable traits. Engineering VNPs for useful applications typically requires chemical modifications and ligations. Standard conjugation techniques target residue side-chains and peptide termini, but remain limited by native-residue availability and lack of site-directed control of conjugation. Fortuitously, recent expansion of the peptide code has allowed for the site-directed addition of noncanonical amino acids that contain bioorthogonal reactive moieties as side chains. By marrying the VNPs with noncanonical amino acid incorporation, we have been able to circumspectly engineer VNPs as tunable platforms for ligand conjugation. Specifically, we have demonstrated the site-directed high-density decoration of Qβ bacteriophage-based VNPs with fluorescent probes and polyethylene glycols. Conjugation sites were designed for both external and internal decoration and the characteristics of each type were observed. Imminent applications of these particles will be in areas of bioimaging and single-particle tracking with promising future applications such as targeted drug delivery, biocatalysis, and nanometric biosensing.

(804h) A Smart Phone Fluorescent Basic Metabolic Panel

Awqatty, B.,
Samaddar, S.,
Cash, K. J., Northeastern University
Clark, H. A., Northeastern University
Dubach, J. M., Massachusetts General Hospital/Harvard Medical School
Originally presented on: 11/8/2013 14:36:00 - 14:54:00

Mobile, point-of-care diagnostics can be used to more efficiently and effectively deliver health care both in and outside the hospital. We have developed fluorescent particle sensors in combination with a handheld detection device and demonstrate the potential as a diagnostic system through measuring the complete basic metabolic panel. Miniaturization of the particle sensors to the nanoscale reduces the volume of needed sample and increases the response time. Fluorescence detection eliminates the need for electrical connection to the sensors and can be used easily with improving optical systems. The fluorescence sensors are based on optode technology and produce dual wavelength emission for quantitative measurements. Sensors to directly measure each of the ions in the panel were developed. Biosensors for the small molecules in the panel were created through a combination of optode particle sensors and enzymes specific for the small molecule of interest. The response of the sensors was used to measure the concentration of the basic metabolic panel components with a custom iPhone case that converts the phone into a dual wavelength fluorescence imager. Future development of this system could be used for a simple point-of-care diagnostic device.

(806a) Measuring Cell Mechanics By Optical Alignment Compression Cytometry

Roth, K. B., Colorado School of Mines
Eggleton, C., University of Maryland Baltimore County
Neeves, K. B., Colorado School of Mines
Marr, D. W. M., Colorado School Of Mines
Originally presented on: 11/8/2013 12:30:00 - 12:48:00

Cell mechanical properties are a useful indicator of cell health. Differences in cell deformability between disease states have been demonstrated in cases such as cancer, malaria, and sickle cell anemia using a variety of bulk and individual-cell measurement techniques. While bulk approaches provide average properties, small cell subpopulations go undetected. Individual cell testing techniques on the other hand, including micropipette aspiration and AFM, are capable of identifying single diseased cells but are time consuming. To increase throughput, we have developed a non-destructive cell deformability testing method, optical alignment compression (OAC) cytometry, that combines an extensional flow geometry with a linear optical trap to induce collisions between cells. To demonstrate this approach, we direct red blood cells (RBCs) using combined hydrodynamic and optical forces toward the geometric stagnation point within a microfluidic platform. Cell deformation occurs at the stagnation point during cell-cell collision due to hydrodynamic drag. We employ the Kelvin-Voigt viscoelastic model to quantify cell response and find expected elastic and viscous parameters for untreated RBCs of 8.62 ± 3.10 μN/m and 1.21 ± 0.69 μN s/m. Measured values of 17.1 ± 7.55 μN/m and 1.15 ± 0.75 μN s/m for glutaraldehyde fixed cells are also consistent with the expected increase in stiffness. To validate our technique, mixed populations were also tested, resulting in measured distributions fully consistent with expectations.

(806b) Quantifying the Contribution of Structural Proteins to Cell Mechanics With a Live Cell Monolayer Rheometer

Elkins, C.,
Shen, W. J.,
Khor, V.,
Kraemer, F.,
Fuller, G. G., Stanford University
Originally presented on: 11/8/2013 12:48:00 - 13:06:00

Cell mechanical properties depend primarily on the cytoskeleton, a multifunctional network which plays a key role in cell shape, motility, and intracellular transport. Additionally, the cytoskeleton is one of the first cell components to be modified by disease. Therefore a number of biological fields would benefit greatly from a device capable of rapidly measuring cell mechanical properties. However, most existing technologies for analyzing cell mechanics are either low throughput single-cell measurements or complicated by the presence of a three-dimensional cell culture gel. Our research addresses this need through the construction of a device capable of rapidly measuring the average mechanical properties of an entire cell monolayer. In this device, a confluent monolayer of cells cultured on collagen-coated glass is gently compressed by a coverslip lowered down to contact the monolayer. After a wait period to allow attachment of the cells to the top coverslip, this top plate is sheared while a sensitive force transducer measures the applied stress, thereby allowing rapid measurement of average mechanical properties across the cell monolayer. The entire apparatus is mounted on a DIC microscope, allowing visualization of the cell deformation during these mechanical measurements. We have applied this cell monolayer rheometer to study the relative impact of several cytoskeletal structural proteins on undifferentiated and differentiated adipocyte cell mechanics. Step-strain experiments using this device show differences in relaxation modulus in cells with disrupted microtubules and actin filaments, as well as cells knocked out for the intermediate filament vimentin. These results indicate that the cell monolayer rheometer may be used to quantify the contribution of various cytoskeletal structural proteins to cell mechanical properties in a way that is relatively high throughput and allows for cells to maintain cell-cell contacts and their adherent morphology throughout the experiment.

(806c) Modes of Cellular Tension Generation in Soft 3D Fibrin Matrices

Adhikari, A. S., Stanford University
Owen, L.,
Gupta, L., Stanford University
Leijnse, N., Stanford University
Dunn, A. R., Stanford University
Originally presented on: 11/8/2013 13:06:00 - 13:24:00

Modes of cellular tension generation in soft 3D fibrin matrices

Arjun S. Adhikari, Leanna M. Owen, Luv Gupta, Natascha Leijnse and Alexander R. Dunn

The ability of living cells to generate and detect mechanical forces plays a central role in diverse biological processes including wound healing, growth and development, and cancer metastasis. At present the vast majority of work characterizing cellular migration and force generation has examined cells on two-dimensional (2D) substrates. As a result, cellular mechanisms for both migration and extracellular matrix (ECM) remodeling in more realistic, three-dimensional (3D) environments remain poorly understood. Here we present a quantitative analysis of how primary fibroblasts, a canonical cell type that plays a central role in wound healing, mechanically deform 3D fibrin hydrogels. The fibrin matrix serves as a simple model for the ECM present at a wound site, and moreover is elastic up to 50% deformations, allowing us to equate matrix displacements with spatiotemporal changes in strain energies.

We used live-cell confocal imaging to simultaneously track cell migration and fibrin matrix distortion by human foreskin fibroblasts (HFFs) on timescales ranging from seconds to hours. Consistent with previous studies of fibroblastic cell types, HFFs generate spindle-like protrusions and contract the fibrin matrix in a dynamic manner. Our preliminary data suggest that myosin localization in the protrusion precedes local contraction of the matrix, and that protrusions deform the matrix in successive contract and release cycles, consistent with local mechanical feedback (Figure 1). Our data additionally yield the surprising observation that changes in cell volume correlate with global deformations of the surrounding matrix: 60% of cells (n = 18) underwent 10% or greater changes in volume in ~ 3 hours, and both increases and decreases in cell volume coincided with matrix deformation (Figure 2). Regulated changes in cell volume may thus constitute a presently overlooked mechanism for cell-driven matrix remodeling in 3D matrices and tissues. Ongoing work examines the role of mechanically activated signal transduction pathways in mediating matrix deformation, both at cell protrusions and in conjunction with cell volume regulation.

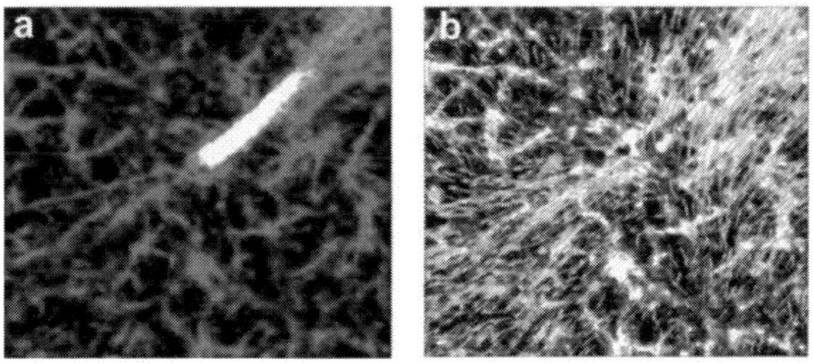

Figure 1. Matrix deformations in a fibrin gel due to protrusive activity of a fibroblast. (**a**) A fibroblast protrusion (GFP actin) in a fibrin matrix (red). (**b**) Vectors showing the matrix deformation. (image size: 33 x 29 µm)

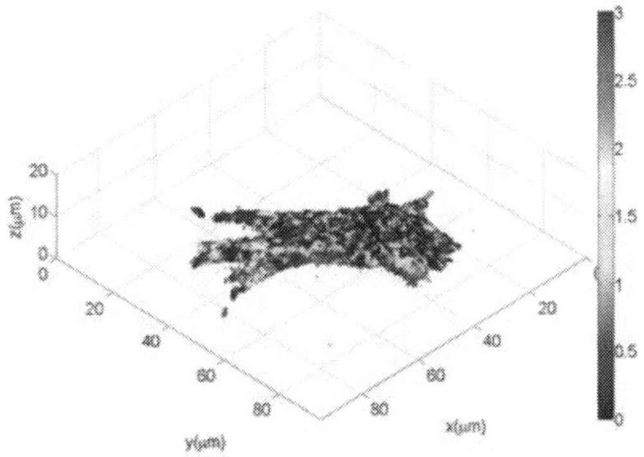

Figure 2. Matrix deformation due to volume change of a fibroblast embedded in a 3D fibrin matrix. The volume of the cell decreased by 10% over 90 minutes. The 3D deformations of the fibrin matrix were quantified using digital volume correlation (DVC) and mapped on the cell surface. Heat map shows the displacement of the matrix in microns.

(806d) Nuclear Structure and Reorganization in Endothelial Cell Responsiveness to VEGF

Dahl, K. N., Carnegie Mellon University
Spagnol, S. T., Carnegie Mellon University
Booth-Gauthier, E.,
Originally presented on: 11/8/2013 13:24:00 - 13:42:00

Mechanical force is found to be increasingly important during development and for proper homeostatic maintenance of cells and tissues. There are many known genome regulators in cells including a myriad of chemical signaling pathways. However, short lived signaling events are not sufficient to explain both the long time responses and permanent changes of cells. We are interested in chromatin organization in endothelial cells and the role of reorganization in enhancing angiogenesis. Our central hypothesis is that chromatin organization is related to gene expression and reorganization is correlated with changes in cellular phenotype.

We image changes in genome structures and quantify movements of the chromatin in live cells. We have decoupled impacts on chromatin movements induced by cellular motors from chromatin decondensation using well-established biological treatments on endothelial cells. We also observe changes in genome organization and movement in the presence of the angiogenic chemical vascular endothelial growth factor (VEGF). We find that early nuclear responses to VEGF appear to be motor induced, followed by chromatin decondensation.

Endothelial cells also respond to extracellular shear stress. The genome within the nucleus is a complex, self-assembled structure which has unique, shear-thinning viscoelastic properties capable of responding proportionally to both the magnitude and duration of applied stress. We observe a statistical change in nuclear rheology at 30 minutes after the application of shear stress. We find that the effects of VEGF and shear stress on the nucleus are additive, suggesting a functionality at the subcellular level.

Combined, these results provide insight into the mechanical regulation of gene expression through altered gene location. We suggest that chemical and mechanical signals are transduced through the cell in multiple ways and cellular structure can enhance chemical responses.

(806e) The Intracellular Mechanical Reponse of Mesenchymal Stem Cells to Oxidative Stress Used to Compare Vertebrate Animal Models of Stem Cell Homing

Rodriguez, K., Georgia Institute of Technology
Khambhati, N., Georgia Institute of Technology
Dawson, M. R., Georgia Institute of Technology
Originally presented on: 11/8/2013 13:42:00 - 14:00:00

THE INTRACELLULAR MECHANICAL REPONSE OF MESENCHYMAL STEM CELLS TO OXIDATIVE STRESS USED TO COMPARE VERTEBRATE ANIMAL MODELS OF STEM CELL HOMING

Kevin Rodriguez, Niti Khambhati, Michelle Dawson

Human mesenchymal stem cells (hMSCs) show great potential as cellular therapeutics. These adult stem cells can be isolated from bone marrow by adherence to plastic then differentiated into a variety of tissues. Natively, MSCs migrate in response to a gradient of soluble factors to sites of inflammation, such as wounds or tumors, where they engraft to aid in regeneration. Systemic infusion of engineered MSCs offers a minimally invasive method of delivering therapeutic genes to diseased tissues; however, transport limitations, which develop with extended culture, genetic modification, and stem cell expansion, reduce the homing efficiency of stem cell-based therapeutics. Extended culture may also result in spontaneous differentiation and cellular aging (senescence)[1]. In previous studies, the intracellular mechanical response of murine and human MSCs to soluble growth factors was characterized with multiple particle tracking microrheology [1, 2]. The cytoskeleton of murine MSCs homogeneously stiffened in response to these factors; whereas, human MSCs underwent a more local stiffening response, which was lost after a few passages. This inability of human MSCs to respond mechanically to soluble growth factors was correlated with loss of other stem cell functions. Murine MSCs are often used in characterizing the homing of MSC-based therapeutics to wound and tumor tissues; however, fundamental differences in mouse and human MSCs may limit their usefulness in these applications.

The aim of this study was to examine the intracellular mechanical response of MSCs isolated from different vertebrate animals to hypoxic conditions, used to stimulate the secretion of soluble growth factors that mediate cell migration. For these studies, MSCs were isolated from mouse, rat, pig, rabbit, and human bone marrow and cultured in hypoxic (2% oxygen) or normoxic (21% oxygen) conditions for at least 24 hours. Multiple particle tracking microrheology was used to characterize the intracellular mechanical properties. Immunostaining and fluorescence microscopy was used to characterize actin stress fiber densities and cell and nuclear shape factors. Colony forming assays were used to characterize MSC regenerative capacity, and differentiation assays were used to characterize multipotency. In normoxic conditions, rat, rabbit and pig MSCs displayed mechanical and morphological properties that were similar to human MSCs; however, only pig MSCs underwent the local stiffening response seen in human MSCs. For all species, reduced oxygen concentrations resulted in cytoskeletal stiffening, which was correlated with increased actin stress fiber densities and cell and nuclear elongation. These changes in cytoskeletal mechanics and morphology may be important in facilitating the migration of MSCs through in vivo transport barriers. These results may contribute to the development of new models for studying stem cell homing.

1. McGrail, D.J., K.M. McAndrews, and M.R. Dawson, *Biomechanical analysis predicts decreased human mesenchymal stem cell function before molecular differences.*Exp Cell Res, 2012.

2. McGrail, D.J., et al., *Differential mechanical response of mesenchymal stem cells and fibroblasts to tumor-secreted soluble factors.* PLoS One, 2012. 7(3): p. e33248.

(806f) Zyxin Morphology, Concentration and Arrangement Modulates Force Magnitude and Directionality in Nanofibrous Environment

Ng, C., Virginia Polytechnic Institute and State University
Nain, A. S., Virginia Tech
Koons, B., Virginia Tech
Originally presented on: 11/8/2013 14:00:00 - 14:18:00

Department of Mechanical Engineering, Virginia Tech, Blacksburg, VA

Figure SEQ Figure * ARABIC 1 shows two fibroblasts deflecting a fiber. A large variety of focal adhesion morphologies occur with their various attachments. (Scale Bar = 20micorns)

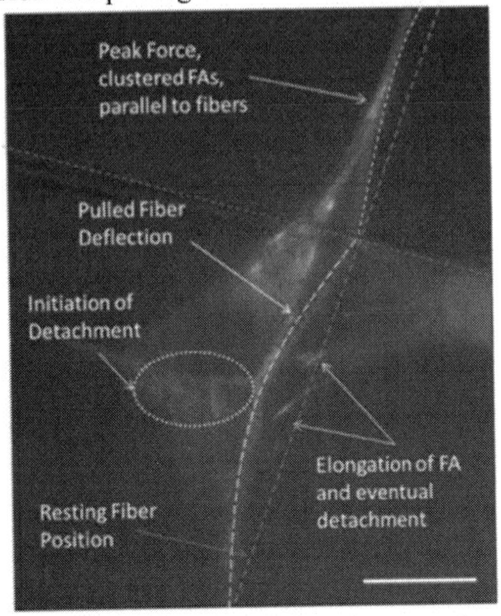

Figure 1 shows two fibroblasts deflecting a fiber. A large variety of focal adhesion morphologies occur with their various attachments. (Scale Bar = 20micorns)

Introduction: In the human body, cells attach to and receive biophysical cues from the surrounding fibrous microenvironment known as the extracellular matrix (ECM). These biophysical cues play an active role in cell morphogenesis, migration, adhesion, proliferation, differentiation and complex inflammatory cascades. However, the exact mechanism by which the cell interacts and responds to the ECM mechanistic environment is not well understood. Recent cellular force measurement techniques including pillars, cantilevers, gels etc., have shown unique focal adhesion morphologies with different force generating behaviors. In this study, we further this understanding by demonstrating the role of zyxin arrangement in force modulation in a suspended nanofibrous environment.

Materials and Methods: Using STEP (Spinneret based Tunable Engineered Parameters) fiber manufacturing platform, we are capable of measuring the forces of migratory cells along suspended fibers ranging from sub 100nm ? micron in diameter. Using an in-house solvent setup, multiple layer fibers are fused together at intersecting points, thus, creating a customized nanonet of fiber networks. In this system, the forces are easily measured by recording the deflection of fiber and using Euler beam mechanics in the elastic limit. Focal adhesion structure and arrangement are studied through time-lapse imaging of NIH3T3 fibroblasts tagged with zyxin and its relation to force.

Results: As shown in the Figure 1, a variety of unique focal adhesion arrangements are observed during force modulation of NIH3T3 fibroblast. Maximum force is exerted when the zyxin proteins align themselves with the deflecting fiber into highly concentrated clusters. Detachment is initiated when the adhesions disperse and unaligned themselves with the fiber. This continues until only a few elongated adhesions remain and eventually completely detach from the fiber.

Conclusion: Zyxin morphology, concentration and arrangement fuctions in determining force directionality and magnitude in a migratory cell in a suspended fibrous environment. Utilizing this knowledge of zyxin's special and temporal role in force modulation can be used to develop diagnostic and drug testing platforms.

(806g) The Dynamic Interplay Between Cell and Substrate Mechanical Properties

Kourouklis, A., Umass, Amherst
Lerum, R. V., University of Massachusetts Amherst
Bermudez, H., University of Massachusetts Amherst
Originally presented on: 11/8/2013 14:18:00 - 14:36:00

The engineering of polymer surfaces or matrices that are capable of controlling cell adhesion has been widely explored. In nearly all of these works, the polymer chains (and ligands) are chemically attached to the underlying substrate, and therefore these systems are inherently static. By contrast, cellular environments such as the extracellular matrix are dynamic and remodeled by biochemical reactions and biophysical forces. Borrowing this concept from Nature, we created polymer films by an interfacial self-assembly process, whereby individual (ligand-modified) chains exhibit lateral mobility. Furthermore, we can simultaneously tune the viscoelastic properties of the films. NIH 3T3 fibroblasts seeded on these films exhibit biphasic responses in spreading and adhesion strength as a function of ligand lateral mobility, with a minimal response for intermediate mobility values. These results establish that ligand mobility is a significant, but thus far overlooked, factor in material-induced cell responses. At the same time, cell spreading showed a weak decreasing response against the film storage modulus, which is in contrast to prior works using densely crosslinked hydrogel networks. We believe that these different trends reflect differences in molecular properties (lateral mobility) and continuum properties (storage and loss moduli), but must also consider cellular feedback response caused by the material itself. To investigate material-induced changes in cell behavior, we are currently exploring the effects of blocking either avb3 or a5b1 integrins and the consequences on cell spreading and focal adhesion (FA) dynamics. We are also studying cell mechanical properties by pharmacologically-induced de-adhesion assays. Previous works have shown a good correlation between de-adhesion and the viscoelastic properties of cells. Our work introduces the fabrication of novel materials whose physical properties more closely represent the dynamic interaction between cells and their extracellular environment. Combined characterization of both molecular and continuum properties, of both cells and their substrates, will give new insight as to how cells sense and respond to their ever-changing surroundings.

(806h) Insights Into Cellular Mechanics of Embryonic Stem Cells

Lim, M., Imperial College London
Teo, A.,
Weihs, D.,
Originally presented on: 11/8/2013 14:36:00 - 14:54:00

Embryonic stem cells (ESCs) have the ability to differentiate in response to mechanical properties of its extracellular microenvironment. Several studies have demonstrated this ability, but little is known about its cellular mechanics in response to changes that occur during cell development. Herein we demonstrate a method in traction force microscopy that provided a means for us to quantify and track cellular mechanics during development stages of ESCs such as proliferation, embryoid body formation and differentiation. Polyacrylamide gels prepared on activated coverslips were conjugated with gelatin and had a stiffness of 4200 Pa. Polyacrylamide gels contained 200 nm fluorescence beads that were used for traction analysis. Mouse ESCs were seeded onto PAA gels and monitored after 1, 2 and 3 days. Temporal changes in the cellular mechanics of ESCs during cell attachment, proliferation and embryoid body formation were monitored and quantified. Using this method, we are able to quantify and show both the spatial and temporal variations in cellular forces of ESCs in embryoid body clusters.

(815a) Physical Modeling of Chromosome Segregation in E. Coli Reveals Impact of Force and DNA Relaxation

Lampo, T., Stanford University
Kuwada, N.,
Wiggins, P.,
Spakowitz, A. J., Stanford University
Originally presented on: 11/8/2013 12:30:00 - 12:47:00

The physical mechanism by which *Escherichia coli* segregates its chromosomes for partitioning into daughter cells is unknown. In several other species of bacteria, the action of a segregation force is responsible for pulling the newly replicated chromosomes apart from the nascent strand. This suggests that such a force may play a role in *E. coli* chromosome segregation. In our previous work, we developed a Rouse polymer model in a viscoelastic medium that accurately describes the equilibrium motion of a chromosomal locus. In this work, we extend our model to examine the effects of a segregation force acting on a single monomer on the polymer dynamic behavior. We then compare our predictions for the mean displacement and velocity autocorrelation to the motion of fluorescently labeled *oriC* loci in *E. coli* during segregation. The *oriC* locus is the origin of replication and is among the first regions to have copies moved to opposite sides of the cell. The mean displacement scales as a power law in time as t □, where □≈0.3-0.4, and shows a positive velocity autocorrelation. Our model predicts that the mean displacement of an *oriC* locus scales as $t^{0.35}$, and that the motion has a positive velocity autocorrelation due to the action of the segregation force. Thus, the motion of the *oriC* locus is consistent with the action of a segregation force, though the source of this force is still a mystery.

(815b) Biomog: A Computational Framework for the Modification Or De Novo Generation of Genome-Scale Biomass Compositions

Tervo, C. J., University of Wisconsin - Madison
Reed, J. L., University of Wisconsin-Madison
Originally presented on: 11/8/2013 12:47:00 - 13:04:00

The success of genome-scale modeling is contingent on a model's ability to accurately predict growth behaviors following perturbation. To date, most research aimed at improving model predictions has focused upon the development of mechanisms to modify the cell's metabolic and genetic networks by adding or removing reactions and refining gene to protein to reaction maps (GPRs) while little focus has been directed towards developing systematic methods of proposing, modifying and interrogating an organism's biomass composition. To address this gap, the biomass modification and generation (BioMog) framework was created and used to generate biomass compositions *de novo,* as well as modify predefined biomasses for models of *Escherichia coli* (iJO1366) and of *Shewanella oneidensis* (iSO783) from high throughput growth phenotype and fitness datasets. This is accomplished by determining production blocked metabolites (i.e., metabolites that cannot be produced by a particular cellular mutant in a given minimal medium). By determining these blocked metabolites for each experiment performed, one can populate lists that support a particular metabolic candidate's inclusion or exclusion from a cell's biomass composition. Using this approach, BioMog's *de novo* biomass proposals have been able to independently recapitulate up to seventy percent of the predefined biomass components and outperformed their predefined counterparts at qualitatively predicting growth phenotypes by up to five percent. Additionally, the BioMog procedure allows for the quantitative assessment of a particular metabolite's essentiality to a cell. Such measures facilitate the determination of erroneous experiments, and inaccurate reaction network and GPR structures. Additionally, the BioMog framework contains an experiment generation component that allows for the directed design of informative experiments. Using such an approach, we correct experimental results relating to the essentiality of *thyA* gene to *E. coli*, as well as, perform phenotype experiments supporting the essentiality of protoheme.

(815c) Experimental and in Silico Analysis of Competitive Metabolic Network Structure-Function Relationships

Folsom, J. P., Montana State University
Carlson, R. P., Montana State University
Originally presented on: 11/8/2013 13:04:00 - 13:21:00

Experimental and theoretical analyses of *Escherichia coli* cultured under nutrient limiting conditions were used to evaluate ecologically competitive metabolic network structure-function relationships relevant to designing bioprocesses or controlling problematic pathogens. An *in silico* analysis of central metabolism was used to identify resource allocation strategies that maximize fitness through molecular-level tradeoffs in pathway resource requirements and metabolic efficiency. The in silico predictions were tested using physiological and proteomic data collected from *E. coli* cultures grown using 12 distinct nutrient-limited chemostat conditions. Consistent with theory, experimental data demonstrated that under iron-, nitrogen-, or carbon-limited growth conditions *E. coli* regulates its metabolism to invest the limiting resource competitively at the cost of other resources like for instance electron donor. The economics-based analysis quantified a continuum of resource allocation strategies that permit competitive cellular functioning under gradients of nutrient scarcities. The results provide a theoretical basis for adaptive optimality as well as metabolic simplicity, and additionally suggest engineering blueprints to redirect metabolism for industrial applications.

(815d) Biofilm Formation During Spaceflight: Insights From Mathematical Modeling of Growth in the Microgravity Environment

Kim, W. S., Rensselaer Polytechnic Institute
Plawsky, J. L., Rensselaer Polytechnic Institute
Collins, C. H., Rensselaer Polytechnic Institute
Originally presented on: 11/8/2013 13:21:00 - 13:38:00

There is an urgent need to understand the effects of microgravity on microbial biofilm formation and the factors that affect the potential for biofilm-related infections, biofouling and biocorrosion during spaceflight. Microbial growth and physiology have been studied extensively in spaceflight and simulated microgravity and many changes have been observed. We have recently conducted two sets of spaceflight experiments (Micro-2 and Micro-2A) and shown the biofilm formation by two opportunistic pathogens, *Pseudomonas aeruginosa* and *Staphylococcus aureus*, is affected by spaceflight. However, while *P. aeruginosa* biofilm formation was found to increase, *S. aureus* biofilm formation was found to decrease during spaceflight compared to ground controls conducted on Earth. In order to understand this descrepancy, as well as our observation that the differences can be modulated by nutrient availability, we have constructed a mathematical model of microbial growth and biofim formation during sapceflight. In addition to gravity, which causes the cells to settle in normal gravity or is omitted in microgravity, the model includes Monod type equations for growth of biofilm and planktonic fractions, a Langmuir isotherm-type binding constant that captures the cell's affinity for a particular surface. Through sensitivity analysis we have identified several key parameters that can affect biofilm formation, either positively or negatively, in microgravity.

(815e) The Toxin Antitoxin Persistence (TAP) Model for Revealing Insights Into Persister Cell Formation in Escherichia Coli

McAnulty, M. J., Virginia Tech
Wood, T. K., Pennsylvania State University
Originally presented on: 11/8/2013 13:38:00 - 13:55:00

Toxin-antitoxin (TA) systems consist of a toxin which is always a protein and an antitoxin which is a protein or mRNA. Many TA systems are active in biofilms and serve to convert cells to a dormant state by reducing metabolism. While in this "persister" cell state, bacteria persevere through certain environmental stresses, such as those incurred by the presence of antibiotics, without attaining resistance through an altered genotype. The exact mechanisms behind persister cell formation remain unclear, so a computational model has been constructed in an attempt to reveal key insights behind this process. Two previously described TA systems (MqsR/MqsA and GhoT/GhoS) are included in the model to explore how they interact with each other and their roles in persistence.

The MqsR/MqsA TA system is of the type II class, in which a labile antitoxin protein (MqsA) binds to a more stable toxin protein (MqsR), blocking its toxic effect in the process (*Appl. Microbiol. Biotechnol.* 64:515, 2004; *J. Bacteriol.* 188: 305, 2006; *PLoS Pathogens* 5:e1000706, 2009). The toxin MqsR is an endoribonuclease that specifically cleaves at 5'-GCU sites, resulting in enhanced degradation of the total cellular mRNA pool, with the exception of 14 mRNA lacking 5'-GCU sites (*Environ. Microbiol.*, 2012, on-line). While the antitoxin MqsA binds to MqsR to inhibit its activity, it also represses transcription from its own operon (*mqsR* and *mqsA* are co-transcribed). During conditions of stress, MqsA is degraded more rapidly from the enhanced action of certain proteases such as Lon (*Nat. Chem. Biol.* 7:359, 2011), thus derepressing the *mqsRA* promoter and increasing MqsR concentrations. Furthermore, the *mqsR* portion of the transcript has only one 5'-GCU site, while the *mqsA* portion has three, leading to the preferential cleavage of the *mqsA* portion by MqsR. The one 5'-GCU site found in the *mqsR*portion may give MqsR negative feedback on its own expression, somewhat decreasing the risk for attaining MqsR concentrations that lead to cell death.

Production of MqsR has been shown to increase the total percentage of persister cells in an *Escherichia coli* population, while deletion of *mqsR* has the opposite effect (*Biochem. Biophys. Res. Commun.* 391:209, 2010); this was the first time that inactivation of a toxin reduced persistence. Production of a version of MqsR that was engineered to be more toxic, MqsR 2-1, increases persistence by decreasing cellular fitness (*Microb. Biotech.* 5:509, 2012). Hence, MqsR not only mediates persistence by enhancing the degradation of the total cellular mRNA pool, but also by not cleaving specific mRNA encoding other genes that are also needed for persistence. One of these mRNA's is that encoding *ghoT*, whose translated product is another toxin whose exact role in persistence has been uncertain until now.

The GhoT/GhoS toxin-antitoxin system is the first example of the type V class, in which the antitoxin protein (GhoS) specifically cleaves the toxin-encoding mRNA (for *ghoT*) (*Nat. Chem. Biol.*8:855, 2012). Like MqsR, production of GhoT increases persistence (*Nat. Chem. Biol.*8:855, 2012). Unlike MqsR, GhoT is not an endoribonuclease, but is a small hydrophobic protein that damages the membrane, leading to the formation of "ghost" cells, when it is overproduced (*Nat. Chem. Biol.*8:855, 2012). The *ghoS* portion of *ghoST* mRNA transcript has two 5'-GCU sites, while the *ghoT* portion has none; hence MqsR cleaves only the *ghoS* portion of the transcript, enriching the *ghoT* portion. GhoT concentrations then become heightened during times of elevated MqsR activity. Hence, the GhoT/GhoS TA system is the first example of a TA system controlled by another TA system (MqsR/MqsA) (*Environ. Microbiol.*, 2012, on-line).

The computational model constructed here to simulate the formation of persister cells via the action of the MqsR/MqsA and GhoT/GhoS TA systems is deterministic and includes over 20 ordinary differential equations. Changes in growth rate due to enhanced toxin activity is taken into account, and persistence caused by toxin-antitoxin activities can be seen as reduced metabolic activity. Previously, we demonstrated conclusively that persister cells have reduced metabolism (*Antimicrob. Agents Chemother.* 57:1468, 2013). The model assumes that the overall protein concentration within a cell is constant, so that cells manipulate their growth rate to ensure it stays constant. Heightened MqsR activity leads to a shorter half-life of the combined mRNA pool within the cell. By assuming that total cellular translation activity is proportional to the total mRNA concentration, MqsR reduces the growth rate. GhoT is modeled as reducing translation rates indirectly. It is assumed to have the same toxic effect as that of the TisB toxin, which acts as a pore specific for channeling small anions. By this mechanism, hydroxyl ions can exit the cell at increased rates and react with protons outside of the cell. They thus have the same virtual effect as allowing the diffusion of protons into the cell as the pH of the cytoplasm equilibrates to respond to the loss of hydroxyl anions. Toxins like these then act to decouple the electron transport chain, and at the same time, the cell must find a way to keep its cytoplasmic pH neutral if it is in a more acidic environment, while its buffering capacity is finite. It may be possible that cells use ATP synthases in reverse, along with a variety of other acid stress responses, to combat this problem. The drop in ATP levels leads to a drop in biosynthetic reactions, with the most major being that of translation. This whole effect is complicated by the fact that *E. coli*manipulates glycolytic (and possibly overall catabolic) activity depending on its demand for ATP. The cell does not take up more carbon to refill ATP that has been lost to increased GhoT activity since it has an upper limit for its catabolic activity, especially at the onset of persistence, when its biosynthetic capacities for increasing overall catabolism by regulation of transcription/translation are hindered. With this assumption, ATP concentrations stabilize at a lower level upon heightened activity of GhoT.

The ordinary differential equations presented in this model allow us to predict the concentration of chromosomal DNA bound or unbound by MqsA, of total and specific mRNA pools, of specific proteins, and of energy-related metabolites (NADH and ATP), with the production of each protein depending on the concentration of its mRNA transcripts and also on translation rates. Two other equations are incorporated to indicate changes in growth rate and translation rates. The metabolites simulated, NADH and ATP, are major molecules produced by cellular catabolism, and their concentrations exert an influence on translation rates. The effects of each toxin on each variable (including growth and translation rates) are visualized upon the onset of a stress condition, such as introducing enhanced proteolytic activity for the degradation of MqsA.

By including the effects of two toxin-antitoxin systems that are known to interact with each other, simulations performed by this model suggest why a cell would have more than one toxin-antitoxin system, if the action of just one is needed to send it into a state of persistence. Unlike the TisB/TisA TA system, which does not depend on other TA systems to enhance activity of the toxin, the GhoT/GhoS TA system relies on MqsR levels for enhancing GhoT concentrations under physiological conditions (*Environ. Microbiol.*, 2012, on-line). Simulations have revealed that the action of GhoT is needed to lower translation rates by reducing ATP levels, decreasing the maximum concentration of MqsR reached and reducing the risk of too much MqsR being produced at the onset of persistence. The GhoT/GhoS system thus acts as another negative feedback mechanism for the MqsR/MqsA TA system, allowing the cell to eventually awaken from a persister state instead of being stuck in a comatose condition. Such a finding is non-intuitive, as the action of a toxin is needed to subdue the action of another toxin. Experimental data to verify the model include showing that, due to its enhanced stability, the effect of production of MqsR 2-1 on persistence is even more so dependent on the action of GhoT, compared to the production of MqsR.

(815f) Analyzing Population Dynamics of Embryonic Stem Cell Cycle Transition Through An Integrated Experimental and Modeling Approach

Task, K., University of Pittsburgh
Koubaa, O.,
Barner, S.,
Banerjee, I., University of Pittsburgh
Originally presented on: 11/8/2013 13:55:00 - 14:12:00

Embryonic stem cells (ESC) have the potential to be used in many therapeutic applications due to their ability to expand in an undifferentiated state while being able to differentiate to any cell type under specific conditions. One important characteristic vital to this behavior is the cell cycle. In somatic cells, the cell cycle is characterized by a lengthy G1 (growth) phase and a long doubling time. ESC are typically characterized by a shortened G1 phase, with an increased proportion of cells residing in the S phase (DNA replication). During differentiation, there is a gradual transition from the ESC cycle to mature cell cycle. However, limited work has been done in the area of dynamics of cell cycle transition from undifferentiated to mature cells. Our objective is to better understand this transition by tracking the cell cycle dynamics of a differentiating population of hESC. In this project we are using an integrated experimental and computational approach to investigate cell cycle traits of pluripotent stem cells (PSC), both at the self-renewal stage and during direct differentiation towards pancreatic lineage.

An important component in the analysis of this system is the heterogeneity in cell population and cell cycle state of each individual cell. We developed a stochastic population modeling framework to predict single cell level information from population based information gathered from flow cytometry data. This model is a cellular automaton model, with the state of the cell being defined by the cell cycle phase which the cell is in (G1, S, or G2/M). The residence times for phases are described by probability distributions and their moments. Individual cells are tracked with time, and in this way the population dynamics are simulated. This model was used with experimentally derived cell synchronization data. Two different PSC lines, H1 hESC and Y1-16 iPS, derived from dermal fibroblasts, were synchronized in the G2 phase with Nocodazole and released back into the cell cycle. Information on the subsequent dynamics of cycle phase distribution was collected via DNA stain (propidium iodide) and flow cytometry. Having developed the model for self-renewal of hESC, we extended it to a differentiating population. Specifically, we directed differentiation towards pancreatic lineage, and experimentally analyzed the cell cycle dynamics during differentiation in addition to the synchronization behavior of the pancreatic progenitor phenotype.

Our population model framework is able to accurately predict phase resident time distributions from experimentally synchronized hESC. PSC synchronization revealed low population heterogeneity in cell cycle transition rates as shown by limited desynchronization of the cycle phase distribution. This data was compared to the population model, and through this comparison probabilistic parameters and characteristics associated with ESC were determined. Only one combination of specific distributions and their associated moments, representing phase residence times, were able to recapitulate the experimental data, and these parameters led to very good agreement between the simulations and experimental dynamics. To validate the model, we performed synchronization/BrdU and CFSE to experimentally measure mean phase and doubling times, and show that these are in very good agreement with those predicted in the model. The validated population model was next extended to induced hESC differentiation towards pancreatic lineage. Differentiation results in changes in phase residence time distributions, with the most prominent effect being an increase in the G1 population.

Together, the population model and synchronization experiments are able to extract single cell information on cycle phase times and variability from simple population analyses. Furthermore, the model is able to predict cell cycle information and describe self-renewal and differentiation behavior to high accuracy. The cell cycle information obtained by applying this methodology can aid in better understanding ESC and help in deriving functional, non-tumorigenic mature cells from a progenitor population for use in therapeutic applications, specifically insulin producing β-cells for diabetes treatment.

(815g) Phenotypic Heterogeneity in the Epithelial to Mesenchymal Transition (EMT) Is Controlled By the NFAT and Phosphorylated-Sp1 Transcription Factors

Gould, R. A., Cornell University
Chakrabarti, A., Cornell University
Butcher, J., Cornell University
Varner, J., Cornell University
Originally presented on: 11/8/2013 14:12:00 - 14:29:00

The epithelial to mesenchymal transition (EMT) is important in development and pathological processes such as fibrosis and cancer. Several extracellular signals trigger EMT, central amongst these are soluble transforming growth factor β (TGFβ) family members. TGFβ isoforms are potent inducers of EMT in both embryonic and pathological conditions. Receptor mediated signaling in response to TGFβ triggers a program that ultimately represses the expression of epithelial genes such as E-cadherin, while simultaneously activating the expression of mesenchymal genes such as Vimentin. The EMT signaling program is complex, containing structural features such as redundancy, feedback and crosstalk. While these architectural features promote robustness, they complicate the understanding of signal flow and rational signal reprogramming. In this study, we developed a dynamic network model of TGFβ induced EMT signaling. Our model used mass action kinetics within an ordinary differential equation (ODE) framework to describe the EMT program initiated by TGFβ isoforms. The interaction network contained 995 protein or mRNA components interconnected through 1700 interactions. A family of model parameters (1700 kinetic constants and 56 non-zero initial conditions) was estimated using 41 protein signaling data data sets generated in DLD1 colon carcinoma, MDCKII and A375 melanoma cells using the Pareto optimal ensemble technique (POETs) multiobjective optimization algorithm. POETs identified more than 15,000 likely TGFβ/EMT models, from which we selected a low-correlation population of approximately 1000 models for analysis. Analysis of the model population suggested that both MCF10A and DLD1 cells could exhibit phenotypic heterogeneity if treated simultaneously with TGFβ2 and VEGFA. Furthermore, the model population predicted that the NFAT and phosphorylated Sp1 transcription factors were key elements driving this heterogeneity. This heterogeneity was characterized by a significant fraction of the population being in a "mixed state" having both high E-cadherin and high Vimentin expression. We tested these predictions experimentally using qRT-PCR and flow-cytometry studies in the presence and absence of NFAT inhibitors. These studies verified that upwards of 45% of the cellular population could be put into the mixed state in the presence of both TGFβ2 and VEGF. Moreover, this response depended upon both the MAPK activation of Sp1 and NFAT action (consistent with the molecular signaling predicted by the model). Lastly, we found that the heterogenous population had significantly different functional behavior at the cellular level. For example, the extent of ductal branch formation significantly increased with MCF10A cells in the mixed phenotype, compared to cells treated with VEGFA alone. Together, these results establish a predictive mechanistic model of EMT susceptibility, and reveal a novel signaling axis between phosphorylated Sp1/NFAT which could possibly regulate carcinoma progression through an EMT versus tubulogenesis response.

(815h) Agent Based Models of Heterogeneous Tumor Cell Populations

Yu, J.,
Bagheri, N., Northwestern University
Originally presented on: 11/8/2013 14:29:00 - 14:46:00

Despite significant improvements in treatment and understanding, cancer continues to grow at an alarming pace, accounting for 1 in 8 deaths worldwide. It is important to identify the mechanisms that control cancer growth, understand how these mechanisms interact, and discern how to manipulate these mechanisms in order to develop treatments. Models act as an invaluable tool for this, given the expensive and time-consuming nature of in vivo/in vitro experimentation.

A significant focus of researchers has been on theoretical, mathematical based models. These strongly deterministic models are fundamental in understanding the basic, physical mechanisms of cancer. These theoretical models, however, fail to point to a single, conceptual framework in which data can be fitted. Given the rapid improvement in "wet lab" tools, it is critical to develop a model that can make use of the large amounts of this high-resolution data.

Cellular automata (CA) has been used to describe the growth of tumor cells using the concept of emergence, which suggest that complex behavior can develop from a simple set of fundamental rules. CA is a more realistic model, where changes in the state of a grid cell depend on a basic set of rules and the states of neighboring cells. The biggest drawback of cellular automata is its static nature. CA cells have a fixed position on the grid environment and only change state, not location, over time.

We propose to apply agent based models (ABM) to study emergent behavior of heterogeneous tumor cell populations. ABMs incorporate dynamic agents that can change both state and location over time based on a set of rules. We intend to use ABMs to model tumor growth based on rules suggested by two hypotheses: the cancer stem cell hypothesis and the clonal evolution model. The cancer stem cell hypothesis suggests that there are certain cancer stem cells that can replicate indefinitely and differentiate into all the cell types in a tumor. It is the migration, and subsequent colonization, of these cancer stem cells leads to metastasis. The clonal evolution model suggests that over time, cancer cells acquire various combinations of mutation. Tumor progression is driven by the natural selection of the most aggressive cancer cells. Metastasis results when the tumor cells have "acquired" the mutations needed to become invasive.

We hope that studying the emergent behavior of these two models and their alignment with experimental data will help clarify the mechanisms by which cancer develops and metastasizes.

(815i) Quality Assessment of Ischemic Livers for Transplantation

Perk, S., Center for Engineering in Medicine, Massachusetts General Hospital, Harvard Medical School, Shriners Hospital for Children
Luitje, M., Center for Engineering in Medicine, Massachusetts General Hospital, Harvard Medical School, Shriners Hospital for Children, Boston, MA
Izamis, M. L., Center for Engineering in Medicine, Massachusetts General Hospital, Harvard Medical School, Shriners Hospital for Children
Bruinsma, B., Center for Engineering in Medicine, Massachusetts General Hospital, Harvard Medical School, Shriners Hospital for Children
Uygun, B., Center for Engineering in Medicine, Massachusetts General Hospital, Harvard Medical School, Shriners Hospital for Children
Yarmush, M. L., Rutgers University and Massachusetts General Hospital
Uygun, K., Center for Engineering in Medicine, Massachusetts General Hospital, Harvard Medical School, Shriners Hospital for Children
Originally presented on: 11/8/2013 14:46:00 - 15:03:00

~4,000 people die every year in the United States while waiting for liver transplantation. ~5,000 marginally damaged livers are discarded due to fear of graft failure. Recent studies on liver preservation focus on new and dynamic modalities such as machine perfusion to recover discarded donor organs to a functional state. Machine-perfusion provides oxygen and nutrition enabling recovery of the organ and is an advantageous alternative to the current gold standard 'cold storage'; however, machine-perfusion and its key factors need to be thoroughly studied for it to become a reality in clinical settings. Two such key factors are determination of biomarkers for graft viability during perfusion and prediction of the likelihood of graft survival after transplantation.

In humans, cardiac death frequently occurs in uncontrolled environments (uDCD). In the absence of cardiac output, ischemic damage to the organ increases in severity as a function of time. Hence, without objective metrics of ischemic duration and organ viability, uDCD organs cannot be safely transplanted.

The aim of this work is the development of an approach that determines the biomarkers for the assessment of uDCD ischemic damage and liver viability during perfusion, and the prediction of clinical outcome post-transplantation.

Machine-perfusion is a bio-process that allows real-time sampling from the graft. The samples are analyzed for blood gases, amino acids, glucose, lactate, and urea. The liver quality at the end of perfusion can further be evaluated via performance indicators such as liver function tests, cell isolation or ATP assays. For the identification of biomarkers to assess liver viability from a large dataset with dynamic trends during or at the end of perfusion, a suitable methodology is statistical process monitoring (SPM) based on partial least squares (PLS) and its extension for dynamic processes: multi-way PLS (MPLS).

MPLS constructs a statistical model of liver dynamic response to perfusion, which enables evaluation of the dynamic patterns of graft function during perfusion and comparisons of unknown uDCD livers' qualities to those of known quality and infer about their ischemic conditions in real-time. Furthermore, MPLS' in-silico predictions based on the liver function trajectory during perfusion, provide an estimate for the likelihood of graft success to the surgeon as a decision support system for transplantation.

This work presents a novel statistical approach for the assessment of ischemic damage in uDCD rat livers and prediction of likelihood of graft survival post transplantation, based on samples taken during perfusions of rat livers for 4 hours and were analyzed with HPLC for amino acids and biochemical assays for lactate, urea, glucose, ALT, and AST.

(820a) Isolation of Fully Synthetic Promoters for Gene Expression in Corynebacterium Glutamicum

Jeong, K. J., KAIST
Yim, S. S., KAIST
An, S. J., KAIST

Originally presented on: 11/8/2013 12:30:00 - 12:48:00

Corynebacterium glutamicum is an important industrial organism that is widely used in the production of amino acids, nucleotides and vitamins. To extend its product spectrum and improve productivity, *C. glutamicum* needs to undergo further engineering, including the development of applicable promoter system. Here, we developed new synthetic promoters with various strengths in *C. glutamicum*. First, fully synthetic promoter library was constructed in which 70-bp length of promoter regions were randomized and green fluorescent protein (GFP) was used as a reporter protein. From promoter library, twenty highly fluorescent cells were isolated by FACS and their promoters were characterized through extensive analysis of DNA sequences and mRNA transcripts. Among twenty promoters, six promoters which have different strengths were selected and it was clearly confirmed that the levels of gene expression under 6 promoters have high correlation with the strengths of promoters. This is the first report of synthetic promoters constructed in *C. glutamicum,* and our screening strategy together with the use of synthetic promoters of various strengths will contribute to the future engineering of *C. glutamicum*

(820b) Controlling Transcriptional Regulation of a Secretion System for Increased Recovery of Folded Heterologous Proteins From the Culture Media

Metcalf, K. J., University of California, Berkeley
Finnerty, C., University of California, Berkeley
Tullman-Ercek, D., University of California, Berkeley
Originally presented on: 11/8/2013 12:48:00 - 13:06:00

Secretion of heterologous proteins in a bacterial culture allows for simplified purification, compared to intracellular expression. However, there is often accumulation of the protein targeted for secretion in the cytosol, resulting in lower yields of protein in the culture media. Previous studies indicated that low native gene expression of a type III secretion system (T3SS) may be limiting titers of secreted protein in this system. We report the engineered expression of a transcription factor, HilA, that transcriptionally regulates a native T3SS from *Salmonella enterica* for high titer yields of diverse heterologous proteins. HilA is a transcriptional activator that has direct and downstream control of the expression of over five transcriptional units that are associated with the T3SS. We characterized the changes in gene expression and protein secretion as a result of the tuned expression of HilA and were able to achieve greater than a 10-fold increase in secreted protein titer. These data demonstrate the use of transcriptional control to achieve a desired performance from a native genetic device.

(820c) Human G Protein-Coupled Receptor Signaling â€“ Engineering Yeast As Reporters of Interacting Pathways

Robinson, A. S., University of Delaware
McNeely, P. M., University of Delaware
Originally presented on: 11/8/2013 13:06:00 - 13:24:00

G Protein-Coupled Receptors (GPCRs) act as members of a large signal transduction protein family, and are targets of approximately 40% of all drug discovery efforts. GPCRs bind ligands at the cell surface and transmit that signal to the cell via dissociation of the heterotrimeric G-proteins and initiation of a downstream signaling cascade that can result in a change in protein production or cell metabolism. Multiple receptor subtypes, receptor promiscuity, and common signaling relays following GPCR activation complicate the signaling process in the native mammalian tissue, and convolute our understanding of the action of potential therapeutic ligands. In contrast, the relative simplicity of the yeast proteome allows the study of human GPCR signaling in the yeast cellular environment, providing a eukaryotic context with reduced crosstalk. In contrast to human cells, yeast posses very few GPCRs, yet the signaling pathways are analogous to those of humans. By engineering the native yeast GPCR-modulated mating pheromone response, it is possible to deconstruct human receptor behavior in this less complex environment.

In this study, we have expressed recombinant human GPCRs involved in neurological and cardiological functions, the adenosine A_{2A} and A_{2B} receptors, in an engineered strain of *Saccharomyces cerevisiae.* To facilitate signal measurement, the endogenous mating pheromone promoter was fused to mCherry to enable quantitation of GPCR signaling via fluorescence measurement. In addition, GPCRs were fluorescently tagged to enable studies of co-localization, and effect of A_{2A} and A_{2B} receptor-specific ligands on pathway activation. These results will be presented in light of the growing discussion of heteroligomerization in signal activation.

(820d) Quantitative Intracellular Analysis of the Cucumber Mosaic Virus Chemically Inducible Viral Amplicon Expression System

Rattanaporn, K.,
Hwang, M. S., UC Davis
Falk, B. W., University of California, Davis
McDonald, K. A., University of California, Davis
Originally presented on: 11/8/2013 13:24:00 - 13:42:00

Transient production of heterologous proteins in plant tissues is gaining importance as a commercially-relevant production strategy due to its inherent speed, scalability, and ability to produce complex glycoproteins. Fast and efficient protein expression is essential for cost effective production of biologics in plants. Plant viral based expression systems provide rapid high-level expression through the viral replication mechanism. In the novel chemically inducible viral amplicon expression system referred to as **CMViva** (*Cucumber Mosaic Virus* inducible viral amplicon) the open reading frame of the CMV coat protein is replaced with the plant codon optimized gene for the target product, which along with other modifications of cDNA of the CMV genome, ensure that infectious CMV virions are not generated. Genetically engineered *Agrobacterium tumefaciens* harboring the CMViva expression system is delivered into harvested *Nicotiana benthamiana*leaf tissue via vacuum agroinfiltration. A viral RNA gene silencing suppressor (GSS) driven by the CaMV 35S promoter is co-expressed to improve production of the target.

In this study the target product is alpha-1-antitrypsin (AAT), a labile 52kD human glycoprotein. People with AAT genetic disorder lack functional AAT in their bloodstream and lungs, leading to early onset emphysema which often requires weekly AAT replacement therapy. Plasma derived AAT is currently the only FDA approved source, however due to production constraints, safety concerns and cost, alternative recombinant production methods are being investigated.

The absolute RNA copy numbers of CMV ORFs, AAT and GSS transcripts is determined using Real-Time Quantitative Reverse Transcription Polymerase Chain Reaction (real time qRT-PCR). The levels of CMV RNA 1, RNA 2, RNA 3, CMV 2b, AAT and GSS mRNAs are measured and compared under non-induced and induced conditions. The mRNA and protein expression levels of rAAT are also compared to those obtained with the constitutive CaMV 35S system. Using the CMViva system, the mRNA levels of the CMV ORFs, AAT and GSS mRNAs reached their maximum at 2 days after induction. The maximum mRNA levels of AAT obtained from the CMViva system were 8 fold higher than the constitutive CaMV 35S system. For the study of rAAT production kinetics, the CMViva system increased the ratio of functional rAAT to total rAAT and increased the functional rAAT levels by 33-fold compared with the CaMV 35S system. These intracellular molecular level data support a conceptual model of the CMViva plant-based expression system.

(820e) The Exploitation of Yeast - Elucidating Quality Control & Regulation in the Secretory Pathway

Young, C. L., University of Delaware
Caplan, J., Delaware Biotechnology Institute
Robinson, A. S., University of Delaware
Originally presented on: 11/8/2013 13:42:00 - 14:00:00

Within the secretory pathway of eukaryotic cells, the endoplasmic reticulum (ER) is responsible for maintaining the fidelity of protein synthesis and maturation. A variety of disturbances including nutrient deprivation, pathogenic infection, and chemical treatment, collectively termed 'ER stress', induce quality control mechanisms to facilitate the recovery of cell homeostasis. ER-associated degradation (ERAD), unfolded protein response (UPR), and autophagy are quality control pathways that occur on various timescales, encompass variations in the spatial organization of multiple organelles, and alter select protein concentrations and intracellular localization. Surprisingly, all three pathways are activated in several neurodegenerative and hereditary diseases. In all cases, as a result of ER stress, there is evidence of atypical, intracellular protein distribution during disease manifestation. Yet, how the accumulation of disease-specific proteins is involved in compromising quality control remains elusive.

By implementing DNA recombination strategies combined with high-resolution imaging techniques, we determined that protein redistribution, resultant spatial changes, and organelle modifications are a consequence of the cell's response to ER stress in the yeast, *S. cerevisiae*. In pursuit of a thorough analysis of the protein redistribution at the subcellular level, multiple yeast expression cassettes have been created to test the effects of codon-optimized fluorescent protein (FP) variants [1], small epitope tags (reviewed in [2]), polylinker length for N- and C-terminal tags, and the inclusion of essential retrieval sequences for ER resident proteins [3]. Consequently, our approach enables one to monitor the trafficking effects of ER chaperones and foldases in real time by incorporating H/KDEL retrieval sequences fused to FP variants. To investigate discrete subpopulations of tagged proteins, live-cell imaging methods and super-resolution techniques (e.g. Fluorescence-Photoactivation Localization Microscopy, F-PALM and Structured Illumination Microscopy, SIM) were used to evaluate organelle morphology. Focused-Ion Beam Scanning Electron Microscopy (FIB-SEM) achieved electron microscope resolution, which facilitated entire three-dimensional organelle reconstructions of yeast cells [6]. Interestingly, the improved spatial resolution confirmed organelle connectivity during periods of prolonged stress (e.g. peripheral ER and mitochondria junctions) and inheritance (e.g. lipid droplets derived from perinuclear ER).

Utilizing FP variants as probes, proteins were recombinantly expressed from their native promoter in order to monitor protein trafficking, analyze localization effects of proteins involved in ERAD, and examine organelle dynamics and morphology as a consequence of local perturbations to the system [4].

Consequently, we confirmed the existence of cellular variability following UPR activation and established that endogenous proteins redistribute in the cell – specifically the ER – in order to perform essential functions that maintain cell homeostasis. To establish that the UPR is a global response to localized perturbations, we evaluated the transcriptional effects of UPR activation. As a result, microarray analysis and q-PCR validation have confirmed the novel repression of 206 genes – highly enriched in protein synthesis and metabolic biological functions – following ER stress [5]. Our results expand the characterization of the UPR and reaffirm diverse, global consequences for the cell.

Due to the intrinsic complexity of biological systems, the integration of a broad spectrum of molecular engineering techniques and assays combined with novel imaging analyses is crucial when investigating regulation of the secretory pathway. Collectively, our experimental approaches have led to a better understanding of ER homeostasis and cell regulation, thus facilitating the implementation of new therapeutic modalities aimed at stress reduction.

References

1. C. L. Young, D. Raden, J. Caplan, K. Czymmek, A. S. Robinson *Optimized Cassettes for Live-Cell Imaging of Proteins and High Resolution Techniques in Yeast*, Yeast, 2012 doi:10.1002/yea.2895. [Epub 2012 Apr 4]

2. C. L. Young, Z. T. Britton, A. S. Robinson *Recombinant Protein Expression and Purification: A Comprehensive Review of Affinity Tags and Microbial Applications*, Biotechnology Journal, 7(4), Jan 10 2012 doi:10.1002/biot.201100155. [Epub ahead of print]

3. C. L. Young, D. L. Raden, A. S. Robinson, *Analysis of Endoplasmic Reticulum Resident Proteins in S. cerevisiae: Implementation of H/KDEL Retrieval Sequences*, Traffic, 2013 apr;14(4):365-81. doi:10.111/tra.12041. Epub 2013 Feb 4.

4. C. L. Young, D. L. Raden, J. Caplan, A. S. Robinson *Dynamics of Endoplasmic Reticulum Resident Proteins and Organelle Morphology in S. cerevisiae,* 2013 (in preparation).

5. D. Wei, S. Jacobs, S. Modla, S. Zhang, C. L. Young, R. Cirino, J. Caplan, K. Czymmek *High-resolution three-dimensional reconstruction of a whole yeast cell using focused-ion beam scanning electron microscopy,* BioTechniques, 2012 Jul;53(1):41-8.

6. T. Yuraszeck[Â] , C. Young[Â] , P. Xu, C. A. Gelmi, F. J. Doyle III, A. S. Robinson *Novel down-regulation pathways in the Unfolded Protein Response from S. cerevisiae provide evidence of a complex regulatory response to ER stress,* 2012 (under revision) [Â] co-authorship.

(820f) Process Cost Modeling to Aid Decisions in Early Vaccine Development

Kristopeit, A., Merck & Co.
Christanti, S., Merck & Co.
Wagner, J., Merck & Co., Inc
Laska, M., Merck & Co., Inc.
Originally presented on: 11/8/2013 14:00:00 - 14:18:00

Fundamental decisions made early in vaccine process development, such as expression system selection, can have profound impacts on the eventual manufacturing cost of the vaccine. It is therefore desirable to consider manufacturing cost when making these early decisions; however, most manufacturing logistics are unknown early in development and it is difficult to gauge the impact of process variables on manufacturing cost. In three case studies, process and cost modeling software was applied in order to provide input to decision early in vaccine development. In the first case study, cost modeling was used to determine the minimum fermentation titer required to achieve an acceptable manufacturing cost for a single component vaccine. This titer was then used as a criteria to screen fermentation expression systems. In the second case study, cost modeling was used to compare expression systems for a multi-component vaccine. This comparison demonstrated the importance of a platform expression system for a multi-component vaccine. In a third case study, the cost impact of an affinity resin used in early stage development was evaluated to determine whether transition to more commercially available chromatography media would significantly impact product cost.

(820g) Understanding Single-Chain Antibody Fragment Production in Pichia Pastoris

Royle, K., Imperial College London
Kontoravdi, C., Imperial College London
Leak, D. J., University of Bath
Originally presented on: 11/8/2013 14:18:00 - 14:36:00

Pichia pastoris is a commonly used expression host for heterologous protein production; considerable yields can be achieved from high cell densities in inexpensive culture media. Despite this, the specific productivity is relatively low. Consequently, the full impact of this host on industrial biotechnology has not yet been realised.

The majority of bioengineering studies to increase specific productivity have targeted one factor in isolation, and despite using comparable strategies have variable outcomes. Here, an integrated experimental and computational modelling approach has been taken to understand how the factors interact and develop a global optimisation strategy. We have selected single-chain antibody fragments (scFvs) as a model heterologous protein for this work as they are industrially relevant.

Initially, a deterministic single cell model of protein production in *P. pastoris* was devised incorporating the essential aspects of transcription, translation and folding in the endoplasmic reticulum (ER). Additionally, as heterologous protein production can induce stress responses, the unfolded protein response (UPR) and ER associated degradation pathway (ERAD) have been accounted for.

Preliminary simulations highlighted key regulators of capacity; the experimental picture, however, is much more complex. *P. pastoris* exhibits clonal variation – whereby a culture, derived from a single cell, shows a range of phenotypic behaviour despite being genetically consistent. We aimed to understand how differences in yield arise from genetically identical strains by systematically characterising those key regulators in strains which express a scFv to different degrees.

Consequently, two single chain antibody fragments were cloned into *P. pastoris* and a high and a low secretor of each identified. Time courses of RT Q-PCR carried out targeting scFV transcript levels, the UPR and the ERAD response. LC-MS/MS was used for absolute quantification of the key regulators Kar2 and PDI, both under normal and conditions of over-expression. The latter provided us with a means to validate the model, and derive strains with a higher level of productivity.

The experimental data has been used to expand the preliminary model, revealing an interesting competition between the degradation and folding pathways. Indeed, high yield can be derived from a number of different pathways across the cellular landscape. Despite this, the knowledge of the upper and lower bounds for parameters from the low and high secreting strains has allowed for their optimal values to be predicted, deriving an *in silico* design strategy. This constitutes an integrated modelling and experimental approach to improving productivity in *P. pastoris*.

(820h) Modulation and Modeling of Monoclonal Antibody N-Linked Glycosylation

Villiger, T. K., ETH Zurich
Morbidelli, M., ETH Zurich
Soos, M., ETH Zurich
Originally presented on: 11/8/2013 14:36:00 - 14:54:00

Mammalian cells represent the state-of-the-art expression system for modern monoclonal antibody (mAb) manufacturing processes due to their ability of posttranslational modifications of the desired product such as N-linked glycosylation. Although these modifications are required to obtain a high efficacy, bioavailability or even a proper folding of the protein, the complex and heterogeneous nature of glycosylation has caused problems to identify the most active form(s) and the development of cell culture processes which produce a desired pattern. It has been shown that cell culture conditions such as pH, C-sources and by-products can have a significant impact on N-linked glycosylation and therefore a distinct product quality can be adjusted with optimized process parameters and feeding strategies. However, by-product accumulation, in particular increasing ammonia concentration[1] and a decreasing availability of nucleotide sugar donors influence the intracellular glycosylation mechanism significantly[2].

In order to produce a constant glycosylation pattern during a fed-batch process, experiments with different operating parameters and various feeding strategies were carried out in a micro bioreactor system which produced mAbs with different glycosylation profiles. Measurements of basic metabolites and intracellular nucleotide sugars were used to tune a mathematical model to describe the resulting quality attributes based on a dynamic mathematical model of N-linked glycosylation[3]. This combined experimental and modeling approach could serve not only as a quality by design instrument during bioprocess development but can also be used to define an engineering design space with process boundaries for large scale production.

Literature

(1) Gawlitzek, M.; Ryll, T.; Lofgren, J.; Sliwkowski, M. B. *Biotechnology and Bioengineering* **2000**, *68*, 637–646.

(2) Kochanowski, N.; Blanchard, F.; Cacan, R.; Chirat, F.; Guedon, E.; Marc, A.; Goergen, J.-L. *Analytical Biochemistry* **2006**, *348*, 243–251.

(3) Jimenez Del Val, I.; Nagy, J. M.; Kontoravdi, C. *Biotechnology Progress* **2011**, *44*, 1–44.

AUTHOR INDEX

AUTHOR INDEX

AUTHOR INDEX

AUTHOR INDEX

AUTHOR INDEX

AUTHOR INDEX

AUTHOR INDEX

AUTHOR INDEX

AUTHOR INDEX

AUTHOR INDEX

AUTHOR INDEX

AUTHOR INDEX

AUTHOR INDEX

AUTHOR INDEX

AUTHOR INDEX

AUTHOR INDEX

AUTHOR INDEX

AUTHOR INDEX

AUTHOR INDEX

AUTHOR INDEX

AUTHOR INDEX